Lecture Notes in Artificial Intelligence　　12319

Subseries of Lecture Notes in Computer Science

More information about this series at http://www.springer.com/series/1244

Ricardo Cerri · Ronaldo C. Prati (Eds.)

Intelligent Systems

9th Brazilian Conference, BRACIS 2020
Rio Grande, Brazil, October 20–23, 2020
Proceedings, Part I

 Springer

Editors
Ricardo Cerri ⓘ
Federal University of São Carlos
São Carlos, Brazil

Ronaldo C. Prati ⓘ
Federal University of ABC
Santo Andre, Brazil

ISSN 0302-9743 ISSN 1611-3349 (electronic)
Lecture Notes in Artificial Intelligence
ISBN 978-3-030-61376-1 ISBN 978-3-030-61377-8 (eBook)
https://doi.org/10.1007/978-3-030-61377-8

LNCS Sublibrary: SL7 – Artificial Intelligence

This Springer imprint is published by the registered company Springer Nature Switzerland AG
The registered company address is: Gewerbestrasse 11, 6330 Cham, Switzerland

Preface

The Brazilian Conference on Intelligent Systems (BRACIS) is one of Brazil's most meaningful events for students and researchers in Artificial and Computational Intelligence. Currently, In its 9th edition, BRACIS originated from the combination of the two most important scientific events in Brazil in Artificial Intelligence (AI) and Computational Intelligence (CI): the Brazilian Symposium on Artificial Intelligence (SBIA), with 21 editions, and the Brazilian Symposium on Neural Networks (SBRN), with 12 editions. The conference aims to promote theory and applications of artificial and computational intelligence. BRACIS also aims to promote international-level research by exchanging scientific ideas among researchers, practitioners, scientists, and engineers.

BRACIS 2020 received 228 submissions. All papers were rigorously double-blind peer-reviewed by an International Program Committee (an average of three reviews per submission), followed by a discussion phase for conflicting reports. At the end of the reviewing process, 90 papers were selected for publication in two volumes of the *Lecture Notes in Artificial Intelligence* series, an acceptance rate of 40%.

We are very grateful to Program Committee members and reviewers for their volunteered contribution in the reviewing process. We would also like to express our gratitude to all the authors who submitted their articles, the general chairs, and the Local Organization Committee, to put forward the conference during the COVID-19 pandemic. We want to thank the Artificial Intelligence and Computational Intelligence commissions from the Brazilian Computer Society for the confidence in serving as program chairs for BRACIS 2020.

We are confident that these proceedings reflect the excellent work in the fields of artificial and computation intelligence communities.

October 2020
Ricardo Cerri
Ronaldo C. Prati

Organization

General Chairs

Hélida Salles Santos	Universidade Federal do Rio Grande, Brazil
Graçaliz Dimuro	Universidade Federal do Rio Grande, Brazil
Eduardo Borges	Universidade Federal do Rio Grande, Brazil
Leonardo Emmendorfer	Universidade Federal do Rio Grande, Brazil

Program Committee Chairs

Ricardo Cerri	Federal University of São Carlos, Brazil
Ronaldo C. Prati	Federal University of ABC, Brazil

Steering Committee

Leliane Barros	Universidade de São Paulo, Brazil
Heloisa Camargo	Universidade Federal de São Carlos, Brazil
Flavia Bernardini	Universidade Federal Fluminense, Brazil
Jaime Sichman	Universidade de São Paulo, Brazil
Karina Delgado	Universidade de São Paulo, Brazil
Kate Revoredo	Universidade Federal do Estado do Rio de Janeiro, Brazil
Renata O. Vieira	Pontifícia Universidade Católica do Rio Grande do Sul, Brazil
Solange Rezende	Universidade de São Paulo, Brazil
Ricardo Prudencio	Universidade Federal de Pernambuco, Brazil
Anne Canuto	Universidade Federal do Rio Grande do Norte, Brazil
Anisio Lacerda	Universidade Federal de Minas Gerais, Brazil
Gisele Pappa	Universidade Federal de Minas Gerais, Brazil
Gina Oliveira	Universidade Federal de Uberlândia, Brazil
Renato Tinós	Universidade de São Paulo, Brazil
Paulo Cavalin	IBM, Brazil

Program Committee

Adenilton da Silva	Universidade Federal de Pernambuco, Brazil
Adrião Dória Neto	Universidade Federal do Rio Grande do Norte, Brazil
Albert Bifet	LTCI, Télécom ParisTech, France
Alberto Paccanaro	Royal Holloway, University of London, UK
Alex Freitas	University of Kent, UK
Alexandre Delbem	Universidade de São Paulo, Brazil
Alexandre Ferreira	Universidade Estadual de Campinas, Brazil

Alexandre Plastino	Universidade Federal Fluminense, Brazil
Aline Paes	Universidade Federal Fluminense, Brazil
Alvaro Moreira	Universidade Federal do Rio Grande do Sul, Brazil
Ana Carolina Lorena	Instituto Tecnológico de Aeronáutica, Brazil
Ana Vendramin	Universidade Tecnológica Federal do Paraná, Brazil
Anísio Lacerda	Centro Federal de Educação Tecnológica de Minas Gerais, Brazil
Anderson Soares	Universidade Federal de Goiás, Brazil
André Coelho	Universidade de Ceará, Fortaleza, Brazil
André Carvalho	Universidade de São Paulo, Brazil
André Rossi	Universidade de São Paulo, Brazil
André Grahl Pereira	Universidade Federal do Rio Grande do Sul, Brazil
Andrés Salazar	Universidade Tecnológica Federal do Paraná, Brazil
Anne Canuto	Universidade Federal do Rio Grande do Norte, Brazil
Araken Santos	Universidade Federal Rural do Semi-árido, Brazil
Aurora Pozo	Universidade Federal do Paranáv, Brazil
Bianca Zadrozny	IBM, Brazil
Bruno Nogueira	Universidade Federal de Mato Grosso do Sul, Brazil
Bruno Travencolo	Universidade Federal de Uberlândia, Brazil
Carlos Ferrero	Instituto Federal de Santa Catarina, Brazil
Carlos Silla	Pontifical Catholic University of Paraná, Brazil
Carlos Thomaz	Centro Universitário da FEI, Brazil
Carolina Almeida	State University in the Midwest of Paraná, Brazil
Celia Ralha	Universidade de Brasília, Brazil
Celine Vens	KU Leuven, Belgium
Celso Kaestner	Universidade Tecnológica Federal do Paraná, Brazil
Cesar Tacla	Universidade Tecnológica Federal do Paraná, Brazil
Cleber Zanchettin	Universidade Federal de Pernambuco, Brazil
Daniel Araújo	Federal University of Rio Grande do Norte, Brazil
Danilo Sanches	Universidade Tecnológica Federal do Paraná, Brazil
David Martins-Jr	Universidade Federal do ABC, Brazil
Delia Farias	Universidad de Guanajuato Lomas del Bosque, Mexico
Denis Fantinato	Universidade Federal do ABC, Brazil
Denis Mauá	Universidade de São Paulo, Brazil
Diana Adamatti	Universidade Federal do Rio Grande, Brazil
Diego Furtado Silva	Universidade Federal de São Carlos, Brazil
Eder Gonçalves	Universidade Federal do Rio Grande, Brazil
Edson Gomi	Universidade de São Paulo
Edson Matsubara	Fundação Universidade Federal de Mato Grosso do Sul, Brazil
Eduardo Borges	Universidade Federal do Rio Grande, Brazil
Eduardo Costa	Corteva Agriscience, USA
Eduardo Goncalves	Escola Nacional de Ciências Estatísticas, Brazil
Eduardo Palmeira	Universidade Estadual de Santa Cruz, Brazil
Eduardo Spinosa	Universidade Federal do Paraná, Brazil
Elaine Faria	Federal University of Uberlândia, Brazil

Elizabeth Goldbarg	Universidade Federal do Rio Grande do Norte, Brazil
Emerson Paraiso	Pontificia Universidade Catolica do Paraná, Brazil
Eraldo Fernandes	Universidade Federal de Mato Grosso do Sul, Brazil
Eric Araújo	Universidade Federal de Lavras, Brazil
Erick Fonseca	Instituto de Telecomunicações, Portugal
Fabiano Silva	Universidade Federal do Paraná, Brazil
Fabrício Enembreck	Pontifical Catholic University of Paraná, Brazil
Fabricio França	Universidade Federal do ABC, Brazil
Fábio Cozman	Universidade de São Paulo, Brazil
Federico Barber	Universitat Politècnica de València, Spain
Felipe Meneguzzi	Pontifícia Universidade Católica do Rio Grande do Sul, Brazil
Fernando Osório	Universidade de São Paulo, Brazil
Flavio Tonidandel	Centro Universitario da FEI, Brazil
Francisco Chicano	University of Málaga, Spain
Francisco de Carvalho	Centro de Informática da UFPE, Brazil
Gabriel Ramos	Universidade do Vale do Rio dos Sinos, Brazil
George Cavalcanti	Universidade Federal de Pernambuco, Brazil
Gerson Zaverucha	Federal University of Rio de Janeiro, Brazil
Giancarlo Lucca	Universidade Federal do Rio Grande, Brazil
Gina Oliveira	Universidade Federal de Uberlândia, Brazil
Gisele Pappa	Universidade Federal de Minas Gerais, Brazil
Gracaliz Dimuro	Universidade Federal do Rio Grande, Brazil
Guilherme Derenievicz	Federal University of Santa Catarina, Brazil
Guillermo Simari	Universidad Nacional del Sur, Argentina
Gustavo Batista	Universidade de São Paulo, Brazil
Gustavo Giménez-Lugo	Universidade Tecnológica Federal do Paraná, Brazil
Heitor Gomes	University of Waikato, New Zealand
Helena Caseli	Universidade Federal de São Carlos, Brazil
Helida Santos	Universidade Federal do Rio Grande, Brazil
Huei Lee	Universidade Estadual do Oeste do Paraná, Brazil
Humberto Bustince	Universidad Publica de Navarra, Spain
Humberto Oliveira	Universidade Federal de Alfenas, Brazil
Isaac Triguero	University of Nottingham, UK
Ivandré Paraboni	Universidade de São Paulo, Brazil
Jaime Sichman	Universidade de São Paulo, Brazil
Jesse Read	École Polytechnique, France
Joao Gama	Universidade do Porto, Portugal
João Balsa	Universidade de Lisboa, Portugal
João Bertini	Universidade Estadual de Campinas, Brazil
João Mendes Moreira	Universidade do Porto, Portugal
João Papa	Universidade Estadual Paulista, Brazil
João Xavier-Júnior	Universidade Federal do Rio Grande, Brazil
João Luís Rosa	Universidade de São Paulo, Brazil
Jomi Hübner	Universidade Federal de Santa Catarina, Brazil
Jonathan Silva	Universidade Federal de Mato Grosso do Sul, Brazil

José Antonio Sanz	Universidad Publica de Navarra, Spain
Julio Nievola	Pontifícia Universidade Católica do Paraná, Brazil
Karina Delgado	Universidade de São Paulo
Kate Revoredo	Universidade Federal do Estado do Rio de Janeiro, Brazil
Krysia Broda	Imperial College London, UK
Leandro Coelho	Pontifícia Universidade Católica do Paraná, Brazil
Leliane Barros	Universidade de São Paulo
Leonardo Emmendorfer	Universidade Federal do Rio Grande, Brazil
Leonardo Ribeiro	Technische Universität Darmstadt, Germany
Livy Real	B2W Digital Company, Brazil
Lucelene Lopes	Roberts Wesleyan College, USA
Luciano Barbosa	Universidade Federal de Pernambuco, Brazil
Luis Garcia	Universidade de Brasília, Brazil
Luiz Carvalho	Universidade Tecnológica Federal do Paraná, Brazil
Luiz Coletta	Universidade Estadual Paulista, Brazil
Luiz Merschmann	Universidade Federal de Lavras, Brazil
Luiza Mourelle	State University of Rio de Janeiro, Brazil
Marcela Ribeiro	Universidade Federal de São Carlos, Brazil
Marcella Scoczynski	Universidade Tecnológica Federal do Paraná, Brazil
Marcelo Finger	Universidade de São Paulo, Brazil
Marcilio de Souto	Université d'Orléans, France
Marcos Domingues	Universidade Estadual de Maringá, Brazil
Marcos Quiles	Federal University of São Paulo, Brazil
Marilton Aguiar	Universidade Federal de Pelotas, Brazil
Marley Vellasco	Pontifícia Universidade Católica do Rio de Janeiro, Brazil
Mauri Ferrandin	Universidade Federal de Santa Catarina, Brazil
Márcio Basgalupp	Universidade Federal de São Paulo, Brazil
Mário Benevides	Universidade Federal Fluminense, Brazil
Moacir Ponti	Universidade de São Paulo, Brazil
Murillo Carneiro	Federal University of Uberlândia, Brazil
Murilo Naldi	Universidade Federal de São Carlos, Brazil
Myriam Delgado	Federal University of Technology of Paraná, Brazil
Nádia Felix	Universidade Federal de Goiás, Brazil
Newton Spolaôr	Universidade Estadual do Oeste do Paraná, Brazil
Patricia Oliveira	Universidade de São Paulo
Paulo Cavalin	IBM Research, Brazil
Paulo Ferreira Jr.	Universidade Federal de Pelotas, Brazil
Paulo Gabriel	Universidade Federal de Uberlândia, Brazil
Paulo Quaresma	Universidade de Évora, Portugal
Paulo Pisani	Universidade Federal do ABC, Brazil
Priscila Lima	Universidade Federal do Rio de Janeiro, Brazil
Rafael Bordini	Pontifícia Universidade Católica do Rio Grande do Sul, Brazil
Rafael Mantovani	Federal Technology University of Paraná, Brazil

Rafael Parpinelli	Universidade do Estado de Santa Catarina, Brazil
Rafael Rossi	Federal University of Mato Grosso do Sul, Brazil
Reinaldo Bianchi	Centro Universitario FEI, Brazil
Renato Assuncao	Universidade Federal de Minas Gerais, Brazil
Renato Krohling	Universidade Federal do Espírito Santo, Brazil
Renato Tinos	Universidade de São Paulo, Brazil
Ricardo Cerri	Universidade Federal de São Carlos, Brazil
Ricardo Silva	Universidade Tecnológica Federal do Paraná, Brazil
Ricardo Marcacini	Universidade de São Paulo, Brazil
Ricardo Prudêncio	Universidade Federal de Pernambuco, Brazil
Ricardo Rios	Universidade Federal da Bahia, Brazil
Ricardo Tanscheit	Pontifícia Universidade Católica do Rio de Janeiro, Brazil
Ricardo Fernandes	Federal University of São Carlos, Brazil
Roberta Sinoara	Instituto Federal de Ciência, Educação e Tecnologia de São Paulo, Brazil
Roberto Santana	University of the Basque Country, Spain
Robson Cordeiro	Universidade de São Paulo, Brazil
Rodrigo Barros	Pontifícia Universidade Católica do Rio Grande do Sul, Brazil
Rodrigo Mello	Universidade de São Paulo
Rodrigo Wilkens	University of Essex, UK
Roger Granada	Pontifícia Universidade Católica do Rio Grande do Sul, Brazil

Contents – Part I

Neural Networks, Deep Learning and Computer Vision

Text Mining and Natural Language Processing

Contents – Part II

Knowledge Representation, Logic and Fuzzy Systems

Machine Learning and Data Mining

Multidisciplinary Artificial and Computational Intelligence and Applications

Evolutionary Computation, Metaheuristics, Constrains and Search, Combinatorial and Numerical Optimization

A New Hybridization of Evolutionary Algorithms, GRASP and Set-Partitioning Formulation for the Capacitated Vehicle Routing Problem

André Manhães Machado$^{(\boxtimes)}$ ⓘ, Maria Claudia Silva Boeres ⓘ,
Rodrigo de Alvarenga Rosa ⓘ, and Geraldo Regis Mauri ⓘ

Universidade Federal do Espírito Santo, Vitoria, ES, Brazil
andre.manhaes@gmail.com, boeres@inf.ufes.br
{rodrigo.a.rosa,geraldo.mauri}@ufes.br

Abstract. This work presents a new hybrid method based on the route-first-cluster-second approach using Greedy Randomized Adaptive Search Procedure (GRASP), Differential Evolution (DE), Evolutionary Local Search (ELS) and set-partitioning problem (SPP) to solve well-known instances of Capacitated Vehicle Routing Problem (CVRP). The CVRP consists of minimizing the cost of a fleet of vehicles serving a set of customers from a single depot, in which every vehicle has the same capacity. The DE heuristic is used to build an initial feasible solution and ELS is applied until a local minimum is found during the local search phase of the GRASP. Finally, the SPP model provides a new optimal solution with regard to the built solutions in the GRASP. We perform computational experiments for benchmarks available in the literature and the results show that our method was effective to solve CVRP instances with a satisfactory performance. Moreover, a statistical test shows that there is not significant difference between the best known solutions of benchmark instances and the solutions of the proposed method.

Keywords: CVRP · Differential evolution · GRASP · Evolutionary local search · Set-partitioning problem

1 Introduction

The Capacitated Vehicle Routing Problem (CVRP) lies in minimizing the cost of a fleet of vehicles serving a set of customers from a unique depot, in which every vehicle has an uniform capacity. It has a direct application to the real-world problems in various activities like distribution, waste management and city logistics with a very active research domain in the last two decades [11]. Since the

---

We want to express our thanks to the National Council for Scientific and Technological Development – CNPq (processes 302261/2019-2 and 307797/2019-8) and FAPES (process 75528452/2016) for financial support.

© Springer Nature Switzerland AG 2020
R. Cerri and R. C. Prati (Eds.): BRACIS 2020, LNAI 12319, pp. 3–17, 2020.
https://doi.org/10.1007/978-3-030-61377-8_1

CVRP is classified as NP-complete and NP-hard [14], heuristics methods are usually devised and applied to instances at medium and large sizes. These methods involve, among others, evolutionary algorithms [17], Simulating Annealing (SA), Tabu Search (TS) [20] either with additional diversification strategies or hybridization techniques [19].

The heuristics usually use two mutually exclusive approaches to solve CVRP: the first way is the cluster-first-route-second in which clients are inserted into clusters and each cluster is solved as a Traveling Salesman Problem (TSP). The second manner is the route-first-cluster-second where vehicle capacity is relaxed to build a TSP called giant tour, then the TSP is break into feasible trips using a split function [11]. The route-first-cluster-second was theoretically proposed in [5], but the first results for CVRP only were presented in [15]. Since then, distinct approaches to tackle the CVRP were proposed [1, 16].

Following the spirit of the hybridization of metaheuristics in [11], this study proposes a new hybrid method called G-DE-SPP based on the route-first-cluster-second approach and using Greedy Randomized Adaptive Search Procedure (GRASP), Differential Evolution (DE), Evolutionary Local Search (ELS) and set-partitioning problem (SPP) to solve well-known instances of CVRP in the literature. The DE is used to build an initial feasible solution and ELS is applied until a local minimum is found during the local search phase of the GRASP. Finally, the SPP model provides a new optimal solution with regard to the built solutions in the previous steps.

The main contributions of this study are: the new hybridization of GRASP applying DE and heuristics for TSP as constructive phase and ELS for local phase, to the best of our knowledge, it has not yet been researched for CVRP; unlike [11], a final step using all previously CVRP solutions found as input data for SPP; good solutions found for almost all instances as shown by the statistical test.

The paper is organized as follows. In the next section, we mathematically define the CVRP problem. Our hybridized method is presented in Sect. 3. Section 4 reports the performance of the proposed heuristic using a set of instances available in the literature, followed by our conclusions in Sect. 5.

2   Mathematical Model for the Capacitated Vehicle Routing Problem

The Capacitated Vehicle Routing Problem (CVRP) is described as a graph $G = (N, E)$ with the set of nodes $N = \{0, 1, \cdots, n\}$ and the set of edges $E \subset N \times N$. Each element $i \in N \setminus \{0\}$ represents a customer with a demand q_i whilst $i = 0 \in N$ designates the depot. The edge $(i, j) \in E$ has a weight $c_{i,j} > 0$ indicating the cost of traveling from $i \in N$ to $j \in N$. A set of vehicles $K = \{1, 2, \cdots, k\}$ with maximum capacity Q must start from the depot $(i = 0)$ to serve each customer $i \in N \setminus \{0\}$ and returns back to the depot. The objective is to find a set of routes that meets all customers' demand and minimize the total routing cost.

The Integer Linear Programming Model (ILPM) of the CVRP can be written as [14]:

$$\text{Minimize} \sum_{k}^{K}\sum_{i}^{N}\sum_{j}^{N} c_{i,j}x_{k,i,j} \tag{1}$$

$$\sum_{k=1}^{K}\sum_{i=0,i\neq j}^{N} x_{k,i,j} = 1, \forall j \in N \setminus \{0\} \tag{2}$$

$$\sum_{j=1}^{N} x_{k,0,j} = 1, \forall k \in K \tag{3}$$

$$\sum_{i=0,i\neq j}^{N} x_{k,i,j} = \sum_{i=0}^{N} x_{k,j,i}, \forall j \in N, \forall k \in K \tag{4}$$

$$\sum_{i=0}^{N}\sum_{j=1,i\neq j}^{N} q_j x_{k,i,j} \leq Q, \forall k \in K \tag{5}$$

$$\sum_{k=1}^{K}\sum_{i\in Y}\sum_{j\in Y,i\neq j} x_{k,i,j} \leq |Y| - 1, \forall Y \subset N \tag{6}$$

$$x_{k,i,j} \in \{0,1\}, \forall k \in K, \forall i, j \in N \tag{7}$$

The objective function (1) minimizes the total travel cost. Constraints (2) ensure that each customer is visited by exactly one vehicle. Constraints (3) and (4) are the flow constraints, which guarantee each vehicle can leave the depot only once and the number of the vehicles arriving at every node $i \in N \setminus \{0\}$ is equal to the number of the vehicles leaving. Constraints (5) make sure that the demands of the customers visited in a route is not greater than the capacity of the vehicle $k \in K$. The sub-tour elimination constraints (6) ensure that the solution contains no cycles. Constraints (7) define the domains of the variables.

3   The Proposed G-DE-SPP Method

3.1   Overview

The G-DE-SPP comprises two sequential stages, the first in which the GRASP heuristic hybridized with Differential Evolution and Evolutionary Local Search is employed to create sub-optimal solutions. Then, it ensues the next stage in which the set-partitioning problem is applied using as input the previous solutions created in the first step. To do this, the concept of route-first-cluster-second underpins the whole proposed procedure.

In the remainder of the text we used the following conventions and definitions. Variables denoted as S are a CVRP solution, but when designated as T represent a tour of the Traveling Salesman Problem (TSP). The variable U represents a set of CVRP solutions. Parameters of each procedure are defined using the letter m. We use the notation $r_i \in S$ to denote the i-th route of CVRP. A j-th node in the route r_i is represented as $t_j \in r_i$ (similarly, $t_j \in T$ designates j-th node in T). The edge $[t_k, t_l] \in r_i$ connects the nodes t_k, t_l in r_i. A n-vector v has dimension n.

The next subsections address the description of each component of the G-DE-SPP method.

3.2   Split and Cost Functions

The proposed G-DE-SPP method explores both TSP and CVRP search spaces, S_{TSP} and S_{CVRP}. A tour T, which belongs to the former space, denotes a partial solution represented by a permutation of nodes. A CVRP solution S, that belongs to the latter, denotes a complete solution containing a set of vehicle trips. The solutions in the different spaces S_{TSP} and S_{CVRP} are converted to each other with the $Split(\cdot)$ and $Split^{-1}(\cdot)$ functions as shown in Fig. 1. The function $Split(\cdot)$ takes a solution T in S_{TSP} and converts it to a solution S in S_{CVRP}. Conversely, $Split^{-1}$ does the reverse way.

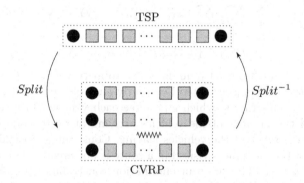

Fig. 1. $Split$ and $Split^{-1}$ functions

The $Split(\cdot)$ function uses as its basis the Bellman's procedure [9] as shown in Algorithm 1. In the outer loop, the nodes $(n_i, n_{i+1}, \cdots, n_j)$ of the tour are check to see if the path $(0, n_i, n_{i+1}, \cdots, n_j, 0)$ is feasible. If so, the path is modeled as an arc-set using the vector P to store the indexes and the vector W to record the trip cost. An optimal splitting of T into feasible trips corresponds to a min-cost path from node 0 to node n in T, where P and W provides the required information to create the path .

The $Split^{-1}(\cdot)$ function is defined as the queuing of the trips $r_0, r_1, \cdots, r_i$ as follows. Let $R = \{r_0, r_1, \cdots, r_i\}$ be the set of routes of S, where $r_k = (n_{k,1}, n_{k,2}, \cdots, n_{k,j_k})$, then $Split^{-1}(S)$ is equal to $(n_{1,1}, n_{1,2}, \cdots, n_{1,j_1}, n_{2,1}, n_{2,2}, \cdots, n_{i,j_i-1}, n_{i,j_i})$.

Algorithm 1. Function $Split(T)$

```
1:  P₀ ← 0;                                23: t ← 0
2:  Pᵢ ← +∞, i = 1, 2, ⋯ , n              24: j ← 0
3:  for i ← 1 to n do                      25: rᵢ ← ∅, i = 1, 2, ⋯ , n
4:      load ← 0                           26: repeat
5:      c ← 0                              27:     t ← t + 1
6:      j ← i                              28:     i ← Wⱼ
7:      repeat                             29:     for k ← i + 1 to j do
8:          load ← load + qᵢ              30:         rᵢ ← rᵢ ∪ {k}
9:          if i = j then                  31:     end for
10:             c ← c₀,ᵢ + cᵢ,₀            32:     j ← i
11:         else                           33: until i = 0
12:             c ← c − cⱼ₋₁,₀ + cⱼ₋₁,ⱼ + cⱼ,₀   34: return {r₀, r₁, ⋯ , rₙ}
13:         end if
14:         if load ≤ Q then
15:             if Pᵢ₋₁ + c < Pⱼ then
16:                 Pⱼ ← Pᵢ₋₁ + c
17:                 Wⱼ ← i − 1
18:             end if
19:             j ← j + 1
20:         end if
21:     until j > n or load > Q
22: end for
```

Cost functions $cost(\cdot)$ and $F_{cost}(\cdot)$, to evaluate T and S solutions, are provided. Given a solution S, $cost(\cdot)$ returns the cost of S when valued by ILPM (Integer Linear Programming Model). The function $F_{cost}(\cdot)$, given a solution T, returns the value $F_{cost}(T) = cost(Split(T))$.

3.3 Differential Evolution (DE)

The DE is an evolutionary algorithm proposed in [13] for optimization problems over a continuous domain. In DE approach, a population of individuals (solutions) evolves throughout a set of iterations (generations) of recombination, evaluation and selection. In the initial stage, this perturbations are large because the individuals are far away from each other. As the evolutionary process advances, the population converges to a small region of the search space. DE has three main parameters of the DE: the scaling factor $m_{de}^f \in [0, \infty]$ that controls the rate at which the population evolves, the crossover probability $m_{de}^c \in [0, 1]$ that determines the ratio of bits that are transferred to the trial vector from its opponent and m_{de}^i that sets the number of iterations.

Due to the its continuous nature, the standard encoding scheme of DE cannot be directly adopted for CVRP. So an indirect representation based on random-key values is imposed [12]. The underlying idea is to represent T as a $n-$vector v of values in $[0, 1]$, and to be able to rebuild T from v using as information the position and its value. Therefore, let T be a TSP solution with n nodes, a n-vector π filled using the function $rand(0, 1)$ (which gives a uniformly distributed random value with the range of 0 and 1) and v a n-vector for DE. Then, for each position i of T and node $t_i \in T$, the position t_i of v is filled with the value of π_i, namely $v_{t_i} = \pi_i$. The function $enc(\cdot)$ is used to encode a solution T as shown in Algorithm 2.

Algorithm 2. $enc(T)$

1: $\pi_i \leftarrow rand(0,1)$, $i = 1$ **to** n
2: $\pi \leftarrow sort(\pi)$
3: **for** $t_i \in T$, $i = 1$ **to** n **do**
4:    $c \leftarrow t_i$
5:    $v_c \leftarrow \pi_i$
6: **end for**
7: **return** v

Conversely, the function $dec(\cdot)$ gets a vector v in DE and converts into a solution T using only the value and its position. The function $dec(\cdot)$ is shown in Algorithm 2.

Algorithm 3. $dec(v)$

1: $p_i = (0,0)$, $i = 1$ **to** n
2: $T = (0)$
3: **for** each position i of v **do**
4:    $p[i] \leftarrow (v[i], i)$
5: **end for**
6: sort p according to the first component of its elements;
7: **for** each pair (a, b) in p **do**
8:    let $c \in N$ be node with index b;
9:    add node c in the last position of T;
10: **end for**
11: **return** T

DE initial population is, in general, random. However, in order to improve the starting point for the algorithm, the initial population was generated using two well-known constructive heuristics: Random Nearest Neighbor Heuristic (RNH) and Random Insertion Heuristic (RIH). Given a partially built tour T, each RNH step consists of checking if the addition of a new unvisited node to T after the last currently inserted, improves its cost. The only difference between RNH e RIH is that, in the latter, the insertion of a new node is possible in any position of T. In both heuristics, the current node to be inserted in the tour is selected randomly among the best m_{ran}^r candidate vertices in each step. More details for both heuristics can be found in [11] and [8].

Algorithm 4 shows the DE algorithm proposed. An initial population composed of tours T is constructed using $RNH(\cdot)$ and $RIH(\cdot)$ procedures. The encoded population obtained using the $enc(\cdot)$ function belongs then to the first generation G_1 of the algorithm. $RNH(\cdot)$ and $RIH(\cdot)$ randomness is calibrated by m_{ran}^r. Then, the population of individuals is iteratively improved throughout m_{de}^i generations, using recombination, mutation and selection tools. In the recombination approach, a candidate solution is added to other population member, based on the weighted difference between two randomly selected individuals and the factor m_{de}^f. In conjunction with selection, the perturbation controlled by the parameter m_{de}^c (mutation) samples around the search space in order to escape of local minimum. The DE algorithm output is a solution S.

Algorithm 4. $DE(m_{ran}^r, m_{de}^{np}, m_{de}^c, m_{de}^f, m_{de}^i)$

1: $half \leftarrow \lfloor m_{de}^{np}/2 \rfloor$
2: $L \leftarrow RNH(m_{ran}^r, half)$
3: $L \leftarrow L \cup RIH(m_{ran}^r, half)$
4: $G_1 = \{enc(T) | T \in L\}$
5: $P = G_1$
6: **for** $i = 1$ **to** m_{de}^i **do**
7:     select three random individuals $\{v_{r1}, v_{r2}, v_{r3}\} \in G_i$
8:     $d_{rand} \leftarrow$ select a random dimension to mutate
9:     **for** each d dimension **do**
10:       **if** $d = d_{rand}$ or $random() \leq m_{de}^c$ **then**
11:         $u_{i,d} \leftarrow v_{r1,d} + m_{de}^f * (v_{r2,d} - v_{r3,d})$
12:       **else**
13:         $u_{i,d} \leftarrow x_{i,d}$
14:       **end if**
15:       **if** $F_{cost}(dec(u_i)) \leq F_{cost}(dec(x_i))$ **then**
16:         add u_i in the offspring G_{i+1}
17:       **else**
18:         add x_i in the offspring G_{i+1}
19:       **end if**
20:     **end for**
21:     $P \leftarrow P \cup \{x | x \in G_{i+1}\}$
22: **end for**
23: $v^* \leftarrow$ best solution in P
24: **return** $Split(dec(v^*))$

3.4 Evolutionary Local Search (ELS)

The ELS heuristic, originally introduced in [18], is inspired on the search algorithms (i.e. hill climbing) and the evolutionary algorithms such as Genetic Algorithms (GA). It uses a population of unitary size that evolves in a loop of applications of the mutation operator, producing a set of offspring solutions. Among all solutions found, the best one is returned as result. The ELS procedure is defined combining the local search $LS(\cdot)$ and mutate $Mutate(\cdot)$ procedure.

Algorithm 5 outlines the local search $LS(\cdot)$ used by $ELS(\cdot)$. The $LS(\cdot)$ applies two classical moves: 2-opt and crossover. For each pair of routes $r_i, r_j \in S$, the 2-opt move repeatedly replaces the edges $[t_{k_1}, t_{k_2}] \in r_i$ and $[t_{k_3}, t_{k_4}] \in r_j$ for $[t_{k_1}, t_{k_3}]$ and $[t_{k_2}, t_{k_4}]$ in S. Likewise, the crossover move repeatedly replaces the edges $[t_{k_1}, t_{k_2}] \in r_i$ and $[t_{k_3}, t_{k_4}] \in r_j$ for $[t_{k_2}, t_{k_3}]$ and $[t_{k_1}, t_{k_4}]$ in S. The 2-opt and crossover moves are only carried out as long as yields a viable and better cost solution. More details regarding 2-opt and crossover moves can be found in

[10] and [11], respectively. Those two moves in $LS(\cdot)$ are applied in an iterative improvement loops until no further enhancement is possible.

Algorithm 5. $LS(S)$

1: $repeat \leftarrow true$
2: **while** $repeat = true$ **do**
3:    $\bar{f} \leftarrow cost(S)$
4:    $S \leftarrow$ 2-opt(S)
5:    $S \leftarrow crossover(S)$
6:    **if** $cost(S) < \bar{f}$ **then**
7:        $\bar{f} \leftarrow cost(S)$
8:    **else**
9:        $repeat \leftarrow false$
10:   **end if**
11: **end while**
12: **return** S

Algorithm 6 shows the $Mutate(\cdot)$ function used by $ELS(\cdot)$, which takes a solution S as input and transform it into a solution T. Under such representation, a classical swap move is performed in T. Finally, the mutated T is converted back into S and it is returned as the result.

Algorithm 6. $Mutate(S)$

1: $T \leftarrow Split^{-1}(S)$
2: Swap randomly two nodes in T
3: $S^{+} \leftarrow Split(T)$
4: **return** S^{+}

Algorithm 7 presents the $ELS(\cdot)$ procedure. Initially, a local search $LS(\cdot)$ is applied to the input solution S and the result is stored as S^{*}. Then, the solution S^{*} is improved in two nested loops. The outer loop set the current best value $\bar{f}$ to infinite so that the first solution generated in the inner loop is always accepted. The inner *for* performs m_{els}^{j} times the mutation and local search procedure over the best solution S^{*}. Therefore, the outer loop allows the inner one to start with a solution out of a local minimum by using the mutate operator. Besides, each new solution built in the ELS is added to set U. The procedure returns the best solution found.

3.5 Greedy Randomized Adaptive Search Procedure (GRASP)

The GRASP algorithm, initially proposed in [7], is an iterative process composed of two phases: construction and local search. In the construction phase,

---

Algorithm 7. $ELS(S, U)$

1: $S^* \leftarrow LS(S);\ U \leftarrow U \cup \{S^*\}$
2: **for** $i \leftarrow 1$ **to** m^i_{els} **do**
3:     $\bar{f} \leftarrow +\infty$
4:     **for** $i \leftarrow 1$ **to** m^j_{els} **do**
5:         $S \leftarrow S^*$
6:         $S \leftarrow Mutate(S)$
7:         $S \leftarrow LS(S)$
8:         $U \leftarrow U \cup \{S\}$
9:         **if** cost(S) $\leq \bar{f}$ **then**
10:             $\bar{S} \leftarrow$ S, $\bar{f} \leftarrow$ cost(S)
11:         **end if**
12:     **end for**
13:     **if** $\bar{f} \leq cost(S^*)$ **then**
14:         $S^* \leftarrow \bar{S}$
15:     **end if**
16: **end for**
17: **return** S^*

---

the objective is to build a feasible solution using a procedure that is random and greedy by definition. In the local search, the objective is to explore the neighborhood of a solution until a local optimal is found. These two phases are repeated by a predetermined number of iterations and the best solution found is returned as the result. The GRASP proposed here is hybridized in both main steps. The construction phase is accomplished by the Differential Evolution (DE) and local phase is performed by the Evolutionary Local Search (ELS). Our GRASP pseudocode is shown in Algorithm 8. The parameter m^i_{grasp} defines its total number of iterations.

Algorithm 8. $GRASP()$

1: $\bar{f} = +\infty$
2: $\bar{S} = \emptyset$
3: $U = \emptyset$
4: **for** $i \leftarrow 1$ **to** m^i_{grasp} **do**
5:     $S \leftarrow$ DE(U)
6:     $S \leftarrow$ ELS(S,U)
7:     **if** cost(S) $\leq \bar{f}$ **then**
8:         $\bar{S} \leftarrow$ S
9:         $\bar{f} \leftarrow$ cost(S)
10:     **end if**
11: **end for**
12: **return** U

3.6   Set-Partitioning Problem (SPP)

The set-partitioning problem (SPP) [2] has as the objective to determine how the items in a given set A can be partitioned into smaller subsets $u \in \mathcal{P}(A)$, where $\mathcal{P}(A)$ is the power set of A. For the CVRP, each subset represents a possible route. However, enumerating each possible route generally is impossible given the number of elements in search space. Therefore, CVRP formulated by SPP usually works with a reduced number of routes.

In this formulation, each route $k \in K$ is represented by a n-binary vector a_k. The value of $a_{k,i}$ is 1 if the node i is visited on the route k, otherwise 0. We associate each vector $k \in K$ with the cost c_k representing the total distance traveled on the route. Also, each route k must be feasible with respect to the capacity constraints. Then, the CVRP can be posed as the following Linear Programming Formulation (LPF):

$$\text{Minimize} \sum_{k \in K} c_k x_k \tag{8}$$

$$\sum_{k \in K} a_{k,i} x_k = 1, i = 1, 2, \cdots, n \tag{9}$$

$$x_k \in \{0, 1\}, \forall k \in K \tag{10}$$

where the binary variable x_k determine if the route represented by column k is in the final solution S.

Using the model LPF for CVRP, we proposed the Algorithm 9. It takes as input the solutions stored in U from which a set of routes K are generated and used it to solve the LPF. The output is the solution S^*.

Algorithm 9. $SPP(U)$

1: $K = \{r | r$ is a route of $S \in U\}$
2: $S^* \leftarrow$ solve LPF for CVRP using K as the set of routes.
3: **return** S^*

3.7   G-DE-SPP Method

The G-DE-SPP method consists of applying the GRASP hybridized with DE and ELS, then use SPP to build a optimal solution based on the routes generated in the GRASP. The Algorithm 10 shows the method proposed.

4   Experiments, Analysis and Results

In this section, the experimental results to analyze the performance of G-DE-SPP are introduced. G-DE-SPP was coded in C++ and the computational tests

Algorithm 10. G-DE-SPP()

1: $U \leftarrow GRASP()$
2: $S \leftarrow SPP(U)$
3: **return** S

were performed on AMD Ryzen 5 2600 Six-Core Processor at 1.5 GHz CPU and 16 GB RAM, running under a Linux 4.4.0×86_64. The proposed method was applied to 75 instances from four standard CVRP benchmarks: A, B, and P in [4] and M in [6]. G-DE-SPP was ran 10 times independently for each tested instance.

DE, ELS and GRASP parameters were determined by tuning G-DE-SPP using A-n32-k5, A-n80-k10 and B-n78-k10 instances. These instances were chosen since they have different sizes and features so they represent in general the instances in the benchmarks. The tuning was executed using the combination of values $m_{grasp}^i \in \{40, 50, \cdots, 70\}$, $m_{els}^i \in \{4, 5, \cdots, 8\}$, $m_{els}^j \in \{100, 200, 300\}$, $m_{de}^f, m_{de}^c \in [0.6, 0.7, \cdots, 0.9]$, $m_{de}^{np} \in [100, 200, 300]$, $m_{de}^i \in [50, 100, 200]$, and $m_{ran}^r \in [0, 0.1, \cdots, 0.5]$. The best value in the tuning was chosen for use in the computational experiments and set as following. For GRASP, $m_{grasp}^i = 40$. For DE, $m_{ran}^r = 0.1$, $m_{de}^f = 0.8$, $m_{de}^c = 0.9$, $m_{de}^i = 100$ and $m_{de}^{np} = 200$. For ELS, $m_{els}^i = 5$ and $m_{els}^j = 200$. The SPP was modeled and solved by CPLEX 12.8, with a maximum execution time of 5 min.

Tables 1 and 2 present the solutions obtained by G-DE-SPP for the A and B benchmarks and P and M benchmarks, respectively. The **Instance** and **BKS** columns indicate the name of the instance and its best known value. The **BFS**, **AVG**, **SD** and **T(s)** columns represent the best solution found, the average solution value, the standard deviation and the average running time for the solutions obtained by G-DE-SPP. The **DEV** column denotes the deviation and it is defined as DEV=100*(AVG-BKS)/BKS. The values of **BFS** and **AVG** columns are in bold when they are equal to **BKS** column.

For benchmark A, in terms of the best solution found, Table 1 reveals that G-DE-SPP always finds BKS (**AVG** column) in 15 out of 27 instances whereas it found at least one BKS (**BFS** column) in 23 out of 27. Furthermore, G-DE-SPP obtains solutions that are very close to BKS with maximum deviation of 0.54% and an average running time lesser than 1 min and 4 s. The quality of the solutions is confirmed by the small value of standard deviation (**SD** column) with values between 0 and 3.55.

For benchmark B, Table 1 reveals that G-DE-SPP always finds BKS (**AVG** column) in 15 out of 23 instances whereas it found at least one BKS (**BFS** column) in 19 out of 23. In addition, G-DE-SPP achieves solutions that are near to BKS with maximum deviation (**DEV** column) of 0.71% and an average running time not exceeding 3 min, except B-n52-k7 and B-n66-k9 instances. In those cases, most of their time is consumed running the SPP formulation and this happens when the gap between the lower and upper bounds is hard to close.

Table 1. Results for Benchmarks A and B.

Instance	BKS	G-DE-SPP				
		BFS	AVG	SD	DEV	T(s)
A-n32-k5	784	784	784	0.00	0.00	6.21
A-n33-k5	661	661	661	0.00	0.00	7.46
A-n33-k6	742	742	742	0.00	0.00	7.44
A-n34-k5	778	778	778	0.00	0.00	8.44
A-n36-k5	799	799	799	0.00	0.00	8.55
A-n37-k5	669	669	669	0.00	0.00	8.89
A-n37-k6	949	949	949	0.00	0.00	9.93
A-n38-k5	730	730	730	0.00	0.00	9.47
A-n39-k5	822	822	822	0.00	0.00	11.61
A-n39-k6	831	831	831.2	0.63	0.02	11.20
A-n44-k6	937	937	937	0.00	0.00	13.00
A-n45-k6	944	944	947.6	2.55	0.38	12.99
A-n45-k7	1146	1146	1146	0.00	0.00	15.97
A-n46-k7	914	914	914	0.00	0.00	14.04
A-n48-k7	1073	1073	1073	0.00	0.00	17.31
A-n53-k7	1010	1010	1010.1	0.32	0.01	20.42
A-n54-k7	1167	1167	1167.6	0.97	0.05	22.12
A-n55-k9	1073	1073	1073	0.00	0.00	18.79
A-n60-k9	1354	1354	1354	0.00	0.00	30.38
A-n61-k9	1034	1035	1035	0.00	0.10	24.61
A-n62-k8	1288	1291	1293.7	1.83	0.44	50.43
A-n63-k10	1314	1314	1316.7	1.06	0.21	32.84
A-n63-k9	1616	1616	1619.4	3.31	0.21	40.93
A-n64-k9	1401	1404	1408.6	2.88	0.54	35.66
A-n65-k9	1174	1177	1177.5	0.53	0.30	28.92
A-n69-k9	1159	1159	1159.3	0.95	0.03	37.58
A-n80-k10	1763	1763	1769.2	3.55	0.35	63.74
B-n31-k5	672	672	672	0.00	0.00	7.19
B-n34-k5	788	788	788	0.00	0.00	10.47
B-n35-k5	955	955	955	0.00	0.00	9.52
B-n38-k6	805	805	805	0.00	0.00	16.04
B-n39-k5	549	549	549	0.00	0.00	10.20
B-n41-k6	829	829	829	0.00	0.00	12.89
B-n43-k6	742	742	742	0.00	0.00	32.84
B-n44-k7	909	909	909	0.00	0.00	15.36
B-n45-k5	751	751	751	0.00	0.00	19.97
B-n45-k6	678	678	678	0.00	0.00	19.70
B-n50-k7	741	741	741	0.00	0.00	18.32
B-n50-k8	1312	1313	1313	0.00	0.08	152.62
B-n51-k7	1032	1032	1032	0.00	0.00	38.11
B-n52-k7	747	747	747	0.00	0.00	274.28
B-n56-k7	707	707	707	0.00	0.00	67.92
B-n57-k7	1153	1154	1161.2	4.78	0.71	20.20
B-n57-k9	1598	1598	1598	0.00	0.00	24.01
B-n63-k10	1496	1496	1502.1	5.45	0.41	32.51
B-n64-k9	861	861	861.2	0.63	0.02	31.14
B-n66-k9	1316	1316	1316.8	0.92	0.06	327.91
B-n67-k10	1032	1033	1034.2	1.03	0.21	40.74
B-n68-k9	1272	1274	1274.6	0.52	0.20	176.04
B-n78-k10	1221	1221	1221.2	0.42	0.02	171.45

Table 2. Results for Benchmarks P and M.

Instance	BKS	G-DE-SPP				
		BFS	AVG	SD	DEV	T(s)
P-n16-k8	**450**	**450**	**450**	0.00	0.00	2.65
P-n19-k2	**212**	**212**	**212**	0.00	0.00	3.26
P-n20-k2	**216**	**216**	**216**	0.00	0.00	3.21
P-n21-k2	**211**	**211**	**211**	0.00	0.00	2.92
P-n22-k2	**216**	**216**	**216**	0.00	0.00	3.35
P-n22-k8	**603**	**603**	**603**	0.00	0.00	3.79
P-n23-k8	**529**	**529**	**529**	0.00	0.00	4.18
P-n40-k5	**458**	**458**	**458**	0.00	0.00	11.56
P-n45-k5	**510**	**510**	**510**	0.00	0.00	15.20
P-n50-k10	**696**	**696**	**696**	0.00	0.00	16.32
P-n50-k7	**554**	**554**	**554**	0.00	0.00	17.55
P-n50-k8	**631**	**631**	632.1	1.60	0.17	16.90
P-n51-k10	**741**	**741**	**741**	0.00	0.00	16.27
P-n55-k10	**694**	**694**	**694**	0.00	0.00	21.90
P-n55-k15	**989**	**989**	**989**	0.00	0.00	18.18
P-n55-k7	**568**	**568**	568.5	0.85	0.09	25.00
P-n60-k10	**744**	**744**	**744**	0.00	0.00	24.89
P-n60-k15	**968**	**968**	**968**	0.00	0.00	22.56
P-n65-k10	**792**	**792**	**792**	0.00	0.00	30.03
P-n70-k10	**827**	**827**	827.1	0.32	0.01	40.61
P-n76-k4	**593**	606	608.6	1.84	2.63	42.52
P-n76-k5	**627**	630	635.8	3.82	1.40	44.96
P-n101-k4	**681**	690	695.7	3.50	2.16	79.50
M-n101-k10	**820**	**820**	**820**	0	0.00	158.12
M-n121-k7	**1034**	1035	1036.7	1.15	0.26	289.52
M-n151-k12	**1015**	1028	1036.9	7.99	2.16	1029.83
M-n200-k17	**1275**	1301	1332.5	23.85	4.51	1662.46

Once more, the maximum standard deviation (**SD** column) of 5.45 asserts the high quality of solutions found.

For benchmark P, with regard to the solutions found, Table 2 shows that G-DE-SPP always finds BKS (**AVG** column) in 17 out of 23 instances whereas it found at least one BKS (**BFS** column) in 20 out of 23. Furthermore, G-DE-SPP get solutions that are near to BKS with maximum deviation (**DEV** column) of 2.63% and an average running time lesser than 80 s. Finally, the maximum standard deviation (**SD** column) obtained was 3.50.

For benchmark M, Table 2 reveals that G-DE-SPP only finds BKS in 1 out of 4 instances. Moreover, G-DE-SPP obtains solutions that are near to BKS with deviation lower or equal to 4.51% and an average time of up to 5 min for the two smaller instances and 30 min or lesser for the greatest instances. However, for the M-n200-k17 instance the standard deviation was 23.85 which shows the algorithm found solutions more spread and not so close the BKS.

To further analyze the solutions of G-DE-SPP, a statistical analysis was conducted using the obtained results. As the tested variables of algorithms (i.e. solutions) are normally not following a distribution of Gaussian type, a non-parametric test should be used such as the Mann–Whitney U test [3]. This test was conducted using the obtained values shown in **BKS** and **AVG** columns. The null hypothesis H_0 is defined as the mean ranks of each pair values of **BKS** and **AVG** are the same. By proceeding this way, the test shows that the p-value is 0.901, which is greater than 0.05. According to this result, the null hypothesis H_0 of Mann-Whitney U test is accepted. Therefore, there are not significant differences between the average solutions of G-DE-SPP and the best known solutions (**BKS**).

5   Conclusion and Future Works

This paper proposed a new hybridization, called G-DE-SPP, of evolutionary algorithms, GRASP and mathematical model to solve Capacitated Vehicle Routing Problem based on the route-first-cluster-second approach. Two types of local search algorithms, 2-opt and crossover, were integrated into the Evolution Local Search (ELS) in local phase of GRASP. Moreover, to enhance the solution's quality and speed up the convergence, the Differential Evolution (DE) was used to create the initial solution of the GRASP. Finally, a set-partitioning problem (SPP) was applied to create a final solution. Furthermore, the experimental results show that our proposed method was effective to solve instances found in the literature.

For future work, our G-DE-SPP method can be applied to solve other VRPs such as CVRP with time windows or Dial-a-ride (DARP). Additionally, G-DE-SPP can be further developed using different split methods and hybridizations.

References

1. Afsar, H.M., Prins, C., Santos, A.C.: Exact and heuristic algorithms for solving the generalized vehicle routing problem with flexible fleet size. Int. Trans. Oper. Res. **21**(1), 153–175 (2014)
2. Agarwal, Y., Mathur, K., Salkin, H.M.: A set-partitioning-based exact algorithm for the vehicle routing problem. Networks **19**(7), 731–749 (1989)
3. Alba, E., Nakib, A., Siarry, P.: Metaheuristics for Dynamic Optimization. Springer, Heidelberg (2013). https://doi.org/10.1007/978-3-642-30665-5
4. Augerat, P., Belenguer, J.M., Benavent, E., Corberán, A., Naddef, D., Rinaldi, G.: Computational results with a branch and cut code for the capacitated vehicle routing problem, vol. 34. IMAG (1995)

5. Beasley, J.E.: Route first-cluster second methods for vehicle routing. Omega **11**(4), 403–408 (1983)
6. Christofides, N., Eilon, S.: An algorithm for the vehicle-dispatching problem. J. Oper. Res. Soc. **20**(3), 309–318 (1969)
7. Feo, T.A., Resende, M.G.C.: A probabilistic heuristic for a computationally difficult set covering problem. Oper. Res. Lett. **8**(2), 67–71 (1989)
8. Gendreau, M., Laporte, G., Vigo, D.: Heuristics for the traveling salesman problem with pickup and delivery. Comput. Oper. Res. **26**(7), 699–714 (1999)
9. Goldberg, A., Radzik, T.: A heuristic improvement of the bellman-ford algorithm. Stanford Univ CA Dept. of Computer Science, Technical report (1993)
10. Irnich, S., Funke, B., Grünert, T.: Sequential search and its application to vehicle-routing problems. Comput. Oper. Res. **33**(8), 2405–2429 (2006)
11. Prins, C.: A grasp× evolutionary local search hybrid for the vehicle routing problem. In: Bio-inspired algorithms for the vehicle routing problem, pp. 35–53. Springer, Heidelberg (2009). https://doi.org/10.1007/978-3-540-85152-3_2
12. Snyder, L.V., Daskin, M.S.: A random-key genetic algorithm for the generalized traveling salesman problem. Eur. J. Oper. Res. **174**(1), 38–53 (2006)
13. Storn, R., Price, K.: Differential evolution-a simple and efficient heuristic for global optimization over continuous spaces. J. Glob. Optim. **11**(4), 341–359 (1997)
14. Toth, P., Vigo, D.: Vehicle Routing: Problems, Methods, and Applications. SIAM (2014)
15. Ulusoy, G., et al.: The fleet size and mix problem for capacitated arc routing. Eur. J. Oper. Res. **22**(3), 329–337 (1985)
16. Ursani, Z., Essam, D., Cornforth, D., Stocker, R.: Localized genetic algorithm for vehicle routing problem with time windows. Appl. Soft Comput. **11**(8), 5375–5390 (2011)
17. Wang, C.H., Lu, J.Z.: A hybrid genetic algorithm that optimizes capacitated vehicle routing problems. Exp. Syst. Appl. **36**(2), 2921–2936 (2009)
18. Wolf, S., Merz, P.: Evolutionary local search for the super-peer selection problem and the p-hub median problem. In: Bartz-Beielstein, T., et al. (eds.) HM 2007. LNCS, vol. 4771, pp. 1–15. Springer, Heidelberg (2007). https://doi.org/10.1007/978-3-540-75514-2_1
19. Zachariadis, E.E., Kiranoudis, C.T.: A strategy for reducing the computational complexity of local search-based methods for the vehicle routing problem. Comput. Oper. Res. **37**(12), 2089–2105 (2010)
20. Zhu, W., Qin, H., Lim, A., Wang, L.: A two-stage tabu search algorithm with enhanced packing heuristics for the 3l-cvrp and m3l-cvrp. Comput. Oper. Res. **39**(9), 2178–2195 (2012)

An Evolutionary Algorithm for Learning Interpretable Ensembles of Classifiers

Henry E.L. Cagnini[1]([⊠]) [iD], Alex A. Freitas[2] [iD], and Rodrigo C. Barros[1] [iD]

[1] School of Technology, PUCRS, Porto Alegre, Brazil
henry.cagnini@edu.pucrs.br, rodrigo.barros@pucrs.br
[2] Computing School, University of Kent, Canterbury, UK
a.a.freitas@kent.ac.uk

Abstract. Ensembles of classifiers are a very popular type of method for performing classification, due to their usually high predictive accuracy. However, ensembles have two drawbacks. First, ensembles are usually considered a 'black box', non-interpretable type of classification model, mainly because typically there are a very large number of classifiers in the ensemble (and often each classifier in the ensemble is a black-box classifier by itself). This lack of interpretability is an important limitation in application domains where a model's predictions should be carefully interpreted by users, like medicine, law, etc. Second, ensemble methods typically involve many hyper-parameters, and it is difficult for users to select the best settings for those hyper-parameters. In this work we propose an Evolutionary Algorithm (an Estimation of Distribution Algorithm) that addresses both these drawbacks. This algorithm optimizes the hyper-parameter settings of a small ensemble of 5 interpretable classifiers, which allows users to interpret each classifier. In our experiments, the ensembles learned by the proposed Evolutionary Algorithm achieved the same level of predictive accuracy as a well-known Random Forest ensemble, but with the benefit of learning interpretable models (unlike Random Forests).

Keywords: Classification · Evolutionary algorithms · Ensemble learning · Machine learning · Supervised learning

1 Introduction

The classification task of machine learning consists of training predictive models for decision-making purposes [31]. Traditionally, classification research has focused mainly on the learned model's predictive accuracy, but model interpretability by users is currently a very active and important topic [6], especially in areas such as medicine, credit scoring, bioinformatics, and churn prediction [12]. Model interpretability is particularly critical in scenarios where models can lead to life-or-death decisions (such as in medicine), or influence decisions that can put several lives at risk, such as the use of recommendation algorithms

© Springer Nature Switzerland AG 2020
R. Cerri and R. C. Prati (Eds.): BRACIS 2020, LNAI 12319, pp. 18–33, 2020.
https://doi.org/10.1007/978-3-030-61377-8_2

in a nuclear power plant [12]. Model interpretability is also often required by law, a major example being the European Union's *Data Protection Regulation*, which includes a "right to explanation" [13].

One field of machine learning that could benefit from interpretability is classification with ensemble learning. Ensembles are sets of classifiers which, when combined, usually perform better than a single strong model, such as when comparing a Random Forest ensemble [4] with a single decision tree learned by C4.5 [2,29]. Ensembles often have hundreds or thousands of models, which greatly hinder their interpretability by human users. Moreover, ensembles often consist of black box base models (e.g. neural networks or support vector machines) that prevent any direct interpretation of their reasoning. Tackling these problems involves learning a small ensemble, consisting of a few directly interpretable models, so that users can interpret each of the models in the ensemble. This is the main problem addressed in this work.

The second problem addressed in this work is that selecting the best setting (or configuration) of hyper-parameters for each base learner in an ensemble is a difficult task *per se* [10,32], which involves testing a very large number of candidate hyper-parameter settings in order to find the best setting for the dataset at hand. Auto-ML (Automated Machine Learning) has recently gained attention due to its capacity of relieving the end user from a manual optimization of algorithms' hyper-parameters, which can be repetitive, tiresome, and often requires advanced domain-specific knowledge [10,28,33].

One way to perform Auto-ML is to employ a population-based algorithm, which explores several regions in the solution space in parallel, and adapts its search depending on the quality of solutions found in those regions. Hence, evolutionary algorithms seem to be a natural choice for the Auto-ML task of optimizing the settings of ensembles' hyper-parameters [14,20,23,33], due to performing a global search in the solution space. Among several types of evolutionary algorithms, we propose an Estimation of Distribution Algorithm (EDA) to evolve an ensemble of interpretable classifiers.

The main difference between EDAs [24] and Genetic Algorithms (GA) [17] is that while GAs implicitly propagate characteristics of good solutions throughout evolution (by carrying on high-quality individuals from one generation to another), EDAs do this explicitly, by encoding those characteristics in a probabilistic graphical model (GM) [16,27].

The rest of this paper is organized as follows. Section 2 describes our proposed method. Sections 3 and 4 present the experimental setup and experimental results, respectively. Section 5 discusses related work. Section 6 presents the conclusions and future research directions.

2   The Proposed Estimation of Distribution Algorithm (EDA) for Evolving Ensembles

EDAs evolve a probabilistic graphic model of candidate solutions, so that candidate solutions (individuals) are sampled from that model and evaluated at each generation. In general, an EDA consists of three stages performed at each generation: (a) sampling of new individuals (candidate solutions) from the probabilistic graphic model; (b) evaluation of the new individuals' performance; and (c) updating of the probabilistic graphic model, based on the best individuals selected from the current generation. Importantly, EDAs avoid the need for specifying genetic operators like crossover and mutation (and their corresponding probabilities of application). That is, instead of generating new individuals by applying genetic operators to selected individuals, they generate new individuals by sampling from the current probabilistic graphic model, which captures the main characteristics of the best individuals selected (based on fitness) along the evolutionary process.

Among several EDA types, we chose PBIL: Probabilistic Incremental Learning [3]. The main characteristic of PBIL is that it assumes independence between variables in the probabilistic graphical model. Although this has the disadvantage of ignoring interactions among variables, it has an important advantage in the context of our task of evolving an ensemble of classifiers: it makes PBIL much more computationally efficient by comparison with other EDA types that consider complex variable interactions – whilst still allowing PBIL to learn ensembles with good predictive accuracy, as shown later.

Another aspect of PBIL is the use of a learning rate α hyper-parameter for updating probabilities in the graphical model, making this process smoother. Take for example two initial probabilities for a binary variable V, $P(V = 0) = 0.5$ and $P(V = 1) = 0.5$, and a learning rate of 0.3. Assume only two individuals are selected to update the graphic model's probabilities, and both have $V = 0$. In this extreme case, an EDA without learning rate would update V so that it would be $P(V = 0) = \frac{2}{2} = 1$ and $P(V = 1) = \frac{0}{2} = 0$ in the next generation. However, using a learning rate, the new probabilities for V are $P(V = 0) = (1 - 0.3) \times 0.5 + 0.3 \times \frac{2}{2} = 0.65$ and $P(V = 1) = (1 - 0.3) \times 0.5 + 0.3 \times \frac{0}{2} = 0.35$. Section 2.3 discusses in more detail how probabilities are updated.

PBIL keeps track of the best individual found so far in a variable φ. At the end of a PBIL run, the returned solution can be the best individual stored in φ or the best individual in the last generation (these two approaches are compared later).

2.1   Individuals (Candidate Solutions)

Each individual is an ensemble, composed of five base models (each learned by a different type of base learner) and an aggregation policy. Regarding base learners, we chose the ones that can generate readily interpretable models [12,18,26]. The recent literature on classification focuses mainly on producing classifiers with ever-increasing predictive performance, with little attention devoted to interpretability [13]. For instance, deep learning classifiers, which have received great attention lately due to obtaining high predictive accuracy in image tasks, are very difficult to interpret [13], with interested researchers shifting the focus from interpreting the models themselves to interpreting their predictions [22].

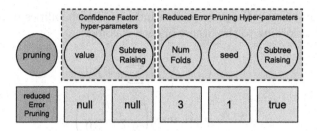

Fig. 1. An example individual in PBIL. Note that, although PBIL assumes probabilistic independence between variables, some values are dependent on others.

The five base learners employed are two decision-tree induction algorithms (C4.5 [29] and CART [5]); two rule induction algorithms (RIPPER [7] and PART [11]); and a Decision Table algorithm [19]. We use these algorithms' implementations in the well-known Weka Toolkit [15]. For the rest of this paper, we will refer to them by their Weka names: J48 for C4.5, SimpleCart for CART, JRip for RIPPER, PART, and Decision Table.

An individual is encoded as an array, where each position denotes a variable, and each value denotes the assigned value for that variable. Some variables may not have any value, because they are not used by an individual. Figure 1 depicts a portion of an individual's array, regarding some variables of its J48 classifier. J48 has three options for tree pruning: *reduced error pruning*, *confidence factor*, and *unpruned*. For this example individual, *reduced error pruning* is used. For this reason, there is no need to set hyper-parameters of the *confidence factor* strategy, which are then set to *null*.

Aggregators. An aggregator is a method responsible for finding a consensus among votes from base models. Consider a three-dimensional probability matrix **P**, of dimensions (B, N, C) – respectively the number of base classifiers in the ensemble, number of instances, and number of classes. The objective of an aggregator is to transform this three-dimensional matrix into a unidimensional array of length N, where each position has the predicted class for each instance.

We use two types of aggregators: majority voting and weighted aggregators. The probabilistic majority voting aggregator uses the fusion function described in [21, p. 150]:

$$\rho_c^{(j)} = \sum_{i=1}^{B} P_c^{(i,j)} \tag{1}$$

$$h(X^{(j)}) = \arg\max_{c \in C} \left(\frac{\rho_c^{(j)}}{\sum_{k=1}^{C} \rho_k^{(j)}} \right) \tag{2}$$

where $\rho_c^{(j)}$ is the sum of the probabilities that the j-th instance has the c-th class, over all B classifiers, and C is the number of classes. The weighted aggregator is similar to majority voting, except that individual probabilities from classifiers are weighted according to the fitness of each classifier:

$$\rho_c^{(j)} = \sum_{i=1}^{B} \psi^{(i)} P_c^{(i,j)} \tag{3}$$

$$h(X^{(j)}) = \arg \max_{c \in C} \left(\frac{\rho_c^{(j)}}{\sum_{k=1}^{C} \rho_k^{(j)}} \right) \tag{4}$$

where $\psi^{(i)}$ is the fitness value of individual $S^{(i)}$.

2.2   Fitness Evaluation

At the start of the evolutionary process, PBIL receives a training set. This set is splitted into five subsets, which are used to compute each individual's fitness by performing an internal 5-fold stratified cross validation (SCV). By keeping the subsets constant throughout all evolutionary process, we allow direct comparisons between individuals from different generations. The fitness function is the Area Under the Receiving Operator Characteristic (ROC) curve (AUC) [8] – a popular predictive accuracy measure.

AUC values are within $[0, 1]$, with the value 0.5 representing the predictive accuracy of random predictions in the case of binary-class problems. In this work, regardless of the number of classes in the dataset, we calculate one AUC for each class, and then average the AUC among all classes. Hence, the fitness of an individual is actually a mean of means: first, the mean AUC among all classes, for a given fold; then, the mean AUC among all five internal folds. Figure 2 depicts the fitness calculation procedure.

```
1: function COMPUTE_FITNESS(X, y, C)
2:     train ← (generate_train_subsets(y))
3:     val ← (generate_validation_subsets(y))
4:     Ψ ← (0|i = 1, 2, ..., |S|)
5:     for i = 1, 2, ..., |S| do
6:         for k = 1, ..., 5 do
7:             S^(i) ← build_model(X^(train^(k)), y^(train^(k)))
8:             P^(i) ← predict(S^(i), X^(val^(k)))
9:             ψ ← 1/5C Σ_{c=1}^{C} AUC(P_c^(i), (y^(j) = c|j ∈ val^(k)))
10:            ψ^(i) ← ψ^(i) + ψ
11:        )
12:    return Ψ
```

Fig. 2. Pseudo-code used for calculating fitness.

2.3 PBIL's Probabilistic Graphical Model

At each generation, new individuals are sampled from the probabilistic graphical model (GM), and the best individuals will update GM's probabilities. Recall that PBIL assumes that the variables in the GM are independent, although we know (as shown in Fig. 1) that there are some dependencies. However, this does not prevent PBIL from finding good-performing solutions; analogously to the overall good performance of the Naïve Bayes classifier, which also assumes that attributes are independent [30].

The sampling procedure is based on hierarchical relationships among the variables representing hyper-parameters in PBIL's GM, as shown in Fig. 3, where the top-level variables are hyper-parameters of base learners that will activate or deactivate the sampling of other variables/hyper-parameters at a lower level. When sampling a new individual, higher-level variables are sampled first, and their descendants are sampled next. Using J48 as example, the variables for this algorithm are *useLaplace, minNumObj, useMDLcorrection, collapseTree, doNotMakeSplitPointActualValue, binarySplits*, and *pruning*. Since none of these variables have any descendent variable, with the exception of *pruning*, the sampling proceeds to choose which type of pruning will be used by J48, and depending on the chosen option, it samples the variables descendent to that option. Unused variables are set to *null*. Once all pertinent variables are sampled, their values are fed to the base classifier constructor, which will in turn generate the model. Figure 3 depicts the variables in PBIL's GM.

Initial Values. There are two types of variables in PBIL's GM: 48 discrete and 2 continuous variables. Discrete variables were first introduced in the original PBIL work [3]. We use the EDA ability of biasing probabilities to increase by 10% the probability to sample values that are the base learner's default in Weka. For all other values, we set uniform probabilities. For instance, for J48's *numFolds*, the default value 3 folds has probability 20%, while each other value in $\{2, 4, 5, 6, 7, 8, 9, 10\}$ has probability 10%. Exceptionally for variable *evaluationMeasure* of Decision Table, value *auc* has a 50% probability of being sampled. We do this to increase the chances that a base learner is using the same metric used as fitness function, which in this work is the AUC.

For continuous variables, we use unidimensional Gaussian distributions. The mean is the default Weka value for the hyper-parameter, and the standard deviation was chosen in a way that borderline values have at least 10% chance of being sampled. Values outside valid range are clipped to the closest valid value. The range of valid values was inferred by inspecting Weka's source code. The list of of variables and its values is present in the source-code of our method[1].

Updating PBIL's GM. The updating of the variables' probabilities is dependent on their type. If a variable is discrete, the update follows the scheme known

---

[1] Available at https://github.com/henryzord/PBIL.

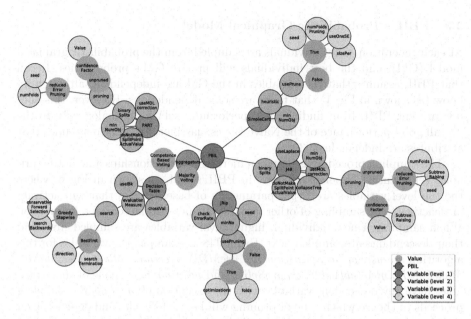

Fig. 3. Graphical Model used by PBIL. Edges denote an implicit correlation, since no probabilistic correlation is designed. Borderless nodes denote values that were sampled from the variable in the above level (i.e. not variables).

as PBIL-iUMDA [33,34], shown in Eq. 5:

$$p_{g+1}(V_j = v) = (1 - \alpha) \times p_g(V_j = v) + \alpha \times p_{\Phi,g}(V_j = v) \qquad (5)$$

where $p_g(V_j = v)$ is the probability that variable V_j assumes discrete value v in the g-th generation (estimated by the proportion of observed occurrences of value v for variable V_j among all individuals in that generation), α is the learning rate, and $p_{\Phi,g}(V_j = v)$ is the proportion of occurrences of value v for V_j in the set of individuals Φ which were selected (based on fitness) at the g-th generation. This process is iterated over all values of a discrete variable. Note that when computing $p_g(V_j = v)$ and $p_{\Phi,g}(V_j = v)$, if some individuals do not have any value set for variable V_j, their *null* values are discarded and do not contribute at all to the updating of probabilities for V_j's values.

Equation 5 was adapted to deal with continuous variables, which encode the mean and the standard deviation of a normal distribution, as follows. The mean of the normal distribution is updated by

$$\mu_{g+1}(V_j) = \mu_g(V_j) + \alpha \times (\mu_g(V_j) - \mu_{\Phi,g}(V_j)) \qquad (6)$$

where $\mu_g(V_j)$ is the mean of the normal distribution of the j-th variable V_j in the g-th generation, α is the learning rate, and $\mu_{\Phi,g}(V_j)$ is the observed mean for the variable V_j in the set of individuals Φ selected at the g-th generation, again considering only individuals where the variable V_j was used.

The standard deviation is decreased as follows:

$$\sigma_{g+1}(V_j) = \sigma_g(V_j) - \frac{\sigma_1(V_j)}{G} \tag{7}$$

where $\sigma_g(V_j)$ is the standard deviation of the j-th variable V_j in the g-th generation, $\sigma_1(V_j)$ the initial standard deviation for variable V_j, and G is the number of generations.

2.4   Early Stop and Termination

If the fitness of the best individual does not improve more than ϵ in ε generations, we assume that PBIL is overfitting the training data, and terminate the run of the algorithm. In our experiments, we use $\epsilon = 5 \times 10^{-4}$ and $\varepsilon = 10$.

At the end of the evolutionary process, we report the best individual from the last generation as the final solution.

2.5   Complexity Analysis

Assume $T(train)$ to be the time to train an ensemble, and $T(fitness)$ to be the time to assert the fitness of said ensemble. At every generation S new ensembles are generated. This process is repeated at most G times (assuming that the early stop mechanism of the previous section is not triggered). This procedure has complexity $GS \times (T(train) + T(fitness))$.

Sampling and updating the graphical model are procedures directly dependent on the number of variables $|V|$. Variables need first to be initialized with default values, for later sampling and update. Variables are sampled S times every generation, and are updated based on the number of fittest individuals, $|\Phi|$. For each variable, we iterate over all of its values, but we assume the number of its values not to be significant – discrete variables have between 2 and 10 values, with 4 as average; continuous variables count as 2 values, i.e. mean and standard deviation of normal distributions. From this analysis we have $|V| \times (1 + G(S + |\Phi|))$. Thus, the overall complexity of training the proposed PBIL is

$$O(GS \times (T(train) + T(fitness)) + |V| \times G(S + |\Phi|)) \tag{8}$$

3   Experimental Setup

3.1   PBIL's Hyper-parameter Optimization

In order to find the best configuration to run PBIL, we perform a grid-search for optimizing five of its hyper-parameters, using eight datasets, hereafter called *parameter-optimization* datasets, described in Table 1. We measure PBIL's AUC on each dataset using a well-known 10-fold cross validation procedure. We emphasize that these datasets were used only for PBIL's hyper-parameter optimization, i.e., they were not used to compare PBIL with baseline algorithms, thus avoiding over-optimistic measures of predictive performance.

We optimize 5 hyper-parameters, with the following range of values: population size: {75, 150}; number of generations: {100, 200}; learning rate: {0.3, 0.5, 0.7}; selection share {0.3, 0.5}; and whether the type of solution returned is the best individual from the last generation or the best individual produced across all generations. Thus, from the 48 combinations of hyper-parameter values, we found the combination that provided the best average AUC across all datasets to be: population size $|S| = 75$, number of generations $G = 200$, learning rate $\alpha = 0.7$, share (proportion) of selected individuals $|\Phi|/|S| = 0.5$, and type of solution returned = best individual from last generation. We use this configuration for conducting further experiments.

3.2   Baseline Algorithms

We compare PBIL with other two algorithms: a baseline ensemble and Random Forest. The baseline ensemble consists of the five base classifiers from PBIL (namely J48, CART, JRip, PART, and Decision Table) with their default hyper-parameter configuration (according to Weka), and a simple majority voting scheme as aggregation policy. The intention of using this baseline algorithm is to check if there is a difference between simply using an ensemble of classifiers, with the simplest voting aggregation policy (i.e. majority voting), and optimizing their hyper-parameter configuration with PBIL.

Random Forest [4] is a well-known ensemble algorithm, and in general it is among the best classification methods regarding predictive performance [9]. A random forest ensemble is solely composed of decision trees. Each decision tree is learned from a different subset of N instances, randomly sampled with replacement from the training set. For each internal node in each tree, a subset of M attributes is randomly sampled without replacement, and the attribute that minimizes the local class impurity is selected as splitting criterion. This process is repeated recursively until no further split improves the impurity metric, when nodes are then turned into leaves.

Random forests usually require a large number of trees in the ensemble to achieve good predictive performance. Also, despite using decision trees, the ensemble as a whole is not directly interpretable, since there are a very large number of trees. Even if the number of trees were small, interpreting each tree would still be problematic due to the large degree of randomness involved in learning each tree. That randomness is necessary to provide diversity to the ensemble, which improves its predictive accuracy, but it hinders interpretability. There are indirect approaches to interpret random forests, using variable importance measures to rank the variables based on their importance in the model, but such measures are out of the scope of this paper.

We also performed a grid-search for optimizing 3 hyper-parameters of Random Forest, with their following ranges of values: number of trees in the forest: {100, 200, 300, 400, 500}; whether to randomly break ties between equally attractive attributes at each tree node, or to simply use the attribute with the smallest index; and maximum tree depth: {0 (no limit), 1, 2, 3, 4}. Hence, Random Forests and PBIL had about the same number of configurations tested by the grid search (50 and 48), and both had their configurations optimized in the same 8 parameter-optimization datasets shown in Table 1, to be fair. We found the best combination to be: number of iterations = 300, do not break ties randomly, and max tree depth = 4. We leave the other two hyper-parameters, the number of instances in the bag for learning each decision tree, and the number of sampled attributes at each tree node, at their default Weka values, respectively N and $\log_2(M-1)+1$.

3.3 Datasets

We use a different set of 9 datasets, described in Table 1, for comparing the predictive performance of the tested algorithms. All datasets, including the ones used for hyper-parameter optimization, were collected from KEEL[2] [1] and the UCI Machine Learning repository[3] [25].

Table 1. Datasets used in this work.

Dataset	Instances	Attributes Total	Categorical	Numeric	Classes
Hyper-parameter optimization datasets					
Australian	690	14	6	8	2
Bupa	345	6	0	6	2
Contraceptive	1473	9	0	9	3
Flare	1066	11	11	0	6
German	1000	20	13	7	2
Pima	768	8	0	8	2
Vehicle	846	18	0	18	4
Wisconsin	699	9	0	9	2
Predictive performance assessment datasets					
Balance-scale	625	4	0	4	3
Blood-transfusion	748	4	0	4	2
Credit-approval	690	15	9	6	2
Diabetic	1151	19	3	16	2
Hcv-Egypt	1385	28	9	19	4
Seismic-bumps	2584	18	4	14	2
Sonar	208	60	0	60	2
Turkiye	5820	32	32	0	13
Waveform	5000	40	0	40	3

---

[2] Available at https://sci2s.ugr.es/keel/datasets.php.
[3] Available at https://archive.ics.uci.edu/ml/datasets.

4   Experimental Results

For each algorithm and each dataset, we run 10 times a 10-fold cross validation procedure, and report the mean unweighted area under the ROC curve among the 10 executions. The results are shown in Table 2.

Table 2. Area under the ROC curve and standard deviations for compared algorithms. Best result for each dataset is shown in bold font.

Dataset	Random forest	Baseline ensemble	The proposed PBIL
Balance-scale	0.8441 ± 0.05	$\mathbf{0.8766 \pm 0.01}$	0.8560 ± 0.00
Blood-transfusion	$\mathbf{0.7354 \pm 0.03}$	0.7335 ± 0.00	0.6742 ± 0.00
Credit-approval	$\mathbf{0.9358 \pm 0.02}$	0.9270 ± 0.00	0.9267 ± 0.00
Diabetic	0.7307 ± 0.04	0.7370 ± 0.01	$\mathbf{0.7674 \pm 0.00}$
Hcv-Egypt	0.5073 ± 0.05	0.4850 ± 0.01	$\mathbf{0.5167 \pm 0.00}$
Seismic-bumps	$\mathbf{0.7823 \pm 0.07}$	0.7715 ± 0.01	0.7553 ± 0.00
Sonar	0.9214 ± 0.08	0.8612 ± 0.01	$\mathbf{0.9356 \pm 0.00}$
Turkiye	$\mathbf{0.8549 \pm 0.01}$	0.8542 ± 0.00	0.8213 ± 0.00
Waveform	0.9562 ± 0.00	0.9502 ± 0.00	$\mathbf{0.9670 \pm 0.00}$

Regarding predictive performance, PBIL and Random forests obtained overall the best results, each with the highest AUC value in 4 datasets. The baseline method obtained the highest value in only one dataset. The largest difference in performance was observed in the blood-transfusion dataset, where the baseline and the Random Forest obtained an AUC value about 6% higher than the AUC of PBIL. In the other datasets, the differences of AUC values among the three methods was relatively small, about 3% or less in general. We believe this is due to the skewed nature of the class distribution in the blood-transfusion dataset.

In addition, the ensembles learned by PBIL and the baseline method have the advantage of consisting of only 5 interpretable base classifiers; so they are directly interpretable by users, unlike Random Forests (as discussed earlier).

Figures 4 and 5 show an ensemble learned by PBIL from the sonar dataset, as an example of such ensembles' interpretability. The models learned by J48 and SimpleCART are both small (with 3 and 13 nodes) and consistently identify Band11 as the most relevant variable in their root nodes. The rule lists learned by JRip and PART are also small, with 5 and 8 rules (most being short rules). The decision table is not so short, with 25 rows, but the fact that all rows refer to the same selected attributes and in the same order (unlike decision trees and rule sets) improves interpretability by users [12].

5   Related Work

Several evolutionary algorithms have been recently proposed for evolving ensembles of classifiers. In [33], another PBIL version was proposed to select the best

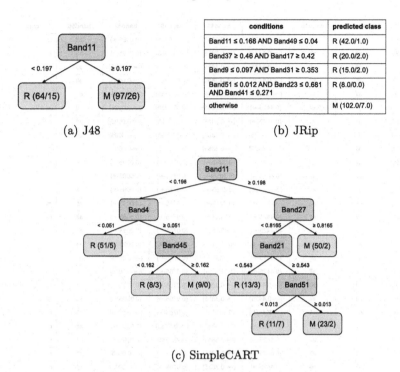

conditions	predicted class
Band11 ≤ 0.168 AND Band49 ≤ 0.04	R (42.0/1.0)
Band37 ≥ 0.46 AND Band17 ≥ 0.42	R (20.0/2.0)
Band9 ≤ 0.097 AND Band31 ≥ 0.353	R (15.0/2.0)
Band51 ≤ 0.012 AND Band23 ≤ 0.681 AND Band41 ≤ 0.271	R (8.0/0.0)
otherwise	M (102.0/7.0)

(a) J48   (b) JRip

(c) SimpleCART

conditions	predicted class
Band11 ≤ 0.198 AND Band52 ≤ 0.0205 AND Band5 ≤ 0.0695 AND Band10 ≤ 0.1665 AND Band7 > 0.0415	R (40.0)
Band47 > 0.063 AND Band37 ≤ 0.48 AND Band18 ≤ 0.914 AND Band49 > 0.0285	M (64.0)
Band54 ≤ 0.0225 AND Band45 > 0.2745 AND Band2 ≤ 0.044	M (9.0)
Band54 ≤ 0.0225 AND Band8 > 0.0655 AND Band27 ≤ 0.846 AND Band28 > 0.3585 AND Band4 ≤ 0.109 AND Band3 ≤ 0.0655	R (25.0)
Band8 > 0.0625 AND Band12 > 0.154 AND Band54 > 0.0105	M (17.0)
Band8 ≤ 0.104	R (14.0)
Band17 > 0.4445	R (11.0/3.0)
otherwise	M (7.0)

(d) PART

Fig. 4. Four of the five base classifiers from the best individual of PBIL for a given run of the sonar dataset. *type* is the class variable (with class labels Rock (R) and Metal (M)), and broadly speaking the features represent the echo returned from hitting rock and metal objects with different frequencies of audio waves.

combination of ensemble method (e.g. bagging, boosting, etc), base learners (e.g. neural networks, SVMs, decision trees, etc.) and their hyper-parameter settings for a given dataset. However, that work focused only on predictive accuracy, so that their learned ensembles are in general non-interpretable (due to being very large and often consisting of non-interpretable classifiers), unlike the ensembles learned in this current work.

band11	band16	band19	band36	band45	band48	band52	band56	type
(0.198-∞)	all	all	(-∞-0.4425]	(-∞-0.3855]	(-∞-0.0755]	(0.0095-∞)	all	m
(-∞-0.198]	all	all	(-∞-0.4425]	(-∞-0.3855]	(-∞-0.0755]	(0.0095-∞)	all	r
(0.198-∞)	all	all	(-∞-0.4425]	(-∞-0.3855]	(0.0755-∞)	(-∞-0.0095]	all	m
(-∞-0.198]	all	all	(-∞-0.4425]	(-∞-0.3855]	(0.0755-∞)	(-∞-0.0095]	all	r
(0.198-∞)	all	all	(-∞-0.4425]	(0.3855-∞)	(-∞-0.0755]	(-∞-0.0095]	all	r
(0.198-∞)	all	all	(0.4425-∞)	(-∞-0.3855]	(-∞-0.0755]	(-∞-0.0095]	all	m
(-∞-0.198]	all	all	(0.4425-∞)	(-∞-0.3855]	(-∞-0.0755]	(-∞-0.0095]	all	r
(0.198-∞)	all	all	(-∞-0.4425]	(-∞-0.3855]	(-∞-0.0755]	(-∞-0.0095]	all	r
(-∞-0.198]	all	all	(-∞-0.4425]	(-∞-0.3855]	(-∞-0.0755]	(-∞-0.0095]	all	r
(-∞-0.198]	all	all	(0.4425-∞)	(0.3855-∞)	(0.0755-∞)	(0.0095-∞)	all	r
(0.198-∞)	all	all	(0.4425-∞)	(0.3855-∞)	(0.0755-∞)	(0.0095-∞)	all	m
(-∞-0.198]	all	all	(-∞-0.4425]	(0.3855-∞)	(0.0755-∞)	(0.0095-∞)	all	m
(0.198-∞)	all	all	(-∞-0.4425]	(0.3855-∞)	(0.0755-∞)	(0.0095-∞)	all	m
(-∞-0.198]	all	all	(0.4425-∞)	(-∞-0.3855]	(0.0755-∞)	(0.0095-∞)	all	r
(0.198-∞)	all	all	(0.4425-∞)	(-∞-0.3855]	(0.0755-∞)	(0.0095-∞)	all	r
(0.198-∞)	all	all	(0.4425-∞)	(0.3855-∞)	(-∞-0.0755]	(0.0095-∞)	all	r
(0.198-∞)	all	all	(0.4425-∞)	(0.3855-∞)	(0.0755-∞)	(-∞-0.0095]	all	m
(-∞-0.198]	all	all	(0.4425-∞)	(0.3855-∞)	(0.0755-∞)	(-∞-0.0095]	all	r
(-∞-0.198]	all	all	(-∞-0.4425]	(-∞-0.3855]	(0.0755-∞)	(0.0095-∞)	all	m
(0.198-∞)	all	all	(-∞-0.4425]	(-∞-0.3855]	(0.0755-∞)	(0.0095-∞)	all	m
(0.198-∞)	all	all	(-∞-0.4425]	(0.3855-∞)	(0.0755-∞)	(-∞-0.0095]	all	r
(0.198-∞)	all	all	(0.4425-∞)	(-∞-0.3855]	(-∞-0.0755]	(0.0095-∞)	all	r
(-∞-0.198]	all	all	(0.4425-∞)	(-∞-0.3855]	(-∞-0.0755]	(0.0095-∞)	all	r
(0.198-∞)	all	all	(0.4425-∞)	(-∞-0.3855]	(0.0755-∞)	(-∞-0.0095]	all	r
(-∞-0.198]	all	all	(0.4425-∞)	(-∞-0.3855]	(0.0755-∞)	(-∞-0.0095]	all	r

Fig. 5. The decision table learned by the best individual from PBIL for a given run of the sonar dataset. This classifier is part of the ensemble composed of classifiers from Fig. 4.

In [20], a Genetic Programming algorithm is used to optimize configurations of ensemble methods (bagging, boosting, etc) and their base learners (logistic regressors, neural networks, etc). In addition, [23] proposes a co-evolutionary algorithm for finding the best combination of hyper-parameters for a set of base classifiers, which might also include the best combination of data pre-processing methods for a given dataset. AUTO-CVE concurrently evolves two populations: a population of base models (using Genetic Programming) and a population of ensembles (using a Genetic Algorithm). In both [20] and [23], again the focus was on predictive accuracy, and those works tend to produce very large ensembles of non-interpretable base classifiers. By contrast, in the current work the learned ensembles are small (with only 5 base classifiers) and consist of interpretable classifiers by design.

6   Conclusion and Future Work

We presented a new evolutionary algorithm (a version of PBIL) for optimizing the configuration of a small ensemble of interpretable classifiers, aiming at maximizing predictive performance on the dataset at hand whilst generating interpretable models by design. The proposed PBIL and Random Forest achieved the best predictive accuracy overall – each was the best in 4 of 9 datasets. The baseline ensemble was the best in one dataset.

Both the proposed PBIL and the baseline ensemble produce interpretable models consisting of only 5 interpretable classifiers, unlike random forest ensembles, which are not directly interpretable as discussed earlier. Note that the baseline ensemble proposed here is not a standard ensemble in the literature, because the literature focuses on large, non-interpretable ensembles. Hence, the results for the baseline ensemble reported here can also be seen as a contribution to the literature, in the sense of being further evidence (in addition to the PBIL's results) that small ensembles of interpretable classifiers can be competitive against large, non-interpretable ensembles.

Future work will involve designing a more advanced version of PBIL encoding dependencies among variables in the graphical model and doing other experiments with more datasets.

Acknowledgment. This study was financed in part by the Coordenação de Aperfeiçoamento de Pessoal de Nível Superior – Brasil (CAPES) – Finance Code 001.

References

1. Alcalá-Fdez, J., et al.: Keel data-mining software tool: data set repository, integration of algorithms and experimental analysis framework. J. Multiple-Valued Logic Soft Comput. **17**, 255–287 (2011)
2. Ali, J., et al.: Random forests and decision trees. Int. J. Comput. Sci. Issues (IJCSI) **9**(5), 272 (2012)
3. Baluja, S., Caruana, R.: Removing the genetics from the standard genetic algorithm. In: Proceedings of the Twelfth International Conference on Machine Learning, pp. 38–46. Elsevier, Tahoe City (1995)
4. Breiman, L.: Random forests. Mach. Learn. **45**(1), 5–32 (2001)
5. Breiman, L., et al.: Classification and Regression Trees. Wadsworth International Group, Belmont (1984)
6. Carvalho, D.V., Pereira, E.M., Cardoso, J.S.: Machine learning interpretability: a survey on methods and metrics. Electronics **8**(8), 832 (2019)
7. Cohen, W.W.: Fast effective rule induction. In: Twelfth International Conference on Machine Learning, pp. 115–123 (1995)
8. Fawcett, T.: An introduction to ROC analysis. Pattern Recogn. Lett. **27**(8), 861–874 (2006)
9. Fernández-Delgado, M., et al.: Do we need hundreds of classifiers to solve real world classification problems? J. Mach. Learn. Res. **15**(1), 3133–3181 (2014)
10. Feurer, M., et al.: Efficient and robust automated machine learning. In: Advances in Neural Information Processing Systems, pp. 2962–2970 (2015)
11. Frank, E., Witten, I.H.: Generating accurate rule sets without global optimization. In: Fifteenth International Conference on Machine Learning, pp. 144–151 (1998)
12. Freitas, A.A.: Comprehensible classification models: a position paper. ACM SIGKDD Explor. Newsl. **15**(1), 1–10 (2014)

13. Fürnkranz, J., Kliegr, T., Paulheim, H.: On cognitive preferences and the plausibility of rule-based models (2018). arXiv preprint arXiv:1803.01316
14. Galea, M., Shen, Q., Levine, J.: Evolutionary approaches to fuzzy modelling for classification. Knowl. Eng. Rev. **19**(1), 27–59 (2004)
15. Hall, M., et al.: The WEKA data mining software: an update. ACM SIGKDD Explor. Newsl. **11**(1), 10–18 (2009)
16. Hauschild, M., Pelikan, M.: An introduction and survey of estimation of distribution algorithms. Swarm Evol. Comput. **1**(3), 111–128 (2011)
17. Holland, J.H.: Adaptation in Natural and Artificial Systems: An Introductory Analysis with Applications to Biology, Control, and Artificial Intelligence. MIT press, Cambridge (1992)
18. Huysmans, J., et al.: An empirical evaluation of the comprehensibility of decision table, tree and rule based predictive models. Decis. Supp. Syst. **51**(1), 141–154 (2011)
19. Kohavi, R.: The power of decision tables. In: Lavrac, N., Wrobel, S. (eds.) ECML 1995. LNCS, vol. 912, pp. 174–189. Springer, Heidelberg (1995). https://doi.org/10.1007/3-540-59286-5_57
20. Kordik, P., Cerny, J., Fryda, T.: Discovering predictive ensembles for transfer learning and meta-learning. Mach. Learn. **107**, 177–207 (2018)
21. Kuncheva, L.I.: Combining Pattern Classifiers: Methods and Algorithms. Wiley, Hoboken (2004)
22. Lapuschkin, S., et al.: Unmasking Clever Hans predictors and assessing what machines really learn. Nat. Commun. **10**(1), 1096 (2019)
23. Larcher, C., Barbosa, H.: Auto-CVE: a coevolutionary approach to evolve ensembles in automated machine learning. In: Proceedings of The Genetic and Evolutionary Computation Conference, pp. 392–400 (2019). https://doi.org/10.1145/3321707.3321844
24. Larrañaga, P., Lozano, J.A.: Estimation of Distribution Algorithms: A new Tool for Evolutionary Computation. Springer, Heidelberg (2001). https://doi.org/10.1007/978-1-4615-1539-5
25. Lichman, M.: UCI machine learning repository (2013). http://archive.ics.uci.edu/ml
26. Luštrek, M., et al.: What makes classification trees comprehensible? Exp. Syst. Appl. **62**, 333–346 (2016)
27. Murphy, K.P.: Machine Learning: A Probabilistic Perspective. MIT Press, Cambridge (2012)
28. Olson, R.S., Urbanowicz, R.J., Andrews, P.C., Lavender, N.A., Kidd, L.C., Moore, J.H.: Automating biomedical data science through tree-based pipeline optimization. In: Squillero, G., Burelli, P. (eds.) EvoApplications 2016. LNCS, vol. 9597, pp. 123–137. Springer, Cham (2016). https://doi.org/10.1007/978-3-319-31204-0_9
29. Quinlan, J.R.: C4.5: Programs for Machine Learning. Morgan Kaufmann Publishers, San Mateo (1993)
30. Rish, I., et al.: An empirical study of the naive Bayes classifier. In: Proceedings of theWorkshop on empirical methods in artificial intelligence, IJCAI 2001, Seattle, USA, vol. 3, pp. 41–46 (2001)
31. Tan, P.N., Steinbach, M., Kumar, V.: Introduction to Data Mining. Pearson, London (2006)
32. Thornton, C., et al.: Auto-WEKA: combined selection and hyperparameter optimization of classification algorithms. In: International Conference on Knowledge Discovery and Data Mining, pp. 847–855. ACM (2013)

33. Xavier-Júnior, J.A.C., et al.: A novel evolutionary algorithm for automated machine learning focusing on classifier ensembles. In: Brazilian Conference on Intelligent Systems. pp. 1–6. IEEE, São Paulo (2018)
34. Zangari, M., et al.: Not all PBILs are the same: unveiling the different learning mechanisms of PBIL variants. Appl. Soft Comput. **53**, 88–96 (2017)

An Evolutionary Analytic Center Classifier

Renan Motta Goulart[1], Saulo Moraes Villela[2],
Carlos Cristiano Hasenclever Borges[2], and Raul Fonseca Neto[2(✉)]

[1] Postgraduate Program in Computational Modeling,
Federal University of Juiz de Fora, Juiz de Fora, Minas Gerais, Brazil
`renan.motta@ice.ufjf.br`
[2] Department of Computer Science, Federal University of Juiz de Fora,
Juiz de Fora, Minas Gerais, Brazil
`cchborges@ice.ufjf.br,{saulo.moraes,raulfonseca.neto}@ufjf.edu.br`

Abstract. Classification is an essential task in the field of Machine Learning, where developing a classifier that minimizes errors on unknown data is one of its central problems. It is known that the analytic center is a good approximation of the center of mass of the version space that is consistent with the Bayes-optimal decision surface. Therefore, in this work, we propose an evolutionary algorithm, relying on the convexity properties of the version space, that evolves a population of perceptron classifiers in order to find a solution that approximates its analytic center. Hyperspherical coordinates are used to guarantee feasibility when generating new individuals and enabling exploration to be uniformly distributed through the search space. To evaluate the individuals we consider using a potential function that employs a logarithmic barrier penalty. Experiments were performed on real datasets, and the obtained results indicate concrete possibilities for applying the proposed algorithm for solving practical problems.

Keywords: Machine learning · Evolutionary algorithm · Version space · Hyperspherical coordinates · Analytic center

1 Introduction

Classification is a Machine Learning task, where it is desired to infer which class a particular instance belongs. If all instances can be separated by a hyperplane in the input space, then the problem is linearly separable. The focus of this work is on finding a linear classifier that achieves a good generalization. It is presented an evolutionary strategy applied to a population of hyperplanes in order to approximate the population's individuals to the analytic center of the version space represented by a compact convex polyhedron bounded by a spherical shell.

Although meta-heuristic and evolutionary computation techniques are widely employed in solving optimization problems, mainly in multi-objective and mixed-integer formulations, their use in finding good classifiers is not much explored.

© Springer Nature Switzerland AG 2020
R. Cerri and R. C. Prati (Eds.): BRACIS 2020, LNAI 12319, pp. 34–48, 2020.
https://doi.org/10.1007/978-3-030-61377-8_3

An extensive search of state-of-the-art research has presented two related works. In [18], the authors present a meta-heuristic technique for training Support Vector Machines (SVM) based on a linear Particle Swarm Optimization (PSO) algorithm. They use a Linear PSO for each decomposed sub-problem involving a reduced number of dual variables. In [13], the authors present a genetic algorithm for the SVM problem. They use binary code to represent the SVM parameters and real code to represent the dual variables. The fitness measure is based on the evaluation of the dual objective function of SVM resulting however in a quadratic complexity in relation to the number of examples of the training set.

In this work, it is important to highlight that the convex properties of the version space guarantee the feasibility of the genetic operators and the use of the system of hyperspherical coordinates makes it possible to carry out an efficient search that explores only the feasible region of the spherical shell. As a fitness measure we choose a potential function [16] that employs a logarithmic barrier penalty, its complexity is linear in relation to the number of examples. This function was previously proposed in [22] on a dual formulation of a projected Newton descent method. In [19], the authors proposed primal, dual, and primal-dual formulations based on interior-point methods for solving the Analytic Center Problem (ACP). They proved the existence and uniqueness of the analytic center solution of a spherical surface when the objective function is strictly convex such as the logarithmic barrier function.

The proposed algorithm can be easily extended to the dual space considering the process of kernelization on the feature space. This is due to the fact that the convex properties and the system of hyperspherical coordinates can also be extended to the dual space.

Experiments were performed on real datasets, and the obtained results were compared with the SVM, Bayes Point Machine (BPM) [8], and the Version Space Reduction Machine (VSRM) [5] results. In order to attest to the accuracy and convergence of the method we verify the fulfillment of the KKT optimality conditions. In relation to the generalization performance we could observe that the proposed method outperforms these baselines algorithms indicating concrete possibilities to be applied for solving practical problems.

The remainder of this paper is structured as follows: initially, Sect. 2 will present the binary classification problem followed by mathematical preliminaries concepts. In Sect. 3, it will be presented the Perceptron Model, and the Analytic Center Problem as well as its primal formulation, properties and optimality conditions. In Sect. 4, the evolutionary classifier is presented, both its implementation details and its operators. Section 5 describes the datasets and experiments performed to validate the algorithm. Finally, Sect. 6 discusses the conclusions of this paper and future work.

2   Binary Classification and Related Concepts

In this section, it will be defined the binary classification problem as well as the related algebraic, geometric, and trigonometric concepts crucial in order to better understand the theoretical foundation of the work.

2.1   Binary Classification Problem

A supervised learning problem can be described as: from a set of known pairs, each consisting of an instance and a related class, correctly infer the label of a new instance of interest. Classification problems are a special case of supervised learning problems where classes are discrete values [12]. If there are only two distinct classes, the classification problem is called binary.

Formally, this problem can be defined as: being Z_m a training set of cardinality m consisting of a set X of dimensionality d called an instance set, and a set of binary values Y called a label set, find a discriminant function that correctly maps a instance X_i to its respective label Y_i and which is able to correctly map instances not yet displayed. The training set Z_m is defined as:

$$Z_m = \{(X_i, Y_i), \ i = 1, \ldots, m\}, \ X_i \in R^d, \ Y_i \in \{-1, 1\}. \tag{1}$$

If we restrict the solution to a linear set of hypotheses, the discriminant function can be considered as a d-dimensional hyperplane that lies in the input space, associated with the decision boundary:

$$X_i : (W \cdot X_i) + b = 0, \tag{2}$$

where W is the normal vector and b is the bias parameter.

The distance between a d-dimensional (W, b) hyperplane and an instance X_i is computed by the functional distance as shown in the following equation:

$$\delta_i = b + \sum_{j=1}^{d} (W_j \cdot X_{i,j}). \tag{3}$$

If we consider it respective label Y_i we can impose for a given instance X_i a classification constraint described by the simple linear inequality:

$$Y_i \cdot (W \cdot X_i + b) \geq 0. \tag{4}$$

In this sense, in order to compute a feasible solution (W, b), our objective is to solve the feasibility problem:

$$Y_i \cdot \delta_i \geq 0, i = 1, \ldots, m. \tag{5}$$

2.2 Version Space

Given a set Z_m of training examples, the set of all possible classifiers that correctly infer the label of each instance is defined as its version space. It can be geometrically visualized as a delimited spherical shell of a unit radius hypersphere considering a unit norm set of W vectors, with hyperplanes passing through it. Each hyperplane is defined by an instance and its label via its respective normal vector $Y_i \cdot X_i$. The hyperspherical surface region that does not violate any classification constraint is the version space. Formally it can be defined as:

$$V_z = (\{W, b\} \ \forall (X_i, Y_i) \in Z_m : Y_i \cdot (W \cdot X_i + b) \geq 0, \ \|W\|_2 = 1). \qquad (6)$$

However, if we consider the unit radius hyperspherical constraint $\|W\|_2 \leq 1$ instead of the unit norm constraint $\|W\|_2 = 1$, the version space can be interpreted as a bounded convex polyhedron where each of its faces is relative to an example of the training set. Particularly, it can be considered as a bounded conical polyhedron or simply a cone, with each hyperplane passing through the origin, if we associate an additional coordinate axis with the bias values.

It is well known from [6] that hypotheses that are near the center of mass of the version space tend to be more efficient. Since computing the center of gravity of a polyhedron in an n-dimensional space is classified as a #P-hard problem [9], approximations that can be efficiently computed, such as the analytic center, are considered.

2.3 Potential Function

The potential function has the property of when approaching the boundary of the feasible region, its value tends to infinity. The logarithmic function fulfills this requirement. More precisely, we take the function $-\ln(|\delta_i|)$ where the argument denotes the distance module related to the instance X_i according to Eq. (3). In this sense, if we consider the hyperplanes with normal vectors $Y_i \cdot X_i$ defining the feasible region of the version space the function values go to infinity as a hypothesis (W, b) approximates the boundary represented by these faces.

2.4 Anaytic Center and Others Approximations

The Support Vector Machine (SVM) [3] is a well-known method for finding the maximum margin hyperplane. It maximizes the distance between both classes in order to provide higher power of generalization on unseen data. The center of the largest hypersphere that can be inscribed on version space is the point relative to the normal vector of the maximum margin hyperplane. However, in cases where the version space is elongated or asymmetric SVM is not a good solution compared to the center of mass or to the polyhedron centroid, see Fig. 1.

It is recognized that the Bayes point consistently outperforms the SVM in the power of generalization. A good approximation for the Bayes point, called the Bayes Point Machine (BPM), was proposed by Herbrich, Graepel, and Campbell [8]. The BPM is a sampling method and consists of using as classifier the

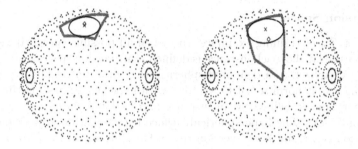

Fig. 1. In red, version spaces for two 3-dimensional problems. The diamond is the center of mass and the cross the center of the largest sphere inscribed. The more elongated the version space is, the more its distance to the center of the sphere. Image taken from [8] page 261. (Color figure online)

average of a set of randomly generated perceptrons consistent with the version space.

Considering that the version space can be defined as the intersection of a finite number of semi-spaces each associated with a classification constraint, it can be represented by a system of linear inequalities defining a polyhedron. The algorithm that accurately computes the volume of an n-dimensional polyhedron and consequently its center of mass was a recursive procedure proposed by Lasserre [10], however, its complexity is exponential, of order $O(d^m)$ with d being the dimension and m the number of inequality constraints.

Aiming to solve the problem related to the prohibitive computational cost to find the center of mass of a polyhedron, the authors [5] presented an algorithm, named Version Space Reduction Machine (VRSM), that approximates the center of mass based on the successive generation of cutting planes. The addition of a cutting plane bisects the version space into two halfspaces and the halfspace to be chosen must reflect the majority opinion of an ensemble formed by a set of randomly generated perceptrons. Thus, after a finite number of iterations, there is a consistent reduction of the version space and convergence to the center of mass of a hypothesis that agrees with the sequential decisions chosen by the ensemble.

Another efficient alternative for approximating the center of mass is by instead calculating its analytical center. It is a well definedness mathematical problem. This point approximates the location of the center of mass of the version space and can be calculated by finding the maximum for a sum of potential functions. For m constraints and considering the use of a logarithmic barrier function, the objective becomes:

$$\min f(W, b) = -\sum_{i=1}^{m} \ln(Y_i \cdot (W \cdot X_i + b)). \tag{7}$$

It is important to mention that the arguments of the logarithmic function must not have negative values. This condition is necessary in order to fulfill the constraint given by Eq. (5) ensuring the consistency of the version space.

2.5   Hyperspherical Coordinates

The spherical coordinate system is a way of representing a vector position that lies in a 3-dimensional space with three values, see Fig. 2a, one for the radius r of a sphere centered at the origin, one for the polar angle ϕ and the other for the azimuthal angle θ, in order to determine a position on such sphere's surface [15]. This coordinate system is used for representing vectors in problems that mainly deal with rotation operations, and therefore is appropriate for representing the W vectors in a version space bounded by a unit radius hypersphere, see Fig. 2b.

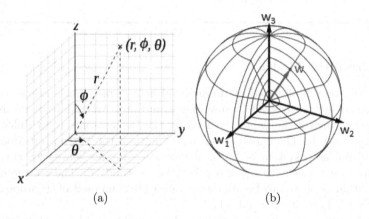

(a)   (b)

Fig. 2. (a) Spherical coordinate system in a 3-dimensional space. (b) Representation of a 3-dimensional unit vector in a unit sphere.

This system can be extended for n-dimensions by increasing the number of values used to represent the polar angles to $n - 1$, where the last angle has a range of $[0, 2\pi]$ and each other angles have a range of $[0, \pi]$. The conversion of an n-dimensional unit vector W of Cartesian coordinates to hyperspherical coordinates, and its inverse procedure, are described by Algorithms 1 and 2.

3   Classifiers

In this section, we introduce the Perceptron Model and the Analytic Center Problem as well as its primal formulation, properties, and optimality conditions.

Algorithm 1: Procedure to convert a Cartesian coordinate vector into a hyperspherical coordinate vector.

Cartesian vector: W;
Dimension of W: d;
Polar angle vector: θ;
$squared\_Sum = 1$;
for $(i = 0; i < d - 2; i + +)$ **do**
$\quad | \quad \phi_i = \arccos\left(\frac{W_i}{\sqrt{squared\_Sum}}\right)$;
$\quad | \quad squared\_Sum = squared\_Sum - W_i \cdot W_i$;
end
if $(W_{d-1} \geq 0)$ **then**
$\quad | \quad \theta = \arccos\left(\frac{W_{d-2}}{\sqrt{squared\_Sum}}\right)$;
else
$\quad | \quad \theta = 2 \cdot \pi - \arccos\left(\frac{W_{d-2}}{\sqrt{squared\_Sum}}\right)$;
end

3.1 Perceptron Model

The Perceptron [20] is based on an artificial neuron model [11], whose connections have weights corrected by a simple procedure. An update of the parameters is performed if the value obtained by the Perceptron output differs from the desired output. The classification constraints of the Perceptron Model can be considered as a system of linear inequations. In this sense, the correction procedure is equivalent to using a relaxation method [1]. The convergence of the algorithm is proven to be guaranteed in a finite number of iterations if the training set is linearly separable [17].

The Perceptron is initialized with random weights and bias. After initialization, the training is performed until all instances can be correctly classified. If an example of the training set is misclassified, that is: $Y_i \cdot (W \cdot X_i + b) < 0$, the correction rule is applied and is defined by:

$$W_{t+1} = W_t + r \cdot Y_i \cdot X_i \tag{8}$$
$$b_{t+1} = b_t + r \cdot Y_i, \tag{9}$$

where W_{t+1} is the updated weight vector, W_t is the weight vector before correction, b_{t+1} is the updated bias parameter, b_t is the bias before correction, r is the learning rate and Y_i is the label associated with the instance X_i.

3.2 Analytic Center Problem

The maximization of the logarithmic barrier function constitutes an efficient approach to approximates the center of mass of the version space because the optimal solution tends to be far away from all constraints or from the polyhedron faces. However, during the optimization process we must guarantee the feasibility

---

Algorithm 2: Procedure to convert a hyperspherical coordinate vector into the Cartesian coordinate vector.

Polar angle vector: ϕ;
Azimuthal angle θ;
Cartesian vector: W;
Dimension of Hyperspherical vector: d_h;
$squared\_Sum = 1$;
for $(i = 0; i < d_h - 1; i++)$ **do**
  | $W_i = \cos(\phi_i) \cdot \sqrt{squared\_Sum}$;
  | $squared\_Sum = squared\_Sum - W_i \cdot W_i$;
end
$W_{d_h-1} = \cos(\theta) \cdot \sqrt{squared\_Sum}$;
$W_{d_h} = |\sqrt{squared\_Sum - W_{d_h-1} \cdot W_{d_h-1}}|$;
if $\theta > \pi$ **then**
  | $W_{d_h} = -1 \cdot W_{d_h}$;
end

---

of the solution ensuring that it remains inside the version space. Another problem is related to the scale of the functional distances for different sample vectors. However, this problem can be overcome by introducing the unit norm constraint. In this sense, the Analytic Center Problem can be formulated as follow:

$$\min_{(W,b)} f(W, b) = -\sum_{i=1}^{m} \ln(Y_i \cdot (W \cdot X_i + b)) \tag{10}$$

$$\text{subject to } Y_i \cdot (W \cdot X_i + b) \geq 0, \ i = 1, \ldots, m \tag{11}$$

$$W^T \cdot W = 1. \tag{12}$$

Although the objective function is strictly convex it is worth mentioning that the feasible set is not due to the existence of the unit norm constraint given by Eq. (12).

3.3 KKT Conditions

To establish the KTT conditions it is necessary to rewrite the optimization problem given by Eqs. (10) to (12) while considering the Lagrangian relaxation of the unit norm constraint. Then, the Lagrangian function becomes:

$$L(W, b, \lambda) = -\sum_{i=1}^{m} \ln(Y_i \cdot (W \cdot X_i + b)) + \lambda \cdot (W^T \cdot W - 1), \tag{13}$$

with the KKT conditions:

$$Y_i \cdot (W \cdot X_i + b) \geq 0, \ i = 1, \ldots, m \tag{14}$$

$$\lambda \geq 0 \tag{15}$$

$$W^T \cdot W = 1 \tag{16}$$

$$\lambda \cdot (W^T W - 1) = 0 \tag{17}$$

$$-\sum_{i=1}^{m} \frac{Y_i \cdot X_i}{Y_i \cdot (W \cdot X_i + b)} + 2 \cdot \lambda \cdot W = 0. \tag{18}$$

In particular, for the bias component we must have:

$$-\sum_{i=1}^{m} \frac{Y_i}{Y_i \cdot (W \cdot X_i + b)} = 0. \tag{19}$$

Observing the KKT conditions we can deduce that Eq. (14) is related to the primal feasibility and then is fulfilled at the optimal solution. Equation (17) is related to the property of complementary slackness and is satisfied because multiplying Eq. (18) by W^T we obtain $2 \cdot \lambda \cdot ||W||_2 \approx m$, which implies that $||W||_2 = 1$ because $\lambda > 0$. Consequently, Eqs. (15) and (16) are fulfilled. Therefore, considering $\lambda = m/2$, the KKT conditions can be simplified to the vector equation:

$$-\sum_{i=1}^{m} \frac{Y_i \cdot X_i}{Y_i \cdot (W \cdot X_i + b)} + m \cdot W = 0. \tag{20}$$

Despite the existence of mathematical programming methods [19, 22] for solving the Analytic Center Problem, we proposed the development of an evolutionary algorithm for this purpose that will be explained in the next section. The major trouble of the mathematical approach for solving the ACP is the difficulty in choosing a starting feasible point, the instability of the interior point methods and the influence of redundant constraints that move way the optimal solution from the true analytical center point. In this sense, the evolutionary approach appears to be a good solution for circumvents these drawbacks.

4   Evolutionary Algorithm

In this section, it is presented a generational evolutionary algorithm that applies recombination and mutation operators [14] to evolve a population of hyperplanes, each represented by their respective weight vector and bias parameter coded in the hyperspherical coordinate system. The algorithm starts its initial main population of individuals by randomly training perceptron classifiers. Then, at each generation, it is created a population of the same size as the main population, composed of child individuals generated by applying the recombination and mutation operators. After this, individuals from the main population are replaced by the child individuals that agree with the replacement criteria.

4.1 Initial Population

The initial population consists of a set of feasible and normalized individuals (hyperplanes) each randomly generated by the perceptron algorithm. Since the initial population is generated though the Cartesian coordinate system, we need to convert each initial individual to the hyperspherical coordinate system, this is done by using Algorithm 1.

4.2 Fitness Measure

The fitness measure used for the evaluation of an individual is based on the objective function of the ACP given by Eq. (7). Therefore, the computational complexity to evaluate an individual (W, b) is linear in relation to the number of examples of the training set. This function is given by:

$$f(W, b) = - \sum_{i=1}^{m} \ln(Y_i \cdot (W \cdot X_i + b)). \tag{21}$$

It is important to highlight that the norm of vector W must remain unitary for every individual of the population and the logarithm function argument can not have negative values for each individual and for each example of the training set.

To perform the inner product we must convert each individual back to the system of Cartesian coordinates, that is done by Algorithm 2. In order to achieve better results the authors [6] recommend the normalization of the training set.

4.3 Recombination Operator

A child individual is initialized by the recombination operator. The parents are two individuals from the population chosen according to a roulette selection process where each individual's probability is relative to its placement on a ordered list sorted by the fitness values of the population. The probability of an individual to be chosen is given by:

$$P_i = \frac{i^2 - (i-1)^2}{n^2}, \tag{22}$$

where P_i is the probability for the i-th individual of the population, and n is the size of the population. Each child has their weights initialized by a convex linear combination of their parents' weights. By considering the convex property of the version space and the feasibility of the initial population these individuals will fulfill the classification constraints. The recombination process using Cartesian coordinates is given by:

$$W_{f,i} = \lambda \cdot W_{p1,i} + (1 - \lambda) \cdot W_{p2,i}, \ \lambda \in [0, 1], \tag{23}$$

where $W_{f,i}$, $W_{p1,i}$ and $W_{p2,i}$ are, respectively, the i-th weights of the child and the parent hyperplanes. The λ parameter is chosen from a uniform probability distribution. It is important to highlight the necessity of normalization of parent individuals before applying recombination in order to preserve the feasibility of the unit radius hyperspherical constraint. Formally, this can be demonstrated as:

Theorem 1. *The recombination operator preserves the feasibility of the unit radius hypersphere constraint of the version space in the Cartesian coordinates representation system if the parents' vectors have unit norm. That is* $||W_f||_2 \leq 1$.

Proof. Let W_{p1} and W_{p2} be the parent vectors with unitary norm $||W_{p1}||_2 = ||W_{p2}||_2 = 1$. Let the child vector be represented by a convex linear combination of W_{p1} and W_{p2}. Then:

$$||\lambda \cdot W_{p1} + (1-\lambda) \cdot W_{p2}||_2 \leq \lambda \cdot ||W_{p1}||_2 + (1-\lambda) \cdot ||W_{p2}||_2 \leq \lambda + (1-\lambda) = 1. \quad (24)$$

The convex linear combination given by Eq. (23) can also be used for computing the children coordinates in the hyperspherical coordinate system represented by the radius, the azimuthal and polar angles. Furthermore, we can prove that if the parents' vectors have unit norm then, after recombination, the children vectors also will have it satisfying the unit norm constraint of the version space. Formally, this can be demonstrated as:

Theorem 2. *The recombination operator preserves the feasibility of the unit norm constraint of the version space in the hyperspherical coordinates representation system if the parents' vectors have unit norm. That is* $r_{hf} = 1$.

Proof. Let r_{hp1} and r_{hp2} be the radius of the parent vectors with unitary norm. Then $r_{hp1} = 1$ and $r_{hp2} = 1$. Therefore, $r_{hf} = r_{hp1} \cdot \lambda + r_{hp2} \cdot (1-\lambda)$, $\lambda \in [0,1] = \lambda + (1-\lambda) = 1$.

Corollary 1. *For any child vector W_f converting from the hyperspherical system to Cartesian system, we have regardless the coordinates values,* $||W_f||_2 = r_{hf} = 1$.

These results show that the vectors generated while in the hyperspherical system preserve the unit norm, therefore it is not necessary to normalize new individuals unlike while using the Cartesian system.

Since each individual represents a direction, the recombination operator can be geometrically interpreted as choosing a new direction between the directions of each parent. Making a convex combination of weight values without this representation would not guarantee a uniform distribution for all possible angles. For example, on a 2-dimensional problem, suppose that the first parent is an angle of $150°$ while the second parent is an angle of $30°$. If λ is set to 0.75, it is expected that the child individual would be the angle of $120°$. However, if the recombination would be applied directly to the weight values, then the resulting angle would be of approximately $136°$.

The recombination operator has the objective of making individuals resemble the best individual in the population, therefore acting as an accelerator by propagating relevant information of old individuals. This can be considered as a primary form of exploitation. On the other hand, the mutation operators, introduced in the following subsection, is responsible for exploring the search space by introducing a random diversity by generating unbiased new individuals [4].

4.4   Mutation Operator

Two mutation operators are used, each makes a change on the weight vector and thus can be interpreted as a rotation of the hyperplane. For this, the hyperspherical coordinate system has two advantages: first, it keeps the vector unit norm since only the angles are being changed, unlike on the Cartesian system; second, if the weight values were to be changed directly, one mutation could interfere with a previous mutation, resulting in the loss of guarantee that the more mutations applied to an individual, the more it will change.

The first mutation operator causes a small rotation on the hyperplane. The intensity of this change does not has a fixed value, instead, it is set as a fraction of the distances from one parent to the other parent. This enables the mutation intensity to scale as the population nears the optimal point of the version space. The first mutation operator procedure can be expressed by the following equation:

$$
\mu = \begin{cases} \tau \cdot \left(\sqrt{(\phi_{hp1} - \phi_{hp2})^2 + \sum (\theta_{hp1} - \theta_{hp2})^2} \right) & \text{if } |\phi_{hp1} - \phi_{hp2}| \leq \pi, \\ \tau \cdot \left(\sqrt{(2\pi - |\phi_{hp1} - \phi_{hp2}|)^2 + \sum (\theta_{hp1} - \theta_{hp2})^2} \right) & \text{otherwise.} \end{cases}
$$

(25)

with μ being the mutation intensity, in radians, and τ the fraction size. Since the ϕ angle is circular, the difference between both parents ϕ angles can be related to the opposite direction they are facing, which is the case if the difference results in a value higher than π, therefore the correct value is its explementary angle.

The second mutation operator also works with the hyperspherical coordinates. It creates a vector by subtracting the hyperspherical coordinates of the parent with lower objective function from the other, after this, the child individual has a 2/3 probability of remaining the same and a 1/3 probability of having this vector added to its hyperspherical coordinates. This operator enables to explore a greater region of the version space while also having a great chance of maintaining feasibility due to its direction being the same as the recombination operator.

4.5   Bias Optimization

The bias parameter is optimized by a binary search procedure on its feasible interval. Since the objective function is a sum of strictly increasing functions, any change to the bias value will increase the sum of potential functions for one class while decreasing the sum of the other. Therefore, the optimal bias value is the point where the sum of derivatives of potential functions for one class is equal to the other. This is solved by considering the minimization of a univariate convex function when we take the set of values of the vector W components. Therefore, for a given individual W_f we have the objective:

$$\min_b f(b) = -\sum_{i=1}^{m} \ln(Y_i \cdot (W_f \cdot X_i + b)). \tag{26}$$

Applying the derivative in Eq. (26) in relation to parameter b and considering m^+ and m^- the cardinality of the respective classes, makes it possible to deduce the first order condition:

$$\sum_{i=1}^{m^+} \frac{1}{Y_i \cdot (W_f \cdot X_i + b)} = \sum_{i=1}^{m^-} \frac{1}{Y_i \cdot (W_f \cdot X_i + b)}, \tag{27}$$

which represents the balance between the inverse of the distances between one class and another in relation to the separating hyperplane. Notice that this is equivalent to the KKT condition given by Eq. (19).

5   Experiments and Results

To validate the proposed method, it was used six linearly separable datasets derived from microarray experiments as shown in Table 1. The datasets are referenced by [2, 7, 21] or [23].

Table 1. Information about the considered datasets.

Set	Features	Samples		
		+1	−1	Total
Prostate	12600	50	52	102
Breast	12625	10	14	24
Colon	2000	22	40	62
Leukemia	7129	47	25	72
DLBCL	5468	58	19	77
CNS	7129	21	39	60

The results were compared to the SVM, BPM and VSRM algorithms by employing a 10×10–10-fold cross-validation, due to the random nature of the methods. The only exception was the SVM, where it was employed a 1×10–10-fold cross-validation. These schemes were adopted in order to reduce the bias of the methods. It was used a stratified cross-validation strategy, where each fold maintains the percentage of data points of each class. For more accurate comparisons it was always selected, for each dataset, the same training and test sets and also the same 10 subsets for cross-validations, preserving the generating seed associated with the random process. The results for SVM, BPM and VSRM were taken from [5]. For the Evolutionary Analytic Center Classifier (EACC), it was used a population of 100 individuals evolved for 50 generations, with the first mutation operator having a 10% probability of being applied on each dimension. Table 2 shows the results obtained from the experiments. The best results for each dataset are highlighted in bold.

Table 2. Comparison of mean classification errors.

Set	SVM		BPM		VSRM		EACC	
Prostate	9.58	1.35	10.46	1.57	9.46	1.77	**8.75**	**1.05**
Breast	20.67	2.49	20.35	3.42	19.83	2.39	**19.33**	**1.54**
Colon	18.69	2.32	15.68	2.35	15.44	2.29	**15.29**	**2.57**
Leukemia	**2.75**	**0.88**	5.50	1.32	3.39	1.30	4.91	1.52
DLBCL	3.81	0.68	3.86	0.79	**3.49**	**0.87**	4.38	1.20
CNS	33.50	1.99	33.33	2.68	32.87	2.15	**32.33**	**3.02**

From the results, we can observe that EACC achieves the best results in 4 of the 6 datasets, while SVM and VSRM in 1 dataset each one.

6 Conclusions and Future Work

An evolutionary algorithm for solving the Analytic Center Problem applied to binary classification tasks was developed. The algorithm evolves a feasible population of hyperplanes in order to approximate the analytic center of the version space. The purpose of the recombination operator is to move the population towards the optimal solution while also preserving the feasibility of each individual. The mutation operator can be interpreted as a rotational motion of a hyperplane. Hyperspherical coordinates were used and proved to be essential for both operators to work correctly. The experiments showed that the Evolutionary Analytic Center Classifier obtained good results competing on accuracy and most of the time overcoming the SVM, BPM, and VSRM classifiers in linearly separable datasets. In this sense, the proposed method presents real possibilities to be applied for solving practical problems. As future work, we plan to develop the dual version of the algorithm for its use in nonlinear classification problems with the introduction of kernel functions, because the convex properties and the system of hyperspherical coordinates can easily be extended to the dual space.

References

1. Agmon, S.: The relaxation method for linear inequalities. Can. J. Math. **6**, 382–392 (1954)

2. Alon, U., et al.: Broad patterns of gene expression revealed by clustering analysis of tumor and normal colon tissues probed by oligonucleotide arrays. Proc. Natl. Acad. Sci. United States Am. **96**(12), 6745–6750 (1999)
3. Cortes, C., Vapnik, V.: Support-vector networks. Mach. Learn. **20**(3), 273–297 (1995)
4. Eiben, A., Schippers, C.: On evolutionary exploration and exploitation. Fundam. Inf. **35**, 35–50 (1998)
5. Enes, K.B., Villela, S.M., Fonseca Neto, R.: Version space reduction based on ensembles of dissimilar balanced perceptrons. In: Proceedings of the Twenty-Fifth International Joint Conference on Artificial Intelligence, pp. 1448–1454 (2016)
6. Gilad-Bachrach, R., Navot, A., Tishby, N.: Bayes and tukey meet at the center point. In: Shawe-Taylor, J., Singer, Y. (eds.) COLT 2004. LNCS (LNAI), vol. 3120, pp. 549–563. Springer, Heidelberg (2004). https://doi.org/10.1007/978-3-540-27819-1_38
7. Golub, T.R., et al.: Molecular classification of cancer: class discovery and class prediction by gene expression monitoring. Science **286**, 531–537 (1999)
8. Herbrich, R., Graepel, T., Campbell, C.: Bayes point machines. J. Mach. Learn. Res. **1**, 245–279 (2001)
9. Kaiser, M.J., Morin, T.L., Trafalis, T.B.: Centers and invariant points of convex bodies. In: Applied Geometry And Discrete Mathematics (1990)
10. Lasserre, J.B.: An analytical expression and an algorithm for the volume of a convex polyhedron in R^n. J. Optim. Theory Appl. **39**, 363–377 (1983)
11. McCulloch, W.S., Pitts, W.: A Logical Calculus of the Ideas Immanent in Nervous Activity, pp. 15–27. MIT Press, Cambridge (1988)
12. Michalski, R.S., Carbonell, J.G., Mitchell, T.M.: Machine Learning: An Artificial Intelligence Approach. Springer, Heidelberg (2013). https://doi.org/10.1007/978-3-662-12405-5
13. MinShu, M.: Genetic algorithms designed for solving support vector classifier. In: International Symposium on Data, Privacy, and E-Commerce, pp. 167–169 (2007)
14. Mitchell, M.: An Introduction to Genetic Algorithms. MIT Press, Cambridge (1998)
15. Moon, P.H., Spencer, D.E.: Field Theory Handbook : Including Coordinate Systems, Differential Equations, and Their Solutions. Springer, Heidelberg (1988). https://doi.org/10.1007/978-3-642-83243-7
16. Nesterov, Y., Nemirovskii, A.S.: Interior point polynomial methods in convex programming: Theory and algorithms (1993)
17. Novikoff, A.: On convergence proofs on perceptrons. In: Proceedings of the Symposium on the Mathematical Theory of Automata, vol. 12, pp. 615–622 (1962)
18. Paquet, U., Engelbrecht, A.: Training support vector machines with particle swarms. In: Proceedings of the International Joint Conference on Neural Networks, vol. 2, pp. 1593–1598 (2003)
19. Raupp, F.M., Svaiter, B.F.: Analytic center of spherical shells and its application to analytic center machine. Comput. Optim. Appl. **43**(3), 329–352 (2009)
20. Rosenblatt: The perceptron, a perceiving and recognizing automaton, Project Para. Cornell Aeronautical Laboratory (1957)
21. Singh, D., et al.: Gene expression correlates of clinical prostate cancer behavior. Cancer Cell **1**(2), 203–209 (2002)
22. Trafalis, T., Malyscheff, A.: An analytic center machine. Mach. Learn. **46**, 203–223 (2002)
23. Zhu, Z., Ong, Y.S., Dash, M.: Markov blanket-embedded genetic algorithm for gene selection. Pattern Recogn. **40**, 3236–3248 (2007)

Applying Dynamic Evolutionary Optimization to the Multiobjective Knapsack Problem

Thiago Fialho de Queiroz Lafetá[(✉)] and Gina Maira Barbosa de Oliveira[(✉)]

Federal University of Uberlândia, Uberlândia, MG, Brazil
`fialhot@gmail.com, gina@ufu.br`

Abstract. Real-world discrete problems are often also dynamic making them very challenging to be optimized. Here we focus on the employment of evolutionary algorithms to deal with such problems. In the last few years, many evolutionary optimization algorithms have investigated dynamic problems, being that some of the most recent papers investigate formulations with more than one objective to be optimized at the same time. Although evolutionary optimization had revealed very competitive algorithms in different applications, both multiobjective formulations and dynamic problems need to apply specific strategies to perform well. In this work, we investigate four algorithms proposed for dynamic multiobjective problems: DNSGA-II, MOEA/D-KF, MS-MOEA and DNSGA-III. The first three were previously proposed in the literature, where they were applied just in continuous problems. We aim to observe the behavior of these algorithms in a discrete problem: the Dynamic Multiobjective Knapsack Problems (DMKP). Our results have shown that some of them are also promising for applying to problems with discrete space.

Keywords: Dynamic multiobjective problems · Evolutionary algorithms · Knapsack problem

1 Introduction

In the last decades, a huge research effort was employed on optimization problems and the algorithms to solve them. Several real-world problems involve optimization. For example, one can want to design a production system in industry which produces better and faster, reducing error rates in processes. Moreover, most of the real-world optimization problems involve several objectives to be simultaneously optimized. More recently, the dynamic nature of the multiobjective optimization problems are also being considered in several investigations [1]. This class of problems is called the Dynamic Multiobjective Optimization Problems (DMOPs). Such problems are characterized by the employment of two or more objectives and by changing the set of objectives and restrictions along the time. DMOPs behavior defines a new challenge to the optimization algorithms since they change the objective space over time [9].

© Springer Nature Switzerland AG 2020
R. Cerri and R. C. Prati (Eds.): BRACIS 2020, LNAI 12319, pp. 49–63, 2020.
https://doi.org/10.1007/978-3-030-61377-8_4

Dynamic optimization is a research topic that has been focused on two decades ago. Nonetheless, the majority of previous investigations had emphasized only on the optimization of a single objective, known as Dynamic Single-objective Optimization Problems (DSOPs) [16]. The field of multiobjective research in dynamic optimization is more recent and has been attracting more and more attention from researchers [1]. Evolutionary algorithms are considered a promising approach for multiobjective and dynamic optimization, to possess the ability to adapt several problems with different objectives and constraints, mainly if these characteristics change over time.

Nevertheless, most of the work related to DMOPs investigates continuous problems and usually handling up to 3 objectives to be optimized. Here we intend to investigate the behavior of some of these previous algorithms in a new scenario. Thus, our focus is to evaluate the performance of evolutionary algorithms (EAs) known in the literature on a discrete dynamic problem with 4 or more objectives to be optimized. Such problems belong to the many-objective class, where the greater the number of objectives the more difficult the search is, since the increase in the number of objectives causes the population to contain only non-dominated solutions, decreasing the selective pressure [15].

In recent years, EAs with different strategies have been proposed aimed at solving DMOPs. The behavior of each strategy is related to the way it reacts to the changes in the environment. Recalling that an environment change refers to the alteration of objective functions and restrictions of the problem over the time. The most basic strategy is to restart the entire population at each environment change, but it is often impractical, as illustrated by Branke [4]. Another way is to use knowledge about previous evolution to speed up the search process after the occurrence of an environment change [16]. Following this way, three types of strategies could be observed in the related literature [1]: (i) introduction of diversity [9]; (ii) prediction [19]; and (iii) memory [6].

Our proposal is to evaluate the performance of different evolutionary algorithms proposed for continuous DMOPs in a discrete many-objective problem. We selected some algorithms found in the literature belonging to each of the three cited strategies. The first algorithm is the DNSGA-II [9], which is based on the previous well-known method NSGA-II [8]. It uses a simple strategy to add diversity in current population after each environment change and was used in several works related to dynamic optimization [2,14,28,29]. MOEA/D-KF [26] was proposed by modifying the multiobjective MOEA/D [31] to a dynamic method that uses a prediction strategy. The algorithm uses the Kalman Filter [17] at each environment change to drive the population for the previous environment towards the Pareto of the next one. The memory-based strategy was also contemplated by investigating the MS-MOEA [33], which presented very promising results, surpassing several algorithms in the literature. The method uses the information stored on the external archive when the change of environment occurs to assist in the evolutionary process in the next environment. Finally, inspired by the DNSGA-II [9], in this work we adapted the well-known

NSGA-III [10], which is widely applied to many-objective problems, proposing the DNSGA-III.

Aiming to evaluate the performance of the selected algorithms, we used a discrete problem well-known in the literature. The knapsack problem (KP) is widely explored in several works [18] because it belongs to the NP-Complete class [5]. Here we use a dynamic multiobjective version of KP called Dynamic Multiobjective Knapsack Problem (DMKP) [11]. It has several dynamic real-life applications, such as cargo loading, budget management, cutting stock, etc. [3]. We generated twelve instances of DMKP by varying the number of objectives (4, 6 and 8), number of items (30 and 50) and number of environmental changes (1 and 2). In each environment change, only the set of objectives varies with the least severity possible, that is, only one objective changes. A small change severity is more appropriate for investigate the strategies mentioned above [4].

2   Problem Formulation

The static version of the problems with multiple objectives to be optimized are known as multiobjective Optimization Problems (MOPs). They involve simultaneous minimization (or maximization) of objectives that satisfy some restrictions. The concept of dominance can be defined, where a solution x dominates another solution y $(x \prec y)$, if x is better than y in at least one objective and at least equal in all the others. All non-dominated solutions in a search space is called the Pareto Optimum (P^*). The goal of solving a MOP is to find P^* or at least a set of solutions close to P^*. Equation 1 formalizes the definition of MOP.

$$min f(x) = \{f_1(x), f_2(x), ..., f_m(x)\}, where \ x \ \epsilon \ \Omega, \ g(x) > 0, \ h(x) = 0 \quad (1)$$

Consider x a search space solution; f_i refers to the value of objective i, where $i = 1, ..., m$; Consider Ω as the representation of the search space (or set of all solutions of problem); The functions g and h represent the set of restrictions of equality and inequality of the problem.

We can see the Dynamic Multiobjective Optimization Problem, or DMOP for short, as an extension of the concept of MOP, where certain characteristics of the problem may change over the time. Whenever there is a change in the MOP it means that the environment of the objective space also changes. The environment change occurs when the set of objectives or restrictions of the problem are modified. The mathematical formulation of DMOPs in given by Eq. 2.

$$min f(x, t) = f_1(x, t), ..., f_m(x, t), \ where \ x \ \epsilon \ \Omega, \ g(x, t) > 0, \ h(x, t) = 0 \quad (2)$$

The same considerations made in Eq. 1 are made in Eq. 2, adding the change of environment that occurs with time t.

A solution x belongs to the Pareto Optimum of time t (P^*_t) if there is no other solution of the search space Ω that dominates it at time t. The most important concept here is that the P^* may vary with time, depending on the environment changes. Therefore, we can assume that any change in the environment can interfere with the dominance relationship between the search space solutions. Two other concepts that we must consider in a dynamic problem are the severity of the change and the frequency of the change.

- The **severity of the change** means how fundamental the changes are in terms of magnitude [27].
- The **frequency of change** determines how often the environment changes, it is usually measured as the number of generations or the number of evaluations of the objective functions [27].

The Knapsack Problem (KP) is described as the challenge of filling a knapsack with available items, where the sum of the values of all items in the knapsack must be maximized. However, the bag has a weight limit, so the sum of the weights of the items in the knapsack cannot exceed this limit. In multiobjective formulation the items have different values and weights for each objective, so the same set of items inside the knapsack must respect the weight restriction of all objective functions and maximize the value of each objective function. This concept can be expressed in mathematical terms by the Eq. 3.

$$\sum_{j=1}^{m} [(max \sum_{i=0}^{n} v_{i,j} * x_i); (\sum_{i=0}^{n} w_{i,j} * x_i \leq Q_j)] \tag{3}$$

Consider a set of items I, where $i = 0, ..., n$ and a set of objectives J, where $j = 1, ..., m$; $v_{i,j}$ and $w_{i,j}$ represents the value and weight of item i in objective function j; x_i is a binary value that receives 1 if item i is in the knapsack and 0 otherwise; Q_j is the maximum weight in function j.

The dynamic characteristic adopted here for this problem is to make a modification in the set of objective functions in each environment change [11]. We also defined that the environment change occurs after a predefined interval of generations. The severity of change should be as low as possible. This recomendation was proposed in for [4], because this is more appropriate for investigate the dynamic multiobjective algorithms. There for, here only one objective function is varied per each environment change.

We generated twelve instances of DMKP by varying the number of objectives (4, 6 and 8), the number of items (30 and 50) and the number of environmental changes (1 and 2). We generate a set of objective functions for each instance. Each objective function contains the value and weight of each item and the maximum weight that the knapsack can support. We generate the value and weight of each item by raffling a number between 0 and 1000, the maximum weight is equal to 60% of the sum of the items weights.

For each environment change, the set of objective varies with the least severity possible, that is, only one objective changes. Before the comparative evaluation of the investigated algorithms, the P^* associated to each instance was approximated by running efficient many-objective algorithms (for static discrete multiobjective problems) using robust parameters (for example, more than 50,000 generations). For each instance, one P^* was needed to be calculated for each environment setting. For example, for the instances with just one environmental change, two P^* were calculated, the one related to the starting environment and the second related to the environment after the single change. In the same way, for the instances with two environmental changes, three P^* were calculated. The size of each approximate P^* varies from 300 to over 100 thousand solutions, as the complexity of the instance increases.

3 Dynamic Multiobjective Evolutionary Algorithms

The Non-dominated Sorting Genetic Algorithm II (NSGA-II) [8] is one of the most-known evolutionary algorithms for solving classical MOPs (static and dealing few objectives). Its basic behavior consists of dividing the population into hierarchical fronts, where a solution from an upper front has a better fitness than another solution from a lower front. The fronts are classified based on the concept of non-dominance, being that a solution that is not dominated by any other would be classified on the uppermost front. They are called the nondominated solutions of the current population and the goal is to approximate this uppermost front to the Pareto Optimum at the end of any arbitrary run. The other solutions are classified in a hierarchy of fronts where the solutions of an upper front dominates the solutions of lower fronts. The metric crowding distance is used to differentiate the solutions of the same front. This distance gives a better fitness for the solutions that are more isolated in the front.

The Dynamic Non-dominated Sorting Genetic Algorithm II (DNSGA-II) [9] is an adaptation of NSGA-II for dynamic problems using a strategy to introduce diversity. The basic evolutionary behavior of the NSGA-II was kept for the DNSGA-II, except in environment change. In such occurrence, a percentage of the current population is modified to promote diversity. Two strategies for this modification was evaluated in DNSGA-II [9] defining two versions of this algorithm. The first version was called DNSGA-II-A, where a percentage of the population is kept intact for the next generation after the environment change (a kind of elitism) and other new individuals are generated at random replacing the worst individuals. The second version was called DNSGA-II-B, where the elitism was also used to keep a percentage of the best solutions while the new individuals that replaces the worst ones are generated by applying mutation over the elite. According to the authors in [9] the performance of both versions are similar, being that DNSGA-II-A is most adequate for DMOPs with a low change frequency. We performed some exploratory experiments with the instances of DMKP investigated in this work and we conclude that the performance of both versions was extremely similar when considering the multiobjective metrics used to compare the evolutionay algorithms. For simplification, we decided to keep just the results of the version with strategy A that we called here just DNSGA-II.

MOEA/D [31] is an evolutionary multiobjective algorithm that decomposes the problem into several subproblems, which are evolved simultaneously. The decomposition is made by a scalarization function, which has the role of transforming each individual in a single scalar value (fitness). The current population is made up of the best solutions for each subproblem. The neighborhood of each subproblem is defined based on the distances between their weighting coefficient vectors. The technique usually used by MOEA/D is the Tchebycheff decomposition [25]. MOEA/D was originally proposed for multiobjective problems with few objectives but it has shown to be a successful algorithm also for many-objective formulations, when 4 or more objetives are used [12,13,24]. Muruganantham proposed an adaptation of MOEA/D for dynamic problems called MOEA/D-KF [26]. This adaptation maintains the evolutionary dynamics of MOEA/D, but modifies the population in the occurrence of an environment change. In such occurrence, part of the population is randomly generated and the other part is generated driven by a Kalman Filter [17].

Proposed by Wang and Li [33] MS-MOEA makes an evolutionary search based on the concept of non-dominance. The algorithm evolves by storing non-dominated solutions as the best solutions. However, as the population has a fixed size, some dominated solutions can be stored to fit this size. Each non-dominated offspring discovered in the current generation replaces another dominated solution in the population. If all solutions stored in the population are also non-dominated, the new non-dominated solution replaces another one at random. Whenever a non-dominated solutions is found, it is stored in an external archive to guarantee that any non-dominated solution is preserved at the end of the run. Since the archive grows, it also participates in the evolutionary process. MS-MOEA uses a memory-based strategy in its dynamic formulation. In any occurrence of an environment change, a part of the population is reinitialized: they are replaced by new individuals, which are generated applying mutation in the non-dominated solutions stored in the archive.

The Non-dominated Sorting Genetic Algorithm III (NSGA-III) is an extension of NSGA-II proposed to deal with (static) many-objectives MOPs [10]. It kept the same basic structure of NSGA-II, the hierarchical dominance fronts, but some adaptations were proposed to manipulate many-objective problems. Instead of using crowding distance, the algorithm applies a niche classification method based in the distance to points traced in a hyperplane, which better represents the difference between two solutions in high-dimensional spaces. It is an application of the reference-point approach previously proposed by Das and Dennis [7]. DNSGA-II was proposed based on NSGA-II and we could not find a similar approach using NSGA-III as a support for dynamic optimization. Therefore, we propose here a dynamic version of NSGA-III based on the strategy employed in DNSGA-II-A. This version is called here DNSGA-III, in which at each environment change a percentage of the population is randomly generated.

4   Major Experiments

Dynamic multiobjective evolutionary algorithms that have already been applied to continuous DMOPs are investigated here aiming to see if these algorithms adapt well to a discrete dynamic many-objective optimization problem. Four algorithms were investigated: DNSGA-II, DNSGA-III, MOEA/D-KF and MS-MOEA. The discrete problem used is the Dynamic Multiobjective Knapsack

Problem (DMKP), in its formulation where only the variation of objectives occurs in each environment change. We investigated instances of 30 and 50 items using formulations with 4, 6 and 8 objectives. Considering the environment change frequency, two types of instances were considered. In the first type, only one environment change happens exactly in the half of the run, whereas in the second two changes of the environment occurs in each third part of the total of generations. As a result, we built twelve instances of DMKP. For each instance, the evolutionary search was performed 100 times for each evaluated algorithm, thus calculating the mean and standard deviation of metrics over the 100 runs. In addition, the execution time of each algorithm was also computed. All algorithms run with the following parameters: 100 generations, 100 individuals, 100% crossover, 10% mutation. Considering these parameters, the number of fitness evaluated is the same for all algorithms. For DNSGA-II and DNSGA-III 20% of the population has been replaced in each environment change. In MS-MOEA, at each change of environment, 20% of the population is made up of individuals from the archive that has been mutated and 80% are random individuals. MOEA/D-KF uses 10 neighbors to form the sub-populations. To evaluate the algorithms performance two multiobjective metrics were used:

Mean Inverted Generational Distance ($\overline{IGD}$) [32]: Based on IGD [20], Wang and Li proposed to use the average of the IGD calculated one step before changing the environment:

$$IGD(P^*, P) = \frac{\sum_{v \in P} d(v, P^*)}{|P|}$$

$$\overline{IGD} = \frac{1}{EC} \sum_{i=1}^{nc} IGD_i$$

where P^* is the Pareto Optimum; P is the Pareto found in the search; $d(v, P^*)$ is the minimum Euclidean distance between v and P^*; EC is the number of environment changes; IGD_i is the value calculated before the occurrence of $(i+1)$th change in the environment. This metric assesses both convergence and diversity, the closer to zero the better the metric value.

Hypervolume Ratio (HVR) [30]: This metric represents the division of Hypervolume [34] of the Pareto found in the search(P) with the Pareto Optimum (P^*):

$$HVR(P^*, x) = \frac{HV(P, x)}{HV(P^*, x)}$$

where $HV(P, x)$ and $HV(P^*, x)$ is the calculation of the Hypervolume of P and P^*, where x is the reference point used by Hyper-volume.

4.1 DMKP Instances with Just One Environment Change (EC = 1)

First, we will analyse the results using instances where the environment change happens once at the starting of generation 51 (half of the run). Figure 1 shows the results of such instances with 30 items. MS-MOEA surpassed all other algorithms followed by DNSGA-II. As the number of objectives increases the supremacy of MS-MOEA becomes more clear. Surprisingly, DNSGA-II surpasses DNSGA-III, nonetheless the static formulation of the later was proposed for many-objective problems. MOEA/D-KF also did not perform as well as MS-MOEA and DNSGA-II, although its performance on HVR becomes higher as the number of objectives increases. Figure 2 shows the results of instances with 50 items and also with just one environment change. One can see that the behavior did not change with the increment on the number of items and the analysis is almost identical to the instances with 30 items.

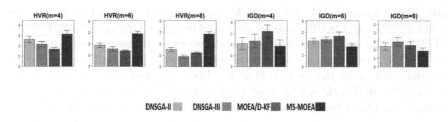

Fig. 1. HVR and $\overline{IGD}$ results for DMKP instances with 30 items (EC = 1).

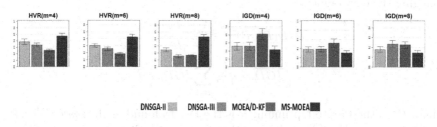

Fig. 2. HVR and $\overline{IGD}$ results for DMKP instances with 50 items (EC = 1).

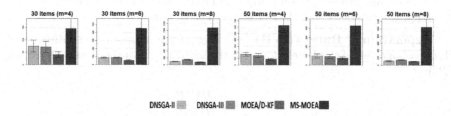

Fig. 3. Processing Time for DMKP instances (EC = 1): 30 and 50 items

Subsequently, the processing time of these experiments is analysed. Figure 3 shows the processing time of each algorithm for each instance previously analysed. MS-MOEA has clearly the longest execution time, being that the increase in items and objectives makes the processing time of MS-MOEA more expensive. MOEA/D-KF is the fastest among the investigated algorithms. DNSGA-II and DNSGA-III spend similar time processing, but we increase the number of objectives, DNSGA-III-A slows down compared to DNSGA-II.

Therefore, although MS-MOEA returned the best performance on the metrics $\overline{IGD}$ and HVR, these improvements cost a huge processing time. On the other hand, DNSGA-II is the second method considering the multiobjective metrics and it is much faster than MS-MOEA. Given the highlight of MS-MOEA and DNSGA-II, we made hypothesis test T [23] with 99% confidence to check if the difference between MS-MOEA and DNSGA-II and the other algorithms was significant. Table 1 shows the results of this test using MS-MOEA as the basis. The green cells indicate the MS-MOEA was significantly higher and the white ones indicate there was no significant difference. Analyzing this table, one can see that MS-MOEA is significantly better in all instances considering both HVR and $\overline{IGD}$. Table 2 shows the results of a similar test using DNSGA-II as the basis (MS-MOEA is ommited here because it was already analysed against NSGA-II in the previous table). Therefore, one can see that in general DNSGA-II is significantly better than DNSGA-III and MOEA/D-KF.

Table 1. Hypothesis test (EC=1). MS-MOEA vs All.

Algorithms	Items	Objectives					
		4		6		8	
		HVR	$\overline{IGD}$	HVR	$\overline{IGD}$	HVR	$\overline{IGD}$
DNSGA-II	30	>	<	>	<	>	<
	50	>	<	>	<	>	<
DNSGA-III	30	>	<	>	<	>	<
	50	>	<	>	<	>	<
MOEA/D-KF	30	>	<	>	<	>	<
	50	>	<	>	<	>	<

Table 2. Hypothesis test (EC=1). DNSGA-II vs All

Algorithms	Items	Objectives					
		4		6		8	
		HVR	$\overline{IGD}$	HVR	$\overline{IGD}$	HVR	$\overline{IGD}$
DNSGA-III	30	>	<	>	<	>	<
	50	>	=	>	=	>	<
MOEA/D-KF	30	>	<	>	<	>	=
	50	>	<	>	<	>	<

4.2   DMKP Instances with Two Environment Changes (EC=2)

The results of the experiments using DMKP instances where the environment changes twice are analysed here. They occur at generations 34 and 67, that is, at each third part of the run. Figures 4 and 5 show the results of the instances with 30 and 50 items respectively. As one can see, the comparative analysis of the four algorithms is very similar to the experiments with just one EC. This observation is corroborated by the results of the Hypothesis T tests shown in Tables 3 and 4 related to the results presented in Figs. 4 and 5. The processing time analysis is also very similar to the previous instances as one can see in 6.

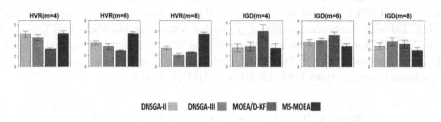

Fig. 4. HVR and $\overline{IGD}$ results for DMKP instances with 30 items (EC = 2).

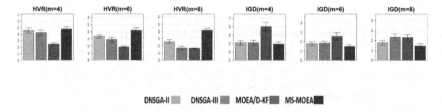

Fig. 5. HVR and $\overline{IGD}$ results for DMKP instances with 50 items (EC = 2).

Table 3. Hypothesis test (EC = 2). MS-MOEA vs All.

Algorithms	Items	Objectives					
		4		6		8	
		HVR	$\overline{IGD}$	HVR	$\overline{IGD}$	HVR	$\overline{IGD}$
DNSGA-II	30	=	=	>	<	>	<
	50	>	<	>	<	>	<
DNSGA-III	30	>	=	>	<	>	<
	50	>	<	>	<	>	<
MOEA/D-KF	30	>	<	>	<	>	<
	50	>	<	>	<	>	<

Table 4. Hypothesis test (EC = 2). DNSGA-II vs All.

Algorithms	Items	Objectives					
		4		6		8	
		HVR	$\overline{IGD}$	HVR	$\overline{IGD}$	HVR	$\overline{IGD}$
DNSGA-III	30	>	=	>	<	>	<
	50	>	=	>	=	>	<
MOEA/D-KF	30	>	<	>	<	>	<
	50	>	<	>	<	>	<

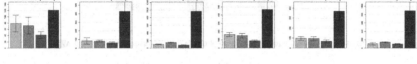

DNSGA-II   DNSGA-III   MOEA/D-KF   MS-MOEA

Fig. 6. Processing Time for DMKP instances (EC = 2): 30 and 50 items

5   Additional Experiments: DNSGA-II and DNSGA-II*

Although MS-MOEA outperforms the other algorithms considering the metrics HVR and $\overline{IGD}$, it takes the longest time to be executed even with the guarantee that the same number of fitness evaluations were used for all the algorithms. On the other hand, DNSGA-II returned a much shorter execution time (specially if one consider the experiments using 6 and 8 objectives) and a good performance on the multiobjective metrics, being that DNSGA-II overcame the other two investigated algorithms (DNSGA-III and MOEA/D-KF). We decided to investigate how the additional time spent using MS-MOEA could improve its metric results and trying to improve DNSGA-II to see if also an extra time could be used for improving its own results. We performed these additional experiments using the instance of 30 items and 8 objectives, because the instances with more objectives returned more discriminant results between MS-MOEA and DNSGA-II and experiments with 30 items are much faster than using 50.

First, we performed some preliminary runs using DNSGA-II by increasing the number of individuals and/or the number of generations. Although both modifications make DNSGA-II spent a longer time (as expected), it was clear that the metrics were more susceptible to an increment in the population size. Therefore, we carried out new runs of DNSGA-II increasing its population size (200, 300, 400 and 500), but fixing the number of generations in 100. Figure 7 shows these results, where we also replicate the results of DNSGA-II and MS-MOEA in the previous experiments with 100 individuals.

One can observe that the successive increase in the population size makes DNSGA-II to improve its metrics towards MS-MOEA results. However, even

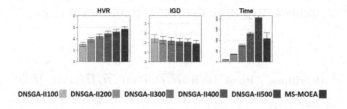

Fig. 7. MS-MOEA vs DNSGA-II results (30 items and 8 objectives)

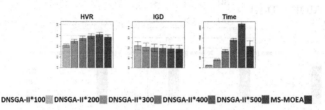

Fig. 8. MS-MOEA vs DNSGA-II* results (30 items and 8 objectives)

using Tp = 500, it was unable to reach the MS-MOEA results. Moreover, starting from 400 individuals, DNSGA-II has a higher computational cost than MS-MOEA. The better explanation we found for this limitation on DNSGA-II compared to MS-MOEA is the employment of an unrestricted external archive in MS-MOEA to store the nondominated solutions found during all generations, while DNSGA-II stores them in its restricted-size current population and some Pareto solutions could be lost over the evolution. In fact, the number of non-dominated solutions in the external archive presented in MS-MOEA runs were around 1000 for the 30-items 8-objectives instance. Therefore, to store so many nondominated solutions, DNSGA-II needs to be executed with Tp=1000 what was not much practical and it will takes much longer time than MS-MOEA. As a promising alternative, we decided to implement a modification over DNSGA-II to also include an external archive in a similar way that it works in MS-MOEA. We call this algorithm version as DNSGA-II* and we carried out additional experiments showed in Fig. 8 using DNSGA-II* by increasing its population size (200, 300, 400 and 500), but with 100 generations. It is possible to observe that there was a significant improvement in all executions of DNSGA-II* comparing it with the standard DNSGA-II. Due to the inclusion of the external file, better metrics (mainly HVR) was achieved with an small increase in processing time.

6   Conclusion

A comparative analysis was carried out concerning the performance of four optimization evolutionary algorithms when they were applied to solve a discrete dynamic problem with many-objectives: the Dynamic Multiobjective Knapsack Problem. Three of them (DNSGA-II, MOEA/D-KF and MS-MOEA) were extracted from the literature being that they were previously applied to optimization problems defined over continuous spaces. DNSGA-III is proposed here based on DNSGA-II by changing the subjacent multiobjective search for the NSGA-III.

Concerning the multiobjective metrics used, HVR and $\overline{IGD}$, MS-MOEA clearly surpassed the other three. However, it takes a much longer processing time. On the other way, DNSGA-II performed well (the second best) in a smaller time. Therefore, both algorithms could be considered efficient for this problem, depending if someone is worried about time. If both the performance of the metrics and the processing time are equally important, the DNSGA-II is the most recommended. We believe that the employment of an external archive is responsible both for MS-MOEA good performance and expensive processing. This information helps the algorithm to adapt to environmental changes but it consumes an extra time to verify the dominance of each new solution against all the others stored in the current archive. This observation led us to propose a modification for DNSGA-II by using an external archive. The resulting algorithm was called here DNSGA-II*, which returned improved results compared to the original one and taking a small extra computational cost. We concluded that the modified DNSGA-II* is the best option among the evaluated algorithms for the investigated DMKP instances. It returns good HVR and IGD+ metrics with a reasonable processing time. Finally, a disappointing result is related to DNSGA-III since we expected that the subjacent NSGA-III search would give competitive advantages to the DNSGA-II in many-objective formulations. On the contrary, its performance decays as the number of objectives is incremented.

For future work, we intend to enlarge the number of scenarios with DMKP by increasing the number of items or objectives to improve our analysis. We also intend to investigate scenarios with more environment changes (from 3 to 50 ECs) to have more dynamic challenges for the algorithms. Such experiments could help us to clarify if the best results obtained with MS-MOEA are due to an interesting behavior for the dynamic problem or they came from its better performance for the "almost static" MKP, with few and sparse changes.

Besides, we are planning to enlarge the number of many-objective evolutionary algorithms to be investigated as the subjacent search. MEANDS [21] and MEANDS-II [22] are natural candidates to this adaptation since they have shown a good performance when applied to discrete static many-objective problems.

Acknowledgment. The authors thank FAPEMIG, CAPES and CNPq.

References

1. Azzouz, R., Bechikh, S., Ben Said, L.: Dynamic multi-objective optimization using evolutionary algorithms: a survey. In: Bechikh, S., Datta, R., Gupta, A. (eds.) Recent Advances in Evolutionary Multi-objective Optimization. ALO, vol. 20, pp. 31–70. Springer, Cham (2017). https://doi.org/10.1007/978-3-319-42978-6_2

2. Azzouz, R., Bechikh, S., Ben Said, L.: Multi-objective optimization with dynamic constraints and objectives: new challenges for evolutionary algorithms. In: Proceedings of the 2015 Annual Conference on Genetic and Evolutionary Computation, pp. 615–622, July 2015
3. Baykasoglu, A., Ozsoydan, F.B.: An improved firefly algorithm for solving dynamic multidimensional knapsack problems. Expert Syst. Appl. **41**(8), 3712–3725 (2014)
4. Branke, J.: Evolutionary optimization in dynamic environments. In: Genetic Algorithms and Evolutionary Computation, vol. 3. Kluwer Academic Publishers, Dordrecht (2001)
5. Branke, J., Orbayı, M., Uyar, Ş.: The role of representations in dynamic knapsack problems. In: Rothlauf, F., et al. (eds.) EvoWorkshops 2006. LNCS, vol. 3907, pp. 764–775. Springer, Heidelberg (2006). https://doi.org/10.1007/11732242_74
6. Cámara, M., Ortega, J., Toro, F.J.: Parallel processing for multi-objective optimization in dynamic environments. In: 2007 IEEE International Parallel and Distributed Processing Symposium, pp. 1–8. IEEE, March 2007
7. Das, I., Dennis, J.E.: Normal-boundary intersection: a new method for generating the Pareto surface in nonlinear multicriteria optimization problems. SIAM J. Optim. **8**(3), 631–657 (1998)
8. Deb, K., Pratap, A., Agarwal, S., Meyarivan, T.A.M.T.: A fast and elitist multi-objective genetic algorithm: NSGA-II. IEEE Trans. Evol. Comput. **6**(2), 182–197 (2002)
9. Deb, K., Rao N., U.B., Karthik, S.: Dynamic multi-objective optimization and decision-making using modified NSGA-II: a case study on hydro-thermal power scheduling. In: Obayashi, S., Deb, K., Poloni, C., Hiroyasu, T., Murata, T. (eds.) EMO 2007. LNCS, vol. 4403, pp. 803–817. Springer, Heidelberg (2007). https://doi.org/10.1007/978-3-540-70928-2_60
10. Deb, K., Jain, H.: An evolutionary many-objective optimization algorithm using reference-point-based nondominated sorting approach, part I: solving problems with box constraints. IEEE Trans. Evol. Comput. **18**(4), 577–601 (2014)
11. Farina, M., Deb, K., Amato, P.: Dynamic multiobjective optimization problems: test cases, approximations, and applications. IEEE Trans. Evol. Comput. **8**(5), 425–442 (2004)
12. França, T.P., de Queiroz Lafetá, T.F., Martins, L.G.A., de Oliveira, G.M.B.: A comparative analysis of moeas considering two discrete optimization problems. In: 2017 Brazilian Conference on Intelligent Systems (BRACIS), pp. 402–407. IEEE, October 2017
13. França, T.P., Martins, L. G., Oliveira, G.M.: MACO/NDS: many-objective ant colony optimization based on non-dominated sets. In: 2018 IEEE Congress on Evolutionary Computation (CEC), pp. 1–8. IEEE, July 2018
14. Goh, C.K., Tan, K.C.: A competitive-cooperative coevolutionary paradigm for dynamic multiobjective optimization. IEEE Trans. Evol. Comput. **13**(1), 103–127 (2008)
15. Ishibuchi, H., Tsukamoto, N., Nojima, Y.: Evolutionary many-objective optimization: a short review. In: 2008 IEEE Congress on Evolutionary Computation (IEEE World Congress on Computational Intelligence), pp. 2419–2426. IEEE, June 2008
16. Jin, Y., Branke, J.: Evolutionary optimization in uncertain environments-a survey. IEEE Trans. Evol. Comput. **9**(3), 303–317 (2005)
17. Kalman, R.E., Bucy, R.S.: New results in linear filtering and prediction theory (1961)
18. Kellerer, H., Pferschy, U., Pisinger, D.: Multidimensional knapsack problems. In: Knapsack Problems, pp. 235–283. Springer, Heidelberg (2004)

19. Koo, W.T., Goh, C.K., Tan, K.C.: A predictive gradient strategy for multiobjective evolutionary algorithms in a fast changing environment. Memetic Comput. **2**(2), 87–110 (2010)

20. Knowles, J., Corne, D.: On metrics for comparing nondominated sets. In: Proceedings of the 2002 Congress on Evolutionary Computation. CEC 2002 (Cat. No. 02TH8600), vol. 1, pp. 711–716). IEEE, May 2002

21. Lafeta, T.F.Q., Bueno, M.L., Brasil, C., Oliveira, G.M.: MEANDS: a many-objective evolutionary algorithm based on non-dominated decomposed sets applied to multicast routing. Appl. Soft Comput. **62**, 851–866 (2018)

22. Lafeta, T.F.Q., Oliveira, G.M.B.: An improved version of a many-objective evolutionary algorithm based on non-dominated decomposed sets (MEANDS-II). In 2019 IEEE International Conference on Systems, Man and Cybernetics (SMC), pp. 3673–3678. IEEE, October 2019

23. Mankiewicz, R.: The story of mathematics. Cassell (2000)

24. Martins, L., França, T., Oliveira, G.: Bio-inspired algorithms for many-objective discrete optimization. In: 2019 8th Brazilian Conference on Intelligent Systems (BRACIS), pp. 515–520. IEEE, October 2019

25. Miettinen, K.: Nonlinear Multiobjective Optimization, vol. 12. Springer Science and Business Media (2012)

26. Muruganantham, A., Tan, K.C., Vadakkepat, P.: Solving the IEEE CEC 2015 dynamic benchmark problems using kalman filter based dynamic multiobjective evolutionary algorithm. Intelligent and Evolutionary Systems. PALO, vol. 5, pp. 239–252. Springer, Cham (2016). https://doi.org/10.1007/978-3-319-27000-5_20

27. Richter, H.: Dynamic fitness landscape analysis. In: Yang, S., Yao, X. (eds.) Evolutionary Computation for Dynamic Optimization Problems, vol. 490, pp. 269–297. Springer, Heidelberg (2013). https://doi.org/10.1007/978-3-642-38416-5_11

28. Roy, R., Mehnen, J.: Dynamic multi-objective optimisation for machining gradient materials. CIRP Ann. **57**(1), 429–432 (2008)

29. Sahmoud, S., Topcuoglu, H.R.: A memory-based NSGA-II algorithm for dynamic multi-objective optimization problems. In: Squillero, G., Burelli, P. (eds.) EvoApplications 2016. LNCS, vol. 9598, pp. 296–310. Springer, Cham (2016). https://doi.org/10.1007/978-3-319-31153-1_20

30. Zitzler, E., Thiele, L.: Multiobjective evolutionary algorithms: a comparative case study and the strength Pareto approach. IEEE Trans. Evol. Comput. **3**(4), 257–271 (1999)

31. Zhang, Q., Li, H.: MOEA/D: a multiobjective evolutionary algorithm based on decomposition. IEEE Trans. Evol. Comput. **11**(6), 712–731 (2007)

32. Wang, Y., Li, B.: Investigation of memory-based multi-objective optimization evolutionary algorithm in dynamic environment. In: 2009 IEEE Congress on Evolutionary Computation, pp. 630–637. IEEE, May 2009

33. Wang, Y., Li, B.: Multi-strategy ensemble evolutionary algorithm for dynamic multi-objective optimization. Memet. Comput. **2**(1), 3–24 (2010)

34. While, L., Bradstreet, L., Barone, L.: A fast way of calculating exact hypervolumes. IEEE Trans. Evol. Comput. **16**(1), 86–95 (2011)

Backtracking Group Search Optimization: A Hybrid Approach for Automatic Data Clustering

Luciano Pacifico[1]([⊠]) and Teresa Ludermir[2]

[1] Departamento de Computacao (DC), Universidade Federal Rural de Pernambuco,
Recife, PE, Brazil
luciano.pacifico@ufrpe.br
[2] Centro de Informatica, Universidade Federal de Pernambuco, Recife, PE, Brazil
tbl@cin.ufpe.br

Abstract. Data clustering is one of the most primitive tasks in pattern recognition, although it is known to be a NP-hard grouping task. Given its complexity, standard clustering methods, such as the partitional data clustering algorithms, are easily trapped in local minimum solutions, due to their lack of good global searching mechanisms. Evolutionary Algorithms (EAs) and Swarm Intelligence (SIs) methods, such as Group Search Optimization (GSO) and Backtracking Search Optimization (BSA), are commonly employed to deal with clustering task, given their capabilities to handle global search problems. In this work, a new hybrid evolutionary algorithm between GSO and BSA is presented, named BGSO, to tackle clustering problem, which combines the best features of GSO and the historical mechanisms of BSA. Also, BGSO is developed in the context of Automatic Clustering approach, which means that it is able to predict the best number of final clusters, so no prior assumption about the data set at hand is required. The proposed approach is compared to standard GSO, BSA and other three EAs and SIs from the literature by means of nine real-world problems, showing promising results considering four clustering metrics.

Keywords: Automatic data clustering · Group search optimization · Backtracking search optimization

1 Introduction

From the past few decades, the amount of data daily produced has increased exponentially. In real life systems, the need for reliable and fast techniques capable of discovering patterns from large data sets is mandatory, given that it is impossible for a human being to analyze massive amounts of data in a short period of time. The process of extracting useful information from raw data is known as Data Mining, and it is a fundamental piece to the process of Knowledge Discovery in Databases (KDD).

© Springer Nature Switzerland AG 2020
R. Cerri and R. C. Prati (Eds.): BRACIS 2020, LNAI 12319, pp. 64–78, 2020.
https://doi.org/10.1007/978-3-030-61377-8_5

Clustering analysis is a field in pattern recognition whose goal is to distribute a set of objects (*patterns*) in categories (*clusters* or *groups*) in such a way that individuals from the same cluster have a high degree of similarity among each other, while individuals from different clusters have a high degree of dissimilarity among each other [23]. The task is performed based only on the information retained in each individual pattern, which makes no prior knowledge required. When applied to a data set, clustering methods are capable of identifying hidden properties present in the data patterns, which makes these techniques very useful for statistical data analysis and exploration in many fields, such as engineering, image understanding, text analysis, engineering, medicine, bioinformatics, and so on [25,41].

The most popular clustering approaches are the partitional clustering algorithms. Partitional algorithms provide a partition of the data set into a prefixed number of clusters (an input parameter for the partitional algorithm). Each cluster is represented by its centroid vector, and the clustering process is driven in an effort to optimize a criterion function iteratively, and, in each step of the execution, all centroids are updated in an attempt to improve the quality of the final solution (best partition found so far). Partitional methods are known for their sensibility to the initial centroid position, what may lead to weak solutions (i.e., the partitional approach may be trapped in a local minimum point) if the algorithm starts in a poor region of the problem space.

From an optimization perspective, clustering is considered as a particular kind of NP-hard grouping problem. Evolutionary Algorithms (EAs) have been increasingly applied to solve a great variety of complex problems, given their capabilities to perform global searches over difficult environments and spaces. In EAs, a population of candidate solutions for the problem at hand is kept and evolved according to a generational process. EAs such as Genetic Algorithm (GA) [22], Differential Evolution (DE) [37] and Backtracking Search Optimization (BSA) [8] perform their search driven by operators that simulate biological processes like mutation, recombination and selection. In this context, Swarm Intelligence (SI) methods are extensions of EAs, in which all the searching operators are employed as attempts to simulate the self-organizing collective behavior of social animals, like swarming, flocking and herding [4]. Examples of SI algorithms are the Ant Colony Optimization (ACO) [14], Particle Swarm Optimization (PSO) [27] and Group Search Optimization (GSO) [20]. Both EAs and SIs searching strategies are guided in an attempt to optimize a criterion function, the fitness function.

EAs and SIs have been successfully adapted from the past years as partitional clustering algorithms [6,35], showing promising results when compared to standard partitional clustering techniques. Another issue concerning the partitional clustering methods is the estimation of the best number of cluster to compose the final partition for a given problem. The manual determination of the optimal number of clusters requires an amount of prior knowledge concerning the problem at hand that may not be available, what limits the application of such algorithms. To avoid that limitation, many Automatic Clustering algorithms have been proposed recently [7,10,16,30,33,38], which try, at the same time, to find the best cluster centroids and to predict the best number of clusters [26] to represent a data set.

In this work, a new hybrid algorithm between Group Search Optimization and Backtracking Search Optimization is proposed, the BGSO, which combines the effective global search mechanisms of GSO with the backtracking features of BSA, in an attempt to combine the best of both worlds to solve clustering problems. BGSO is employed as an Automatic Clustering model that will optimize the number of clusters composing the final partition of the problem and, at the same time, will find the best cluster centroids for each group.

This work is organized as follows. Section 2 presents GSO algorithm, followed by a brief introduction on BSA model (Sect. 3). Next (Sect. 4), the proposed BGSO algorithm is presented. Experimental results are shown in Sect. 5, followed by some conclusions and leads to future works (Sect. 6).

2 Group Search Optimization

Group search optimization is inspired by animal social searching behavior and group living theory. GSO employs the Producer-Scrounger (PS) model as a framework. The PS model was firstly proposed by Barnard and Sibly [3] to analyze social foraging strategies of group living animals. PS model assumes that there are two foraging strategies within groups: *producing* (*e.g.*, searching for food); and joining (*scrounging*, *e.g.*, joining resources uncovered by others). Foragers are assumed to use producing or joining strategies exclusively. Under this framework, concepts of resource searching from animal visual scanning mechanism are used to design optimum searching strategies in GSO algorithm [20].

In GSO, the population G of S individuals is called *group*, and each individual is called a *member*. In a n-dimensional search space, the i-th member at the t-th searching iteration (*generation*) has a current position $\mathbf{X}_i^t \in \Re^n$ and a head angle $\alpha_i^t \in \Re^{n-1}$. The search direction of the i-th member, which is a vector $\mathbf{D}_i^t(\alpha_i^t) = (d_{i1}^t, \ldots, d_{in}^t)$ can be calculated from α_i^t via a polar to Cartesian coordinate transformation:

$$d_{i1}^t = \prod_{q=1}^{n-1} \cos(\alpha_{iq}^t),$$

$$d_{ij}^t = \sin(\alpha_{i(j-1)}^t) \prod_{q=1}^{n-1} \cos(\alpha_{iq}^t)(j = 1, \ldots, n-1),$$

$$d_{in}^t = \sin(\alpha_{i(n-1)}^t) \tag{1}$$

A group in GSO consists of three types of members: producers, scroungers and dispersed members (or *rangers*) [20]. The rangers are introduced by GSO mo-del, extending standard PS framework.

During each GSO search iteration, a group member which has found the best fitness value so far (most promising area form the problem search space) is chosen as the producer ($\mathbf{X}_p$) [9], and the remaining members are scroungers or rangers.

The producer employs a scanning strategy (*producing*) based on its vision field, generalized to a n-dimensional space, which is characterized by maximum pursuit angle $\theta_{max} \in \Re^{n-1}$ and maximum pursuit distance $l_{max} \in \Re$, given by Eq. (2).

$$l_{max} = \|\mathbf{U} - \mathbf{L}\| = \sqrt{\sum_{k=1}^{n}(U_k - L_k)^2} \tag{2}$$

where U_k and L_k denote the upper bound and lower bound of the k-th dimension from the problem space, respectively.

In GSO, at the t-th iteration the producer $\mathbf{X}_p^t$ will scan laterally by randomly sampling three points in the scanning field: one at zero degree (Eq. (3)), one in the right hand side hypercube (Eq. (4)) and one in the left hand side hypercube (Eq. (5)).

$$\mathbf{X}_z = \mathbf{X}_p^t + r_1 l_{max} \mathbf{D}_p^t(\alpha_p^t) \tag{3}$$

$$\mathbf{X}_r = \mathbf{X}_p^t + r_1 l_{max} \mathbf{D}_p^t(\alpha_p^t + \frac{\mathbf{r}_2 \theta_{max}}{2}) \tag{4}$$

$$\mathbf{X}_l = \mathbf{X}_p^t + r_1 l_{max} \mathbf{D}_p^t(\alpha_p^t - \frac{\mathbf{r}_2 \theta_{max}}{2}) \tag{5}$$

where $r_1 \in \Re$ is a normally distributed random number (mean 0 and standard deviation 1) and $\mathbf{r}_2 \in \Re^{n-1}$ is a uniformly distributed random sequence in the range $U(0, 1)$.

If the producer is able to find a better resource than its current position, it will fly to this point; if no better point is found, the producer will stay in its current position, then it will turn its head to a new generated angle (Eq. (6)).

$$\alpha_p^{t+1} = \alpha_p^t + \mathbf{r}_2 \alpha_{max} \tag{6}$$

where $\alpha_{max} \in \Re$ is the maximum turning angle.

If after $a \in \Re$ iterations the producer cannot find a better area, it will turn its head back to zero degree (Eq. (7)).

$$\alpha_p^{k+a} = \alpha_p^k \tag{7}$$

All scroungers will join the resource found by the producer, performing *scrounging* strategy according to Eq. (8).

$$\mathbf{X}_i^{t+1} = \mathbf{X}_i^t + \mathbf{r}_3 \circ (\mathbf{X}_p^t - \mathbf{X}_i^t) \tag{8}$$

where $\mathbf{r}_3 \in \Re^n$ is a uniform random sequence in the range $U(0, 1)$ and $\circ$ is the Hadamard product or the Schur product, which calculates the entrywise product of two vectors.

The rangers will perform random walks through the problem space [21], according to Eq. (9).

$$\mathbf{X}_i^{t+1} = \mathbf{X}_i^t + l_i \mathbf{D}_i^t(\alpha_i^{t+1}) \tag{9}$$

where

$$l_i = a r_1 l_{max} \tag{10}$$

In GSO, when a member escapes from the search space bounds, it will turn back to its previous position inside the search space [13]. GSO algorithm is presented in Algorithm 1.

Algorithm 1. GSO

$t \leftarrow 0$.
Initialize randomly position $\mathbf{X}_i^{(0)}$ and head angles $\alpha_i^{(0)}$ of all members $\mathbf{X}_i^{(0)} \in G$.
Calculate the fitness value $(fitness(\mathbf{X}_i^{(0)}))$ for each member $\mathbf{X}_i^{(0)}$.
while (termination conditions are not met) **do**
 Pick the best group member as the $\mathbf{X}_p^t$ for the current generation.
 Execute producing ($\mathbf{X}_p^t$ only) by evaluating three random points in its visual scanning field, $\mathbf{X}_z^t$ (eq. (3)), $\mathbf{X}_r^t$ (eq. (4)) and $\mathbf{X}_l^t$ (eq. (5)).
 Choose a percentage from the members (but the $\mathbf{X}_p^t$) to perform scrounging (eq. (8)).
 Ranging: The remaining members will perform *ranging* through random walks (eq. (9)).
 Calculate the new fitness value $fitness(\mathbf{X}_i^t)$ for each group member $\mathbf{X}_i^t$.
 $t \leftarrow t + 1$.
end while
Return $\mathbf{X}_p^{t_{max}}$.

---

3   Backtracking Search Optimization

BSA is an EA designed to be a global optimizer [8]. BSA has a single control parameter and a simple structure that is effective and capable of solving different optimization problems. Furthermore, BSA is a population-based method and possesses a memory in which it stores a population from a randomly chosen previous generation for generating the search-direction matrix. BSA is divided in five processes: Initialization, Selection-I, Mutation, Crossover and Selection-II.

In BSA, the population G is randomly initialized on the problem search space. BSA's the historical population $oldG$ is determined as in Eq. (11).

$$oldG^{t+1} = \begin{cases} G^t, & \text{if } rand_i < rand_j \\ oldG^t, & \text{otherwise} \end{cases} \tag{11}$$

where $rand_i$ and $rand_j$ are values obtained from a uniformly distributed random sequence in the range $U(0, 1)$. After that, the order of the individual in $oldG$ is changed (*permuted*) according to Eq. (12).

$$oldG = permuting(oldG) \tag{12}$$

A trial population M is generated using the mutation operator using Eq. (13).

$$M^t = G^t + F \otimes (oldG^t - G^t) \tag{13}$$

where F is a parameter amplitude factor (generally set as $F = 3 \cdot N$, where $N(0,1)$ is the standard normal distribution).

The crossover operator is obtained by a two-steps process: firstly, calculates a binary integer-valued matrix (map) of size $S \times n$ (where S is the population size, and n is the number of dimensions of each population individual) to indicate the mutant individual to be manipulated by using the relevant individual; after that, the relevant dimensions of mutant individual are updated by using the relevant individual. The BSA's crossover operator is described in Algorithm 2.

Algorithm 2. Crossover Operator in BSA

Initiate $map_{1:S,1:n} := 1$.
if $a < b | a, b \in U(0,1)$ **then**
  for i from 1 to S **do**
    $map_{i,u_{(1:\lceil mixrate \cdot rand \cdot n \rceil)}} := 0 | u = permuting(<1,2,\ldots,n>)$.
  end for
else
  for i from 1 to S **do**
    $map_{i,randi(n)} := 0$
  end for
end if-else
$T := M$
for i from 1 to S **do**
  for j from 1 to n **do**
    if $map_{i,j} = 1$ **then**
      $T_{i,j} := G_{i,j}$
    end if
  end for
end for

---

where T is a trial population, $\lceil \rceil$ is the ceiling function, $mixrate$ is the parameter in BSA's crossover process that controls the number of elements of individuals that will mutate in a trial by using $\lceil mixrate \cdot rand \cdot n \rceil$). After crossover, a boundary control mechanism is applied to avoid out-bounded individuals [8].

The Selection-II operator is performed after the evaluation of trial population T: each individual in T^t with a better fitness value than its correspondent individual in G^t will be selected to compose the new population G^{t+1} (Eq. (14)).

$$\mathbf{G}_i^{t+1} = \begin{cases} \mathbf{T}_i^t, & \text{if } fitness(\mathbf{T}_i^t) \text{ is better than } fitness(\mathbf{X}_i^t) \\ \mathbf{X}_i^t, & \text{otherwise} \end{cases} \tag{14}$$

The BSA algorithm is presented in Algorithm 3.

Algorithm 3. BSA

$t \leftarrow 0$.

Initialize randomly position $\mathbf{X}_i^{(0)} \in G$.

Calculate the fitness value ($fitness(\mathbf{X}_i^{(0)})$) for each individual $\mathbf{X}_i^{(0)}$.

while (termination conditions are not met) **do**

    Update and permute the historical population matrix $oldG^t$ using Selection-I operator, according to eq. (11) and eq. (12).

    Execute Mutation operator (eq. (13)).

    Generate the trial population T^t by the application of Crossover operator (Algorithm 2).

    Apply a boundary control mechanism to out-bounded individuals from T^t.

    Calculate the fitness value for each individual $fitness(\mathbf{T}_i^t)$ from T^t.

    Determine the new population G^{t+1} using Selection-II operator (eq. (14)).

    $t \leftarrow t + 1$.

end while

Return $\mathbf{X}_best^{t_{max}}$.

---

4 Proposed Approach: Backtracking Group Search Optimization

This section presents the proposed partitional algorithm for automatic data clustering: the BGSO. BGSO executes a global search through the problem space, while, in parallel, performs an attempt to predict the best number of clusters for the final partition of a given data set.

Formally, consider a partition P_C of a data set with N patterns $\mathbf{x}_j \in \Re^m$ ($j = 1, 2, ..., N$) in at most C_{max} clusters. Each cluster is represented by its centroid vector $\mathbf{g}_c \in \Re^m$ ($c = 1, 2, ..., C_{max}$). Each population individual $\mathbf{X}_i \in \Re^n$ (where $n = C_{max} + C_{max} \times m$) in population G represents C_{max} activation threshold values and C_{max} cluster centroids at the same time, one for each candidate cluster [7,10,16,26], as illustrated in Fig. 1.

At the t-th generation, the $\mathbf{X}_i^t$ individual will be evaluated by considering only its cluster centroids that are active, that is, cluster centroids with a threshold value such that $t_{ic}^t \geq 0.5$. Many functions are commonly adopted as the fitness function in Automatic Clustering applications [26], such as Dunn Index [15], Calinski-Harabasz Index [5] and Davies-Bouldin Index [11]. Such measures seek out the optimization of both the number of clusters and the cluster centroids themselves at the same time.

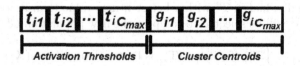

Fig. 1. Individual representation: the first C_{max} features represent activation thresholds for each candidate cluster, while the following $C_{max} \times m$ are the C_{max} m-dimensional candidate cluster centroids.

The initialization process is executed by the random choice of C_{max} patterns from the data set currently in analysis to compose the initial cluster centroids, for each individual $\mathbf{X}_i^{(0)}$, as much as the determination of each activation threshold $t_{ic}^{(0)}$ (where $c = 1, 2, \ldots, C_{max}$) by picking a value from a uniformly distributed random sequence in the range $U(0, 1)$.

After the initialization and the evaluation of $G^{(0)}$ according to the selected fitness function, the generational process begins. In BGSO, there will be three types of members as in GSO (producers, scroungers and rangers). BGSO implements a searching strategy that takes advantage of both GSO and BSA mechanisms. In each generation, half the scroungers will follow the recombination mechanisms of BSA, while the other half will perform scrounging as in standard GSO. Each BGSO generation starts by the execution of Selection-I operator, just like in BSA. After the determination and reorganization of the current historical population matrix $oldG^t$ (see Eq. (11) and Eq. (12)), the current best member $\mathbf{X}_p^t$ is selected to perform producing operator. The remaining members are chosen as scroungers or rangers by a given probability.

In BGSO, if at the t-th generation the i-th member $\mathbf{X}_i^t$ is chosen as a scrounger, it will perform, with a fifty percent probability, scrounging operator as in GSO (Eq. (8)); otherwise, it will perform recombination operator (mutation and crossover) just like in BSA (see Eq. (13) and Algorithm 2). Rangers will execute raging operator as in GSO (Eq. (9)). Both rangers and scroungers will compose the current trial population T^t. After that, GSO boundary control mechanism is applied to the trial population T^t, avoiding out-bounded members. If at any time the i-th $\mathbf{X}_i^t$ member represents less than two active clusters, it will be randomly reinitialized using the same process adopted on initialization step. Finally, the trial population T^t is evaluated and BSA's Selection-II operator (Eq. (14)) is used to determine the new population G^{t+1}.

BGSO algorithm is presented in Algorithm 4.

5   Experimental Analysis

In this section, we test the clustering capabilities of the proposed BGSO, in comparison to five other automatic clustering evolutionary and swarm intelligence algorithms, by means of nine real-world data sets: Banknote Authentication, Breast Cancer Wisconsin, Pima Indians Diabetes, Heart (Statlog), Ionosphere, Iris, Page Blocks Classification, Seeds and Waveform. All real-world data sets are benchmark classification and clustering problems acquired from UCI Machine Learning Repository [2]. The selected real data set features are shown in Table 1, presenting different degrees of difficulties, such as unbalanced and overlapping classes, different number of classes and features, and so on.

For comparison purposes, four clustering measures are employed: the Calinski-Harabasz Index (CH) [5], the Corrected Rand Index (CR) [24], the Davies-Bouldin Index (DB) [11], and the Jaccard Index (JI) [19].

Algorithm 4. BGSO

$t \leftarrow 0$.

Initialization: For each member $\mathbf{X}_i^{(0)} \in G^{(0)}$, pick C_{max} patterns randomly as the initial cluster centroids $\mathbf{g}_{ic}(c = 1, 2, \ldots, C_{max})$. Randomly determine the cluster activation thresholds $t_{ic}^{(0)}$ and head angles $\alpha_i^{(0)}$ of all members $\mathbf{X}_i^{(0)} \in G^{(0)}$. After that, assign each pattern $\mathbf{x}_j$ to its closest active cluster.

Calculate the fitness value $(fitness(\mathbf{X}_i^{(0)}))$ for each member $\mathbf{X}_i^{(0)}$.

while (termination conditions are not met) **do**

    Update and permute the historical population matrix $oldG^t$ using Selection-I operator, according to eq. (11) and eq. (12).

    Execute Mutation operator (eq. (13)).

    Generate the trial population T^t by the application of Crossover operator (Algorithm 2).

    Pick the best group member as the $\mathbf{X}_p^t$ for the current generation.

    Execute producing ($\mathbf{X}_p^t$ only) by evaluating three random points in its visual scanning field, $\mathbf{X}_z^t$ (eq. (3)), $\mathbf{X}_r^t$ (eq. (4)) and $\mathbf{X}_l^t$ (eq. (5)). For each evaluated point ($\mathbf{X}_z^t$, $\mathbf{X}_r^t$ and $\mathbf{X}_l^t$), determine its partition by assigning each data pattern to the active cluster with the nearest centroid.

    Choose a percentage from the members (but the $\mathbf{X}_p^t$) to perform scrounging:

    if $rand < 0.5$ **then**

        Execute scrounging according to eq. (8). Replace the $\mathbf{T}_i^t$ member by the result of scrounging operator in trial population T^t.

    else

        Keep the already determined $\mathbf{T}_i^t$ member in trial population T^t.

    end if-else

    Ranging: The remaining members will perform ranging through random walks (eq. (9)) and replace their corresponding positions in current trial population T^t.

    Apply GSO's boundary control mechanism to the out-bounded members from T^t.

    Reinitialize all members in T^t presenting less than two active clusters.

    Calculate the fitness value for each member $fitness(\mathbf{T}_i^t)$ from T^t.

    Determine the new population G^{t+1} using Selection-II operator (eq. (14)).

    $t \leftarrow t + 1$.

end while

Return $\mathbf{X}_p^{t_{max}}$.

The selected comparison evolutionary and swarm intelligence algorithms are: Genetic Algorithm, Differential Evolution, Particle Swarm Optimization, standard Group Search Optimization and standard Backtracking Search Optimization. The selected approaches are state-of-the-art models from evolutionary computing and data clustering literature, being successfully applied in many applications [1,28,29,31,34–36,39,40]. All EAs and SIs have been adapted to the context of partitional automatic clustering, using the same approach adopted by our proposed BGSO (see Sect. 4 and Algorithm 4). All algorithms use Calinski-Harabasz Index as their fitness function, running in a MATLAB 7.6 environment. Thirty independent tests have been executed for each data set, and all methods have started with the same initial population in each test, obtained by a random

Table 1. Real-world data set features.

Data set	Patterns	Feat	Classes
Banknote Authentication	1372	4	2
Cancer	699	9	2
Diabetes	768	8	2
Heart	270	13	2
Ionosphere	351	34	2
Iris	150	4	3
Page Blocks Classification	5473	10	5
Seeds	210	7	3
Waveform	5000	21	3

Table 2. Hyperparameters for each EA.

Algorithm	Parameter	Value
All EAs and SIs	t_{max}	200
	S	100
	C_{max}	20
GA	crossover rate	0.8
	mutation rate	0.1
	selection rate	0.8
DE	F	0.8
	crossover rate	0.9
PSO	c_1	2.0
	c_2	2.0
	w	0.9 to 0.4
GSO and BGSO	scroungers rate	0.8
	θ_{max}	π/a^2
	α_0	$\pi/4$
	α_{max}	$\theta_{max}/2$
BSA and GBSO	$mixrate$	1
	F	$3N(0,1)$

process, as explained in Sect. 4. Table 2 presents the hyperparameters for each EA and SI models.

The evaluation criterion includes an empirical analysis and a rank system employed through the application of Friedman test [17,18] for all the comparison clustering measures. The Friedman test is a non-parametric hypothesis test that ranks all algorithms for each data set separately. If the null-hypothesis (all ranks are not significantly different) is rejected, Nemenyi test [32] is adopted as the

Table 3. Experimental results for the real-world data sets (average ± standard deviation).

Data set	Algorithm	$CH^\uparrow$	$CR^\uparrow$	$DB^\downarrow$	$JI^\uparrow$	C
Banknote authentication	GA	1423.4 ± 0.202	0.0487 ± 0.0015	0.8709 ± 0.0012	0.3803 ± 0.0008	2 ± 0
	DE	1423.6 ± 0.153	0.0485 ± 0.0006	0.8704 ± 0.0009	0.3804 ± 0.0006	2 ± 0
	PSO	1387.4 ± 107.6	**0.0647 ± 0.0420**	0.8863 ± 0.0378	0.3573 ± 0.0522	2.7667 ± 2.063
	BSA	1423.5 ± 0.2572	0.0491 ± 0.0010	0.8708 ± 0.0008	**0.3805 ± 0.0007**	2 ± 0
	GSO	**1423.7 ± 0.0486**	0.0487 ± 0.0003	**0.8702 ± 0.0004**	**0.3805 ± 0.0003**	2 ± 0
	BGSO	**1423.7 ± 0.0485**	0.0486 ± 0.0003	**0.8702 ± 0.0004**	**0.3805 ± 0.0002**	2 ± 0
Cancer	GA	1038.9 ± 1.979	0.8320 ± 0.0090	0.7618 ± 0.0006	0.8599 ± 0.0067	2 ± 0
	DE	1038.9 ± 2.572	0.8344 ± 0.0084	0.7618 ± 0.0006	0.8618 ± 0.0062	2 ± 0
	PSO	1029.3 ± 65.95	0.8372 ± 0.0121	0.7873 ± 0.1429	0.8633 ± 0.0116	2.0333 ± 0.1826
	BSA	1040.1 ± 1.165	0.8351 ± 0.0083	0.7615 ± 0.0003	0.8623 ± 0.0062	2 ± 0
	GSO	**1041.4 ± 0.0918**	0.8385 ± 0.0029	**0.7612 ± 0.0001**	0.8647 ± 0.0022	2 ± 0
	BGSO	**1041.4 ± 0.2197**	**0.8391 ± 0.0028**	**0.7612 ± 0.00004**	**0.8651 ± 0.0021**	2 ± 0
Diabetes	GA	1139.1 ± 2.102	0.0443 ± 0.0036	0.6646 ± 0.0042	0.3789 ± 0.0041	3 ± 0
	DE	1140.0 ± 2.251	0.0450 ± 0.0025	0.6651 ± 0.0032	0.3793 ± 0.0026	3 ± 0
	PSO	996.14 ± 187.6	**0.0501 ± 0.0164**	0.8084 ± 0.2226	**0.3806 ± 0.0050**	4.6667 ± 2.928
	BSA	1136.5 ± 3.586	0.0453 ± 0.0046	**0.6638 ± 0.0037**	**0.3806 ± 0.0050**	3 ± 0
	GSO	1141.8 ± 2.930	0.0451 ± 0.0010	0.6673 ± 0.0044	0.3783 ± 0.0017	3 ± 0
	BGSO	**1142.0 ± 1.117**	0.0450 ± 0.0012	0.6679 ± 0.0015	0.3781 ± 0.0012	3 ± 0
Heart	GA	206.95 ± 0.0036	0.0295 ± 0.0012	0.9875 ± 0.0006	0.3606 ± 0.0009	2 ± 0
	DE	206.95 ± 0	**0.0302 ± 0**	0.9871 ± 0	**0.3611 ± 0**	2 ± 0
	PSO	206.84 ± 0.0995	0.0250 ± 0.0037	**0.9871 ± 0.0014**	0.3591 ± 0.0012	2 ± 0
	BSA	206.95 ± 0	**0.0302 ± 0**	0.9871 ± 0	**0.3611 ± 0**	2 ± 0
	GSO	206.95 ± 0.0041	0.0301 ± 0.0005	0.9873 ± 0.0007	0.3610 ± 0.0006	2 ± 0
	BGSO	**206.95 ± 0.0025**	0.0300 ± 0.0008	0.9874 ± 0.0006	0.3608 ± 0.0006	2 ± 0
Ionosphere	GA	115.65 ± 1.198	0.1464 ± 0.0132	1.5341 ± 0.0111	0.4190 ± 0.0064	2 ± 0
	DE	115.48 ± 1.601	0.1427 ± 0.0158	1.5367 ± 0.0143	0.4175 ± 0.0074	2 ± 0
	PSO	116.13 ± 9.484	**0.1791 ± 0.0214**	1.5375 ± 0.0895	**0.4317 ± 0.0084**	2.0667 ± 0.258
	BSA	117.27 ± 0.9134	0.1564 ± 0.0151	1.5206 ± 0.0094	0.4233 ± 0.0075	2 ± 0
	GSO	118.43 ± 0.3889	0.1697 ± 0.0091	1.5158 ± 0.0052	0.4298 ± 0.0043	2 ± 0
	BGSO	**118.76 ± 0.1395**	0.1737 ± 0.0053	**1.5127 ± 0.0024**	0.4315 ± 0.0027	2 ± 0
Iris	GA	561.58 ± 0.256	0.7302 ± 0.0001	0.6622 ± 0.0013	0.6958 ± 0.0003	3 ± 0
	DE	**561.63 ± 0**	0.7302 ± 0	**0.6620 ± 0**	**0.6959 ± 0**	3 ± 0
	PSO	560.80 ± 2.540	0.7301 ± 0.0004	0.6636 ± 0.0047	0.6956 ± 0.0007	3 ± 0
	BSA	**561.63 ± 0**	0.7302 ± 0	**0.6620 ± 0**	**0.6959 ± 0**	3 ± 0
	GSO	561.37 ± 0.8113	**0.7316 ± 0.0040**	0.6627 ± 0.0023	0.6971 ± 0.0037	3 ± 0
	BGSO	**561.63 ± 0**	0.7302 ± 0	**0.6620 ± 0**	**0.6959 ± 0**	3 ± 0
Page blocks classification	GA	14395.2 ± 567.3	0.0070 ± 0.0154	**0.5250 ± 0.0342**	0.6044 ± 0.0893	5.5 ± 0.509
	DE	**16343.2 ± 778.1**	0.0003 ± 0.0129	0.6159 ± 0.0311	0.5195 ± 0.0745	7.5 ± 0.861
	PSO	13372.5 ± 1920.9	0.0109 ± 0.0059	0.5307 ± 0.0318	0.6634 ± 0.0300	**4.7000 ± 0.8769**
	BSA	15007.2 ± 529.4	0.0057 ± 0.0156	0.5626 ± 0.0523	0.6031 ± 0.0735	5.9667 ± 0.8087
	GSO	12456.9 ± 1436.9	**0.0110 ± 0.0108**	0.5364 ± 0.0267	0.6667 ± 0.0232	4.3667 ± 0.5561
	BGSO	12837.6 ± 1623.7	0.0096 ± 0.0075	0.5357 ± 0.0388	**0.6673 ± 0.0246**	4.5333 ± 0.6288
Seeds	GA	375.31 ± 0.7548	**0.7178 ± 0.0086**	0.7535 ± 0.0010	**0.6827 ± 0.0081**	3 ± 0
	DE	372.38 ± 2.3840	0.7106 ± 0.0209	0.7564 ± 0.0041	0.6763 ± 0.0194	3 ± 0
	PSO	**375.73 ± 0.2892**	0.7159 ± 0.0028	**0.7535 ± 0.0007**	0.6808 ± 0.0026	3 ± 0
	BSA	370.66 ± 5.5973	0.6988 ± 0.0274	0.7603 ± 0.0081	0.6656 ± 0.0243	3 ± 0
	GSO	375.68 ± 0.3881	0.7153 ± 0.0040	**0.7535 ± 0.0007**	0.6803 ± 0.0037	3 ± 0
	BGSO	375.05 ± 2.0970	0.7163 ± 0.0118	0.7543 ± 0.0040	0.6814 ± 0.0110	3 ± 0
Waveform	GA	2518.7 ± 11.88	0.3473 ± 0.0112	1.3783 ± 0.0036	0.4374 ± 0.0067	2 ± 0
	DE	2544.2 ± 8.190	0.3597 ± 0.0057	1.3734 ± 0.0021	0.4450 ± 0.0035	2 ± 0
	PSO	2552.6 ± 58.11	0.3669 ± 0.0213	**1.3705 ± 0.0047**	0.4480 ± 0.0209	2.0333 ± 0.1826
	BSA	2536.2 ± 7.4291	0.3537 ± 0.0093	1.3745 ± 0.0027	0.4413 ± 0.0056	2 ± 0
	GSO	2546.3 ± 10.52	0.3608 ± 0.0065	1.3733 ± 0.0027	0.4456 ± 0.0040	2 ± 0
	BGSO	**2559.4 ± 2.4081**	**0.3686 ± 0.0026**	**1.3704 ± 0.0007**	**0.4505 ± 0.0017**	2 ± 0

post-hoc test. According to Nemenyi test, the performance of two algorithms are considered significantly different if the corresponding average ranks differ by at least the critical difference

$$CD = q_a \sqrt{\frac{n_{alg}(n_{alg}+1)}{6n_{data}}} \tag{15}$$

where n_{data} represents the number of data sets, n_{alg} represents the number of compared algorithms and q_a are critical values based on a Studentized range statistic divided by $\sqrt{2}$ [12]. Since CH, CR and JI are *maximization metrics* (indicated by ↑), the best methods will obtain higher ranks for the Friedman test, while for DB (a *minimization metric*, indicated by ↓), the best methods will find lower average ranks for the Friedman test.

The experimental results are presented in Table 3. As we can observe in an empirical analysis, the proposed BGSO was able to find the best values for the fitness function (CH) for most cases (seven out of nine problems). Almost all algorithms (except for PSO) have been able to predict the exact estimated number of final clusters for six out of nine data sets, which is a good result, compatible with many works from the literature [38]. Even for Diabetes, Waveform and Page Blocks Classification, the best number of clusters found by the EAs and SIs is not very much distant from the expected values, what is quite acceptable, given the different degrees of separability among the original classes in such data sets.

Table 4. Overall Evaluation: Average Ranks for the Friedman Test for each metric, with $CD = 2.5132$.

Algorithm	$CH^{\uparrow}$	$CR^{\uparrow}$	$DB^{\downarrow}$	$JI^{\uparrow}$
GA	63.7167	73.5815	103.5000	77.3593
DE	85.4296	78.5759	108.3000	77.4833
PSO	101.1148	100.7630	**73.6870**	95.4185
BSA	72.6685	86.1889	101.6130	88.0852
GSO	105.6963	99.4519	81.8444	99.7370
BGSO	**114.3741**	**104.4389**	**74.0556**	**104.9167**

Considering the overall evaluation performed through the application Friedman hypothesis tests (Table 4), BGSO obtained the best rank values for all four evaluation metrics, showing its robustness. GSO have been able to achieve the second best rank values for CH and JI (followed by PSO), and PSO have reached the best rank for DB (with no statistically significant difference from BGSO), and the second best rank value for the CR (with no statistically significant difference from GSO).

6   Conclusions

In this work, a new hybrid evolutionary algorithm is presented, which combines Group Search Optimization and Backtracking Search Optimization: the Backtracking Group Search Optimization (BGSO). In BGSO, the historical information of BSA is aggregated to the searching process of GSO, keeping the best features of both models. The proposed hybrid algorithm is adapted to the context of automatic data clustering as a partitional clustering model.

In an attempt to validate BGSO, five state-of-the-art algorithms are adapted as partitional automatic clustering approaches and compared to BGSO: GA, DE, PSO, GSO and BSA. Nine real-world data sets are employed, and four clustering metrics are used for comparison purposes. The experimental evaluation included an empirical analysis and a hypothesis test (Friedman test).

The experiments showed that BGSO is able to find better solutions than standard GSO and BSA algorithms in most cases, and in an overall evaluation, BGSO has been able to outperform all comparison approaches in relation to the selected clustering indices.

As future works, we intend to extend our analysis on the behavior of BGSO by employing controlled scenarios obtained through the use of synthetic data sets, so we can understand the best features and limitations of the proposed model on different clustering problems. Also, we intend to evaluate the influence of the fitness function on BGSO performance.

References

1. Akbari, M., Izadkhah, H.: GAKH: a new evolutionary algorithm for graph clustering problem. In: 2019 4th International Conference on Pattern Recognition and Image Analysis (IPRIA), pp. 159–162. IEEE (2019)
2. Asuncion, A., Newman, D.: Uci machine learning repository (2007)
3. Barnard, C., Sibly, R.: Producers and scroungers: a general model and its application to captive flocks of house sparrows. Anim. Behav. **29**(2), 543–550 (1981)
4. Bonabeau, E., Dorigo, M., Theraulaz, G.: Swarm Intelligence: From Natural to Artificial Systems, vol. 4. Oxford University Press, New York (1999)
5. Caliński, T., Harabasz, J.: A dendrite method for cluster analysis. Commun. Stat. Theory Methods **3**(1), 1–27 (1974)
6. Chen, C.Y., Ye, F.: Particle swarm optimization algorithm and its application to clustering analysis. In: 2004 IEEE International Conference on Networking, Sensing and Control, vol. 2, pp. 789–794. IEEE (2004)
7. Chen, G.B., et al.: Automatic clustering approach based on particle swarm optimization for data with arbitrary shaped clusters. In: 2016 Seventh International Conference on Intelligent Control and Information Processing (ICICIP), pp. 41–48. IEEE (2016)
8. Civicioglu, P.: Backtracking search optimization algorithm for numerical optimization problems. Appl. Math. Comput. **219**(15), 8121–8144 (2013)
9. Couzin, I.D., Krause, J., Franks, N.R., Levin, S.A.: Effective leadership and decision-making in animal groups on the move. Nature **433**(7025), 513–516 (2005)
10. Das, S., Abraham, A., Konar, A.: Automatic clustering using an improved differential evolution algorithm. IEEE Trans. Syst. Man Cybern. Part A Syst. Hum. **38**(1), 218–237 (2007)
11. Davies, D.L., Bouldin, D.W.: A cluster separation measure. IEEE Trans. Pattern Anal. Mach. Intell. **2**, 224–227 (1979)
12. Demšar, J.: Statistical comparisons of classifiers over multiple data sets. J. Mach. Learn. Res. **7**, 1–30 (2006)

13. Dixon, A.: An experimental study of the searching behaviour of the predatory coccinellid beetle adalia decempunctata (l.). J. Animal Ecol., pp. 259–281 (1959)
14. Dorigo, M., Maniezzo, V., Colorni, A.: Ant system: optimization by a colony of cooperating agents. IEEE Trans. Syst. Man Cybern. Part B: Cybern. **26**(1), 29–41 (1996)
15. Dunn, J.C.: A fuzzy relative of the isodata process and its use in detecting compact well-separated clusters (1973)
16. Elaziz, M.A., Nabil, N., Ewees, A.A., Lu, S.: Automatic data clustering based on hybrid atom search optimization and sine-cosine algorithm. In: 2019 IEEE Congress on Evolutionary Computation (CEC), pp. 2315–2322. IEEE (2019)
17. Friedman, M.: The use of ranks to avoid the assumption of normality implicit in the analysis of variance. J. Am. Stat. Assoc. **32**(200), 675–701 (1937)
18. Friedman, M.: A comparison of alternative tests of significance for the problem of m rankings. Ann. Math. Stat. **11**(1), 86–92 (1940)
19. Halkidi, M., Batistakis, Y., Vazirgiannis, M.: Cluster validity methods: part i. ACM Sigmod Record **31**(2), 40–45 (2002)
20. He, S., Wu, Q.H., Saunders, J.R.: Group search optimizer: an optimization algorithm inspired by animal searching behavior. IEEE Trans. Evol. Comput. **13**(5), 973–990 (2009)
21. Higgins, C.L., Strauss, R.E.: Discrimination and classification of foraging paths produced by search-tactic models. Behav. Ecol. **15**(2), 248–254 (2004)
22. Holland, J.H.: Genetic algorithms. Sci. Am. **267**(1), 66–72 (1992)
23. Hruschka, E.R., Campello, R.J., Freitas, A.A., et al.: A survey of evolutionary algorithms for clustering. IEEE Trans. Syst. Man Cybern. Part C (Applications and Reviews) **39**(2), 133–155 (2009)
24. Hubert, L., Arabie, P.: Comparing partitions. J. Classif. **2**(1), 193–218 (1985)
25. Jain, A.K., Murty, M.N., Flynn, P.J.: Data clustering: a review. ACM Comput. Surv. (CSUR) **31**(3), 264–323 (1999)
26. José-García, A., Gómez-Flores, W.: Automatic clustering using nature-inspired metaheuristics: a survey. Appl. Soft Comput. **41**, 192–213 (2016)
27. Kennedy, J., Eberhart, R.: Particle swarm optimization. In: Proceedings of the IEEE International Conference on Neural Networks, 1995, vol. 4, pp. 1942–1948. IEEE (1995)
28. Latiff, N.A., Malik, N.N.A., Idoumghar, L.: Hybrid backtracking search optimization algorithm and k-means for clustering in wireless sensor networks. In: 2016 IEEE 14th Intl Conf on Dependable, Autonomic and Secure Computing, 14th Intl Conf on Pervasive Intelligence and Computing, 2nd Intl Conf on Big Data Intelligence and Computing and Cyber Science and Technology Congress (DASC/PiCom/DataCom/CyberSciTech). pp. 558–564. IEEE (2016)
29. Li, L., Liang, Y., Li, T., Wu, C., Zhao, G., Han, X.: Boost particle swarm optimization with fitness estimation. Nat. Comput. **18**(2), 229–247 (2019)
30. Liu, Y., Wu, X., Shen, Y.: Automatic clustering using genetic algorithms. Appl. Math. Comput. **218**(4), 1267–1279 (2011)
31. Mortezanezhad, A., Daneshifar, E.: Big-data clustering with genetic algorithm. In: 2019 5th Conference on Knowledge Based Engineering and Innovation (KBEI), pp. 702–706. IEEE (2019)
32. Nemenyi, P.: Distribution-free multiple comparisons. In: Biometrics, vol. 18, p. 263. International biometric soc 1441 I ST, NW, SUITE 700, Washington, DC 20005–2210 (1962)

33. Omran, M., Salman, A., Engelbrecht, A.: Dynamic clustering using particle swarm optimization with application in unsupervised image classification. In: Fifth World Enformatika Conference (ICCI 2005), Prague, Czech Republic, pp. 199–204 (2005)
34. Pacifico, L.D.S., Ludermir, T.B.: Hybrid k-means and improved self-adaptive particle swarm optimization for data clustering. In: 2019 International Joint Conference on Neural Networks (IJCNN), pp. 1–7. IEEE (2019)
35. Pacifico, L.D., Ludermir, T.B.: Hybrid k-means and improved group search optimization methods for data clustering. In: 2018 International Joint Conference on Neural Networks (IJCNN), pp. 1–8. IEEE (2018)
36. Souza, E., Santos, D., Oliveira, G., Silva, A., Oliveira, A.L.: Swarm optimization clustering methods for opinion mining. Natural Comput., 1–29 (2018)
37. Storn, R., Price, K.: Differential evolution–a simple and efficient adaptive scheme for global optimization over continuous spaces. international computer science institute, berkeley. Technical report, CA, 1995, Tech. Rep. TR-95-012 (1995)
38. Tam, H.H., Ng, S.C., Lui, A.K., Leung, M.F.: Improved activation schema on automatic clustering using differential evolution algorithm. In: 2017 IEEE Congress on Evolutionary Computation (CEC), pp. 1749–1756. IEEE (2017)
39. Toz, G., Yücedağ, İ., Erdoğmuş, P.: A fuzzy image clustering method based on an improved backtracking search optimization algorithm with an inertia weight parameter. J. King Saud Univ.-Comput. Inf. Sci. 31(3), 295–303 (2019)
40. Wang, H., Zuo, L., Liu, J., Yi, W., Niu, B.: Ensemble particle swarm optimization and differential evolution with alternative mutation method. Natural Comput., 1–14 (2018)
41. Xu, R., Wunsch, D., et al.: Survey of clustering algorithms. IEEE Trans. Neural Netw. 16(3), 645–678 (2005)

Dynamic Software Project Scheduling Problem with PSO and Dynamic Strategies Based on Memory

Gabriel Fontes da Silva, Leila Silva, and André Britto$^{(\boxtimes)}$

Department of Computing, Federal University of Sergipe, São Cristovão, Brazil
{gabriel.silva,leila,andre}@dcomp.ufs.br

Abstract. The Software Project Scheduling Problem (SPSP) aims to allocate employees to tasks in the development of a software project, such that the cost and duration, two conflicting goals, are minimized. The dynamic model of SPSP, called DSPSP, considers that some unpredictable events may occur during the project life cycle, like the arrival of new tasks, which implies on schedule updating along the project. In the context of Search-Based Software Engineering, this work proposes the use of dynamic optimization strategies, based on memory, together with the particle swarm optimization algorithm (PSO) to solve the DSPSP. The results suggest that the addition of these dynamic strategies improves the quality of the solutions in comparison with the application of the PSO algorithm only.

Keywords: Software project scheduling problem · Search-based software engineering · Particle swarm optimization · Dynamic strategy

1 Introduction

The Software Engineering field contains many problems with conflicting goals; they are optimization problems. Search-Based Software Engineering (SBSE) [1] is a domain of Software Engineering that applies search based techniques to find optimal or near optimal solutions to these problems. Metaheuristics such as bio inspired algorithms [2] or genetic algorithms [3] are commonly used in this area.

A relevant problem with conflicting goals, when developing a software project, is to define which tasks each employee should develop. This problem is known as Software Project Scheduling Problem (SPSP), whose aim is to find the best project schedule with duration and cost minimized. The duration tends to increase when the cost decreases, and vice versa. For example, the project cost increases when new employees are hired to develop the project more quickly; on the other hand, the project duration increases when there are few employees working on the project, and thus, less development costs.

There are two well-known models for the SPSP: **Static** [4], which does not consider dynamic events like task effort uncertainty, and a new task or employee arrival; and **Dynamic** [5] (abbreviated by DSPSP), that considers these dynamic

© Springer Nature Switzerland AG 2020
R. Cerri and R. C. Prati (Eds.): BRACIS 2020, LNAI 12319, pp. 79–94, 2020.
https://doi.org/10.1007/978-3-030-61377-8_6

events. In the case of SPSP, only one schedule is established for the project, whereas in DSPSP, the schedule set up at the beginning of the project, is updated after the occurrence of dynamic events.

Problems like DSPSP, whose search space changes occasionally, tend to be hard to be solved. The metaheuristic applied must be able to identify differences in the search space, what it is not so simple. Nguyen et al. [7] present some approaches to handle problems that have dynamic events, like multi-population, prediction, the introduction of diversity, and memory, i.e. historical data, of previous solutions.

Few works incorporate dynamic optimization strategies in metaheuristics to solve the DSPSP. Shen et al. [5] propose an ϵ-domination based Multi-Objective Evolutionary Algorithm (MOEA), denoted as dϵ-MOEA, to solve DSPSP and incorporate some dynamic strategies to construct the initial population of the algorithm after a dynamic event, such as: usage of information from previous schedules in the current one; incorporation of random individuals, to introduce diversity; and proactive repair of the schedule, to preserve the solution stability. The use of an initialization strategy in dϵ-MOEA produces better results than other strategies investigated in [5]. In [8] strategies used in [5] are applied to a Q-learning-based mechanism that chooses appropriated search operators for different scheduling environments. These works show, for the DSPSP, that the application of dynamic optimization strategies together with metaheuristics can improve the results achieved by the application of the metaheuristic only.

This paper aims to investigate the Particle Swarm Optimization (PSO) [2] metaheuristic's performance for solving the DSPSP, in three scenarios: the application of the algorithm without dynamic strategies; the inclusion of a dynamic strategy based on historical solutions, similar to [5]; and the inclusion of a new strategy based on memory, here proposed, that stores the best solutions after each new rescheduling. We are not aware of any other approach that uses PSO for solving the DSPSP, with and without dynamic strategies, and thus, we intend to contribute to a larger investigation of this problem in the context of SBSE. To evaluate the proposed approach we use six DSPSP's benchmark instances, with a distinct number of tasks and employees.

This paper is organized as follows. Related works are discussed in Sect. 2. In Sect. 3 the formulation of the DSPSP as an optimization problem is described. In Sect. 4 the algorithms are briefly described. The experiments are discussed in Sect. 5. At last, Sect. 6 presents conclusions and directions for future work.

2   Related Works

There are many studies considering the SPSP static model in the context of SBSE. Rezende et al. [6] provide a relevant systematic review in this subject. Nevertheless, few studies have investigated the DSPSP. Seminal works for DSPSP are the works of Shen et al. [5,8], mentioned in Sect. 1. Besides, Rezende [9] proposes an extension of the model of [5], by considering two more dynamic events and the influence of the team experience. In this work, the author compares the performance of the Cooperative Multiobjective Differential Evolution (CMODE) algorithm to the Non-dominated Sorting Genetic Algorithm III (NSGA-III).

The use of dynamic strategies in metaheuristics to solve DSPSP also have been few investigated. In [5] strategies like historical solutions, diversity introduction, and self-adaptive mechanisms are used in dϵ-MOEA, whereas [8] adopt a predicting mechanism in MOTAMAQ. Furthermore, Rezende [9] applies proactive repair and historical solutions to the CMODE and NSGA-III algorithms.

The PSO algorithm was applied to solve some software engineering problems, in the context of SBSE. For example, Bardsiri et al. [10] propose a hybrid estimation model based on a combination of a PSO and Analogy-based estimation (ABE) to increase the accuracy of software development effort estimation. Andrade et al. [11] propose a hyper-heuristic, denoted as GE-SPSP, to configure the metaheuristic Speed-Constrained PSO (SMPSO), based on Grammatical Evolution, to solve the SPSP.

Differently from those works, this work proposes the application of the metaheuristic SMPSO [12] for the DSPSP, by incorporating two distinct dynamic strategies, one new and the other based on the work of [5] for the dϵ-MOEA.

3   Dynamic Software Project Scheduling Problem

The model of DSPSP adopted in this work is due to Shen et al. [5], which is an extension of the work of [4]. A brief description of this model is given in what follows.

3.1   Employees

The DSPSP considers a set of employees E, where e_i ($i = 1, 2, ..., |E|$) denotes each employee in the project. Employees have a salary and a maximum dedication to the project, ranging from 0 to 1. Furthermore, each employee has a set of skills, which is denoted by $skills_i$. All the necessary skills required by the project, for the set of employees, are represented by the set $S = \{sk_1, sk_2, ..., sk_{|S|}\}$. Figure 1 (a) shows an example of the DSPSP, where skills like Technical leadership, UML modeling, Programming, Database, and Web design are in S. Table 1 presents each variable associated to employees, where t_l represents a rescheduling point.

3.2   Tasks

Tasks are executed by employees and each task requires a skill set for its execution and an estimated effort expressed in person-month. There is a precedence relationship between each task, represented by an acyclic directed graph $G(T, A)$, where $T = \{\tau_1, ..., \tau_n\}$ is the vertex set, representing the set of tasks in the project and A is an edge set representing tasks dependency, i.e., the edge $(\tau_i, \tau_j) \in A$ means that task τ_i needs to be concluded before that task τ_j starts. This graph is called Task Precedence Graph (TPG) and an example is showed in Fig. 1 (b). Table 2 presents each variable associated to tasks.

Table 1. Employee parameters in DSPSP.

Name	Description
e_i	An employee in the project
e_i^{skills}	Indicates how much an employee masters a task in the project, and is defined as $e_i^{skills} = \{prof_i^1, prof_i^2, ..., prof_i^S\}$, where $prof_i^k \in [0, C]$ $(k = 1, 2, ..., |S|)$ expresses the e_i proficiency to the kth skill. If $prof_i^k = 0$, then e_i does not have this skill; and if $prof_i^k = C$, then e_i masters the kth skill
$skills_i$	The skill set of the employee and defined as $skills_i = \{k | prof_i^k > 0, k = 1, 2, ..., |S|\}$
$e_i^{max\_ded}$	Maximum dedication of e_i to the project that is the monthly workday percentage that e_i can support. If $e_i^{max\_ded} > 1$, then e_i is able to do overtime at work
$e_i^{hab\_sal}$	The e_i monthly salary in a regular workday
$e_i^{ext\_sal}$	The e_i salary, if he/she does overtime at work
$e_i^{aval}(t_l)$	Binary variable that indicates whether e_i is available or not in t_l
$e\_aval(t_l)$	A set that indicates all available employees in t_l, i. e., $e\_aval(t_l) = \{e_i | e_i^{aval}(t_l) = 1, i = 1, 2, ..., |E|\}$

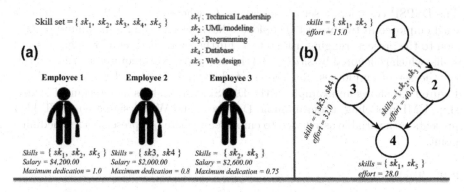

Fig. 1. Examples of (a) employees properties; (b) TPG; and (c) a dedication matrix.

Table 2. Task parameters in DSPSP.

Name	Description
τ_j	A task in the project
τ_j^{skills}	Required skill set to τ_j. Defined as $\tau_j^{hab} = \{sk_j^1, sk_j^2, ..., sk_j^k | k = 1, 2, ..., |S|\}$, where $sk_j^k = 1$ means that the kth skill is required in τ_j, and $sk_j^k = 0$ not
req_j	A specific skill set to τ_j, where is defined as $req_j = \{k | sk_j^k = 1, k = 1, 2, ..., |S|\}$
$\tau_j^{tot\_eff\_est}$	The initial estimated effort required to complete τ_j in person-month. The uncertainess of this value follow a normal distribution $\phi(\mu_j, \sigma_j)$
$\tau_j^{unfinished}(t_l)$	Binary variable that indicates whether τ_j is finished or not in t_l
$\tau_j^{aval}(t_l)$	Binary variable that indicates whether τ_j is available or not in t_l
$\tau\_aval(t_l)$	A set that indicates all available tasks in t_l, i. e., $\{\tau_j | \tau_j^{aval}(t_l) = 1, j = 1, 2, ..., |T| + |T_{new}(t_l)|\}$, where $T_{new}(t_l)$ is the set of new tasks that arrives in t_l
TPG	An acyclic directed graph $G(T, A)$, where the vertices represent the tasks, and the edges indicate the precedences between each task

3.3 Solution Representation

A solution to DSPSP is expressed by a dedication matrix M_{ij} of size $|E| \times |T|$, where the rows represent the employees and columns the tasks. Thus, all cells in the dedication matrix determine the dedication of an employee in some tasks, which is defined as $e_{ij}^{prof} = \prod_{k \in req_j} \frac{prof_i^k}{C}$. For example, in Fig. 1 (c), the employee e_1 has dedicated to the task τ_1 a total of 41% of his/her workday. If the cell m_{ij} has value 0, this means that employee e_i is not allocated to the task τ_j.

3.4 Dynamic Events

The DSPSP proposed in [5] incorporates one uncertainty and three dynamic events to the model in [4], which are listed in what follows.

- **Uncertainness about task efforts:** early in the project life cycle, a software model, such as COCOMO [13] is used to estimate the effort of each task. However, due to changes in task specifications or erroneous estimates, task efforts need to be modified. Therefore, it is assumed that the effort value follows a normal distribution $\phi(\mu, \sigma)$, in such a way that the mean μ and the standard deviation σ values is assigned to each task. The value of the initial effort assumes μ.

– **New tasks arrival:** during the project life cycle it is common that new tasks arrive. They can be urgent or regular tasks. When urgent tasks appear, all running tasks must pause, so that these urgent tasks can begin. On the other hand, when regular tasks appear, they can be scheduled in a normal way, without priority.
– **Employee leaves or returns the project:** an employee can leave the project due to many reasons. He/She got sick, being part of multiple projects, or becoming a father/mother. Likewise, the employee may return to the project, and this event is denoted by the time since the employee leaves until the moment he/she returns.

3.5   Objective Functions

The DSPSP considers four objective functions to be optimized, as shown in the Table 3. Such objectives arise at any time t_l ($t_l > t_0$), where there is the following information about the project: (1) a set of available employees $e\_aval(t_l)$; (2) a set of available tasks $\tau\_aval(t_l)$ with the rest of the estimated efforts; and (3) the TPG $G(T(t_l), A(t_l))$ that updates in t_l.

Table 3. DSPSP objective functions.

Name	Description
$f_1(t_l)$	Remaining project duration in t_l
$f_2(t_l)$	Initial project cost for available tasks in t_l
$f_3(t_l)$	It denotes the robustness of the project by measuring how sensitive scheduling is to task efforts uncertainess
$f_4(t_l)$	The stability between schedulings in t_l and t_{l-1}, where $l \geq 1$. Measures how much the schedule in t_l differs from the previous schedule

The optimization of these objectives is expressed by:

$$\min \mathbf{F}(t_l) = [f_1(t_l), f_2(t_l), f_3(t_l), f_4(t_l)] \tag{1}$$

where $f_1(t_l)$, $f_2(t_l)$, $f_3(t_l)$ and $f_4(t_l)$ are defined as *duration*, *cost*, *robustness*, and *stability*, respectively.

The objective function $f_1(t_l)$ gives us the maximum required time to complete the remaining efforts of each task in t_l. Thus, the model of [5] defines that $\tau_j^{fin\_eff}(t_l)$ is the finalized effort from t_0 to t_l for each $\tau_j \in \tau\_aval(t_l)$, and $\tau_j^{est\_rem\_eff}(t_l)$ is the estimated remaining effort of a task τ_j in t_l, such that $\tau_j^{est\_rem\_eff}(t_l) = \tau_j^{tot\_eff\_est} - \tau_j^{fin\_eff}(t_l)$. If $\tau_j^{est\_rem\_eff}(t_l) \leq \tau_j^{fin\_eff}(t_l)$ and $\tau_j^{unfinished}(t_l) = 1$, then the initial estimated effort was erroneous, and $\tau_j^{tot\_eff\_est}$ needs to be reestimated many times by a value B from the normal distribution $\phi(\mu_j, \sigma_j)$ until the condition $B > \tau_j^{fin\_eff}$ be true. Hence, Eq. 2 defines $f_1(t_l)$, where I is the initial scenario considering the remaining effort $\tau_j^{est\_rem\_eff}(t_l)$ as the required effort to complete τ_j in t_l; $\tau_j^{start}(t_l)$ is the time, in months, wherein τ_j started to be processed, from the generated rescheduling after t_l; and τ_j^{end} is the time wherein τ_j ends.

$$f_1(t_l) = duration_I = \max_{\{j | \tau_j \in \tau\_aval(t_l)\}} (\tau_j^{end}(t_l)) - \min_{\{j | \tau_j \in \tau\_aval(t_l)\}} (\tau_j^{start}(t_l)) \quad (2)$$

The function $f_2(t_l)$ defines the project cost based on the salary of each employee e_i available in t_l. Equation 3 denotes the initial project cost, where $t' \geq t_l$ and $e\_cost_i^{t'}$ represents how much was paid to e_i in t'.

$$f_2(t_l) = cost_I = \sum_{t' \geq t_l} \sum_{e_i \in e\_aval(t_l)} e\_cost_i^{t'}. \quad (3)$$

There are two ways to obtain the value of $e\_cost_i^{t'}$, as shown in Eq. 4 and Eq. 5, where $\tau\_active(t') = \{\tau_j | \tau_j^{unfinished}(t') = 1\}$. If $\sum_{j \in \tau\_aval(t')} m_{ij}(t') \leq 1$, then

$$e\_cost_i^{t'} = e_i^{hab\_sal} \cdot t' \cdot \sum_{j \in \tau\_active(t')} m_{ij}(t'), \quad (4)$$

otherwise if $1 < \sum_{j \in \tau\_aval(t')} m_{ij}(t') \leq e_i^{max\_ded}$, then

$$e\_cost_i^{t'} = e_i^{hab\_sal} \cdot t' \cdot 1 + e_i^{ext\_sal} \cdot t' \cdot \left(\sum_{j \in \tau\_active(t')} m_{ij}(t') - 1 \right) \quad (5)$$

To compute *robustness*, Eq. 6 is applied. In this equation the values of $f_1(t_l)$ and $f_2(t_l)$ are estimated, which defines the set $\{\theta_q | q = 1, 2, ..., \gamma\}$, where θ_q is the qth effort sample of the task, and γ is a input parameter set to be 30. To generate θ_q is necessary estimate the total cost of the sample θ_q, defined as $\tau_j^{tot\_cost_q}$, many times by normal distribution $\phi(\mu_j, \sigma_j)$, until the condition $\tau_j^{tot\_cost_q} > \tau_j^{fin\_eff}(t_l)$ be true. Thereby, is assumed that $\theta_q = \{\tau_j^{rem\_eff_q}(t_l) | \tau_j^{rem\_eff_q}(t_l) = \tau_j^{tot\_cost_q} - \tau_j^{fin\_eff}(t_l), \tau_j \in \tau\_aval(t_l)\}$, where $\tau_j^{rem\_eff_q}(t_l)$ represents the qth remaining effort sample of τ_j. λ is a weight parameter that determines the relative importance of cost over duration, which is set to be 1.

$$f_3(t_l) = robustness$$

$$= \sqrt{\frac{1}{\gamma} \sum_{q=1}^{\gamma} \left(\max\left(0, \frac{duration_q(t_l) - duration_I(t_l)}{duration_I(t_l)} \right) \right)^2}$$

$$+ \lambda \sqrt{\frac{1}{\gamma} \sum_{q=1}^{\gamma} \left(\max\left(0, \frac{cost_q(t_l) - cost_I(t_l)}{cost_I(t_l)} \right) \right)^2} \quad (6)$$

The *stability* function (see Eq. 7) is the weighted sum of the dedication deviations, and it is calculated for all available tasks in t_l ($l \geq 1$).

$$f_4(t_l) = stability =$$
$$\sum_{\{i|e_i \in e\_aval(t_{l-1}) \cap e\_aval(t_l)\}} \sum_{\{j|\tau_j \in \tau\_aval(t_{l-1}) \cap \tau\_aval(t_l)\}} \omega_{ij}|m_{ij}(t_l) - m_{ij}(t_{l-1})|$$
$$(7)$$

In Eq. 7 ω_{ij} is the weight parameter and contains three possible values, as shown in Eq. 8. When $\omega_{ij} = 2.0$, an employee e_i does not work in task τ_j in t_{l-1}, but works in t_l, which can decrease the productivity of the employee, because may be necessary extra training and time to get familiar with the task. Differently of the first case, when $\omega_{ij} = 1.5$, the employee e_i works in task τ_j in t_{l-1}, but not in t_l. In the last case, there is no penalty, because e_i remains on τ_j in t_l.

$$\omega_{ij} = \begin{cases} 2.0; & \text{if } m_{ij}(t_{l-1}) = 0 \text{ and } m_{ij}(t_l) > 0, \\ 1.5; & \text{if } m_{ij}(t_{l-1}) > 0 \text{ and } m_{ij}(t_l) = 0, \\ 1.0; & \text{otherwise.} \end{cases} \qquad (8)$$

3.6   Constraints

The DSPSP considers three type of constraints, where the first two are hard constraints, and the third one is a soft constraint. These constraints are listed below and the constraints handling are detailed in [5].

(*i*) **No Overwork Constraints.** The total dedication of an available employee to the running tasks $e\_work_i^{t'}$, in any time $t' \geq t_l$, must not exceed the maximum employee dedication, i. e.:

$$\forall e_i \in e\_aval(t'), \forall t' \geq t_l, \text{ s.t. } e\_work_i^{t'}$$
$$= \sum_{j \in \tau\_active(t')} m_{ij}(t') \wedge e\_work_i^{t'} \leq e_i^{max\_ded}. \qquad (9)$$

(*ii*) **Task skill constraints.** The skill set of each employee scheduled to task τ_j must be in the task required skill set req_j, and for each running task, at least one available employee must be part of the team, as shown Eq. 10.

$$\forall \tau_j \in \tau\_aval(t_l) \text{ s.t. } req_j \subseteq \bigcup_{e_i \in e\_aval(t_l)} \{skills_i | m_{ij}(t_l) > 0\}. \qquad (10)$$

(*iii*) **Maximum headcount constraint.** There is a limit of employees working together in any available task, which is denoted as $\tau_j^{max\_emp}$ and is estimated by the formula $\tau_j^{max\_emp} = \max\{1, round(2/3(\tau_j^{tot\_eff\_est})^{0,672})\}$ [14]. However, if the number of employees in the task cannot be reduced, without violating the constraint (ii), then the effort task is penalized. Equation 11 shows the constraint, where $\tau_j^{team\_size}(t_l)$ is the team size of τ_j in t_l, and $\tau_j^{min\_emp}(t_l)$ is the minimum team size in τ_j to satisfy the constraint ii.

$$\forall \tau_j \in \tau\_aval(t_l), \tau_j^{team\_size} \leq \max\left(\tau_j^{max\_emp}, \tau_j^{min\_emp}(t_l)\right). \tag{11}$$

4 Speed-Constrained PSO Applied to DSPSP

The PSO [2] is a metaheuristic inspired by social behavior of bird flocking or fish schooling. The set of possible solutions is a set of particles, called a swarm, which moves in the search space, in a cooperative search procedure. These movements are performed by the velocity operator that is guided by a local and a social component. In Multi-Objective Optimization, Multi-Objective Particle Swarm Optimization (MOPSO), the Pareto dominance relation is adopted to establish preferences among solutions to be considered as leaders. By exploring the Pareto dominance concepts, each particle in the swarm could have different leaders, but only one may be selected to update the velocity. This set of leaders is stored in an external archive (or repository) that contains the best non-dominated solutions found so far. Normally, this archive is bounded and has a maximum size. So, two important features of PSO are the method to archive the solutions in the repository and how each particle will choose its leader (leader's selection). The basic steps of a MOPSO algorithm are the initialization of the particles, computation of the velocity, position update, and update of the leader's archive.

The SMPSO algorithm [12] is a basic MOPSO algorithm that uses the procedure that limits the velocity of each particle. The velocity of the particle is limited by a constriction factor χ, which varies based on the values of C_1 and C_2 that are specific parameters which control the effect of the local and global best particles, respectively. Besides, the SMPSO introduces a mechanism that links or constraints the accumulated velocity of each variable j (in each particle). Also, after the velocity of each particle has been updated a mutation operation is applied. A polynomial mutation is applied in 15% of the population, randomly selected. Furthermore, SMPSO uses the Crowding Distance archive.

Algorithm 1 shows the pseudocode of the SMPSO, adapted from [12], such that: lines 1–3 initialize the position, best and velocity of the particles, respectively; line 4 initializes the leaders archive with the non-dominated solutions in the swarm; in sequence, the loop updates every particle in the swarm at each step, until some stop condition is reached; and finally, at line 13, the archive of leaders is returned.

4.1   Dynamic Optimization Strategies

The application of the SMPSO to the DSPSP is straightforward since the solution to the problem, the dedication matrix, can be represented by a vector of real numbers. Thus, the dedication matrix represents the particle position. The problem constraints are current handled by DSPSP modeling, described in Sect. 3.6. Hence, although SMPSO can be applied in the problem solution, the incorporation of strategies to deal with the dynamic features of DSPSP may improve the SMPSO results.

In order to try the best performance of the SMPSO for the DSPSP, two differents kinds of dynamic strategies based on historical solutions were applied, i.e, solutions from previous schedules will influence later schedules. The use of memory strategy is typical when the solutions return to regions near their previous locations in the search space, after the occurrence of a change in the dynamic problem [7]. In DSPSP, e.g., when an employee e_i is available in the project, the feasible solutions in some time t_l can be reused in a time t_k $(k > l)$, when the employee returns to the project after leaving it for some period of time.

Historical Solutions. The first strategy applied is based on the work of [5], in which a percentage of historical solutions' usage n is adopted to construct the initial population after the occurrence of a dynamic event. Shen et al. [5] added this strategy to the dϵ-MOEA algorithm with $n = 20$, whose repaired and random solutions strategies compose the remaining 80%. In our experiments, n assumes the values 25, 50, 75, and 100, and random solutions complete the remaining portion of the population, to analyze the impact of varying the percentage of historical solutions in SMPSO. The pseudocode of this algorithm is similar to Algorithm 1. The main difference is the swarm initialization in line 1, which includes the mentioned strategy. We denote this variant as SMPSO-HP.

Memory Approach. The second strategy modifies the SMPSO algorithm by including a new archive, called *prevLeaders* (see Algorithm 2). This archive still uses as Crowding distance as filter selection, but is updated with the best solutions in *leaders* (see line 16). Moreover, at line 6 *leaders* receive the best solutions of previous schedules. The input parameter *schedule* indicates in which rescheduling point the project simulation is situated. The condition in line 5 is to prevent solutions with different number of objectives in *schedule* = 0 and *schedule* = 1, as stability is not considered when *schedule* = 0. We called this variant as SMPSO-MA.

<table>
<tr><td>

Algorithm 1: SMPSO pseudocode

---

1 Initialize the swarm;
2 Initialize the best solution (*Best*);
3 Initialize the speed vector (*speed*);
4 Initialize leaders archive (*leaders*);
5 **while** *a stop condition is not reached* **do**
6   update *speed*;
7   update swarm position;
8   swarm mutation;
9   evaluate the swarm;
10   update *leaders*;
11   update *Best*;
12 **end**
13 **return** *leaders*;

</td><td>

Algorithm 2: SMPSO-MA pseudocode

---

input: Reschedule indicator of the project *schedule*
1 Initialize the swarm;
2 Initialize the best solution (*Best*);
3 Initialize the speed vector (*speed*);
4 Initialize leaders archive (*leaders*);
5 **if** *schedule* > 1 **then**
6   update *leaders* with *prevLeaders*
7 **end**
8 **while** *a stop condition is not reached* **do**
9   update *speed*;
10   update swarm position;
11   swarm mutation;
12   evaluate the swarm;
13   update *leaders*;
14   update *Best*;
15 **end**
16 update *prevLeaders* with *leaders*
17 **return** *leaders*;

</td></tr>
</table>

5 Experiments

To evaluate our hypothesis, the following algorithms are executed 10 times for each instance: SMPSO, SMPSO-HP25, SMPSO-HP50, SMPSO-HP75, SMPSO-HP100, and SMPSO-MA. In what follows the parameters of algorithms, the DSPSP instances, the metrics, and statistical tests used, as well as the results achieved, are described.

5.1 Parameterization

We use the DSPSP model of [5] implemented in Java available in the repository of our research group[1]. The SMPSO and variants are implemented using the jMetal framework. In all experiments, each algorithm evaluates 6000 times the objective function in a reschedule, allowing a fair comparative analysis between all metaheuristics. Since there is a great variation in the number of dynamic events in each simulation, we delimit only the first 100 reschedule events for each experiment; the events always occur in the same sequence for all algorithms. Table 4 presents all parameters applied in the metaheuristics. The first four parameters are the same for all algorithms. The parameter historical solutions' usage expresses the percentage of historical solutions used when a new population is updated in the SMPSO-HPn and is used only for this algorithm.

---

[1] https://github.com/rodrigoamaral/spsp-jmetal.

Table 4. Algorithms parameterization used in the experiments.

Number of reschedule events	100
Stop criteria	< 10000
Mutation	Polynomial mutation
Archive type	Crownding distance
Archive size	100
Population size	100
Historical solutions' usage (%)	$\{25, 50, 75, 100\}$

5.2 Instances

We used six artificial instances extracted from the work of [5]. These instances simulate real projects and consider different numbers of employees, skills, tasks, and new tasks arrival in each scenario analyzed. The occurrence of dynamic events is associated with the instances and follows the same sequence in each execution. We choose the instances regarding the number of tasks and employees, which simulates small and medium-sized projects. Table 5 presents the parameters of all instances used in our experiments, with their respective identifiers, where T_{new} is the set of new tasks that will arrive. The parameters $skills_i^+$ and $skills_i^-$ specify the maximum and minimum values of the employee's skills, respectively.

Table 5. Instances identifiers and their respective parameters used in experiments.

Instance ID	$|E|$	$|skills_i^-|$	$|skills_i^+|$	$|T|$	$|T_{new}|$
I_1	5	4	5	10	10
I_2	10	4	5	10	10
I_3	15	4	5	10	10
I_4	5	4	5	20	10
I_5	10	4	5	20	10
I_6	15	4	5	20	10

5.3 Metrics and Statistical Tests

We use hypervolume to compare the results of each algorithm, commonly used in multiobjective optimization works. Hypervolume is a metric that measures the quality from a non-dominated solution set $P = \{p^1, p^2, \ldots, p^n\}$, by calculating the polytopus area formed between the set P and a reference point r, which is a input parameter [15]. So, when a solution set has high values of hypervolume, this means a better schedule set, considering all objectives computed. In our experiments, the program produces a file with the objective evaluation values after a rescheduling. These values are normalized and used in hypervolume calculation, which ranges from 0 to 1. As the stability objective requires the information of two schedules, it is not optimized at the first schedule. Thus, the initial reference point is $(1.1, 1.1, 1.1)$. In the following reschedules, where stability is considered, the reference point is $(1.1, 1.1, 1.1, 1.1)$.

To compare the results, we consider an accurate statistical level by the non-parametric hypothesis test of Kruskal-Wallis and Tukey as post-hoc test [16], in a multiple comparisons $(n \times n)$. We use a 5% as significance level to assert whether there was a statistical difference between the results of the algorithms.

5.4 Results and Discussion

Table 6 shows the results achieved. The proposed approaches were confronted with the basis SMPSO algorithm. Since each algorithm was compared with SMPSO, the symbols $\approx$, $+$ and $-$ indicate if the variant is significantly equal, better or worse than SMPSO, respectively. If the p-value obtained from the statistical test is greater than 0.05, then we consider significantly equal; otherwise, if p-value < 0.05, then it is significantly better. We regard worse when the variant presents values of hypervolume smaller. Furthermore, in this table, a cell-shaded in gray indicates the best value of hypervolume, for a given instance.

Table 6. Means and standard deviations values of the hypervolumes for each instance.

Algorithm	Hypervolume Mean (Hypervolume standart deviation)		
	I_1	I_2	I_3
SMPSO	0.887 (5.81E-01)	1.186 (8.54E-02)	1.169 (1.46E-01)
SMPSO-HP25	0.884 (5.80E-01)$-$	1.188 (9.16E-02)$\approx$	1.177 (1.46E-01)$\approx$
SMPSO-HP50	0.886 (5.80E-01)$\approx$	1.187 (8.36E-02)$\approx$	1.175 (1.49E-01)$\approx$
SMPSO-HP75	0.889 (5.82E-01)$\approx$	1.187 (8.80E-02)$\approx$	1.170 (1.42E-01)$\approx$
SMPSO-HP100	0.879 (5.76E-01)$-$	1.201 (9.15E-02)$+$	1.167 (1.55E-01)$-$
SMPSO-MA	1.392 (5.66E-02)$+$	1.364 (5.96E-02)$+$	1.362 (1.33E-01)$+$
	I_4	I_5	I_6
SMPSO	1.171 (3.06E-01)	1.115 (8.50E-02)	1.075 (7.45E-02)
SMPSO-HP25	1.154 (3.02E-01)$-$	1.112 (8.48E-02)$-$	1.072 (7.53E-02)$-$
SMPSO-HP50	1.163 (3.04E-01)$-$	1.115 (8.75E-02)$\approx$	1.068 (7.83E-02)$-$
SMPSO-HP75	1.156 (3.03E-01)$-$	1.126 (8.11E-02)$\approx$	1.078 (7.57E-02)$\approx$
SMPSO-HP100	1.161 (3.03E-01)$-$	1.121 (8.85E-02)$\approx$	1.076 (7.48E-02)$\approx$
SMPSO-MA	1.383 (6.60E-02)$+$	1.367 (7.48E-02)$+$	1.356 (7.66E-02)$+$

Considering the instance I_1, that has the fewest employees and tasks, the variant SMPSO-HP25 and SMPSO-HP100 were worse than SMPSO, presenting lower values of hypervolume. For the variants SMPSO-HP50 and SMPSO-HP75, the statistical test of Kruskal-Wallis did not indicate statistical difference to the SMPSO, thus, they are significant equals. Nonetheless, the SMPSO-MA presents the highest value of hypervolume, and p-value < 0.05 when compared to the SMPSO. For this fact, the SMPSO-MA obtained the best performance for this instance. By adding more employees to the project and preserving the number of tasks, i.e., the instances I_2 and I_3, the variants SMPSO-HP25, SMPSO-HP50, and SMPSO-HP75 had performance significantly equals to the SMPSO, in both instances. The SMPSO-HP100 presented best results in I_2, however, in I_3, it was significantly worse to the SMPSO, while the SMPSO-MA maintained the best values of hypervolume and, consequently, was significantly better than the SMPSO.

When the project contains a large number of tasks, but few employees, i.e., the instance I_4, note that the variant SMPSO-HPn was significantly worse to the SMPSO in all n values. Differently, the SMPSO-MA once again presented the best performance compared to the other variants, obtaining a p-value $<$ 0.05. Nonetheless, when we consider the instances I_5 and I_6, which regard more employees to the project, the SMPSO-HP25 remains worse than SMPSO. This does not happen with the SMPSO-HP75 and SMPSO-HP100, as even having better values of hypervolume, however, by the Kruskal-Wallis test, are significant equals to the SMPSO. The SMPSO-HP50 remains worse in I_6 and is equal in I_5 when compared to SMPSO. The SMPSO-MA remains better to the SMPSO and the other variants in both instances.

The results suggest that, when comparing the SMPSO with the SMPSO-HPn, there is no gain in extending SMPSO with the dynamic optimization strategy based on the work of [5]. Also, the variation of the percentage of historical solutions does not reveal any significant impact on the quality of the solutions achieved. Nevertheless, when comparing SMPSO-MA with the SMPSO and the other variants, the results show a significant difference; the SMPSO-MA achieves the best values of hypervolume in all instances and presents a significant difference. Thus, the proposed memory approach was able to improve the results of SMPSO, when solving a dynamic optimization problem.

6 Conclusion

This paper aims to investigate whether the inclusion of two distinct dynamic strategies based on historical solutions improve the performance of the SMPSO when this algorithm is applied to DSPSP. For this purpose, ten independent executions of each algorithm were accomplished, by considering six instances that simulate real project scenarios. The results suggest that the SMPSO-MA is better than the SMPSO and the SMPSO-HPn, for all value of n. Future works may explore issues, such as: **a)** to evaluate the algorithms in more instances of DSPSP; **b)** to increase the number of objective evaluations in each algorithm, which may influence the result precision; **c)** to apply the memory approach in other metaheuristics and compare their performance with the SMPSO-MA.

Acknowledgement. This study was financed in part by CAPES, Brazil - Finance Code 001 and by the Universal CNPq grant, project number 425861/2016-3.

References

1. Harman, M., Jones, B.F.: Search-based software engineering. Inf. Softw. Technol. **43**(14), 833–839 (2001)
2. Eberhart R., Kennedy J.: A new optimizer using particle swarm theory. In: Proceedings of the Sixth International Symposium on Micro Machine and Human Science, pp. 39–43. IEEE (1995)
3. Holland, J.H.: Adaptation in Natural and Artificial Systems: An Introductory Analysis with Applications to Biology, Control, and Artificial Intelligence, 1st edn. MIT Press, Cambridge (1992)
4. Alba, E., Chicano, J.F.: Software project management with GAs. Inf. Sci. **177**(11), 2380–2401 (2007)
5. Shen, X., Minku, L.L., Bahsoon, R., Yao, X.: Dynamic software project scheduling through a proactive-rescheduling method. IEEE Trans. Softw. Eng. **42**, 658–686 (2016)
6. Rezende, A.V., Silva, L., Britto, A., Amaral, R.: Project scheduling problem in the context of search-based software engineering: a systematic review. J. Syst. Softw. **155**, 43–56 (2019)
7. Nguyen, T.T., Yang, S., Branke, J.: Evolutionary dynamic optimization: a survey of the state of the art. Swarm Evol. Comput. **6**, 1–24 (2012)
8. Shen, X., Minku, L.L., Marturi, N., Guo, Y., Han, Y.: A Q-learning based memetic algorithm for multi-objective dynamic software project scheduling. Inf. Sci. **428**, 1–29 (2018)
9. Rezende, A.V.: Otimização com muitos objetivos por evolução diferencial aplicada ao escalonamento dinâmico de projeto de software. Master thesis, Federal University of Sergipe (2019)
10. Bardsiri, V.K., Jawawi, D.N.A., Hashim, S.Z.M., Khatibi, E.: A PSO-based model to increase the accuracy of software development effort estimation. Softw. Qual. J. **21**(3), 501–526 (2013)
11. de Andrade, J., Silva, L., Britto, A., Amaral, R.: Solving the software project scheduling problem with hyper-heuristics. In: Rutkowski, L., Scherer, R., Korytkowski, M., Pedrycz, W., Tadeusiewicz, R., Zurada, J.M. (eds.) ICAISC 2019. LNCS (LNAI), vol. 11508, pp. 399–411. Springer, Cham (2019). https://doi.org/10.1007/978-3-030-20912-4_37
12. Nebro, A.J., Durillo, J.J., García-Nieto, J., Coello, C.A.C., Luna, F., Alba, E.: SMPSO: a new PSO metaheuristic for multi-objective optimization. In: IEEE Symposium on Computational Intelligence in Multi-criteria Decision-Making, pp. 66–73 (2009)
13. Boehm, B.W.: Software engineering economics. IEEE Trans. Softw. Eng. **10**(1), 4–21 (1984)
14. Chang, C.K., Jiang, H., Di, Y., Zhu, D., Ge, Y.: Time-line based model for software project scheduling with genetic algorithms. Inf. Softw. Technol. **50**(11), 1142–1154 (2008)

15. Fonseca, C.M., Paquete, L., López-Ibánez, M.: An improved dimension-sweep algorithm for the hypervolume indicator. In: 2006 IEEE International Conference on Evolutionary Computation, pp. 1157–1163. IEEE (2006)
16. Kruskal, W., Wallis, W.: Use of ranks in one-criterion variance analysis. J. Am. Stat. Assoc. 47(260), 583–621 (1952)

Evaluation of Metaheuristics in the Optimization of Laguerre-Volterra Networks for Nonlinear Dynamic System Identification

Victor O. Costa$^{(\boxtimes)}$ and Felipe M. Müller

Department of Applied Computing, Federal University of Santa Maria,
Santa Maria 97105-900, Brazil
vcosta.acad@gmail.com, felipe@inf.ufsm.br

Abstract. The main objective of nonlinear dynamic system identification is to model the behaviour of the systems under analysis from input-output signals. To approach this problem, the Laguerre-Volterra network architecture combines the connectionist approach with Volterra-based processing to achieve good performance when modeling high-order nonlinearities, while retaining interpretability of the system's characteristics. In this research we assess the performances of three metaheuristics in the optimization of Laguerre-Volterra Networks using synthetic input-output data, a task in which only the simulated annealing metaheuristic was previously evaluated.

Keywords: Nonlinear dynamic systems · Laguerre-Volterra networks · Metaheuristics

1 Introduction

Nonlinear dynamic system identification from input-output (IO) data using the Volterra series is a broadly studied field [5,16,18,20,29]. The functional power series, as coined by Volterra in [37], can be seen as a Taylor series with memory, in the sense that it can approximate stable nonlinear functions with dynamic behaviour. This functional expansion representation allows for the extraction of kernels that characterize the system dynamics for any given nonlinear order, which can be used to interpret characteristics of the systems under analysis. For example, the first order Volterra kernel of a system is akin to the convolution kernel that describes a linear time-invariant (LTI) system, and represents the linear component of a given nonlinear system.

In practice, the estimation of high-order Volterra kernels implies in the computation of a large number of coefficients, resulting in heavy computational burdens and in the need for large amounts of input-output data [2]. These difficulties often lead to the use of low-order models, which cannot properly represent high-order nonlinear behaviour and are therefore biased [21]. To mitigate the

© Springer Nature Switzerland AG 2020
R. Cerri and R. C. Prati (Eds.): BRACIS 2020, LNAI 12319, pp. 95–109, 2020.
https://doi.org/10.1007/978-3-030-61377-8_7

forementioned problems, it is possible to convolve the input signal with a filter bank and then expand the Volterra series in terms of the bank's outputs. This representation assumes that the filter bank properly represents the system's dynamic characteristics and is capable of reducing the number of parameters in the series. The search for adequate filter banks often lead to the use of orthonormal basis functions, and in [38] authors use a discretized version of the continuous Laguerre functions with this purpose for the first time. Subsequently, [19] applies the Laguerre functions proposed in [27] for this task, which are built to be orthonormal in discrete time.

As the next step towards a decrease in the number of parameters when modeling Volterra systems, [1] proposes a connectionist model which applies the Laguerre functions proposed in [27] as a filter bank for the input signal and then propagates the bank's outputs through a layer of polynomial activation functions. This model is referred to as the Laguerre-Volterra network (LVN), and there is a straightforward relationship between the parameters of the network and the corresponding Volterra kernels. The LVN has been applied to model electrical activity in the dentate gyrus [1] and cerebral hemodynamics [23,24], both using backpropagation (BP) to optimize the continuous parameters of the network. Although BP is successful in the optimization of various connectionist models (e.g. MLP, CNN, RNN), [8] argues the presence of nonmonotonic activation functions in the LVN architecture makes BP-based optimization prone to local minima issues. Hence, [8] proposes the use of the simulated annealing metaheuristic [15] to optimize the LVN continuous parameters. Subsequently, [13] uses the same metaheuristic in this task.

This research is focused on the practical issue of optimizing the parameters of LVN models from IO data with metaheuristic algorithms, and contributes to the literature with a comparison of three metaheuristics for this purpose, namely simulated annealing (SA), particle swarm optimization (PSO) and ant colony optimization for continuous domains ($ACO_{\mathbb{R}}$). This comparison is relevant because although the literature evidences the success of the metaheuristic paradigm in the training of LVNs, SA is the only algorithm from this family ever employed for the task. All source code used to collect the results reported in this paper are publicly available at a Git repository [6] under the GNU GPL v3.

The rest of the paper is organized as follows. Section 2 presents relevant literature in the application of metaheuristics for Volterra-based system identification. Section 3 develops on the Laguerre-Volterra network architecture and on the metaheuristics under analysis. Section 4 explains the generation of synthetic input-output data from two simulated nonlinear dynamic systems. In Sect. 5, we discuss the results obtained for each metaheuristic, and finally, Sect. 6 presents concluding remarks and indications for possible future research topics.

2   Metaheuristics in Volterra Models

Although the use of metaheuristics to optimize continuous parameters of the LVN architecture has only recently been proposed, there is plenty of literature relating metaheuristics to Volterra-based system identification. Here we present some of the existing research to offer a perspective on some ways in which these fields may interact.

In [40], authors propose using the first and second order terms of the Volterra series as inputs of a three-layer feedforward artificial neural network, in which each node computes the same exponential activation function. The continuous parameters of this architecture consist of activation function thresholds; and weights between input and hidden layers, and between hidden and output layers. All network weights and thresholds are optimized using PSO with the mean squared error (MSE) as cost function, which is shown to achieve smaller error rates when compared to backpropagation, and to be robust to noisy train signals. This approach combines the Volterra modeling and connectionist paradigm in such a way that the final model has more free parameters than the second order Volterra series itself.

The authors of [41] reorganize the Volterra series equations to a state-space representation, and the resulting parameters are estimated using the Kalman smoother (KS) adaptive filter. This approach limits the method to off-line processing, since the KS equations are not causal. The optimization of the smoother parameters with gradient-based methods is considered too complex for the proposed system, encouraging the use of metaheuristics for the task. In this way, the metaheuristics artificial bee colony (ABC), PSO and genetic algorithm (GA) are applied in the second level of inference, i.e. in the search for the parameters of a system which in its turn optimizes the parameters of another system. The presented results show that metaheuristics based on the swarm intelligence paradigm [39], i.e. PSO and ABC, achieved significantly lower errors than GA and standalone KS. These two metaheuristics achieve error rates lower than 1% in noiseless case and lower than 3% when considering the presence of input-additive Gaussian white noise (GWN), which results in a 20 dB signal-to-noise ratio (SNR).

Considering synthetic IO signals from simulated systems, [11] compares state transition algorithm (STA), real coded genetic algorithm (RCGA) and covariance matrix adapted evolution strategy (CMA-ES) metaheuristics in the Volterra series identification task. The research also compares the same metaheuristics with the design of multivariate PID controllers. In the experiments, CMA-ES had the lowest overall MSE for Volterra series optimization, while STA presented the best performance for PID controllers design.

A novel adaptive version of the ABC metaheuristic is presented in [43], with an inertia weight parameter varying nonlinearly according to the number of past iterations. This algorithm is then applied to search the coefficients of a single-input single-output second-order Volterra series, and the reported results show that the adaptive mechanism enhances the identification of simulated nonlinear systems.

Different procedures for initialization of hierarchical populations in GA are compared in [3], which optimizes Volterra systems considering that for each identification tasks only a subset of the Volterra series coefficients are active.

According to the results, the hierarchical population scheme reduces the overall computational burden of the system identification task. The proposed procedure is also shown to be robust to three noise levels, with SNRs of 10, 15 and 20 dB.

The non-structural parameters of the LVN are optimized with SA for the first time in [8], which also proposes the use of ℓ_1 regularization in the cost function. In this way, a sparse representation is achieved by guiding insignificant parameters towards near-zero values in the search process, implicitly solving the structure selection problem. A recurrent sparse version of the LVN architecture is also proposed, and applied to model the Hodgkin-Huxley neuronal firing equations [12]. Subsequently, [13] employs the LVN in the context of large scale synapse simulations, with the network's continuous parameters optimized using a fusion between SA and linear equation solving with the Moore-Penrose pseudoinverse. In [9,10], SA is used to optimize the free parameters of the proposed neuronal mode network (NMW) architecture for system identification from spike-train IO data. This architecture is not Volterra-equivalent itself but stems from this framework.

3   Theoretical Background

As the LVN architecture is described in this section, we highlight how its particularities determine the search space of the metaheuristics presented later.

3.1   Laguerre-Volterra Network Structure

In this subsection we use the notation from [21], Sects. 2.1, 2.3.2 and 4.3, only changing indexing parenthesis for brackets to reinforce that the processing occurs in discrete time. The LVN architecture, proposed in [1], is displayed in Fig. 3. It is composed of a filter bank in which each filter is a discrete Laguerre functions (DLFs) as proposed in [27], cascaded with a layer of static polynomial activation functions, whose outputs are summed with an offset value to give the network's output.

A bank composed of L LTI filters has L outputs, and each output can be computed by convolving the input signal with the corresponding impulse responses. The Laguerre finite impulse response (FIR) of order j, b_j, is defined in (1) as a function of the discrete-time index m. This relation shows that the filter characteristics depend directly on the α parameter and on the filter order j. Figure 1 shows that the number of zero-crossings of the Laguerre filters' impulse response depends on the order j. The order also determines the spread of significant values for each impulse response over time, in such a way that low order filters will rapidly get to near-zero values.

$$b_j[m] = \alpha^{(m-j)/2}(1-\alpha)^{1/2}\sum_{k=0}^{j}(-1)^k \binom{m}{k}\binom{j}{k}\alpha^{j-k}(1-\alpha)^k \tag{1}$$

where $0 < \alpha < 1$, $0 \le j \le L-1$ and $0 \le m \le M-1$. In this FIR implementation, the memory extent of the system, M, is explicitly represented.

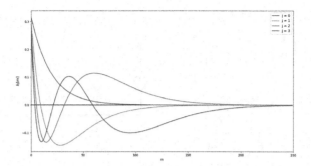

Fig. 1. Finite impulse response of each DLFs for a filter bank with $L = 4$ and $\alpha = 0.9$.

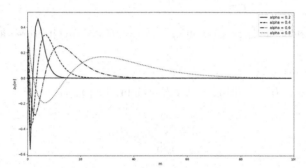

Fig. 2. Finite impulse response of DLFs with order $j = 2$, for a range of α values.

Increasing the α parameter produces a similar effect to the increment of the order j with respect to the filter's spread of significant values, as shown in Fig. 2, directly influencing its memory characteristic. Although the FIR realization of the Laguerre filter bank allows us to examine the forementioned characteristics, it is possible obtain the bank's outputs with reduced computational complexity by using a recursive, IIR, implementation of the Laguerre functions proposed by Ogura [27]. This approach dismisses the convolution operation, and the output of each DLF is defined in (2) and initialized by (3) for an input signal $x[n]$.

$$v_j[n] = \sqrt{\alpha}v_j[n-1] + \sqrt{\alpha}v_{j-1}[n] - v_{j-1}[n-1] \tag{2}$$

$$v_0[n] = \sqrt{\alpha}v_0[n-1] + T\sqrt{1-\alpha}x[n], \tag{3}$$

where T is the sampling period of the discrete input signal $x[n]$.

The input of each of the H nonlinear nodes is a weighted sum of the Laguerre filter bank outputs, with unique weights for each node, as shown in (4). Each hidden node h is a zero-memory polynomial function of nonlinear order Q, with output defined in (5).

$$u_h[n] = \sum_{j=0}^{L-1} w_{h,j}v_j[n], \tag{4}$$

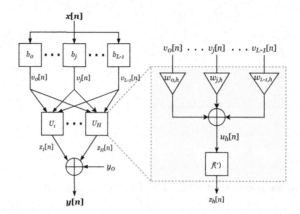

Fig. 3. Structure of the Laguerre-Volterra network (LVN) architecture, adapted from [21]

where $h = 1, 2, ..., H$, resulting in $L * H$ weights in $\{w_{h,j}\}$.

$$z_h[n] = \sum_{q=1}^{Q} c_{h,q} u_h^q[n],$$ (5)

with $H * Q$ polynomial coefficients in $\{c_{h,q}\}$. When the equivalent Volterra model of the network is extracted, its nonlinear order is also Q.

The network output is a summation of hidden layer outputs and offset term, following (6). In the three-layer perceptron architecture the connections between hidden and output layers are weighted, while in the LVN the output layer computes a simple sum. This is because weighting these connections would give the same results as multiplying the polynomial coefficients of a given node by the connection's weight. Also, the polynomial functions in the hidden layer do not need zeroth order terms due to the trainable output offset y_o. These facts contribute to the compactness of the network's parameters. Considering the Laguerre parameter α, hidden layer input weights $\{w_{h,j}\}$, polynomial coefficients $\{c_{h,q}\}$ and output offset y_0, optimizing the network's parameters is a problem of dimensionality $H * (L + Q) + 2$. In this way, the LVN architecture is more compact when compared to both the original Volterra series representation and the Volterra model expanded with Laguerre basis functions [1].

$$y[n] = \sum_{h=1}^{H} z_h[n] + y_0$$ (6)

All relations shown in this subsection depend directly on the LVN structure (L, H and Q). Among the procedures proposed in the literature to select the structure of LVN models, the selection using Bayesian information criterion (BIC) is discouraged by [21], which favours ascending order model selection with a statistical stopping criterion. More recently, [8] starts with a roof high-dimensional model and leverages the sparsity of parameters introduced by ℓ_1 regularization to reduce the structure via model pruning. Since the present research focus in the optimization of continuous network parameters rather than structure selection, in the rest of the paper we assume this step was already performed.

3.2 Metaheuristics

Simulated Annealing. The connection between the fields of multivariate combinatorial optimization and condensed matter physics was analyzed initially in [15], showing how it is possible to optimize cost functions by combining the relationship proposed by Metropolis in [22], used to simulate a collection of atoms at equilibrium according to the Boltzmann distribution, and the annealing procedure from metallurgy. In this way, the SA metaheuristic treats the solution under optimization and its cost value as a configuration and the associated energy of this configuration for a given system, respectively. Although it was initially proposed in the context of combinatorial optimization, SA was later adapted to optimize over continuous variables [4,17,35].

For every local iteration, one variable x_i of the current solution $\mathbf{x}$ is randomly chosen to be perturbed, generating a new value for the variable as given in $x_i' = x_i + \Delta x$, where Δx is the chosen perturbation. This new solution with one modified variable then has its cost evaluated. For the sake of comparison, here we use a constant step-size of random sign Δx for all variables as done in the other researches that use SA to train LVNs [8,13]. With the cost function treated as the system's energy, E, the configuration generated by a perturbation in the solution is accepted according to the probabilistic Metropolis equation, in which $\delta E = E(x') - E(x)$ is the difference in terms of costs between the new and current solutions, defined in (7). According to this relation, a newly generated solution is automatically accepted when its costs is smaller than the current cost, and otherwise the probability of acceptance depends on how close the new solution's cost is to the current cost, and on a temperature T. It is noteworthy that due to the behaviour of SA, the best solution until the moment has non-zero probability of being substituted by a solution with worst fitness. For this reason, the current best solution must be stored apart from the current solution.

$$P(\Delta E) = \begin{cases} 1 & \text{if } \Delta E \leq 0, \\ e^{-\Delta E/T} & \text{otherwise.} \end{cases} \qquad (7)$$

The annealing is applied when at every global iteration, T is reduced according to some cooling schedule and the expected number of accepted solutions is also reduced because of (7). The authors of [26] compare cooling procedures and favours the use of the exponential decay cooling schedule, defined in (8). In this way, considering n_{iter}^{local} local iterations, the total number of function evaluations (FE) for SA is $n_{iter}^{global} * n_{iter}^{local}$.

$$T(t) = \eta T(t-1), \quad t = 1, 2, 3, ..., n_{iter}^{global} \qquad (8)$$

where n_{iter}^{global} is the number of global iterations.

Particle Swarm Optimization. Inspired by the behaviour of swarms in nature, [14] proposes the PSO metaheuristic with the intent of extracting computational intelligence from simple social interactions. In the proposal, the swarm is composed of m particles, each of which represents a different solution. These solutions are initialized randomly, and then iteratively updated. Each particle i is associated with three D-dimensional vectors, where D is the dimensionality of the search space. The vectors are velocity $\mathbf{v_i}$, current position $\mathbf{x_i}$ and personal best position $\mathbf{pbest_i}$. At every iteration, each particle has the cost of its position evaluated, and if it is the lowest cost found by the particle until the moment, these coordinates are stored in the $\mathbf{pbest_i}$ vector. If it is also the lowest cost ever found in the entire swarm, the coordinates are stored in the $\mathbf{gbest}$ vector. The particles' velocity vectors are then updated stochastically, taking into account the best coordinates visited by the particle itself and by the entire swarm according to (9), while their position vectors are updated according to (10). When updating velocities, the c_p and c_g parameters are used as weights for the personal and global best positions, respectively.

$$\mathbf{v_i}(t) = \mathbf{v_i}(t-1) + rand(0,1) * c_p(\mathbf{pbest_i} - \mathbf{x_i}(t-1)) \\ + rand(0,1) * c_g(\mathbf{gbest} - \mathbf{x_i}(t-1)) \tag{9}$$

$$\mathbf{x_i}(t) = \mathbf{x_i}(t-1) + \mathbf{v_i}(t), \tag{10}$$

where t is the current iteration of the algorithm and $rand(0,1)$ is an uniform random number.

As in [42], previous research indicate that PSO is vulnerable to stagnation due to low swarm convergence. Beyond that, there may be uncontrolled increment of the velocities, leading to instability in the algorithm [30]. To alleviate these problems, [31] proposes the parameter ω which acts as a forgetting factor over past velocities of a given particle, modifying the velocity update relationship to (11).

$$\mathbf{v_i}(t) = \omega * \mathbf{v_i}(t-1) + rand(0,1) * c_p(\mathbf{pbest_i} - \mathbf{x_i}(t-1)) \\ + rand(0,1) * c_g(\mathbf{gbest} - \mathbf{x_i}(t-1)) \tag{11}$$

In this way, at each iteration all particles are stochastically guided to search space regions considered promising by both the particle itself and the whole swarm. Considering a total of n_{iter} iterations, the objective function is evaluated $m * n_{iter}$ times.

Ant Colony Optimization for Continuous Domains. The ant colony optimization (ACO) family of metaheuristics is inspired by the foraging behaviour of ants [33], and guides the search probabilistically using the cost landscape captured as pheromone representation. Although the first algorithms of this framework were proposed in the context of combinatorial optimization, metaheuristics for continuous domains such as $ACO_\mathbb{R}$ [32] were proposed later.

$ACO_\mathbb{R}$ keeps information about the search history using an archive that keeps the best k solutions found so far. In the optimization process, all solutions of the archive are initialized randomly and the cost function is evaluated for each solution. The archive is then sorted in ascending order of cost, in such a way that the best solutions are located in the first positions of the archive, and each of these positions is associated with a weight ω_j according to (12).

$$\omega_j = \frac{1}{qk\sqrt{2\pi}}e^{\dfrac{-[rank(j)-1]^2}{2q^2k^2}}, \tag{12}$$

in which rank(j) is the archive position and q is a parameter of the algorithm. Hence, the best solution of the archive has the highest weight.

At each iteration, each ant of the population chooses a solution from the archive, with the probability of solution l being selected given by the weight of l divided by a normalizing constant, as displayed in (13).

$$p_l = \frac{\omega_l}{\sum_{j=1}^{k}\omega_j} \tag{13}$$

The selected solution is then used as the center μ of the probability density function of (14), with standard deviation defined in (15) as the average absolute distance from the chosen solution to the entire archive, from which a new solution is then sampled.

$$g_i(x;\mu_i,\sigma_i) = \frac{1}{\sigma_i\sqrt{2\pi}}e^{-\dfrac{(x-\mu_i)^2}{2\sigma_i^2}} \tag{14}$$

$$\sigma_i = \xi\sum_{j=1}^{k}\frac{|s_j^i - \mu_i|}{k-1}, \tag{15}$$

where s_j^i is the value of the ith continuous variable of the jth solution in the archive. The random variable x is the domain of the PDF.

All new solutions then have their costs evaluated, and are appended to the achieve. The entire achieve is then sorted and only the best k solutions are kept. Considering a total of n_{iter} iterations, $ACO_{\mathbb{R}}$ evaluates a total of $k + m * n_{iter}$ objective functions during the optimization process.

4   Synthetic Signals

The two pairs of input-output signals used in this research are synthetic, being extracted from simulated systems. One of these systems has finite nonlinear order, while the other one has infinite order in the Taylor and Volterra senses. The finite order system can be considered ideal because it uses the LVN architecture with structure $L = 5$, $H = 3$, $Q = 4$, and all continuous parameters are uniformly randomized. The output of this system is defined in (6).

The second system is defined by a linear filter cascaded by an infinite-order static nonlinearity. The linear filter is composed of a sum of three recursive exponentially weighted moving averages (EWMA) with randomized smoothing constants, defined in (16) and (17).

$$e_i[n] = \beta_i e_i[n-1] + (1 - \beta_i)x[n] \tag{16}$$

$$p[n] = \sum_{i=1}^{3} e_i[n], \tag{17}$$

where β_i, $e_i[n]$ and $p[n]$ are the smoothing constant, the output of the ith EWMA and the final filter output, respectively.

The static nonlinearity is a composite function of exponential following sine, as given in (18). Each of these functions is of infinite order, and the resulting composition is expected increase the difficulty of the modeling task.

$$y_{inf}[n] = e^{sin(p[n])} \tag{18}$$

To enable a rich display of non-linear interactions among multiple frequency components of the presented systems, the input signals must be broadband [21]. Thus, although the input spectrum is not required to be white, we use GWN signals as input for both simulated systems, and we consider that the IO signals were sampled 25 Hz.

To evaluate the generalization performance of the LVN optimization task, for each of the defined systems we extract two pairs of IO signals, namely the train and test signals. While the train signals are used to define a cost function for the search process, the test signals allows us to assess the resulting models on data unseen during the optimization. In many cases, such as for some physiological signals, the acquisition of long signals is intractable [21]. Therefore, we use short train IO signals with only 1024 data points to take this intractability into account. The test signals, on the other side, are 2048 points long. In this way, the search for parameters is guided with limited data, while the resulting models are evaluated using more data.

5   Evaluation and Discussion

To compare the presented metaheuristics in the optimization of LVN models, we use both systems from Sect. 4. To guide the search, the cost of a given set of parameters is defined as the normalized mean squared error (NMSE) between the expected output and the LVN model output. All experiments detailed in this section were executed in Python 3.6.8 with the NumPy module [28] for vector operations, and the SciPy [36] and scikit-posthoc [34] modules for statistical significance analysis. The environment is composed of an octa-core Intel Xeon ®:E-5405 processor, an 8 GiB RAM, running the Ubuntu 18.04.3 operating system.

Due to the inherent stochastic behaviour of metaheuristics, we perform each experiment 30 times and present the resulting statistics. Since we deal with two

Table 1. Parameters used for each metaheuristic.

SA	PSO	$ACO_{\mathbb{R}}$
$n_{iter}^{local} = 100$	$m = 20$	$m = 10$
$\eta = 0.99$	$w = 0.99$	$k = 50$
$\Delta x = 10^{-2}$	$c_g = 2$	$q = 0.01$
$T_0 = 10$	$c_p = 2$	$\xi = 0.85$

distinct systems, for each of them we use a different LVN structure. The LVN structures for the presented finite and infinite order systems are ($L = 5$, $H = 3$, $Q = 4$) and ($L = 2$, $H = 3$, $Q = 5$), respectively. The parameters used for each metaheuristic are presented in Table 1, none of which were fine tuned for the application.

Tables 2 and 3 show that the PSO metaheuristic rapidly converges to low error rates, but does not improve in later search stages. Oppositely, the first iterations of SA achieve high error rates, but with enough iterations it consistently reaches lower rates than PSO. Although SA never exhibits lower train errors compared to $ACO_{\mathbb{R}}$, it has lower test errors for later stages considering the infinite order system.

Table 2. Average errors achieved by each metaheuristic when optimizing the *finite order system*, with standard deviations in parenthesis.

	Train NMSE			Test NMSE		
FE	SA	PSO	$ACO_{\mathbb{R}}$	SA	PSO	$ACO_{\mathbb{R}}$
1×10^2	19.151	0.304	0.381	19.481 (14.759)	0.316 (0.012)	0.379 (0.063)
1×10^3	11.840	0.301	0.242	12.052 (10.541)	0.313 (0.004)	0.249 (0.018)
5×10^3	1.032	0.301	0.235	1.056 (2.796)	0.313 (0.005)	0.250 (0.018)
1×10^4	0.724	0.301	0.234	0.745 (2.652)	0.313 (0.005)	0.251 (0.019)
1×10^5	0.245	0.301	0.233	0.264 (0.032)	0.313 (0.005)	0.252 (0.022)

These differences between train and test average errors reveal the presence of the overfitting phenomenon in our experiments, in which the minimization of train cost does not imply in generalization capability. The phenomenon can be seen when, for both systems, $ACO_{\mathbb{R}}$ starts with relatively low train and test error rates, and consistently improves the train errors along the iterations. However, the test error rates do not follow this improvement and are subject to stagnation or even mild worsening. This also happens to SA when optimizing the infinite order system, for which it reaches competitively low error rates, when a great decrease in the average train error (from 1×10^4 to 1×10^5 function evaluations) actually hurts the test performance.

Table 3. Average errors achieved by each metaheuristic when optimizing the *infinite order system*, with standard deviations in parenthesis.

FE	Train NMSE			Test NMSE		
	SA	PSO	ACO$_\mathbb{R}$	SA	PSO	ACO$_\mathbb{R}$
1×10^2	1.100	0.401	0.447	1.112 (0.596)	<u>0.403</u> (0.037)	0.413 (0.033)
1×10^3	0.945	0.397	0.310	0.946 (0.495)	0.402 (0.036)	<u>0.382</u> (0.023)
5×10^3	0.536	0.397	0.303	0.489 (0.140)	0.401 (0.036)	<u>0.380</u> (0.013)
1×10^4	0.429	0.397	0.300	<u>0.368</u> (0.029)	0.401 (0.036)	0.379 (0.015)
1×10^5	0.306	0.397	0.299	<u>0.378</u> (0.018)	0.401 (0.036)	0.380 (0.014)

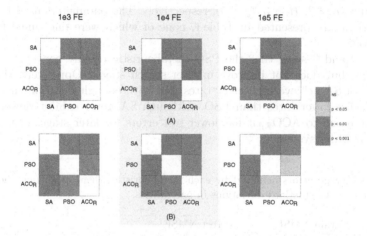

Fig. 4. Heat map representation of the Nemenyi test for a range of function evaluations, considering finite (A) and infinite (B) order systems. Results below the 95% confidence threshold are considered non-significant, and there are different levels of significance.

Before we make recommendations from the average errors, it is necessary to assess the statistical significance of the stated results. Considering a confidence threshold of 99%, the Friedman test for multiple repeated measures [7] rejects the null hypothesis that the distributions of test errors are identical for all metaheuristics. Therefore, we are able to use the Nemenyi post-hoc test [25] to compare the significances between algorithms, in such a way that it is possible to assess the significance of specific wins and losses of each metaheuristic. To ease interpretation, we represent the Nemenyi tests as heat maps in Fig. 4, which show that for a small number of FE (1×10^3) both ACO$_\mathbb{R}$ and PSO significantly outperform SA in the 99.9% confidence threshold. When considering more FEs, the wins of ACO$_\mathbb{R}$ over SA stop being significant. The SA wins over ACO$_\mathbb{R}$ in late search stages, however, are not consistently significant because of the overfitting issues discussed previously. In this way, the best practice to date in LVN optimization seems to be using $ACO_\mathbb{R}$ with a relatively small number of function evaluations (e.g. 5×10^3 FE).

6   Conclusion

The results on metaheuristics optimizing the parameters of Laguerre-Volterra networks show that the $ACO_\mathbb{R}$ metaheuristic achieves significantly smaller error rates within few iterations, when compared to the other two metaheuristics. When the number of iterations is greater, however, SA is able to achieve similar or slightly better performance. We identify the overfitting phenomenon as a major issue when metaheuristics are able to achieve low train errors, hurting the generalization capacity of the LVNs. Future research ideas include searching in the wider space of multi-input multi-output network models with a more diverse group of metaheuristics, and the use of real-world input-output data.

References

1. Alataris, K., Berger, T., Marmarelis, V.: A novel network for nonlinear modeling of neural systems with arbitrary point-process inputs. Neural Networks **13**(2), 255–266 (2000)
2. Allgöwer, F., Zheng, A.: Nonlinear Model Predictive Control, vol. 26. Birkhäuser, Basel (2012)
3. de Assis, L.S., Junior, J.R.d.P., Tarrataca, L., Haddad, D.B.: Efficient Volterra systems identification using hierarchical genetic algorithms. Appl. Soft Comput. **85**, 105745 (2019)
4. Brooks, D.G., Verdini, W.A.: Computational experience with generalized simulated annealing over continuous variables. Am. J. Math. Manag. Sci. **8**(3–4), 425–449 (1988)
5. Chon, K.H., Holstein-Rathlou, N.H., Marsh, D.J., Marmarelis, V.Z.: Comparative nonlinear modeling of renal autoregulation in rats: Volterra approach versus artificial neural networks. IEEE Trans. Neural Networks **9**(3), 430–435 (1998)
6. Costa, V.O.: Metaheuristics for Laguerre-Volterra networks optimization (2020). https://github.com/vctrop/metaheuristics_for_Laguerre-Volterra_networks_optimization/tree/BRACIS2020
7. Friedman, M.: The use of ranks to avoid the assumption of normality implicit in the analysis of variance. J. Am. Stat. Assoc. **32**(200), 675–701 (1937)
8. Geng, K., Marmarelis, V.: Methodology of recurrent Laguerre-Volterra network for modeling nonlinear dynamic systems. IEEE Trans. Neural Netw. Learn. Syst. **28**, 1–13 (2016)
9. Geng, K., et al.: Mechanism-based and input-output modeling of the key neuronal connections and signal transformations in the CA3-CA1 regions of the hippocampus. Neural Comput. **30**(1), 149–183 (2018)
10. Geng, K., et al.: Multi-input, multi-output neuronal mode network approach to modeling the encoding dynamics and functional connectivity of neural systems. Neural Comput. **31**(7), 1327–1355 (2019)
11. Gurusamy, S.: Metaheuristic algorithms for the identification of nonlinear systems and multivariable PID controller tuning. Ph.D. thesis, Kalasalingam Academy of Research and Education, Krishnankoil, India (2017)

12. Hodgkin, A.L., Huxley, A.F.: A quantitative description of membrane current and its application to conduction and excitation in nerve. J. Physiol. **117**(4), 500 (1952)

13. Hu, E.Y., Yu, G., Song, D., Bouteiller, C.J.M., Berger, W.T.: Modeling nonlinear synaptic dynamics: a Laguerre-Volterra network framework for improved computational efficiency in large scale simulations. In: 40th Annual International Conference of the IEEE Engineering in Medicine and Biology Society (EMBC), pp. 6129–6132 (2018)

14. Kennedy, J., Eberhart, R.: Particle swarm optimization. In: Proceedings of ICNN 1995 – International Conference on Neural Networks, vol. 4, pp. 1942–1948. IEEE (1995)

15. Kirkpatrick, S., Gelatt, C.D., Vecchi, M.P.: Optimization by simulated annealing. Science **220**(4598), 671–680 (1983)

16. Korenberg, M.J., Hunter, I.W.: The identification of nonlinear biological systems: Volterra kernel approaches. Ann. Biomed. Eng. **24**(2), A250–A268 (1996). https://doi.org/10.1007/BF02648117

17. Locatelli, M.: Simulated annealing algorithms for continuous global optimization. In: Pardalos, P.M., Romeijn, H.E. (eds.) Handbook of Global Optimization. NOIA, vol. 62, pp. 179–229. Springer, Boston (2002). https://doi.org/10.1007/978-1-4757-5362-2_6

18. Marmarelis, P.Z., Marmarelis, V.Z.: Analysis of physiological signals. In: Analysis of Physiological Systems. CBM, pp. 11–69. Springer, Boston (1978). https://doi.org/10.1007/978-1-4613-3970-0_2

19. Marmarelis, V.Z.: Identification of nonlinear biological systems using Laguerre expansions of kernels. Ann. Biomed. Eng. **21**(6), 573–589 (1993). https://doi.org/10.1007/BF02368639

20. Marmarelis, V.Z.: Modeling methology for nonlinear physiological systems. Ann. Biomed. Eng. **25**(2), 239–251 (1997). https://doi.org/10.1007/BF02648038

21. Marmarelis, V.Z.: Nonlinear Dynamic Modeling of Physiological Systems, vol. 10. Wiley, Hoboken (2004)

22. Metropolis, N., Rosenbluth, A.W., Rosenbluth, M.N., Teller, A.H., Teller, E.: Equation of state calculations by fast computing machines. J. Chem. Phys. **21**(6), 1087–1092 (1953)

23. Mitsis, G.D., Poulin, M.J., Robbins, P.A., Marmarelis, V.Z.: Nonlinear modeling of the dynamic effects of arterial pressure and co2 variations on cerebral blood flow in healthy humans. IEEE Trans. Biomed. Eng. **51**(11), 1932–1943 (2004)

24. Mitsis, G.D., Zhang, R., Levine, B., Marmarelis, V.: Modeling of nonlinear physiological systems with fast and slow dynamics. II. Application to cerebral autoregulation. Ann. Biomed. Eng. **30**(4), 555–565 (2002). https://doi.org/10.1114/1.1477448

25. Nemenyi, P.: Distribution-free multiple comparisons. Ph.D. thesis, Princeton University, New Jersey, United States (1963)

26. Nourani, Y., Andresen, B.: A comparison of simulated annealing cooling strategies. J. Phys. A Math. Gen. **31**(41), 8373 (1998)

27. Ogura, H.: Estimation of wiener kernels of a nonlinear system by means of digital Laguerre filters and their fast algorithm. In: Nonlinear Signal and Image Processing, pp. 855–858. IEEE (1995)

28. Oliphant, T.E.: A Guide to NumPy, vol. 1. Trelgol Publishing, New York (2006)

29. Palm, G., Pöpel, B.: Volterra representation and wiener-like identification of nonlinear systems: scope and limitations. Q. Rev. Biophys. **18**(2), 135–164 (1985)

30. Poli, R., Kennedy, J., Blackwell, T.: Particle swarm optimization. Swarm Intell. **1**(1), 33–57 (2007). https://doi.org/10.1007/s11721-007-0002-0

31. Shi, Y., Eberhart, R.: A modified particle swarm optimizer. In: 1998 IEEE International Conference on Evolutionary Computation Proceedings. IEEE World Congress on Computational Intelligence, pp. 69–73 (1998)
32. Socha, K., Dorigo, M.: Ant colony optimization for continuous domains. Eur. J. Oper. Res. **185**(3), 1155–1173 (2008)
33. Stützle, T., et al.: Parameter adaptation in ant colony optimization. In: Hamadi, Y., Monfroy, E., Saubion, F. (eds.) Autonomous Search, pp. 191–215. Springer, Heidelberg (2011). https://doi.org/10.1007/978-3-642-21434-9_8
34. Terpilowski, M.: scikit-posthocs: Pairwise multiple comparison tests in Python. J. Open Source Softw. **4**(36), 1169 (2019). https://doi.org/10.21105/joss.01169
35. Vanderbilt, D., Louie, S.G.: A Monte Carlo simulated annealing approach to optimization over continuous variables. J. Comput. Phys. **56**(2), 259–271 (1984)
36. Virtanen, P., et al.: SciPy 1.0: fundamental algorithms for scientific computing in Python. Nat. Methods **17**, 261–272 (2020)
37. Volterra, V.: Theory of Functionals and of Integral and Integro-Differential Equations. Blackie & Son, London (1930)
38. Watanabe, A., Stark, L.: Kernel method for nonlinear analysis: identification of a biological control system. Math. Biosci. **27**(1), 99–108 (1975)
39. Yang, X., Deb, S., Fong, S., He, X., Zhao, Y.: From swarm intelligence to metaheuristics: nature-inspired optimization algorithms. Computer **49**, 52–59 (2016)
40. Yang, Y.S., Chang, W.D., Liao, T.L.: Volterra system-based neural network modeling by particle swarm optimization approach. Neurocomputing **82**, 179–185 (2012)
41. Yazid, E., Liew, M.S., Parman, S., Kurian, V.J.: Improving the modeling capacity of Volterra model using evolutionary computing methods based on Kalman smoother adaptive filter. Appl. Soft Comput. **35**, 695–707 (2015)
42. Ye, H., Luo, W., Li, Z.: Convergence analysis of particle swarm optimizer and its improved algorithm based on velocity differential evolution. Comput. Intell. Neurosci. **2013**, 384125 (2013)
43. Zhang, Y., Rong, Y., Yan, C., Liu, J., Wu, X.: Kernel estimation of Volterra using an adaptive artificial bee colony optimization and its application to speech signal multi-step prediction. IEEE Access **7**, 49048–49058 (2018)

EvoLogic: Intelligent Tutoring System to Teach Logic

Cristiano Galafassi[1]([✉]), Fabiane F. P. Galafassi[2], Eliseo B. Reategui[1],
and Rosa M. Vicari[1]

[1] Centro de Estudos Interdisciplinares em Novas Tecnologias da Educação,
Universidade Federal do Rio Grande do Sul (UFRGS), Porto Alegre, RS 90040-060, Brazil
cristianogalafassi@gmail.com, eliseoreategui@gmail.com,
rosa@inf.ufrgs.br
[2] Universidade Federal do Pampa (UNIPAMPA),
Campus Itaqui, Itaqui, RS 970650-000, Brazil
fabiane.penteado@gmail.com

Abstract. This article presents the cognitive model of the EvoLogic Intelligent Tutoring System, developed to assist in the teaching-learning process of Natural Deduction in Propositional Logic. EvoLogic consists of 3 agents, among which, the Pedagogical agent (treated here as the student model) and the Specialist agent (based on a Genetic Algorithm) compose the cognitive model. This cognitive model allows an efficient model tracing mechanism to be developed, which will follow each student's step during the theorem proof. The purpose of the article, in addition to presenting the EvoLogic, is to analyze the efficiency of the ITS in a known exercise that has already been studied in the literature (applied to 57 students). The results show that the EvoLogic obtained all the solutions presented by the students, allowing it to follow the student's steps, providing real-time feedback, based on the steps that the students are taking (in real time), known as model tracing.

Keywords: Natural Deduction in Propositional Logic · Genetic algorithms · Cognitive model · Model tracing

1 Introduction

In recent years, with the advancement of the processing capacity of computers, Artificial Intelligence (AI) has been used in several fields. The advances in AI associated with the emergence of new technologies ended up impacting another field of research, the field of education.

Among these technologies we have: Intelligent Tutoring Systems (ITS), Serious Games, Affective Intelligent Tutoring Systems, Learning Management Systems, Educational Intelligent Robotics and Massive Online Open Courses. Each of these applications, however, makes use of AI technologies in different ways.

R. Cerri and R. C. Prati (Eds.): BRACIS 2020, LNAI 12319, pp. 110–121, 2020.
https://doi.org/10.1007/978-3-030-61377-8_8

In the field of Education, AI has been used, primarily, in ITS with the objective of expanding access to knowledge, as well as favoring the personalization of the teaching-learning process. The emergence of these tools is the result of the researchers' interest in understanding and increasingly simulating the teaching and learning process in order to improve the quality of teaching, so that it is possible to achieve better levels of proficiency.

ITS are computer programs designed to incorporate AI techniques commonly used in education. A characteristic of AI and education is the use of intelligence to reason about teaching and learning, representing what, when and how to teach certain contents. This is aligned with Nwana's [1] statement, that the production of a well-designed teaching-learning system has three aspects: 1) They know what they teach; 2) They know how to teach; and 3) Observe how students are learning.

Still in the educational context, the present work addresses the theme of logic. Logic is a component found in curricular matrices of computing and informatics courses [2], being typically taught in the first or second semester of these courses. The course has as basic contents: propositional logic - propositions, formulas and truth tables (propositions and logical operators, material implication and logical equivalence, formulas and precedence, construction of truth tables for propositional formulas, among others) and Natural Deduction in Propositional Logic (NDPL) - arguments, inference rules and evidence (valid arguments, truth table for arguments, formal demonstrations, natural deduction rules, among others).

Among the various support tools for teaching logic there are: proof assistants [3, 4], theorem prover [5–10] and ITS [11–15].

In this sense, the present work aims to present the cognitive model of ITS EvoLogic, developed to deal with problems of Natural Deduction in Propositional Logic (NDPL). This cognitive model composes the EvoLogic model tracing, used to follow the individual steps of each student to provide important feedback during the performance of an exercise. This article presents a study in which several solutions obtained by Evologic to Propositional Logic problems were contrasted with exercises carried out by 57 students in a previous work [16], highlighting the ability to track students by different lines of reasoning previously identified.

2 Theoretical Framework

Among the tools found in the literature that support the teaching of Logic, some ITS are worth highlighting: Logic-ITA [11, 12], P-Logic Tutor [13] and AProS [14], in addition to the Heráclito environment [15, 16].

Logic-ITA [11, 12] is a web-based teaching/learning assistant for propositional logic. Its field of application is the construction of formal proofs in logic. The system acts as an intermediary between the teacher and students: first, it provides students with an environment to practice formal proofs with feedback and, on the other hand, allows teachers to monitor progress and errors of the class. The system is adapted for two different types of use: for students, it is an autonomous ITS, while for teachers, it includes the functionality to configure learning levels, adjust the parameters to progress through these levels, monitor class progress and collect data. The purpose of P-Logic Tutor [13] is to teach students fundamental concepts of propositional logic and theorem-proofing techniques.

The P-Logic Tutor plays a dual role as an educational tool and a research environment. It introduces students to the fundamental concepts of propositional logic and also provides practice in solving theorems. The software also provides an environment in which it is possible to track learning, explore the cognitive issues of problem-solving, and to investigate the possibilities of learning. The AProS project [14] started in 2006 seeking to cover the contents of logic. However, the project has expanded and currently has several tools that interconnect making it possible to search for exercises, solve theorems and accompany students in the process of proving the theorem. Among them is the Proof Tutor that bridges the gap between the tools Proof Lab and Truth Lab test. It allows students who are struggling and receive suggestions dynamically obtained from other proofs that have already been generated.

Heráclito is an ITS for teaching logic, allowing students to solve exercises from truth table even on proof of arguments by the rules of Natural Deduction. To do so, it offers one Electronic Logic Workbook - LOGOS that lets the student to create and edit formulas, truth tables and proofs of propositional logic [15].

In addition to ITS, there are proof assistants, such as: Coq [3] and HOL [4], which provide specification languages based on advanced logics (high order logics), capable of offering sophisticated support for the construction of formal proofs in these Logics. Prover9/Mace4 [5], EProver [6] and Classic SPASS [7] are automatic theorem provers of first order logic; in addition to the editors/proofreaders JAPE [8], Pandora [9] and Isabelle [10].

Among ITS presented (Logic-ITA, P-Logic Tutor, AProS and Heráclito), all make use of a formal demonstration that provides an appropriate symbolic structure to monitor the process of teaching and learning in the Logical deduction. However, only AProS and Heráclito use demonstrations similar to those used by teachers and students in the classroom. Still with regard to the demonstration, only Heráclito presents the possibility of continuing the proof of the theorem where the student left off, as well as providing the next step (based on the current proof). In this context, EvoLogic is similar to Heráclito, for having all these characteristics. However, the difference is in the process in how the solution is generated. While Heráclitus continues the proof from the point which has been submitted (new exercise or partly solved by the student), generating one possible solution among the various existing, unlike EvoLogic. By owning a Genetic Algorithm (GA) as a specialist, the EvoLogic gets numerous solutions for the same exercise at the time the student starts it. These different solutions can lead to different lines of reasoning, allowing the model tracing to provide feedbacks (suggesting new steps to the student) according to the behavior.

3 Materials and Methods

Among the materials, we mention:

- NDPL exercises: consist of a sequence of steps (application of deduction rules) that seek to prove a theorem. These exercises are typically taught in the early semesters of computer courses, with the aim of developing students' logical thinking. The NDPL exercises covered here are based on the book written by Gluz and Py, similar to Pospesel [16].

- Genetic Algorithm: EvoLogic's specialist agent consists of an GA adapted to solve NDPL problems. This agent follows the classic operation of GA with adaptations in the representation and interpretation of the solution, crossing and mutation operators that must avoid the generation of unviable solutions and in the fitness function.

As for the method, this research is classified as exploratory, as it seeks to identify how students behave when solving NDPL exercises, analyzing the cognitive model of EvoLogic.

4   EvoLogic

ITS EvoLogic consists of a multi-agent system that seeks to follow the student in the process of solving a problem. More specifically, the focus of the ITS is on the resolution of NDPL problems, where it is necessary to conduct out multiple steps to obtain a viable solution.

EvoLogic is presented as an architecture composed of 3 agents: Interface, Pedagogical and Specialist. In this work, the focus is on the cognitive model, composed of the Pedagogical and Specialist agents. It should be noted that the traditional pedagogical

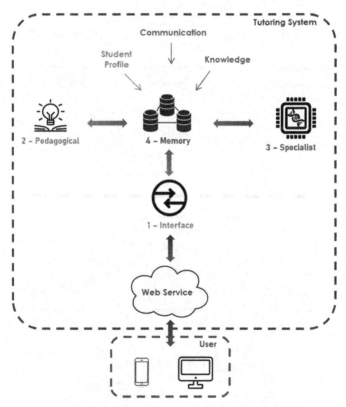

Fig. 1. ITS EvoLogic multiagent architecture.

model in an ITS comprises: teaching-learning strategies, pedagogical tactics and the student model. Within the scope of this work, the focus of the Pedagogical agent is on the student model. Communication between agents takes place through the exchange of simple messages, supported by memory structures, which also store the student model and the knowledge about the exercises solved by the Specialist agent. The organization from the point of view of a multi-agent system can be seen in Fig. 1, which highlights the elements that compose the system and the web service, used to receive messages from the student's interaction with his work environment.

The Specialist Agent consists of a GA adapted to solve NDPL problems, inspired by the classic models of literature [17]. In this way, here the differential elements are highlighted, which make the Specialist Agent capable of dealing with the resolution of the problems in question: interpretation of the solution, representation of the individual, crossover and mutation operators and fitness function.

The interpretation of the solution consists of identifying whether the solution to the problem is correct, in other words, whether the theorem can be proved. It also evaluates the steps that were taken to obtain the solution. Figure 2 shows two correct solutions for a simple exercise, identifying the steps that have been taken.

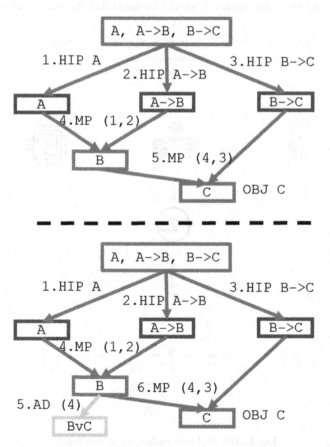

Fig. 2. Example of two solutions for an NDPL exercise.

At the top of Fig. 2, it is possible to observe a solution composed of 5 steps, all leading directly to objective C. By contrast, at the bottom, it also has a correct solution, reaching goal C, but it can be seen that was held an extra step (5.AD (4)). This step does not make the solution unviable, since the Addition rule can be applied during the process. However, these steps can indicate that the student is having some difficulty, and intervention by the tutor may be convenient. It is worth mentioning that in some exercises, the student may choose to use only basic rules, avoiding the use of derived rules. In such cases, it is important that the tutor clearly identifies the difference between the lines of reasoning and the extra steps.

The representation of the individual, in the GA, is done through a two-dimensional vector, where the applied rule and its dependents are stored. An example can be seen in Table 1 where the index was added so that it was possible to view the dependencies of steps 4 and 5.

Table 1. Representation of the individual.

Index	0	1	2	3	4	5
Steps	OBJ C	HIP A	HIP A->B	HIP B->C	MP B	MP C
Dependency	–	–	–	–	1,2	4,3

The crossover operator was developed to prevent unviable solutions from being generated (*i.e.*, solutions that contain errors). To generate a new individual, one parent's objective and hypotheses are copied (this information will always be the same for all individuals) and one of the other steps of that same father is randomly selected, copying it together with his dependencies to the new individual. Then, a random step from the second parent is selected and copied along with its dependents to the new individual. The process is repeated until the new individual is complete. Finally, the mutation operator replaces a step with a new one or includes a new step in the current solution. Fitness function is given by Eq. 1.

$$f = (1 - x) \times t_i + (x \times (E_i \times c)) \tag{1}$$

Where:

- x is an integer variable assuming a value of 1 if individual i solves the problem and 0 otherwise;
- t_i is the size of individual i (i.e., the number of genes);
- E_i is the efficiency of individual i, calculated only when $x = 1$ through Eq. 2;
- c is a constant that determines the weight given to a solution that solves the problem. In this work, $c = 25$, will be used, 25 being the maximum size of an individual.

Using $(1 - x)$ and x, it is possible to zero a part of the equation, prioritizing individuals who have more genetic characteristics (larger individuals) or more efficient individuals. This allows individuals who have a greater genetic load to pass on part of their genes, in addition to allowing individuals who present valid solutions to have a prominent place in the population by calculating the efficiency of the solution (Eq. 2).

$$E_i = t'_i/t_i \tag{2}$$

Where:

- t_i is the size of the individual i;
- t'_i is the size of the individual i considering only the characteristics that compose the solution to the problem.

Equation 2 shows the relationship between the size of the solution and the part that composes the solution to the problem. In several situations, an individual can represent the solution to a problem that requires several steps. Some of these steps can be simplified, without having to appear in the individual. In this context, only the steps that are directly linked to the solution are considered in the variable t'_i.

Still in Eq. 1, the constant c acts as a bias to prioritize correct solutions to the detriment of solutions that do not solve the problem. It is worth mentioning that all individuals are acceptable during the evolutionary process since the problem addressed in this work is not trivial and demands that a comprehensive exploration of the solution space is performed.

Finally, the Pedagogical agent, regarding the cognitive model, is responsible for the student model, following the steps taken by the student and categorizing it by its quality and by the line of reasoning followed during the resolution of the exercise.

5 Results

In 2019 [15] a new student model was proposed for the Heráclito Environment, in the context of NDLP, where the results about the resolution of 10 exercises were presented, so that all the steps performed by the students were stored. Therefore, it is possible to reproduce the steps of each student, individually, in EvoLogic, seeking to identify the lines of reasoning of each student. In order to highlight the characteristics of the cognitive model, we chose to use data related to exercise: $A<->Q$, $F<->R$, $A \wedge R$ $| - F \wedge Q$ where the hypotheses are $A<->Q$, $F<->R$ and $A \wedge R$ and the goal is $F \wedge Q$. Among the 57 students who participated in the experiment, 44 correctly solved the exercise while the rest (13 students) started the resolution, but did not complete it successfully (either due to difficulty or lack of time). When starting the exercise, both Heráclito and EvoLogic, solve the exercise in order to identify if it has a viable solution. Given the operating characteristic of the Heráclito, only one solution is obtained at the beginning (where new solutions are generated based on the student steps). The solution obtained by Heráclito is shown in Table 2, where OBJ defines what wanted to prove, HIP represents the Hypotheses of the problem and -EQ, SP, MP and CJ represent the Elimination of Equivalence, Simplification, Modus Ponens and Conjunction, respectively. The first column identifies the index while the second column shows the steps taken by the student, where there is the rule applied, the result of applying the rule and the lines that were used to obtain the given result.

Table 2. Solution obtained by the heráclito environment

Heráclito	
0	OBJ F∧Q
1	HIP A<->Q
2	HIP F<->R
3	HIP A∧R
4	-EQ R->F (2)
5	-EQ A->Q (1)
6	SP A (3)
7	SP R (3)
8	MP Q (6, 5)
9	MP F (7, 4)
10	**CJ F∧Q (9, 8)**

On the other hand, EvoLogic, by using an evolutionary mechanism as a specialist, may obtain several solutions to the same problem over the generations of the population. When applying it in the exercise in question, 8 different solutions were obtained (by removing the extra steps), the students followed by 5 of them, shown in the Tables 3, 4 and 5, which also shows the population in which they were obtained. The parameters used to obtain these results were: Population Size 50, Mutation Rate 5% and Total Population Generation 500 (stopping criterion).

Table 3. Solutions 1 and 2 obtained by EvoLogic.

EvoLogic – 1 (135)		EvoLogic – 2 (112)	
0	OBJ F∧Q	0	OBJ F∧Q
1	HIP A<->Q	1	HIP A<->Q
2	HIP F<->R	2	HIP F<->R
3	HIP A∧R	3	HIP A∧R
4	-EQ R->F (2)	4	-EQ A->Q (1)
5	-EQ A->Q (1)	5	-EQ R->F (2)
6	SP A (3)	6	SP A (3)
7	SP R (3)	7	SP R (3)
8	MP Q (6, 5)	8	MP F (5, 7)
9	MP F (7, 4)	9	MP Q (4, 6)
10	**CJ F∧Q (9, 8)**	10	**CJ F∧Q (8, 9)**

Table 4. Solutions 3 and 4 obtained by EvoLogic.

EvoLogic – 3 (193)		EvoLogic – 4 (189)	
0	OBJ F∧Q	0	OBJ F∧Q
1	HIP A<->Q	1	HIP A<->Q
2	HIP F<->R	2	HIP F<->R
3	HIP A∧R	3	HIP A∧R
4	-EQ A->Q (1)	4	SP R (3)
5	SP A (3)	5	SP A (3)
6	SP R (3)	6	-EQ A->Q (1)
7	MP Q (4, 5)	7	MP Q (6, 5)
8	-EQ R->F (2)	8	-EQ R->F (2)
9	MP F (8, 6)	9	MP F (8, 4)
10	**CJ F∧Q (8, 7)**	**10**	**CJ F∧Q (9, 7)**

Table 5. Solution 5 obtained by EvoLogic.

EvoLogic – 5 (203)	
0	OBJ F∧Q
1	HIP A<->Q
2	HIP F<->R
3	HIP A∧R
4	SP R (3)
5	-EQ R->F (2)
6	MP F (4, 5)
7	SP A (3)
8	-EQ A->Q (1)
9	MP Q (7, 8)
10	**CJ F∧Q (6, 9)**

From the point of view of the GA, the last solution obtained was in the 203 generation (considering the 8 obtained), although 500 populations were generated. This phenomenon may occur due to a lack of genetic variability, preventing GA to get new solutions. In this context, whenever the student presents a line of reasoning that is not contemplated, a new instance of the GA is performed, initializing the population with characteristics similar to those presented by the student. It is noteworthy that in this exercise it was not necessary to perform this procedure, since all the solutions presented by the students were obtained in the first execution of the GA.

Based on the steps taken by the 57 students, a simulation was carried out, using the cognitive model of EvoLogic, where the individual steps of each student were performed (in a simulated way).

Considering solutions 1 and 2, the students first chose to manipulate the Hypotheses and then combine the results and reach the solution. It can be seen that the first steps (4 to 7) applied rules directly to the hypotheses and, only in the following steps (8 and 9) seek to derive the propositions to reach the solution (step 10).

In the case of solution 5, it is possible to verify a different behavior, where the student applies the rules in two Hypotheses (steps 4 and 5) and already manipulates them (step 6). Then, it returns to the hypotheses (steps 7 and 8) and again manipulates them (step 9), reaching the solution.

Comparing the two strategies, there are two different lines of reasoning that need separate monitoring. In the first case (solutions 1 and 2), of the 28 students who presented this solution, 5 of them took extra steps, which may indicate that they were exploring the possibilities to understand how to proceed with the resolution. In the second case (solution 5), 4 students followed the steps in the same order, indicating that they could have identified the solution line before starting to interact with the environment (none of these students took extra steps).

Besides, the other 12 students followed the steps of solutions 3 and 4, with 11 of them taking one or more extra steps. Analyzing these solutions, it can be seen that there is an alternation between the application of rules in the hypotheses and their derivatives. This alternation differs from that presented in solution 5 by the way they occur as well as by the number of extra steps. These steps indicate that students have identified that some rules can be applied, however not one rule leads directly to the resolution of the exercise.

It is worth mentioning that all the steps taken by the students were followed and categorized, both in terms of quality (whether or not it is part of a direct solution) and in terms of their line of reasoning.

5.1 Discussion

The difference between Heráclito and EvoLogic is in the way of dealing with exercise solutions. While Heráclito performs the proof from the point where the student is, EvoLogic already has the solution and only follows the student's reasoning through the solution.

In some cases, in the work of [15], participants stressed out that they ended up getting confused. During the resolution of the exercises, some students were following a reasoning line that would require 3 steps to complete the test, while Heráclito pointed out that they needed only 2 steps (message sent as a form of incentive to the student). Since Heráclito continued the student's partial proof, a path with fewer steps was identified. This may have occurred because Heraclitus completed the proof of the theorem using some derived rule, while the student preferred to use only basic rules (obtaining the same result with more steps). It is important to note that both solutions are correct and that the difference is in the line of reasoning followed by the student. Unlike Heraclitus, EvoLogic could have verified that there was still more than one possible path, starting from where the student was, preventing such a message from being sent inaccurately.

This subtle difference opens the possibility to explore more precise pedagogical mediation strategies, as well as the creation of an automated model tracing mechanism, which estimates which line of reasoning the student is following, providing feedback directed to his profile.

Based on the students' steps, it was found that some of them went objectively to the solution while others ended up by taking some extra steps. These extra steps may indicate that, for the student, the resolution process is unclear. The identification mechanism of the student's line of reasoning may use several characteristics such as possible continuations from the partial solution, history of the student's behavior (how and which steps he has been taking), estimates of knowledge of rules that involve each path (student model) and also the probability of knowing a certain solution (also based on the student model).

6   Conclusions

This paper presented the Intelligent Tutoring System EvoLogic, in particular, its cognitive model, which consists of 2 agents: a Pedagogical and a Specialist agent. The Pedagogical agent represents the student model, while the Specialist is responsible for solving the NDPL exercises started by the students. The cognitive model aims to follow each step of the student, identifying when he or she is showing some difficulty and providing relevant feedback.

To evaluate the capabilities of the Specialist agent, a known exercise, applied to 57 students and recorded step by step, was given to the EvoLogic, in a simulated process. Each step taken by the students was simulated at the ITS, where the solutions were analyzed and compared.

EvoLogic obtained 8 direct solutions to the problem (excluding the extra steps) while 44 students that successfully proved the theorem, presented 5 of them. It was possible to observed, that the students took different approaches in these solutions, choosing to explore all the hypotheses before considering their derivations, or deriving the results to the maximum and returning the hypotheses when necessary. In this process, it was found that in some lines of reasoning (solutions 3 and 4), students presented several extra steps. This may indicate that students were having difficulties, providing support so that pedagogical mediation strategies aimed at each behavior can be developed.

In this way, the objectives of this study are met, opening a range of possible strategies for future applications. Therefore, as future work, a more detailed analysis of the Specialist agent is sought, especially with regard to problems with more complex deduction rules, as well as the formalization of an automated model tracing, supported by the solutions generated by the agent Specialist.

References

1. Nwana, H.S.: Intelligent tutoring systems: an overview. Artif. Intell. Rev. **4**, 251–277 (1990). https://doi.org/10.1007/BF00168958
2. Resolução CNE/CES n° 5, de 16 de novembro de 2016. http://portal.mec.gov.br/component/content/article?id=12991. Accessed 01 May 2020
3. The Coq Proof Assistant. Disponível em. https://coq.inria.fr/. Accessed 01 May 2020

4. HOL - Interactive Theorem Prover. https://hol-theorem-prover.org/. Accessed 05 May 2020
5. Prover9 and Mace4. https://www.cs.unm.edu/~mccune/prover9/. Accessed 01 May 2020
6. The E Theorem Prover. https://wwwlehre.dhbw-stuttgart.de/~sschulz/E/E.html. Accessed 01 May 2020
7. Classic SPASS: An Automated Theorem Prover for First-Order Logic with Equality. https://www.mpi-inf.mpg.de/departments/automation-of-logic/software/spass-workbench/classic-spass-theorem-prover/. Accessed 01 May 2020
8. JAPE. http://japeforall.org.uk/. Accessed 01 May 2020
9. PANDORA - Proof Assistant for Natural Deduction using Organised Rectangular Areas. http://www.doc.ic.ac.uk/pandora/newpandora/quick_start.html. Accessed 01 May 2020
10. Isabelle proof assistant. https://isabelle.in.tum.de/. Accessed 01 May 2020
11. Lesta, L., Yacef, K.: An intelligent teaching assistant system for logic. In: Cerri, S.A., Gouardères, G., Paraguaçu, F. (eds.) ITS 2002. LNCS, vol. 2363, pp. 421–431. Springer, Heidelberg (2002). https://doi.org/10.1007/3-540-47987-2_45
12. Yacef, K.: The logic-ITA in the classroom: a medium scale experiment. Int. J. Artif. Intell. Educ. (IJAIED) **15**, 41–62 (2005)
13. Lukins, S., Levicki, A., Burg, J.: A tutorial program for propositional logic with human/computer interactive learning. In: Proceedings of the 33rd SIGCSE Technical Symposium on Computer Science Education, pp. 381–385 (2002)
14. Sieg, W.: The APros project: strategic thinking & computational logic. Log. J. IGPL **15**(4), 359–368 (2007)
15. Galafassi, F.F.P., Galafassi, C., Vicari, R.M., Gluz, J.C.: Identifying knowledge from the application of natural deduction rules in propositional logic. In: Demazeau, Y., Matson, E., Corchado, J.M., De la Prieta, F. (eds.) PAAMS 2019. LNCS (LNAI), vol. 11523, pp. 66–77. Springer, Cham (2019). https://doi.org/10.1007/978-3-030-24209-1_6
16. Pospesel, H.: Introduction to Logic: Propositional Logic, Revised Edition, 3rd edn. Pearson, London (1999)
17. Holland, J.H.: Adaptation in Natural and Artificial Systems. University of Michigan Press, Ann Arbor (1975)

Impacts of Multiple Solutions on the Lackadaisical Quantum Walk Search Algorithm

Jonathan H. A. de Carvalho[1]([✉]) [ID], Luciano S. de Souza[2] [ID],
Fernando M. de Paula Neto[1] [ID], and Tiago A. E. Ferreira[2,3] [ID]

[1] Centro de Informática, Universidade Federal de Pernambuco, Recife, Brazil
{jhac,fernando}@cin.ufpe.br
[2] Departamento de Estatística e Informática, Universidade Federal Rural de
Pernambuco, Recife, Brazil
{luciano.serafim,tiago.espinola}@ufrpe.br
[3] Harvard School of Engineering and Applied Sciences, Harvard University,
Cambridge, USA
taef@seas.harvard.edu

Abstract. The lackadaisical quantum walk is a graph search algorithm
for 2D grids whose vertices have a self-loop of weight l. Since the technique depends considerably on this l, research efforts have been estimating the optimal value for different scenarios, including 2D grids with
multiple solutions. However, specifically two previous works have used
different stopping conditions for the simulations. Here, firstly, we show
that these stopping conditions are not interchangeable. After doing such
a pending investigation to define the stopping condition properly, we
analyze the impacts of multiple solutions on the final results achieved by
the technique, which is the main contribution of this work. In doing so,
we demonstrate that the success probability is inversely proportional to
the density of vertices marked as solutions and directly proportional to
the relative distance between solutions. These relations presented here
are guaranteed only for high values of the input parameters because,
from different points of view, we show that a disturbed transition range
exists in the small values.

Keywords: Quantum computing · Quantum walk · Search algorithm

1 Introduction

As the classical random walks, the quantum walks are divided into discrete-time
and continuous-time models, which were introduced in [1] and [8], respectively.
Since these two quantum models are not equivalent when analyzed in detail [18],
a wide range of researches has been trying to find the more efficient one, mostly
in spatial search problems.

The algorithm proposed by Grover [10] can successfully search for a single
element within a disordered database of N items in $O(\sqrt{N})$ steps, which is a

© Springer Nature Switzerland AG 2020
R. Cerri and R. C. Prati (Eds.): BRACIS 2020, LNAI 12319, pp. 122–135, 2020.
https://doi.org/10.1007/978-3-030-61377-8_9

quadratic speedup over the classical counterparts. However, Benioff [4] showed that a quantum robot using Grover's algorithm is no more efficient than a classical robot because both require $O(N \log \sqrt{N})$ steps to search 2-dimensional spatial regions of size $\sqrt{N} \times \sqrt{N}$.

Childs and Goldstone [7] addressed this 2D spatial search problem using a continuous-time quantum walk but failed to provide substantial speedup. Ambainis et al. [3] proposed an algorithm capable of finding the solution after $O(\sqrt{N} \log N)$ steps using a discrete-time model, outperforming the previous works. However, Childs and Goldstone [6] achieved this same runtime later using a continuous-time model.

Over time, quantum walks for other graph structures have been developed, such as for hypercubes and complete graphs [14,22,24], but attempts to improve the search on 2D grids continued as well. In particular, the lackadaisical quantum walk (LQW) developed in [26] has been drawing attention because it improved the 2D spatial search by making a simple modification to the algorithm proposed in [3]. The modification was to attach a self-loop of weight l at each vertex of the space. When this l is optimally adjusted, the approach can find the solution in $O(\sqrt{N \log N})$ steps, which is an $O(\sqrt{\log N})$ improvement over that loopless version presented in [3].

That improvement was achieved by fitting the self-loop weight to $l = 4/N$, where N is the total number of vertices. This optimal value, although, is only one instance of a general observation about the LQW searching vertex-transitive graphs with $m = 1$ solutions. For these cases, the optimal self-loop weight l equals the degree of the graph without loops, which is 4 for 2D lattices, divided by N [20]. An analytical proof of this conjecture is given in [11] using the fact that the LQW can be approximated by the quantum interpolated walk.

However, that conjecture does not hold when the number of solutions m in the search space is higher than 1. Thus, another adjustment of l is required. Saha et al. [21] showed that $l = \frac{4}{N(m+\lfloor \sqrt{m}/2 \rfloor)}$ is the optimal value when m solutions are arranged as a block of $\sqrt{m} \times \sqrt{m}$ within the search space. In contrast, Nahimovs [15] demonstrated that this new fit of l is not optimal for arbitrary placements. Rather, two other adjustments were proposed, both in the form $l = \frac{4(m-O(m))}{N}$. After, Giri and Korepin [9] showed that one of these m solutions can be obtained with sufficiently high probability in $O(\sqrt{\frac{N}{m} \log \frac{N}{m}})$ steps.

In this paper, we are not focused on adjusting the weight of l for different scenarios than the ones addressed by previous works. We observed that Wong [26] and Nahimovs [15] found their results using different stopping conditions for the simulations. This naturally raised a question about the interchangeability of the conditions for use in subsequent researches. However, our results showed that the stopping condition used in [15] is satisfied prematurely since the probability of measure a solution continues improving in the next iterations. Only the condition used in [26] pursues the highest amplitude amplification of the solutions during the system evolution and, thus, should be chosen in the works from now on.

Choosing the appropriate stopping condition was an issue to be solved before we numerically investigate the impacts of multiple solutions, considering two high-level properties, on the success probability achieved at the end. Here, we show that the success probability is inversely proportional to the density of solutions and directly proportional to the relative distance between solutions. However, these relations are only valid for high values of the input parameters, which are the total number of vertices N and the number of solutions m. From different points of view, we show that disturbed behaviors exist in a transition between small to high values of both N and m. All of these impacts of multiple solutions can determine the appropriateness of the technique for future practical applications. Thus, the motivation here is in the direction of developing the LQW and better understanding its limitations from numerical experiments.

This paper is organized as follows: Sect. 2 presents some theoretical background about the task of search on 2D grids by the LQW. Here, the reader is expected to be familiar with the basics of quantum computing. If it is not the case, knowledge from the basic to the advanced levels can be obtained in [12,13,17,27], to mention just a few. In Sect. 3, the different stopping conditions used in previous works are compared. After that, Sect. 4 relates the impacts on the success probability to both the density of solutions and the relative distance between solutions, although a kind of transition range exists. Finally, Sect. 5 presents concluding remarks.

2   Search with the Lackadaisical Quantum Walk

The classical random walk is a probabilistic movement in which a particle jumps to its adjacent positions based on the outcome of a non-biased random variable at each step [18]. Generally, this random variable is a fair coin that has one degree of freedom for each possible direction of movement in the space at hand.

The quantum walk, in turn, is a generalized concept in comparison with the classical random walk. That high-level idea of conditioned movements remains, but quantum operations are responsible for evolving the system. In this context, quantum properties allow the quantum walk to spread quadratically faster than the classical one [18]. Consequently, this advantage can be used strategically to develop faster search algorithms.

2.1   Spatial Search with a Quantum Walk

Ambainis et al. [3] proposed a quantum walk algorithm to search a single vertex, also called the marked vertex, in the 2-dimensional grid of $L \times L = N$ vertices. In that work, the process evolved on the Hilbert space $\mathcal{H} = \mathcal{H}_C \otimes \mathcal{H}_P$, where $\mathcal{H}_C$ is the 4-dimensional coin space, spanned by $\{|\uparrow\rangle, |\downarrow\rangle, |\leftarrow\rangle, |\rightarrow\rangle\}$, and $\mathcal{H}_P$ represents the N-dimensional space of positions, spanned by $\{|x,y\rangle : x,y \in [0,\ldots,L-1]\}$.

Firstly, the coin toss is accomplished by the operator C presented in Eq. 1, which combines the coin operators C_0 and C_1 in such a way that C_1 is applied only to the marked state $|v\rangle$, whereas C_0 is applied to the others. Also, C_0 was defined as the Grover diffusion coin $C_0 = 2|s\rangle\langle s| - I_4$, where $|s\rangle = \frac{1}{2}(|\uparrow\rangle + |\downarrow\rangle +$

$|\leftarrow\rangle + |\rightarrow\rangle)$, and C_1 was defined as $-I_4$. In turn, I_4 denotes the 4-dimensional identity operator.

$$C = C_0 \otimes (I_4 - |v\rangle\langle v|) + C_1 \otimes |v\rangle\langle v| \qquad (1)$$

Then, the flip-flop shift operator S_{ff} is applied to move the quantum particle while inverting the coin state, as presented in Eq. 2. This shift works mod $\sqrt{N} = L$ because the grid has periodic boundary conditions. Finally, the quantum walk is a repeated application of the operator $U = S_{ff} \cdot C$ to the quantum system $|\psi\rangle$, which begins in the state $|\psi(0)\rangle = \frac{1}{\sqrt{N}} \sum_{x,y=0}^{\sqrt{N}-1} |s\rangle \otimes |x,y\rangle$.

$$\begin{aligned}
S_{ff}|\rightarrow\rangle|x,y\rangle &= |\leftarrow\rangle|x+1,y\rangle \\
S_{ff}|\leftarrow\rangle|x,y\rangle &= |\rightarrow\rangle|x-1,y\rangle \\
S_{ff}|\uparrow\rangle|x,y\rangle &= |\downarrow\rangle|x,y+1\rangle \\
S_{ff}|\downarrow\rangle|x,y\rangle &= |\uparrow\rangle|x,y-1\rangle
\end{aligned} \qquad (2)$$

As a result, the marked vertex can be obtained at the measurement with a probability $O(1/\log N)$ after $T = O(\sqrt{N \log N})$ steps. To achieve a success probability near to 1, it was applied amplitude amplification [5], which implied additional $O(\sqrt{\log N})$ steps. Hence, the total running time of this quantum walk based search algorithm is $O(\sqrt{N} \log N)$.

2.2   Improved Running Time by the Lackadaisical Quantum Walk

The LQW search algorithm [26] is an approach strictly based on that algorithm designed in [3]. The main modification is to attach a self-loop of weight l at each vertex of the 2D grid, which implies other changes in the loopless technique. First, $\mathcal{H}_C$ is spanned now by $\{|\uparrow\rangle, |\downarrow\rangle, |\leftarrow\rangle, |\rightarrow\rangle, |\circlearrowleft\rangle\}$ because of the new degree of freedom. However, no changes are required for $\mathcal{H}_P$.

Regarding the coin operator, C_0 was defined as the Grover diffusion coin for weighted graphs [25], so $C_0 = 2|s_c\rangle\langle s_c| - I_5$, where $|s_c\rangle$ is the non-uniform distribution presented in Eq. 3. Also, better results were found when $C_1 = -C_0$, outperforming that choice of $C_1 = -I$ used in [3]. About the shift operator S_{ff}, it works like an identity operator when applied to $|\circlearrowleft\rangle|x,y\rangle$. Finally, the quantum system $|\psi\rangle$ begins in a uniform distribution between all vertices with their edges in the weighted superposition $|s_c\rangle$ instead of the uniform $|s\rangle$.

$$|s_c\rangle = \frac{1}{\sqrt{4+l}}(|\uparrow\rangle + |\downarrow\rangle + |\leftarrow\rangle + |\rightarrow\rangle + \sqrt{l}|\circlearrowleft\rangle) \qquad (3)$$

As a result, the LQW with $l = 4/N$ finds the marked vertex with a success probability close to 1 after $T = O(\sqrt{N \log N})$ steps. This is an $O(\sqrt{\log N})$ improvement over the loopless algorithm. More sophisticated approaches have also achieved this improvement in running time [2,19,23], but the LQW is a significantly simpler and equally capable technique. Moreover, the success probability converges closer and closer to 1 if the number of vertices N increases when using that optimal l.

These numerical results were found by simulations that stopped when the first peak in the success probability occurred. For that, the stopping condition monitored the success probability at each step. When the current value was smaller than the immediately previous one for the first time, the simulation stopped, and this immediately previous result was reported as the maximum found.

2.3 Lackadaisical Quantum Walk with Multiple Solutions

If there are multiple marked vertices, the results for the case with only one do not hold. Significant research efforts have focused on adjusting the weight l optimally for multiple solution cases. Firstly, Saha et al. [21] addressed m marked vertices arranged as a block of $\sqrt{m} \times \sqrt{m}$ within a $\sqrt{N} \times \sqrt{N}$ grid, as already studied for the loopless version in [16].

In this scenario, the LQW can produce success probabilities that exceed 0.95 for large values of N with the optimal weight value of $l = \frac{4}{N(m+\lfloor\sqrt{m}/2\rfloor)}$. However, these results do not hold if the solutions are randomly sampled, as Nahimovs [15] demonstrated. A new choice of l is required.

Thereby, Nahimovs [15] searched for new optimal values in the form $l = \frac{4}{N} \cdot a$. Thus, l was adjusted as a factor of the optimal value for $m = 1$ reported in [26], which was $l = 4/N$. As a result, two adjustments were proposed: $l = \frac{4m}{N}$, for small values of m, and $l = \frac{4(m-\sqrt{m})}{N}$, for large values of m. To find these optimal values of l, the m solutions were arranged following the M_m set presented in Eq. 4. However, random placements of solutions yielded similar results.

$$M_m = \{(0, 10i) \mid i \in [0, m-1]\} \tag{4}$$

Regarding the simulation, a different stopping condition was used rather than monitoring the success probability at each step. Alternatively, the inner product $|\langle\psi(t)|\psi(0)\rangle|$ was monitored until its minimum is achieved, so the simulation stopped when this inner product became close to 0 for the first time since the process is periodic.

3 Comparison Between Different Stopping Conditions

Two different stopping conditions have been used in previous works without concerning the interchangeability issue. Therefore, the interchangeability between these conditions became an open question. Before conducting further experimental analysis, it is necessary to verify whether or not those conditions converge to the same points from equal initial settings. Here, we made such an investigation.

The experiment setup was equal to the one used in [15], i.e., a space of 200×200 vertices with the m solutions following the M_m set. Under this scheme, a maximum of 20 solutions could be defined in that space, since $M_{20} = \{(0,0),(0,10),\ldots,(0,180),(0,190)\}$. A different organization other than the M_m set is required for m values higher than 20 on the 200×200 grid.

As a result, the stopping conditions converged to the same points for $l = \frac{4m}{N}$. However, it is not the case for $l = \frac{4(m-\sqrt{m})}{N}$, as presented in Table 1, which contrasts the results obtained monitoring the marked vertices against the ones obtained monitoring the inner product $|\langle\psi(t)|\psi(0)\rangle|$. As can be seen, the results tend to converge to the same points as m increases. Nevertheless, the conditions were not equivalent because each one was satisfied at a different step T, which implied different success probabilities Pr as well. Also, in all cases investigated, monitoring the inner product $|\langle\psi(t)|\psi(0)\rangle|$ generated inferior probabilities of measure a solution at the end of the simulation.

Table 1. The convergence step T and the final success probability Pr, as the number of solutions m increases, for different stopping conditions used in previous works.

	Stopping conditions			
	Marked vertices		$\|\langle\psi(t)\|\psi(0)\rangle\|$	
m	T	Pr	T	Pr
1	399	0.140828	420	0.138489
5	409	0.878178	288	0.593276
10	297	0.867440	249	0.704010
15	290	0.835395	254	0.747045
20	288	0.818635	268	0.778724

To investigate the divergence more deeply, Fig. 1 shows the system evolution step by step for the $m = 5$ case, which had the most significant results. The black line represents the condition that monitors the marked vertices, whereas the blue dashed line represents the one that monitors the inner product $|\langle\psi(t)|\psi(0)\rangle|$. As can also be seen in Table 1, the condition that monitors the inner product in absolute value is satisfied prematurely at the step $T = 288$ since the success probability continues increasing until $T = 409$. After the step $T = 288$, although, the curves have a similar growth damping, which raised a question about monitoring the real value of the inner product, rather than its absolute value.

Figure 2 shows the system evolution during 1000 steps while monitoring both the marked vertices, represented again by the black line, and the inner product without calculating its absolute value, represented by the green dashed line. In this way, one curve tracks the other throughout the entire evolution.

From these experimental results, it is possible to conclude that the stopping conditions used in previous works are equivalent if, and only if, the inner product is considered without calculating its absolute value. Nevertheless, all results that will be discussed in this work from now on were found using the stopping

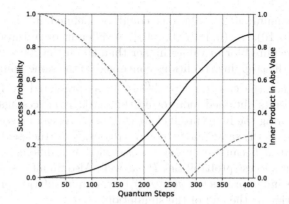

Fig. 1. System evolution step by step until the condition that monitors the marked vertices is satisfied. The black line represents the monitoring of that condition, whereas the blue dashed line represents the monitoring of the inner product in absolute value. (Color figure online)

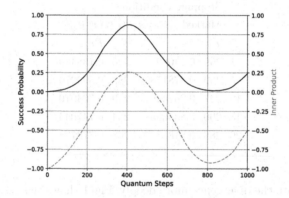

Fig. 2. System evolution step by step during 1000 steps. The black line represents the monitoring of the marked vertices, whereas the green dashed line represents the monitoring of the inner product without calculating its absolute value. (Color figure online)

condition that monitors the marked vertices. Since the goal is to measure the quantum system when the maximum amplification in the success probability occurs, monitoring the marked vertices is the most natural choice for simulations.

4   Solution Setups Affecting the Success Probability

In the last investigation, the evolution of the LQW was analyzed in order to verify if the different stopping conditions would be interchangeable. After choosing the stopping condition properly, the next step was to address factors that affect the success probability achieved at the end. Previous works have already demonstrated the considerable dependence on the weight l. Expanding the analysis, we addressed the density of solutions and the relative distance between solutions.

4.1 Previous Evaluations About Densities of Solutions and a Complementary Experiment

Previous works have already performed experiments that can reveal relations between the success probability and the density of solutions ρ, which is $\rho = \frac{m}{N}$. However, such works did not make links between the density of solutions and the success probability achieved at the end, at least not from this point of view. Here, we discuss these previous experiments briefly but aiming to identify the impacts of ρ in the final results. Finally, a complementary experiment was made.

Firstly, Wong [26] investigated the impacts of adding more unmarked vertices in a space with only one solution. Actually, that experiment evaluated how density decreasing could affect the success probability. As a result, the success probability tends to improve, even though some unknown behavior for the first values of N exists.

After, Nahimovs [15] inserted more and more solutions in the 200×200 grid when adjusting the value of l for multiple marked vertices. Since the space size was fixed, that experiment increased the density with each addition of a new solution. However, the probability of finding a marked vertex was smaller when m increased, which would be a counter-intuitive idea in a classical environment, but in the quantum world is different.

The following explanation is based on [18]. Consider $|\omega\rangle$ as the state where the total energy of the quantum system is equally distributed between the marked states, so the success probability is 1. The goal of Grover's algorithm is to rotate the state of the system $|\psi\rangle$ to get as close to $|\omega\rangle$ as possible. However, this is an iterative process in which $|\psi\rangle$ rotates at each step by an angle θ that is proportional to m. If $N \gg m$, increasing m implies fewer steps T, even though $|\psi\rangle$ gets less close to $|\omega\rangle$ at the end, resulting in smaller success probabilities as well. On the other hand, θ is inversely proportional to N, so increasing N implies more steps T, but $|\psi\rangle$ gets more close to $|\omega\rangle$ at the measurement.

Thus, these previous experiments suggest that the success probability is inversely proportional to the density of solutions of the search space. In this work, a complementary experiment was made to fill the gap not addressed by those previous works. Thereby, we investigated how density decreasing, like in [26], can affect the quantum walk with multiple marked vertices, like in [15]. While we conducted that experiment, we also searched for the optimal value of l in the form $l = \frac{4}{N}a$, like in [15] again.

Figure 3 shows the peaks in the success probability, represented by the black line, and the optimal a values that generated these peaks, represented by the brown dashed line, both as functions of N. The density of solutions decreased in this case because m was always equal to 10, so N increased by the addition of unmarked vertices. Moreover, the optimal a values were searched with steps of size 0.5 and N varied from 10^4 to 10^6 with the $m = 10$ solutions following the M_{10} set.

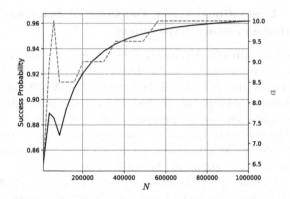

Fig. 3. Peaks in the success probability and the respective optimal a values, both as functions of N, which varied from 10^4 to 10^6, with the $m = 10$ solutions located in the 2D grid according to the M_m set.

Again, there is an unknown behavior for the first N values. After that, the success probability tends to 1 as well as the optimal a value tends to the number of solutions $m = 10$. Hence, the construction $l = \frac{4(m-O(m))}{N}$ proposed in [15] is a way of adjusting for the cases where density is not small enough, because a is equal to m in the best cases of density and, consequently, $l = \frac{4m}{N}$.

4.2 A New Set of Solutions Increasing Relative Distances

In the last experiment, the $m = 10$ solutions were located always at the points $\{(0,0),(0,10),\ldots,(0,90)\}$, following the M_{10} set. That distribution of solutions did not take advantage of the increment on the total number of vertices N. If the solutions were located farther away from each other, it would be possible to continue evaluating how the success probability depends on the density of solutions but also on the relative distance between solutions.

Thus, we propose an alternative to the M_m set that is the $P_{L,m}$ set presented in Eq. 5. Following this new set, the m solutions are located depending on the number of vertices in each dimension L so that the size of the search space is better used. For example, $m = 10$ solutions on the 200×200 grid would be located at the points $\{(0,0),(20,20),\ldots,(180,180)\}$, following the $P_{200,10}$ set. As a consequence, the solutions are father apart using the $P_{L,m}$ set than using the M_m set.

$$P_{L,m} = \left\{ \left(\left\lfloor \frac{L}{m} \right\rfloor i, \left\lfloor \frac{L}{m} \right\rfloor i \right) \ \middle| \ i \in [0, m-1] \right\} \tag{5}$$

Then, our complementary experiment that evaluated density decreasing with $m = 10$ solutions was redone, but using the $P_{L,m}$ set this time for localizing the solutions farther from each other. The results obtained with this new set of solutions are contrasted in Table 2 with the ones obtained previously, which used the M_m set. These results for the M_m set are exactly the success probabilities already presented in Fig. 3 but in terms of L now because L is the variable used to define the $P_{L,m}$ set, even though $L^2 = N$.

Table 2. The number of steps and the success probability as L increases for different sets of $m = 10$ solutions in the 2D grid.

L	M_m		$P_{L,m}$	
	T	Pr	T	Pr
100	147	0.849178	109	0.902339
200	293	0.889219	223	0.927680
300	511	0.871665	342	0.940301
400	747	0.908749	460	0.948348
500	965	0.930714	581	0.953927
600	1181	0.943863	700	0.958288
700	1407	0.951843	822	0.961646
800	1623	0.956787	941	0.964317
900	1857	0.959761	1063	0.966613
1000	2097	0.961896	1187	0.968522

For all values of L, the set of solutions $P_{L,m}$ generated better results because the success probability was higher and with a smaller number of steps. Besides this, that unknown behavior for the first values of L did not appear in the results with the new set of solutions. Finally, it is possible to conclude from these numerical results that the success probability is directly proportional to the relative distance between solutions.

Regardless of whether an unknown behavior exists or does not exist for the first L values, the success probability had an asymptotic and growing behavior for higher values of L in all previous cases. However, that is not true for all values of m. Table 3 shows the same investigation of density decreasing with the $P_{L,m}$ set again, but for $m = \{3, 4, 5\}$, and not for $m = 10$ as before.

The qualitative behaviors found for these m values are not equal to the behaviors for both $m = 1$, as reported in [26], and $m = 10$, as shown in Fig. 3 and Table 2. In those $m = 1$ and $m = 10$ cases, an unknown behavior existed during a transition from small to high values of L and, then, the success probability improved continuously. However, $m = \{3, 4, 5\}$ can be seen as a kind of transition from small to high values but from the perspective of the number of solutions m. It suggests that the asymptotic and growing behavior for the success probability is only guaranteed for values high enough of both L and m.

Table 3. Values of m that do not have asymptotic and growing behaviors for the success probability as the density of solutions decreases.

L	Pr		
	$m = 3$	$m = 4$	$m = 5$
100	0.991433	0.986119	0.981772
200	0.988165	0.992391	0.990397
300	0.985744	0.993451	0.992697
400	0.984221	0.993206	0.993754
500	0.983418	0.992604	0.994283
600	0.983100	0.991933	0.994585
700	0.983081	0.991282	0.994717
800	0.983252	0.990683	0.994677
900	0.983548	0.990138	0.994557
1000	0.983927	0.989644	0.994392

4.3   Evaluation of Density Increasing with the New Set of Solutions

In fact, the density of solutions in the search space affects the success probability achieved by the LQW. We already analyzed the density decreasing by the addition of unmarked states using both the M_m and $P_{L,m}$ sets. Regarding increases in density, we complement this analysis here using the $P_{L,m}$ set since results using the M_m set are found in [15]. Increases in density occur by adding more solutions in a space of fixed size $L \times L$.

Figure 4 shows the success probability achieved as a function of the number of solutions m for 2D grids with different numbers of vertices per dimension L. The colored lines represent the results for spaces with L varying from 100 to 1000 and with the number of solutions $m = \{1, 2, \ldots, 10\}$ following the $P_{L,m}$ set.

As expected, because of the inversely proportional relation, the success probability decreases as the density of solutions increases by the addition of more solutions to the search space. However, these results with the $P_{L,m}$ set also had that transitory phenomenon. For all cases analyzed, there are intervals where a disturbed behavior exists and, then, the success probability tends to decrease continuously.

This was one more different perspective showing some uncertainty about the behavior of the LQW with small values of some input parameter. Thus, it is more confident to apply the LQW search algorithm in real scenarios where the input parameters are higher to avoid those unknown behaviors presented in this work.

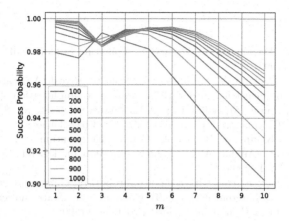

Fig. 4. The success probability as density increases by having more solutions in the spaces with the number of vertices per dimension L varying from 100 to 1000. Each colored line represents the results for a different space with the number of solutions $m = \{1, 2, \ldots, 10\}$ following the $P_{L,m}$ set.

5 Conclusions

In this work, first, we demonstrated that different stopping conditions used in previous works are not interchangeable. Calculating the absolute value of the inner product $\langle \psi(t)|\psi(0) \rangle$ implies prematurely stops. Rather, the real value must be used. After choosing the stopping condition properly, we demonstrated that the success probability is inversely proportional to the density of solutions and directly proportional to the relative distance between solutions. However, those relations are guaranteed only for high values of the input parameters. From different points of view, a transitory phenomenon existed between small to high values.

Here, we took a step towards establishing the limitations of the technique from numerical data. Future works should mathematically define upper and lower bounds considering the impacts of multiple solutions stated in this paper, which will enable better comparisons between the LQW and other quantum algorithms for 2D searches. However, the analysis needs to occur in controlled schemes because information about the solution setups might not be available in real applications. Also, that transitory phenomenon between small to high values should be studied in future works, as well as the limits of proximity between solutions.

Acknowledgments. Science and Technology Support Foundation of Pernambuco (FACEPE) Brazil, Brazilian National Council for Scientific and Technological Development (CNPq), and Coordenação de Aperfeiçoamento de Pessoal de Nível Superior - Brasil (CAPES) - Finance Code 001 by their financial support to the development of this research.

References

1. Aharonov, Y., Davidovich, L., Zagury, N.: Quantum random walks. Phys. Rev. A **48**(2), 1687 (1993)
2. Ambainis, A., Bačkurs, A., Nahimovs, N., Ozols, R., Rivosh, A.: Search by quantum walks on two-dimensional grid without amplitude amplification. In: Iwama, K., Kawano, Y., Murao, M. (eds.) TQC 2012. LNCS, vol. 7582, pp. 87–97. Springer, Heidelberg (2013). https://doi.org/10.1007/978-3-642-35656-8_7
3. Ambainis, A., Kempe, J., Rivosh, A.: Coins make quantum walks faster. In: Proceedings of 16th Annual ACM-SIAM Symposium on Discrete Algorithms, pp. 1099–1108. SIAM (2005)
4. Benioff, P.: Space searches with a quantum robot. In: Quantum Computation and Information (Washington, D.C., 2000), Contemporary Mathematics, vol. 305, pp. 1–12. American Mathematical Society, Providence (2002)
5. Brassard, G., Hoyer, P., Mosca, M., Tapp, A.: Quantum amplitude amplification and estimation. Contemp. Math. **305**, 53–74 (2002)
6. Childs, A.M., Goldstone, J.: Spatial search and the Dirac equation. Phys. Rev. A **70**(4), 042312 (2004)
7. Childs, A.M., Goldstone, J.: Spatial search by quantum walk. Phys. Rev. A **70**(2), 022314 (2004)
8. Farhi, E., Gutmann, S.: Quantum computation and decision trees. Phys. Rev. A **58**(2), 915 (1998)
9. Giri, P.R., Korepin, V.: Lackadaisical quantum walk for spatial search. Mod. Phys. Lett. A **35**(08), 2050043 (2019)
10. Grover, L.K.: Quantum mechanics helps in searching for a needle in a haystack. Phys. Rev. Lett. **79**(2), 325 (1997)
11. Høyer, P., Yu, Z.: Analysis of lackadaisical quantum walks. arXiv preprint arXiv:2002.11234 (2020)
12. McMahon, D.: Quantum Computing Explained. Wiley, Hoboken (2007)
13. Mermin, N.D.: Quantum Computer Science: An Introduction. Cambridge University Press, New York (2007)
14. Meyer, D.A., Wong, T.G.: Connectivity is a poor indicator of fast quantum search. Phys. Rev. Lett. **114**(11), 110503 (2015)
15. Nahimovs, N.: Lackadaisical quantum walks with multiple marked vertices. In: Catania, B., Královič, R., Nawrocki, J., Pighizzini, G. (eds.) SOFSEM 2019. LNCS, vol. 11376, pp. 368–378. Springer, Cham (2019). https://doi.org/10.1007/978-3-030-10801-4_29
16. Nahimovs, N., Rivosh, A.: Quantum walks on two-dimensional grids with multiple marked locations. Int. J. Found. Comput. Sci. **29**(04), 687–700 (2018)
17. Nielsen, M.A., Chuang, I.L.: Quantum Computation and Quantum Information, 10th anniversary edn. Cambridge University Press, New York (2010)
18. Portugal, R.: Quantum Walks and Search Algorithms. Springer, New York (2013). https://doi.org/10.1007/978-1-4614-6336-8
19. Portugal, R., Fernandes, T.D.: Quantum search on the two-dimensional lattice using the staggered model with Hamiltonians. Phys. Rev. A **95**(4), 042341 (2017)
20. Rhodes, M.L., Wong, T.G.: Search on vertex-transitive graphs by lackadaisical quantum walk. arXiv preprint arXiv:2002.11227 (2020)
21. Saha, A., Majumdar, R., Saha, D., Chakrabarti, A., Sur-Kolay, S.: Search of clustered marked states with lackadaisical quantum walks. arXiv preprint arXiv:1804.01446 (2018)

22. Shenvi, N., Kempe, J., Whaley, K.B.: Quantum random-walk search algorithm. Phys. Rev. A **67**(5), 052307 (2003)
23. Tulsi, A.: Faster quantum-walk algorithm for the two-dimensional spatial search. Phys. Rev. A **78**(1), 012310 (2008)
24. Wong, T.G.: Grover search with lackadaisical quantum walks. J. Phys. A Math. Theor. **48**(43), 435304 (2015)
25. Wong, T.G.: Coined quantum walks on weighted graphs. J. Phys. A Math. Theor. **50**(47), 475301 (2017)
26. Wong, T.G.: Faster search by lackadaisical quantum walk. Quantum Inf. Process. **17**(3), 1–9 (2018). https://doi.org/10.1007/s11128-018-1840-y
27. Yanofsky, N.S., Mannucci, M.A.: Quantum Computing for Computer Scientists. Cambridge University Press, New York (2008)

Multi-objective Quadratic Assignment Problem: An Approach Using a Hyper-Heuristic Based on the Choice Function

Bianca N. K. Senzaki⬥, Sandra M. Venske⬥, and Carolina P. Almeida(✉)⬥

Midwestern Parana State University - UNICENTRO, Guarapuava, Paraná, Brazil
biancanamie@gmail.com, {ssvenske,carol}@unicentro.br

Abstract. The *Quadratic Assignment Problem* (QAP) is an example of combinatorial optimization problem and it belongs to NP-hard class. QAP assigns interconnected facilities to locations while minimizing the cost of transportation of the flow of commodities between facilities. Hyper-Heuristics (HH) is a high-level approach that automatically selects or generates heuristics for solving complex problems. In this paper is proposed the use of a selection HH to solve the multi-objective QAP (mQAP). This HH is based on the MOEA/DD (Evolutionary Many-Objective Optimization Algorithm Based on Dominance and Decomposition) and Choice Function strategy. The heuristics selected by HH correspond to the operators that generate new solutions in an iteration of the multi-objective evolutionary algorithm. IGD metric and statistical tests are applied in order to evaluate the algorithm performances in 22 mQAP instances. The effectiveness of the proposed method is shown and it is favorably compared with three other evolutionary multi-objective algorithms: IBEA, SMS-EMOA e MOEAD/DRA.

Keywords: Combinatorial problems · Hyper-heuristics · Multi-objective approach

1 Introduction

Combinatorial optimization problems are noted in various applications, including communications network design, VLSI (Very Large-Scale Integration) design, machine vision, airline crew scheduling, corporate planning, computer-aided design and manufacturing, data-base query design, cellular telephone frequency assignment, constraint directed reasoning, and computational biology [1].

The *Quadratic Assignment Problem* (QAP), introduced by Koopmans and Beckmann [2], is an example of combinatorial optimization problem and it belongs to NP-hard class [3]. QAP was initially derived as a mathematical model of assigning a set of economic activities to a set of locations [2]. Afterwards, a large variety of other applications of the QAP is known including such areas

© Springer Nature Switzerland AG 2020
R. Cerri and R. C. Prati (Eds.): BRACIS 2020, LNAI 12319, pp. 136–150, 2020.
https://doi.org/10.1007/978-3-030-61377-8_10

as scheduling, wiring problems in electronics, parallel and distributed computing, statistical data analysis, design of control panels and typewriter keyboards, sports, chemistry, archeology, balancing of turbine runners, computer manufacturing, and transportation [1].

For NP-hard optimization problems in the real world, there are several heuristics being proposed, many find efficient solutions for a given problem, but the heuristic being good for a specific problem does not mean that it will get good results for all [4]. Hyper-Heuristics (HH) is a high-level approach that automatically selects or generates heuristics for solving complex problems [5]. A difference between hyper-heuristics and metaheuristics is the search space where they operate: HHs do a search in the space of heuristics while metaheuristics do a search in the space of solutions [6]. The HHs can be divided into generation or selection, in generation, new heuristics are built from various components, and selection methods are used to choose the heuristic to be used in each iteration.

In this work is proposed the use of a selection HH to solve the multi-objective QAP (mQAP), based on the MOEA/DD (Evolutionary Many-Objective Optimization Algorithm Based on Dominance and Decomposition) [7] and Choice Function (CF) strategy [8]. In the proposed approach, named MOEA/DD$_{CF}$, the heuristics selected by HH correspond to the operators that generate new solutions in an iteration of the multi-objective evolutionary algorithm. The effectiveness of the proposed method is shown by comparison with the original algorithm, without using HH. The performance of the algorithm with hyper-heuristic is compared with three other evolutionary multi-objective algorithms: IBEA [9], SMS-EMOA [10] e MOEAD/DRA [11].

The contribution of this work is in the analysis of the behavior of an algorithm based on hyper-heuristics for multi-objective QAP. As far as the authors are aware, there are no works on this in the literature.

Section 2 provides the background for Multi-objective Quadratic Assignment Problem, Hyper-Heuristics and Choice Function. Related works are shown in Sect. 2.3. The proposed approach is described in Sect. 3. In the Sect. 4 experiments are presented and discussed. Finally, the conclusions are presented in Sect. 5.

2 Background

This section presents some basic information of some topics directly related to the proposed approach: Multi-objective Quadratic Assignment Problem, Hyper-Heuristics, and Choice Function. Some works related to this proposal are also briefly commented.

2.1 Multi-objective Quadratic Assignment Problem (mQAP)

QAP assigns n interconnected facilities to n locations while minimizing the cost of transportation of the flow of commodities between facilities [2]. The Multi-objective Quadratic Assignment Problem (mQAP) [12] models situations where two or more types of flows exist. Given a square location matrix of order n A = $\{a_{ij}\}$ and m square flow matrices of order n $B_k = \{b_{rs}^k\}$, $k = 1, \ldots, m$, mQAP can be formulated as [12]:

$$Minimize\ \vec{C}(\pi) = \{C^1(\pi), C^2(\pi), \ldots, C^m(\pi)\} \qquad (1)$$
$$\pi \in P(n)$$

where $\vec{C}^k(\pi) = \sum_{i=1}^n \sum_{j=1}^n a_{ij} b_{\pi_i \pi_j}^k$ $1 \leq k \leq T$, a_{ij} is the distance between locations i and j, b_{ij}^k is the k-th flow between facilities i and j and π_i gives the location of facility i in permutation $\pi \in P(n)$, where $P(n)$ is the set of all permutations of $\{1, 2, \ldots, n\}$.

2.2 Hyper-Heuristics (HH)

Hyper heuristics are high-level methodologies developed for the optimization of a wide range of NP-Hard problems. They can choose in the search space whats the option that brings a good overall result in the moment to fulfill their goals using heuristic components to create heuristics (generation heuristics) or methods to choose the best one in each situation (selection heuristics), each of them can be subdivided in construction or perturbation heuristics, the first create new solutions and the second modify an existing solution. HH's can learn in two ways, online, where the learning takes place during the execution of the program with the help of a feedback provided by the program, or offline when it already comes with the rules defined before the execution of the problem [5].

A selection type HH with online learning was chosen due to the ability to adapt for the situation, seeking the best way to achieve its goal.

Choice Function (CF). Choice Function is a high-level strategy based on the selection type HH that seeks in a space of low-level heuristics, who's is the most efficient crossover and mutation to solve the problem. Every time a heuristic is used, it receives a reward for the performance in the algorithm. In the CF Eq. 2, h is the operator and lh is the last operator used, $f1$ is the performance of an operator, $f2$ is the performance considering an pair of operators used in order, $f3$ shows how long the operator has not been used. α, β and γ indicate the weights that each function receives, and at each iteration their values are modified [8,13].

$$CF(h) = \alpha * f1(h) + \beta * f2(lh, h) + \gamma * f3(h) \qquad (2)$$

In this work, the search space is composed of four operators that uses two parents to generate their children, and two mutations who take an the generated child and modify it. The operators are Cyclic Crossover [14], Permutation Two Points Crossover [15], Order Crossover [14] and Partially Matched Crossover [14]. The mutations are the Insertion mutation [16] and the Swap mutation [16].

2.3   Related Works

Quadratic assignment problem is considered a challenge for many researchers mainly due to its complexity and multiple applications in real time. There are many works in the literature that approach the problem as mono-objective [17–24]. However, few works apply the multi-objective [25–28] or many-objective [29] approach to QAP. As far as the authors are aware, there are no works using the hyper-heuristic approach to the mQAP.

In [25] are conducted experiments to approach the multi-objective QAP using a hybrid multi-objective version of extremal optimisation, named Hybrid Multi-objective Extremal Optimisation (HMEO). HMEO consists of a multi-objective extremal optimisation framework, for the coarse-grain search, which contains a novel multi-objective combinatorial local search framework for the fine-grain search. HMEO is applied to solve a group of eight different mQAPs considering two objectives by Knowles and Corne [30]. The results obtained show that the HMEO is able to obtain competitive results to SPEA2 and NSGA-II.

In [26] is investigated hybrid algorithms combining Transgenetic Algorithms and Evolutionary Multi-objective Optimization (EMO) frameworks to deal with mQAP. The authors compare the ability of EMO algorithms based on Pareto dominance with those based on decomposition to deal with the mQAP. Thus, two hybrid algorithms are proposed to deal with the mQAP: NSTA (TA + NSGA-II) and MOTA/D (TA + MOEA/D). The proposed algorithms are compared with NSGA-II and MOEA/D in 126 instances of the mQAP considering two and three objectives: Knowles and Corne [30], and Paquete and Stützle [31].

In [28], the authors characterize and study the performance of different memory strategies applied on memetic algorithms for solving different types of instances of the Bi-objective Quadratic Assignment Problem. The memetic approach is tested in a set of 26 instances, which were generated by the instance generator tool.

In order to scale-up optimisation in many-objective search spaces, [29] uses cartesian product of scalarization functions to reduce the number of objectives of the search space. The author use a stochastic a local search algorithm with product functions to evaluate solutions within a local search run with the goal of generating the entire Pareto front. The performance of algorithm is compared using several many-objective QAP instances.

As far we know, this is the first time that mQAP has been treated by a Hyper-Heuristic. HHs, including those based on Choice Function, have been successfully applied to several complex optimization problems [32–34].

3   Proposed Approach

MOEA/DD is a unified paradigm which combines the dominance- and decomposition-based technique. Our approach, named MOEA/DD$_{CF}$, uses the MOEA/DD framework coupled with a HH based on choice function in order to choose between pairs of genetic operators within the evolutionary process. There are eight combinations, of the following operators: Cyclic Crossover (CX), Permutation Two Points Crossover (2PX), Order Crossover (OX) and Partially Matched Crossover (PMX), Swap mutation and Insertion mutation.

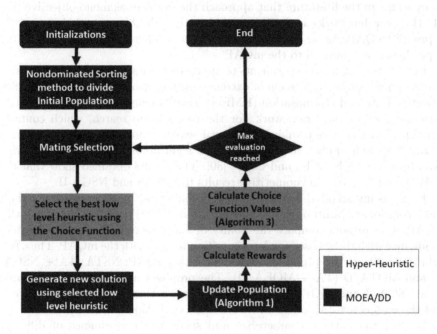

Fig. 1. Steps of MOEA/DD$_{CF}$.

The Fig. 1 shows the steps of the MOEA/DD$_{CF}$. It starts initializing and evaluating the population with random permutations. Then it is created a weight vector, according to [7,35], for each individual and assign them in their respective neighborhoods. Next, it's used a fast nondomination sorting method [36] in the generated population to organize it in fronts, according to the nondomination level. The first front isn't dominated by anybody, from the second front onward, it withdraw the already used ones and search for the nondominated individual until it identifies all fronts. After the nondominated sorting stage, the mating selection chooses the parents that will suffer the operators' action. First, it is determined the scope from which k solutions (parents) will be randomly selected. There is a probability λ of choosing the parents from the neighborhood of the current subregion and a probability 1.0- λ of selecting them from the whole population.

In sequence, the selection by the choice function starts after the use of every operator in the search space of the CF, then it chooses the operator with the highest CF value found. The selected operator is used with the parents chosen by the mating selection, to generate new solutions with the possibility to be accepted in the population.

The population update process is based on a procedure which is conducted in a hierarchical manner, sequentially depending on Pareto dominance, local density estimation and scalarization functions. The Algorithm 1 shows how the process is made. It starts by finding the subregion of the new solution y, adding y and the population on a new temporary population P' and updating its non-domination levels. If P' only have one level, the worst solution in P' will be removed. If it has more levels, then the F_{level}, that shows how many solutions are in the last level of nondomination, will be used. If this level has only one solution, the result depends on how many solutions exist in the associated sub-region, if is not the only one in the subregion then remove it, but if is the only one, then the worst solution in P' will be deleted. If P' have more then one solution in the F_{level}, will search in its most crowded subregion associated with solutions at that level, if has more then one solution then remove for the worst inside this subregion or if is the only solution, then delete for the worst in P'. So, the population nondomination level update is done. To search for the worst solution of the entire population, is used the Algorithm 2.

The rewards for the CF is calculated using the difference of the mean value of the individual fitness of the worst parent and the best offspring. The Algorithm 3 shows how each variable is updated every time it is used. The reward is send to the CF and all of the operators values are updated. In lines 1 to 4, the $f1$ is added with the reward of the used operator s, $f2$ is added with the sum of the reward of s and the one of the *lastOperator*, $f3$ receives the current time. In lines 4 to 14, the α, β and γ are updated according to the reward, if positive, then everyone return to a fixed predetermined value, and if negative, α and β will decrease and γ will increase while γ is less then 0.9. This give a better chance to use the less used operators. Then, the *lastReward* and *lastOperator* are updated for the next time it will be called, and in lines 17 to 19, every operator will be updated with the new values.

The termination criteria is the quantity of evaluations used, when it reaches the maximum number of evaluations showed in Table 2 it ends and the relevant data for the tests are saved. The algorithm finishes and outputs the set of nondominated solutions when the maximum number of evaluations is reached.

4 Experiments and Discussion

Our proposed approach is configured to run 30 test trials varying by random seed. The parameters considered in this study are given in Table 2.

In experiments, we first intend to evaluate the impact of HH method. We compared the proposed algorithm with HH (MOEA/DD$_{CF}$) with the classic version (MOEA/DD). MOEA/DD uses the most selected operators by the choice function, they are: cycle crossover and swap mutation.

All of algorithms were applied to solve the bi- and 3-objective quadratic assignment problem, in a experimentation considering 22 benchmark instances with 10, 20 and 30 locations [30]. See Table 1 for the characteristics of the test suite problems [37].

Algorithm 1. Update Population

    Input: Population P^{gen}, offspring solution **y**
    Output: Population P^{gen}

1: Find the subregion associated with y
2: $P' = P^{gen} \cup \{\mathbf{y}\}$
3: Update the nondomination level structure of P'
4: **if** $level == 1$ **then** //$level$ is the last nondomination level
5:     $\mathbf{x}' = \text{LOCATE\_WORST}(P')$
6:     $P^{gen} = P' \setminus \mathbf{x}'$
7: **else**
8:     **if** $|F_{level}| == 1$ **then** //F_{level} has only one solution, $\mathbf{x^r}$
9:         **if** $|\Phi^r| > 1$ **then** //$|\Phi^r|$ is the subregion associated with the solution $\mathbf{x^r}$
10:           $P^{gen} = P' \setminus \mathbf{x^r}$
11:         **else**//Φ^r has only one solution that is important for diversity
12:           $\mathbf{x}' = \text{LOCATE\_WORST}(P')$
13:           $P^{gen} = P' \setminus \mathbf{x}'$
14:         **end if**
15:     **else**
16:         Identify the most crowded subregion Φ^h associated with solutions in F_{level}
17:         **if** $|\Phi^h| > 1$ **then**
18:           //Find the worst solution associated with Φ^h
19:           $\mathbf{x^h} = argmax_{\mathbf{x} \in \Phi^r} g^{ws}(\mathbf{x}|\boldsymbol{\lambda}^h, \mathbf{z}*)$
20:           $P^{gen} = P' \setminus \mathbf{x^h}$
21:         **else**//Φ^h has only one solution that is important for diversity
22:           $\mathbf{x}' = \text{LOCATE\_WORST}(P')$
23:           $P^{gen} = P' \setminus \mathbf{x}'$
24:         **end if**
25:     **end if**
26: **end if**
27: Update the nondomination level structure of P^{gen}
28: Return P^{gen}

Algorithm 2. Find the Worst Solution in the Population (LOCATE_WORST)

    Input: Population P'
    Output: The worst solution $\mathbf{x}'$

1: Identify the most crowded subregion Φ^h in P', ties are broken using the following equation: $h = argmax_{i \in S} \sum_{\mathbf{x} \in \Phi^i} g^{ws}(\mathbf{x}|\boldsymbol{\lambda}^i)$
2: Find the solution set R, which is a subset of Φ^h that belongs to the worst nondomination level
3: Find the worst solution $\mathbf{x}' = argmax_{\mathbf{x} \in R} g^{ws}(\mathbf{x}|\boldsymbol{\lambda}^h)$
4: Return $\mathbf{x}'$

In comparison, we use the IGD (Inverted Generational Distance) [38] and Hypervolume [39] metrics. The IGD and Hypervolume measures assess different properties of a nondominated solution set, and provide a single performance value for the set. Inverted Generational Distance indicates how far the approximation front is from a reference set. Lower values of IGD represents a better

Algorithm 3. Pseudocode of the Choice Function

    Intput: Operator used s, reward of the operator *reward*

1: f1[s] $\leftarrow$ f1[s] + reward;
2: f2[lastOperator][s] $\leftarrow$ f2[lastOperator][s] + reward + lastReward;
3: f3[s] $\leftarrow$ currentTime();
4: **if** reward $\geq$ 0 **then**
5:     $\alpha \leftarrow$ 0.49;
6:     $\beta \leftarrow$ 0.49;
7:     $\gamma \leftarrow$ 0.02;
8: **else**
9:     **if** $\gamma \leq$ 0.9 **then**
10:       $\alpha \leftarrow \alpha$ - 0.01;
11:       $\beta \leftarrow \beta$ - 0.01;
12:       $\gamma \leftarrow$ 1 - ($\alpha + \beta$);
13:     **end if**
14: **end if**
15: lastReward $\leftarrow$ reward;
16: lastOperator $\leftarrow$ s;
17: **for** i $\leftarrow$ 0 ; i < 8 ; i ++ **do**
18:     CF[i] $\leftarrow$ (α * f1[i]) + (β * f2[s][i]) + (γ * (f3[i] - currentTime()));
19: **end for**

Table 1. Test suite.

Test name	Instance category	Number of locations	Number of flows
KC10-2fl-[1,...5]rl	Real-like	10	2
KC10-2fl-[1,2,3]uni	Uniform	10	2
KC20-2fl-[1,...5]rl	Real-like	20	2
KC20-2fl-[1,2,3]uni	Uniform	20	2
KC30-3fl-[1,2,3]uni	Uniform	30	3
KC30-3fl-[1,2,3]rl	Real-like	30	3

performance. The hypervolume measures the hypervolume portion of the objective space that is weakly dominated by an approximation set A. The higher the hypervolume dominated by an approximation set, the better it is. The reference set is constructed considering the nondominated solutions of the union of the approximation sets obtained by all the algorithms being compared. The results are ranked according to the Kruskall-Wallis statistical test with Dunn-Sidak's post-hoc test [40]. All tests are applied with significance level of 5%.

The hyper-heuristic effect is presented in Tables 3 and 4. It present mean and standard deviation of IGD and Hypervolume values for MOEA/DD$_{CF}$ and MOEA/DD, respectively. Both metrics indicate advantages for MOEA/DD$_{CF}$ in instances with 2 objectives and a certain disadvantage in instances with 3 objectives. According to IGD MOEA/DD$_{CF}$ significantly outperforms MOEA/DD in

Table 2. Parameters used in the experiments.

MOEA/DD parameters		
Name	Values	Description
NP	100	Population size
MAX-EV	200000 for bi-objective	Maximum number of evaluations
	300000 for tri-objective	
C	20	Neighborhood size
δ	0.9	Scope selection probability

7 of the 22 instances considered, obtaining a better performance for the smaller instances. MOEA/DD is the best algorithm for 2 instances. The performance of both algorithms are statistically equivalent for 10 of the 22 instances. Hypervolume shows that MOEA/DD$_{CF}$ significantly outperforms MOEA/DD in 6 of the 22 instances considered, obtaining a better performance for the smaller instances, while MOEA/DD is the best algorithm for 5 instances. The performance of both algorithms are statistically equivalent for 8 of the 22 instances.

The Empirical Attainment Function (EAF) provides a graphical description of the distribution of a Pareto front approximation set, using the notion of goal-attainment [41]. This metric can be used to identify which regions of the objective space one approximation set is better than the other and what is the probability of this happening. In this sense the attainment function is a more robust metric than the others, but has a high computational cost [41]. Figure 2 presents the EAF for the graphical comparison of MOEA/DD$_{CF}$ and MOEA/DD in instance KC10-2fl-5rl. By EAF it is possible to observe that the MOEA/DD$_{CF}$ surpass the MOEA/DD, specially at the right end of the objective space.

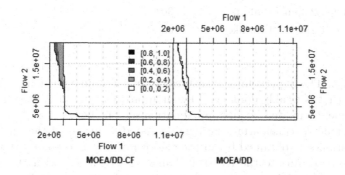

Fig. 2. Empirical attainment function for instance KC10-2fl-5rl.

An analysis of the dynamics of operator selection by HH is shown in Fig. 3. The preferred combinations are PMX and CX crossover with Swap mutation and the least preferred is OX crossover with insertion mutation.

Table 3. The IGD statistics based on 30 independent runs for MOEA/DD$_{CF}$ and MOEA/DD. Dark gray cells emphasize the best indicator value and light gray cells indicate its statistically equivalent results according to the statistical tests.

	MOEA/DD$_{CF}$	MOEA/DD
KC10-2fl-1rl	$3.51e-04_{8.8e-05}$	$4.01e-04_{6.8e-05}$
KC10-2fl-1uni	$2.42e-03_{6.6e-04}$	$2.24e-03_{5.6e-04}$
KC10-2fl-2rl	$1.89e-03_{5.2e-04}$	$2.15e-03_{3.2e-04}$
KC10-2fl-2uni	$6.13e-02_{1.1e-02}$	$6.44e-02_{1.4e-02}$
KC10-2fl-3rl	$5.11e-04_{2.3e-04}$	$6.20e-04_{2.2e-04}$
KC10-2fl-3uni	$1.03e-04_{3.7e-05}$	$1.18e-04_{3.5e-05}$
KC10-2fl-4rl	$0.00e+00_{0.0e+00}$	$1.10e-06_{4.1e-06}$
KC10-2fl-5rl	$5.04e-04_{2.5e-04}$	$1.32e-03_{8.6e-04}$
KC20-2fl-1rl	$7.65e-04_{1.6e-04}$	$7.98e-04_{1.8e-04}$
KC20-2fl-1uni	$8.97e-04_{2.8e-04}$	$9.34e-04_{2.1e-04}$
KC20-2fl-2rl	$6.96e-04_{1.5e-04}$	$7.25e-04_{1.9e-04}$
KC20-2fl-2uni	$9.06e-03_{2.8e-03}$	$9.55e-03_{3.4e-03}$
KC20-2fl-3rl	$2.79e-04_{6.1e-05}$	$2.71e-04_{6.2e-05}$
KC20-2fl-3uni	$1.82e-04_{3.2e-05}$	$1.66e-04_{2.4e-05}$
KC20-2fl-4rl	$8.27e-04_{1.1e-04}$	$8.80e-04_{1.2e-04}$
KC20-2fl-5rl	$5.97e-04_{1.5e-04}$	$5.44e-04_{1.4e-04}$
KC30-3fl-1rl	$4.97e-04_{8.4e-05}$	$4.55e-04_{5.9e-05}$
KC30-3fl-1uni	$4.82e-04_{5.3e-05}$	$4.48e-04_{6.1e-05}$
KC30-3fl-2rl	$5.48e-04_{8.0e-05}$	$5.18e-04_{6.9e-05}$
KC30-3fl-2uni	$1.31e-03_{2.1e-04}$	$1.21e-03_{2.0e-04}$
123KC30-3fl-3rl	$4.13e-04_{4.3e-05}$	$3.97e-04_{3.5e-05}$
KC30-3fl-3uni	$4.16e-04_{3.2e-05}$	$4.03e-04_{4.9e-05}$

Table 4. The hypervolume statistics based on 30 independent runs for MOEA/DD$_{CF}$ and MOEA/DD. Dark gray cells emphasize the best indicator value and light gray cells indicate its statistically equivalent results according to the statistical tests.

	MOEA/DD$_{CF}$	MOEA/DD
KC10-2fl-1rl.dat	$7.82e-01_{5.2e-03}$	$7.79e-01_{2.3e-03}$
KC10-2fl-1uni.dat	$8.55e-01_{1.2e-02}$	$8.38e-01_{3.1e-02}$
KC10-2fl-2rl.dat	$9.37e-01_{0.0e+00}$	$9.37e-01_{7.7e-03}$
KC10-2fl-2uni.dat	$1.00e+00_{0.0e+00}$	$1.00e+00_{0.0e+00}$
KC10-2fl-3rl.dat	$8.32e-01_{2.6e-03}$	$8.29e-01_{8.6e-03}$
KC10-2fl-3uni.dat	$6.17e-01_{1.8e-03}$	$6.15e-01_{4.7e-03}$
KC10-2fl-4rl.dat	$7.41e-01_{0.0e+00}$	$7.41e-01_{0.0e+00}$
KC10-2fl-5rl.dat	$9.38e-01_{8.4e-04}$	$9.30e-01_{2.8e-02}$
KC20-2fl-1rl.dat	$7.04e-01_{4.2e-02}$	$7.09e-01_{3.9e-02}$
KC20-2fl-1uni.dat	$6.53e-01_{2.8e-02}$	$6.55e-01_{3.7e-02}$
KC20-2fl-2rl.dat	$8.17e-01_{5.0e-02}$	$8.06e-01_{3.4e-02}$
KC20-2fl-2uni.dat	$5.28e-01_{1.6e-01}$	$5.74e-01_{1.4e-01}$
KC20-2fl-3rl.dat	$7.88e-01_{1.2e-02}$	$7.89e-01_{2.3e-02}$
KC20-2fl-3uni.dat	$6.10e-01_{1.6e-02}$	$6.12e-01_{1.4e-02}$
KC20-2fl-4rl.dat	$8.37e-01_{3.3e-02}$	$8.47e-01_{3.1e-02}$
KC20-2fl-5rl.dat	$7.73e-01_{4.6e-02}$	$7.82e-01_{4.2e-02}$
KC30-3fl-1rl.dat	$6.19e-01_{3.0e-02}$	$6.65e-01_{3.6e-02}$
KC30-3fl-1uni.dat	$5.26e-01_{2.8e-02}$	$5.73e-01_{3.9e-02}$
KC30-3fl-2rl.dat	$6.87e-01_{7.6e-02}$	$6.97e-01_{8.9e-02}$
KC30-3fl-2uni.dat	$4.55e-01_{9.2e-02}$	$5.32e-01_{7.9e-02}$
KC30-3fl-3rl.dat	$6.73e-01_{3.7e-02}$	$6.93e-01_{2.6e-02}$
KC30-3fl-3uni.dat	$5.10e-01_{3.6e-02}$	$5.69e-01_{2.3e-02}$

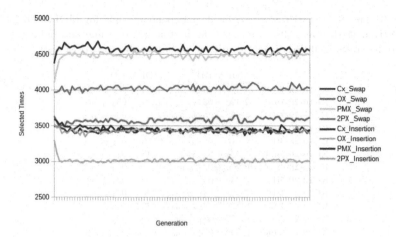

Fig. 3. Choice function operator selection throughout evolution process for instance KC10-2fl-1rl.

Tables 5 and 6, respectively, present mean and standard deviation of IGD and Hypervolume metric values for MOEA/DD$_{CF}$, SMS-EMOA, IBEA, and MOEA/D-DRA. Based on the results of Table 5, IGD values, MOEA/DD$_{CF}$ obtains statistically better mean values than the other algorithms in 11 of 22

Table 5. The IGD statistics based on 30 independent runs for MOEA/DD$_{CF}$, SMS-EMOA, IBEA and MOEA/D-DRA. Dark gray cells emphasize the best indicator value and light gray cells indicate its statistically equivalent results according to the statistical tests.

	MOEADD$_{CF}$	SMS-EMOA	IBEA	MOEA/D-DRA
KC10-2fl-1rl	$2.05e-03_{6.3e-06}$	$2.05e-03_{2.6e-05}$	$2.17e-03_{4.9e-05}$	$1.53e-03_{1.2e-04}$
KC10-2fl-1uni	$3.28e-03_{9.5e-05}$	$3.31e-03_{1.7e-04}$	$3.43e-03_{1.6e-04}$	$2.32e-03_{4.2e-04}$
KC10-2fl-2rl	$3.64e-03_{1.5e-04}$	$3.59e-03_{1.9e-04}$	$3.68e-03_{1.3e-04}$	$1.82e-03_{5.0e-04}$
KC10-2fl-2uni	$5.70e-03_{1.5e-04}$	$5.34e-03_{4.0e-04}$	$5.46e-03_{3.4e-04}$	$3.41e-03_{5.3e-04}$
KC10-2fl-3rl	$2.17e-03_{1.7e-05}$	$2.16e-03_{2.4e-05}$	$2.22e-03_{3.9e-05}$	$1.63e-03_{2.2e-04}$
KC10-2fl-3uni	$5.49e-04_{5.9e-06}$	$5.57e-04_{1.3e-05}$	$5.99e-04_{1.6e-05}$	$9.43e-04_{1.2e-04}$
KC10-2fl-4rl	$3.17e-03_{7.1e-11}$	$3.16e-03_{5.5e-05}$	$3.18e-03_{2.2e-05}$	$2.09e-03_{5.5e-04}$
KC10-2fl-5rl	$2.17e-03_{6.1e-05}$	$2.16e-03_{5.6e-05}$	$2.63e-03_{6.8e-05}$	$1.52e-03_{1.2e-04}$
KC20-2fl-1rl	$2.86e-03_{1.2e-04}$	$2.88e-03_{8.5e-05}$	$2.96e-03_{1.1e-04}$	$3.16e-03_{2.1e-04}$
KC20-2fl-1uni	$2.94e-03_{8.0e-05}$	$2.94e-03_{6.6e-05}$	$2.99e-03_{6.2e-05}$	$3.29e-03_{1.4e-04}$
KC20-2fl-2rl	$3.12e-03_{9.3e-05}$	$3.14e-03_{1.4e-04}$	$3.23e-03_{7.2e-05}$	$3.38e-03_{3.5e-04}$
KC20-2fl-2uni	$5.14e-03_{2.2e-04}$	$5.01e-03_{2.8e-04}$	$5.05e-03_{2.3e-04}$	$4.83e-03_{4.4e-04}$
KC20-2fl-3rl	$2.03e-03_{3.8e-05}$	$2.12e-03_{7.8e-05}$	$2.11e-03_{2.4e-05}$	$2.49e-03_{1.9e-04}$
KC20-2fl-3uni	$8.07e-04_{2.2e-05}$	$8.42e-04_{2.1e-05}$	$8.50e-04_{2.3e-05}$	$1.47e-03_{8.9e-05}$
KC20-2fl-4rl	$2.80e-03_{1.4e-04}$	$2.88e-03_{1.1e-04}$	$2.98e-03_{1.2e-04}$	$3.09e-03_{2.6e-04}$
KC20-2fl-5rl	$2.61e-03_{6.6e-05}$	$2.71e-03_{1.1e-04}$	$2.71e-03_{7.3e-05}$	$2.93e-03_{2.9e-04}$
KC30-3fl-1rl	$2.00e-03_{8.3e-05}$	$2.50e-03_{8.8e-05}$	$2.39e-03_{4.2e-05}$	$3.58e-03_{1.8e-04}$
KC30-3fl-1uni	$1.89e-03_{5.8e-05}$	$2.33e-03_{6.8e-05}$	$2.25e-03_{7.4e-05}$	$3.43e-03_{1.7e-04}$
KC30-3fl-2rl	$2.03e-03_{8.4e-05}$	$2.57e-03_{9.7e-05}$	$2.44e-03_{7.5e-05}$	$3.52e-03_{2.4e-04}$
KC30-3fl-2uni	$3.66e-03_{1.2e-04}$	$3.96e-03_{1.3e-04}$	$4.07e-03_{1.4e-04}$	$5.11e-03_{2.4e-04}$
KC30-3fl-3rl	$1.66e-03_{5.3e-05}$	$2.23e-03_{9.3e-05}$	$2.06e-03_{6.2e-05}$	$2.85e-03_{1.6e-04}$
KC30-3fl-3uni	$1.30e-03_{4.5e-05}$	$2.15e-03_{2.5e-04}$	$1.66e-03_{8.4e-05}$	$2.66e-03_{1.3e-04}$

Table 6. The hypervolume statistics based on 30 independent runs for MOEA/DD$_{CF}$, SMS-EMOA, IBEA and MOEA/D-DRA. Dark gray cells emphasize the best indicator value and light gray cells indicate its statistically equivalent results according to the statistical tests.

	MOEADD$_{CF}$	SMS-EMOA	IBEA	MOEA/D-DRA
KC10-2fl-1rl.dat	$8.73e-01_{2.2e-03}$	$8.71e-01_{7.0e-03}$	$8.68e-01_{3.6e-03}$	$5.86e-01_{2.7e-02}$
KC10-2fl-1uni.dat	$9.36e-01_{4.4e-03}$	$9.21e-01_{1.2e-02}$	$9.18e-01_{1.4e-02}$	$5.46e-01_{5.8e-02}$
KC10-2fl-2rl.dat	$9.87e-01_{2.3e-03}$	$9.81e-01_{5.4e-03}$	$9.82e-01_{4.6e-03}$	$6.95e-01_{6.0e-02}$
KC10-2fl-2uni.dat	$9.94e-01_{2.2e-02}$	$9.31e-01_{6.8e-02}$	$9.52e-01_{5.6e-02}$	$4.85e-01_{6.0e-02}$
KC10-2fl-3rl.dat	$9.04e-01_{2.7e-03}$	$9.00e-01_{5.2e-03}$	$8.93e-01_{9.7e-03}$	$6.02e-01_{3.4e-02}$
KC10-2fl-3uni.dat	$6.12e-01_{2.5e-03}$	$6.13e-01_{2.8e-03}$	$6.06e-01_{4.5e-03}$	$4.59e-01_{1.4e-02}$
KC10-2fl-4rl.dat	$9.63e-01_{0.0e+00}$	$9.61e-01_{1.3e-02}$	$9.58e-01_{1.4e-02}$	$6.12e-01_{5.9e-02}$
KC10-2fl-5rl.dat	$9.70e-01_{1.3e-03}$	$9.69e-01_{5.4e-03}$	$9.65e-01_{6.4e-03}$	$7.82e-01_{3.9e-02}$
KC20-2fl-1rl.dat	$8.35e-01_{2.6e-02}$	$8.52e-01_{2.2e-02}$	$8.41e-01_{3.0e-02}$	$3.52e-01_{1.9e-02}$
KC20-2fl-1uni.dat	$7.70e-01_{1.9e-02}$	$7.81e-01_{2.2e-02}$	$7.73e-01_{2.3e-02}$	$2.72e-01_{1.4e-02}$
KC20-2fl-2rl.dat	$9.26e-01_{2.2e-02}$	$9.22e-01_{2.3e-02}$	$9.15e-01_{2.4e-02}$	$3.97e-01_{3.0e-02}$
KC20-2fl-2uni.dat	$8.95e-01_{3.1e-02}$	$8.68e-01_{4.0e-02}$	$8.63e-01_{4.7e-02}$	$2.64e-01_{3.4e-02}$
KC20-2fl-3rl.dat	$8.41e-01_{1.4e-02}$	$7.95e-01_{1.7e-02}$	$8.63e-01_{1.5e-02}$	$4.05e-01_{2.2e-02}$
KC20-2fl-3uni.dat	$5.99e-01_{9.4e-03}$	$6.06e-01_{8.2e-03}$	$6.11e-01_{7.2e-03}$	$3.51e-01_{1.2e-02}$
KC20-2fl-4rl.dat	$8.98e-01_{2.0e-02}$	$9.18e-01_{2.1e-02}$	$9.19e-01_{2.5e-02}$	$4.33e-01_{2.9e-02}$
KC20-2fl-5rl.dat	$8.97e-01_{2.4e-02}$	$8.92e-01_{2.4e-02}$	$9.00e-01_{2.1e-02}$	$4.29e-01_{2.8e-02}$
KC30-3fl-1rl.dat	$7.15e-01_{1.7e-02}$	$6.04e-01_{2.7e-02}$	$7.62e-01_{2.1e-02}$	$1.45e-01_{1.1e-02}$
KC30-3fl-1uni.dat	$5.82e-01_{2.5e-02}$	$5.78e-01_{2.9e-02}$	$6.61e-01_{2.2e-02}$	$8.89e-02_{8.8e-03}$
KC30-3fl-2rl.dat	$7.63e-01_{3.2e-02}$	$6.20e-01_{4.4e-02}$	$7.95e-01_{3.5e-02}$	$1.69e-01_{1.1e-02}$
KC30-3fl-2uni.dat	$6.65e-01_{3.3e-02}$	$7.49e-01_{3.9e-02}$	$7.40e-01_{3.0e-02}$	$7.75e-02_{9.1e-03}$
KC30-3fl-3rl.dat	$6.67e-01_{1.9e-02}$	$5.10e-01_{2.2e-02}$	$6.99e-01_{2.1e-02}$	$1.70e-01_{1.0e-02}$
KC30-3fl-3uni.dat	$5.00e-01_{2.0e-02}$	$4.58e-01_{2.4e-02}$	$5.95e-01_{1.7e-02}$	$1.19e-01_{8.0e-03}$

instances and is equivalent to the best algorithm in two instances. MOEA/DD$_{CF}$ achieves better performance for larger instances. The second best algorithm was the MOEA/D-DRA, significantly better in six instances, followed by the SMS-EMOA. The worst results were obtained by the IBEA.

According to Hypervolume results, Table 6, MOEA/DD$_{CF}$ and IBEA obtain similar performance, being statistically better mean values than the other algorithms in 6 of 22 instances and are equivalent to the best algorithm in 5 instances. Both algorithms are followed by the SMS-EMOA. The worst results were obtained by the MOEA/D-DRA.

5   Conclusions

In this work we proposed the application of the HH selection to solve the multi-objective QAP, based on the MOEA/DD algorithm and using choice function strategy. The proposed algorithm, named MOEA/DD$_{CF}$, were applied using the multi-objective bi- and 3-objective quadratic assignment problem and 22 benchmark instances were considered for the experiments.

The contribution of this work is in the analysis of the behavior of an algorithm based on hyper-heuristic for multi-objective QAP. In the tests, we first evaluated the impact of the HH method, comparing the proposed algorithm using HH with the canonical version. Then, we compared our approach with other multi-objective algorithms: SMS-EMOA, IBEA and MOEA/D-DRA.

All analyzes took into account two quality indicators and non-parametric statistical tests. According to the results of IGD and hypervolume, MOEA/DD$_{CF}$ significantly outperforms MOEA/DD in most of instances considered, specially in smaller instances. MOEA/DD$_{CF}$ is also favorably compared other multi-objective algorithms. For the IGD metric, MOEA/DD$_{CF}$ competes with MOEA/D-DRA and it is better in large instances. According to hypervolume indicator, MOEA/DD$_{CF}$ competes with IBEA and SMS-EMOA and achieves better performance especially for small instances.

QAP is one of the most difficult problems in NP-hard class and MOEA/DD$_{CF}$' results point to promising studies using hyper-heuristics for this problem.

In the future, we will extend our approach in the direction of test other selection heuristics instead of choice function. Additionally, we intend to test MOEA/DD$_{CF}$ on other benchmark instances considering many objective optimization, for example. We also intend to analyze more deeply, different characteristics of the mQAP test instances and the hyper-heuristics.

References

1. Çela, E.: The Quadratic Assignment Problem: Theory and Algorithms. Combinatorial Optimization. Springer, New York (1998). https://doi.org/10.1007/978-1-4757-2787-6
2. Koopmans, T.C., Beckmann, M.J.: Assignment problems and the location of economics activities. Econometrica **25**(1), 53–76 (1957)
3. Sahni, S., Gonzales, T.: P-complete approximation problems. J. Assoc. Comput. Mach. **23**(3), 555–565 (1976)
4. Drake, J.H., Kheiri, A., Özcan, E., Burke, E.K.: Recent advances in selection hyper-heuristics. Eur. J. Oper. Res. **285**(2), 405–428 (2020)
5. Burke, E., et al.: Hyper-heuristics: a survey of the state of the art. J. Oper. Res. Soc. **64**, 1695–1724 (2013). https://doi.org/10.1057/jors.2013.71
6. Chakhlevitch, K., Cowling, P.: Hyperheuristics: recent developments. In: Cotta, C., Sevaux, M., Sörensen, K. (eds.) Adaptive and Multilevel Metaheuristics. SCI, vol. 136, pp. 3–29. Springer, Heidelberg (2008). https://doi.org/10.1007/978-3-540-79438-7_1
7. Li, K., Deb, K., Zhang, Q., Kwong, S.: An evolutionary many-objective optimization algorithm based on dominance and decomposition. IEEE Trans. Evol. Comput. **19**(5), 694–716 (2015)
8. Kendall, G., Soubeiga, E., Cowling, P.: Choice function and random hyperheuristics. In: Proceedings of the Fourth Asia-Pacific Conference on Simulated Evolution And Learning, SEAL, pp. 667–671. Springer (2002)
9. Zitzler, E., Künzli, S.: Indicator-based selection in multiobjective search. In: Yao, X., et al. (eds.) PPSN 2004. LNCS, vol. 3242, pp. 832–842. Springer, Heidelberg (2004). https://doi.org/10.1007/978-3-540-30217-9_84
10. Beume, N., Naujoks, B., Emmerich, M.: SMS-EMOA: multiobjective selection based on dominated hypervolume. Eur. J. Oper. Res. **181**(3), 1653–1669 (2007)

11. Zhang, Q., Li, H.: MOEA/D: a multi-objective evolutionary algorithm based on decomposition. IEEE Trans. Evol. Comput. **11**(6), 712–731 (2007)
12. Knowles, J., Corne, D.: Towards landscape analyses to inform the design of hybrid local search for the multiobjective quadratic assignment problem. In: Abraham, A., del Solar, J.R., Koppen, M. (eds.) Soft Computing Systems: Design, Management and Applications, pp. 271–279. IOS Press, Amsterdam (2002)
13. Maashi, M., Özcan, E., Kendall, G.: A multi-objective hyper-heuristic based on choice function. Expert Syst. Appl. **41**(9), 4475–4493 (2014)
14. Otman, A., Abouchabaka, J.: A comparative study of adaptive crossover operators for genetic algorithms to resolve the traveling salesman problem. Int. J. Comput. Appl. **31**, 03 (2012)
15. Talbi, E.-G.: Metaheuristics: From Design to Implementation. Wiley Publishing, Hoboken (2009)
16. Larrañaga, P., Kuijpers, C., Murga, R., Inza, I., Dizdarevic, S.: Genetic algorithms for the travelling salesman problem: a review of representations and operators. Artif. Intell. Rev. **13**, 129–170 (1999). https://doi.org/10.1023/A:1006529012972
17. Shukla, A.: A modified bat algorithm for the quadratic assignment problem. In: 2015 IEEE Congress on Evolutionary Computation (CEC), pp. 486–490, May 2015
18. Lv, C.: An improved particle swarm optimization algorithm for quadratic assignment problem. In: 2012 Fourth International Conference on Multimedia Information Networking and Security, pp. 258–261, November 2012
19. Gunawan, A., Ng, K.M., Poh, K.L., Lau, H.C.: Hybrid metaheuristics for solving the quadratic assignment problem and the generalized quadratic assignment problem. In: 2014 IEEE International Conference on Automation Science and Engineering (CASE), pp. 119–124, August 2014
20. Ahmed, Z.H.: An improved genetic algorithm using adaptive mutation operator for the quadratic assignment problem. In: 2015 38th International Conference on Telecommunications and Signal Processing (TSP), pp. 1–5, July 2015
21. Montero, A.R., López, A.S.: Ant colony optimization for solving the quadratic assignment problem. In: 2015 Fourteenth Mexican International Conference on Artificial Intelligence (MICAI), pp. 182–187, October 2015
22. Abderrahim, I.A., Loukil, L.: Hybrid PSO-TS approach for solving the quadratic three-dimensional assignment problem. In: 2017 First International Conference on Embedded Distributed Systems (EDiS), pp. 1–5, December 2017
23. Ma, Z., Wang, L., Lin, S., Zhong, Y.: Comparative study of inhomogeneous simulated annealing algorithms for quadratic assignment problem. In: 2018 11th International Congress on Image and Signal Processing, BioMedical Engineering and Informatics (CISP-BMEI), pp. 1–7, October 2018
24. Okano, T., Katayama, K., Kanahara, K., Nishihara, N.: A local search based on variant variable depth search for the quadratic assignment problem. In: 2018 IEEE 7th Global Conference on Consumer Electronics (GCCE), pp. 60–63, October 2018
25. Gómez-Meneses, P., Randall, M., Lewis, A.: A hybrid multi-objective extremal optimisation approach for multi-objective combinatorial optimisation problems. In: IEEE Congress on Evolutionary Computation, pp. 1–8, July 2010
26. de Almeida, C.P., Gonçalves, R.A., Goldbarg, E.F., Goldbarg, M.C., Delgado, M.R.: Transgenetic algorithms for the multi-objective quadratic assignment problem. In: 2014 Brazilian Conference on Intelligent Systems, pp. 312–317, October 2014

27. Rajeswari, M., Jaiganesh, S., Sujatha, P., Vengattaraman, T., Dhavachelvan, P.: A study and scrutiny of diverse optimization algorithm to solve multi-objective quadratic assignment problem. In: 2016 International Conference on Communication and Electronics Systems (ICCES), pp. 1–5, October 2016

28. Sandoval-Soto, R., Villalobos-Cid, M., Inostroza-Ponta, M.: Tackling the biobjective quadratic assignment problem by characterizing different memory strategies in a memetic algorithm. In: 2017 36th International Conference of the Chilean Computer Science Society (SCCC), pp. 1–12, October 2017

29. Drugan, M.M.: Stochastic pareto local search for many objective quadratic assignment problem instances. In: 2015 IEEE Congress on Evolutionary Computation (CEC), pp. 1754–1761, May 2015

30. Knowles, J., Corne, D.: Instance generators and test suites for the multiobjective quadratic assignment problem. In: Fonseca, C.M., Fleming, P.J., Zitzler, E., Thiele, L., Deb, K. (eds.) EMO 2003. LNCS, vol. 2632, pp. 295–310. Springer, Heidelberg (2003). https://doi.org/10.1007/3-540-36970-8_21

31. Paquete, L., Stutzle, T.: A study of stochastic local search algorithms for the biobjective QAP with correlated flow matrices. Eur. J. Oper. Res. **169**(3), 943–959 (2006). http://ideas.repec.org/a/eee/ejores/v169y2006i3p943-959.html

32. de Almeida, C.P., Gonçalves, R.A., Venske, S., Lüders, R., Delgado, M.: Hyperheuristics using multi-armed bandit models for multi-objective optimization. Appl. Soft Comput., 106520 (2020). http://www.sciencedirect.com/science/article/pii/S1568494620304592

33. Gómez, R.H., Coello, C.A.C.: A hyper-heuristic of scalarizing functions. In: Proceedings of the Genetic and Evolutionary Computation Conference, GECCO 2017, pp. 577–584. ACM. New York (2017)

34. Gonçalves, R.A., de Almeida, C.P., Venske, S.M.G.S., Delgado, M.R.d.B.d.S., Pozo, A.T.R.: A new hyper-heuristic based on a restless multi-armed bandit for multi-objective optimization. In: 2017 Brazilian Conference on Intelligent Systems (BRACIS), pp. 390–395, October 2017

35. Das, I., Dennis, J.E.: Normal-boundary intersection: a new method for generating the pareto surface in nonlinear multicriteria optimization problems. SIAM J. Optim. **8**(3), 631–657 (1998)

36. Deb, K., Pratap, A., Agarwal, S., Meyarivan, T.: A fast and elitist multiobjective genetic algorithm: NSGA-II. IEEE Trans. Evol. Comput. **6**(2), 182–197 (2002)

37. Coello, C., Lamont, G.: Applications of Multi-objective Evolutionary Algorithms. Advances in Natural Computation. World Scientific, Singapore (2004)

38. Santos, T., Xavier, S.: A Convergence Indicator for Multi-Objective Optimisation Algorithms. TEMA (São Carlos), vol. 19, pp. 437–448, December 2018

39. Zitzler, E., Knowles, J., Thiele, L.: Quality assessment of Pareto set approximations. In: Branke, J., Deb, K., Miettinen, K., Słowiński, R. (eds.) Multiobjective Optimization. LNCS, vol. 5252, pp. 373–404. Springer, Heidelberg (2008). https://doi.org/10.1007/978-3-540-88908-3_14

40. Conover, W.J.: Practical Nonparametric Statistics, 3rd edn. Wiley, Hoboken (1999)

41. Grunert da Fonseca, V., Fonseca, C.M., Hall, A.O.: Inferential performance assessment of stochastic optimisers and the attainment function. In: Zitzler, E., Thiele, L., Deb, K., Coello Coello, C.A., Corne, D. (eds.) EMO 2001. LNCS, vol. 1993, pp. 213–225. Springer, Heidelberg (2001). https://doi.org/10.1007/3-540-44719-9_15

On Improving the Efficiency of Majorization-Minorization for the Inference of Rank Aggregation Models

Leonardo Ramos Emmendorfer$^{(\boxtimes)}$ [iD]

Center for Computational Sciences, Federal University of Rio Grande,
Rio Grande 96203-900, Brazil
`leonardoemmendorfer@furg.br`

Abstract. The Plackett-Luce model represents discrete choices from a set of items and it is often applied to rank aggregation problems. The iterative majorization-minorization method is among the most relevant approaches for finding the maximum likelihood estimation of the parameters of the Plackett-Luce model, but its convergence might be slow. A noninformative initialization is usually adopted which assumes all items are equally relevant at the first step of the iterative inference process. This paper investigates the adoption of approximate inference methods which could allow a better initialization, leading to a smaller number of iterations required for the convergence of majorization-minorization. Two alternatives are adopted: a spectral inference method from the literature and also a novel approach based on a Poisson probabilistic model. Empirical evaluation is performed using synthetic and real-world datasets. It was revealed that initialization provided by an approximate method can lead to statistically significant reductions in both the number of iterations required and also in the overall computational time when compared to the scheme usually adopted for majorization-minorization.

Keywords: Majorization-minorization · Plackett-Luce · Poisson · Rank aggregation

1 Introduction

The aggregation of pairwise comparisons and partial rankings is a relevant task in the context of belief function theory with applications in many areas [14] such as econometrics [15], psychometrics [2,21], sports ranking [5,18] and multiclass classification [8].

The Luce's axiom of choice [13] states that the probability of choosing an item should not depend on the specific set of items from which the choice is made. This assumption is also denoted as the independence from irrelevant alternatives (IIA). Let S be the set of m items $\{1, 2, \cdots, m\}$ which will be chosen by an agent, or might represent competitors in a contest. The probability of selecting an item i from the options in S is given by the Plackett-Luce (PL) model [13,18] as:

© Springer Nature Switzerland AG 2020
R. Cerri and R. C. Prati (Eds.): BRACIS 2020, LNAI 12319, pp. 151–165, 2020.
https://doi.org/10.1007/978-3-030-61377-8_11

$$\mathcal{P}(i \text{ is the preferred item in } S) = \mathcal{P}(i \succeq S) = \frac{\gamma_i}{\sum_{j \in S} \gamma_j} \tag{1}$$

Each parameter γ_i represents a positive-valued utility associated with an item i. That value relates to the extent to which i is superior to other items, which translates to the item being selected more frequently than others. As a concrete example, consider the items to be sports teams, where γ_i represents the overall strength of team i [9].

The model allows the computation of the probability of any specific ranking κ with w_κ items $Q = \{s_1, s_2, \cdots, s_{w_\kappa}\}$, $Q \subseteq S$, as the successive application of (1):

$$\mathcal{P}(s_1 \succ s_2 \succ \cdots \succ s_{w_\kappa}) = \prod_{r \in \{1, 2, \cdots, w_\kappa\}} \frac{\gamma_{s_r}}{\sum_{j \in A_r^\kappa} \gamma_j} \tag{2}$$

where $a \succ b$ represents that item $a \in Q$ was ranked better than item $b \in Q$ in a contest κ; the item i which is the best ranked among all competitors receives $rank(i) = 1$, the second is ranked as 2 and so on. A_r^κ is the set of options available from Q at each rank r as $A_r^\kappa = \cup_{f=r}^{w_\kappa} \{s_f\} = \{s_r, s_{r+1}, \cdots, s_{w_\kappa}\}$, for a contest κ.

Similarly, the Bradley–Terry [2] (BT) model for pairwise comparisons describes the probabilities of the possible outcomes when individuals are judged against one another in pairs [9]:

$$\mathcal{P}(a \text{ is preferred over } b) = \mathcal{P}(a \succ b) = \frac{\gamma_a}{\gamma_a + \gamma_b} \tag{3}$$

Other variants and extensions of that type of choice model have been proposed. The Rao-Kupper (RK) model [19], for instance, extends the BT model to the case where a comparison between two items can result in a tie.

The first iterative algorithm for maximum likelihood estimation (MLE) of parameters for a model of pairwise comparisons was proposed in 1929 [22], which was later shown to be a special case of the majorization-minorization (MM) approach [9]. The algorithm was later extended to the general case of the Plackett-Luce model. Besides MM, other iterative algorithms from the literature such as Newton-Raphson among others [3,6] are adopted for the MLE. The MM algorithm still provides an attractive approach, however, it was shown to present slow convergence in some cases [9]. This prevents the approach to be adopted in a wider range of cases, such as real-time applications which demand faster responses.

This work evaluates two approximate methods for the estimation of parameters $\gamma = \{\gamma_1, \gamma_2, \cdots, \gamma_m\}$ of the Placket-Luce model, which are applied to obtain better initial values for the MM iterative process and would allow a reduction in the number of iterations until convergence. A method from the literature is adopted, along with a novel approximate approach.

The paper is organized as follows. Section 2 revises exact and approximate methods for the MLE of parameters of the Plackett-Luce model. Section 3 presents the proposed approximate method. The experimental evaluation is described in Sect. 4. Section 5 concludes the paper.

2 Maximum Likelihood Estimation for the Placket-Luce Model

The log-likelihood of the parameters γ for model (2) given rankings of the form $Q = \{s_1, s_2, \cdots, s_{w_\kappa}\}$, $Q \in \mathcal{D}$ is:

$$log\mathcal{L}(\gamma|\mathcal{D}) = \sum_{\kappa \in \{1,2,\cdots,n\}, r \in \{1,2,\cdots,w_\kappa\}} (log\gamma_{s_r} - log \sum_{j \in A_r^\kappa} \gamma_j) \qquad (4)$$

The function in (4) is strictly concave and the model admits a unique MLE $\gamma^\star = \{\gamma_1^\star, \gamma_2^\star, \cdots, \gamma_m^\star\}$ [14]. Several approaches were adopted for finding $\gamma^\star$ by minimization of (4), but the majorization-minorization approach can be considered as a standard for the MLE in the context of Luce's model [9,14].

Algorithm 1 MM($\mathbf{x}^{(0)}$)

Input: $\mathbf{x}^{(0)}$
Output: $\mathbf{x}^\star$ which minimizes $f(\mathbf{x})$
1. $k \leftarrow 0$
2. **repeat**
3. $\mathbf{x}^{(k+1)} \leftarrow argmin_\mathbf{x} g(\mathbf{x}|\mathbf{x}^{(k)})$
4. $\delta \leftarrow |\mathbf{x}^{(k+1)} - \mathbf{x}^{(k)}|$
5. $k \leftarrow k + 1$
6. **until** $\delta < \tau$, where τ is a predefined thereshold
7. **return** $\mathbf{x}^{(k)}$

The MM approach was studied under various names for over 30 years [9,12]. The iterative process adopted by MM for finding the $\mathbf{x}^\star$ which minimizes $f(\mathbf{x})$ starts from an initial value $\mathbf{x}^{(0)}$. A surrogate function g that majorizes f such that $f(\mathbf{x}) \leq g(x|x^{(k)})$ is adopted at each iterative step of MM. The next $x^{(k+1)}$ is obtained from $x^{(k+1)} = argmin_\mathbf{x} g(\mathbf{x}|\mathbf{x}^{(k)})$. The successive minimization of g, which might be a more tractable function, leads to the minimization of the original function f. Algorithm 1 illustrates the adoption of MM for a given pair of functions $f(\mathbf{x})$ and $g(\mathbf{x})$. The approach is quite general and covers the widely adopted Expectation-Maximization algorithm [4] as special case [12].

Theoretical analysis of MM as an MLE has been proposed in the literature. In [7] the number of samples enough to drive the mean-square error to down to zero is analyzed. It was also noticed that MM might present slow convergence in some cases [9] yet the scale of the issue and its apparent unpredictability is surprising [14].

Other estimators have been proposed, both for the PL and BT models. The Rank Centrality proposed in [16] builds a graph where the nodes are the items and the transition probability is constructed from the comparisons between outcomes. This spectral approach has been a building block for several ranking algorithms [17]. In [14] the connection between the spectral approach of Rank

Centrality and the MLE, providing a unifying view to the problem. The Accelerated Spectral Ranking algorithm [1] efficiently finds the parameters of PL models and also achieves optimal sample complexity under certain conditions. In [11] the inference Luce's model parameters from data is reformulated by breaking rankings into pairwise comparisons and finding the transition matrix of the corresponding Markov chains [14].

2.1 Luce Spectral Ranking

The Luce spectral ranking algorithm [14] is adopted for approximate inference of PL models by building a Markov chain from an expression derived from the optimality condition of the log-likelihood.

Algorithm 2 LSR($\mathcal{D}$)

Input: rankings $\mathcal{D}$ as in (2)
Output: estimation $\hat{\gamma}^{LSR}$ of the parameters of a PL model
1. $\lambda \leftarrow 0_{m \times m}$
2. **for all** $(i, A^{\kappa}_{rank(i)})$
3.   **for** $j \in A^{\kappa}_{rank(i)} - \{i\}$
4. $\lambda_{j,i} \leftarrow \lambda_{j,i} + \frac{m}{|A^{\kappa}_{rank(i)}|}$
5.   **end for**
6. **end for**
7. $\hat{\gamma}^{LSR} \leftarrow$ stationary distribution of the Markov chain corresponding to λ

The transition probabilities are computed by rewarding each item i in a contest by a fixed amount of incoming rate that is evenly split across the alternatives $A^{\kappa}_{rank(i)}$ corresponding to the rank of i in each contest κ. Algorithm 2 illustrates the pseudocode of LSR. Once the transition matrix is built, the algorithm gives its corresponding stationary distribution as the result for the estimation of $\hat{\gamma}^{LSR}$. The algorithm has running time $\mathcal{O}(T + S)$ where $T = \sum_{l,\kappa} |A^{\kappa}_l|$ and S is the cost associated with finding the stationary distribution of a $m \times m$ transition matrix [14]. Since a single iteration of MM for the problem considered here is $\mathcal{O}(T)$, the approximate computation performed by LSR is expected to be obtained at a much smaller cost.

3   Proposed Approximate Model

An ordered choice of items from a set S is called the result of a contest. Let us define that an item i "failures" when it is overlooked in favor of another competing item j, or $rank_{\kappa}(i) > rank_{\kappa}(j)$ in a contest κ. Let us also define the random variable $X_i \geq 0$ as the degree of failure of an item in a given contest. For instance, X_i can be computed as $rank_{\kappa}(i) - 1$. Let us assume the probability that the degree of failure for each X_i follows the Poisson model as:

$$P(X_i = x) = \frac{\lambda_i^x e^{-\lambda_i}}{x!} \tag{5}$$

Input data comprises ranking results from n contests $c_1, c_2, \cdots, c_n$, with sizes $q_1 = |c_1|, q_2 = |c_2|, \cdots, q_n = |c_n|$, respectively. Each contest c_κ contains rankings for a subset of items from S. The parameter $\hat{\lambda}_i$ which represents the expected value $E(X_i)$ is inferred from data as the MLE:

$$\hat{\lambda}_i = \frac{\hat{R}_i}{W} \tag{6}$$

where $\hat{R}_i = \frac{\sum_{\{\kappa|i\in c_\kappa\}}(rank_k(i)-1)}{|\{\kappa|i\in c_\kappa\}|}$ is sample mean over all $rank_\kappa(i) - 1$ of each item i, considering contests $c_\kappa \in \{c_i, c_2, \cdots, c_n\}$ which contain item i and W is a reference number of opponents adopted for all items.

The set of strength values $\boldsymbol{\gamma} = \{\gamma_1, \gamma_2, \cdots, \gamma_m\}$ as adopted in the PL model (1) will be interpreted from (5) as the probability of i to be the first item selected, or the item better ranked, independently from the ranking of the opponents. Each $\hat{\gamma}_i$ is therefore defined as the probability that respective item i has no failures as:

$$\hat{\gamma}_i := P(X_i = 0) = \frac{\hat{\lambda}_i^0 e^{-\hat{\lambda}_i}}{0!} = e^{-\hat{\lambda}_i} = e^{-\frac{\hat{R}_i}{W}} = e^{-\frac{\sum_{\{\kappa|i\in c_\kappa\}}(rank_k(i)-1)}{W|\{\kappa|i\in c_\kappa\}|}} \tag{7}$$

The reference number of opponents W remains to be set. It should be valid for all items since the strength values obtained from (7) must refer to the same scale, to be useful as measures for the relative strength of the respective items when compared to each other under diverse types of contests. However, since it is not feasible to set a single value for W which represents contests of all sizes, we propose to adopt a weighted average over a range of contest sizes $2, 3, \cdots, \bar{q}$ where $\bar{q} = max\{|c_1|, |c_2|, \cdots, |c_n|\}$ is the largest number of items per contest as obtained from data. A default value for W can, therefore, be computed as:

$$W = \frac{\sum_{w=1}^{\bar{q}-1} w(\bar{q}-w)}{\sum_{w=1}^{\bar{q}-1}(\bar{q}-w)} \tag{8}$$

which is the weighted mean among all possible amounts of opponents in the interval $w \in \{1, 2, \cdots, \bar{q} - 1\}$, where the weight of each w is given by $(\bar{q} - w)$.

The weighted average in (8) is designed as an attempt to represent the relevance of each contest size, where smaller sizes received greater weigh. Notice that the highest weight $\bar{q} - 1$ is set to the smallest contest, which corresponds to a single competitor, or a pairwise comparison. The greater relevance of smaller contest sizes relates to the form every full or partial ranking is composed in (2), which always includes smaller contest sizes. This supports the assumption that smaller contest sizes should be emphasized as in (8) even when input data comprises greater contests majorly.

Algebraic manipulation from (8) leads to:

$$W = \frac{\bar{q}+1}{3} \tag{9}$$

Substituting (9) into (7) leads to the expression for the strength of each item i under the proposed Poisson model:

$$\hat{\gamma}_i^{Poisson} = e^{-\frac{3\sum_{\{c_k|i\in c_k\}}(rank_k(i)-1)}{(\bar{q}+1)|\{c_k|i\in c_k\}|}} \tag{10}$$

The strength values estimated from (10) can be adopted in the Luce model (1, 2) for computing the probability of each item i to be selected by substituting each γ_i for the respective $\hat{\gamma}_i$, which leads to:

$$\mathcal{P}^{Poisson}(i \succeq S) = \frac{\hat{\gamma}_i^{Poisson}}{\sum_{j\in S}\hat{\gamma}_j^{Poisson}} \tag{11}$$

Since the computation of each $\hat{\gamma}_i^{Poisson}$ in (10) is performed in time $\mathcal{O}(n)$, the method proposed has running time $\mathcal{O}(m \times n)$.

4   Empirical Evaluation

Empirical evaluation is performed using synthetic and real-world datasets. For both cases, the error metric is the same as adopted by [14] and [7], where a log-transformation is applied for the parameters. The root-mean-squared-error E_{RMS} between the exact $\gamma^\star$ and approximate $\hat{\gamma}$ is computed as:

$$E_{RMS} = \sqrt{\frac{\sum_{i=1}^m (\hat{\theta}_i - \theta^\star)^2}{m}} \tag{12}$$

where $\hat{\theta}_i = \log(\hat{\gamma}_i - \frac{\sum_{i=1}^m \log \hat{\gamma}_i}{m})$ and $\theta^\star_i = \log(\gamma^\star_i - \frac{\sum_{i=1}^m \log \gamma^\star_i}{m})$, which leads to $\sum_{i=1}^m \hat{\theta}_i = 0$ and $\sum_{i=1}^m \theta^\star_i = 0$. The approach adopted for computing parameters θ conforms to the random utility formulation adopted in the logit model [15].

4.1   Evaluation on Synthetic Datasets

Initially, synthetic datasets are generated and grouped according to the number of items considered. The parameters adopted for the generation of synthetic datasets were adapted from an experiment in [7] which was planned in the context of a statistical efficiency assessment procedure. Several values for the number of items were investigated, with $m \in \{2, 4, 8, 16, 32, 64, 128, 256, 512, 1024\}$. For each m, a total of 100 synthetic datasets were generated. The parameters $\gamma\star_i$ for each dataset were obtained randomly as:

$$\gamma\star_i \sim e^{[-2,2]} \tag{13}$$

For each dataset obtained from (13), $n = 64$ rankings were generated where each ranking contains all m items (full rankings with $q = m$), also as in [7].

The experiments for each group of datasets are performed as follows. The reference values for the number of iterations and computational time of each dataset are obtained from the execution of MM given an initial uniform uninformative set of parameters $\gamma^0 = [\frac{1}{m}, \frac{1}{m}, \cdots, \frac{1}{m}]$, as usually adopted in the literature. Two alternative treatments correspond to the execution of MM given as input the result from approximate inference methods, which are the Poisson model proposed in (10) and the LSR algorithm, which are called Poisson+MM and LSR+MM respectively. The effects on both the number of iterations and the total computational time are assessed. For each group, Wilcoxon signed-rank tests are performed for evaluating the statistical significance of an eventual difference between the reference and each of the alternative approaches, at a 0.05 significance level.

Table 1 shows the results from the evaluations. Both methods Poisson+MM and LSR+MM were able to achieve significantly lower number of MM iterations when compared to the reference MM approach for most groups considered, at the significance level considered. The average number of iterations for Poisson+MM and LSR+MM is very similar for all m considered, while sample standard deviation $\hat{\sigma}$ decreases consistently as a function of m, for both alternative treatments and also for the reference approach. The exception is the case corresponding to $m = 2$, where a reduction could not be achieved since the number of iterations of MM does not vary over all the 100 datasets considered, independently from the initialization adopted. This prevented the Wilcoxon signed-rank tests to be applied. Figure 1 illustrates the number of iterations of MM as a function of m, where a trend can be perceived and one might infer that similar conclusions would arise for other itemset sizes m above the range considered here.

Table 1. Average number of iterations of MM $\pm$ sample standard deviation $\hat{\sigma}$ using a usual uniform initialization (denoted as MM) as the reference and two alternative initialization approaches (LSR+MM, Poisson+MM) computed from 1,000 synthetic datasets equally distributed over 10 groups according to the itemset size m. Statistically significant improvements at a significance level of $\alpha = 0.05$ are shown in bold.

Number of items (m)	MM avg. $\pm$ σ	Poisson+MM avg. $\pm$ σ	p-value	LSR+MM avg. $\pm$ σ	p-value
2	2.0 ± 0.0	2.0 ± 0.0	–	2.0 ± 0.0	–
4	39.9 ± 30.1	**36.6 ± 27.6**	<0.0001	**37.0 ± 27.6**	<0.0001
8	37.3 ± 20.6	**32.7 ± 18.1**	<0.0001	**33.5 ± 17.7**	<0.0001
16	38.9 ± 9.0	**33.0 ± 7.8**	<0.0001	**33.5 ± 7.7**	<0.0001
32	37.3 ± 5.9	**30.5 ± 4.8**	<0.0001	**31.0 ± 5.0**	<0.0001
64	38.3 ± 3.7	**30.3 ± 2.9**	<0.0001	**30.9 ± 2.8**	<0.0001
128	38.8 ± 2.1	**29.7 ± 1.6**	<0.0001	**30.3 ± 1.7**	<0.0001
256	38.3 ± 1.6	**28.3 ± 1.2**	<0.0001	**28.9 ± 1.1**	<0.0001
512	38.8 ± 1.2	**27.9 ± 1.0**	<0.0001	**28.3 ± 0.8**	<0.0001
1024	39.9 ± 1.0	**27.1 ± 0.7**	<0.0001	**27.7 ± 0.7**	<0.0001

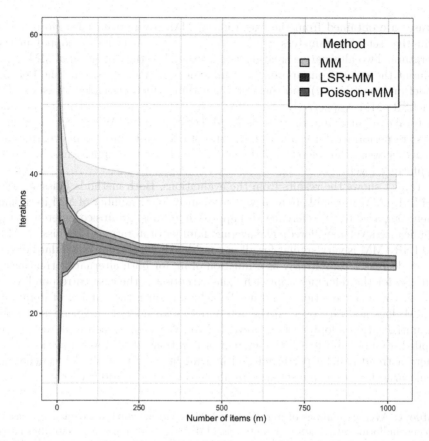

Fig. 1. Average number of iterations of MM $\pm$ sample standard deviation $\hat{\sigma}$ using a usual uniform initialization (denoted as MM) as the reference and two alternative initialization approaches (LSR+MM, Poisson+MM) computed from 1,000 synthetic datasets equally distributed over 10 groups according to the itemset size m.

The computational time resulting from approximate computation of initial values for PL parameters using Poisson or LSR and the subsequent execution of MM are summed up in both cases in order to be compared to the respective reference value from the execution of MM with a uniform input, for each dataset. Results from this evaluation are shown in Table 2. LSR+MM was not able to achieve improvements in computational time, although the number of iterations is actually better than the reference MM approach. This results from the payoff corresponding to the execution of LSR, which prevents the total computational time to be improved. A diverse result arises when Poisson+MM is considered. The average computational time of Poisson+MM is inferior to MM for all groups considered. Statistically significant improvements, however, are only detected for greater itemset sizes, for $m = 512$ and 1024. Observable trends for the computa-

Table 2. Average computational time $\pm$ sample standard deviation $\hat{\sigma}$ of MM using a usual uniform initialization (denoted as MM) as the reference and computational time of MM summed with the computational time of an alternative initialization approach (LSR+MM, Poisson+MM) computed from 1,000 synthetic datasets equally distributed over 10 groups according to the itemset size m. Statistically significant improvements at a significance level of $\alpha = 0.05$ are shown in bold.

Number of items (m)	MM avg. $\pm$ σ	Poisson+MM avg. $\pm$ σ	p-value	LSR+MM avg. $\pm$ σ	p-value
2	0.003 $\pm$ 0.006	0.002 $\pm$ 0.005	0.5297	0.006 $\pm$ 0.008	0.0572
4	0.005 $\pm$ 0.009	0.004 $\pm$ 0.008	0.6444	0.019 $\pm$ 0.015	0.0003
8	0.005 $\pm$ 0.007	0.004 $\pm$ 0.008	0.4644	0.047 $\pm$ 0.012	<0.0001
16	0.006 $\pm$ 0.009	0.008 $\pm$ 0.010	0.2860	0.124 $\pm$ 0.018	<0.0001
32	0.008 $\pm$ 0.010	0.007 $\pm$ 0.008	0.5941	0.380 $\pm$ 0.024	<0.0001
64	0.010 $\pm$ 0.010	0.008 $\pm$ 0.008	0.2081	1.360 $\pm$ 0.039	<0.0001
128	0.012 $\pm$ 0.009	0.009 $\pm$ 0.008	0.1667	5.263 $\pm$ 0.328	<0.0001
256	0.021 $\pm$ 0.009	0.019 $\pm$ 0.009	0.2120	22.376 $\pm$ 1.559	<0.0001
512	0.046 $\pm$ 0.013	**0.032 $\pm$ 0.010**	0.0005	91.357 $\pm$ 0.354	<0.0001
1024	0.085 $\pm$ 0.013	**0.063 $\pm$ 0.012**	<0.0001	363.467 $\pm$ 7.580	<0.0001

tional time as a function of m (not shown) suggest that significant improvements in computational time would be obtained with even greater values for m.

4.2   Evaluation on Real-World Datasets

Two real-world datasets are used for the evaluation of the performance of Poisson+MM and LSR+MM when compared to the reference execution MM with uniform initialization. The same datasets were adopted in [14] for the empirical evaluation of LSR. The number of items m, the number of rankings n, as well as the size of the greatest contest size $\bar{q}$ for each dataset, are given in Table 3.

The NASCAR[1] [9] dataset contains multiway partial rankings from 36 automobile races for the 2002 NASCAR season in the United States of America. Each of the races involved 43 drivers, with some drivers participating in all 36 races and some participating in only one. Altogether, 83 different drivers participated in at least one race [14].

The Sushi[2] dataset [10] contains multiway partial rankings provided by sushi consumers after performing a sensory test. It was designed so that the more frequently supplied types of sushi in restaurants were more frequently shown to respondents. The dataset comprises 5000 contests with sizes varying from 2 to 10 items.

---

[1] https://rdrr.io/cran/PLMIX/man/d_nascar.html.
[2] http://www.kamishima.net/sushi/.

Table 3. Datasets adopted in the empirical evaluation.

Dataset	m	n	$\overline{q}$
NASCAR	83	36	43
Sushi	100	5000	10

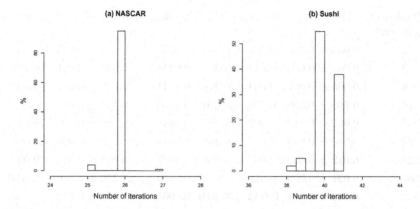

Fig. 2. Distribution of the resulting number of iterations required by MM for the inference of the parameters of the Plackett-Luce model for each datasets (a) NASCAR and (b) Sushi, considering 100 runs where each run is performed given random initializations for the parameters $\gamma^0{}_i \sim e^{[-2,2]}$ as input.

Initially, the sensitivity of MM to the initial values γ^0 is investigated. A total of 100 random realizations for γ^0 are generated as in (13) from the distribution $\gamma^0{}_i \sim e^{[-2,2]}$, and MM is executed given each random vector as input. Figure 2 illustrates the results for both datasets considered. The distribution of the number of iterations until convergence is highly concentrated at 26 and 40 respectively for the NASCAR and Sushi datasets. Similar conclusions concerning to the sensitivity to random initialization values were already pointed out in [9].

The effectiveness of each approximate algorithm on improving the convergence of MM are shown in Table 4. Both algorithms were able to reduce the number of iterations until convergence down to a value that was not achieved by the random variation of the parameters, as illustrated in the previous experiment. This represents the effectiveness of both Poisson and LSR to reduce the number of iterations required by MM for both datasets considered. The computational time obtained by the summation of LSR and the subsequent execution of MM leads to a value above the computational time of the adoption of MM with an uninformative input. The adoption of the proposed Poisson model, by the other side, leads to better computational times for both datasets.

Table 4. Number of iterations required by the reference MM method and computational time of MM summed with the computational time of an approximate inference method (Poisson+MM and LSR+MM).

Dataset	Method					
	MM		Poisson+MM		LSR+MM	
	MM iterations	Total time (s)	MM iterations	Total time (s)	MM iterations	Total time (s)
NASCAR	26	0.006	21	0.005	21	0.349
Sushi	40	0.410	32	0.349	32	2.557

4.3 Evaluation of the Proposed Method as an Approximate Inference Approach

In order to better understand how the proposed method compares to other approximate inference methods adopted for Plackett-Luce models and also to provide insights for future research, results obtained directly from both approximate inference methods adopted are evaluated and compared. The E_{RMS} values obtained from the evaluation of the results of the Poisson method were 0.342 relative to the Poisson method and 1.396 from LSR when the NASCAR dataset is considered. For the Sushi dataset $E_{RMS} = 0.567$ was obtained from the Poisson method and 0.931 from LSR. The values obtained from our implementation of LSR differ slightly from the lower values reported in [14] for the same datasets. The approximate algorithm GMM-F proposed in [20] is also evaluated as reported in [14], which achieved $E_{RMS} = 0.751$ for the NASCAR dataset and 0.130 for the Sushi dataset.

The computational time of the Poisson method in this evaluation is also lower when compared to LSR for both datasets considered. The Poisson method required less than 0.001 s for computing an approximate solution for the NASCAR dataset while LSR took 0.344 s. When the Sushi dataset is considered the computational times of the algorithms are 0.005 s and 2.221 s respectively from Poisson and LSR. Computational times reported for the GMM-F algorithm in [14] are 0.06 s and 0.19 s for the NASCAR and Sushi dataset respectively. Those results for GMM-F should be considered carefully, since the computational architecture used in [14] is not the same as the one adopted here.

The E_{RMS} results from the Poisson method could be further improved. The adoption of a standard reference number of opponents W is defined by an ad-hoc rule from a weighted average over a range of values. One can obtain from (9), under the assumptions made, the relation $\frac{W}{\bar{q}+1} = \frac{1}{3}$. Although this issue might be subject to further theoretical and empirical investigation, we should here illustrate some possible alternatives which could be explored. Suppose that W could be a parameter of the novel method such that $\frac{W}{\bar{q}+1} \in [0, 1]$ instead of the constant value $\frac{1}{3}$ which resulted from (8).

Figure 3 illustrates the result of the variation of W on the E_{RMS} of the results from the Poisson model in (6), when compared to the exact values of

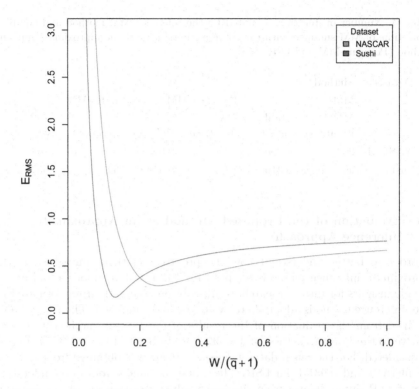

Fig. 3. E_{RMS} computed from the results of the Poisson approximate inference method (6) applied to two datasets from the literature, as a function of $\frac{W}{\overline{q}+1}$ with varying W.

the parameters, as obtained from MM. The value of $\frac{W}{\overline{q}+1}$ corresponding to the optimal E_{RMS} is actually close to the theoretical reference $\frac{1}{3}$ as predicted in (9), at least for the NASCAR dataset. The optimal $E_{RMS} = 0.291$ for the NASCAR dataset under the Poisson model corresponds to $\frac{W}{\overline{q}+1} = 0.26$, while the theoretical $\frac{W}{\overline{q}+1} = \frac{1}{3}$ leads to an $E_{RMS} = 0.342$. When the Sushi dataset is considered, a single numerical global optimum for the E_{RMS} of the Poisson model is also evident. However, the optimal $E_{RMS} = 0.166$ corresponds to a relation $\frac{W}{\overline{q}+1} = 0.12$, farther apart from the theoretically predicted value in (9), which leads to a much higher $E_{RMS} = 0.567$, as already shown.

The same evaluation is performed for the groups of randomly generated synthetic datasets. The average E_{RMS} was computed from the Poisson models resulting from 100 synthetic datasets built for each itemset size considered ($m \in \{2, 4, 8, 16, 32, 64, 128, 256, 512, 1024\}$). Average optimal E_{RMS} for each group resulted values ranging from 0.017 to 0.135 while the average corresponding $\frac{W}{\overline{q}+1}$ resulted values ranging from 0.14 to 0.17, also farther apart from the theoretically predicted value $\frac{1}{3}$. Figure 4 illustrates the results for a subset of the groups for better visualization. Similar curves arise from the remaining groups.

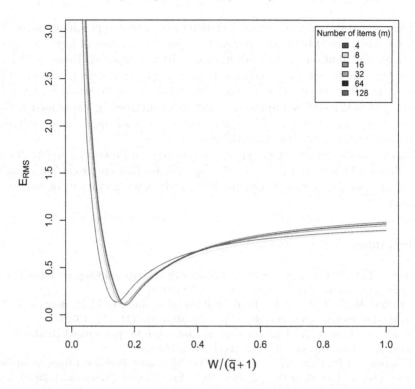

Fig. 4. Average E_{RMS} computed from the results of the Poisson approximate inference method (6) applied to synthetic datasets grouped by itemset size $m \in \{4, 8, 16, 32, 64, 128\}$, as a function of $\frac{W}{\bar{q}+1}$ with varying W.

5   Discussion and Conclusions

This work evaluated the adoption of two approximate methods for the estimation of initial values provided to the MM method when applied to the inference of parameters of the PL model. The candidate approaches are both compared to the adoption of the usual MM initialization which assumes that all items are equally strong. A novel approximate method which is based on the adoption of a simple Poisson probabilistic model was able to achieve statistically significant reduction in both the number of iterations required by MM and also in the computational time.

Although the Poisson model proposed was adopted for the approximate estimation of parameters of PL models, the assumptions made before the development of the model in (10) are somewhat stronger than IIA assumption made by PL. While the PL model assumes that the preference for an item a over b should not depend on other items $i \in S$, the model proposed in (10) assumes, in some sense, that the preference for an item a should not depend on the preference for any other item $i \in S$.

The computational time records relative to the other approximate algorithm adopted (LSR) could be much improved, majorly as a result of how exactly the stationary distribution of the transition matrix is computed. However, for the sake of this evaluation, this would not raise major concerns since the aim of the paper is not to compare LSR to the novel method but to verify whether the efficiency of MM could be improved, which was confirmed by experiments. Conceivably, the same results could be obtained using other approximate inference methods, which is to be further investigated.

Further work should better explain the theoretical properties and/or limitations that might arise from the model proposed. Further empirical investigation should also be performed, which might include a wider variety of real-world datasets.

References

1. Agarwal, A., Patil, P., Agarwal, S.: Accelerated spectral ranking. In: International Conference on Machine Learning, pp. 70–79 (2018)
2. Bradley, R.A., Terry, M.E.: Rank analysis of incomplete block designs: I. The method of paired comparisons. Biometrika **39**(3/4), 324–345 (1952)
3. Caron, F., Doucet, A.: Efficient Bayesian inference for generalized Bradley-Terry models. J. Comput. Graph. Stat. **21**(1), 174–196 (2012)
4. Dempster, A.P., Laird, N.M., Rubin, D.B.: Maximum likelihood from incomplete data via the EM algorithm. J. Roy. Stat. Soc. Ser. B (Methodol.) **39**(1), 1–22 (1977)
5. Elo, A.E.: The Rating of Chessplayers, Past & Present. Arco, New York (1978)
6. Guiver, J., Snelson, E.: Bayesian inference for Plackett-Luce ranking models. In: Proceedings of the 26th Annual International Conference on Machine Learning, pp. 377–384 (2009)
7. Hajek, B., Oh, S., Xu, J.: Minimax-optimal inference from partial rankings. In: Advances in Neural Information Processing Systems, pp. 1475–1483 (2014)
8. Hastie, T., Tibshirani, R.: Classification by pairwise coupling. Ann. Stat. **26**, 451–471 (1998)
9. Hunter, D.R., et al.: MM algorithms for generalized Bradley-Terry models. Ann. Stat. **32**(1), 384–406 (2004)
10. Kamishima, T., Akaho, S.: Efficient clustering for orders. In: Zighed, D.A., Tsumoto, S., Ras, Z.W., Hacid, H. (eds.) Mining Complex Data. Studies in Computational Intelligence, vol. 165, pp. 261–279. Springer, Heidelberg (2009). https://doi.org/10.1007/978-3-540-88067-7_15
11. Kumar, R., Tomkins, A., Vassilvitskii, S., Vee, E.: Inverting a steady-state. In: Proceedings of the Eighth ACM International Conference on Web Search and Data Mining, pp. 359–368 (2015)
12. Lange, K., Hunter, D.R., Yang, I.: Optimization transfer using surrogate objective functions. J. Comput. Graph. Stat. **9**(1), 1–20 (2000)
13. Luce, R.D.: Individual Choice Behavior: A Theoretical Analysis. Wiley, New York (1959)
14. Maystre, L., Grossglauser, M.: Fast and accurate inference of Plackett-Luce models. In: Advances in Neural Information Processing Systems, pp. 172–180 (2015)

15. McFadden, D., et al.: Conditional logit analysis of qualitative choice behavior. In: Zarembka, P. (ed.) Frotiers in Econometrics, pp. 105–142. Academic Press, Cambridge (1973)
16. Negahban, S., Oh, S., Shah, D.: Iterative ranking from pair-wise comparisons. In: Advances in Neural Information Processing Systems, pp. 2474–2482 (2012)
17. Negahban, S., Oh, S., Thekumparampil, K.K., Xu, J.: Learning from comparisons and choices. J. Mach. Learn. Res. **19**(1), 1478–1572 (2018)
18. Plackett, R.L.: The analysis of permutations. J. Roy. Stat. Soc. Ser. C (Appl. Stat.) **24**(2), 193–202 (1975)
19. Rao, P., Kupper, L.L.: Ties in paired-comparison experiments: a generalization of the Bradley-Terry model. J. Am. Stat. Assoc. **62**(317), 194–204 (1967)
20. Soufiani, H.A., Chen, W., Parkes, D.C., Xia, L.: Generalized method-of-moments for rank aggregation. In: Advances in Neural Information Processing Systems, pp. 2706–2714 (2013)
21. Thurstone, L.L.: The method of paired comparisons for social values. J. Abnorm. Soc. Psychol. **21**(4), 384 (1927)
22. Zermelo, E.: Die berechnung der turnier-ergebnisse als ein maximumproblem der wahrscheinlichkeitsrechnung. Math. Z. **29**(1), 436–460 (1929)

On the Multiple Possible Adaptive Mechanisms of the Continuous Ant Colony Optimization

Victor O. Costa[✉][iD] and Felipe M. Müller[iD]

Department of Applied Computing, Federal University of Santa Maria,
Santa Maria 97105-900, Brazil
vcosta.acad@gmail.com, felipe@inf.ufsm.br

Abstract. Among the existing techniques to improve the performance of metaheuristics in optimization problems, adaptive parameter control consists in varying one or more parameters of a given metaheuristic according to some indicator of the search conditions. This approach allows metaheuristics to change algorithmic behaviour during the search, and is particularly relevant for the optimization of dynamic problems. In this research we theoretically analyse in which ways the parameters of the ant colony optimization for continuous domains metaheuristic can be adapted, regarding how each parameter influences exploration and exploitation characteristics of the algorithm. Our experimental contributions include validating the colony success rate as a search condition estimator and choosing suitable maps from this estimator to the parameters q and ξ of the algorithm. Beyond that, we compare the performances of three proposed adaptive versions of the base metaheuristic and show the benefits of simultaneously adapting multiple parameters.

Keywords: Parameter adaption · Ant colony optimization · Continuous domains

1 Introduction

Optimization problems are often characterized by infeasibility in the computation of exact solutions. For some of these problems, often referred to as NP-hard [19], the existence of algorithms capable of computing exact solutions in polynomial time is unknown. Metaheuristic algorithms provide general search methodologies [3] and have been widely used for hard optimization problems without requiring deep knowledge about the function under optimization [10].

Although metaheuristics are considered to be general, their behaviours are often governed by some set of parameters [7,9]. The choice of values for these parameters can be accomplished by means of parameter tuning and/or control [6]. In tuning, parameters are chosen according to performance measures in a set of problem instances and remain static over new searches, while in control the adjustment of parameters happens over time according to some rule

© Springer Nature Switzerland AG 2020
R. Cerri and R. C. Prati (Eds.): BRACIS 2020, LNAI 12319, pp. 166–178, 2020.
https://doi.org/10.1007/978-3-030-61377-8_12

or evolutionary mechanism. Regarding control, [20] emphasizes that updating parameters online, depending on the search stage, may improve the robustness of metaheuristics. Parameter control methods can be classified as *prescheduled* or *adaptive* [20] depending on whether they respond to the course of time or to the search conditions, respectively. Adaptive parameter control is specially useful when the function under optimization dynamically alters its fitness landscape [11].

Ant colony optimization (ACO) is a well-established family of metaheuristics inspired by the foraging behaviour of ants [20] which guides the search for near-optimal solutions probabilistically, taking advantage of the objective function landscape captured in pheromone representation. Although initially proposed in the context of categorical optimization, ACO was later adapted to continuous domains (ACO$_\mathbb{R}$) in [19]. ACO$_\mathbb{R}$ keeps track of the search history via a solution archive which stores the most promising solutions found during the search.

In this research, we investigate the multiple ways in which the parameters of the ACO$_\mathbb{R}$ metaheuristic can be adapted during the search for solutions with minimum cost. As a study case, we experiment using an evolutionary state estimator (ESE) drawn from swarm intelligence literature [15] to adapt parameters of ACO$_\mathbb{R}$, and show the benefits of adapting multiple parameters. While simultaneous adaption of more than one parameter is present in the literature of ACO algorithms for categorical optimization, e.g. [4,12], we present the first results for this task using ACO$_\mathbb{R}$ to the best of our knowledge. All source code used to collect the results reported in this paper are publicly available at a Git repository [5] under the GNU GPL v3 license.

The rest of this paper is organized as follows. Section 2 reviews the functioning of ACO$_\mathbb{R}$. Section 3 presents a theoretical analysis of in which ways this same metaheuristic can be adapted to the search conditions, along with a review of the related literature. In Sect. 4 we propose the use of a success rate strategy to estimate search conditions, and define three adaptive versions of ACO$_\mathbb{R}$ based on this strategy. The description of experiments and discussion of results is given in Sect. 5. Finally, Sect. 6 concludes the paper with a summary of the conducted research, along with the description of some possible future directions.

2   Ant Colony Optimization for Continuous Domains

In the optimization process of ACO$_\mathbb{R}$ [19], the entire solution archive is initialized randomly and the cost function of each random solution is evaluated. The solutions are then sorted in ascending order of cost, in such a way that the best solutions are located in the first positions of the archive, and each of the k solutions in the archive is associated with a weight w_j according to (1).

$$w_j = \frac{1}{qk\sqrt{2\pi}} e^{\dfrac{-[rank(j) - 1]^2}{2q^2k^2}}, \tag{1}$$

in which $rank(j)$ is the archive position, and q is a parameter of the algorithm. Hence, most promising solutions of the archive have the highest weight.

A central concept to $ACO_\mathbb{R}$ is that the archive ultimately represents a multidimensional Gaussian kernel, and each of its solutions define the center of a multidimensional Gaussian function [19]. The realization of the kernel takes place when, for every iteration, each of the m ants in the population chooses a solution l from the archive with probability defined in (2) to use as the center, μ, of the Gaussian probability density function (PDF) defined in (3).

$$p_l = \frac{\omega_l}{\sum_{j=1}^{k} \omega_j} \tag{2}$$

The standard deviation of this Gaussian function is described in (4) as the average absolute distance from the chosen solution to the entire archive, multiplied by a parameter ξ. Each of the ants then uses its own PDF to sample a new solution, which has its cost evaluated and is appended to the archive. Finally, the archive is sorted and the worst m solutions are removed at the end of each iteration, such that the archive remains with the best k solutions.

$$g_i(x; \mu_i, \sigma_i) = \frac{1}{\sigma_i \sqrt{2\pi}} e^{-\frac{(x - \mu_i)^2}{2\sigma_i^2}} \tag{3}$$

$$\sigma_i = \xi \sum_{j=1}^{k} \frac{|s_j^i - \mu_i|}{k - 1}, \tag{4}$$

where s_j^i is the value of the ith continuous variable of the jth solution in the archive. The random variable x is the domain of the PDF.

3   Possible Parameter Adaptions for $ACO_\mathbb{R}$

To change the parameters of metaheuristics dynamically during the search, a measure of the search conditions must be defined (i.e. an ESE), and then mapped to parameter values in such a way that it improves exploration and exploitation capabilities of the metaheuristic (also referred as diversification and intensification, respectively). To this end, we now analyse how each parameter of $ACO_\mathbb{R}$ affects the search and which adaption opportunities these parameters offer.

As described in Sect. 2, each of the ants must select a solution from the archive as the center of a Gaussian distribution prior to sampling a new solution. Considering (1) and (2), when q approaches zero ants have a strong tendency to select top-ranked solutions as their centers, while the selection probabilities are less biased for higher values of this parameter [19]. Therefore we can say that low values of q favor elitism and exploitation of promising search regions, while high values favor exploration of alternative regions.

In [22], authors modify $ACO_\mathbb{R}$ for multimodal optimization problems using niching methods. For each niche, the ESE is a function of maximum and minimum fitness values of each niche and also of the entire archive. A different value of q is used for each niche, which is defined by an exponential function of the

ESE. The proposed algorithm is then enhanced with mutation and local search operators and evaluated against state-of-the-art multimodal optimization algorithms, evidencing the superiority of the proposed method to handle multimodal problems.

The dispersion of the PDF used by an ant to sample a solution is defined in (4) by taking into account the archived solutions and a parameter ξ. High dispersion values lead to high diversity in the search by increasing the probability of newly generated solutions to be far from the centers of the PDFs from which they were sampled. As the dispersion is directly proportional to ξ, the higher the value for this parameter, the slower the convergence speed of the algorithm, since the solution generation process will be less biased towards regions already present in the archive. In ACO literature, this parameter is related to the pheromone evaporation rate [19].

Instead of only assessing the search conditions to guide the adaption of the parameter q, [17] computes the population diversity measure defined in [18] and imposes a linear decay for this measure. The parameter is then adapted in the attempt to ensure that the diversity measure follows the scheduled decay. The value of q is stochastically chosen, being sampled from a distribution with mean defined by the adaption mechanism. Meanwhile, this work assigns a completely random variable to the parameter ξ.

In [1], authors use an array of 14 predefined values for ξ. Each of these values has an associated probability of being selected to influence the solution generation mechanism, and this probability is defined by the past success of each of the values in sampling of good solutions.

Since in $ACO_\mathbb{R}$ each solution present in the archive defines the center of a different Gaussian function, the archive size k determines the complexity of the resulting Gaussian kernel PDF [19]. Large archive sizes favor exploration for allowing less promising solutions to be memorized, when compared with a small archive. Beyond that, when k is increased the elitism of the center selection procedure is reduced according to (1). To take advantage of this relationship, [13] proposes a growing solution archive. The growing mechanism by itself would be characterized as prescheduled parameter control, but after a certain number of iterations with relatively low fitness improvement the archive is restarted to its initial size. By doing this, the algorithm adapts its archive size to the search conditions.

When the optimization problem is constrained with a fixed budget of time or number of objective function evaluations, the population size parameter m determines how this budget is used over time in the search. When compared to large populations, small populations lead to a greater number of iterations for the same budget. In the most extreme case of a population with a single ant, the archive is updated to each new solution generated. With a large population, a number solutions from the same non-updated archive will be selected at every iteration, and therefore we reason that the use of large population sizes results in greater exploration capabilities. This is the only parameter of $ACO_\mathbb{R}$ parameter which has not been adapted in the literature, and therefore this is an open research topic to the best of our knowledge.

4   Colony Success Rate

The search conditions estimator used in this research is drawn from the idea of success rate, a concept which has already been explored to guide the adaption of evolutionary [2] and swarm-based metaheuristics [15]. When, at iteration t, an an ant selects a solution s_l from the archive with probability as in (2) and uses it to sample a new solution with the PDF defined in (3), the success of the given ant depends on whether the newly sampled solution improved over the center of the PDF from which it was sampled. This description of the ant's success, su_{ant}, is also expressed in (5).

$$su_{ant}(t) = \begin{cases} 1 \text{ if } cost(s_{ant}, t) < cost(s_l, t), \\ 0 \qquad \qquad \text{otherwise.} \end{cases} \tag{5}$$

After the success of every solution generated by the population at a given iteration is evaluated, the colony success rate (CSR) is computed as the percentage of solutions which improved over their respective selected solutions, as in (6).

$$sr_{pop}(t) = \frac{\sum_{i=1}^{m} su_i(t)}{m} \tag{6}$$

A high success rate indicates that the selected solutions are probably closer to local optima regions than to some global optimum, since in the first case it would be easier for ants to improve over the selected solutions. Beyond the perspective it provides about the search landscape, the CSR is easy to compute and restricted to the range $0 \leq sr_{pop}(t) \leq 1$. These characteristics make it a promising estimator.

Once an ESE is defined, it is mapped to the values of parameters under adaption using algebraic functions, fuzzy-based or entropy-based control [23]. In this research we explore linear, exponential and sigmoid real-valued functions to map the CSR to parameter values. The expressions for these functions and the definition of their constants (concerning maximum and minimum values of a given parameter when the domain is constrained to the $[0, 1]$ interval) are displayed in Table 1.

With the CSR as search condition estimator, we define three adaptive versions of $ACO_{\mathbb{R}}$ depending on which parameter is dynamically controlled:

- $AELACO_{\mathbb{R}}$: Adaptive elitism level $ACO_{\mathbb{R}}$, which maps sr_{pop} to q.
- $AGDACO_{\mathbb{R}}$: Adaptive generation dispersion $ACO_{\mathbb{R}}$, which maps $1 - sr_{pop}$ to ξ.
- $BAACO_{\mathbb{R}}$: Bi-adaptive $ACO_{\mathbb{R}}$, which applies adaptions of both $AELACO_{\mathbb{R}}$ and $AGDACO_{\mathbb{R}}$.

Table 1. Functions evaluated in the map from ESE to parameter values.

Name	Expression	Constants
Linear	$p(x) = Ax + B$	$A = p_{max} - p_{min}$
		$B = p_{min}$
Exponential	$p(x) = Ae^{Bx}$	$A = p_{min}$
		$B = \ln\left(\dfrac{p_{max}}{p_{min}}\right)$
Sigmoid	$p(x) = \dfrac{2}{1 + Qe^{-Bx}}$	$Q = \dfrac{2 - p_{min}}{p_{min}}$
		$B = \ln\left(\dfrac{p_{max}}{2 - p_{max}} \cdot \dfrac{2 - p_{min}}{p_{min}}\right)$

The rest of this paper is dedicated to evaluate how well the functions in Table 1 map the ESEs to parameter values, and to assess the performance of each proposed modification. In this way, to the best of our knowledge, we conduct the first research of the literature with simultaneous adaption of multiple parameters in $ACO_{\mathbb{R}}$.

5   Experimental Study

5.1   Settings and Description of Experiments

All experiments detailed in this section were executed using Python 3.6.8 with the NumPy module [16] for vector operations, and the DEAP module [8] for benchmark functions. The the environment is composed of an octa-core Intel Xeon ®:E-5405 processor, an 8 GiB RAM, running the Ubuntu 18.04.3 operating system. Because the algorithms under analysis are stochastic, we run each experiment 100 times and report the resulting statistics.

To select which functions from Table 1 will be used to map the ESE to parameter values in $AELACO_{\mathbb{R}}$ and $AGDACO_{\mathbb{R}}$, we split the synthetic objective functions of Table 2 in train and test partitions. Train functions are used to evaluate the adequacy of each map for each version of the algorithm, while test functions are used to evaluate the performance of the proposed metaheuristics.

In the performance evaluation, we take into account the basic version of $ACO_{\mathbb{R}}$, all three proposed modifications and adaptive versions of the particle swarm optimization (PSO) and simulated annealing (SA) metaheuristics from the literature. The adaptive inertia weight particle swarm optimization (AIW-PSO) algorithm [15] relies on the swarm success rate as ESE, which is mapped linearly to the values of the inertia weight parameter. Beyond the inertia weight, AIWPSO is also governed by its population size (m) and by the weights that govern the trade-off between social and individual behaviour (c_g and c_p). In [14], authors propose a version of SA in which the perturbations used to generate new solutions are random variables following the Bates distribution, and the adaption takes place by modifying the dispersion of the distribution though the

Table 2. Objective functions used in this research, with minimization as objective and global optima at $f(x) = 0$.

Partition	Function	d	Range	Expression
Train	(1) Ackley	3	$[-15, 30]^d$	$f_1(\mathbf{x}) = 20 - 20e^{-0.2\sqrt{\frac{1}{d}\sum_{i=1}^d x_i^2}}$ $+e - e^{\frac{1}{d}\sum_{i=1}^d cos(2\pi x_i)}$
	(2) Griewank	3	$[-600, 600]^d$	$f_2(\mathbf{x}) = \frac{1}{4000}\sum_{i=1}^d x_i^2$ $- \prod_{i=1}^d cos\left(\frac{x_i}{\sqrt{i}}\right) + 1$
	(3) Rosenbrock	3	Unbounded	$f_3(\mathbf{x}) = \sum_{i=1}^{d-1} 100(x_i^2 - x_{i+1})^2$ $+(1 - x_i)^2$
	(4) Schwefel	3	$[-500, 500]^d$	$f_4(\mathbf{x}) = \alpha \cdot d - \sum_{i=1}^d x_i sin(\sqrt{|x_i|})*$ $*\alpha = 418.9828872724339$
Test	(5) Bohachevsky	3	$[-100, 100]^d$	$f_5(\mathbf{x}) = \sum_{i=1}^{d-1} 0.7 + x_i^2 + 2x_{i+1}^2$ $-0.3cos(3\pi x_i) - 0.4cos(4\pi x i_{i+1})$
	(6) Cigar	3	Unbounded	$f_6(\mathbf{x}) = x_0^2 + 10^6 \sum_{i=1}^d x_i^2$
	(7) Himmelblau	2	$[-6, 6]^d$	$f_7(\mathbf{x}) = (x_1^2 + x_2 - 11)^2$ $+(x_1 + x_2^2 - 7)^2$
	(8) Rastrigin	3	$[-5.12, 5.12]^d$	$f_8(\mathbf{x}) = 10d$ $+ \sum_{i=1}^d x_i^2 - 10cos(2\pi x_i)$
	(9) Schaffer	3	$[-100, 100]^d$	$f_9(\mathbf{x}) = \sum_{i=1}^{d-1}(x_i^2 + x_{i+1}^2)^{0.25}$ $\cdot[sin^2(50 \cdot (x_i^2 + x_{i+1}^2)^{0.10}) + 1]$
	(10) Sphere	3	Unbounded	$f_{10}(\mathbf{x}) = \sum_{i=1}^d x_i^2$

so-called crystallization factor. Since in the proposal each dimension has its own crystallization factor, the search happens in an anisotropic manner. The other parameters that influence this metaheuristic are the number of local iterations ($local_{itr}$), the initial temperature (T_0) and the cooling constant (η). The proposed SA modification is hereinafter called the adaptive crystallization factor simulated annealing (ACFSA). Here we implement only the mechanisms proposed to adapt parameters of the original algorithms, and leave aside additional mechanisms unrelated to parameter adaption.

To avoid offline parameter tuning, we employed values already present in the literature for the constant parameters of all algorithms, and also for minimum and maximum values of the adaptive parameters. These values are displayed in Table 3.

Table 3. Parameters used for each metaheuristic.

$ACO_\mathbb{R}$	$AELACO_\mathbb{R}$	$AGDACO_\mathbb{R}$	$BAACO_\mathbb{R}$	AIWPSO	ACFSA
$m = 10$	$m = 10$	$m = 10$	$m = 10$	$m = 20$	$local_{itr} = 100$
$k = 50$	$k = 50$	$k = 50$	$k = 50$	$w \in [0,1]$	$\eta = 0.99$
$q = 0.01$	$q \in [0.01, 1]$	$q = 0.01$	$q \in [0.01, 1]$	$c_g = 2$	$T_0 = 50$
$\xi = 0.85$	$\xi = 0.85$	$\xi \in [0.1, 0.85]$	$\xi \in [0.1, 0.85]$	$c_p = 2$	

5.2 Results and Discussion

As mentioned earlier, here we compare the performances of three different functions to map the CSR estimator to the values of q (in $AELACO_\mathbb{R}$ and $BAACO_\mathbb{R}$) and ξ (in $AGDACO_\mathbb{R}$ and $BAACO_\mathbb{R}$).

Regarding the performance of each function from Table 1, the average costs reported in Tables 4 and 5 for a total of 5E3 objective function evaluations on the train partition evidence that nonlinear maps prevail over linear ones for the proposed algorithms. More specifically, the exponential function was the best map to parameter q for all functions, while the sigmoid was the best map to ξ for most of the functions, in the context of $AELACO_\mathbb{R}$ and $AGDACO_\mathbb{R}$, respectively. We note, however, that the exponential and sigmoid functions have only minor performance differences in most cases. Henceforth, the proposed parameter adaptions are always associated with the forementioned maps.

Table 4. Average costs of $AELACO_\mathbb{R}$ for each map from ESE to q.

Function	Linear	Exponential	Sigmoid
1	7.919E−14	0.000E+00	3.553E−17
2	5.112E−02	4.470E−02	4.486E−02
3	3.200E−01	2.238E−01	2.793E−01
4	2.458E+02	2.232E+02	2.323E+02

Table 5. Average costs of $AGDACO_\mathbb{R}$ for each map from ESE to ξ.

Function	Linear	Exponential	Sigmoid
1	2.120E−02	3.908E−16	3.553E−16
2	4.427E−02	4.019E−02	3.815E−02
3	5.958E−01	2.747E−01	3.453E−01
4	1.669E+02	1.509E+02	1.363E+02

To evaluate how the proposed algorithms perform against the basic non-adaptive $ACO_\mathbb{R}$ and other adaptive algorithms, we collect results of $ACO_\mathbb{R}$,

AELACO$_\mathbb{R}$, AGDACO$_\mathbb{R}$, BAACO$_\mathbb{R}$, AIWPSO and ACFSA in the test partition of objective functions, considering a budget of $1e5$ function evaluations. With BAACO$_\mathbb{R}$ we use the best performing map for each parameter, i.e. the exponential map for q and the sigmoidal one for ξ. The average cost values shown in Table 6 evidence the advantages of the proposed adaptions. We note that, for most functions, the basic ACO$_\mathbb{R}$ already outperforms the adaptive versions of other metaheuristics in terms of average minimum cost, but it is mostly outperformed by its adaptive versions. Apart from instances 5, 6 and 10, for which multiple algorithms reach perfect convergence, BAACO$_\mathbb{R}$ has the lowest average cost when compared to all other metaheuristics. The corresponding standard deviations in Table 7 show that the dispersion of cost values is similar among ant-based algorithms. Nevertheless, AGDACO$_\mathbb{R}$ and BAACO$_\mathbb{R}$ have zero standard deviation for an instance in which ACO$_\mathbb{R}$ does not. When considering all metaheuristics, AIWPSO has the lowest standard deviation for instances 8 and 9, but its average cost for these instances is relatively high. Therefore, the low standard deviation values of AIWPSO for these instances cannot be considered evidence of superior robustness.

Table 6. Average cost of metaheuristics for each test function.

Func.	ACO$_\mathbb{R}$	AELACO$_\mathbb{R}$	AGDACO$_\mathbb{R}$	BAACO$_\mathbb{R}$	AIWPSO	ACFSA
5	1.239E−02	4.129E−03	0.000E+00	0.000E+00	0.000E+00	7.572E−01
6	0.000E+00	0.000E+00	0.000E+00	0.000E+00	0.000E+00	1.038E−05
7	3.313E−31	2.367E−31	2.445E−31	1.578E−31	1.642E−19	1.059E−08
8	1.363E+00	1.522E+00	1.234E+00	1.094E+00	2.896E+00	2.882E+01
9	2.676E−03	2.490E−03	1.322E−03	1.215E−03	5.624E+00	1.695E+01
10	0.000E+00	0.000E+00	0.000E+00	0.000E+00	0.000E+00	2.629E−07

Table 7. Standard deviations of cost values found by metaheuristics for each test function.

Func.	ACO$_\mathbb{R}$	AELACO$_\mathbb{R}$	AGDACO$_\mathbb{R}$	BAACO$_\mathbb{R}$	AIWPSO	ACFSA
5	7.044E−02	4.109E−02	0.000E+00	0.000E+00	0.000E+00	6.473E−01
6	0.000E+00	0.000E+00	0.000E+00	0.000E+00	0.000E+00	1.017E−−04
7	3.893E−31	3.615E−31	3.648E−−31	3.155E−31	1.634E−18	1.048E−07
8	1.031E+00	1.505E+00	1.016E+00	9.999E−01	4.768E−01	1.464E+01
9	5.696E−03	5.846E−03	3.594E−03	3.211E−03	1.776E−15	3.377E+00
10	0.000E+00	0.000E+00	0.000E+00	0.000E+00	0.000E+00	2.615E−06

To analyse the statistical significance of the wins and losses of the proposed adaptive metaheuristics against the basic ACO$_\mathbb{R}$, we use the Wilcoxon signed-rank test for paired data [21] with the costs present in Table 6. Table 8 displays the p-values between each proposed modification and $ACO_\mathbb{R}$ for the set of test functions. Although the average costs of $AELACO_\mathbb{R}$ and $AGDACO_\mathbb{R}$ are lower when compared to the basic algorithm in many functions, most of these wins are of little statistical significance. On the other side, $BAACO_\mathbb{R}$ has the best overall average cost among test functions and is characterized by the most significant wins. If we consider the common confidence threshold of 95%, $BAACO_\mathbb{R}$ significantly outperforms $ACO_\mathbb{R}$ in instances 7 and 9, while all other algorithms have performance similar to $ACO_\mathbb{R}$. The performance of $BAACO_\mathbb{R}$ is a consequence of how the simultaneous adaption of parameters creates a balance between exploration and exploitation during the search. For high values of the CSR, when promising solutions are most probably in local minima regions, the parameters q and ξ of $BAACO_\mathbb{R}$ are expected to have high and low values, respectively. In this setting, due to q ants are not elitist and have a similar probability of selecting any solution from the archive to define the centers of the distributions from which they will sample new solutions, but because of ξ this distribution has low dispersion, i.e. the sampled solutions are likely to be close to the centers of the distributions. This results in exploratory behaviour through exploitation of regions which are not considered promising. When the success rate is low, higher quality solutions are preferably chosen, but the dispersion of the sampling is also high, which results in the conservation of solution diversity even when exploiting top-ranked regions.

Table 8. Statistical significance of adaptive versions against $ACO_\mathbb{R}$. Green-colored boxes represent wins against the base algorithm, while orange boxes indicate losses.

Function	$AELACO_\mathbb{R}$	$AGDACO_\mathbb{R}$	$BAACO_\mathbb{R}$
5	0.5775	0.1025	0.1025
6	–	–	–
7	0.0641	0.0934	0.0005
8	0.6675	0.3437	0.0724
9	0.7991	0.0597	0.0167
10	–	–	–

While Table 6 displays the performances of algorithms after a large number of iterations, Fig. 1 offers a perspective of how each algorithm behaves over time. Each of the plotted curves is the average cost history considering 100 runs, with a granularity of 100 function evaluations. Among all presented algorithms, AIWPSO is the one which converges faster to good search regions, but in later search stages it is mostly outperformed by $ACO_\mathbb{R}$-based algorithms. On the other side, the curves for ACFSA were not displayed for most of the functions due to the late convergence of this algorithm. When comparing $ACO_\mathbb{R}$-based algorithms, there is no clear winner in terms of convergence speed.

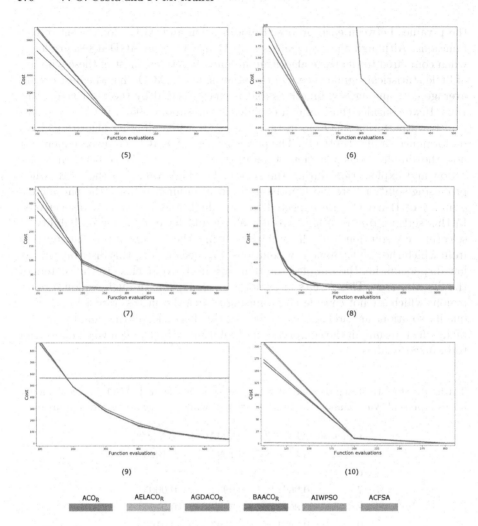

Fig. 1. Cost history of each metaheuristic for each test function, considering the average costs over 100 runs of the experiments.

6   Conclusion

Based on the parameter control literature, we have analysed in which ways the parameters of the $ACO_\mathbb{R}$ metaheuristic can be adapted according to a search condition estimator. The conducted experimental study shows that the colony success rate is an adequate estimator of the search conditions, leading to algorithms with good performance when mapped by nonlinear functions to the parameters under adaption. Beyond that, our experiments evidence the benefits of combining multiple adaptions when their effects on diversification and intensification behaviours are known and well explored, which resulted in the promising performance of BAACO$_\mathbb{R}$. An obvious future work direction is to evaluate BAACO$_\mathbb{R}$ against a larger pool of adaptive metaheuristics for a greater number of objective functions. With the benefits of the simultaneous adaption of multiple ACO$_\mathbb{R}$ parameters, applying this approach with sets of parameters not considered here is a promising research field. We also identify that the effectiveness of adapting of the population size parameter in ACO$_\mathbb{R}$ is as an open research question.

References

1. Abdelbar, A.M., Salama, K.M.: Parameter self-adaptation in an ant colony algorithm for continuous optimization. IEEE Access **7**, 18464–18479 (2019)
2. Bartz-Beielstein, T.: Evolution strategies and threshold selection. In: Blesa, M.J., Blum, C., Roli, A., Sampels, M. (eds.) HM 2005. LNCS, vol. 3636, pp. 104–115. Springer, Heidelberg (2005). https://doi.org/10.1007/11546245_10
3. Boussaïd, I., Lepagnot, J., Siarry, P.: A survey on optimization metaheuristics. Inf. Sci. **237**, 82–117 (2013)
4. Chusanapiputt, S., Nualhong, D., Jantarang, S., Phoomvuthisarn, S.: Selective self-adaptive approach to ant system for solving unit commitment problem. In: Proceedings of the 8th Annual Conference on Genetic and Evolutionary Computation, pp. 1729–1736 (2006)
5. Costa, V.O.: Bank of metaheuristics (2020). https://github.com/vctrop/bank_of_metaheuristics/tree/BRACIS2020
6. Eiben, Á.E., Hinterding, R., Michalewicz, Z.: Parameter control in evolutionary algorithms. IEEE Trans. Evol. Comput. **3**(2), 124–141 (1999)
7. Favaretto, D., Moretti, E., Pellegrini, P.: On the explorative behavior of MAX–MIN ant system. In: Stützle, T., Birattari, M., Hoos, H.H. (eds.) SLS 2009. LNCS, vol. 5752, pp. 115–119. Springer, Heidelberg (2009). https://doi.org/10.1007/978-3-642-03751-1_10
8. Fortin, F.A., De Rainville, F.M., Gardner, M.A., Parizeau, M., Gagné, C.: DEAP: evolutionary algorithms made easy. J. Mach. Learn. Res. **13**, 2171–2175 (2012)
9. Hoos, H.H.: Automated algorithm configuration and parameter tuning. In: Hamadi, Y., Monfroy, E., Saubion, F. (eds.) Autonomous Search, pp. 37–71. Springer, Heidelberg (2011). https://doi.org/10.1007/978-3-642-21434-9_3
10. Huang, C., Li, Y., Yao, X.: A survey of automatic parameter tuning methods for metaheuristics. IEEE Trans. Evol. Comput. **24**, 201–216 (2019)
11. Karafotias, G., Hoogendoorn, M., Eiben, Á.E.: Parameter control in evolutionary algorithms: trends and challenges. IEEE Trans. Evol. Comput. **19**(2), 167–187 (2014)
12. Li, Z., Wang, Y., Yu, J., Zhang, Y., Li, X.: A novel cloud-based fuzzy self-adaptive ant colony system. In: 2008 Fourth International Conference on Natural Computation, vol. 7, pp. 460–465. IEEE (2008)
13. Liao, T., Montes de Oca, M.A., Aydin, D., Stützle, T., Dorigo, M.: An incremental ant colony algorithm with local search for continuous optimization. In: Proceedings of the 13th Annual Conference on Genetic and Evolutionary Computation, pp. 125–132 (2011)
14. Martins, T., Sato, A., Tsuzuki, M.: Adaptive neighborhood heuristics for simulated annealing over continuous variables. INTECH Open Access Publisher (2012)
15. Nickabadi, A., Ebadzadeh, M.M., Safabakhsh, R.: A novel particle swarm optimization algorithm with adaptive inertia weight. Appl. Soft Comput. **11**(4), 3658–3670 (2011)

16. Oliphant, T.E.: A Guide to NumPy, vol. 1. Trelgol Publishing, USA (2006)
17. Omran, M., Polakova, R.: A memetic and adaptive continuous ant colony optimization algorithm. In: Aliev, R.A., Kacprzyk, J., Pedrycz, W., Jamshidi, M., Babanli, M.B., Sadikoglu, F.M. (eds.) ICSCCW 2019. AISC, vol. 1095, pp. 158–166. Springer, Cham (2020). https://doi.org/10.1007/978-3-030-35249-3_20
18. Poláková, R., Tvrdík, J., Bujok, P.: Adaptation of population size according to current population diversity in differential evolution. In: 2017 IEEE Symposium Series on Computational Intelligence (SSCI), pp. 1–8. IEEE (2017)
19. Socha, K., Dorigo, M.: Ant colony optimization for continuous domains. Eur. J. Oper. Res. **185**(3), 1155–1173 (2008)
20. Stützle, T., et al.: Parameter adaptation in ant colony optimization. In: Hamadi, Y., Monfroy, E., Saubion, F. (eds.) Autonomous Search, pp. 191–215. Springer, Heidelberg (2011). https://doi.org/10.1007/978-3-642-21434-9_8
21. Wilcoxon, F., Katti, S., Wilcox, R.A.: Critical values and probability levels for the wilcoxon rank sum test and the wilcoxon signed rank test. Sel. Tables Math. Stat. **1**, 171–259 (1970)
22. Yang, Q., et al.: Adaptive multimodal continuous ant colony optimization. IEEE Trans. Evol. Comput. **21**(2), 191–205 (2016)
23. Zhang, J., et al.: A survey on algorithm adaptation in evolutionary computation. Front. Electr. Electron. Eng. **7**(1), 16–31 (2012)

A Differential Evolution Algorithm for Contrast Optimization

Artur Leandro da Costa Oliveira[1] and André Britto[1,2(✉)]

[1] Post-Graduation Program in Computer Science,
Federal University of Sergipe, Sergipe, Brazil
arturleandrodacosta@gmail.com
[2] Departament of Computing, Federal University of Sergipe, Sergipe, Brazil
andre@dcomp.ufs.br

Abstract. Image Enhancement is one of the most important phases of the image processing system. Contrast Enhancement plays a key role in this step. Histogram Equalization (HE) is one of the main tools used to improve the contrast of an image. However, the use of HE causes an increase in the natural brightness of the image, which is not desirable in many types of applications such as consumer electronics products. To solve these limitations, it is proposed in this paper a variation of the Differential Evolution metaheuristic algorithm for Contrast Optimization called DECO. The results obtained were statistically compared with other techniques and metaheuristic algorithms. The results showed that DECO is competitive compared with other techniques.

Keywords: Differential Evolution · Contrast Optimization

1 Introduction

Despite the evolution of photo camera sensors and processing algorithms in capturing the image, it is still a challenge to obtain images that present a good contrast quality. Image pre-processing is one of the key steps in the image processing area. Its main function is to transform the image obtained to present an improvement for the subsequent processing step, such as detection and identification [7].

Image enhancement is a technique widely used in this step. The main objective of image enhancement is to make changes to the image's attributes so that its use in a given task is more appropriate for a specific observer [11]. For example, image enhancement is essential in medical diagnostics, or iris and fingerprint enhancement systems for biometric recognition systems. It is usually done by increasing the contrast or by suppressing the noise [11].

A wide number of algorithms are used to enhance the image quality. Histogram Equalization (HE) is considered one of the most common techniques used to enhance the contrast of grayscale images [7]. Its goal is to use an evenly distributed histogram with a cumulative density function [3]. The HE can cause

© Springer Nature Switzerland AG 2020
R. Cerri and R. C. Prati (Eds.): BRACIS 2020, LNAI 12319, pp. 179–194, 2020.
https://doi.org/10.1007/978-3-030-61377-8_13

problems related to the image brightness level, mainly in consumer electronics applications. In this case, preserving the brightness is essential for the success of the application [3]. Different methodologies based in HE are found to fix the problem of preservation of brightness. However, these methods fail to produce, with low cost, natural quality images. CLAHE [16] (Contrast Limited Adaptive Histogram Equalization) is a well known adaptive method of HE in the literature. Its improvement allows it to add restrictions to the HE and preserve certain characteristics of the image.

In recent years, with the emergence of several quality metrics for improving images, it has been possible to use metaheuristic algorithms to perform this step. Contrast optimization aims to improve image contrast by defining it as a restricted nonlinear optimization problem. Several works have been developed for this purpose. Evolutionary Genetic Algorithms [9,13,18], Particle Swarm Optimization (PSO) [1] and Ant Colony Optimization (ACO) [10] have good results for the problem. In [9] a new representation of the solution widely used in Contrast Optimization is presented. The genetic algorithm proposed obtains excellent results in improving the contrast of images. However, the defined objective function can be further improved to maximize its results. The algorithm performance is questionable concerning the execution time, because few iterations were used in its tests. They also do not implement statistical testing procedures, which make it difficult to compare the divergent results of experiments between researchers.

This paper presents a metaheuristic algorithm implementation called DECO (Differential Evolution for Contrast Optimization). Its implementation has singularities for the contrast improvement problem, in addition, an objective function with additional parameters is used, based on in [14] that improves the results of the final images. The main contribution of contrast optimization algorithm DECO is related to your high convergence speed for the better solution, besides its simple structure, versatility, and robustness. Our implementation is based on Differential Evolution (DE) developed by Storn and Price [20] which is one of the most superior evolutionary algorithms.

In addition, DECO is evaluated through a comparative analysis between standard methods of contrast improvement and Contrast Optimization techniques. In those set of experiments, a statistical comparison based on [5] is performed between different algorithms. The performance of the algorithms is analyzed using three different criteria. The adapted fitness function proposed by [9], the *Peak Signal-to-Noise Ratio* (PSNR) image quality criterion, and the time that the algorithm took to find the best solution. The results found demonstrate that DECO achieves positive results when solving the problem of improving image quality.

The remainder of the paper is organized in: Sect. 2 details the problem to be addressed. Section 3 presents the algorithm and its specifications, while the experiments and results are provided in Sect. 4. The conclusion is reported in Sect. 5.

2   Background

2.1   Problem Formulation

Exists two categories of approaches for solving the problem of image enhancement: filtering techniques and contrast enhancement methods. Filtering consists of replacing the gray level of a given pixel with another calculated through the neighborhood. The contrast enhancement is the operation that maps the gray levels of the image creating a new set of gray levels distributing more homogeneously. The contrast enhancement can be divided into two types: global or local. Global techniques are based on mapping all gray levels of the image and making changes to that mapping to change the contrast of the whole image. Unlike global techniques, local ones use different functions in different areas of the image to make local changes [1].

Global techniques are suitable for general image enhancement. However, they often fail to adapt to local characteristics of the image, such as brightness. Because gray levels with high frequency often tend to dominate those with low frequency. Local techniques are able to contour with the problems presented by global techniques. However, they have a high cost of performance and tend to bring a high level of improvement in only certain portions of the image and generally increase the noise in the image.

The contrast enhancement problem can be solved using optimization techniques with approaches to global image processing techniques. The representation of the solutions is based on the same used in [9] that manipulates it to represent the chromosome in the Genetic Algorithm. The desired solution can be seen in the upper part of Fig. 1. An ordered vector of size D of integers values that vary in the range of [0,255]. The size D represents the number of gray levels of the input image. Each level of gray found in the image is indicated in an index of the vector. The first index of the solution means the lowest level of gray found in the image. The second index is the second smallest, and so on.

To remap the image the value of the first gray level of the generated solution is used instead of the first gray level value of the original image. In this way, different histograms are generated trough modifications in the solution. Given the example in Fig. 1 the transformation that will be applied to the input image has the following description: $f(40) = 0$, $f(88) = 42$, $f(97) = 79$, $f(121) = 113$, $f(143) = 188$ and $f(201) = 255$.

To convert the solution into an image, it is necessary to use a specific data structure for the problem. Such structure maintains the coordinates of each gray level of the original image. Thus it is possible to change the levels of the original image according to each gray level of the transformed solution. So it is possible to observe the image generated and perform a qualitative analysis of it.

The objective function, also known as fitness, has the role of evaluating and measuring the quality of the generated solution. For the contrast enhancement problem, a grayscale image with good contrast include many intensive edges [18]. The function fitness adopted in [9] is shown in Eq. 1:

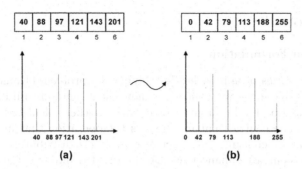

Fig. 1. (a) Histogram of the input image and (b) Histogram of the output image.

$$fitness(s) = log(log(E(I(s)))) * ne(I(s)) \tag{1}$$

Where *fitness(s)* represents the objective function of a solution *s* and *I(s)* represents the image generated after remapping the solution. *E(I(s))* is the sum of the intensities of the generated image edges. This sum is calculated using the following expression [4]:

$$E(I(s)) = \sum_x \sum_y \sqrt{h(x,y)^2 + v(x,y)^2} \tag{2}$$

On Eq. 2, *h(x,y)* and *v(x,y)* represent, respectively, the horizontal and vertical gradient for the point (x, y) of the image. These values are obtained from the convolution of the Sobel edge detector [7]. The *log(log(E(I(s))))* is used in the sum of the intensities to prevent the production of unnatural images [9]. The number of edges detected in the image is represented by *ne(I(s))* using the convolution result and an automatic threshold value [17].

To evaluate the quality of the solution in our algorithm, a new value will be used that improves the Eq. 1 by adding the entropy of the image. The formula demonstrated in Eq. 3, used by [1,6,8,14,19], has the entropy of the image *H(I(s))* as more one component of the calculation. The higher the entropy value of the image, the more quality it has.

$$fitness(s) = log(log(E(I(s)))) * \frac{ne(I(s))}{PH * PV} * H(I(s)) \tag{3}$$

In addition to entropy, two new values appear in the equation: *PH* and *PV*. These values represent respectively the number of horizontal and vertical pixels in the image.

2.2 Related Work

Different approaches for improving contrast using metaheuristics are found in the literature. The first methods that provided the basis for contrast optimization used evolutionary strategies and genetic algorithms as an example in [12,15]. In those works, the standard selection approaches in the genetic algorithm were not performed. Instead, a technique based on user experience was implemented to select the best individuals. The tests in [15] obtained good results on magnetic resonance images and in [12] on satellite images. The work on [18] is one of the pioneers to completely automate the optimization process by metaheuristics. The proposed solution removes the human need by declaring an automatic objective function. The function is performed based on the sum of the edges intensities. Their implementation based on a genetic algorithm uses a Prewitt operator [7] to calculate the objective function. The method proved to be effective in the experiments as it presented results that improved and kept natural the contrast of the tested images.

Modern approaches have come to address the use of different optimization algorithms to solve the problem of image contrast. Different ways of measuring the objective function are adopted, such as entropy, the number of pixels in the image, and the intensity of these pixels. In [19] the Differential Evolution (DE) algorithm with different chaotic sequence approaches is proposed to solve the problem of contrast optimization. Their results prove to be satisfactory when comparing to the different DE approaches proposed. A hybrid algorithm that uses the join of Ant Colony (ACO), Simulated Annealing (SA), and a genetic algorithm is proposed in [10]. The results are compared with the algorithms of [1,13] and it shows to be superior.

The work on [9] proposes a spatial approach to represent the solution of the problem. This methodology is used as a reference by several articles in the literature. Both for its behavior in mapping the solution, and in the comparison of its results. His approach, previously discussed, is the same used in this paper described in the previous section. Their genetic algorithm has specific selection, crossover and mutation operations that lead to excellent results in image contrast and PSNR in the most divergent types of images. The results compared to other optimization algorithms and contrast improvement techniques of the epoch demonstrated superiority. In [6] an algorithm based on Artificial Bee Colony (ABC) is proposed. This proposal uses the concept of [9] to represent the solution, however, it uses a more robust fitness function. Their results surpass the genetic algorithm of [9] in 4 out of 5 cases.

The proposal in [2] diverges from the previous works because it does not represent the solution for optimization. It uses the Artificial Bee Colony to find the best parameters for the Incomplete Beta Function (IBF), which has proven to be effective in enhancing image contrast. Most related works present qualitative and quantitative tests to validate their results. However, as verified in the tests presented in [2,6,9], no statistical methodologies are used to validate the results found. Some works [6,9] has a small number of iterations and is not reported the number of times that the algorithms were executed in order to obtain a diversity of cases for analysis.

3   Differential Evolution for Contrast Optimization - DECO

The algorithm proposed in this paper called DECO (Differential Evolution for Contrast Optimization) is based on the Differential Evolution (DE) algorithm proposed by [21]. The main motivation for choosing this algorithm as a base is that it is one of the evolutionary algorithms that present high consistency and a high degree of performance when compared to other evolutionary algorithms. Its global search capability, effective handling of restrictions, reliable performance, and low need for information mean that the algorithm has a high potential to solve several problems in image processing [19].

3.1   Differential Evolution (DE)

The standard approach of the DE algorithm needs four different parameters for its initialization. A maximum value t_{max} representing the number of iterations performed or the number of generations. The size of the population to be used: *popsize*. The mutation rate factor F used to create the differential vector and CR the probability of performing the crossover.

Mutation Operation (or Differential Operation). DE has variations in its implementation forms. The general convention for the nomenclature of these variations is given by $DE/\alpha/\beta/\gamma$. The α represents how the differential vector will be disturbed. The β is the number of differential vectors used in the perturbation and the γ represents the type of crossover. In this paper, was implemented the standard form of DE, also known as $DE/rand/1/bin$ [21] represented by Eq. 4:

$$v_{i,G+1} = x_{r1,G} + F \cdot (x_{r2,G} - x_{r3,G}) \tag{4}$$

The *rand* means that the vector will be randomly selected and the *bin* means that the crossover is performed in "Binomial" scheme. A variation of the algorithm was also implemented, which was called DECO-BEST. The proposed algorithm has the form $DE/best/2/bin$ as base. The *best* means that the best vector will be selected. This form is a high benefit method that deserves special attention [20]. It mutate individuals according to the following Eq. 5:

$$v_{i,G+1} = x_{best,G} + F \cdot (x_{r1,G} + x_{r2,G} - x_{r3,G} - x_{r4,G}) \tag{5}$$

In the above equations, the value x_i represents an individual from the population of size NP. The elements r_1, r_2, r_3, and r_4 are random individuals selected from the population. G represents the generation which that individual belongs. The mutation factor is described by $F \in [0,2]$ and $v_{i,G+1}$ represents the differential vector generated to perform the crossover operation with the selected individual. In Eq. 5 the value described by $x_{best,G}$ represents the best fitness solution of that generation.

Crossover Operation. The crossover is applied after the mutation operation to mix individuals with the differential vectors resulting from the operation, increasing the diversity of individuals. The resulting individual is generated by:

$$u_{ji,G+1} = \begin{cases} v_{ji,G+1} & \text{if (randb(j)} \leq \text{CR) or j = rnbr(i)}, \\ x_{ji,G} & \text{if (randb(j)} > \text{CR) and j} \neq \text{rnbr(i)}. \end{cases} \tag{6}$$

On Eq. 6, j represents each position of the individual's vector. The value of *randb(j)* is the jth random value generated in the range of [0,1]. *rnbr(i)* is also a random value of $i \in [1,D]$ it guarantees that at least one position of the vector will be changed.

Selection Operation. The selection step has as main objective to select which individual will be chosen for the next generation, $G + 1$. Among the choices are the individual from that iteration $x_{i,G}$ and the individual generated by the mutation and crossover $u_{i,G+1}$. The choice is based on the fitness value of the individuals. The one with the highest fitness value will be chosen for the next generation.

3.2 DECO Implementation

The pseudocode in Algorithm 1 describes the main steps that have been implemented for the proposed contrast enhancement algorithm. First, the algorithm receives an image and the same parameters of the DE. Through the input image it is possible to extract basic structures that will serve as support for the rest of the algorithm. After that the steps follow the same structure as the standard implementation of the DE algorithm [21].

An initial population of size *NP* is created with random values. From t_{max} iterations, changes are made to the populations to improve the objective function. The change is based on, for each p of the population a new individual q is generated by adding and subtracting vector from p with 3 other individuals in the population: a, b, and c. The new son q will be compared with the father p and it will be added to the population if it has a greater fitness value, otherwise it will be discarded. Because we are dealing with a maximization problem, higher values of the objective function indicate optimization in the final result. The *best* individual found is maintained and updated in each generation. Thus, at the end of the optimization, the data for the experiments will be extracted from the *best* solution and the image generated through it.

***ExtractMapper* Function.** This function, referenced on line 2, is responsible for generating the M structure that will be used to maintain the data from the original input image and perform the solution conversion. A scan of the original image is performed to extract the standard solution and the size of this solution D which represents the number of gray levels of the individuals that will be generated. It is important to note that this value must be ordered so that there is

---

Algorithm 1: DECO

---

input : An image I and the DE parameters: t_{max}, *popsize*, F, CR
output: The optimized image O

1 **begin**
2 $\quad$ $M \longleftarrow ExtractMapper(I)$;
3 $\quad$ $P \longleftarrow InitializePopulation(M, popsize)$;
4 $\quad$ **repeat**
5 $\quad\quad$ $Q \longleftarrow \emptyset$;
6 $\quad\quad$ **for** $p \in P$ **do**
7 $\quad\quad\quad$ $a \longleftarrow$ a copy of an individual other than p, randomly chosen;
8 $\quad\quad\quad$ $b \longleftarrow$ a copy of an individual other than p and a, randomly chosen;
9 $\quad\quad\quad$ $c \longleftarrow$ a copy of an individual other than p, a and b, randomly chosen;
10 $\quad\quad\quad$ $d \longleftarrow GenerateDifferential(a, b, c, F)$;
11 $\quad\quad\quad$ $q \longleftarrow CrossOver(d, p, CR)$;
12 $\quad\quad\quad$ **if** $Fitness(q, M) < Fitness(p, M)$ **then**
13 $\quad\quad\quad\quad$ $q \longleftarrow Copy(p)$
14 $\quad\quad\quad$ **if** $Fitness(best, M) < Fitness(q, M)$ **then**
15 $\quad\quad\quad\quad$ $best \longleftarrow Copy(q)$
16 $\quad\quad\quad$ $Q.add(q)$;
17 $\quad\quad$ $P \longleftarrow Copy(Q)$;
18 $\quad$ **until** *reach the maximum number of iterations*: t_{max};
19 $\quad$ $O \longleftarrow GenerateImage(M, best)$;

---

a reference between the original solution and the new individuals. Also, a coordinate map of the original solution is kept in memory. Such map is responsible for transform individuals into images. Each gray level of the original image is changed in its different coordinates, for each gray level position changed in the new individual.

***InitializePopulation* Function.** With the solution size D found in the previous step, it is possible to initialize, line 3, a population of size NP of individuals through a random process. An individual in the population is represented by a vector of size D with values in the range of [0,255]. To maximize the range of gray levels, a value of 0 is assigned to the first element of the vector and a value of 255 to the last one [9]. The vector created is increasingly ordered in order to comply with the requirements of the proposed solution.

***GenerateDifferential* Function.** Also known as the Mutation Operation (or Differential Operation), this function is demonstrated in line 10. Its task is to generate the individual d through the operation defined in the Eq. 4, in the case of DECO-BEST the Eq. 5. It is important to note that the values generated in this operation can escape the domain of the integers in the range of [0,255]. Therefore, it is important to perform rounding and apply the range limits. If a value generated is not within the bounds, a new random value that does not belong to the individual is generated.

***Crossover* Function.** In this function, referenced on line 11, the individual q is generated by mixing the individual from the differential operation d and the individual p from the current iteration. The function implementation follows the steps of the Eq. 6 where there will be at least one change in values between individuals. At the end of the process, it is necessary to reorder the vector of the individual q because the exchange of values in the crossover can generate anomalies in the resulting image for that individual.

Fitness Function. It is necessary to transform the solution into an image to calculate the objective function based on Eq. 3. The structure in M must be used to perform the conversion. With the resulting image it is possible to apply the Sobel edge detector to extract the gradient according to Eq. 2 and extract the remaining values that result in fitness. The fitness calculation of each individual is stored in memory internally in the individual's object to avoid excessive-performance expenses. Because the steps taken to reach fitness proved to be costly in large size images.

GenerateImage Function. After performing the optimization step, the best fitness individual named *best* is chosen. It will be possible to obtain a final image of this individual, on line 19, by remapping *best* with the data kept in M for the entry image I and the initial solution that is kept in memory.

4   Experimental Results

In this section the two DECO and DECO-BEST implementations are applied to a validation process. Different types of grayscale images were used that cover different levels of contrast indicating different scenarios of use of the algorithms. Several algorithms and metrics are used, through a statistical comparison, to obtain validity in the tests performed.

4.1   Algorithms and Parameters

The experiments were performed with algorithms widely used in the literature to improve the contrast, such as Histogram Equalization [7] and CLAHE [16]. Also were included methods of contrast optimization based on metaheuristics, such as the genetic algorithm proposed in [9] and the ABC demonstrated in [6], which use the same representation of the solution and obtained good results in their experiments. First, it is necessary to carry out initial isolated tests on DECO and DECO-BEST to find initial parameters that would bring relevant results to the work.

For DECO, DECO-BEST algorithms, and the genetic proposed by [9] the t_{max} value for the number of iterations it was set to $1,000$, and the value NP for the population size is set to 20. In DECO and DECO-BEST algorithms, the following parameter values were used: $F = 1$, and $CR = 0.9$. Such values obtained good results in initial isolated tests with different types of images. In the genetic algorithm, the crossover rate and mutation rate values were respectively 0.8 and 0.1. As indicated in the tests carried out in [9]. In the ABC algorithm [6] were used the same parameters proposed in the paper: number of solutions 25, both for the number of employed bees and onlookers. And the threshold of activation of the scouts limit is set to 5. Only the number of iterations t_{max} has been changed to the value of 400 to balance the number of executions of the objective function with the other algorithms.

4.2 Images and Metrics

A set of 5 images in grayscale with different sizes, taken from the literature and related works, were used for the experiments. The images are named by: *Airplane* (size: 512×512), *Cameraman* (size: 256×256), *Crowd* (size: 512×512), *Messier83* (size: 640 × 640), and *Chest X-Ray* (size: 482 × 551). *Airplane* and *Cameraman* presents a good variation of pixel intensities and frequencies. The *Messier83* and *Chest X-Ray* images have many pixels with low intensity. *Messier83* also has a high region with low frequency of pixels intensity. The *Crowd* presents the highest variation of pixels intensity between the images, which results in a more equalized image. These images will make it possible to evaluate differences in the execution of the algorithms in different image sizes and characteristics.

The performance of the algorithms is analyzed using three different criteria. The fitness value of the best individual in each algorithm is used to analyze which optimization achieved the best performance. With this comparison it is possible to evaluate which algorithm can reach a higher number of intensive edges. The *Peak Signal-to-Noise Ratio* (PSNR) image quality criterion, which measures the noise ratio between the original image and the enhanced images. And the execution time that the algorithm took to find the best solution.

Metrics evaluation will be performed using non-parametric tests with multiple comparisons [5]. Each algorithm was executed 20 times and its values of μ (mean) and σ (standard deviation) for each metric are counted for each algorithm. It is necessary to apply a multiple comparison between the algorithms to know if they are equivalent or not. In this paper the Kruskal–Wallis test was used to verify if one algorithm is significantly different from the other. It is necessary to reach a *p-value* < 0.05 for a $\alpha = 0.05$. This value indicates the lowest level of significance that results in the rejection of the null hypothesis [5].

4.3 Results

In the first step, the algorithms were statistically evaluated for their performance over the fitness value of the best solution. Only metaheuristic algorithms participate in this phase because they are directly influenced by the value of fitness. The values of μ (mean) and σ (standard deviation) for each algorithm are shown in Table 1. It is shown in Table 2 the values of *p-value* resulting from the Kruskal–Wallis test to compare the results of fitness in the algorithms. The values in bold in this table represent that the algorithms are comparable to each other. It is possible to notice that the DECO and DECO-BEST algorithms had comparable results in all images. The Kruskal-Wallis test of these two algorithms results in a *p-value* > 0.05.

Table 1. *Fitness* mean and standard deviation.

Test image	[9]	[6]	DECO	DECO-BEST
Airplane	$\mu = 19.79$	$\mu = 20.46$	$\mu = 19.98$	$\mu = 19.99$
	$\sigma = 0.16$	$\sigma = 0.10$	$\sigma = 0.06$	$\sigma = 0.05$
Cameraman	$\mu = 17.40$	$\mu = 17.79$	$\mu = 17.45$	$\mu = 17.44$
	$\sigma = 0.20$	$\sigma = 0.09$	$\sigma = 0.03$	$\sigma = 0.04$
Crowd	$\mu = 21.22$	$\mu = 21.75$	$\mu = 21.37$	$\mu = 21.40$
	$\sigma = 0.10$	$\sigma = 0.09$	$\sigma = 0.02$	$\sigma = 0.03$
Messier83	$\mu = 5.51$	$\mu = 5.59$	$\mu = 5.56$	$\mu = 5.56$
	$\sigma = 0.03$	$\sigma = 0.09$	$\sigma = 0.01$	$\sigma = 0.01$
Chest X-Ray	$\mu = 16.89$	$\mu = 16.19$	$\mu = 16.29$	$\mu = 16.28$
	$\sigma = 0.28$	$\sigma = 0.15$	$\sigma = 0.03$	$\sigma = 0.03$

Table 2. Kruskal-Wallis test on *fitness*.

Comparison	*Airplane*	*Cameraman*	*Crowd*	*Messier83*	*Chest X-Ray*
[9] versus [6]	5e−14	5.3e−07	1.0e−13	0.0014	4.6e−09
[9] versus DECO	0.00049	**0.99**	**0.02987**	0.0019	0.00044
[9] versus DECO-BEST	0.00014	**1.00**	0.00034	0.0027	3.0e−05
[6] versus DECO	0.00014	3.0e−06	5e−06	**0.9998**	**0.12570**
[6] versus DECO-BEST	0.00049	3.3e−07	0.00133	**0.9980**	**0.38305**
DECO versus DECO-BEST	**0.99068**	**0.98**	**0.58495**	**0.9996**	**0.93247**

$$\alpha = 0.05$$

In the images *Airplane* and *Crowd* it is possible to observe a better result of algorithm [6] despite its high standard deviation value. At *Cameraman* the results are very similar, but for Kruskal-Wallis, algorithm [6] differs from the others. In *Messier83* only the algorithm [9] obtained a lower result than the rest of the algorithms, differently from *Chest X-Ray* where it obtained the best result. It is notable in all images that the standard deviation of DECO and DECO-BEST are lower than their competitors, demonstrating a standard of reliability in the execution of the algorithm. And despite some draws, the algorithm [6] was more successful in maximizing the objective function.

In the second step, the algorithms are compared using the PSNR. The purpose of this test is to analyze the noise ratio of the generated image compared to the original image. The HE and CLAHE techniques participate in these experiments, but the mean value and standard deviation for these algorithms are not shown because several executions of these techniques arrive at the same result. The test data can be found in Table 3. The results of *PSNR* shows that the HE and CLAHE algorithms add less noise to the images, only in *Messier83* that their results are worse than the metaheuristic algorithms, mainly the HE result. Among the metaheuristic algorithms it is possible to see in Table 4 that there were several equivalences between them. The algorithm in [9] added less noise to the image only in the *Airplane* and in *Cameraman*, where there was equivalence with all the other algorithms. [6] was superior in isolation in the image *Crowd* and the *Cameraman*, in *Messier83* it was equivalent to the others. DECO and DECO-BEST were superior in the image *Chest X-Ray* and equivalent in *Cameraman* and *Messier83*. When matching the data of these two tests, it is interesting to note, for example, that in the image *Chest X-Ray* the algorithms DECO and DECO-BEST despite being comparable with [6] in fitness, maintain a better quality in the final image. In the image *Crowd* the algorithm [6] is superior in isolation in both metrics.

Table 3. *PSNR* mean and standard deviation.

Test image	HE	CLAHE	[9]	[6]	DECO	DECO-BEST
Airplane	21.87	23.64	$\mu = 20.70$	$\mu = 19.43$	$\mu = 19.70$	$\mu = 19.49$
			$\sigma = 0.51$	$\sigma = 0.89$	$\sigma = 0.43$	$\sigma = 0.47$
Cameraman	21.58	22.86	$\mu = 18.33$	$\mu = 18.19$	$\mu = 18.06$	$\mu = 18.46$
			$\sigma = 0.56$	$\sigma = 0.72$	$\sigma = 0.69$	$\sigma = 0.35$
Crowd	15.39	21.38	$\mu = 11.26$	$\mu = 12.60$	$\mu = 11.74$	$\mu = 11.92$
			$\sigma = 0.28$	$\sigma = 0.48$	$\sigma = 0.27$	$\sigma = 0.34$
Messier83	3.68	19.78	$\mu = 20.60$	$\mu = 21.69$	$\mu = 21.67$	$\mu = 21.54$
			$\sigma = 0.45$	$\sigma = 1.32$	$\sigma = 0.36$	$\sigma = 0.38$
Chest X-Ray	14.36	18.54	$\mu = 16.82$	$\mu = 16.98$	$\mu = 17.54$	$\mu = 17.50$
			$\sigma = 0.41$	$\sigma = 0.80$	$\sigma = 0.59$	$\sigma = 0.47$

Table 4. Kruskal-Wallis test on *PSNR*.

Comparison	*Airplane*	*Cameraman*	*Crowd*	*Messier83*	*Chest X-Ray*
[9] versus [6]	5.5e−06	**0.875**	5.3e−11	0.00129	**0.8994**
[9] versus DECO	0.00047	**0.66**	0.01403	6.4e−06	0.0051
[9] versus DECO-BEST	3.8e−06	**0.75**	0.00030	0.00014	0.0046
[6] versus DECO	**0.76870**	**0.98**	0.00078	**0.63272**	0.0437
[6] versus DECO-BEST	**0.99985**	**0.28**	0.02817	**0.94988**	0.0405
DECO versus DECO-BEST	**0.72528**	**0.14**	**0.72528**	**0.91446**	**1.0000**

$\alpha = 0.05$

The third step is characterized by accounting for the total execution time of the algorithms, demonstrated in Table 5. As expected, the HE and CLAHE techniques obtain execution time values in milliseconds, due to the algorithm technique, so only the results of the metaheuristic algorithms will be demonstrated. DECO and DECO-BEST had better results compared to other algorithms, reaching execution times well below expectations, demonstrating the performance power that DE carries.

Table 5. *Execution time* mean and standard deviation.

Test Image	[9]	[6]	DECO	DECO-BEST
Airplane	$\mu = 46.67s$	$\mu = 177.59s$	$\mu = 28.66s$	$\mu = 29.27s$
	$\sigma = 1.16s$	$\sigma = 8.59s$	$\sigma = 2.49s$	$\sigma = 0.03s$
Cameraman	$\mu = 11.34s$	$\mu = 30.21s$	$\mu = 6.92s$	$\mu = 7.07s$
	$\sigma = 0.25s$	$\sigma = 0.82s$	$\sigma = 0.60s$	$\sigma = 0.04s$
Crowd	$\mu = 51.54s$	$\mu = 208.24s$	$\mu = 31.63s$	$\mu = 32.39s$
	$\sigma = 1.65s$	$\sigma = 13.11s$	$\sigma = 2.65s$	$\sigma = 0.84s$
Messier83	$\mu = 69.87s$	$\mu = 213.72s$	$\mu = 42.49s$	$\mu = 44.49s$
	$\sigma = 1.79s$	$\sigma = 11.22s$	$\sigma = 3.73s$	$\sigma = 0.58s$
Chest X-Ray	$\mu = 44.28s$	$\mu = 190.63s$	$\mu = 27.55s$	$\mu = 27.87s$
	$\sigma = 1.04s$	$\sigma = 136.64s$	$\sigma = 2.57s$	$\sigma = 0.03s$

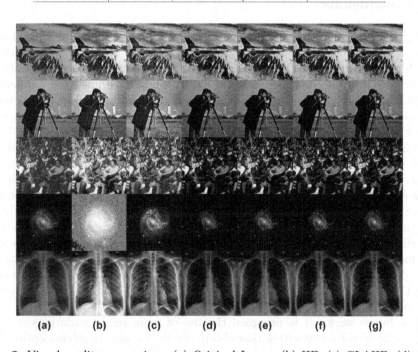

(a)   (b)   (c)   (d)   (e)   (f)   (g)

Fig. 2. Visual quality comparison: (a) Original Image, (b) HE, (c) CLAHE, (d) [9], (e) [6], (f) DECO and (g) DECO-BEST. From top to bottom are the respective images: *Airplane, Cameraman, Crowd, Messier83* and *Chest X-Ray*

The last step of the experimental process is characterized by performing a visual analysis of the images generated by the algorithms. Figure 2 shows the comparison of all final solutions taken from the tests of each algorithm for each image. It is noticed that the HE and CLAHE algorithms, despite having obtained excellent results in the PSNR and runtime tests, change the characteristics of the images. Brightness is the most affected feature in these algorithms, the biggest effect of this loss is seen in *Messier83*. That image has a high concentration of pixel frequency in low-level intensities, which caused HE to have difficulties in improving the contrast. Therefore, the use of optimization algorithms is validated to obtain images as close as possible to the original, with enhancement only in contrast levels. It is visually noticeable that there is a lower loss in the natural brightness of the images in the results of the metaheuristic algorithms. The visual differences between the metaheuristic algorithms themselves are difficult to be noticed, but in certain cases there are greater amplitudes of contrasts in certain specific locations of the images. The noise is also little perceived visually, despite the obtained values of *PSNR* which was lower than HE and CLAHE in some cases.

5   Conclusions

In this work, an image contrast improvement based on metaheuristics was implemented using two well-known strategies of the Differential Evolution algorithm which resulted in the DECO and DECO-BEST algorithms. The representation of the optimization problem was based on [9]. The algorithms were compared qualitatively and statistically with other contrast improvement algorithms and techniques: Histogram Equalization (HE), CLAHE, the genetic algorithm proposed in [9], and the ABC proposed in [6].

The results found were promising, the resulting images showed to have a good quality standard, with low noise level, and a visual contrast improvement. Regarding the objective function, the DECO and DECO-BEST algorithms achieved similar results. They demonstrated to be competitive in comparison to the other algorithms and their standard deviation of the obtained fitness values was much smaller than the others which demonstrates a high convergence in the search for the solution. The results obtained in PSNR were equivalent between the metaheuristic algorithms which indicates a small deformation of the image caused by the use of this solution representation. In terms of execution time, DECO and DECO-BEST resulted in much shorter times than the metaheuristic algorithms.

For future works, it is proposed the use of more recent variations of the differential evolution algorithm that obtain excellent results in solving optimization problems. A greater variety of contrast improvement techniques can also be compared in the experiments to improve the results found. Another path to be explored is to change the representation of the problem by mapping it as a multi-objective problem. The motivation is to improve not only edge strength but also PSNR and other combined quality criteria, such as SSIM and IFC. In this way, the final result of the image quality will be improved.

Acknowledgements. This study was financed in part by the Coordenação de Aperfeiçoamento de Pessoal de Nível Superior - Brasil (CAPES) - Finance Code 001.

References

1. Braik, M., Sheta, A.F., Ayesh, A.: Image enhancement using particle swarm optimization. In: World Congress on Engineering, vol. 1, pp. 978–988 (2007)
2. Chen, J., Yu, W., Tian, J., Chen, L., Zhou, Z.: Image contrast enhancement using an artificial bee colony algorithm. Swarm Evol. Comput. **38**, 287–294 (2018)
3. Chen, S.D., Ramli, A.R.: Contrast enhancement using recursive mean-separate histogram equalization for scalable brightness preservation. IEEE Trans. Consum. Electron. **49**(4), 1301–1309 (2003)
4. DaPonte, J.S., Fox, M.D.: Enhancement of chest radiographs with gradient operators. IEEE Trans. Med. Imaging **7**(2), 109–117 (1988)
5. Derrac, J., García, S., Molina, D., Herrera, F.: A practical tutorial on the use of nonparametric statistical tests as a methodology for comparing evolutionary and swarm intelligence algorithms. Swarm Evol. Comput. **1**(1), 3–18 (2011)
6. Draa, A., Bouaziz, A.: An artificial bee colony algorithm for image contrast enhancement. Swarm Evol. Comput. **16**, 69–84 (2014)
7. Gonzalez, R.C., Woods, R.E., et al.: Digital Image Processing. Prentice Hall, Upper Saddle River (2002)
8. Gorai, A., Ghosh, A.: Gray-level image enhancement by particle swarm optimization. In: 2009 World Congress on Nature & Biologically Inspired Computing (NaBIC), pp. 72–77. IEEE (2009)
9. Hashemi, S., Kiani, S., Noroozi, N., Moghaddam, M.E.: An image contrast enhancement method based on genetic algorithm. Pattern Recogn. Lett. **31**(13), 1816–1824 (2010)
10. Hoseini, P., Shayesteh, M.G.: Hybrid ant colony optimization, genetic algorithm, and simulated annealing for image contrast enhancement. In: IEEE Congress on Evolutionary Computation, pp. 1–6. IEEE (2010)
11. Maini, R., Aggarwal, H.: A comprehensive review of image enhancement techniques. arXiv preprint arXiv:1003.4053 (2010)
12. Munteanu, C., Lazarescu, V.: Evolutionary contrast stretching and detail enhancement of satellite images. In: Proceedings of MENDEL 1999, pp. 94–99 (1999)
13. Munteanu, C., Rosa, A.: Evolutionary image enhancement with user behavior modeling. ACM SIGAPP Appl. Comput. Rev. **9**(1), 8–14 (2001)
14. Munteanu, C., Rosa, A.: Gray-scale image enhancement as an automatic process driven by evolution. IEEE Trans. Syst. Man Cybern. Part B Cybern. **34**(2), 1292–1298 (2004)
15. Poli, R., Cagnoni, S.: Evolution of pseudo-colouring algorithms for image enhancement with interactive genetic programming. Cognitive Science Research Papers-University OF Birmingham CSRP (1997)
16. Reza, A.M.: Realization of the contrast limited adaptive histogram equalization (CLAHE) for real-time image enhancement. J. VLSI Signal Process. Syst. Signal Image Video Technol. **38**(1), 35–44 (2004)
17. Rosin, P.L.: Edges: saliency measures and automatic thresholding. Mach. Vis. Appl. **9**(4), 139–159 (1997)
18. Saitoh, F.: Image contrast enhancement using genetic algorithm. In: IEEE SMC 1999 Conference Proceedings. 1999 IEEE International Conference on Systems, Man, and Cybernetics (Cat. No. 99CH37028), vol. 4, pp. 899–904. IEEE (1999)
19. dos Santos Coelho, L., Sauer, J.G., Rudek, M.: Differential evolution optimization combined with chaotic sequences for image contrast enhancement. Chaos, Solitons Fractals **42**(1), 522–529 (2009)

20. Storn, R., Price, K.: Minimizing the real functions of the ICEC'96 contest by differential evolution. In: Proceedings of IEEE International Conference on Evolutionary Computation, pp. 842–844. IEEE (1996)
21. Storn, R., Price, K.: Differential evolution-a simple and efficient heuristic for global optimization over continuous spaces. J. Global Optim. **11**(4), 341–359 (1997)

Neural Networks, Deep Learning and Computer Vision

A Deep Learning Approach for Pulmonary Lesion Identification in Chest Radiographs

Eduardo Henrique Pais Pooch$^{(\boxtimes)}$, Thatiane Alves Pianoschi Alva, and Carla Diniz Lopes Becker

Universidade Federal de Ciências da Saúde de Porto Alegre (UFCSPA), Rua Sarmento Leite, 245, Porto Alegre, Rio Grande do Sul, Brazil
{edupooch,thatiane,carladiniz}@ufcspa.edu.br

Abstract. Radiography is a primary examination used to diagnose chest conditions, as it is fast, low cost, and widely available. If the physician cannot conclude de diagnosis with the radiography, a computed tomography scan may be required. However, this exam is expensive and has low availability, mainly in the public health system of developing countries and low-income locations, which can delay the treatment and cause complications to the patient's health condition. Computer-aided diagnosis systems provide more resources for medical diagnostic decision-making, increasing the accuracy of the assessment of the patient's clinical condition. The main objective of this work is to develop a deep-learning-based approach that performs an automatic analysis of digital images of chest radiographs to aid the detection of pulmonary nodules and masses, aiming to extract sufficient relevant information from the image, optimizing the initial phase of the diagnosis of lung lesions. The developed approach uses neural networks in a dataset of 8,178 annotated chest radiographs extracted from a public dataset. Half of it is of images annotated with "nodule" or "mass", and the other half is of images with "no findings". We implemented and tested convolutional neural networks and data preprocessing techniques to create a classification model. A model with five convolution layers that achieved 0.72 accuracy, 0.75 sensitivity, and 0.68 specificity. The proposed approach achieved results comparable to state of the art for lesion identification using limited computational power and can assist radiological practice as a second opinion, which can improve the rates of early diagnosed cancer.

Keywords: Deep learning · Chest x-ray · Image processing

1 Introduction

Radiography was the first modality used in medical imaging, and even with the invention of new modalities like computed tomography (CT), magnetic resonance imaging, and others, it remains a significant modality to evaluate the chest [7].

© Springer Nature Switzerland AG 2020
R. Cerri and R. C. Prati (Eds.): BRACIS 2020, LNAI 12319, pp. 197–211, 2020.
https://doi.org/10.1007/978-3-030-61377-8_14

The x-ray exam can provide useful information and is a primary examination used to diagnose chest conditions, as it is fast, low cost, and widely available. If the physician can not define the diagnosis with chest radiography, a CT scan is required. However, a CT scan exposes the patient to more radiation, is high-priced, and has low availability, especially in the public health system and in low-income locations.

In most countries, lung cancer is the leading cause of cancer death [2], and the global 5-year survival rates vary between 10% and 20%. Its poor prognosis is mainly caused by the lack of effective early detection methods and the inability to cure metastatic diseases [10]. In the early stages, the 5-year survival rate is about 60%, while in later stages, it can be less than 5% [17]. Missed lung cancer is an important concern among radiologists, and 90% of the times, the misdiagnosis occurs on chest radiographs, mostly because of observer error [5]. In some stated cases, the radiography presents signs of early-stage cancer and is undetected by a radiologist when the cancer is still resectable [19]. The inability to recognize an abnormal chest radiograph may also lead to a failure in requesting a potentially valuable CT examination [7], which has a higher sensitivity for finding pulmonary nodules [6].

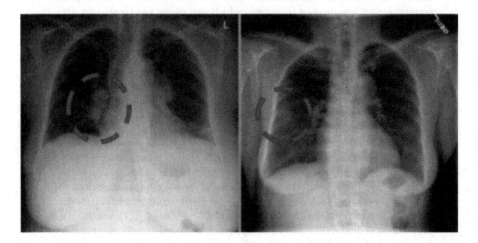

Fig. 1. Example of a pulmonary mass (left) and nodule (right) appearance on a chest radiography. Image source: Wang *et al.* [22]

A pulmonary lesion appearance on chest radiography may indicate lung cancer. A nodule is a lesion smaller than 30 mm, and it is most likely benign. If the lesion diameter is larger than 30 mm, it is described as a mass and has a higher chance of malignity [11]. Developing automated methods for identifying these signs might improve early detection rates of lung cancer, which leads to a better prognosis [17].

Computer-aided diagnosis (CAD) systems provide more resources for medical diagnostic decision-making. CAD systems use computer vision techniques to perform automated analysis and interpretation of medical images. Deep learning methods can automate a series of tasks, and it is the method of choice for most computer vision problems, including medical image analysis [14], the most successful method being convolutional neural networks (CNNs).

The first use of CNNs in medical imaging was for lung nodule detection in 1995 [15]. Due to computer limitations at the time, the method used image blocks of 16×16 pixels subsampled from blocks of 32×32 pixels by averaging the values of every 4 pixels in each 2×2 block. The used CNN had two hidden layers, each one with 12 nodes, and finally, a fully connected layer with 10 nodes, which outputs a label *yes* or *no* for nodule presence. This approach used a total of 55 chest images, and only 25 contained nodules. The authors recognized that the method needed more data to be validated.

Recently, a series of studies have defined that CNNs are the current standard for image exam classification [14]. In classification problems, the models receive an input image and output a diagnosis label. CNNs have been explored for nodule detection and analysis both in chest radiographs and chest CT. Even though chest radiographs are the most common radiological exam, a higher number of studies use CT images for nodule detection [14], probably because a CT exam offers more information and has a higher sensitivity for nodule identification and characterization [16].

Litjens *et al.* [14] reviewed 308 papers on deep learning in medical image analysis and described the key aspects of successful deep learning methods. Even though choosing the right CNN architecture is important, some other aspects outside of the network may be critical in finding a good solution, such as preprocessing techniques and data augmentation. Another important aspect is model hyper-parameter optimization, but it still is a rather empirical practice, since there is no clear methodology.

A problem with applying deep learning techniques is gathering the data for training deep neural networks without overfitting the model. The radiological practice was one of the first medical fields to be completely digitized, and it has been using picture archiving and communication systems (PACS) for almost ten years, which made the creation of large datasets of medical images possible. A study by Wang et al. [22] presented a chest x-ray database comprising more than $100,000$ annotated chest radiographs extracted from the hospital's PACS, which enabled the development of deep learning tools to identify diseases on radiographs. In the same work, the authors used CNNs for supervised image classification, which could achieve an area under the curve (AUC) of 0.69 in identifying masses and AUC of 0.66 in identifying nodules.

This work proposes an automated approach using CNNs to identify pulmonary nodules and masses in chest radiographs. A CNN architecture was built based on LeNet, tested, and improved based on its results. We compare the built CNN with Inception-v3 [21], the most popular architecture used in medical image analysis recently [14]. We trained the LeNet-based model from scratch, while the Inception-v3 model was previously trained on ImageNet aiming to transfer learning from the ImageNet dataset to the chest x-ray one. Therefore, we make a comparison in training a small model from scratch to fine-tuning a large pre-trained network.

The proposed model can assist radiologists to evaluate the chest condition using radiographs in a CAD system for lesion detection and serve as a second opinion, which could improve diagnostic accuracy, making it possible to detect cancer at earlier stages, and to conclude false-positive diagnosis with radiographs, reducing the demand for costly exams like CT.

This study is organized into four sections. In Sect. 2, we present the suggested methodology. Section 3 presents the results of the dataset simulations, comparison, and discussion of the efficiency of the methods. Finally, in Sect. 4, conclusions about the paper and indications of possible works are presented.

2   Background

Deep learning software attempts to simulate brain processes using a large array of connected neurons in an artificial neural network (ANN). The ANN can learn, with high-level precision, to recognize patterns in digital representations of images and other data and became popular models in the past due to its ability to represent non-linearities and approximate mostly any function. Because of computers processing capacity evolution and improvements in mathematical approach, engineers and computer scientists can now model many more layers in ANNs than ever before, creating deep neural networks (DNN). The DNNs trained using a high amount of data, usually have high generalization capacity.

CNN is a type of DNN that has layers of convolution operations that extract features of spatial information. CNNs exploit natural signals' spatially-local correlation by using local connections between neurons of adjacent layers and shared weights [13], which means that the model can learn to detect the same object at different positions of the image. Being a location invariant method makes it a very suitable method for lesion identification in radiographs since the patient positioning might vary, and a lesion can happen in multiple parts of the lungs.

Input Layer. A CNN for image classification receives an image as input. The input layer holds the image that will be classified, so it has to be of the same height, width, and depth as the image. The size of the input layer affects how much memory is necessary to train the network and how many pooling layers the network will support since it reduces the dimensionality of the input.

Convolution Layer. The convolution layer's parameters consist of a set of K learnable filters and biases. Every filter is small spatially (along width and height). In the convolution layer, each filter W is convolved across the width and height of the input image X and added with the bias b_k. As the filter slides over the width and height of the input image, it forms a feature map X_k that represent some types of image features, such as edges. The mathematical equation of the convolution neuron (see Eq. 1) is similar to the multilayer perceptron (MLP) one since it has an operation between the weights and the input and it adds a bias b_k. However, instead of a multiplying W and X the CNN applies a convolution operation.

$$\mathbf{X}_k^l = \mathbf{W}_k^{l-1} * \mathbf{X}^{l-1} + b_k^{l-1} \tag{1}$$

For the convolution layer, we can define some hyperparameters that can improve the network results on different datasets. The stride size defines the number of pixels that shift over each iteration of the convolution operation. We can apply a padding operation to fill the image matrix with zeros so that the output matrix will be the same size as the input.

Activation Layers. Activation layers are composed of activation functions that operate on the output of the previous layer, mapping these values into the desired range. The rectified linear unit (ReLU) insert non-linearities in the network, so it can easily obtain sparse representations and model complex functions. ReLU's introduction [8] made the evolution of deep learning methods possible since it is more computationally efficient than previous functions used in neural networks [13], which optimized the training process. The ReLU function (see Eq. 2) returns 0 for any $x \leq 0$ and x for $x > 0$, what modifies the feature map like shown on Eq. 3.

$$R(x) = max(0, x) \tag{2}$$

$$\mathbf{X}_k^l = R\left(\mathbf{W}_k^{l-1} * \mathbf{X}^{l-1} + b_k^{l-1}\right) \tag{3}$$

The last layer consists of a function that returns the probabilities of the input image being of each class. The softmax function (see Eq. 4) takes the previous layer output y and returns a value in the interval $(0, 1)$, and every added value sums to 1.

$$softmax(y_i) = \frac{e^{y_i}}{\sum_j e^{y_j}} \tag{4}$$

Pooling Layers. The pooling layer is a way to apply a dimensionality reduction technique in the network architecture, which reduces the input size but keeps relevant information by merging semantically similar features into one [13]. The network architectures of this work were constructed using max-pooling layers of size 2×2, which takes the largest element each different 2×2 block, reducing the size of the input image by half.

Fully-Connected Layers. Fully-connected (FC) or dense layers are regular MLP layers. FCs are composed of nodes that take a 1D array as input and returns a value based on the defined function of weights and biases of each neuron. As the input of a convolutional neural network is a matrix, the input must be transformed in an array to apply the FC layer functions. This operation is called flattening and combines all the features from the previous layers to output the values for the last activation function.

Metrics. Cross-entropy is an error function that indicates how likely the samples are correctly classified based on their inferred probabilities. Since the problem of lesion identification is binary, the cross-entropy calculation is as described in Eq. 5, in which n is the number of samples, y_i is the predicted label for sample i, being 0 for normal or 1 for lesion presence, and p_i is the probability of the input sample i being of the lesion class, which is the value returned by the softmax function in the final layer.

$$\mathcal{L}(\theta) = -\frac{1}{n} \sum_{i=1}^{n} y_i \log{(p_i)} + (1 - y_i) \log{(1 - p_i)} \tag{5}$$

The model training occurs by adjusting the values of the weights W and bias b of each neuron to minimize the binary cross-entropy. This adjustment is made by testing the network in a subset of the data and searching for the minimum of the function using a method called stochastic gradient descent. Each iteration adjusts the weights and bias based on the learning rate. The number of training examples used to create the subset is called the batch size. Each iteration passes the batch size through the network. An epoch is the passing of all the training samples.

We split the data into two sets, train and validation. The train set is the one that the network uses to adjust its weights so that it can model the distribution of the samples on the training set. We use the validation set to verify that the model can be generalized to a set that the network has not seen during training.

If the model infers that a positive sample is positive for lesion, it is a true positive (TP) outcome. Otherwise, if it was a normal sample, it is a false positive (FP). If the model outputs "normal" for a normal sample, it is a true negative (TN), and if it was a "lesion" sample, it is a false negative (FN). The metrics used to evaluate performance are accuracy, sensibility, and specificity. Accuracy (Eq. 6) is the number of true outcomes over the total of samples. Sensitivity (Eq. 7) is the number of TP over every positive sample, and specificity (Eq. 8) is the number of TN over every negative sample.

$$accuracy = \frac{TP + TN}{TP + TN + FP + FN} \tag{6}$$

$$sensitivity = \frac{TP}{TP + FN} \tag{7}$$

$$specificity = \frac{TN}{TN + FP} \tag{8}$$

3   Materials and Methods

3.1   Data Description

The NIH Chest X-ray [22] is an annotated dataset composed of 112,120 images of 30,805 patients. Each image has disease labels extracted from the radiological reports using natural language processing (NLP). The labels are common thoracic radiological findings such as atelectasis, cardiomegaly, effusion, infiltration, mass, nodule, pneumonia, and pneumothorax. We show an example of radiography with pulmonary mass and nodule appearance in Fig. 1. One or more labels could be present in a single image. For normal exams, the label is "No Finding". Besides the clinical data, some patient data, like age and gender, and exam data, like pixel spacing and patient positioning, are also available.

From the NIH ChestX-ray dataset (112,120), we extracted a subset (8,178) of images with nodules or masses and a selection of images with no findings using a Python script. The script selected entries with posteroanterior (PA) positioning, that contained the strings "Mass" or "Nodule" or "No Findings" in the "Finding Labels" field. Images labeled with "Mass" or "Nodule" were assigned to the class "Lesion" the images labeled with "No Findings" went to the "Normal" class.

Considering the dataset contained multiple images from the same patient, only one image per patient was selected for each class, to avoid very similar images separated between the train and test sets, which could bias the validation metrics—resulting in 4,089 images of radiographs with lesions and 22,459 images with no findings. Then, we sampled 4,089 "Normal" samples to balance the dataset, therefore making a total of 8,178 images to train and test the CNN models, split in a 50/50 relation between the two classes.

The age and gender distributions between the two classes are similar, although there are more male patients presenting a lesion than male patients with no findings, probably because of the higher incidence of lung cancer in male patients [20]. The label frequency is the number of samples that present the string in the finding labels field; it is higher for nodules (2,527 samples) than for masses (1,855 samples), with 293 samples labeled with both nodules and masses. Figure 2 shows an analysis of the selected samples' age, gender, and label distribution.

3.2   Model Development

The described models were built and trained using the Python 3.6 programming language and tools available in Keras 2.2.2 [4], a high-level Python deep learning library, which runs on top of TensorFlow [1], a software library for numerical computation using the graphics processing unit (GPU). The used version of TensorFlow is 1.10.0. The adopted technologies are all open-source. The hardware used to develop and test this is an Intel Xeon e3 central processing unit, 3.50 GHz, 16 GB of system random access memory (RAM), Nvidia Quadro K620 GPU with 2 GB DDR3 video RAM and Windows 10 Education operational system.

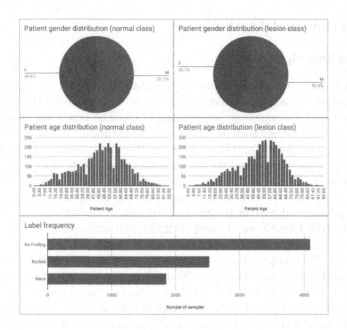

Fig. 2. Dataset distribution of patients age and gender separated by class and number of times a finding appeared in the labels of the samples (label frequency).

At first, we created CNet-2, a CNN architecture based on LeNet [12]. It consists of a sequential model with 12 layers, starting with an input layer of size 128 × 128. The first is a convolution layer with 20 5 × 5 filters, followed by a ReLU activation layer, and a max-pooling with 2 × 2 strides and pool size. The next three layers are another set of convolution, ReLU activation, and max pooling, but with 50 filters. Then a flatten layer and an FC layer of 500 nodes and a ReLU activation layer, followed by a dense layer with 2 nodes, which is the number of classes, and a softmax activation layer, which returns the classes probabilities.

Then we augmented CNet-2 in order to improve the results. The CNet-3 is similar to the CNet-2 model, but instead of having two sets of convolution, ReLU, and max pooling, it has three, and the new convolution layer has 100 filters. CNet-4 has four sets, the first three are like the ones in CNet-3, and the fourth has 200 filters. CNet-5 has one more set than CNet-4, but the fourth set with 150 filters and the fifth with 200. CNet-6 is the same as CNet-5 but with one additional set of layers with 250 convolution filters. Table 1 shows the number of layers and parameters for each architecture.

A batch size of 64 samples was used during training to avoid resource exhaustion since a higher value of batch size requires more memory space. The initial learning rate of the models is 10^{-3}. We trained the models for 120 epochs, which we empirically observed to be the number of epochs that reduced overfitting and increased validation accuracy.

3.3 Inception-v3

Another tested model was the Inception-v3 [21], developed by Szegedy *et al.*, which is a very popular architecture in medical image analysis [14]. The model used is the one available on Keras library, with weights pre-trained on ImageNet, which can transfer the learning from another dataset into the proposed application.

Table 1. Layers and number of parameters of each of the network architectures used in this work. A reduced number of parameters in models with more layers is due to our increase in sets of Convolution, ReLU, and Pooling layers, reducing the final number of parameters needed for the fully-connected layer.

Network	Layers	Parameters
CNet-2	12	25,628,072
CNet-3	15	12,856,212
CNet-4	18	6,636,412
CNet-5	21	2,061,562
CNet-6	24	1,411,812
Inception-v3 [21]	159	21,802,784

3.4 Preprocessing

Histogram equalization is an image processing technique to improve structure visualization. It spreads the gray levels of an image, highlighting the differences between similar pixels. Global equalization can be useful in images where most of the gray levels are confined in a specific range of the histogram, but if they are already distributed, some information may be lost in gray levels near the edges (pixels next to black or white). A better approach is to use an adaptive equalization [18], which splits the image into small blocks and applies the histogram equalization in each block separately. A contrast limit is also applied to avoid amplifying noise in those regions. This technique is called contrast limited adaptive histogram equalization (CLAHE). Figure 3 shows the results of the CLAHE application in a sample image of the dataset, in which it is possible to observe that the darker lung areas became closer to black, improving the contrast between structures of similar shades on the radiography (More details about the method available on [18]).

As a preprocessing method, CLAHE was applied to the images of the dataset before training and testing the CNNs. CLAHE was implemented using OpenCV 3.4 [3] for Python. The threshold for contrast limit is of 2.0, and the tile grid size is 4×4.

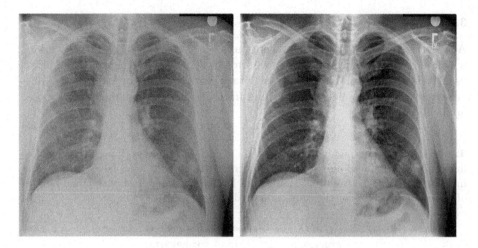

Fig. 3. Comparison of original image (left) of a chest radiography and after CLAHE application (right). (Original image source: Wang et al. [22])

4   Results and Discussion

The models were trained on 80% of the samples of the dataset and validated on the remaining 20% of it. Figure 4 shows the training and validation metrics during 120 epochs. The training of the models aims at minimizing the training loss value, which should also minimize validation loss and maximize training and validation accuracy. When the training and validation metrics are distant, it means that the model is not well generalized to unknown data and is presenting overfitting of the training samples.

The results of each experiment are described in Table 2, which shows the accuracy, sensitivity, and specificity of each model. As shown, adding more layers does not necessarily improve the network performance. Some models had a considerable difference between sensitivity and specificity. Most models have lower sensitivity, which makes them not a good fit for a real-world application, since it performs poorly on its main objective, identifying nodules and masses.

The model which achieved the best overall performance without preprocessing was the CNet-5, which scored higher sensitivity for lesion detection. We decided to perform the CLAHE preprocessing on this architecture only to reduce the total number of experiments. Applying CLAHE to the dataset images improved the accuracy, sensitivity, and specificity of the CNet-5 model. It confirms what was stated by Litjens *et al.* [14] that the performance of the same network architecture can be improved based on data preprocessing and other aspects outside of the network. The results are shown in Table 3.

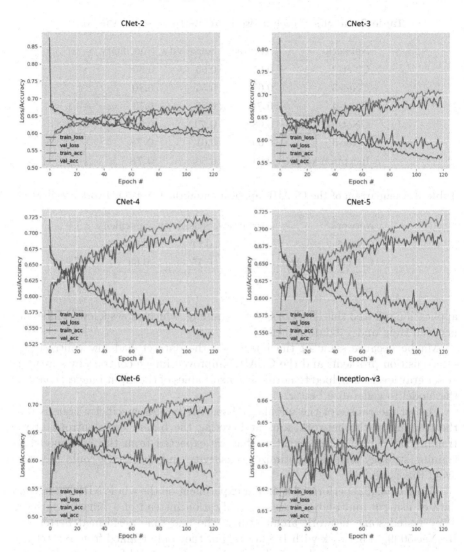

Fig. 4. Training and validation metrics during each of the 120 epochs for the six developed models. The graphs show each model's training loss (red), validation loss (blue), training accuracy (purple) and validation accuracy (gray) values (vertical axis) at the end of an epoch (horizontal axis). (Color figure online)

It seems that the equalization technique reduced overfitting during the training of the CNet-5, as seen in Fig. 5, which shows accuracy and loss during training and validation. On the CNet-5 training without CLAHE, after epoch number 30, the loss and accuracy metrics on the train and validation sets are distant, which is caused by overfitting the model to the training set, undermining generalization and causing a performance reduction on unknown data. The CLAHE

Table 2. Results of each network architecture used in this work.

Network	Accuracy	Sensitivity	Specificity
CNet-2	0.65	0.65	0.65
CNet-3	0.67	0.53	0.80
CNet-4	**0.70**	0.65	0.75
CNet-5	0.68	**0.71**	0.64
CNet-6	0.69	0.61	**0.78**
Inception-v3 [21]	0.64	0.50	**0.78**

Table 3. Comparison of the CLAHE application on the CNet-5 network architecture.

Network	Accuracy	Sensitivity	Specificity
CNet-5	0.68	0.71	0.64
CNet-5 + CLAHE	0.72	0.75	0.68

application reduced this effect in CNet-5 training, which shows a lower difference between train and validation metrics.

Maybe because some of the images on the dataset had low contrast due to acquisition problems and the CLAHE improved image contrast by spreading closer gray levels, it helped to normalize pixel values of the input images, enabling the model to generalize better.

One of the causes of the unbalanced sensitivity and specificity issue is that the training of the network aims at lowering the binary cross-entropy of the training set, which is not exactly what we expected from the final model. A training strategy that also optimizes sensitivity and specificity could might to avoid this issue.

Wang *et al.* [22] performed a similar experiment on the whole NIH ChestX-ray dataset using a multi-label disease classification instead of a binary approach. They tested some network architectures, the one that performed better was ResNet-50 [9], a network with 168 layers, but they only trained from scratch the transition and prediction layers. They achieved 0.66 AUC for nodule detection and 0.69 AUC for masses, which were some of the lower rates, as stated by the authors because these pathologies contain small objects.

The receiver operating characteristic (ROC) curve for each of the seven approaches is in Fig. 6. Our best approach achieved 0.72 AUC for lesion detection. However, there is some difference in both approaches to make a statistically significant quantitative comparison of AUCs, since our work uses both nodules and masses in one class against normal chest radiographs on the other with a balanced number of samples, used only PA positioning, and discarded repeated samples from the same patients.

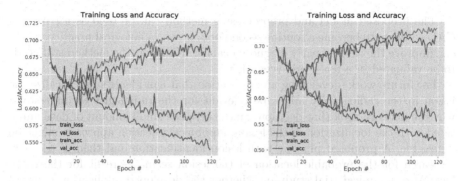

Fig. 5. Training and validation loss and accuracy metrics during the 120 epochs of the CNet-5 model. Training with the original images (left) and using images preprocessed with CLAHE (right).

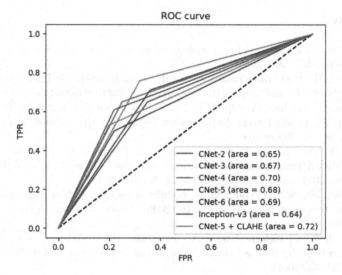

Fig. 6. ROC Curve and AUC (area) of each developed approach. Vertical axis shows true positive rate (TPR) and the horizontal axis shows the false positive rate (FPR).

5   Conclusions

In this work, we presented a strategy to identify pulmonary lesions in chest radiographs using CNN in an image classification approach. Our best model had 5 convolution layers, a reduced number of parameters, and a total depth of 21 layers. It was fully trained, and achieved results close to state of the art using limited hardware.

Even with thousands of images, this model can still present overfitting. A preprocessing strategy by applying histogram equalization on the example images was tested and evaluated and improved the network's results by reducing overfitting.

The proposed model and preprocessing method can be used as a part of a CAD software to become a routine second opinion on radiological practice, which can improve the rates of early diagnosed cancer. However, the method still needs to be validated in a real-world scenario.

For future work, developing an application that can identify and localize the lesion in the radiography might be relevant for the radiologist to understand why the network took that decision. Since there is a small amount of data annotated with the localization of the lesion, an object detection approach might be unfeasible. A proposed pipeline with lesion identification and then, if positive, to search for the probable location of the lesion seems more fit for the problem. Also, a clinical trial study on whether the developed application is useful to increase diagnostic accuracy of radiologists with the achieved sensitivity and specificity values might be relevant.

References

1. Abadi, M., et al.: TensorFlow: a system for large-scale machine learning. In: OSDI, vol. 16, pp. 265–283 (2016)
2. Araujo, L.H., et al.: Lung cancer in Brazil. Jornal Brasileiro de Pneumologia **44**(1), 55–64 (2018). https://doi.org/10.1590/s1806-37562017000000135
3. Bradski, G., Kaehler, A.: OpenCV. Dr. Dobb's J. Softw. Tools **3** (2000)
4. Chollet, F., et al.: Keras: deep learning library for theano and tensorflow **7**(8) (2015). https://keras.io/k
5. del Ciello, A., Franchi, P., Contegiacomo, A., Cicchetti, G., Bonomo, L., Larici, A.R.: Missed lung cancer: when, where, and why? Diagn. Interv. Radiol. **23**(2), 118–126 (2017). https://doi.org/10.5152/dir.2016.16187
6. Diederich, S.: Solitary pulmonary nodule: detection and management. Cancer Imaging **6**(Special Issue A), S42–S46 (2006). https://doi.org/10.1102/1470-7330.2006.9004
7. Gibbs, J.M., Chandrasekhar, C.A., Ferguson, E.C., Oldham, S.A.: Lines and stripes: where did they go?–From conventional radiography to CT. Radiographics **27**(1), 33–48 (2007)
8. Glorot, X., Bordes, A., Bengio, Y.: Deep sparse rectifier neural networks. In: Proceedings of the Fourteenth International Conference on Artificial Intelligence and Statistics, pp. 315–323 (2011)
9. He, K., Zhang, X., Ren, S., Sun, J.: Deep residual learning for image recognition. In: Proceedings of the IEEE Conference on Computer Vision and Pattern Recognition, pp. 770–778 (2016)
10. Hirsch, F.R., Franklin, W.A., Gazdar, A.F., Bunn, P.A.: Early detection of lung cancer: clinical perspectives of recent advances in biology and radiology. Clin. Cancer Res. **7**(1), 5–22 (2001)
11. Larici, A.R., et al.: Lung nodules: size still matters. Eur. Respir. Rev. **26**(146), 170025 (2017). https://doi.org/10.1183/16000617.0025-2017
12. LeCun, Y., Bottou, L., Bengio, Y., Haffner, P.: Gradient-based learning applied to document recognition. Proc. IEEE **86**(11), 2278–2324 (1998)
13. LeCun, Y., Bengio, Y., Hinton, G.: Deep learning. Nature **521**(7553), 436 (2015)
14. Litjens, G., et al.: A survey on deep learning in medical image analysis. Med. Image Anal. **42**, 60–88 (2017). https://doi.org/10.1016/j.media.2017.07.005

15. Lo, S.C., Lou, S.L., Lin, J.S., Freedman, M., Chien, M., Mun, S.: Artificial convolution neural network techniques and applications for lung nodule detection. IEEE Trans. Med. Imaging **14**(4), 711–718 (1995). https://doi.org/10.1109/42.476112

16. Marchiori, E., Irion, K.L.: Avanços no diagnóstico radiológico dos nódulos pulmonares. Jornal Brasileiro de Pneumologia **34**(1), 2–3 (2008)

17. Mountain, C.F.: Revisions in the international system for staging lung cancer. Chest **111**(6), 1710–1717 (1997). https://doi.org/10.1378/chest.111.6.1710

18. Pizer, S.M., et al.: Adaptive histogram equalization and its variations. Comput. Vis. Graph. Image Process. **39**(3), 355–368 (1987)

19. Shah, P.K., et al.: Missed non-small cell lung cancer: radiographic findings of potentially resectable lesions evident only in retrospect. Radiology **226**(1), 235–241 (2003). https://doi.org/10.1148/radiol.2261011924

20. Siegel, R.L., Miller, K.D., Jemal, A.: Cancer statistics, 2017. CA Cancer J. Clin. **67**(1), 7–30 (2017). https://doi.org/10.3322/caac.21387

21. Szegedy, C., Vanhoucke, V., Ioffe, S., Shlens, J., Wojna, Z.: Rethinking the inception architecture for computer vision. arXiv preprint arXiv:1512.00567 (2015)

22. Wang, X., Peng, Y., Lu, L., Lu, Z., Bagheri, M., Summers, R.M.: Chestx-ray8: hospital-scale chest x-ray database and benchmarks on weakly-supervised classification and localization of common thorax diseases. CoRR abs/1705.02315 (2017)

A Pipelined Approach to Deal with Image Distortion in Computer Vision

Cristiano Rafael Steffens[(✉)] [iD], Lucas Ricardo Vieira Messias[iD],
Paulo Lilles Jorge Drews-Jr[iD], and Silvia Silva da Costa Botelho

Centro de Ciências Computacionais, Universidade Federal do Rio Grande – FURG,
Rio Grande, Brazil
cristianosteffens@furg.br
http://nautec.c3.furg.br

Abstract. Image classification is a well-established problem in computer vision. Most state-of-the-art models rely on Convolutional Neural Networks to achieve near-human performance in that task. However, CNNs have shown to be susceptible to image manipulation, which undermines the trustability of perception systems. This property is critical, especially in unmanned systems, autonomous vehicles, and scenarios where light cannot be controlled. We investigate the robustness of several Deep-Learning based image recognition models and how the accuracy is affected by several distinct image distortions. The distortions include ill-exposure, low-range image sensors, and common noise types. Furthermore, we also propose and evaluate an image pipeline designed to minimize image distortion before the image classification is performed. Results show that most CNN models are marginally affected by mild miss-exposure and Shot noise. On the one hand, the proposed pipeline can provide significant gain on miss-exposed images. On the other hand, harsh miss-exposure, signal-dependent noise, and impulse noise, incur in a high impact on all evaluated models.

Keywords: Computer vision · Data preprocessing · Image distortion

1 Introduction

Computer vision has become an important component of many robotic and autonomous systems. Vision systems combine elements of the camera, hardware, and computer algorithms to process visual data into useful information. Domestic and assistive robots [9,18], autonomous cars [2], harvesting robots [29], automated visual inspection [17,23], and surveillance robots [14] are some practical uses of computer vision systems applied to robotics.

This study was financed in part by the Coordenação de Aperfeiçoamento de Pessoal de Nível Superior – Brasil (CAPES) – Finance Code 001.

© Springer Nature Switzerland AG 2020
R. Cerri and R. C. Prati (Eds.): BRACIS 2020, LNAI 12319, pp. 212–225, 2020.
https://doi.org/10.1007/978-3-030-61377-8_15

Image Recognition is one of the most mature computer vision tasks. Recent advancements in Deep Learning (DL) based classification have attained near human level accuracy in various datasets. Among the state-of-the-art Convolutional Neural Networks (CNN) we highlight VGG [22], DenseNet [8], Inception-ResNet-v2 [25], MobileNet-v2 [21], NASNetLarge [35], NASNetMobile [35], Inception-v3 [5], and ResNeXt [33].

Large and varied training datasets are crucial for machine learning classifiers to achieve good performance [30]. Most of the DL image classification models are developed to maximize accuracy on one of the following datasets: Imagenet ILSVRC [20], MS-COCO [12], CIFAR [11], and PASCAL VOC [4]. These datasets have provided support for the development of modern CNN based image recognition models, with extensive and diverse labeled training data. They provide a truthful benchmark to evaluate and compare within distinct models. Nevertheless, we usually quantify the generalization capabilities of any trained model by measuring its performance on a held-out test set.

Image recognition models have been under hard scrutiny [10,19]. Nevertheless, an open issue seems to be overlooked: 'Are classification CNNs ready to deal with ill-exposed, noisy, or over-compressed images?'. While these conditions are commonplace in any computer vision pipeline, we notice that their impact in the final prediction accuracy lacks a deeper assessment. The contribution of this paper is threefold. First, we propose a theoretical framework to support the investigation on the impact of several image distortions on state-of-the-art image recognition models. Then, using a comprehensive set of image recognition models, we evaluate their robustness against image distortions that are common to most computer-vision and machine-vision applications. Finally, we propose an alternative pipeline and discuss the impact of image pre-processing restoration in the classification accuracy.

2 Methodology

We divide the methodology into two main parts. First, we measure the impact of image distortion on several DL based image recognition models using synthetically generated samples. Then, we propose and evaluate a pipeline-based approach which intends to mitigate the image distortions, and, therefore, improve classification results. The remaining of this section provides the fundamentals to reproduce our experiments.

2.1 Part I: Assessing the Impact

To assess the robustness of image recognition models towards miss-exposure and noise we used pre-trained models on the ImageNet ILSRVC Challenge [20] dataset. All CNN recognition models are used 'as is'. We keep the same input shapes, weights and biases, and pooling layers provided by the original authors, without any fine-tuning.

The ILSRVC Challenge rules state that, for each image, the image classifier should generate a list of up to five labels, rated by confidence. The quality of the classifier is evaluated based on the label that best matches the ground-truth label for the image. Given this assessment strategy, an algorithm can identify multiple objects in one image without being penalized in the case one of the objects identified was not included in the ground truth.

For each model, the evaluation methodology is organized in 5 sequential steps as follows:

1. Data Loading – All JPEG compressed images are loaded using the Python Scikit-Image [32] library with an 8-bit unsigned integer data type. Image size and aspect ratio may vary among files.
2. Distortion – The images are distorted with the following distortions: Gamma power transformation, percentile-wise truncation, Salt and Pepper noise, Gaussian noise, Poisson noise, and Speckle noise. A sample showing the original image as well as the distorted and restored versions is shown in Fig. 1. Details on the distortions are presented in Sect. 2.1.
3. Pre-processing – In order to fit the input size of the classification network the images need to be cropped and reshaped. We use first-order spline interpolation for resizing. For down-scaling, we use a Gaussian filter with $\sigma = \frac{s-1}{2}$ as anti-aliasing strategy, where s is the scaling factor. Further adjustments are made for each model as to fit their particularities.
4. Inference – Once pre-processed, the content of each image is labeled by the classification model. After the inference is performed, the outcomes are stored to allow further evaluation.
5. Evaluation – In the evaluation, we take in account five popular classifier metrics: Top-1, Top-3, and Top-5 Accuracy, Precision, and F1-Score. Precision and F1-Score provide relevant information once they consider the number of true instances for each label. For the sake of brevity, only Top-1 accuracy results will be tabulated in the body of the present text.

Distortions. Figure 1 shows the impact of several distortions applied to the images and a restored counterpart (which will be discussed later in Sect. 2.2). The details about distortion models are discussed in the upcoming sections.

Gamma Power Transformation is a nonlinear operation used to encode and decode luminance values in image systems [27]. It is used to adjust and compensate the response of some luminance levels in the input image. We use Gamma Power Transformation to mimic the conditions observed in under-exposed and overexposed images as $\hat{I} = I^\gamma$. The power transformation is followed by min-max normalization in order to adjust pixel values to a valid representation range. This transformation results in lost data in dark regions, when $\gamma > 1$, or bright and washed-out regions, when $\gamma < 1$. For simulation purposes, we used $\gamma = [\frac{1}{4}; \frac{1}{6}; \frac{1}{8}; 4; 6; 8]$.

Additive White Gaussian Noise (AWGN) is randomly added to the input image. The random noise follows a normal distribution, defined by:

$$p(z) = \frac{1}{\sqrt{2\pi\sigma^2}} e^{-(z-\bar{z})^2/2\sigma^2}, \tag{1}$$

where z represents intensity, $\bar{z}$ is the mean value of z and σ its standard deviation. In this work, we used $\sigma = 23.55$, which results in severely damaged images.

Shot Noise also know as Photon or Poisson Noise [31], it is a data-dependent noise model. A Poisson model of noise may be more appropriate than a Gaussian model for low light conditions where the noise is due to low photon counts [27]. Talbot *et al.* [28] claims that image sensor noise is dominated by Poisson statistics, even at high illumination level, this being a typical effect in images captured by robots.

Salt and Pepper Noise (S&P) is an impulse noise, added to an image by setting white (pixel value equals 255 in an 8-bit per color color-space) and black pixels (pixel value equals 0) with a probability per pixel P. In our experiments, we use $P = 0.3$. In real applications, Salt & Pepper noise is often associated with dead pixels the camera's sensor array.

Speckle Noise is originated from coherent processing of back-scattered signals from multiple distributed points [31]. The Speckle interference of an image I is expressed as $\hat{I} = I + (n \times I)$, where n is a uniform noise (with $\mu = 0$ and $\sigma^2 = 1$). Speckle noise in real applications is often related to environmental conditions that affect the imaging sensor during image acquisition. It is also common in medical images, as well as active Radar images [16].

Classification Models. Image classification models are built to predict the classes of objects present in an image. The remaining of this paper explores CNN based classification models, which have been adjusted for the ImageNet Large Scale Visual Recognition Challenge (ILSVRC). Convolutional networks have recently enjoyed great success in this task. Among these, we highlight the following popular models: *i.* VGG, by Simonyan *et al.* [22], which obtained both first and second place in the ILSVRC-2014; *ii.* ResNet, by [5], which obtained first place in the ILSVRC-2015; *iii.* Inception-v3, by [26], which introduces factorized convolutions and aggressive regularization; *iv.* Inception-ResNet-v2, by [25], which combines residual connections; *v.* MobileNetV1, by [7], which includes depthwise separable convolutions between the regular convolutions layers; *vi.* DenseNet, by [8], where each layer obtains additional inputs from all preceding layers and passes on its own feature-maps to all subsequent layers; *vii.* NASNet, by [35], NASNet which automates network design using information acquired on a small dataset.

It is important to highlight that many state of the art object localization, semantic segmentation, instance segmentation, image recognition, object localization, object tracking, and visual odometry models, as well as feature-based

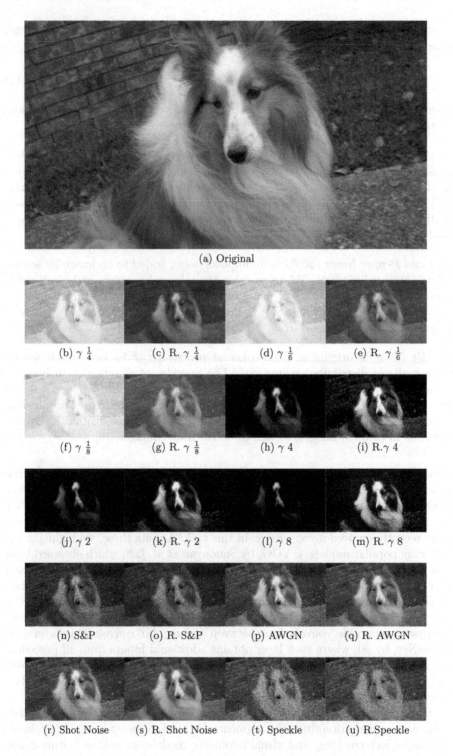

(a) Original

(b) $\gamma \frac{1}{4}$ (c) R. $\gamma \frac{1}{4}$ (d) $\gamma \frac{1}{6}$ (e) R. $\gamma \frac{1}{6}$

(f) $\gamma \frac{1}{8}$ (g) R. $\gamma \frac{1}{8}$ (h) $\gamma 4$ (i) R.$\gamma 4$

(j) $\gamma 2$ (k) R. $\gamma 2$ (l) $\gamma 8$ (m) R. $\gamma 8$

(n) S&P   (o) R. S&P   (p) AWGN   (q) R. AWGN

(r) Shot Noise   (s) R. Shot Noise   (t) Speckle   (u) R.Speckle

Fig. 1. Example of the image distortions and their restored versions.

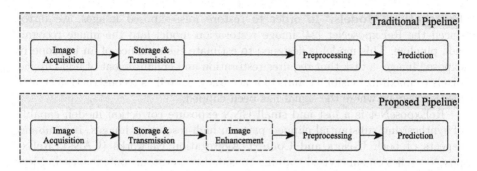

Fig. 2. Traditional deep-learning based computer vision pipeline versus the proposed approach

loss functions, are, in fact, built upon these popular models. Therefore, one can assume that many of them may present a similar performance degradation when submitted to less than optimal input data.

2.2 Part II: An Alternative Image Recognition Pipeline

Preliminary results, which will be discussed in Sect. 3 have provided some insights and instigated further developments. In this direction, the question that we tried to answer is "How can we prevent ill-exposure and noise in real robotics applications (weakly controlled illumination, self-driving cars)?".

In robotics and autonomous systems, where vision-based perception is required, the usual pipeline is designed as follows. First, a sensor acquires the RAW data. Then, the RAW data is transformed into an sRGB image and compressed. In this step, most cameras include some basic denoising and exposure compensator. Next, the compressed data is sent to a computer vision algorithm. After that, the algorithm extracts the useful data. Finally, the extracted data is used to actuate the autonomous system behavior.

Figure 2 presents an overview of a typical deep-leaning based passive computer vision pipeline. We can break this pipeline in four broad activities: Image Acquisition, Storage and Transmission, Preprocessing, and Prediction. The first steps are often built into the hardware itself. While most manufacturers provide some level of configuration, they hardly provide full control of the image acquisition settings. The same holds for storage and transmission, where the settings are usually restricted to popular image compression algorithms and settings to balance quality vs time constraints.

In order to minimize the undesirable impacts of noise and miss-exposure, we modify the traditional pipeline by introducing an image restoration step, which operates in the RGB colorspace, after the image has already been compressed and transmitted. The restoration takes place immediately before the computer vision algorithm. The models used for restoration are described in Sect. 2.2.

Restoration Models. In order to restore miss-exposed images, we introduced the ReExposeNet [24] image restoration model into the image recognition pipeline. This model is designed to estimate the radiance of an improperly exposed image, a task that requires restoration and enhancement of non-clipped pixels to maximize visibility and color accuracy, as well as reconstruction strategies for regions where the signal has been clipped.

ReExposeNet is a fast and small CNN exposure correction model, capable of synthesizing substantial clipped parts in high-resolution images. It combines aspects of both U-Nets and Context Aggregation Networks (CANs). ReExposeNet relies on supervised training considering a custom content-based objective function to maximize restoration and reconstruction in clipped areas. It has been adjusted considering both synthetic and real miss-exposed images in three different datasets. ReExposeNet is released as a one-size-fits-all solution, which can be consistently applied on a wide range of image miss-exposure levels. For the present work, we used the model as released by its authors, without further fine-tuning.

To restore images damaged by noise, we used the DnCNN-3 model [34]. DnCNN-3 is a very deep feed-forward denoising convolutional neural network. It relies on residual learning and batch normalization to speed up the training process as well as boost the denoising performance. Zhang *et al.* claims to provide a single DnCNN model to tackle several general image denoising tasks, such as blind Gaussian denoising, single image super-resolution, and JPEG image deblocking. The authors show that the DnCNN model can not only exhibit high effectiveness in several general image denoising tasks but also be efficiently implemented by benefiting from GPU computing, which makes it adequate for real-time applications.

3   Results

We conducted a comprehensive evaluation of the robustness of state-of-the-art image recognition neural networks towards image distortion. The provided metrics relate to the performance obtained on the ImageNet ILSRVC Challenge validation subset. All results were computed on the Imagenet ILSRVC [20] Validation subset, which consists of 1000 distinct categories with 50 images each. We using 32-bit floating-point precision for the inference in all networks. We notice the Top-1 Accuracy in all evaluated models is slightly different from the official reports. This may be related to the image processing libraries, re-scaling and interpolation strategies as well as image cropping strategies.

Table 1 shows the impact of damaged images and the effects of restoration on the evaluated models model. We highlight each result in the presented tables according to the following strategy. Results for undamaged images are shown in black. Conditions that worsened the accuracy by up to 10% are marked in green (low impact). Conditions that worsened the performance of the model by any value between 10% and 30% are shown in orange (moderate impact). Conditions that worsened the network outcomes by more than 30% are shown in red (critical impact). This scale allows for fast visualization of the models' robustness in the light of its original performance.

Results for VGG-16 [22] indicate that this model is highly susceptible to image distortion. Except for Shot noise, all image distortions resulted in an accuracy drop larger than 10%. Nevertheless, we notice that the pipeline, including restoration, provides an expressive gain even under extreme miss-exposure. Otherwise, for noisy images, we notice that the inclusion of the denoising model worsened the results on both AWGN and Shot noise. For S&P and Speckle noise, the restoration offered marginal improvements.

Table 1. Top-1 Accuracy for each model on distorted images.

Classification Model	[22]	[5]	[25]	[26]	[8]	[35] *	[35]**	[21]
Original Images	0.612	0.668	0.747	0.773	0.663	0.806	0.693	0.600
Gamma 4	0.401	0.459	0.591	0.626	0.458	0.691	0.486	0.376
Gamma 4 R.	0.553	0.611	0.697	0.730	0.633	0.761	0.631	0.563
Gamma 6	0.261	0.313	0.454	0.493	0.314	0.580	0.339	0.237
Gamma 6 R.	0.435	0.498	0.611	0.650	0.527	0.692	0.525	0.438
Gamma 8	0.175	0.217	0.342	0.385	0.224	0.467	0.239	0.157
Gamma 8 R.	0.429	0.491	0.586	0.628	0.506	0.665	0.507	0.425
Gamma 1/4	0.455	0.501	0.645	0.683	0.445	0.745	0.551	0.295
Gamma 1/4 R.	0.622	0.665	0.723	0.754	0.560	0.781	0.650	0.479
Gamma 1/6	0.330	0.376	0.553	0.598	0.313	0.680	0.425	0.186
Gamma 1/6 R.	0.625	0.662	0.722	0.753	0.548	0.781	0.649	0.456
Gamma 1/8	0.236	0.280	0.469	0.516	0.222	0.612	0.324	0.130
Gamma 1/8 R.	0.618	0.647	0.716	0.746	0.526	0.776	0.636	0.419
Gauss	0.508	0.534	0.687	0.716	0.588	0.765	0.607	0.436
Gauss R.	0.497	0.559	0.660	0.714	0.585	0.749	0.603	0.450
Poisson	0.586	0.626	0.732	0.759	0.651	0.796	0.671	0.567
Poisson R.	0.445	0.506	0.608	0.672	0.545	0.714	0.553	0.377
S&P	0.143	0.126	0.362	0.405	0.191	0.527	0.243	0.070
S&P R.	0.141	0.131	0.322	0.370	0.203	0.469	0.225	0.092
Speckle	0.081	0.069	0.261	0.311	0.145	0.423	0.167	0.041
Speckle R.	0.091	0.086	0.259	0.310	0.161	0.402	0.174	0.057

Resnet [5] shows to be robust against mild exposure variations ($\gamma = [\frac{1}{4}; 4]$) and Poisson noise. Pixel value truncation, however, shows to have a higher impact dropping Top-1 accuracy levels from 0.668 to 0.603, for truncation in the bright part, and 0.593, for truncation in the darker pixels. Coarse miss-exposure, Gaussian noise, impulse noise, and Speckle noise also have high impact in the metric.

From Inception-v3 network [25] the results show that, in general, distorted images have a significant interference on the classification performance. This model appears to be robust when applied in distortions as Gamma power transformation with $\gamma = [\frac{1}{4}; \frac{1}{6}; 4; 6]$, quantile-based truncation with both values, Gaussian and Poisson noise. However, when applied heavier distortions as $\gamma = [\frac{1}{8}; 8]$, S&P, and speckle, the model's accuracy decays significantly.

The Inception-ResNet-v2 [26] model is able to achieve Top-5 accuracy larger than 0.5 in all but speckle-noise conditions. As in the models discussed above, we observe that the models are robust towards slightly miss-exposed images resulting from $\gamma = [\frac{1}{4}; 4]$ or quantile-wise pixel value truncation, which make the accuracy drop by a small percentage. We also observe the model's robustness in the face of Poisson and Gaussian noise.

The results for DenseNet201 [8] indicate that this model is little affected by minor disturbances in pixel intensity ($\gamma = \frac{1}{4}$ and $\gamma = 4$), pixel value truncation at Q_1, and Poisson noise. Gaussian noise shows to have a moderate impact. Gross miss-exposure generated through power transformations with $\gamma = [\frac{1}{6}; \frac{1}{8}; 6; 8]$, however, has shown to have expressive impact in the classification accuracy. The metrics show a mirrored effect, displaying similar results for both dark and bright images. Nevertheless, none of the above seems to affect classification accuracy as much as the impulse noise and speckle noise.

NASNetLarge [35]* is the model that offers the best accuracy, precision, and F1-Score among all models considered in this study. We find it to be robust towards a wide range of ill exposure levels, gross Gaussian and Poisson noise. Nevertheless, it also suffers under salt and pepper and speckle noise. NASNetMobile [35]** is a smaller version (in terms of trainable parameters) of NASNet. This model struggles considerably more under harsh conditions of miss-exposure, Salt & Pepper noise, and Speckle interference as compared to NASNetLarge. Light under-exposure and light over-exposure, as well as Poisson noise, tend to have little effect on the accuracy and precision of image recognition.

Lastly, the results obtained by MobileNetV2 [21], a model conceived to be embedded in low capacity devices, sacrificing accuracy for feasibility. On MobileNetV2 only the original images, $\gamma = 2$, quantile-wise truncated images, and images affected by Poisson noise resulted in Top-1 accuracy higher than 0.5. Power transformation and noise resulted in expressive accuracy drops. S&P and Speckle noise result in fluky predictions.

Figure 3 provides an overview for all evaluated models, showing their performance under optimal conditions, with distorted images, and with restored images. Considering the results as a whole, the top-3 models with the best performance are Inception-ResNet-v2 [25] and NASNetLarge [35] and Inception-v3 [26]. Overall, the NASNetLarge model was the best model in terms of accuracy with expressive gains over the second place. However, the Xception model showed to be cost-effective with a moderated size (in terms of trainable parameters) and a small difference from the best accuracy.

Robustness towards typical image distortions seems to be strongly linked with the amount of trainable parameters. On one end, NASNetLarge [35], Inception-ResNet-v2 [25], and Inception-v3 [3] are less affected by noise and harsh miss-exposure. On the other end, MobileNetV2 [21], DenseNet201 [8], and NAS-NetMobi [35] have shown to be fragile in dealing with distorted images.

Poisson noise seems to have little impact on most of the evaluated models. We believe this to be related to the fact that most digital images already present this type of distortion due to sensor and electronics properties. It is plausible that the models were able to learn and avoid it during training time. Even in extreme conditions, Gaussian noise shows little impact on image recognition. This may be attributed to data augmentation techniques used during training time, which often include Gaussian noise as an alternative to generating synthetic samples.

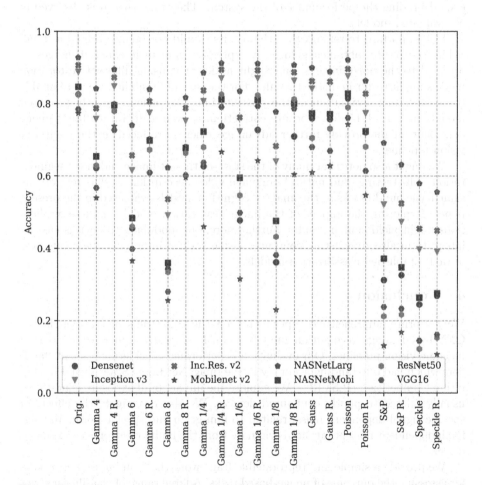

Fig. 3. Top-1 Accuracy overview for several models under distinct noise conditions and after image restoration

We notice that the majority of the evaluated models show a small deterioration in accuracy when applied to slightly miss-exposed images. Nevertheless, we might recall that most of the recent advances in image recognition provide marginal improvements over previously existing state-of-the-art image recognition models. Looking in perspective, even a modest reduction in accuracy undermines years of refinement.

For overexposed images, the ReExposeNet model was able to restore the performance of the object recognition model to the same levels observed in undamaged images. For dark images, we also notice that the restoration model contributed significantly to reduce the number of misclassifications. Yet for noisy images, DnCNN-3, even though restored images are aesthetically pleasing, classification results show it is unable to improve the prediction accuracy, sometimes even degrading the performance of the system. The same pattern is observed in all evaluated models.

It is important to remark that the results should not be interpreted as final. While the results obtained evince an expressive gain in ill-exposed images as opposed to the noisy images, one might notice that using different restoration models might result in distinct results. For the sake of feasibility, we were unable to perform such evaluation at the present time. We believe recently proposed models such as [1,15] for exposure correction, as well as [6,13] for blind image denoising, might provide further advantage when introduced in the computer vision pipeline.

We might also note that many of the state-of-the-art image and video segmentation, self-steering, and natural language video description models rely on the same basic build blocks as the image recognition models evaluated in the present paper. Exploring the impact of the image restoration models in these applications might yield similar results. Furthermore, our modular approach is easy to integrate in already existing computer vision setups, improving the robustness of well established classification models.

4   Conclusion

We perform comprehensive experiments on the robustness of state-of-the-art CNN-based image recognition models towards several common conditions in machine vision pipelines. Our evaluation is performed considering a set of classifiers which had outstanding accuracy in the ImageNet Large Scale Visual Recognition Competition (ILSVRC). We explore the impact of ill exposure conditions, usually related to poorly configured devices. We also investigate the effects of specific signal-dependent noise, related to film-grain and speckle noise, and signal independent noise, that often originate from the faulty electronics and sensor defects.

We provide a simple and reproducible framework that can be used as a basis for assessing the outcome of image-based tasks. A large range of classifier metrics allows for sensitivity and specificity assessment, which offers a succinct representation of the performance of the classifier. Such metrics are equally relevant and provide a straightforward indication of their applicability in robotics and autonomous systems.

Our proposal is highly inspired on approaches that are already proven in other research fields. Tests from many models of object recognition indicate that existing CNNs are little affected by slight miss-exposure or saturated pixel values. Poisson noise and AWGN also have limited effect on the accuracy in most of the tested classification models. Otherwise the models appear vulnerable to image distortions caused by severe miss-exposure and signal-independent noise. For miss-exposed images, the improved pipeline, including an additional image enhancement step, has shown to provide an expressive gain in terms of robustness.

Many relevant research questions arise from the proposed approach and the experimental results presented in this paper. One especially fruitful avenue for further investigation is to target segmentation, mapping, and localization systems to analyze the robustness of these models in light of the growing number of robotics and automation applications that include scene understanding and labeling as a fundamental building block.

References

1. Afifi, M., Derpanis, K.G., Ommer, B., Brown, M.S.: Learning to correct overexposed and underexposed photos. arXiv preprint arXiv:2003.11596 (2020)
2. Chen, C., Seff, A., Kornhauser, A., Xiao, J.: DeepDriving: learning affordance for direct perception in autonomous driving. In: The IEEE International Conference on Computer Vision (ICCV), December 2015
3. Chollet, F.: Xception: deep learning with depthwise separable convolutions. In: Proceedings of the IEEE Conference on Computer Vision and Pattern Recognition, pp. 1251–1258 (2017)
4. Everingham, M., Van Gool, L., Williams, C.K., Winn, J., Zisserman, A.: The pascal visual object classes (VOC) challenge. Int. J. Comput. Vision 88(2), 303–338 (2010)
5. He, K., Zhang, X., Ren, S., Sun, J.: Deep residual learning for image recognition. In: Proceedings of the IEEE Conference on Computer Vision and Pattern Recognition, pp. 770–778 (2016)
6. Hou, Y., et al.: NLH: a blind pixel-level non-local method for real-world image denoising. IEEE Trans. Image Process. 29, 5121–5135 (2020)
7. Howard, A.G., et al.: MobileNets: efficient convolutional neural networks for mobile vision applications. arXiv preprint arXiv:1704.04861 (2017)
8. Huang, G., Liu, Z., Van Der Maaten, L., Weinberger, K.Q.: Densely connected convolutional networks. In: Proceedings of the IEEE Conference on Computer Vision and Pattern Recognition, pp. 4700–4708 (2017)
9. Iocchi, L., Holz, D., Ruiz-del Solar, J., Sugiura, K., Van Der Zant, T.: RoboCup@Home: analysis and results of evolving competitions for domestic and service robots. Artif. Intell. 229, 258–281 (2015)
10. Karim, R., Islam, M.A., Mohammed, N., Bruce, N.D.: On the robustness of deep learning models to universal adversarial attack. In: 2018 15th Conference on Computer and Robot Vision (CRV), pp. 55–62. IEEE (2018)
11. Krizhevsky, A., Nair, V., Hinton, G.: The CIFAR-10 dataset. http://www.cs.toronto.edu/kriz/cifar.html 55 (2014)
12. Lin, T.-Y., et al.: Microsoft COCO: common objects in context. In: Fleet, D., Pajdla, T., Schiele, B., Tuytelaars, T. (eds.) ECCV 2014. LNCS, vol. 8693, pp. 740–755. Springer, Cham (2014). https://doi.org/10.1007/978-3-319-10602-1_48

13. Liu, D., Wen, B., Jiao, J., Liu, X., Wang, Z., Huang, T.S.: Connecting image denoising and high-level vision tasks via deep learning. IEEE Trans. Image Process. **29**, 3695–3706 (2020)
14. Lopez, A., Paredes, R., Quiroz, D., Trovato, G., Cuellar, F.: Robotman: a security robot for human-robot interaction. In: 2017 18th International Conference on Advanced Robotics (ICAR), pp. 7–12, July 2017. https://doi.org/10.1109/ICAR. 2017.8023489
15. Lv, F., Lu, F.: Attention-guided low-light image enhancement. arXiv preprint arXiv:1908.00682 (2019)
16. Maity, A., Pattanaik, A., Sagnika, S., Pani, S.: A comparative study on approaches to speckle noise reduction in images. In: 2015 International Conference on Computational Intelligence and Networks, pp. 148–155. IEEE (2015)
17. Molina, M., Frau, P., Maravall, D.: A collaborative approach for surface inspection using aerial robots and computer vision. Sensors **18**(3), 893 (2018)
18. Piyathilaka, L., Kodagoda, S.: Human activity recognition for domestic robots. In: Mejias, L., Corke, P., Roberts, J. (eds.) Field and Service Robotics. STAR, vol. 105, pp. 395–408. Springer, Cham (2015). https://doi.org/10.1007/978-3-319-07488-7_27
19. Recht, B., Roelofs, R., Schmidt, L., Shankar, V.: Do imagenet classifiers generalize to imagenet? arXiv preprint arXiv:1902.10811 (2019)
20. Russakovsky, O., et al.: ImageNet large scale visual recognition challenge. Int. J. Comput. Vision **115**(3), 211–252 (2015). https://doi.org/10.1007/s11263-015-0816-y
21. Sandler, M., Howard, A., Zhu, M., Zhmoginov, A., Chen, L.C.: MobileNetV2: inverted residuals and linear bottlenecks. In: Proceedings of the IEEE Conference on Computer Vision and Pattern Recognition, pp. 4510–4520 (2018)
22. Simonyan, K., Zisserman, A.: Very deep convolutional networks for large-scale image recognition. arXiv preprint arXiv:1409.1556 (2014)
23. Soares, L.B., et al.: Seam tracking and welding bead geometry analysis for autonomous welding robot. In: 2017 Latin American Robotics Symposium (LARS) and 2017 Brazilian Symposium on Robotics (SBR), pp. 1–6. IEEE (2017)
24. Steffens, C.R., Huttner, V., Messias, L.R.V., Drews, P.L.J., Botelho, S.S.C., Guerra, R.S.: CNN-based luminance and color correction for ill-exposed images. In: 2019 IEEE International Conference on Image Processing (ICIP), pp. 3252–3256, September 2019. https://doi.org/10.1109/ICIP.2019.8803546
25. Szegedy, C., Ioffe, S., Vanhoucke, V., Alemi, A.A.: Inception-v4, Inception-ResNet and the impact of residual connections on learning. In: Thirty-First AAAI Conference on Artificial Intelligence (2017)
26. Szegedy, C., Vanhoucke, V., Ioffe, S., Shlens, J., Wojna, Z.: Rethinking the inception architecture for computer vision. In: Proceedings of the IEEE Conference on Computer Vision and Pattern Recognition, pp. 2818–2826 (2016)
27. Szeliski, R.: Computer Vision: Algorithms and Applications. TCS. Springer, London (2010). https://doi.org/10.1007/978-1-84882-935-0
28. Talbot, H., Phelippeau, H., Akil, M., Bara, S.: Efficient Poisson denoising for photography. In: 2009 16th IEEE International Conference on Image Processing (ICIP), pp. 3881–3884. IEEE (2009)
29. Taqi, F., Al-Langawi, F., Abdulraheem, H., El-Abd, M.: A cherry-tomato harvesting robot. In: 2017 18th International Conference on Advanced Robotics (ICAR), pp. 463–468, July 2017. https://doi.org/10.1109/ICAR.2017.8023650

30. Therrien, R., Doyle, S.: Role of training data variability on classifier performance and generalizability. In: Medical Imaging 2018: Digital Pathology, vol. 10581, p. 1058109. International Society for Optics and Photonics (2018). https://doi.org/10.1117/12.2293919

31. Verma, R., Ali, J.: A comparative study of various types of image noise and efficient noise removal techniques. Int. J. Adv. Res. Comput. Sci. Softw. Eng. **3**(10) (2013)

32. van der Walt, S., et al.: The scikit-image contributors: Scikit-image: image processing in Python. PeerJ **2**, e453 (2014). https://doi.org/10.7717/peerj.453

33. Xie, S., Girshick, R., Dollár, P., Tu, Z., He, K.: Aggregated residual transformations for deep neural networks. In: Proceedings of the IEEE Conference on Computer Vision and Pattern Recognition, pp. 1492–1500 (2017)

34. Zhang, K., Zuo, W., Chen, Y., Meng, D., Zhang, L.: Beyond a Gaussian denoiser: residual learning of deep CNN for image denoising. IEEE Trans. Image Process. **26**(7), 3142–3155 (2017)

35. Zoph, B., Vasudevan, V., Shlens, J., Le, Q.V.: Learning transferable architectures for scalable image recognition. In: Proceedings of the IEEE Conference on Computer Vision and Pattern Recognition, pp. 8697–8710 (2018)

A Robust Automatic License Plate Recognition System for Embedded Devices

Lucas S. Fernandes[4], Francisco H. S. Silva[1,2], Elene F. Ohata[1,2],
Aldísio Medeiros[1,2], Aloísio V. Lira Neto[3], Yuri L. B. Nogueira[5],
Paulo A. L. Rego[4,5], and Pedro Pedrosa Rebouças Filho[1,2(✉)]

[1] Laboratório de Processamento Digital de Imagens, Sinais e Computação Aplicada,
Instituto Federal de Federal de Educação, Ciência e Tecnologia do Ceará, Fortaleza,
Ceará, Brazil
pedrosarf@ifce.edu.br,
{herculessilva,eleneohata,aldisio.medeiros}@lapisco.ifce.edu.br
[2] Programa de Pós-Graduação em Engenharia de Teleinformática,
Universidade Federal do Ceará, Fortaleza, Ceará, Brazil
[3] Polícia Rodoviária Federal, Rio Largo, Brazil
aloisio.lira@prf.gov.br
[4] Group of Computer Networks, Software Engineering and Systems (GREat),
Universidade Federal do Ceará, Fortaleza, Brazil
lucasfernandes42@lia.ufc.br, pauloalr@ufc.br
[5] Department of Computing, Federal University of Ceará, Fortaleza, Ceará, Brazil
yuri@dc.ufc.br

Abstract. Automatic License Plate Recognition (ALPR) systems are
used in many real-world applications, such as road traffic monitoring
and traffic law enforcement, and the use of deep learning can result in
efficient methods. In this work, we present an ALPR system efficient
for edge computing, using a combination of MobileNet-SSD for vehicle
detection, Tiny YOLOv3 for license plate detection and OCR-net for
character recognition. This method was evaluated in two datasets on a
NVIDIA Jetson TX2 system, obtaining 96.87% of accuracy and 8 FPS of
framerate in a public real-world scenario dataset and achieving 90.56%
of accuracy and 11 FPS of framerate in a private dataset of traffic monitoring images, considering the recognition of at least six characters. It
is faster than related works with similar deep learning approaches, that
achieved at most 2 FPS, and slightly inferior in accuracy, with less than
10% of difference in the worst scenario. This shows the proposed method
is well balanced between accuracy and speed, thus, suitable for embedded
devices.

Keywords: ALPR · OCR · Deep learning · Edge computing ·
Embedded systems · Jetson TX2

© Springer Nature Switzerland AG 2020
R. Cerri and R. C. Prati (Eds.): BRACIS 2020, LNAI 12319, pp. 226–239, 2020.
https://doi.org/10.1007/978-3-030-61377-8_16

1   Introduction

The Automatic License Plate Recognition (ALPR) problem consists in detecting and reading one or more license plates (LP) in images. The ALPR systems are used in a relevant number of real-world applications, such as road traffic monitoring, automatic toll collection, traffic law enforcement, and parking lot access control [3].

In general, ALPR systems are composed of the following stages: the vehicle's LP region is located and cropped from the input image, LP's characters are segmented and classified, thus, reading the LP. However, recent approaches first detect the regions of the vehicle before the LP detection in order to reduce false positives and the processing time [7,9,10,16,17]. Also, recent works replace character recognition and classification for character detection, combining both stages [10,16,17] while some works use completely segmentation-free approaches [2,5,6].

In each country or region, LPs obey some established patterns. In Brazil, for example, most LPs are composed of a white background with black characters or a red background with white characters. Besides considering the possible differences between LPs, ALPR systems have to deal with the variation of the quality of the images, which may differ in illumination, shadows, blur, inclinations, and other kinds of distortions.

Therefore, many recent approaches have used robust deep learning techniques [9,10,16,17], as it has improved the state-of-the-art of object detection, speech recognition, among others [11]. These methods achieve good efficiency on a high-end Graphic Processing Unit (GPU) but may be too costly for local processing in weaker devices, being more suitable for cloud computing deployment. This approach using heavy deep learning methods incurs significant latency, energy, and financial overheads and also raises privacy concerns [13], thus, limiting the possibilities of real-world applications.

In this context, a better approach is to use edge computing, i.e., computing the data locally on small and low powered edge devices, as this approach is more attractive to several applications, such as robotics, drone-based surveillance, and autonomous driving [13].

In this work, we propose a complete ALPR system for Brazillian LPs based on the combination of deep learning object detection techniques and efficient enough for embedded system execution. Our main goal is to achieve a balance between accuracy and timing using convolutional neural networks (CNNs) to detect and read an LP aiming a suitable performance in realistic scenarios, allowing the extension of real-world applications through edge computation.

The remainder of this paper is organized as follows. We review related works in Sect. 2. The materials used in our experiments are presented in Sect. 4. The details of the proposed system are described in Sect. 3. In Sect. 5, we report and discuss the experimental results. Finally, conclusions and future works are given in Sect. 6.

2 Related Works

In this section, we will present several works proposed for the different stages of an ALPR system, Vehicle Detection, and LP Detection and Recognition (LPDR). Bringing performance data in precision and execution time of the methods, since the focus of our work is on-time processing real, we must pay attention not only to the efficiency of the method but also to the response time.

2.1 Vehicle Detection

The first stage of an ALPR system is the detection of the vehicle since the LP must be attached to its body. The system's hit rate is highly dependent on the quality of the vehicle's detection, and once the detection method returns a closed image of the vehicle where the LP is cut, or even no LP appears, the system will not be able to recognize all characters and no characters belonging to that vehicle's LP. Next, we will discuss vehicle detection methods proposed by different authors.

Wang et al. [18] proposed a new structure called Envolving Boxes, which determines and refines the object boxes through different representations of attributes of each object. A gyro-fine network (FTN) is responsible for the refinement of the boxes. The method was evaluated using Faster R-CNN in the DETRAC benchmark, where it achieved an improvement of 9.5% mAP, running at 9–13 FPS on an Nvidia Titan X GPU.

Sang et al. [15] proposed a CNN based on Yolov2, Yolov2_Vehicle. During the training of the network, the K-means ++ algorithm groups the bounding and anchoring boxes. Other improvements were imposed, such as the normalization of object boxes to improve the method of calculating losses, removal of repeated convolutional layers, and the merging of attributes from different layers in order to improve the extraction of attributes. The Yolov2_Vehicle was tested in the vehicle dataset of the Beijing Institute of Technology (BIT), reaching 94.78% mAP and running on 4 Nvidia Tesla K80 GPUs.

The Faster R-CNN with Envolving Boxes proposed by Wang et al. [18] proved unable to process real-time images (30 FPS, for example) on a medium-performance GPU. While Yolov2_Vehicle, proposed by Sang et al. [15], required 4 GPUs for use in training and validation. Therefore, both methods may not achieve a satisfactory framerate in limited systems used in embedded applications, such as Jetson TX2.

2.2 License Plate Detection and Recognition

After detecting the vehicle, there are two more crucial steps for the operation of the ALPR, and these are the Detection and Recognition of LP characters (LPDR). Below we describe works that proposed LPDR methods, which used classical attribute extractors in Computer Vision, Convolutional Neural Networks (CNNs), as well as Machine Learning algorithms.

Bulan *et al.* [2] proposed a method for recognizing LP where the first step consists of identifying flaws in the detection of the LP through the legibility classification of the characters present in the LP—using the Transfer Learning technique, with CNN AlexNet as feature extractor and Linear kernel Support Vector Machine (SVM) as classifier. The character recognition stage, on the other hand, used the HOG and LeNet extractors in conjunction with the SVM-Linear classifier. The method achieved more than 99% accuracy, running on an Nvidia GTX 570 GPU, but with a frame rate of 0.5 FPS.

Björklund *et al.* [1] used LP synthetic images from the European Union (EU) to train LP detection and recognition CNNs. The detection task validation was performed in the AOLP dataset, where it performed an accuracy of 99.30%, as well as the recognition task that reached 99.80%. The methods together required an average of 845 ms to process each 640×480 image on a Jetson TX1 embedded system, while on an Nvidia GTX GPU 1080, the same procedure took 25.5 ms.

The method proposed by Bulan *et al.* [2] is not feasible for a real-time application, since its framerate during the validation process was 0.5 FPS. Björklund *et al.* [1] used synthetic images for training its detection and reconnaissance networks, but does not concern itself with vehicle detection, and its validation was performed with images with and without vehicles, but all with license plates, which does not match the purpose of our method.

2.3 Complete ALPR Systems

Some studies have proposed the complete ALPR system, from vehicle detection, through detection to LP recognition. This type of system receives an image with vehicles present, returning the license plate characters of each vehicle in the image. Some real-time applications use this type of system, such as parking lots, speed cameras, and police vehicles. We cite some examples of Complete ALPR Systems below.

Laroca *et al.* [10] implemented a complete Automatic Plate Recognition (ALPR) system composed of three versions based on the YOLO (You Only Look Once) architecture: Yolov2 used in vehicle detection, Fast-Yolov2 responsible for plate detection on a given vehicle, and CR-Net which is a version of YOLO adapted for the detection and recognition of license plate characters. This method reached 95.90% accuracy among the license plates present in Dataset UFPR-ALPR [9], also proposed by the author. The experiment ran at 73 FPS on a high-capacity GPU.

Silva and Jung [17] proposed an ALPR method for LP images in different conditions of visibility, perspectives, and projections. This method uses a network architecture called Warped Planar Object Detection Network (WPOD-NET), which detects the LP and performs a perspective readjustment to assist in character recognition, which is the next step to LP detection. This method was evaluated in OpenALPR Datasets (types BR and EU), SSIG, and AOLP (RP), as well as in the Dataset proposed in their work, the CD-HARD, which contains LP at different angles and distances, presenting greater difficulty for systems of ALPR. The method reached 93.52% on OpenALPR-US, 91.23% on

OpenALPR-BR, LP datasets in frontal perspectives, while on CD-HARD, it reached 75.00% LP accuracy at higher angles and distances, running at 5 FPS on an Nvidia GPU Titan X.

In our work, we propose a complete ALPR system for embedded applications in real-time, since, every day, the need for recognition of LP in security systems and public security in general, both for storing information and in parking with access control, as well as consultations carried out by security agents and also record of infractions in the house of speed cameras. We will compare our method with those previously mentioned, Laroca *et al.* [10] and Silva *et al.* [17], both using a server machine and a device created for embedded applications, since both validated their methods based on LP from Brazil, which is the type that we will validate our work.

3   Proposed Methodology

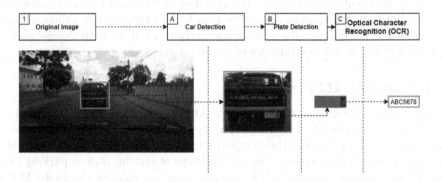

Fig. 1. Proposed methodology flow chart (we covered the LP to maintain the vehicle's anonymity).

The proposed pipeline for LP recognition illustrated in Fig. 1, and is composed of: (A) car detection, (B) LP detection, and (C) LP character recognition. The first step is the vehicle detection using MobileNet-SSD; the detected vehicles are isolated from the rest of the input image. These isolated vehicles are the input for the LPD-net, because detecting the vehicle first may decrease the number of false positives and result in better images with larger and easier LPs for posterior detection. The LPD-net identifies LPs for each vehicle image. The next step consists of the LP character recognition using the OCR-net. Finally, the characters are replaced according to Brazilian LPs patterns.

3.1   Vehicle Detection

Since vehicles are common objects in pre-trained weights of usual deep learning object detection approaches, we decided not to train new weights from scratch.

The SSD-300 [12] with MobileNet [8] as the backbone and PASCAL-VOC [4] pre-trained weights is fast and accurate enough for this approach, even without any additional changes or training in the model. So, the MobileNet-SSD detects the vehicles in the input image, and then they are isolated in separate images that will be used in the next stages.

3.2   License Plate Detection

For each vehicle image, the LPs must be detected and isolated. In this work, we use the Tiny YOLOv3 architecture [14] changing the last layer for one class detection, resulting in the License Plate Detection Network (LPD-net), as shown in Table 1. The network is small and, thus, its speed should be efficient for most of the systems, including embedded systems.

Table 1. License Plate Detection Network (LPD-net): a modification of Tiny YOLOv3 for one class output

	Layer	Filters	Size	Input shape	Output shape
0	conv	16	$3 \times 3/1$	$414 \times 414 \times 3$	$414 \times 414 \times 16$
1	max		$2 \times 2/2$	$414 \times 414 \times 16$	$207 \times 207 \times 16$
2	conv	32	$3 \times 3/1$	$207 \times 207 \times 16$	$207 \times 207 \times 32$
3	max		$2 \times 2/2$	$207 \times 207 \times 32$	$104 \times 104 \times 32$
4	conv	64	$3 \times 3/1$	$104 \times 104 \times 32$	$104 \times 104 \times 64$
5	max		$2 \times 2/2$	$104 \times 104 \times 64$	$52 \times 52 \times 64$
6	conv	128	$3 \times 3/1$	$52 \times 52 \times 64$	$52 \times 52 \times 128$
7	max		$2 \times 2/2$	$52 \times 52 \times 128$	$26 \times 26 \times 128$
8	conv	256	$3 \times 3/1$	$26 \times 26 \times 128$	$26 \times 26 \times 256$
9	max		$2 \times 2/2$	$26 \times 26 \times 256$	$13 \times 13 \times 256$
10	conv	512	$3 \times 3/1$	$13 \times 13 \times 256$	$13 \times 13 \times 512$
11	max		$2 \times 2/1$	$13 \times 13 \times 512$	$13 \times 13 \times 512$
12	conv	1024	$3 \times 3/1$	$13 \times 13 \times 512$	$13 \times 13 \times 1024$
13	conv	256	$1 \times 1/1$	$13 \times 13 \times 1024$	$13 \times 13 \times 256$
14	conv	512	$3 \times 3/1$	$13 \times 13 \times 256$	$13 \times 13 \times 512$
15	conv	18	$1 \times 1/1$	$13 \times 13 \times 512$	$13 \times 13 \times 18$
16	yolo				
17	route	13			
18	conv	128	$1 \times 1/1$	$13 \times 13 \times 256$	$13 \times 13 \times 128$
19	upsample		$2\times$	$13 \times 13 \times 128$	$26 \times 26 \times 128$
20	route	198			
21	conv	256	$3 \times 3/1$	$26 \times 26 \times 384$	$26 \times 26 \times 256$
22	conv	18	$1 \times 1/1$	$26 \times 26 \times 256$	$26 \times 26 \times 18$
23	yolo				

For training the LPD-net, we used a private dataset, which is presented in Sect. 4.2. The four corners of each LP were manually labeled and no data augmentation techniques were required. The training of the network used the follow parameters: 416 × 416 for input size; 50k iterations of mini-batches containing 64 images; learning rate of 0.001 in the first 25k iterations and 0.0001 in the rest of them.

3.3   Optical Character Recognition

Eventually, since the LP image is isolated, the characters can be recognized using an Optical Character Recognition network (OCR-net) [16]. We decided not to train a model from scratch since there are pre-trained weighs with satisfying results [17]. Also, the vast majority of Brazilian LPs have a pattern of three letters, followed by four numbers in a uniform background color. Thus, some heuristics are applied to replace digits and letters when it is needed, as shown in Table 2.

Table 2. Replacing heuristics for correcting the recognized text

Letter		Digit
O	⇔	0
Q	⇒	0
D	⇒	0
I	⇔	1
Z	⇒	2
A	⇔	4
S	⇔	5
G	⇔	6
T	⇔	7
B	⇔	8

4   Materials

This section provides information about the Jetson TX2, and the datasets evaluated in this paper.

4.1   Jetson TX2

Jetson TX2 is a power-efficient computing device. It has a powerful processor that helps to bring artificial intelligence processing power to end products. The Jetson TX2 is composed of a GPU with Nvidia Pascal and a dual-core 64 bit

ARM processor. Furthermore, it has an 8 GB RAM with a speed of 59.7 GB/s. It has standard connections for cameras, displays, mouse, and keyboard, as well as GPIO pins, which allow for fast prototyping.

4.2 Datasets

In this paper, we used two datasets of Brazilian LP images to evaluate the proposed methodology. The first one is a private dataset composed of 1988 images of cars obtained from traffic monitoring cameras from Brazil. The resolution of the images is 752×540 and they were captured during the daytime and also during nighttime, resulting in some black and white pictures. This dataset was split into 1331 images for training and 657 for testing. Samples of the private dataset are presented in Fig. 2. Since this is a private dataset, we omitted characteristics from the vehicle that can identify it, including the license plate.

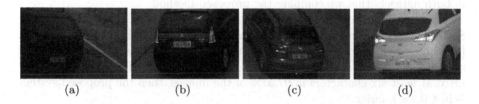

(a)   (b)   (c)   (d)

Fig. 2. Samples of the private dataset of car images

The second dataset, which is called UFPR-ALPR dataset, is from real-world scenarios, where a camera was placed inside a moving vehicle. Three different cameras were used on the images acquisition; for each camera, approximately 1,500 images with 1920×1080 pixels of size were captured, totaling 4,500 images, 150 vehicles, and over 30,000 characters. This dataset was split in 40% for training, 40% for testing, and the remaining for validation. The UFPR-ALPR dataset can only be used for academic research. Because of the private dataset has only images of cars, we filtered the UFPR-ALPR validation set in order to have only car images. Samples of the UFPR-ALPR dataset are shown in Fig. 3. We can observe that this dataset has a proper perspective for an embedded system.

(a)   (b)   (c)   (d)

Fig. 3. Samples of the UFPR-ALPR of car images.

5   Experimental Results

In this section, we evaluate the proposed ALPR system in two steps. The first step consists of experiments using the UFPR-ALPR dataset using a computer composed by Nvidia GTX 1070 as GPU, 8 GB of RAM, with Ubuntu 16.04 LTS as the operating system. In the second step, we evaluated the proposed ALPR system in an embedded platform, which is the Jetson TX2.

Since the LPD-net was trained only with Brazilian plates and cars, we compare our results with the works of Silva and Jung [17] and Laroca et al. [10]. Both papers proposed a complete ALPR system, and used a larger dataset than ours; Silva and Jung [17] affirmed to be tuned for Brazilian plates, and Laroca et al. [10] used Brazilian plates in its plate recognition training.

We performed ten runs on the validation set of each dataset on each system. All runs resulted in the same value. The final results of the proposed system are presented in Table 3 in bold. This table also presents the results for the methods in both datasets that we compare the proposed method.

For the UFPR-ALPR dataset, our system recognized all seven characters in 85.27% of the images, resulting in an improvement of 6.24% when compared to [17], but a reduction of 9.52% when compared to [10]. However, the proposed system identified at least six characters in 96.87% of the LPs, while [10] recognized at least six characters in 97.57% of the images, then the proposed system is just 0.7% inferior.

For the private dataset, we can note that the proposed approach achieved a recognition rate 3.51% inferior than [10] on all characters correct, but obtained a result 2.13% better on at least six characters correct. This result can indicate that the dataset used on the training stage influences the results. In both datasets, the OCR-net could be responsible for the big difference between a completely correct LP and a correct six-characters plate.

In Fig. 4, we can observe that the proposed method surpass the method proposed by [17] in both datasets, and is better than [10] in the private dataset. In addition, even though the proposed method does not have samples of the UFPR-ALPR dataset in its training, it has similar results to the method proposed by [10].

Table 3. Recognition rates of the proposed ALPR system in the Nvidia GTX 1070 and Jetson TX2.

Dataset	ALPR	≥6 characters	All correct
UFPR-ALPR cars	Silva and Jung [17]	87.57%	79.03%
	Laroca et al. [10]	97.57%	94.79%
	Proposed	**96.87%**	**85.27%**
Private dataset	Silva and Jung [17]	86.76%	75.34%
	Laroca et al. [10]	88.43%	81.74%
	Proposed	**90.56%**	**78.23%**

5.1   Evaluation Using Nvidia GTX 1070

In Table 4, we present an average time required for processing the proposed ALPR system in the UFPR-ALPR dataset using an Nvidia GTX 1070 as GPU divided into stages. The LPD-net runs at 105 FPS, making it feasible for embedded systems. The slowest step is the vehicle detection, running at 22 FPS. This speed is due to the MobileNet-SSD with Pascal-VOC weights, which contains 20 classes, then one image can have additional classes, making the time increase.

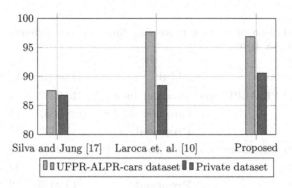

Fig. 4. Graphical representation of the results for at least six characters correct.

Table 5 shows a comparison of the average of the processing times between different ALPR systems using an Nvidia GTX 1070 as GPU. For the UFPR-ALPR-cars dataset, we can observe that [17] achieved only 2 FPS, and [10] reached 5 FPS, meaning that the proposed system is three times faster than the system proposed by [10]. In the private dataset, the proposed system stood out again, reaching about fives faster than the compared approaches. All three systems had a better FPS in the private dataset because it has only one car per image.

Table 4. Results of average time required for processing the ALPR system in **UFPR-ALPR dataset** using a Nvidia GTX 1070.

ALPR Stage	Time (ms)	FPS
Vehicle detection	45.2018	22
LP detection	9.5340	105
Character recognition	10.2778	97
ALPR	65.0137	15

5.2   Evaluation Using Jetson TX2

Since the results on Nvidia GTX 1070 were promising, we decided to embed the system on a Jetson TX2. Table 6 exhibits the average time required for processing the proposed ALPR system in the UFPR-ALPR dataset using a Jetson TX2 divided into stages. We can note that the complete system took approximately 122 ms to execute the three stages per image. Despite the time being double than the Nvidia GTX 1070, this time still is efficient for an embedded platform, considering that we have a complete ALPR-system.

Table 5. Comparison of average processing time between different ALPR systems using a Nvidia GTX 1070.

Dataset	ALPR	Time (ms)	FPS
UFPR-ALPR cars	Silva and Jung [17]	413.9287	2
	Laroca et al. [10]	198.5857	5
	Proposed	65.0137	15
Private dataset	Silva and Jung [17]	79.2450	13
	Laroca et al. [10]	90.9189	11
	Proposed	19.2940	52

Table 7 presents a comparison of the average processing time between different ALPR systems using a Jetson TX2. For the UFPR-ALPR-cars dataset, we can observe that the proposed system is significantly faster than [17] and [10]. For the private dataset, all systems improved their times due to the characteristics of the dataset, highlighting the proposed system that is approximately five times faster than the other approaches.

In Fig. 5, we can observe that the proposed system is faster than the other approaches in both datasets used in the experiments, processing more frames per seconds. Thus, the proposed system is feasible and can be applied as a real-time application in a Jetson TX2.

Table 6. Results of average time required for processing the ALPR system in **UFPR-ALPR dataset** using a Nvidia Jetson TX2.

ALPR stage	Time (ms)	FPS
Vehicle detection	63.3129	16
LP detection	40.9666	24
Character recognition	18.5270	54
ALPR	122.8065	8

Table 7. Comparison of average processing time between different ALPR systems using a Nvidia Jetson TX2

Dataset	ALPR	Time (ms)	FPS
UFPR-ALPR cars	Silva and Jung [17]	755.3228	1
	Laroca et al. [10]	647.4782	1
	Proposed	122.8065	8
Private dataset	Silva and Jung [17]	467.7464	2
	Laroca et al. [10]	436.7330	2
	Proposed	93.9716	11

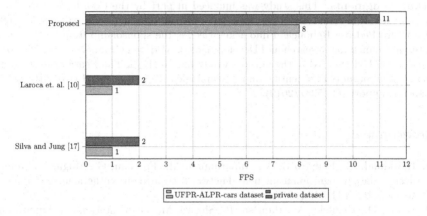

Fig. 5. Chart comparing the average framerate of the ALPR systems in the UFPR-ALPR-cars and private dataset using Jetson TX2.

6   Conclusion and Future Works

In this paper, we proposed a complete ALPR system for Brazilian LPs. We used two existing CNN networks: MobileNet-SSD with Pascal-VOC weights for the detection of the cars, and OCR-net for character recognition. In addition, we created the LPD-net, a CNN network modified from the Yolov3 Tiny for the plate detection.

In order to evaluate the proposed system, we used two datasets: one private dataset and one public dataset, the UFPR-ALPR dataset. Furthermore, we assessed the results in two platforms, Nvidia GTX 1070 and Nvidia Jetson TX2. This comparison was made, mainly because many papers published make the assumption that there is unlimited computing power. However, this is not the case when dealing with mobile or portable systems.

When considering the complete ALPR system and the recognition of at least six characters, the proposed approach achieved 96.87% in the UFPR-ALPR dataset and 90.56% in the private dataset. Besides, the proposed system accomplished the best processing times in both datasets in both platforms, Nvidia GTX

1070 and Nvidia Jetson TX2; in Nvidia GTX 1070, the system obtained 65.01 ms and 19.29 ms for the UFPR-ALPR dataset and the private dataset, respectively. In Jetson TX2, the system reached the times 122.81 ms and 93.97 ms for the UFPR-ALPR dataset and the private dataset, respectively. Thus, these results indicated to be efficient and feasible for embedded systems.

For future works, first, we aim to implement the system as a real-time application inside a car, connecting cameras to the Jetson TX2. Also, we intend to create a CNN network to detect the cars so we can speed up this step and, consequently, the overall process. We also want to expand the proposed system for motorcycles and other types of vehicles.

Acknowledgments. This study was financed in part by the Coordenação de Aperfeiçoamento de Pessoal de Nível Superior - Brasil (CAPES) - Finance Code 001. Also Pedro Pedrosa Rebouças Filho acknowledges the sponsorship from the Brazilian National Council for Research and Development (CNPq) via Grants Nos. 431709/2018-1 and 311973/2018-3. Also, the authors would like to thank The Ceará State Foundation for the Support of Scientific and Technological Development (FUNCAP) for the financial support (6945087/2019).

References

1. Björklund, T., Fiandrotti, A., Annarumma, M., Francini, G., Magli, E.: Robust license plate recognition using neural networks trained on synthetic images. Pattern Recogn. **93**, 134–146 (2019)
2. Bulan, O., Kozitsky, V., Ramesh, P., Shreve, M.: Segmentation-and annotation-free license plate recognition with deep localization and failure identification. IEEE Trans. Intell. Transp. Syst. **18**(9), 2351–2363 (2017)
3. Du, S., Ibrahim, M., Shehata, M., Badawy, W.: Automatic license plate recognition (ALPR): a state-of-the-art review. IEEE Trans. Circuits Syst. Video Technol. **23**(2), 311–325 (2012)
4. Everingham, M., Van Gool, L., Williams, C.K., Winn, J., Zisserman, A.: The pascal visual object classes (VOC) challenge. Int. J. Comput. Vision **88**(2), 303–338 (2010)
5. Gonçalves, G.R., Diniz, M.A., Laroca, R., Menotti, D., Schwartz, W.R.: Real-time automatic license plate recognition through deep multi-task networks. In: 2018 31st SIBGRAPI Conference on Graphics, Patterns and Images (SIBGRAPI), pp. 110–117. IEEE (2018)
6. Gonçalves, G.R., Diniz, M.A., Laroca, R., Menotti, D., Schwartz, W.R.: Multi-task learning for low-resolution license plate recognition. In: Nyström, I., Hernández Heredia, Y., Milián Núñez, V. (eds.) CIARP 2019. LNCS, vol. 11896, pp. 251–261. Springer, Cham (2019). https://doi.org/10.1007/978-3-030-33904-3_23
7. Gonçalves, G.R., Menotti, D., Schwartz, W.R.: License plate recognition based on temporal redundancy. In: 2016 IEEE 19th International Conference on Intelligent Transportation Systems (ITSC), pp. 2577–2582. IEEE (2016)
8. Howard, A.G., et al.: MobileNets: efficient convolutional neural networks for mobile vision applications. arXiv preprint arXiv:1704.04861 (2017)
9. Laroca, R., et al.: A robust real-time automatic license plate recognition based on the YOLO detector. In: 2018 International Joint Conference on Neural Networks (IJCNN), pp. 1–10. IEEE (2018)

10. Laroca, R., Zanlorensi, L.A., Gonçalves, G.R., Todt, E., Schwartz, W.R., Menotti, D.: An efficient and layout-independent automatic license plate recognition system based on the YOLO detector. arXiv preprint arXiv:1909.01754 (2019)
11. LeCun, Y., Bengio, Y., Hinton, G.: Deep learning. Nature **521**(7553), 436 (2015)
12. Liu, W., et al.: SSD: single shot multibox detector. In: Leibe, B., Matas, J., Sebe, N., Welling, M. (eds.) ECCV 2016. LNCS, vol. 9905, pp. 21–37. Springer, Cham (2016). https://doi.org/10.1007/978-3-319-46448-0_2
13. Mittal, S.: A survey on optimized implementation of deep learning models on the NVIDIA Jetson platform. J. Syst. Architect. **97**, 428–442 (2019)
14. Redmon, J., Farhadi, A.: YOLOv3: an incremental improvement. arXiv preprint arXiv:1804.02767 (2018)
15. Sang, J., et al.: An improved YOLOv2 for vehicle detection. Sensors **18**(12), 4272 (2018)
16. Silva, S.M., Jung, C.R.: Real-time Brazilian license plate detection and recognition using deep convolutional neural networks. In: 2017 30th SIBGRAPI Conference on Graphics, Patterns and Images (SIBGRAPI), pp. 55–62. IEEE (2017)
17. Silva, S.M., Jung, C.R.: License plate detection and recognition in unconstrained scenarios. In: Ferrari, V., Hebert, M., Sminchisescu, C., Weiss, Y. (eds.) ECCV 2018. LNCS, vol. 11216, pp. 593–609. Springer, Cham (2018). https://doi.org/10.1007/978-3-030-01258-8_36
18. Wang, L., Lu, Y., Wang, H., Zheng, Y., Ye, H., Xue, X.: Evolving boxes for fast vehicle detection. In: 2017 IEEE International Conference on Multimedia and Expo (ICME), pp. 1135–1140. IEEE (2017)

Assessing Deep Learning Models for Human-Robot Collaboration Collision Detection in Industrial Environments

Iago R. R. Silva[1($\boxtimes$)], Gibson B. N. Barbosa[1], Carolina C. D. Ledebour[1], Assis T. Oliveira Filho[1], Judith Kelner[1], Djamel Sadok[1], Silvia Lins[2], and Ricardo Souza[2]

[1] Universidade Federal de Pernambuco, Recife, Brazil
{iago.silva,gibson.nunes,carolina.cani,assis.tiago,jk,jamel}@gprt.ufpe.br
[2] Centro de Inovações, Ericsson, Brazil
{silvia.lins,ricardo.s.souza}@ericsson.com

Abstract. The increasing adoption of industrial robots to boost production efficiency is turning human-robot collaborative scenarios much more frequent. In this context, technical factory workers need to be safe at all times from collisions and prepare for emergencies and potential accidents. Another trend in industrial automation is the usage of machine learning techniques - specifically, deep learning algorithms - for image classification. Following these tendencies, this work evaluates the application of deep learning models to detect physical collision in human-robot interactions. Security camera images are used as the primary information source for intelligent collision detection. Unlike other proposed approaches in the literature that apply sensors like Light Detection And Ranging (LIDAR), Laser Range Finder (LRF), or torque sensors from robots, this work does not consider extra sensors, using only 2D cameras. Results show more than 99% of accuracy in the evaluated scenarios, revealing that approaches adopting deep learning algorithms could be promising for human-robot collision avoidance in industrial scenarios. The proposed models may support safety in industrial environments and reduce the impact of collision accidents.

Keywords: Deep learning · Collision detection · Human-robot collaboration

1 Introduction

The United States Bureau of Labor Statistics on Occupational Accidents in Industry recently reported as many as 2.8 million cases of workplace injuries and illnesses. These indicators are decreasing over the years [9]. One of the factors behind this reduction is the introduction of technological innovations. Nonetheless, the occurrence of accidents remains considerably high.

The use of collaborative robots in the industry represents a growing market as they increase productivity and efficiency at lower costs. Robots can also make

© Springer Nature Switzerland AG 2020
R. Cerri and R. C. Prati (Eds.): BRACIS 2020, LNAI 12319, pp. 240–255, 2020.
https://doi.org/10.1007/978-3-030-61377-8_17

factories safer when used in conjunction with machine learning techniques to prevent accidents. The trend report in [1] estimates that by 2020 there will be a 60% increase in collaborative robot sales over those taking place in 2019. Since the purpose of these robots is to work together with humans within proximity, safety measures must be carefully weighted in order to guarantee people's health and well-being.

To this end, several systems apply intelligent algorithms in the detection of risk situations. Machine learning algorithms have been used in several real applications [4]. These algorithms can determine with an acceptable level of accuracy a classification task. The use of machine learning is applicable for collision detection [15]. Among the used techniques, we can especially highlight Deep Learning. Note that it has achieved widespread use in many scientific domains. When a database composed of images is used for classification, the Convolutional Neural Networks (CNN) are better suited for such task [10]. They are deep learning architectures known to extract the best image features and classify them automatically.

In the industry scenario, with a high rate of occupational accidents and the increasing number of collaborative robots present in the market, it is necessary to take special care regarding workers' safety when interacting with these robots. To deal with this problem, it is necessary to identify collision between humans and robots, preferably before taking place. In an attempt to solve this, the works of Gecks [5] and Henrich [6] use various static cameras to construct a 3D model of collision of objects in the environment, which can be applied to motion planning. Pan et al. [13] use LIDAR (a scanning laser range finder) and a stereo camera to generate a point cloud data. This is a three-dimensional representation of the environment generated by sensors. They then apply a probabilistic collision detection machine learning algorithm that computes contacts between robot and objects.

The objective of our work is to detect collision and non-collision situations in human-robot collaboration work. For this purpose, we create a human-robot collision database for the classification task. We used a UR5 robotic arm as a study object. We created two databases that we use in two test cases: in the first one, the worker wears casual clothes without color contrast to the robot. In the second test case, the worker uses Personal Protective Equipment (PPE) contrasts with the robotic arm. The PPE behaves like markers that facilitate the classification task. In both test cases, the images are obtained from a security camera. We resized the images, and we inserted them into the learning models. We applied the CNN-32-64-128, VGG-16, VGG-19 and MobileNet network architectures to both databases. These networks are based on deep learning. We used the metrics accuracy, sensitivity, specificity and loss to evaluate the four models.

Note that this work seeks to evaluate the models generated from the training, that can detect a collision using a single camera point of view with images extracted from the security camera. So, in the future, they can be applied for real-time classification in a factory. However, in this work, real-time classification performance is not evaluated.

Unlike what was reported in some previous works, we just used cameras to make our approach. The cameras are sensors commonly present in industrial environments. It is a advantage once we do not use extra sensors. The use of deep learning to this end is a contribution too, once these architectures based on transfer learning can provide more accurate results. According to the results, the proposed models in this work may support the detection of collision situations in the context of human-robot collaboration in industrial environments. The contributions previously cited may support or improve the safety in industrial environments and reduce the impact of collision accidents.

We organize the rest of this work as follows. Section 2 presents the related works of collision detection in human-robot collaboration environments. Section 3 provides the methods and materials used in this work such as scenario setup, creation of dataset, machine learning models, and their details. Section 4 describes the experiments and their obtained results. Section 5 shows the paper conclusions and future works.

2   Related Works

Collision detection or collision avoidance in industrial environments in the context of human-robot collaboration has been the goal of several research initiatives, as shown in the literature. The adopted approaches to this problem fall into two main classes: (I) using sensors with mathematical modeling (without machine learning) or (II) using machine learning algorithms. In this section, we present some works targeted for collision detection in an industrial environment.

Lee et al. [11,12] modeled a system for collision detection based on the motor and the joint friction torque. The authors made a mathematical modeling to classify the situations. They did not use extra external sensors to achieve the objective. A machine learning approach was not proposed in this paper, but it is possible to make a mathematical classification in the proposed scenario. The authors performed experiments using a robot arm. This robotic arm is usually used for human-robot collaboration in industrial environments. The results show that the proposed method may provide a low-cost solution for detection in collision situations.

The works proposed by Sharkawy et al. [14,15] provide a system based on classic neural network approach. The KUKA LWR manipulator and joint torque sensors from the robot give the data trained in the neural network. It is a proprietary manipulator. The neural network architecture used was the multilayer perceptron (MLP). In the model evaluation, the classification model provided 84% accuracy on collision situations, with 8% of error in the positive class and 16% of error in the negative class.

Takiguchi et al. [18] adopted a sensor known as Laser Range Finder (LRF) and odometry to estimate the position and speed of objects, respectively. The collected information is applied to a deep learning algorithm to define safe paths in which the robot does not encounter obstacles. LRF data is captured and classified according to the distance between the points, forming a clustered point cloud. These points are inferred as the object, and its speed is estimated from that. Another work based on this type of sensor, Das and Yip [2] uses machine learning to detect collision for robotics applications. The authors propose a system for achieving faster robot movement planning response time, considering the objects that obstruct the path. For this, the authors make a three-dimensional environment mapping. The evaluation metrics used were in respect to the model size and training time.

In some of the analyzed works, it is necessary to use sensors to feed the algorithm's training. Thus, they fail to preserve the characteristic of flexibility, since not all devices benefit from these types of sensors. In industrial environments, we need to consider the limitation of use of some sensors. These industrial environments generally provide the camera system monitoring that may provide a lot of information about the scenario. These cameras, in most cases, are two-dimensional, so they do not give depth notions. Our work aims to develop a system to be applied in the industrial environment where it is not necessary to use extra external sensors. In this paper, we use security cameras to capture the images. In addition, this work focuses on identifying the accident, so that its consequences are reduced since it was identified quickly, allowing emergency services to be notified promptly. Note that most of the other contributions are focused on movement planning. Another contribution that we can cite is the use of transfer and deep learning for better decision support. We aim to evaluate the applicability of our proposed models for accident collision detection in human-robot collaboration, differently from the reported works.

3 Materials and Methods

In this section, we describe the adopted materials and methods. We report the steps taken to achieve the work results, such as scenario setup, image retrieval, database definition, and deep learning models.

3.1 Scenario Setup

The scenario consists of a UR5 robot, a person, a security camera, and a computer. We describe the equipment specification in Table 1. Initially, the worker will interact with the scenario to simulate several possible work situations in the industrial environment, in direct contact, or near the robot (Fig. 1).

A single camera is responsible for capturing the images of interactions. The goal is to verify if the camera can retrieve images that can lead to some evidence of the possible collision. Moreover, we adopted the restriction of the absence of depth information in the camera images, once most security cameras do not provide this kind of information, even knowing that this negatively affects the neural network training. The network will classify uncertain cases (e.g. a blind spot, or overlap situations of human-robot) as collisions. In this paper we chose

Table 1. Configurations of the equipment used in the experiments. In the left is the attribute, in the right is its value.

Equipment	Value
Robot	UR5 robot arm
Camera	Intelbras VHD 1120 D G3
PC Processor	Core i7-2600 CPU - 3.4 GHz
PC RAM	8 GB
PC System type	64 bits
PC Video card	GeForce GTX 1060 6 GB

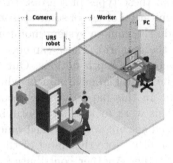

Fig. 1. Schematic of the experiments scenario.

to label these unresolved cases as collisions in order to avoid confusion in the neural network classification.

The scenario simulation consists of a person interacting with the robot in the following cases:

– human outside robot's workspace;
– human robot collaboration with object handover;
– accidental collision caused by the human;
– accidental collision caused by the robot;
– near collision situation;
– intentional blind spot to cause a point of view confusion.

The previous described situations provide data to train our proposed deep learning models. To strongly affirm this, we need to evaluate our models and check their accuracy. It is important to define different situations that faithfully represent real human-robot collaboration to teach the deep learning models to make accurate predictions and can differentiate these situations.

We simulated the cases listed above in two different situations: In the first situation, the person is wearing casual clothes, so he did not wear any special industrial clothing. In the second set of experiments, he wore PPE, including reflexives red shirt and pants and safety orange helmet.

3.2 Image Retrieval and Database Definition

We captured the simulated scenario videos and saved these in the AVI format. We extracted the frames to a folder using a specific public domain software, creating JPEG files. Next, it was necessary to separate the images into two classes: with and without collision. We perform the labeling based on the human body overlap in the robot according to the camera viewpoint. As a result, we obtained two folders: collision/overlap (based on 2D viewpoint) and collision-free image sets. We performed the interaction cases described in Sect. 3.1 with casual clothes and with PPE. Figure 2 shows several examples of images taken from our database (in this case, the worker with and without special clothes).

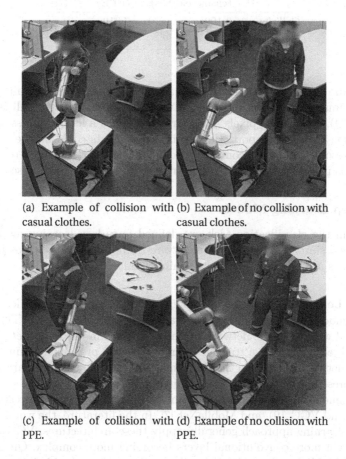

(a) Example of collision with casual clothes.

(b) Example of no collision with casual clothes.

(c) Example of collision with PPE.

(d) Example of no collision with PPE.

Fig. 2. Example of images in the created database for collision and no collision and the worker with and without PPE.

After the databases creation, the next step was to obtain several images for both databases in both classes. Table 2 shows the number of images in each database and for each class. As result according to the rule previously defined (occlusion/overlap cases labeled as collision), the final dataset was balanced the data generated, then the class's quantity became as similar. Deep learning models usually require thousands of images for the training to offer sufficient generalization in the learning. Our database meets this requirement.

Table 2. The number of images contained in each database. Despite the difference in the number of images in each class, there is a sufficient number of images for the database to be considered balanced.

	Database A (casual clothes)	Database B (PPE)
Collision	8,109	7,810
No collision	7,391	6,515

We made changes to the database images to use in the deep learning models. First, the image resolution initially captured by the camera is in full High Definition (HD), 1080 × 720 pixels. This format requires a lot of processing, which the machine may not support. Hence, the experiments could not be performed with this resolution. To overcome this problem, we resized the image to 138 × 178 pixels. This procedure was sufficient to represent the experimental scenario.

3.3   Deep Learning Models

To perform the classification task, we defined four deep learning models. The first one is a simple model based on *not pre-trained weights*, whereas the other three models are based on transfer learning architectures.

Transfer Learning Architecture. In the models that use transfer learning, we defined a module that contains all pre-trained network architecture. The module contains all weights of the network initially obtained in the *Imagenet Challenge* [3]. Our objective is to use transfer learning and use previous learning through the Imagenet database. Then these weights are not trainable in our architecture. Figure 3 presents the base network architecture used in this work.

We defined the transfer learning module by the VGG architectures [17] (VGG-16 and VGG-19), and MobileNet [7]. In the literature, the works that use a transfer learning approach generally apply these architectures. VGG architectures contain more convolutional layers becoming more complex. On the other hand, the MobileNet is used for mobile application or when it is necessary to use low-cost processing. In this work, we propose the use of these commonly used architectures aiming to evaluate in our collision detection application. We need to implement our system as a real-time application in the future. For this reason, it is necessary to evaluate the deep neural network complexity and its impacts on the final classification results.

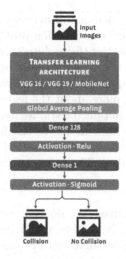

Fig. 3. The used networks architecture. The difference of each experiment is the transfer learning architecture used. Then after there are some common trainable layers.

VGG architectures use only 3 × 3 convolutional layers. The VGG-16 and VGG-19 contain 16 and 19 layers, respectively. The MobileNet is very fast to train compared to other architectures. MobileNet is the simplest compared to the others, with a smaller number of parameters. On the other hand, there are more trainable parameters. Table 3 presents a comparison of these architectures (it compares the number of all parameters in each network).

Table 3. Number of total parameters (trainable and non-trainable) for each architectures.

Architecture	Total	Trainable	Non-trainable
VGG-16	14,780,481	65,793	14,714,688
VGG-19	20,090,177	65,793	20,024,384
MobileNet	3,409,729	180,865	3,228,864

As shown in Fig. 3 there are other layers after the transfer learning. These layers are the trainable part of our architecture and are common to all transfer learning models. First, we performed a data reduction using the pooling layer. Single values replace a complete nearest region of values. In this pooling layer we used the global average function. It can provide a feature mapping and reduces overfitting.

We defined a dense layer after the data reduction. This layer contains 128 neurons with adjustable weights in each training network process. This is the fine tuning process, where there is an adaptation of the learn based on the new samples. We selected 128 neurons of neurons empirically. We believe that a simple fine-tuning to a binary classification problem may be sufficient due to the robust transfer learning part of the architecture.

As a result, the network architecture may use previously acquired knowledge unified with the knowledge acquired in the neural network training process. This forces deep networks learning more about the feature generated by the transfer learning architecture. For the obtained data, we apply the Rectified Linear Unit (ReLU) activation. Then the data moves to the fully connected layer. In this layer, we defined a neuron and a sigmoid activation to determine the probability of the data belonging to the class. Then the classification is performed through this probability.

CNN-32-64-128 Architecture. We do not use pre-trained weights in this architecture. All layers and their weights are changed and improved along with the training. We used this deep learning architecture from the literature [16]. This architecture was used to another machine learning application, but in this work, we aim to evaluate the results in the collision detection application empirically. Figure 4 shows the CNN architecture used.

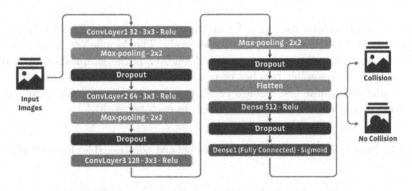

Fig. 4. The used networks architecture. The difference between each experiment is the transfer learning architecture used. Then there are some common trainable layers.

The network's first part is composed by three convolutional layers with kernel size 3×3. The first convolutional layer contains 32 filters and the other convolutional layers have 64 and 128 filters respectively. After each convolution, there are pooling layers containing the $max()$ function. Succeeding each pooling, we defined a dropout rate, aiming to obtain less overfitting in the training model. Finally, we established a dense layer following a fully connected with a sigmoid function (binary classification).

4   Experiments and Results

In this section we present the experimental settings and the results obtained with the experiments.

4.1   Experimental Setup

In the experiments was set some configurations in the neural network architectures and in the data. We will detail these configurations in the next subsections.

General Settings. For a fair comparison, we set the same configuration hyperparameters for all architectures. Table 4 presents these parameters and their values.

Table 4. The network architectures hyperparameters. On the left are the hyperparameters, and on the right their values.

Parameters	Values
Optimizer	Adam
Loss function	Binary cross entropy
Epochs	100
Batch size	24

The Adam optimizer [8] is a combination of RMSprop and stochastic gradient descendant optimizer. We choose this optimizer because it is very fast for the training process, thanks to requiring low memory space. It works well with large databases using deep neural networks. We set up the loss function binary cross entropy because our approach is based on a classification task and the last layer uses the sigmoid as activation function. Usually the number of epochs should be selected when it is sufficient to stagnate the accuracy increase and loss decrease. Empirically we selected 100 epochs and it was sufficient to attend this conditions.

The batch size is the number of samples propagated in the neural networks. It is important select a number observing the memory capacity of the PC (Table 1). We defined 24, because it is the maximum possible to train the network without PC crash.

We adopted the K-fold (cross-validation) as the model validation method, with $K = 5$. The 5-fold provide us 5 executions containing different classification results. Finally we calculate the average of results for all defined metrics.

Metrics. For model evaluation, we selected some representative metrics commonly used to evaluate machine learning models and prediction results. These metrics are based on the amount of correctness of the positive (TP) and negative (TN) classes. We represent the positive samples as collision/overlap, and the negative samples as non-collision. The errors of the positive (FP) and negative (FN) classes are also considered. The metrics to evaluate the prediction results are: accuracy (ACC), sensitivity (SEN), specificity (SPE), and the F1-score. The Eqs. (1), (2), and (3), and (4) show the formulas that correspond to the first three metrics:

$$ACC = \frac{TP + TN}{TP + TN + FP + FN} \cdot 100 \tag{1}$$

$$SEN = \frac{TP}{TP + FN} \cdot 100 \tag{2}$$

$$SPE = \frac{TN}{TN + FP} \cdot 100 \tag{3}$$

$$F1 - score = \frac{2TP}{2TP + FP + FN} \cdot 100 \tag{4}$$

We may consider the SEN as the collision rate, because the metric calculates the acceptance of the collision class. Equally we may consider the SPE as the non-collision rate, where it is calculated the non-collision class rate.

About the loss, it is necessary to know the predicted probability observation (p_c) of currently sample, and the boolean indicator (y) that indicates if the value was classified correctly, and the number of classes (M). The Eq. (5) depicts the metric formula:

$$LOSS = - \sum_{c=1}^{M} y \cdot log(p_c) \tag{5}$$

4.2   Results and Discussion

After performing the experiments, we collected the classification results. First, we present the accuracy (on train and test set) and the LOSS for each model. Tables 5 and 6 show these results for the databases A and B, respectively. We show all results using the mean±standard deviation for each metric.

Table 5. Accuracy and LOSS results obtained using casual clothes.

Model	Train ACC	Test ACC	LOSS
VGG-16	96.73 ± 0.24	90.92 ± 0.03	0.10 ± 0.01
VGG-19	95.65 ± 0.30	88.80 ± 1.30	0.27 ± 0.02
MobileNet	86.80 ± 9.42	58.68 ± 3.75	1.30 ± 0.26
CNN-32-64-128	98.40 ± 0.90	95.02 ± 1.07	0.08 ± 0.08

As shown in the Tables 5 and 6 the VGG-16 outperforms the other transfer learning-based models, but CNN-32-64-128 outperforms the VGG-16. First, for the training and testing set accuracy results are close. This phenomenon shows that the models has not overfitted. The behavior of VGG-19 was similar to that of the VGG-16. The model has not overfitted for both databases. The accuracy of the CNN-32-64-128 model in both databases was very close. The model has not

Table 6. Accuracy results obtained using PPE.

Model	Train ACC	Test ACC	LOSS
VGG-16	98.40 ± 0.89	96.00 ± 0.07	0.06 ± 0.01
VGG-19	96.40 ± 0.74	95.05 ± 0.14	0.07 ± 0.02
MobileNet	93.80 ± 5.21	75.41 ± 5.02	0.51 ± 0.19
CNN-32-64-128	99.00 ± 0.70	98.85 ± 0.80	0.02 ± 0.04

overfitted in both databases. The initial results show the efficacy of the model in detecting overlap and non-overlap.

Different from previous discussed models, the MobileNet model has overfitted. The training set accuracy outperformed on large-scale the testing set accuracy. This is an evidence of model overfitting, the model is only used to the training set, and has difficulty to classify a different set. Consequently, it show us that the model cannot be applied in our detection system because of its low accuracy rate. In the transfer learning models we can note that the VGG models outperform the MobileNet.

About the results obtained in the database A (Table 5), due to the accuracy of the VGG models are very close it was not possible to determinate with precision who is better. Observing the LOSS metric, we can note that the values follow the same behavior of the accuracy. While the accuracy increases, the loss decreases. The LOSS of the MobileNet is greater than the VGG models. In the database B (Table 6), analyzing the accuracy metric is possible to note that the VGG models continue outperforming the model MobileNet. Just like the experimental results in A database, the VGG models accuracy remains close in the B database. We can see that the accuracy increases when compared with the experimental results for database A. It also reflects the LOSS metric. The LOSS metric presents better results and it follows the accuracy results.

The accuracy metric is not sufficient to determine the machine learning model's robustness. We show in the Tables 7 and 8 the metrics and results, including SEN, SPE and LOSS.

Table 7. Sensitivity and specificity results obtained using the database A (casual clothes).

Model	SEN	SPE	F1-score
VGG-16	90.01 ± 00.25	89.46 ± 00.10	89.86 ± 00.05
VGG-19	84.93 ± 00.32	97.00 ± 00.34	84.32 ± 00.23
MobileNet	53.06 ± 05.48	90.02 ± 05.31	52.80 ± 05.67
CNN 32-64-128	98.63 ± 01.42	94.31 ± 01.21	97.30 ± 01.50

About the results for the database A (Table 7) obtained in other metrics, SEN and SPE, we can note some considerations. First, all models could provide a high acceptance in the positive classes (SEN, when the collision/overlap happens), except the MobileNet model. This means that VGG models and the CNN-32-64-128 model are capable to detect when the collision/overlap happens, the F1-score confirm this hypothesis, indicating the high acceptance of the positive classes. The VGG-19 model outperformed the VGG-16 with almost 0.08 difference. These mentioned models could guarantee this acceptance rate (SPE) superior to 0.95. The results show that the CNN-32-64-128 outperformed the transfer learning models in all metrics. It occurs because this model has greater complexity compared to the other models. In the transfer learning models there is just a dense layer to be trained while in the CNN-32-64-128 there are another convolutional layers to be trained. This suggests that the configuration chosen empirically may not be the best one. There is a possibility of improving transfer learning results increasing the number of neurons or trainable layers.

Table 8. Sensitivity and specificity results obtained using the database B (PPE).

Model	SEN	SPE	F1-score
VGG-16	96.43 ± 00.22	96.21 ± 00.19	96.21 ± 00.23
VGG-19	94.81 ± 00.37	96.64 ± 00.31	94.90 ± 00.26
MobileNet	69.21 ± 05.42	73.42 ± 05.10	64.19 ± 05.36
CNN 32-64-128	99.10 ± 01.50	98.11 ± 01.24	98.75 ± 01.20

The experimental results for the database B are in the Table 8. We can note a high impact in the MobileNet results. The SEN rate was not well as it got in the database A. But it got a better SPE rate, making the same comparison. These results can explain the improvement in relation to MobileNet model in the accuracy rate in the second database. About the VGG models, the VGG-19 got a low decrease in the SEN and F1-score metrics, but got a high increase in the SPE metric. The VGG-16 got an increase of 0.06–0.08 in these metrics and outperform the VGG-19 model. Again the CNN-32-64-128 outperformed the transfer learning models in this database. The same explanation used previously should be used now to explain this phenomenon.

Three of four proposed models presented efficient results. The exception is the MobileNet transfer learning model. The bad results for this model can be explained by the low complexity of this model in relation to the others used in this work. As we showed in the Table 3 this model contains less trainable parameters compared to others models that use transfer learning. With the lower complexity of the model, it tends to present inferior results compared to the other models. The higher complexity of the other models provided better ranking results on all metrics that we analyze in this paper. One thing that could also help in the increasing of models accuracy is the generalization of the data. It can facilitate the classification process as the database can be easily separated by straight lines or curves defined by the neural network in the arbitrary space.

Another point to highlight is about the special PPE clothing use in the experiments. When the worker wore casual clothes, the results in the SEN and SPE metrics showed an imbalance in some models. This has been observed with the results obtained through the VGG-19 and MobileNet models. With the use of PPE, the results for these metrics increased by about 10%. Overall accuracy results also increased considerably with the use of PPE in all models. The results show that the use of special clothes increased the database capacity generalization. It provided better results for learning models classification. Thus, we can affirm that the special clothes provided better results for collision detection and non-collision, a result that is very encouraging. The standard deviation of the results obtained with casual clothes and PPE indicate that there is no overlap in the accuracy rates obtained. Therefore, we can infer that for the model validation method adopted with 5 folds, it is not necessary to carry out statistical tests to ratify the conclusions obtained.

Overall, we can consider the results obtained as satisfactory. The paper's goal was to propose machine learning models capable of identifying collisions in industrial environments. The detection occurs without the inclusion of external sensors, that can change the common factory environment, differently of some previously published works. We only used a two-dimensional camera and special clothing that workers commonly wear in their daily work. It was sufficient to provide good accuracy, higher than 0.9 (or 90%). With this achieved accuracy ratio, we can guarantee reliability in the collision detection system to be implemented. It is an application that detects accidents, a good accuracy rate is required for the system to actually be implemented in a real time application. We can affirm that when this system is deployed, there is evidence that it can help the rapid relief and reduction of consequences caused by the accident in industrial environments. The evidence is pointed out throughout the obtained results. Finally, to solve the problem of data labeling (overlap/collision), it is necessary to use more 2D cameras in different points of view to have a detailed description of the scene. With that, it is necessary to make tests to identify overlap/collision in each one of them for final inference about collision in the scene. With this, the real rate of TP can be improved.

5   Conclusion

In this paper, we presented a comparison of deep learning models for human-robot collaboration collision detection in industrial environments. The first three models are based on transfer learning, while the last one is a convolutional architecture without the pre-trained weights. The model with the best classification results was the CNN-32-64-128, which obtained an accuracy result of 0.9932, with the worker wearing PPE. It was possible to get well results of collision and non-collision detection.

With every experimental process, we were able to create a balanced database containing collision and non-collision data. About the transfer learning based models, only MobileNet did not achieve good results, leaving the highlight to the VGG models. Another critical point is that the use of PPE clothing provided an accuracy increase and balance rates comparing to the SEN and SPE metrics. In both databases, the data provided a good generalization. It was further ratified in database B, in which the worker dressed PPE. The results demonstrate that it was possible to obtain a good collision detection accuracy without the use of external or internal robotic sensors. These results were obtained only by using standard tools in industrial environments such as PPE clothing, a two-dimensional camera, and a computer to perform the data processing.

As future work, for reducing occlusions effects and improve the real time classification, more cameras may be introduced to take other points of view and to build ensemble of classifiers with the overlap classification results. This ensemble will be intended to support the final decision whether the collision actually occurred at any given time in real time. Then we should evaluate the developed model by currently work in real-time application with network metrics. Finally, we also look in to detect other risk situations in industrial environments by applying deep learning models.

Acknowledgments. This work was supported by the Ericsson Research (Brazil), Conselho Nacional de Desenvolvimento Científico e Tecnológico (CNPq), Coordenação de Aperfeiçoamento de Pessoal de Nível Superior (CAPES), and the Fundação de Amparo a Ciência e Tecnologia de Pernambuco (FACEPE).

References

1. 365 Microsoft Dynamics: 2019 manufacturing trends report (2019). Accessed 09 Sep 2019
2. Das, N., Yip, M.: Learning-based proxy collision detection for robot motion planning applications. arXiv preprint arXiv:1902.08164 (2019)
3. Deng, J., Dong, W., Socher, R., Li, L.J., Li, K., Fei-Fei, L.: ImageNet: a large-scale hierarchical image database. In: CVPR 2009 (2009)
4. Ferreira, J.L.C., Aloise, A.F., Matter, V.K., Barbosa, J.L.V., Rigo, S.J., de Oliveira, K.S.F.: A model for predicting disapproval of apprentices in distance education using decision tree. In: Proceedings of the XV Brazilian Symposium on Information Systems, SBSI 2019, Aracaju, Brazil, 20–24 May 2019, pp. 14:1–14:8 (2019). https://doi.org/10.1145/3330204.3330223
5. Gecks, T., Henrich, D.: Multi-camera collision detection allowing for object occlusions. In: 37th International Symposium on Robotics (ISR 2006)/4th German Conference on Robotics (Robotik 2006) (2006)
6. Henrich, D., Gecks, T.: Multi-camera collision detection between known and unknown objects. In: 2008 Second ACM/IEEE International Conference on Distributed Smart Cameras, pp. 1–10. IEEE (2008)
7. Howard, A.G., et al.: MobileNets: efficient convolutional neural networks for mobile vision applications. arXiv preprint arXiv:1704.04861 (2017)

8. Kingma, D.P., Ba, J.: Adam: a method for stochastic optimization. In: 3rd International Conference on Learning Representations, ICLR 2015, San Diego, CA, USA, 7–9 May 2015, Conference Track Proceedings (2015). http://arxiv.org/abs/1412.6980

9. Bureau of Labor Statistics: Employer-reported workplace injuries and illnesses - 2017 (2018)

10. LeCun, Y., Bottou, L., Bengio, Y., Haffner, P., et al.: Gradient-based learning applied to document recognition. Proc. IEEE **86**(11), 2278–2324 (1998)

11. Lee, S., Kim, M., Song, J.: Sensorless collision detection for safe human-robot collaboration. In: 2015 IEEE/RSJ International Conference on Intelligent Robots and Systems (IROS), pp. 2392–2397, September 2015. https://doi.org/10.1109/IROS.2015.7353701

12. Lee, S., Song, J.: Collision detection for safe human-robot cooperation of a redundant manipulator. In: 2014 14th International Conference on Control, Automation and Systems (ICCAS 2014), pp. 591–593, October 2014. https://doi.org/10.1109/ICCAS.2014.6987848

13. Pan, J., Chitta, S., Manocha, D.: Probabilistic collision detection between noisy point clouds using robust classification. In: Christensen, H.I., Khatib, O. (eds.) Robotics Research. STAR, vol. 100, pp. 77–94. Springer, Cham (2017). https://doi.org/10.1007/978-3-319-29363-9_5

14. Sharkawy, A.N., Aspragathos, N.: Human-robot collision detection based on neural networks. Int. J. Mech. Eng. Robot. Res **7**(2), 150–157 (2018)

15. Sharkawy, A.-N., Koustoumpardis, P.N., Aspragathos, N.A.: Manipulator collision detection and collided link identification based on neural networks. In: Aspragathos, N.A., Koustoumpardis, P.N., Moulianitis, V.C. (eds.) RAAD 2018. MMS, vol. 67, pp. 3–12. Springer, Cham (2019). https://doi.org/10.1007/978-3-030-00232-9_1

16. Silva, I.R.R., Silva, G.S.L., Souza, R.G., Santos, W.P., Fagundes, R.A.A.: Model based on deep feature extraction for diagnosis of Alzheimer's disease. In: Proceedings of the 2019 International Joint Conference on Neural Networks (IJCNN), vol. 1 (2019)

17. Simonyan, K., Zisserman, A.: Very deep convolutional networks for large-scale image recognition. In: 3rd International Conference on Learning Representations, ICLR 2015, San Diego, CA, USA, 7–9 May 2015, Conference Track Proceedings (2015). http://arxiv.org/abs/1409.1556

18. Takiguchi, T., Lee, J.H., Okamoto, S.: Collision avoidance algorithm using deep learning type artificial intelligence for a mobile robot. In: Proceedings of the International MultiConference of Engineers and Computer Scientists, vol. 1 (2018)

Diagnosis of Apple Fruit Diseases
in the Wild with Mask R-CNN

Ramásio Ferreira de Melo[1,2]($\boxtimes$) (iD), Gustavo Lameirão de Lima[1] (iD),
Guilherme Ribeiro Corrêa[1] (iD), Bruno Zatt[1] (iD),
Marilton Sanchotene de Aguiar[1] (iD), Gilmar Ribeiro Nachtigall[3] (iD),
and Ricardo Matsumura Araújo[1] (iD)

[1] Graduate Program in Computer Science (PPGC),
Federal University of Pelotas - UFPel, Pelotas, Brazil
{ramasio.melo,gustavolameirao,gcorrea,zatt,marilton,
ricardo}@inf.ufpel.edu.br
[2] Federal Institute of Tocantins - IFTO, Computer Science, Araguatins, Brazil
ramasiomelo@ifto.edu.br
[3] EMBRAPA Grape & Wine, Vacaria, Brazil
gilmar.nachtigall@embrapa.br

Abstract. A major challenge in image classification tasks using Machine Learning, and in particular when using deep neural networks, is domain shifting in deployment. This happens when images during usage are capture in different conditions from those used during training. In this paper, we show that despite previous works on the diagnosis of apple tree diseases using standard Convolutional Neural Networks displaying high accuracy, they do so only when no domain shift is present. When the trained model is asked to classify photos of apples taken in the wild, a 22% reduction in F1 score is observed. We propose to treat the task as a segmentation problem and test two different approaches, showing that using Mask R-CNN allows not only to improve performance in the original domain by 3%, but also significantly reduce losses in the new domain (only 6% reduction in F1 score). We establish segmentation as an important alternative towards improving diagnosis of apple tree diseases from photos.

Keywords: Deep learning · Instance segmentation · Apple fruits

1 Introduction

The quick diagnosis of diseases and disorders in cultivars is essential for higher production yields in agriculture. Many disorders present themselves by changing visual characteristics of the plant, such as the presence of discoloration or well-defined spots. In these cases, an expert can correctly diagnose by visual inspection, drawing on previous knowledge of the disorders. Such expertise, however, can be expensive to obtain, taking many hours of training, and not be widely available.

One solution that has considerable adoption is the creation of printed quick guides that helps a non-expert to reach a diagnosis by comparing the affected cultivar with labeled photos. Such method can be slow and prone to errors due to the similarity of many disorders and the difficulty of printing the wide range of ways that a disorder may affect a plant. Machine Learning (ML) solutions have been proposed for such scenario, automating the diagnosis process using models trained on labeled photos.

Convolutional Neural Networks (CNN) are often used as the underlying model, providing improved performance compared to previous approaches using more traditional models [9,19]. When using CNNs, images are labeled by experts and used to train a network – in this case, the whole image is fed to the model without any additional information.

However, a major issue with deploying a system based on CNNs is domain shifting in the wild. Images taken in the wild are subject to angular and luminosity variations, partial and complete occlusions, brightness, contrast and texture changes, the presence of external objects, among other interference [14]. This shift may hinder the model's performance when deployed, reducing its reliability.

In this context, we consider the possibility of training a model using segmented images and evaluate this approach using a dataset of apple fruits containing images taken both in controlled settings and in the wild after harvest. Fruits in our dataset may have Alternaria or Scabies disease, two common conditions that cause spots and changes in the structure of fruits [7,19].

We evaluate two different segmentation - one that roughly segments the whole fruit and another where each spot in each fruit is segmented. Our hypothesis is that such granularity allow for better generalization of the model, allowing it to better focus on parts of the image that are relevant for the task. In order to test this hypothesis, we conduct extensive experiments that, in particular, evaluate the performance when training a network using only images taken in controlled settings and testing it on images taken in the wild.

The main contributions of this work are:

(i) A dataset of annotated images for segmentation of diseases of the type Alternaria and Scabies in apple fruits;
(ii) An evaluation of the Mask R-CNN method applied to the diagnosis of Alternaria and Scabies diseases in apple fruits;
(iii) A comparison of the approach based on segmentation to conventional CNN based models, showing and improved performance when using the former.

The rest of the paper is organized as follows: Sect. 2 presents an overview of related works. Section 3 presents the main goals and the methodology of this work. Section 4 addresses the instances segmentation Mask R-CNN architecture. Section 5 details the dataset. Section 6 describes the experiments carried out and the training process, and the Mask R-CNN model adopted. Section 7 evaluates the results of the tests performed. Finally, Sect. 8 presents the limitations, proposals for future activities and conclusions of this work.

2   Related Works

Machine learning models have been used successfully in analyzing large data sets of apple images taken in controlled settings, such as disease classification in apple fruits or leaves [1,3,7,19], detection of defective apple fruits [13].

Nachtigall et al. [19] developed a CNN model for diagnosing apple tree disorders using a dataset of approximately 10,000 images of leaves and fruits divided into 12 classes. They reported 97% accuracy in apple leaf's classification and 91% accuracy in apple fruit classification using a simple AlexNet architecture. All images were obtained with fruits and leaves in controlled settings.

The issue of domain shifting received ample reporting in the literature (see e.g. [26] for a survey). Domain shifting happens when training and test data differ in their joint probability distribution – e.g. when one trains a model to classify objects taken with a particular camera, but then uses the trained model to classify the same objects with photos taken with a very different camera. This often results in poor performance and several techniques for Domain Adaptation exists, most requiring access to at least some examples in the target domain [8,20]. Zero-shot learning [16] is an instance of domain adaptation where no examples from the target domain is made available during training.

Image segmentation is the task of segmenting an image into different parts, aiming at e.g. detecting objects' locations in images or counting objects. Deep neural network approaches have proven very successful at this task, in particular semantic segmentation – i.e. pixel-level classification [6,23]. See [17] for an extensive survey on image segmentation.

3   Goals and Methodology

The main objective of this paper is to evaluate the performance of training a deep neural network for Instance segmentation of apples' diseases in photos and using the network for diagnosing these diseases in unseen photos that may have suffered domain shifting – i.e. were taken in very different settings compared to those used during training. We establish as specific goals:

(i) To evaluate whether training a model for segmentation improves generalization to unseen domain-shifted images, when compared to conventional CNN's trained for whole-image classification;
(ii) To evaluate different segmentation granularity, one encompassing the whole fruit and another focusing on the characteristic visual manifestation of each disease;

In order to do so, we leverage two datasets containing labeled images of apple fruits displaying symptoms of diseases. In one, photos of the fruits are taken under controlled settings using a semi-professional camera (we will refer to this dataset as the Lab dataset); in the other, photos are taken using a simple smartphone, after harvesting (we will refer to this dataset as the Wild dataset).

Using the two diseases (Alternaria and Scabies) that intersect both datasets, we labeled using a segmentation tool several photos from the Lab dataset. Experiments with two types of segmentation are conducted: one enclosing the whole fruit and another segmenting each individual visual manifestation of the diseases (often discolored/distorted patches or spots).

For each segmentation type, we then train a Mask R-CNN [15], a deep neural network aimed at learning instance segmentation for images, using the labeled images and then use the trained model to infer the disease from both unseen Lab images and Wild images, measuring its performance using several metrics.

In addition, we also train a more standard CNN using solely images from the Lab dataset and also evaluate its performance on unseen images from the Lab and Wild datasets.

Our hypothesis is that training a model for segmentation allows it to focus on parts of the image that are relevant to the task and that this leads to better generalization when the domain shifts – i.e. we hypothesize that Mask R-CNN will perform better in the Wild dataset than the CNN counterpart.

In what follows, we detail each part of this methodology.

4 Mask R-CNN Architecture

Mask R-CNN[15] is a deep neural network developed in 2017, evolved from Faster R-CNN [22] for object detection and instance segmentation in images.

Mask R-CNN is based on the concept of Regions of Interest (ROI). It consists of using a Features Pyramid Network (FPN) to generate the features maps of the images. Then, a Region Proposal Network (RPN) analyzes the features maps generated by the FPN to find possible regions of the image (anchors) capable of representing an object [21, 22, 24].

Anchors can form regions of interest of varying sizes, so ROI Align operation resizes them to a single size. In the sequence, the convolutional and fully-connected layers are responsible for making the prediction.

The Mask R-CNN has three outputs: the class, the bounding box, and the segmentation mask [4, 14, 15]. Figure 1 details the Mask R-CNN architecture adopted in this work.

Mask R-CNN uses a multi-task loss function and weights can be assigned to loss functions during training to adjust the model [5, 21]. The Equation of multi-task loss function is given by:

$$Total\ Loss = Class\ Loss + Bbox\ loss + Mask\ Loss \tag{1}$$

Details on the adjustments of the loss function used in the proposed model are discussed in Sect. 6. Additionally, Mask R-CNN can use different architectures such as ResNet, VGG, MobileNet, among others [4, 14].

5 Dataset Preparation

We use part of a dataset of apple tree disorders that affect fruits and leaves, provided and validated by researchers at EMBRAPA. The dataset includes more

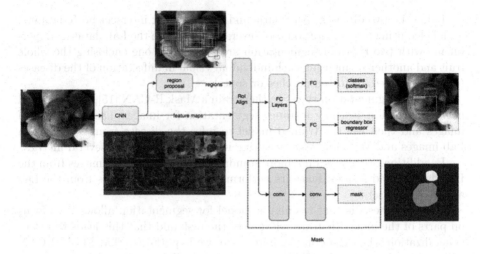

Fig. 1. The original Mask R-CNN architecture adapted from [14]

than 7000 labeled images of apple fruits divided into 14 classes (11 disorders and 3 types of healthy fruits). All photos were taken under controlled settings, using a single camera, white background and controlled lighting. More details about the dataset can be found in [19].

A second dataset, also provided by EMBRAPA for this project, contains photos of apple fruits taken during harvest in the wild. This dataset contains samples of only two diseases and photos are labeled as either having a fruit with Alternaria or Scab. Photos were taken using a smartphone and the photographer was only instructed to keep the fruit of interest centered and with the symptoms, when present, visible. Since photos were taken during harvest, most photos contain other apples in the background – apples are not attached to the tree anymore.

Apple Scab is a disease caused by the fungus *Venturia Inaequalis*. It is present in all apple-producing regions in the world. It causes small circular lesions, isolated or scattered. In an advanced state, it provokes dark-colored cracks and cancers in fruits and leaves [9,19]. Alternaria attacks several cultivars such as cotton, rice, beans, corn, among others. It can occur in leaves and fruits of apple trees. The disease starts with circular brown spots, can cause leaf necrosis, and spread through the fruits, giving an appearance of rot [7,9]. Figure 2 shows examples of each disease.

We used the VGG Image Annotator (VIA) [12] to annotate 1629 images available in the original dataset, equally divided between Alternaria and Scab classes[1]. VIA is an open-source web tool used for graphic annotation in images, audios, and videos.

---

[1] The dataset will be made available after publication.

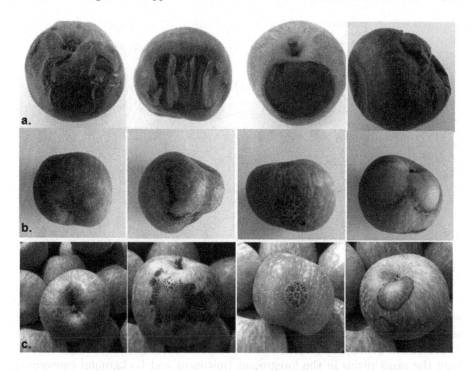

Fig. 2. Dataset preparation: a. Alternaria lab; b. Sarna lab; c. Scab wild

The annotations of the images must be accurate at the pixel level. However, it is common that the polygons extracted in these annotation tools do not accurately represent the object's limits, compromising the training and evaluation of segmentation techniques [10,11,14].

In order to address this, two different approaches were used in the annotation (see Fig. 3): i) circles annotation which demarcates the whole fruit to the edges, ii) polygons annotation which demarcates regions that correspond to visual disease symptoms.

6   Model Adjustment

6.1   Loss Function

As previously described in Sect. 4, the Mask R-CNN loss function combines losses for classification, location of the bounding box (bbox), and mask.

In summary, five loss calculations are used to compose the total loss function of the Mask R-CNN model. The first loss functions of the class and bounding box are applied during the training of the RPN which is responsible for finding the ROI's in the images. In this step, the RPN class loss is computed for cross-entropy, and the RPN bbox loss is calculated via regression function. The best-rated ROI's per class is then passed to ROI Align through non-maximum

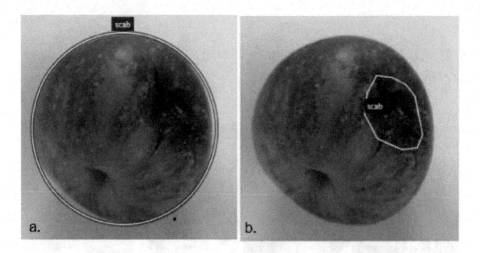

Fig. 3. Image annotation approaches: a. Whole fruit; b. Symptoms

suppression (NMS). The Mask R-CNN loss functions are then applied to the selected ROI's. The mrcnn class and mrcnn bbox loss functions were used in the former model Faster R-CNN. Mrcnn mask loss uses binary cross-entropy to evaluate the mask pixels in the foreground (instance) and background concerning the class label [15,22].

The weights of the Mask R-CNN loss function were adjusted in training performed only with Lab images and prioritized the classification task. Therefore, the weights of the loss functions assigned to the class were adjusted together. For example, higher weights were attributed only to classification losses up to a maximum value of 6. The loss function weights of the Mask R-CNN model used for training are as follows:

- RPN class loss: 2.0;
- RPN bbox loss: 1.0;
- MR-CNN class loss: 2.0;
- MR-CNN bbox loss: 1.0;
- MR-CNN mask loss: 1.0;

6.2   Training

Transfer learning was employed to train the Mask R-CNN model. We adopted the pre-trained weights from the COCO challenge dataset [18] to initialize the network. COCO provides a large dataset of annotated images for object detection and segmentation tasks and includes a generic apple class.

We modified the Mask R-CNN output to fit the task and preserved the knowledge of the upper layers. This step is essential for learning the coordinates of the bounding boxes, masks, and classes of the model (see Fig. 4).

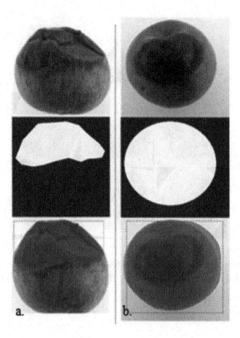

Fig. 4. First row: original images. Second row: Masks. Third row: Instances and bounding boxes. First column (a) shows the symptoms (polygon) approach while (b) shows whole-fruit (circles) approach.

Additionally, the task of segmenting apple disease symptoms is a complicating factor when considering small datasets, since ground truth masks do not have a defined pattern or format and can vary widely depending on the stage of the disease. We applied data-augmentation using geometric transformations in the training images to improve diversity of examples. Details on the applied data-augmentation transformations are presented in Fig. 5.

The Mask R-CNN was developed using the Matterport implementation [2], TensorFlow and Keras API. The model is trained for 40 epochs and 100 steps per epoch, with a learning rate of 0.001, momentum rate of 0.9 and weight decay of 0.0001. Backbones and optimizers are evaluated in the Sect. 7. The size of the RPN's anchors varies between 16 and 256 and generates 200 ROI's for each image during training. The best ROI evaluated by class are filtered by non-maximum suppression (NMS). A maximum number of 100 instances can be detected per image, in inference mode. The model uses input image sizes of 256 × 256. We used Google Colab[2] on a single Tesla K80 GPU.

---

[2] The source code for the Mask R-CNN model will be made available after publication.

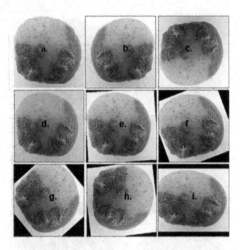

Fig. 5. Data augmentation in training images: a. Original image; b. Horizontal flip; c. Vertical flip; d. Multiply; e. Rotation 15°; f. Rotation 45°; g. Rotation 90°; h. Rotation 270°; i. Scale;

6.3 Evaluation

We used two sets of validation data to evaluate the performance of the Mask R-CNN model. A dataset consisting of Lab images and another consisting of images in the wild.

We evaluate the model both for segmentation and classification performance. As metrics, we use Average Accuracy (mAP), Average Recovery (mAR) and F1 Measure (F1). We adopted IoU threshold values of (0.5), (0.75) and (0.5: 0.95) and mean Intersection over Union (mIoU) to assess the quality of the segmentation. We compared the predictive ability of the Mask R-CNN model with a traditional CNN model, in both datasets.

Intersection over the union (IoU) is a measure that assesses the overlap of bounding boxes. The IoU value is calculated by the intersection between the predicted area and the ground-truth area to analyze whether the detection is valid (True Positive) or not (True Negative). It is given by:

$$IoU = \frac{Area\ Overlap}{Area\ Union} \tag{2}$$

An IoU threshold greater than or equal to 0.5 is used to output 1 as prediction and 0 otherwise. We calculate the mAP, mAR, and F1-measure on the IoU threshold equal to 0.5 because it is considered a standard measure of segmentation quality [4,14].

7 Experimental Results and Discussion

The results of the Mask R-CNN model are organized into three parts. The first step analyzes the performance of the model for the classification of apple fruit diseases. The second step assesses the quality of the segmentation generated by the model. Finally, the Mask R-CNN model best evaluated in the two previous steps is compared to a traditional CNN classifier.

The experiments cover two architectures, ResNet-50 and ResNet-101, and two optimizers, SGD and ADAM. The images of train and validation are partitioned with an 80:20 ratio. The same set of validation of images in the wild was used for evaluating the models. All experiments went through the same training steps, discussed in the previous section.

Table 1 shows the results of all experiments performed for the classification task. The values of mAP, mAR, and F1 shown in the table are calculated for the IoU threshold of 0.5 and the best-obtained results are presented in bold.

Table 1. Summary of the experiments applied in the apple fruit disease classification task. Tests 1–4: the training and validation data set contains laboratory images (1099 and 366, respectively). In tests 5–8: the training set consists of laboratory images (1099), the validation set contains images in the wild (164). In tests 9 to 12: the training set contains laboratory images (310), the validation set contains images in the wild (164)

Experiments	Dataset Annotation	Classification task results				
		Backbone	Optimizer	mAP	mAR	F1
Test 1	Whole-fruit (1099/366)	ResNet-50	ADAM	0.99	0.99	0.99
Test 2		**ResNet-50**	**SGD**	**0.99**	**0.99**	**0.99**
Test 3		ResNet-101	ADAM	0.99	0.99	0.99
Test 4		ResNet-101	SGD	0.98	0.99	0.98
Test 5	Whole-fruit (1099/164)	ResNet-50	ADAM	0.91	0.92	0.91
Test 6		**ResNet-50**	**SGD**	**0.93**	**0.93**	**0.93**
Test 7		ResNet-101	ADAM	0.88	0.91	0.89
Test 8		ResNet-101	SGD	0.92	0.92	0.92
Test 9	Symptoms (310/164)	ResNet-50	ADAM	0.87	0.92	0.89
Test 10		**ResNet-50**	**SGD**	**0.91**	**0.92**	**0.91**
Test 11		ResNet-101	ADAM	0.78	0.92	0.84
Test 12		ResNet-101	SGD	0.78	0.92	0.84

All tests that used the ResNet-50 backbone and SGD optimizer combined obtained better results in the evaluation, both for the laboratory image validation dataset and for the Wild image validation dataset in the two annotation approaches.

The R-CNN mask model achieved high predictive performance when trained and tested on Lab images, reaching 99% precision and recall rates. Test 6 achieved the best prediction results in the Wild image validation dataset with the whole-fruit annotation approach and obtained 93% precision and recall and

F1-measure rates. It surpassed the tests using the symptom annotation approach.

Using the ResNet-50 backbone led to the highest values of mAP and mAR, with an emphasis on Test 6, which uses the SGD optimizer and proved to be the best solution for the classification task in both validation datasets (Lab and Wild).

The higher the IoU threshold, the better correspondence between the prediction and the ground truth. Thus, the values of mAP@0.5 (VOC PASCAL metrics), mAP@0.75, and mAP@.5:.95 (both, standard COCO metrics) and mIoU are compared below to assess the quality of segmentation of Mask R-CNN model.

The mAP@.5:.0.95 is the standard measure of the COCO detection challenge that measures mAP achieved for different IoU thresholds, from 0.5 to 0.95 [0.5: 0.05: 0.95].

The results presented in the Table 2 summarizes the mAP rates achieved for IoU thresholds. Again, the tests which use the ResNet-50 backbone performed better in the segmentation task, in general. For the Lab validation set, the ADAM classifier was able to provide better results for the instances segmentation as it is possible to observe in Tests 1 and 3. These tests' results are practically equivalent, with a small advantage for test 3, which reached a precision rate of 75% for IoU threshold of 0.5:.95 and mean IoU rate of 85%.

In this case, the choice of the appropriate solution can consider issues such as processing, storage capacities, training, and inference times, that were not analyzed here and depend on the specific application. The tests performed on the Wild validation set that obtained the best results used the SGD classifier and maintained the same pattern as the classification task. the best solution was achieved in Test 6 which reached higher the values of mAP@0.5, mAP@0.7, and mAP@.5:.95 with rates of 93%, 71%, and 62%, respectively, and mIoU rate of 78%.

Table 2. Evaluating the quality of the segmentation for IoU thresholds

Segmentation task results											
Annotation	Whole-fruit				Whole-fruit				Symptoms			
Tests	1	2	3	4	5	6	7	8	9	10	11	12
mAP (IoU@0.5)	**0.99**	0.99	0.99	0.98	0.91	**0.93**	0.88	0.92	0.87	**0.91**	0.78	0.78
mAP (IoU@0.75)	**0.91**	0.90	0.90	0.90	0.65	**0.71**	0.65	0.65	0.70	**0.71**	0.55	0.58
mAP (IoU@.5:.95)	0.74	0.71	**0.75**	0.66	0.56	**0.62**	0.61	0.62	0.58	**0.55**	0.50	0.45
mIoU Score	0.84	0.84	**0.85**	0.80	0.77	**0.78**	0.77	0.77	0.80	**0.80**	0.78	0.79
Train Lab	1099				1099				310			
Validation Lab.	366				–				–			
Validation Wild	–				164				164			

In all experiments, when using as backbone the ResNet-101 accuracy dropped considerably, probably due to insufficient training data. In contrast, the lower number of ResNet-50 parameters helps prevent overfitting, reduces training time, and allows the model to be used on less expensive hardware, such as mobile or embedded devices.

Test 6 details the best performance solution found for both tasks. These results are compared to a CNN assessed on the same data set of Lab and Wild images. We performed approximately the same number of tests on the proposed CNN that followed the same pattern as the Mask R-CNN model. We tested in an ad hoc fashion different backbones, optimizers and hyper-parameters and show only the best found solution.

The proposed CNN model uses ResNet-50 as the backbone and ADAM optimizer and it was trained for 80 epochs with a learning rate of 0.001, weight decay of 0.0001, and a mini-batch size of 32. The results achieved in the validation set composed of Lab images are comparable to those found in previous work [9,19,25]. The results are detailed in the Table 3.

Table 3. Comparing apple disease classification models

Models	Dataset Images			Classification task results					
	Train	Validation		Laboratory			Wild		
	Lab	Lab	Wild	mAP	mAR	F1	mAP	mAR	F1
CNN	1099	366	164	0.97	0.96	0.96	0.81	0.76	0.75
Mask R-CNN	1099	366	164	**0.99**	**0.99**	**0.99**	**0.93**	**0.93**	**0.93**

Therefore, it is possible to conclude that the Mask R-CNN model, trained with Lab images, performed well in the task of classifying apple diseases in both Lab and Wild images and surpassed the results achieved by a CNN method in both cases, with emphasis on the 12% increase in the average precision rate on images in the wild.

A CNN typically analyzes the entire image to label an object, which does not work well when subjected to the classification of images with a lot of interference, as in the case as in the case of images in the wild. The main advantage of using the Mask R-CNN to classify apple fruit diseases in the wild is that the use of ROIs makes the network focus only on the regions of the image that potentially best represent the object. Our results show that it can decrease the interference in the images and allow for better generalization.

Figure 6 shows examples of inference (Test 6) when applied to unseen images, using a minimum confidence value of 0.95. It can be seen that the model was effective in solving the classification and segmentation tasks even with diverse visual manifestation of the symptoms.

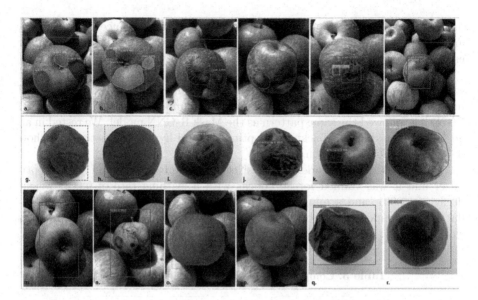

Fig. 6. Viewing images of apples in inference mode: Scabies (a–f, i–p); Alternaria: (g–k, q, r); Masks and bounding boxes: (a, b, i, k, o, p); Masks: (c, d, h, i); Refined bounding boxes: (e, f, j, k, m, n, q, r)

8   Conclusion

In this work we tackled the issue of domain shifting when using deep neural network models to diagnose apple tree diseases from photos.

We showed that a standard CNN trained on images taken in a controlled environment suffers a 22% drop in F1 score when applied to images in the wild. By training a Mask R-CNN to segment input images we were able to reduce this drop to 6% while improving performance in the original domain by 3%.

This result was obtained by using a segmentation where the whole-fruit is annotated using a circle. This simpler segmentation outperformed a finer-grained one where individual visual symptoms were annotated. This was unexpected but also welcome since annotating the whole-fruit is much easier.

The main contribution of this work is establishing segmentation as an important alternative towards improving diagnosis of apple tree diseases from photos taken in different contexts.

As future work, we aim at annotating segments for more disorders and diseases and more images, in order to allow for a direct comparison against models trained on all available classes of the original dataset. There is also opportunities to use different backbones, including more modern ones such as DenseNet.

Acknowledgment. This work was supported by the Conselho Nacional de Desenvolvimento Científico e Tecnológico - CNPq (Edital Universal 407780/2016-5) and by the Coordenaçño de Aperfeiçoamento de Pessoal de Nível Superior - CAPES (Finance

Code 001). We also gratefully acknowledge the support of NVIDIA Corporation with the donation of the GPU used for this research.

References

1. Abd El-aziz, A.A., Darwish, A., Oliva, D., Hassanien, A.E.: Machine learning for apple fruit diseases classification system. In: Hassanien, A.-E., Azar, A.T., Gaber, T., Oliva, D., Tolba, F.M. (eds.) AICV 2020. AISC, vol. 1153, pp. 16–25. Springer, Cham (2020). https://doi.org/10.1007/978-3-030-44289-7_2
2. Abdulla, W.: Mask R-CNN for object detection and instance segmentation on Keras and TensorFlow (2017). https://github.com/matterport/Mask_RCNN
3. Agarwal, A., Sarkar, A., Dubey, A.K.: Computer vision-based fruit disease detection and classification. In: Tiwari, S., Trivedi, M.C., Mishra, K.K., Misra, A.K., Kumar, K.K. (eds.) Smart Innovations in Communication and Computational Sciences. AISC, vol. 851, pp. 105–115. Springer, Singapore (2019). https://doi.org/10.1007/978-981-13-2414-7_11
4. Allehaibi, K.H.S., Nugroho, L.E., Lazuardi, L., Prabuwono, A.S., Mantoro, T., et al.: Segmentation and classification of cervical cells using deep learning. IEEE Access 7, 116925–116941 (2019)
5. Anantharaman, R., Velazquez, M., Lee, Y.: Utilizing mask R-CNN for detection and segmentation of oral diseases. In: 2018 IEEE International Conference on Bioinformatics and Biomedicine (BIBM), pp. 2197–2204. IEEE (2018)
6. Badrinarayanan, V., Kendall, A., Cipolla, R.: SegNet: a deep convolutional encoder-decoder architecture for image segmentation. IEEE Trans. Pattern Anal. Mach. Intell. 39(12), 2481–2495 (2017)
7. Ballester, P., Correa, U.B., Birck, M., Araujo, R.: Assessing the performance of convolutional neural networks on classifying disorders in apple tree leaves. In: Barone, D.A.C., Teles, E.O., Brackmann, C.P. (eds.) LAWCN 2017. CCIS, vol. 720, pp. 31–38. Springer, Cham (2017). https://doi.org/10.1007/978-3-319-71011-2_3
8. Ballester, P., Correa, U.B., Araujo, R.M.: Lateral representation learning in convolutional neural networks. In: 2018 International Joint Conference on Neural Networks (IJCNN), pp. 1–8. IEEE (2018)
9. Dewliya, S., Singh, M.P.: Detection and classification for apple fruit diseases using support vector machine and chain code. Int. Res. J. Eng. Technol. (IRJET) 2(04), 2097–2104 (2015)
10. Dias, P.A., Tabb, A., Medeiros, H.: Multispecies fruit flower detection using a refined semantic segmentation network. IEEE Robot. Autom. Lett. 3(4), 3003–3010 (2018)
11. Dias, P.A., Medeiros, H.: Semantic segmentation refinement by Monte Carlo region growing of high confidence detections. In: Jawahar, C.V., Li, H., Mori, G., Schindler, K. (eds.) ACCV 2018. LNCS, vol. 11362, pp. 131–146. Springer, Cham (2019). https://doi.org/10.1007/978-3-030-20890-5_9
12. Dutta, A., Zisserman, A.: The VGG image annotator (VIA). arXiv preprint arXiv:1904.10699 (2019)
13. Fan, S., et al.: On line detection of defective apples using computer vision system combined with deep learning methods. J. Food Eng. 286, 110102 (2020)
14. Gonzalez, S., Arellano, C., Tapia, J.E.: Deepblueberry: quantification of blueberries in the wild using instance segmentation. IEEE Access 7, 105776–105788 (2019)
15. He, K., Gkioxari, G., Dollár, P., Girshick, R.: Mask R-CNN. In: Proceedings of the IEEE International Conference on Computer Vision, pp. 2961–2969 (2017)

16. Lampert, C.H., Nickisch, H., Harmeling, S.: Learning to detect unseen object classes by between-class attribute transfer. In: 2009 IEEE Conference on Computer Vision and Pattern Recognition, pp. 951–958. IEEE (2009)
17. Lateef, F., Ruichek, Y.: Survey on semantic segmentation using deep learning techniques. Neurocomputing **338**, 321–348 (2019)
18. Lin, T.-Y., et al.: Microsoft COCO: common objects in context. In: Fleet, D., Pajdla, T., Schiele, B., Tuytelaars, T. (eds.) ECCV 2014. LNCS, vol. 8693, pp. 740–755. Springer, Cham (2014). https://doi.org/10.1007/978-3-319-10602-1_48
19. Nachtigall, L.G., Araujo, R.M., Nachtigall, G.R.: Use of images of leaves and fruits of apple trees for automatic identification of symptoms of diseases and nutritional disorders. Int. J. Monit. Surveill. Technol. Res. (IJMSTR) **5**(2), 1–14 (2017)
20. Oquab, M., Bottou, L., Laptev, I., Sivic, J.: Learning and transferring mid-level image representations using convolutional neural networks. In: Proceedings of the IEEE Conference on Computer Vision and Pattern Recognition, pp. 1717–1724 (2014)
21. Ramcharan, A., et al.: Assessing a mobile-based deep learning model for plant disease surveillance. arXiv preprint arXiv:1805.08692 (2018)
22. Ren, S., He, K., Girshick, R., Sun, J.: Faster R-CNN: towards real-time object detection with region proposal networks. In: Advances in Neural Information Processing Systems, pp. 91–99 (2015)
23. Ronneberger, O., Fischer, P., Brox, T.: U-Net: convolutional networks for biomedical image segmentation. In: Navab, N., Hornegger, J., Wells, W.M., Frangi, A.F. (eds.) MICCAI 2015. LNCS, vol. 9351, pp. 234–241. Springer, Cham (2015). https://doi.org/10.1007/978-3-319-24574-4_28
24. Shi, J., Zhou, Y., Zhang, W.X.Q.: Target detection based on improved mask RCNN in service robot. In: 2019 Chinese Control Conference (CCC), pp. 8519–8524. IEEE (2019)
25. Sujatha, P.K., Sandhya, J., Chaitanya, J.S., Subashini, R.: Enhancement of segmentation and feature fusion for apple disease classification. In: 2018 Tenth International Conference on Advanced Computing (ICoAC), pp. 175–181. IEEE (2018)
26. Wang, M., Deng, W.: Deep visual domain adaptation: a survey. Neurocomputing **312**, 135–153 (2018)

Ensemble of Algorithms for Multifocal Cervical Cell Image Segmentation

Geovani L. Martins[1]([⊠]) , Daniel S. Ferreira[1,2] , Fátima N. S. Medeiros[2] ,
and Geraldo L. B. Ramalho[1]

[1] Instituto Federal de Educação, Ciência e Tecnologia do Ceará,
Fortaleza, Ceará, Brazil
geovani.martins@lapisco.ifce.edu.br
[2] Departamento de Engenharia de Teleinformática, Universidade Federal do Ceará,
Fortaleza, Ceará, Brazil

Abstract. One of the main challenges for cell segmentation is to separate overlapping cells, which is also a challenging task for cytologists.
Here we propose a method that combines different algorithms for cervical cell segmentation of Pap smear images and searches for the best
result underlying the maximization of a similarity coefficient. We carried
out experiments with three state-of-the-art segmentation algorithms on
images with clumps of cervical cells. We extracted features such as coefficient of variation and overlapping ratios for each cell grouping and
selected the most appropriate algorithm to segment each cell clump. For
decision criterion, we identified the cell clumps of the training dataset
and calculated the mentioned features. We segmented each clump by
the algorithms and reckoned the Dice measure from each segmentation.
Finally, we used the kNN classifier to predict the best algorithm among
neighboring k-clumps by choosing the one with the largest number of
wins. We validated our proposal on multifocal cervical cell images and
obtained an average Dice around 76.6% without using a threshold value.
These results demonstrated that the proposed ensemble of segmentation
algorithms is promising and suitable for cervical cell image segmentation.

Keywords: Cervical cells · Cell segmentation · Ensemble of
algorithms · Multifocal cytology

1 Introduction

Cervical cancer is a chronic degenerative disease with a high degree of lethality
and morbidity, however, it has a great possibility of cure if diagnosed early.
According to the latest worldwide estimate, 570,000 new cervical cancer cases
were diagnosed in 2018, and 311,000 women died from this disease. About 85%
of the deaths occurred in underdeveloped countries [2].

The most commonly used method for cervical cancer screening is the Pap
(Papanicolaou) smears [16]. Nevertheless, this examination is based on human

© Springer Nature Switzerland AG 2020
R. Cerri and R. C. Prati (Eds.): BRACIS 2020, LNAI 12319, pp. 271–286, 2020.
https://doi.org/10.1007/978-3-030-61377-8_19

visual analysis, which can lead to misinterpretation, resulting in false negative results [5]. Besides the subjectivity of the diagnosis, the successive evaluation of several slides can cause eye strain, contributing to the occurrence of errors. Thus, the use of computational techniques improves the quality of the exam and works as a pre-screening to the cytopathologist, increasing accuracy and decreasing the waiting time of the results.

By processing images, it is possible to extract information to highlight details, which helps and provides means for human interpretation. The first step to perform image analysis is commonly the segmentation. This procedure consists of defining automatic image cut-offs to identify objects or regions of interest to obtain a semantic interpretation of the scene according to the application. Segmentation subdivides an image into its constituent parts or objects [7] and is one of the key tools in medical image analysis.

Computational approaches for automatic cervical cell pre-diagnosis have been a frequent subject in recent years [3, 26]. It is relevant to emphasize the existence of factors that damage the performance of these algorithms, such as intrinsic aspects of the slides or even of images, such as capture conditions, noise, and resolution. Furthermore, the number of cells sampled per frame, the overlap between these cells, the low contrast of the cell cytoplasm, and the presence of mucus and blood [6] also lead to a considerable variation of the results.

Since the computational systems for cervical cell pre-diagnosis usually need reliable information about the cell features, the performance of segmentation methods plays an important role in accurate systems. As observed in [29], hand-crafted cell features are valuable for abnormal cell detection when there exists reliable segmentation. Recently, there are methods for cervical cell nuclei segmentation based on deep learning [25], which are highly efficient for this task. These methods commonly require high computational cost and a large number of images for the training stage.

In images of Pap smears, we can extract different information through segmentation, as shown in Fig. 1. From the frame in Fig. 1a, it is possible to extract nuclei (Fig. 1b), the cytoplasmic mass of each cell (Fig. 1c) and the cytoplasmic mass of the cell clump (Fig. 1d). The segmentation result is represented as observed in Fig. 1e, in which the nuclei present yellow borders, and the cytoplasms have distinct colors identifying the different elements detected in the frame.

In general, we can divide the algorithms that deal with cervical cell segmentation into four approaches: (i) methods that only segment nuclei, (ii) methods that segment images containing a single cell, (iii) methods that segment nuclei and the boundaries of cell clumps in multiple cell images and (iv) methods that segment nuclei and regions of each cytoplasm with several cells [14, 26].

Nuclei detection and segmentation are important steps in cancer diagnosis. The most traditional methods of the literature segment the nuclei from isolated or partially overlapping cells. In [19, 20], the authors estimate the contours of the nuclei initially through morphological operations and, finally, apply the snake algorithm to find the final contours. In [9], the researchers introduce an approach

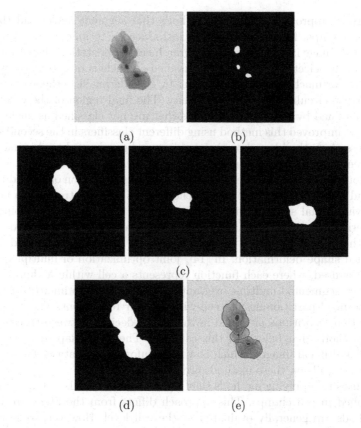

Fig. 1. Results of cell components of a Pap smear image. (a) Original image, (b) nuclei, (c) individual cytoplasm of cells, (d) cell clump, (e) final segmentation.

based on a Bayesian classifier to separate an overlapped or clumped nuclei. Various schemes using curvature information have been investigated to separate the overlapped nuclei, such as in [4,11,27]. In [8], the authors published an interesting nuclei segmentation review.

The second category of cell segmentation methods processes the nucleus and cytoplasm into images with only one cell. In [3], they proposed a method to segment the nucleus and cytoplasm of cervical cells by using the fuzzy C-means (FCM) clustering technique. The pre-processed image is segmented into seven groups using through FCM technique. Then each group is identified as a nucleus, cytoplasm, or background image based on two thresholds. In [13], the authors introduced a snake-based method to obtain the regions of the nucleus and cytoplasm, called radiating gradient vector flow (RGVF), which requires for initialization the initial outline of the object of interest and the map of edges of the image. The initial contours for the nucleus and cytoplasm are obtained by dividing the image into three regions using the k-means method.

The third approach consists of methods that segment nuclei and the mass of the cell clumps. In [10], the authors introduced a technique with two steps. Initially, the image is divided using a hierarchical segmentation algorithm. Then, the previously selected segments are classified as nucleus or cytoplasm with the support vector machines (SVM) classifier by considering size, the average intensity of pixels, circularity, and homogeneity. The final region of the cytoplasmic mass is obtained by joining all segments that are not classified as nuclei. In [6], the authors improved this method using different classifiers in the second stage of the process. A method based on thresholding and graph cut-based segmentation was presented to segment images containing several cervical cells [28].

The fourth approach focuses on the complete segmentation of individual cytoplasm and nuclei of overlapping cells. The authors in [15] introduced a new continuous variational segmentation framework with star-shape prior using directional derivatives to segment overlapping cervical cells in Pap smear images. In [23], the authors proposed an approach based on superpixel-based features and guided shape deformation. In [14], joint optimization of multiple level set functions is used, where each function represents a cell within a clump. In [12], the authors segmented multiple overlapping cervical cells in microscopic images with superpixel partitioning and cell-wise contour refinement. The ensemble of segmentation algorithms proposed in this paper decides between three state-of-the-art methods that belong to this category. They are adapted to work with stacks of multifocal images, which is a more informative dataset for segmentation. More details on these methods are given in Sect. 2.3.

Scenarios for applying methods that segment cervical cells in Pap smears can be identified in cell clumps. This approach differs from the literature because the methods are generally evaluated at the cell level. However, some features obtained at the level of cell clump can better characterize the complexity of the segmentation. Thus, we propose a methodology that combines state-of-the-art algorithms in the segmentation of cervical cells into multifocal images according to characteristics extracted directly from cell clumps.

The remainder of this paper is structured as follows. Section 2 presents the methodology used to implement the ensemble approach. The comparative results of the evaluated methods are exposed and discussed in Sect. 3. Finally, Sect. 4 discusses the conclusions and future works.

2   Materials and Methods

The ensemble of segmentation algorithms methodology shown in Fig. 2 can be divided into two main important flows. The dashed flow illustrates the data training to support the ensemble approach decision of which algorithm is more appropriate to the test samples, represented by the solid flow.

Figure 2 shows the training flow performed based on a training dataset provided by the *IEEE International Symposium on Biomedical Imaging* (ISBI 2015). The training images are pre-processed in Step 2 to detect cell clumps. The estimation of cell clumps is performed from the ground truth (GT) provided by the

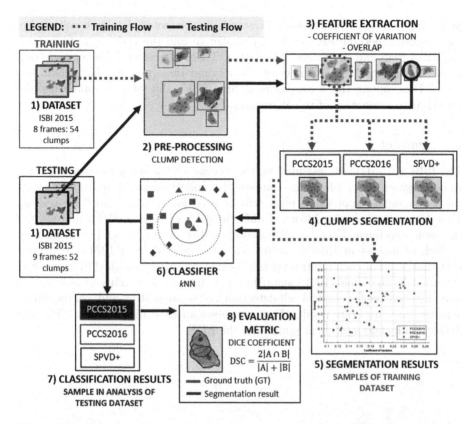

Fig. 2. Methodology of the ensemble of segmentation algorithms for multifocal cervical images.

ISBI 2015 dataset. Then, Step 3 extracts the overlap and coefficient of variation from the detected clumps. We discarded the samples with isolated cells because the overlap in these cases is equal to zero, which is not significant to analyze the algorithm performance. Thus, 54 cell clumps were obtained. Step 4 uses three different methods (PCCS2015, PCCS2016, and SPVD+) to segment each sample. Finally, Step 5 evaluates which algorithm offers the best segmentation for each cell clump according to the Dice coefficient (see Subsect. 2.4). The ensemble of segmentation algorithms predicts the best segmentation method for test samples based on a scatter diagram that relates the extracted features and winner method.

The testing dataset contains nine images available by ISBI 2015. The first three steps in Fig. 2 are similar for both training and testing flows. The difference between them consists of the number of cell clumps, which are 52 samples to the testing dataset. After the feature extraction in Step 3, the testing flow proceeds to Step 6. In this stage, the segmentation of each clump is performed by the ensemble of segmentation algorithms, which chooses the most appropriate method according to the characteristics extracted. For each cell clump, the proposed method uses the kNN classifier to find the training samples with similar overlap and coefficient of variation. Then, the ensemble of segmentation algorithms segments the cells using the method that presents the highest values of Dice among the k neighbors in Step 7. Finally, Step 8 approaches the quantitative evaluation of the algorithms.

2.1   Dataset

The ISBI 2015 dataset provided by the *Second Overlapping Cervical Cytology Image Segmentation Challenge* consists of a collection of 17 multi-layer cervical cell volumes, from which eight will be used for training and 9 for testing. This dataset was obtained from 17 different fields of view (FOV) acquired from the same specimen. Each sample is composed of 20 images or layers, forming a stack of images multifocal, with each layer defined by a 1024×1024 8-bit PNG file [14]. For each multi-layer cell volume, an image obtained by a one-pass extended depth of field (EDF) algorithm [1] is also provided. This set of images is more informative for cervical cell detection and segmentation and, consequently, achieves more accurate results. Figure 3 shows an EDF image of the ISBI 2015 dataset and some multifocal samples of the EDF.

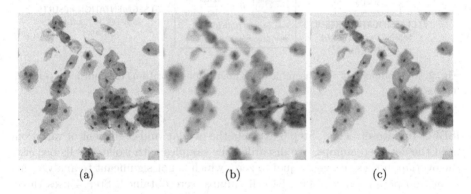

(a)   (b)   (c)

Fig. 3. Sample of ISBI 2015 dataset. (a) EDF, (b)–(c) multifocal samples.

2.2   Feature Extraction

Two main features are analyzed in our study, the overlap and coefficient of variation, which are presented in more detail in the following.

Coefficient of Variation (CV). The coefficient of variation, also known as the relative standard deviation (RSD), is a standardized measure of the dispersion of a probability distribution or frequency distribution. It is defined as the ratio of the standard deviation σ to the mean μ.

$$CV = \frac{\sigma}{\mu}. \tag{1}$$

We considered the coefficient of variation as a parameter of the roughness of the cell clumps, measuring the pixel intensity variation.

Overlap (O). We propose this metric to estimate the overlap in cell clumps. Based on each cell cytoplasm annotations, it is possible to estimate the overlap of a cell with its neighbors according to the following equation:

$$O_{cell_i} = \frac{1}{Q} \left[1 - \sum_{j=0;j\neq i}^{Q} \frac{A_i - A_j}{A_i} \right], \tag{2}$$

where Q is the number of cells overlapping with cell i, A_i stands for the area of the cell i, and A_j corresponds to the area of the cell j, which overlaps with cell i. Then, we calculate the overlap in the cell clump O_{clump}, according to:

$$O_{clump} = \frac{1}{Q_c} \left[\sum_{j=0}^{Q_c} O_{cell_j} \right], \tag{3}$$

where Q_c is the number of cells in a clump. Our overlap metric results zero for the single cells. Therefore, we discarded these cases of our experiments.

2.3   Ensemble

We describe the adopted segmentation algorithms and the proposed decision criterion of our ensemble method in the following subsections.

Segmentation Methods. We choose methods based on nuclei and cytoplasm segmentation on multifocal images with several cervical cells to compose our ensemble algorithm. In [17], the researchers proposed the first algorithm. Here, we named it PCCS2015. This algorithm ranked first in the *Second Segmentation of Overlapping Cervical Cells from Extended Depth of Field Cytology Image Challenge* that is held under the auspices of the IEEE International Symposium on Biomedical Imaging (ISBI 2015). It processes cervical images based on three steps: detection of nuclei using an iterative thresholding approach, segmentation of cell clumps with Gaussian mixture and morphological operations, and segmentation of each cell cytoplasm dividing the depth images into grid squares and classifying them based on gradient features.

The second algorithm of our proposal is PCCS2016. This algorithm corresponds to an improvement of PCCS2015 and was proposed in [18]. The PCCS2016 method implemented a new algorithm to identify regions of potential nuclei, and segments cell clumps by learning a Gaussian mixture model using the

EM (Expectation-Maximization) algorithm. It also implements a new distance measure to approximate the limit of the cytoplasm of each detected nucleus.

The third algorithm (SPVD+) was introduced in [22] and won the second place in the ISBI 2015 challenge. This method is an improved version of SPVD [24]. It consists of three main steps: rough segmentation of subcellular compartments using superpixel combined to Voronoi diagrams, structural refinement of the cytoplasm boundary through the calculus of variations, and morphological reconstruction combined to optimization methods to determine the minimum enclosing ellipse.

Decision Criterion. The k-nearest neighbor (kNN) classifier was chosen to support the ensemble of segmentation algorithms in the decision task of selecting among all algorithms, the most suitable to segment a cell clump. This classifier is defined as follows for a given training set:

$$T = \{< x_1, y_1 >, < x_2, y_2 >, ..., < x_n, y_n >\}. \tag{4}$$

The kNN classifier searches for a subset $S \subset T$ of k samples closer to x. In this way, the label assigned to x is the one with the highest frequency among the samples in S. A distance function calculates the distance between all x and all training samples. Our experiments were performed with the *Cityblock, Euclidean* and *Mahalanobis* distances, and the values of $k = 3, 5, 7$, and 9.

As shown in Step 5 of the training flow, each cell clump is represented by the algorithm with the highest value of the Dice metric calculated from its corresponding segmentation result. Thus, this algorithm represents the sample at a point in the scatter diagram related to the extracted features. Then, the selection of the algorithm that should segment a given sample in the testing flow is the one that wins among the k-nearest neighbors defined by the classifier.

2.4   Evaluation Metric

The Zijdenbos Similarity Index (ZSI) [30], also known as Dice similarity coefficient (DSC), is defined as:

$$DSC = \frac{2|A \cap B|}{|A| + |B|}, \tag{5}$$

where A denotes a reference region, and B is a segmentation result of an image region. A and B are measured in pixels in the two-dimensional space. This measure indicates the segmentation accuracy. According to [21], the "good" cell segmentation result occurs when $DSC > 0.7$. The false negative rate (FNR) can be obtained from the proportion of cells with $DSC \leq 0.7$. Our approach does not establish a specific threshold for the Dice values, as suggested in [21]. Therefore, the FNR metric is discarded from our analysis.

3   Experimental Results and Discussion

The results of the training and testing flows are shown and discussed in Subsects. 3.1 and 3.2, respectively.

3.1   Preliminary Data Analysis

Figure 4 shows the histograms of the segmentation results of each state-of-art method on the training dataset. We also present the segmentation results for our approach, where the best algorithm for each cell clump was chosen. Therefore, the histogram of the ensemble of algorithms corresponds to the possible maximization of the Dice metric for the training dataset.

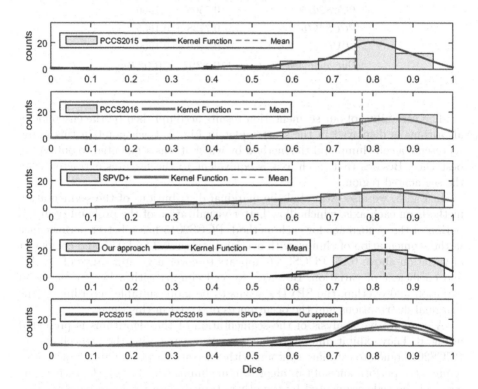

Fig. 4. Histograms of the Dice values for single segmentation methods and our proposed ensemble algorithm.

The histograms in Fig. 4 reveal the occurrence of methods that did not perform a good segmentation in certain cell clumps. For instance, PCCS2015 and PCCS2016 presented samples with Dice values equal to zero.

The probability density function curves show the data trend. Moreover, it demonstrates that the use of the ensemble of algorithms is suitable and promising for cell segmentation since the Dice values did not obtain values below 0.6, and its corresponding curve is concentrated above this value.

The dashed vertical lines in Fig. 4 represent the mean of the Dice values. The ensemble of segmentation algorithms achieved the best average Dice. In absolute terms, we observed $DSC = 0.8297$ and the lowest standard deviation in comparison to the state-of-the-art methods. This demonstrates a low variation for the Dice metric in this experiment. Table 1 shows the absolute values of the mean Dice of the results for comparison purposes.

Table 1. Segmentation results (Training flow).

Methods	Dice
PCCS2015	0.7569 ± 0.1566
PCCS2016	0.7738 ± 0.1656
SPVD+	0.7169 ± 0.1900
Our approach (best case)	**0.8297 ± 0.0849**

The variability of our segmentation results accomplished by the ensemble of algorithms is displayed in the scatter plot in Fig. 4. Each point in this graph represents a cell clump and the method by which it was segmented to obtain the best Dice. Besides that, each axis is equivalent to the features extracted from the segmented region.

In Fig. 5, we also estimated the marginal distributions of the segmentation methods on each axis, which allow better visualization of the probability distribution of the occurrence for each method. PCCS2015 has a better performance in the segmentation of clumps with overlap between 0.3 and 0.75 and coefficient of variation above 0.15. PCCS2016 appears to cover a greater range of overlap and coefficient of variation. However, for cell clumps with a low coefficient of variation and overlap, the SPVD+ seems to be more suitable according to the marginal distributions and the occurrences observed in this range.

A comparative analysis of the segmentation of the algorithms is presented in Fig. 6. They exhibit different behaviors when segmenting the two-cell clump. PCCS2015 tends to estimate contours with a serrated aspect. This characteristic reduces the performance of this algorithm in clumps with low overlap, where the edges of the cells segmented by the other studied algorithms are more regular. The edge segmentation of PCCS2016 has a smoother aspect due to its refining algorithm. Cell clumps with moderate overlap and coefficient of variation can achieve good results with the PCCS2016 algorithm since it defines the cell contour based on the intensity of the pixels in radial lines that depart from the nuclei. Excessive variation in pixel intensity can affect the definition of the contour in these lines. SPVD+ estimates the contour of the cytoplasm by connecting candidate stretches of edge, resulting from a high-pass filter, which approaches an ellipse. As the SPVD+ traces the ellipse based on the Voronoi diagram from

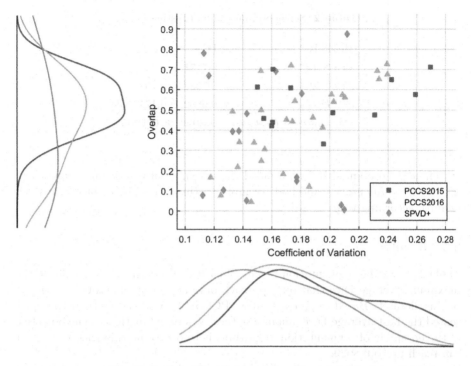

Fig. 5. Scatter plot of the training flow samples.

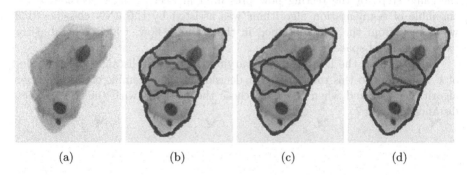

(a)   (b)   (c)   (d)

Fig. 6. Comparative analysis of the segmentation of methods. (a) Cell clump, (b) PCCS2015, (c) PCCS2016 and (d) SPVD+. The blue lines correspond to the cytoplasm annotations of each cell in a clump. The red lines represent the segmentation of the methods. (Color figure online)

the nuclei, clumps with high overlap can generate smaller ellipses, distancing themselves from the true cytoplasm. The cell clump used in this analysis has overlap and coefficient of variation equals to 0.4215 and 0.1599, respectively. The PCCS2015 algorithm obtained the best segmentation with $DSC = 0.9160$.

Table 2. Segmentation results (Testing flow).

Methods	Dice
PCCS2015	0.7171 ± 0.1541
PCCS2016	0.7588 ± 0.1683
SPVD+	0.6952 ± 0.1718
Our approach (best case)	**0.8029 ± 0.1005**

The data presented in Fig. 5 aided the ensemble of algorithms approach to decide the segmentation method to apply to the new samples of cervical cell clumps, according to testing flow.

3.2   Results

Table 2 shows the segmentation results of each method and the estimate of the maximum average Dice that our approach obtained in an experiment with the testing dataset. Based on these results, PCCS2016 was the method that presented the best average Dice among the three evaluated methods. The objective of the ensemble of segmentation algorithms is to overcome this Dice value that can reach up to 0.8029.

Given the promising results of Table 2, we performed a case study following the same steps of the testing flow presented in Sect. 2. The decision of the ensemble of segmentation algorithms was assisted by the kNN classifier that we trained from the data obtained in the training flow, which generated the two-dimensional space of Fig. 5.

Table 3 shows the Dice result by varying the type of distance and the value of k. According to this table, the best decision of the ensemble of segmentation algorithms occurred when the kNN chose the methods with the *Mahalanobis* distance and $k = 7$.

Table 3. Segmentation results using the kNN classifier (Testing flow).

		$k = 3$	$k = 5$	$k = 7$	$k = 9$
Distance	*Cityblock*	0.7526 ± 0.1384	0.7591 ± 0.1291	0.7635 ± 0.1284	0.7566 ± 0.1399
	Euclidean	0.7522 ± 0.1384	0.7607 ± 0.1278	0.7626 ± 0.1283	0.7439 ± 0.1741
	Mahalanobis	0.7517 ± 0.1378	0.7586 ± 0.1294	$\mathbf{0.7658 \pm 0.1262}$	0.7489 ± 0.1732

Table 4 shows the confusion matrix between the single segmentation algorithm and the predicted method by the kNN, considering all testing cell clumps. Since our ensemble approach uses the predicted single algorithms to perform the cell segmentation in each clump, our proposal achieves the best performance if we observed a diagonal confusion matrix. However, there were cases in which some samples were classified incorrectly. They were not segmented by the appropriate method, which could provide the best segmentation. Thus, we found quantities outside the main diagonal. Our approach was able to identify 25 out of 31 cell clumps that PCCS2016 would be more efficient. Only 6 samples were classified by methods that did not optimize the Dice value. The kNN provided the worst predictions for SPVD+ clumps, where 8 out of 9 were incorrectly addressed to other segmentation methods.

Table 4. Confusion matrix in predicting the clump cell segmentation method by the our best kNN(k) setup: $k = 7$ and *Mahalanobis* distance (Testing flow).

Methods	Predicted methods		
	PCCS2015	*PCCS2016*	*SPVD+*
PCCS2015	**5**	5	2
PCCS2016	1	**25**	5
SPVD+	1	7	**1**

Thus, given the conditions of this experiment, the case study using kNN resulted in $DSC = 0.7658$ for the ensemble of segmentation algorithms. This value was greater than the Dice values of state-of-the-art methods, such as PCCS2016, and can improve its results for up to DSC $= 0.8029$.

Our experiments demonstrated that our approach optimized the final segmentation result compared to the results of each isolated method. The preliminary analysis justifies this result when it presents state-of-the-art methods that do not segment all the cell clumps of the training flow, according to the Dice histograms. Furthermore, the scatter plot on Fig. 5 shows ranges of the greater and lesser probability of segmentation results for each method. This is important to analyze these regions to decide the segmentation approach of each cell clump.

4   Conclusion

Here, we proposed an ensemble of algorithms to segment overlapping cells of Pap smears using multifocal images from the ISBI 2015 database. We compare our proposal to state-of-the-art cervical cell segmentation methods, and we state that it is capable of improving cell detection performance. Our method was applied to clumps of overlapping cells, from which features were extracted and used to select the most suitable segmentation algorithms to process different regions in the same cell image. In addition, we introduced a measure to calculate the overlap ratio in cell clumps from the ground truth of the cells. Further work will investigate new features that can improve the characterization of the cell clumps as well as other decision-making models. We will also evaluate other cross-validation procedures that better apply to the size of the dataset.

Acknowledgment. This work was supported by CAPES/CNPq-PVE (401442/2014-4) and CNPq.

References

1. Bradley, A.P., Bamford, P.C.: A one-pass extended depth of field algorithm based on the over-complete discrete wavelet transform. In: Image and Vision Computing New Zealand (IVCNZ), pp. 279–284 (2004)
2. Bray, F., Ferlay, J., Soerjomataram, I., Siegel, R.L., Torre, L.A., Jemal, A.: Global cancer statistics 2018: globocan estimates of incidence and mortality worldwide for 36 cancers in 185 countries. CA Cancer J. Clin. **68**(6), 394–424 (2018)
3. Chankong, T., Theera-Umpon, N., Auephanwiriyakul, S.: Automatic cervical cell segmentation and classification in Pap smears. Comput. Methods Programs Biomed. **113**(2), 539–556 (2014)
4. Cong, G., Parvin, B.: Model-based segmentation of nuclei. Pattern Recogn. **33**(8), 1383–1393 (2000)
5. Gay, J., Donaldson, L., Goellner, J.: False-negative results in cervical cytologic studies. Acta Cytol. **29**(6), 1043–1046 (1985)
6. Gençtav, A., Aksoy, S., Önder, S.: Unsupervised segmentation and classification of cervical cell images. Pattern Recogn. **45**(12), 4151–4168 (2012)
7. Gonzalez, R.C., Woods, R.E.: Digital Image Processing, vol. 3. Prentice Hall, Upper Saddle River (2008)
8. Irshad, H., Veillard, A., Roux, L., Racoceanu, D.: Methods for nuclei detection, segmentation, and classification in digital histopathology: a review-current status and future potential. IEEE Rev. Biomed. Eng. **7**, 97–114 (2014)
9. Jung, C., Kim, C., Chae, S.W., Oh, S.: Unsupervised segmentation of overlapped nuclei using Bayesian classification. IEEE Trans. Biomed. Eng. **57**(12), 2825–2832 (2010)
10. Kale, A., Aksoy, S.: Segmentation of cervical cell images. In: Proceedings of 20th International Conference on Pattern Recognition, pp. 2399–2402. IEEE Computer Society, Istanbul (2010)
11. Kumar, S., Ong, S.H., Ranganath, S., Ong, T.C., Chew, F.T.: A rule-based approach for robust clump splitting. Pattern Recogn. **39**(6), 1088–1098 (2006)

12. Lee, H., Kim, J.: Segmentation of overlapping cervical cells in microscopic images with superpixel partitioning and cell-wise contour refinement. In: IEEE Conference on Computer Vision and Pattern Recognition (CVPR) Workshops, pp. 63–69 (2016)

13. Li, K., Lu, Z., Liu, W., Yin, J.: Cytoplasm and nucleus segmentation in cervical smear images using radiating GVF snake. Pattern Recogn. **45**(4), 1255–1264 (2012)

14. Lu, Z., Carneiro, G., Bradley, A.P.: An improved joint optimization of multiple level set functions for the segmentation of overlapping cervical cells. IEEE Trans. Image Process. **24**(4), 1261–1272 (2015)

15. Nosrati, M.S., Hamarneh, G.: Segmentation of overlapping cervical cells: a variational method with star-shape prior. In: IEEE International Symposium on Biomedical Imaging (ISBI), pp. 186–189 (2015)

16. Papanicolaou, G.N.: A new procedure for staining vaginal smears. Science **95**(2469), 438–439 (1942)

17. Phoulady, H.A., Goldgof, D.B., Hall, L.O., Mouton, P.R.: An approach for overlapping cell segmentation in multi-layer cervical cell volumes. In: IEEE International Symposium on Biomedical Imaging (ISBI). IEEE (2015)

18. Phoulady, H.A., Goldgof, D.B., Hall, L.O., Mouton, P.R.: A new approach to detect and segment overlapping cells in multi-layer cervical cell volume images. In: IEEE International Symposium on Biomedical Imaging (ISBI), pp. 201–204. IEEE (2016)

19. Plissiti, M.E., Charchanti, A., Krikoni, O., Fotiadis, D.I.: Automated segmentation of cell nuclei in Pap smear images. In: Proceedings IEEE International Special Topic Conference on Information Technology in Biomedicine, Greece, pp. 26–28 (2006)

20. Plissiti, M.E., Nikou, C., Charchanti, A.: Automated detection of cell nuclei in Pap smear images using morphological reconstruction and clustering. IEEE Trans. Inf Technol. Biomed. **15**(2), 233–241 (2011)

21. Radau, P., Lu, Y., Connelly, K., Paul, G., Dick, A., Wright, G.: Evaluation framework for algorithms segmenting short axis cardiac MRI. MIDAS J. Cardiac MR Left Ventricle Segmentation Challenge **49** (2009)

22. Ramalho, G.L., Ferreira, D.S., Bianchi, A.G., Carneiro, C.M., Medeiros, F.N., Ushizima, D.M.: Cell reconstruction under voronoi and enclosing ellipses from 3D microscopy. In: IEEE International Symposium on Biomedical Imaging (ISBI). IEEE (2015)

23. Tareef, A., et al.: Automatic segmentation of overlapping cervical smear cells based on local distinctive features and guided shape deformation. Neurocomputing **221**, 94–107 (2017)

24. Ushizima, D.M., Bianchi, A.G., Carneiro, C.M.: Segmentation of subcellular compartments combining superpixel representation with voronoi diagrams. In: IEEE International Symposium on Biomedical Imaging (ISBI), pp. 1–2. IEEE (2014)

25. Wan, T., Xu, S., Sang, C., Jin, Y., Qin, Z.: Accurate segmentation of overlapping cells in cervical cytology with deep convolutional neural networks. Neurocomputing **365**, 157–170 (2019)

26. Wang, P., Wang, L., Li, Y., Song, Q., Lv, S., Hu, X.: Automatic cell nuclei segmentation and classification of cervical Pap smear images. Biomed. Signal Process. Control **48**, 93–103 (2019)

27. Zhang, C., Sun, C., Su, R., Pham, T.D.: Clustered nuclei splitting via curvature information and gray-scale distance transform. J. Microsc. **259**(1), 36–52 (2015)

28. Zhang, L., Kong, H., Chin, C.T., Liu, S., Chen, Z., Wang, T., Chen, S.: Segmentation of cytoplasm and nuclei of abnormal cells in cervical cytology using global and local graph cuts. Comput. Med. Imaging Graph. **38**(5), 369–380 (2014)

29. Zhao, L., Li, K., Yin, J., Liu, Q., Wang, S.: Complete three-phase detection framework for identifying abnormal cervical cells. IET Image Proc. **11**(4), 258–265 (2017)
30. Zijdenbos, A.P., Dawant, B.M., Margolin, R.A., Palmer, A.C.: Morphometric analysis of white matter lesions in MR images: method and validation. IEEE Trans. Med. Imaging **13**(4), 716–724 (1994)

Improving Face Recognition Accuracy for Brazilian Faces in a Criminal Investigation Department

Jones José da Silva Júnior$^{(\boxtimes)}$ (ID) and Anderson Silva Soares (ID)

Universidade Federal de Goiás, Goiânia, GO 74690-900, Brazil
jjsjunior@gmail.com, anderson@inf.ufg.br
https://www.inf.ufg.br/

Abstract. This work addresses a critical problem in the use of the Face Recognition (FR) task by a police department state of Brazil. FR is a valuable crime-fighting tool that can help the police service prevent and detect crime, preserve public safety, and bring offenders to justice. Although significant advances have been shown in the last years, the works are based on large labeled datasets and supervised training. But with this approach, the lack of representative data distribution is an issue, known as data bias, mainly according to some aspects that makes FR harders: gender and race. Recent works have suggested that these two aspects may cause a significant accuracy drop. Thus, the paper is concerned over the FR data bias problem for Brazilian faces. Using pre-trained models learned from public datasets, we demonstrate that even in the small training dataset, it is possible to improve the accuracy of Brazilian faces with simple yet effective implementation tricks in fine-tuning. Two important conclusions wast obtained from this study using a non-public police dataset. First, there is a strong suggestion of data bias concerning ethnicity when evaluating models trained with public datasets on Brazilian faces, and second, the fine-tuning task implemented over non-public police dataset showed a relevant improvement to minimize the dataset bias problem.

Keywords: Face recognition · Deep learning · Real-world application

1 Introduction

The growing attention in Face Recognition (FR) system is due to the need for identity verification in the digital world and the need for face analysis and modeling techniques in multimedia data management and computer entertainment. The last years showed significant progress in this area, owing to advances in deep learning algorithms. Currently, the algorithms with the highest accuracy rates are based on Deep Convolutional Neural Networks, DCNN, trained in a supervised manner by specific cost functions for this task, exceeding human identification capacity [20].

© Springer Nature Switzerland AG 2020
R. Cerri and R. C. Prati (Eds.): BRACIS 2020, LNAI 12319, pp. 287–301, 2020.
https://doi.org/10.1007/978-3-030-61377-8_20

Although research in automatic face recognition has been conducted, although decades, there are critical challenges to address, however, regarding FR applied to law enforcement, two relevant tasks can be enumerated: Live Face Recognition (LFR) and Face Identification (FI). The LFR consists of monitoring live digital video cameras spread across public spaces and verify if each detected person is on a blacklist of wanted people. Another relevant task on law enforcement is Face Identification, generally called as Face Recognition, which consists of search on a gallery with thousands or millions of images to find potential matches to a probe image. This task is an open-set protocol problem, and none of the gallery and probe images identities are present at the training set. A probe image is usually a person of interest obtained during a criminal investigation from several sources such as local crime CCTVs, false identification documents, and fake profiles in social networks.

In this Face Recognition scenario, DCNNs are used to extract encoded and compact representations from facial images, or embeddings. These embeddings are compared by some distance metric, which aims to be equivalent to facial similarity. Embeddings from more similar faces should have shorter distances. Therefore, it is a metric distance learning (DML) problem. A work published on 2014 called *Facenet* [16] adopted the loss function *Triplet Loss* in conjunction with online *triplets* generation with semi-hard samples. The model reached state-of-art with an accuracy of 99.63% in the Labeled Faces in the Wild (LFW) benchmark. It is a proposal based on a model that learns to directly map facial images to a Euclidean space, where the distances correspond directly to the degree of similarity between the faces. Subsequent work achieved results very close to *FaceNet* using training data with a lower number of samples - 500 times smaller. One of these works is known as *AM-Softmax* [19], which consists of a modified version of Softmax loss to supervise the training of DCNN models and drive feature embeddings learning with an emphasis on the angular separation between those of the same and different identities. This work obtained an accuracy of 99.12% in LFW.

All these states of art proposals share the common property of been based on supervised training. Thus, all these models are subjected to the data bias issue, which contributes to the depreciation of accuracy in people with facial features not equally represented in the training data [7,8,10,13,20]. Due to this factor, models using this approach have shown higher error rates when evaluated in certain demographic subgroups, such as afro-descendant. The data bias problem leads to several Civil rights groups to raise alerts, alleging the possibility of unfair treatment, making this a critical subject to law enforcement agencies. On July-2019 a false positive identification from a LFR system in Rio de Janeiro, Brazil, caused an innocent person to be taken to a police station by Military Police [14]. Although the scientific investigation is required to find out whether this case is a data bias problem, this situation warned about the possibility of prejudice in FR systems.

We could not find works digging into data bias problem in the context of law enforcement in Brazil. Face recognition is a important tool on criminal investigations, since there's a regular scenario where detectives obtain face images of crime suspects, from CCTV and fake profiles social media, and need to get the real identification of this person of interest [18]. Also, there is a lack of methods and datasets to evaluate the real accuracy of FR algorithms trained on foreign faces datasets, mainly when applied on real-world criminal investigation scenarios, with probe match search on large gallery sets.

In this work, we investigate how to reduce data bias in models trained in public data collection, composed by celebrities worldwide, when used in Brazilian, facial images. The main objective is to verify, through training in fine-tuning mode, the best cost function to improve, for the context of Brazilian faces, a pre-trained Convolutional Neural Network architecture in these collections is no adequate distribution of images regarding ethnicity.

2 Background

2.1 Data Bias

The term Bias has been widely used in machine learning and statistics with somewhat different meanings. In this work we adopt the definition of [9] where the author defines bias as any basis for choosing one generalization over another, other than strict consistency with the instances. In the context data, this definition can be understood to any preference for choosing one hypothesis explaining the data over other (equally acceptable) hypotheses, where such preference is based on extra information independent of the data.

The origin can be several problems such as a non-distributed training data in a representative way. In this case, two scenarios can be presented:

- the features of each sample are not sufficiently captured by the model during training (or are not available);
- the training set does not contain representative examples of the problem to be addressed.

According to [8] the bias problem is not caused by Artificial Intelligence itself, but the methods of training models. For facial recognition to work as expected, with high accuracy and fairness, the training set must have balance and comprehensiveness in its samples, with enough representative diversity to reflect all possible ways in which faces can inherently be different.

Many public available datasets for training and testing, such as *Casia-Webface* [1], *MS-celeb-1M* [4] and *VGGFace2* [2] are collected on the internet with majority of photos of celebrities and famous people in good lightning, pose, face expression and mostly caucasian race. These datasets are made up of 84.4%, 76.3% and 74.2% caucasian people respectively, and The LFW bencharmark LFW, 69.9% [21].

According to [10] the American Civil Liberties Union, ACLU, carried out the following test with the company *Amazon*[1] facial recognition tool, known

---

[1] Amazon.com, Inc.

as *Rekognition*: using wanted criminals images, the organization compared to photos of US deputies and 28 of them were falsely recognized as being one of the criminals. This article also found that the error rate was 100% higher in Afro-descendant representatives, thus exposing, in addition to the false positive problem, a higher propensity to make mistakes in specific ethnic groups.

A gender and skin color audit algorithm was proposed in [13] and named *Gender Shades*, which used a dataset labeled by gender and skin color called *Pilot Parliaments Benchmark* with equal distribution by theses subgroups, was used to evaluate commercial facial recognition solutions.

The evaluated face recognition products were *Face++*, *MSFT*, *IBM*, *Amazon* and *Kairos*. Performance results are listed in Table 1:

Table 1. Error rate in Pilot Parliaments Benchmark. This table shows de error rates (%) for PPB groups and subgroups. FW = Femela white skin; MW = Male white skin; FD = Female dark skin; MD = Male dark skin.

Company	Total	Female	Male	Dark skin	White skin	FW	MW	FD	MD
Face++	1,6	2,5	0,9	2,6	0,7	4,1	1,3	1,0	0,5
MSFT	0,48	0,90	0,15	0,89	0,15	1,52	0,33	0,34	0,0
IBM	4,41	9,36	0,43	8,16	1,17	16,97	0,63	2,37	0,26
Amazon	8,86	18,73	0,57	15,11	3,08	**31,37**	1,26	7,12	0,00
Kairos	6,60	14,10	0,60	11,10	2,80	22,50	1,30	6,40	0,00

According to results in Table 1, all products achieved lower error rates in the group of males in the gender category and white in skin color category. Conversely, dark-skinned women are mistakenly classified in the highest number. The *Amazon* has the highest error rate in the group of black women, 31.37%, followed by *Kairos* with 22.5%. These results are quite high values compared to men of white skin, which *Amazon* reached 0%, suggesting that ethnic and gender data bias can result in unfairness false positives for specific subgroups and need to be better studied and overcome.

In [10] authors performed cross check experiments, in a controlled way, the leverage of training samples genders on the accuracy rate regarding to white and black individuals.

Two models *LightCNN-9* [24] was trained from scratch using training samples with only one race each and evaluated by two test datasets separated by race: black and white individuals. This experiment was accomplished using the following datasets: *CMU Multi-PIE* [12], Craniofacial Longitudinal Morphological *Morphological (MORPH) Album-2* [5] and *Racial Faces in-the-Wild*, RFW [21]

The accuracy of these models for each test dataset is:

– Model trained with only white people images achieved 79,23% accuracy on white individuals test dataset and 34,31% on the dataset with black individuals.

– Model trained with only black people images achieved 82,36% accuracy on black individuals test dataset and 28,89% on the dataset with white individuals.

These results suggest that models learn embeddings with extraction of specific features for each race, demonstrating the need of using train datasets with samples closer to the final use regarding the demography.

Since miscegenation is one of the main characteristics of the Brazilian population, it is quite complex and subjective to carry out any classification according to this criterion, in addition to being a very controversial topic. IBGE[2] itself adopted self-classification as a race definition criteria in the 2010 census [3]. In this research, 47.73% declared themselves as prados, a criterion that encompasses mulattos, cablocos, cafuzos, mamelucos or mestizos of black people with another races.

Due to these factors, this work considers Brazilian faces as belonging to a unique and own ethnicity, so that in this way it is possible to evaluate the behavior of the models in relation to this categorization. There is a lack of works digging more precisely the data bias problems regarding to public security. Specially when we consider Brazil realities, where we can fit on gender categorization.

2.2 Transfer Learning

Transfer Learning is a technique that aims to improve the accuracy performance of a model by transferring the knowledge from a pre-trained model with a related domain. Due to the dependence on large dataset data of DCNN, this technique can be used to get betters results when the amount of data train available is not enough to train from scratch. Furthermore, according to [26], transfer learning can also be employed to correct the unequal distribution data problem.

Fine-Tuning. Fine-tuning is a particular form of transfer learning and has been used with multiple purposes on Face Recognition [20]. On [2] this technique was used to find an optimal performance by training first on MS-Celeb-1M then finetuning on VGGface2 supervised by *Softmax Loss* function. The model fine-tuned achieved higher accuracy on IJB-A [6] than the others training with each one of this data set alone.

3 Method

This work uses known cost functions to fine-tune a pre-trained state-of-art FR model [15]. This reference model is used to improve performance accuracy on Brazilian data set in the context of law enforcement. Our approach focuses on the face recognition step, where each previously detected and cropped image face with the same size is submitted to a model to extract a compact and discriminate face representation in vector-column shape, i.e., embedding. So, the

---

[2] Brazilian Institute of Geography and Statistics.

embedding is a encoded representation of the face in an feature space, where distances corresponds to face similarities. The embedding is matched to each other according to metric distance enforced by the cost function during training, so that an indicator, or distance measure, indicates the probability of being from the same person or not.

In face recognition research field the most relevant performance improvements has been achieved by new cost functions [20], since they supervise models to generate for discriminate embeddings. The cost functions employed on this work are *FaceNet* [16] and *AM-Softmax* [19].

Facenet was chosen because achieved the second highest accuracy in the LFW collection (99.67%) [20]. Several later works were based on modified versions of *Softmax loss* and also reached the state-of-art, as *AM-Softmax*, which achieved accuracy only 0.5% lower than *FaceNet* using *Casia-Webface* dataset, which is 100 times smaller than *Facenet*'s private dataset. Therefore, there are two proposals with close results and with very different training methods, the first being based on triplets and the second a classifier.

3.1 FaceNet

The *FaceNet* generated embedding, represented by $f(x) \in \mathbb{R}^d$, encodes an input image x in a euclidean space of d dimensions. These embeddings are normalized such that $\|f(x)\|_2 = 1$. The cost function used in *FaceNet* for DCNN training is called *Triplet Loss*, inspired by [22]. The embedding are generated from a x image, such that the quadratic distance between all faces, regardless of the conditions of the image, is small for the same identity and large for images of different identities. The function *Triplet Loss* supervises the training of the model so that the embedding generated from the anchor image sample x_i^a of a given person must be spatially closer to the embedding from another image of the same identity, the positive sample x_i^p, than the embedding from another person's identity, negative example x_i^n, in a space of features $\mathbb{R}^d$.

The cost function *Triplet Loss* is define by Eq. 1:

$$L = \sum_{n=1}^{n} [\|f(x_i^a) - f(x_i^p)\|_2^2 - \|f(x_i^a) - f(x_i^n)\|_2^2 + \alpha] \tag{1}$$

where α is the margin that separates embeddings from different identities samples.

The *FaceNet* authors also proposed an online triplet generation method within a *mini-batch* that violates the constraint in 1 for faster convergence. Thus, the positive sample is select such that $\arg \max_{x_i^n} \|f(x_i^a) - f(x_i^p)\|_2^2$ and the negative such that $\arg \min_{x_i^n} \|f(x_i^a) - f(x_i^n)\|_2^2$. To avoid the local minima another criteria was proposed to select the negative in 2:

$$L = \|f(x_i^a) - f(x_i^p)\|_2^2 < \|f(x_i^a) - f(x_i^n)\|_2^2 \tag{2}$$

and defined as **semi-hard** negatives.

3.2 AM-Softmax

The *AM-Softmax* cost function [19] is an adaptation from *Softmax Loss* and defined in Eq. 3:

$$L_s = \frac{1}{n} \sum_{i=1}^{n} \log \frac{e^{W_{y_i}^T f_i}}{\sum_{j=1}^{c} e^{W_j^T f_i}} \tag{3}$$

where f_i is the input of last fully connected layer associated to from the i-th sample, $W_{y_i}^T$ is the weight vector associated to the y class and W_j the j-th weight vector from the last fully connected layer. After successive transformations using properties like cosine similarity the cost function *AM-Softmax* is defined by 4:

$$L_s = \frac{1}{n} \sum_{i=1}^{n} \log \frac{e^{s.(W_{y_i}^T f_i - m)}}{e^{s.(W_{y_i}^T f_i - m)} + \sum_{j=1, j \neq y_i}^{c} e^{s W_j^T f_i}} \tag{4}$$

where m is the angular margin separating interclass samples and s is an hyper-parameter to scale de cosine values of the angle between weight vector and embedding. Suggested values are $s = 30$ e $m = 0.35$.

3.3 Reference Model

This work uses as reference the *Inception-Resnet-v1* [17] model available at [15], trained with *CASIA-Webface* dataset [1] and generates 128 dimensional (128d) embeddings. This model achieved 98.5% accuracy in the LFW benchmark.

The training was supervised by the cost function *Center Loss* proposed in [23] for joint supervision with *Softmax*. This function aims to learn a center, in a Euclidean space $\mathbb{R}^d$, of dimension d, for each identity and penalize the model when the intraclass embeddings are far from the center, forcing greater intraclass compression.

The cost function *Center loss* is defined by Eq. 5 :

$$L_c = \frac{1}{2} \sum_{i=1}^{m} \|x_i - c_{y_i}\|_2^2 \tag{5}$$

where x_i is i-th embedding sample, y_i is the class, or identity, of this sample and c_{y_i} is the center of all embeddings from samples of y_i class, so $\{c_{y_i}, x_i\} \in \mathbb{R}^d$. The expression $\|x_i - c_{y_i}\|_2^2$ computes the distance between the embedding x_i and its center c_{y_i}.

As it is a joint supervision, the total error calculated by the two cost functions is defined by the Eq. 6:

$$L = L_s + \lambda L_c \tag{6}$$

where λ is an scalar parameter to balance the leverage of *Center Loss* on total error.

4 Experiments

Detection and face alignment was performed using the method developed by [25]. The aligned images have shape of 160 × 160 pixels. The reference model is used to choose the probe set by using itself for finding mismatch identification on the Brazilian dataset, and to serve as a pre-trained model for the fine-tuning training proposed in this work.

Brazilian Dataset. The Brazillian dataset used in the experiments was built from databases of private photos of the police department state (omitted just for the double-blind review). This dataset is composed of 61,221 images of 27,653 identities, or classes, as displayed in the Table 4.

Table 2. Dataset statistics. This Table shows the Brazillian dataset statistics.

Total identities	Total samples	Median	Standard Dev	Most samples identity	Less samples identities
27.653	61.221	2,21	1,23	32	2

This dataset was then split into training and testing sets. The test suite, in turn, consists of the gallery, probe, and validation samples. The Probes set those samples in which the embeddings generated by the reference model has intraclass distance greater than the interclass. That is, scenarios in which there was an error in face identification. The classes of the mismatched samples were also kept out of the training set and added to the gallery in order to guarantee the same hard scenarios (Table 2).

Figure 1 shows the samples selection scheme for the probe set, as well the identity split in the mismatch occurrences.

The validation set was created from the random selection of 20% of the remaining identities, totaling 5,000. For each class, a sample was also randomly selected to compose the validation set, which will be compared against the gallery. The remaining samples of each class were moved to the gallery as match images. This set of data aims to evaluate the generalization of the models, so that it is possible to guarantee that they, after training, have improved the accuracy in the most difficult examples, collection of tests, but also maintain the accuracy rate in the other cases.

A structure of the test collection can be seen in Fig. 2. The embedding of probe and validation samples are matched to the gallery using the distance metric consistent with each model, determined by the function used.

In Table 4 we enumerate the splitted dataset with the number of samples and classes. The sum of the sub-totals of the samples from the data collections exceeds the total number of samples and classes due to the fact that there is an

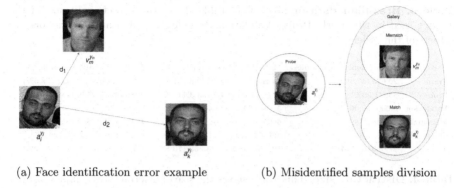

(a) Face identification error example   (b) Misidentified samples division

Fig. 1. (a) Face identification error, since the distance d_1 between examples of different classes is less than the distance d_2 between samples of the same identity. (b) Samples from classes with misidentification are moved to gallery, in order to maintain the same scenario in later model evaluations.

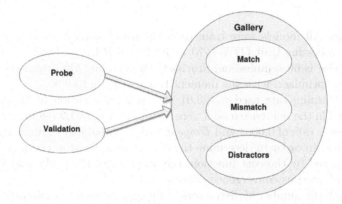

Fig. 2. This Figure shows the dataset structure for models evaluation performance. Each probe and validation sample is compared to the gallery. The gallery is formed by: 1) **Match** The match samples of probe and validation identities; 2) **Distractors**, which are samples with no correspondence and aims to increase identification difficulty; **Mismatch** Samples that were mistakenly consider closer, by reference model, to a sample than their matches.

overlap of samples between the probe and matche images, since a given sample can be in both bins.

Table 4 show the number of disctrators grouped by genre.

Fine-Tuning Models. The models were fined-tuned, which implies that all the weights of the reference model were loaded before the training started, and thus all the features previously learned provides an advanced starting point (Table 3).

In order to compare the performance of cost functions in the most isolated way, the same values were used for all common hyper-parameters of the

Table 3. Brazillian dataset size. This Table shows the subsets split size of the dataset used in this work. Probe, Validation, Mismatch, Match and Distractors are subsets of the gallery.

	Total (identities/samples)	Male (identities/samples)	Female (identities/samples)
Train	20,672/45,698	14,268 (69.1%)/32,666 (71.4%)	6,404 (30.9%)/13,032 (28.6%)
Probe	838/1,252	477 (56,9%)/706 (56,3%)	361 (43,0%)/546 (43,6%)
Validation	5,000/5,000	3,464 (69,2%)/3,464 (69.2%)	1,536 (30,7%)/1,536 (30,7%)
Mismatch	1,143/2,782	669 (58,5%)/1808 (64.9%)	474 (41,4%)/974 (35,0%)
Match	5,838/7,340	3,941 (67.5%)/5,185 (70.6%)	1,897 (32.5%)/2155 (29.3%)
Total	27,653/61,221	18,878 (68.2%)/43,359 (70.8%)	8,887 (31.7%)/17,862 (29.2%)

Table 4. Distractors size. This Table shows the size of disctractores samples, as well the rates of males on females.

	Total	Male	Female
Disctrators	208,187	132,576 (63.67%)	75,618 (36.3%)

architecture. All models were trained on the same computer, equipped with a Graphics Processing Unit *GTX-1070* with 8 GB of *RAM* and 1920 *CUDA* cores. The *batch* size is 60 samples and Stochastic Descending Gradient with *Momentum* [11] as optimizer, with the moment term γ equals to 0.9.

The η learning rate used was 0.01 and as regularization techniques, those already used in the reference model were maintained, with **L2** regularization with a weight decay rate of 0.0001 and *Dropout* with a probability of maintaining each neuron of 0.8. In order to determine the correct time for stop training, the error curve generated by the cost function, the accuracy of the probe and validation sets were adopted as convergence criteria.

Although the number of distractors in the gallery used to evaluate the results is more than 208,000 individuals, during the training 5,400 were used. This reduction aims to reduce the computational cost and time of training.

The Fig. 3 shows the training/test framework. The embeddings are collected from the output of the last Fully Connected Layer. Then, the distance between them are evaluated using de similarity metric.

5   Results

The *Triplet Loss* supervised model was trained with 135 epochs. The stop training criteria was not increasing the validation and Probe accuracy after ten epochs, which happened after 125 epochs. Furthermore, as shown in Fig. 4, the value of triplet loss stayed steady after 130 epochs.

The model using cost function *AM-Softmax* was trained for 6 h, 178 epochs in total. As a criterion for convergence, accuracy in the training set was also considered. As can be seen in Fig. 5a, the accuracy in the training set reached a value above 99% from epoch 150, keeping this level until the end. However, the accuracy of the validation set in Fig. 5b stabilized around 96.5% between

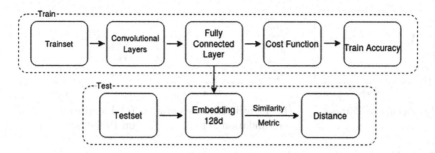

Fig. 3. This figure show train and test framework main modules adopted on this work. The cost function is only used training, then is dismissed for testing, hence the face embeddings are collected as a output of the last Fully Connected Layer and matched by a metric distance.

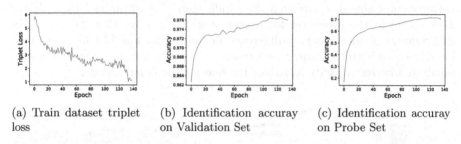

(a) Train dataset triplet loss

(b) Identification accuray on Validation Set

(c) Identification accuray on Probe Set

Fig. 4. This shows the evolution of (a) Triplet loss on train dataset; (b) Identification accuracy on validation set; (c) Identification accuracy on probe set. The training was interrupted after the loss stopped decreasing and identification on validation and probe set stoped increasing.

Table 5. Rank-1 results. This Table shows validation and test results regarded to genders.

Função Custo	Validação			Teste		
	Feminino (1536)	Masculino (3464)	Total (5000)	Feminino (546)	Masculino (706)	Total (1252)
Baseline	88,15%	92,58%	91,22%	0,18%	0,14%	0,16%
AM-Softmax	90,89%	94,60%	93,46%	32,42%	44,05%	38,98%
Triplet Loss	**94,79%**	**96,62%**	**96,06%**	**50,18%**	**54,53%**	**52,64%**

periods 60 and 179, not exceeding 96.7%. In probe set, as shown in Fig. 5c, the model reached 58% in epoch 120, with a maximum value of 58.7%, oscillating around 58.5% until the end training.

Rank-1. Table 5 enumerates Rank-1 identification accuracy for the three evaluated models and full set of distractors.

Figure 6 shows the *Rank-1* accuracy split by gender, both probe and validation set. Our implementation using *Triplet Loss* model not only achieved

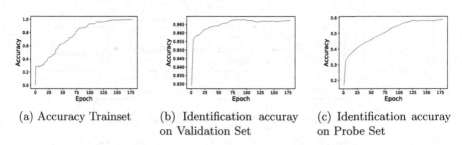

(a) Accuracy Trainset   (b) Identification accuray on Validation Set   (c) Identification accuray on Probe Set

Fig. 5. This shows the evolution of (a) train dataset accuracy; (b) Identification accuracy on validation set; (c) Identification accuracy on probe set. The training was interrupted after all theses indicators stop increasing, meaning that the model converged.

better results but also narrowed the gender gap. The accuracy of the *Triplet Loss* model in the probe set for women was 4.15% lower than for men. The *AM-Softmax* gender accuracy difference in the results was 11.63%. Therefore, in addition to better accuracy performance in both genres, our implementation achieved a better capacity to reduce the bias for the female gender.

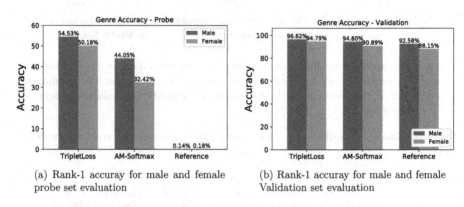

(a) Rank-1 accuray for male and female probe set evaluation   (b) Rank-1 accuray for male and female Validation set evaluation

Fig. 6. These graphs shows Rank-1 accuracy with respect of gender. Triplet Loss achieved higher accuracy for both genders in Validation and Probe sets.

To analyze the variation in *Rank-1* accuracy according to the increase of the number of samples, and evaluate the scalability of each model, Fig. 7 shows the accuracy with different quantity of distractors. The model trained with supervision of *TripletLoss* achieved better results in all variations in the number of distractors.

CMC Courve. The CMC courves in Fig. 8 were drawn to assess the accuracy of models regarding to variation of Rank metric, from Rank-1 to Rank-30. *Triplet Loss* model surpassed other models performance for all values of Rank-N.

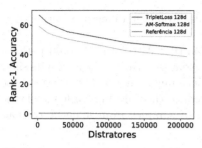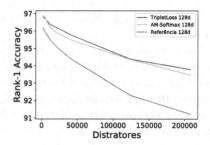

(a) Probe set Rank-1 accuracy under a increase in number of distractors

(b) Validation set Rank-1 accuracy under a increase in number of distractors

Fig. 7. Rank-1 accuracy considering various amounts of distractors. Triplet Loss model performed better for all distractor quantities.

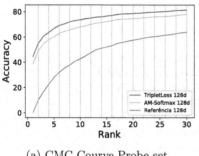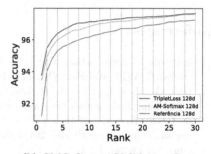

(a) CMC Courve Probe set

(b) CMC Courve Validation Set

Fig. 8. CMC courves of performance models from Rank-1 to Rank-30. Triplet Loss model achieved higher accuracy in Probe set for every value of Rank.

6 Concluding Remarks

Contributions. The main contribution of this work is to point out *Triplet loss* as the cost function with the best performance for fine tuning with datasets with restricted size. The online generation of triplets provides a greater exploitation of the limited data, driving the model to a better extraction of discriminative features.

This work also demonstrates that it is possible to improve the performance of already state-of-art pre-trained models in massive datasets when applied on subjects with different demographic distribution than the trainset. There is a strong suggestion for the existence of data bias concerning ethnicity when evaluating models trained with public datasets on Brazilian faces.

We also presented an accuracy comparison of state-of-the-art cost functions for finetuning training. In the original work, the FaceNet *FaceNet*, the authors performed supervised training using Triplet Loss, reporting 99.63% accuracy in LFW, a training dataset consisting of 200 million samples from 8 million individuals. Whereas a model trained using Am-Softmax reported 99,12% accuracy using a training dataset of 0.41M samples from 10k different identities*Am-Softmax*.

These numbers may lead to the conclusion that AM-Softmax could perform better than FaceNet when trained on the same dataset. However, the results show that Triplet Loss performed better in this finetuning scenario. These results also demonstrate that Triplet Loss was able to reduce the data bias in female faces, decreasing the proportion of errors relative to samples from male individuals compared to AM-Softmax.

Future Directions. In research should gather a wider and more representative dataset to train and evaluate FR models on Brazillian faces. As demonstrated in this work, high accuracy on a specific benchmark does not mean similar performance on different demographic subgroups.

References

1. Institute of Automation, C.A.o.C.: CASIA WebFAce. http://www.cbsr.ia.ac.cn/english/CASIA-WebFace-Database.html. Acessado em abril de 2019
2. Cao, Q., Shen, L., Xie, W., Parkhi, O.M., Zisserman, A.: VGGFace2: a dataset for recognising faces across pose and age (2017)
3. de Geografia e Estatística, I.B.: Censo demográfico do brasil. https://sidra.ibge.gov.br/Tabela/#resultado (2010). Acessado em 23 Mar 2020
4. Guo, Y., Zhang, L., Hu, Y., He, X., Gao, J.: MS-Celeb-1M: a dataset and benchmark for large-scale face recognition. In: Leibe, B., Matas, J., Sebe, N., Welling, M. (eds.) ECCV 2016. LNCS, vol. 9907, pp. 87–102. Springer, Cham (2016). https://doi.org/10.1007/978-3-319-46487-9_6
5. Rawls, A.W., Ricanek, K.: MORPH: development and optimization of a longitudinal age progression database. In: Fierrez, J., Ortega-Garcia, J., Esposito, A., Drygajlo, A., Faundez-Zanuy, M. (eds.) BioID 2009. LNCS, vol. 5707, pp. 17–24. Springer, Heidelberg (2009). https://doi.org/10.1007/978-3-642-04391-8_3
6. Klar, B.F., et al.: Pushing the frontiers of unconstrained face detection and recognition: iarpa janus benchmark A (2015)
7. Nicholls, M.E., Churches, O., Loetscher, T.: Perception of an ambiguous figure is affected by own-age social biases. Sci. Rep. **8**, 12661 (2018)
8. Merler, M., Ratha, N., Feris, R.S., Smith, J.R.: Diversity in faces (2019)
9. Mitchell, T.M.: The need for biases in learning generalizations. Laboratory for Computer Science Research, Department of Computer Science (1980)
10. Nagpal, S., Singh, M., Singh, R., Vatsa, M., Ratha, N.: Deep learning for face recognition: pride or prejudiced? (2019)
11. Qian, N.: On the momentum term in gradient descent learning algorithms. Neural Netw. Official J. Int. Neural Netw. Soc. **12**(1), 145–151 (1999)
12. Gross, R., Matthews, I., Cohn, J., Kanade, T., Baker, S.: Multi-PIE. Image Vis. Comput. **28**, 807–813 (2010)
13. Raji, I.D., Buolamwini, J.: Actionable auditing: investigating the impact of publicly naming biased performance results of commercial AI products. In: AAAI ACM Conference on AI Ethics and Society (2019)

14. Rio, G.: Sistema de reconhecimento facial da pm do rj falha, e mulher é detida por engano (2019). https://g1.globo.com/rj/rio-de-janeiro/noticia/2019/07/11/sistema-de-reconhecimento-facial-da-pm-do-rj-falha-e-mulher-e-detida-por-engano.ghtml. Accessed 27 May 2020
15. Sandberg, D.: Face recognition using Tensorflow. https://github.com/davidsandberg/facenet. Acessado em 01 Apr 2020
16. Schroff, F., Kalenichenko, D., Philbin, J.: FaceNet: a unified embedding for face recognition and clustering (2015)
17. Szegedy, C., Ioffe, S., Vanhoucke, V., Alemi, A.: Inception-v4, inception-resnet and the impact of residual connections on learning (2016)
18. The New York Times: How the police use facial recognition, and where it falls short (2020). https://www.nytimes.com/2020/01/12/technology/facial-recognition-police.html. Acessado em 04 Apr 2020
19. Wang, F., Liu, W., Liu, H., Cheng, J.: Additive margin softmax for face verification (2018)
20. Wang, M., Deng, W.: Deep face recognition: a survey. arXiv preprint arXiv:1804.06655 (2019)
21. Wang, M., Deng, W., Hu, J., Tao, X., Huang, Y.: Racial faces in-the-wild: reducing racial bias by information maximization adaptation network (2018)
22. Weinberger, K.Q., Blitzer, J., Saul., L.K.: Distance metric learning for large margin nearest neighbor classification. MIT Press (2011)
23. Wen, Y., Zhang, K., Li, Z., Qiao, Yu.: A discriminative feature learning approach for deep face recognition. In: Leibe, B., Matas, J., Sebe, N., Welling, M. (eds.) ECCV 2016. LNCS, vol. 9911, pp. 499–515. Springer, Cham (2016). https://doi.org/10.1007/978-3-319-46478-7_31
24. Wu, X., He, R., Sun, Z., Tan, T.: A light CNN for deep face representation with noisy labels (2015)
25. Zhang, K., Zhang, Z., Li, Z., Qiao, Y.: Joint face detection and alignment using multitask cascaded convolutional networks. IEEE Sig. Process. Lett. **23**(10), 1499–1503 (2016). https://doi.org/10.1109/LSP.2016.2603342
26. Zhuang, F., et al.: A comprehensive survey on transfer learning (2019)

Neural Architecture Search in Graph Neural Networks

Matheus Nunes[(✉)] and Gisele L. Pappa

Universidade Federal de Minas Gerais, Belo Horizonte, Minas Gerais, Brazil
{mhnnunes,glpappa}@dcc.ufmg.br

Abstract. Performing analytical tasks over graph data has become increasingly interesting due to the ubiquity and large availability of relational information. However, unlike images or sentences, there is no notion of sequence in networks. Nodes (and edges) follow no absolute order, and it is hard for traditional machine learning (ML) algorithms to recognize a pattern and generalize their predictions on this type of data. Graph Neural Networks (GNN) successfully tackled this problem. They became popular after the generalization of the convolution concept to the graph domain. However, they possess a large number of hyperparameters and their design and optimization is currently hand-made, based on heuristics or empirical intuition. Neural Architecture Search (NAS) methods appear as an interesting solution to this problem. In this direction, this paper compares two NAS methods for optimizing GNN: one based on reinforcement learning and a second based on evolutionary algorithms. Results consider 7 datasets over two search spaces and show that both methods obtain similar accuracies to a random search, raising the question of how many of the search space dimensions are actually relevant to the problem.

Keywords: Graph Neural Networks · Neural architecture search · Evolutionary algorithms · Reinforcement learning

1 Introduction

Performing analytical tasks over graph[1] data has become increasingly interesting due to the ubiquity and large availability of relational information. Predicting interaction between proteins, classifying users in social networks and recommending movies to users are some classical examples of such tasks [24]. However, unlike images (formed by a grid of pixels) and sentences (formed by a string of ordered words), there is no notion of sequence in networks. Nodes (and edges) follow no absolute order, so it is hard for traditional machine learning (ML) algorithms, which were built to handle data stored in tensors, to recognize a pattern and generalize their predictions on this type of data [23].

---

[1] In this work we use the terms "graph" and "network" interchangeably. When referring to "neural networks" we will use NN or "neural network".

© Springer Nature Switzerland AG 2020
R. Cerri and R. C. Prati (Eds.): BRACIS 2020, LNAI 12319, pp. 302–317, 2020.
https://doi.org/10.1007/978-3-030-61377-8_21

Due to the success of convolutional neural networks (CNNs) for tasks such as image classification [12], object identification [14] and semantic segmentation [1], a large body of work began to re-define the concept of convolution to the graph domain. Following the work of Gori et al. [10] and Scarselli et al. [17] on Graph Neural Networks (GNNs), the concept of spectral-based graph convolution function was defined by Bruna et al. [3] and later refined by Defferrard et al. [5]. In this approach, unlike traditional neural networks where the architecture is composed by fully connected layers of neurons, graph neural networks follow the graph structure itself [17]. Forward propagation is done on the nodes of the graph, which pass information onto the next layer by aggregating information from the neighborhood and applying an activation function to the result.

Since the concept of convolution was adapted to the context of graphs, a plethora of GNN models were proposed, including GraphSAGE [11], Graph Attention Networks (GAT) [20], Graph Isomorphism Network (GIN) [22] and many others. These methods achieve state-of-the-art results on tasks such as node classification and link prediction. However, the design and optimization of GNN architectures is currently hand-made, based on heuristics or empirical intuition, which makes it an ineffective and error prone task [22].

Automated Machine Learning (AutoML) appears as a solution to this problem, as it aims to automate the process of building and optimizing machine learning pipelines, relieving users from that burden [7]. Neural Architecture Search (NAS) is considered the current challenge in automating machine learning algorithms [8]. Its methods are composed by a search space of possible architectures, a search method to explore this space and an evaluation framework for the generated architectures.

To the best of our knowledge, there were few attempts in the literature to employ NAS for GNNs [9,25]. In these works, reinforcement learning methods are used to explore similar search spaces. The NAS literature poses two main types of methods as the most effective to solve the problem: reinforcement learning (RL) and evolutionary algorithms (EAs) [8]. The second type of technique has been so far overlooked in the context of GNNs.

This work employs an EA previously proposed for NAS in the context of image classification [16] to optimize GNNs and performs a comparative analysis of the method with reinforcement learning and random search in terms of model accuracy and runtime. It also conducts a study of the characteristics of the previously proposed search spaces for GNNs in order to identify opportunities for performance improvement on GNN NAS algorithms. Results show that both RL and EA are able to find equivalent models in terms of accuracy, with EA being faster in some cases, which corroborates previous findings for image classification. Furthermore, following the already discussed problems of large search spaces – such as those required in the case of GNNs – with many low effective dimensions [2], we show a Random Search is able to find architectures with equivalent accuracy while being faster. We discuss these results in the light of previous works that discuss this problem.

The remainder of this work is organized as follows. Section 2 introduces background on GNNs and Sect. 3 discusses related work. Section 4 describes the methodology followed to apply the tested methods in GNN search spaces, while Sect. 5 presents the results. Finally, Sect. 6 draws conclusions and discusses directions of future work.

2   Background

In this work, we assume as input a graph composed of a set of nodes and edges, $G = (N, E)$. Each node $n_i \in N$ is attached to a feature/attribute vector $x_i \in X$, and a label $l_i \in L$. The presence of node labels indicates that we are assuming a **supervised learning** situation. We define by $\mathcal{N}(i)$ the neighborhood of a node i, i.e., the set of nodes connected to i by an edge. The primary concept behind GNNs is that each node in the graph represents an abstract concept, and edges represent the relationship between these concepts. Therefore, the node's features should correlate with its neighboring features, defining a state (or hidden node representation) $h_i \in \mathcal{H}_N$ for each node [17].

Traditionally, each GNN layer is composed of a function that aggregates information from the neighborhood of each node $\mathcal{N}(i)$, forming an intermediate vector $h_{\mathcal{N}(i)}$, and a second function that combines this value with the current node representation h_i, which in turn goes through an activation function before being output [11,17]. Formally, this process can be defined as:

$$h_{\mathcal{N}(i)}^{(k)} = \text{aggregate}(h_j^{k-1} : j \in \mathcal{N}(i)) \tag{1}$$

$$h_i^{(k)} = \text{activate}(\text{combine}(h_i^{(k-1)}, h_{\mathcal{N}(i)}^{(k)})) \tag{2}$$

By convention, the first hidden representation of each node is its feature vector, $h_i^{(0)} = x_i$ [13]. Figure 1 shows how the structure of a GNN is generated. Given the graph represented in part (a) of the figure, which has 4 nodes n_i and a feature vector x_i associated to each of them, an intermediate representation is generated ((b) in the figure). In this representation, for each node, the neighborhood information generates the intermediate vectors h_i according to the process described in Eq. 1. The third part of the picture (c) shows the GNN itself, where each layer corresponds to an update of the state of the feature vectors of the current node.

In this work we consider undirected graphs and a one-hop neighborhood for each node, which means that only features from a node's direct neighbors are considered in aggregation. There are many options of aggregation and activation functions, and other mechanisms can also be added to this standard GNN architecture. These components choices are the main subject of this paper, as detailed in Sect. 4.1.

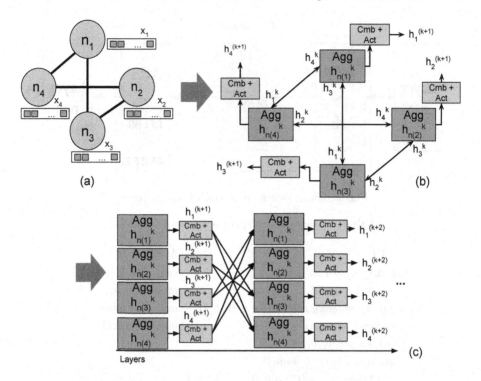

Fig. 1. Structure of a GNN, adapted from Scarselli et al. [17]

3   Related Work

NAS is considered the current challenge in automating machine learning algorithms, after the success of automated feature engineering [8]. Famous NAS works can be roughly split into two categories: Reinforcement Learning (RL) [4,26] and Evolutionary Algorithms (EA) [16]. It has been shown that both types of methods are able to find models that perform better than hand-crafted engineered ones, but Real et al. presents empirical proof that EA-based and RL-based methods are able to find equally well-suited models in terms of performance, with EA-based methods finding less complex models in less overall time [8,16]. Our idea is to adapt and employ NAS methods to the task of finding a good GNN model for large-scale graph embedding, whereas in previous works, the tasks of interest were mostly image classification and object detection.

To the best of our knowledge, NAS has not yet been largely explored in the context of GNNs. GraphNAS [9] is one of the few that uses RL to find feasible architectures for the node classification task. The authors define a search space composed of sampling, aggregation and gated functions, which can be extended to account for hyperparameters. Auto-GNN [25] follows the same line of work, exploring RL and a similar search space to GraphNAS.

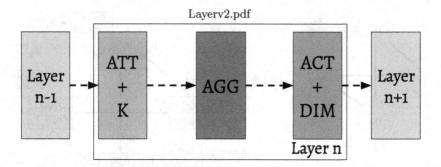

Fig. 2. Macro search space GNN layer example

Table 1. Macro search space options for 5 actions.

ATT	AGG	ACT
const, $e_{ij} = 1$	sum	tanh
gcn, $e_{ij} = 1/d_i d_j$	mean	linear
gat, $e_{ij} = leaky\_relu((W_l h_i + W_r h_j))$	max	softplus
sym-gat, $e_{ij} = e_{ji} + e_{ij}$	mlp	sigmoid
cos, $e_{ij} =< W_l h_i, W_r h_j >$		elu
linear, $e_{ij} = tanh(sum(W_l h_j))$		relu
gen_linear, $e_{ij} = W_a tanh(W_l h_i + W_r h_j)$		relu6
		leaky_relu
K	2^i, $i \in \{1, ..., 6\}$	
DIM	2^i, $i \in \{2, ..., 8\}$	

4   Methodology

The problem of NAS in GNNs can be formally defined as follows. Given a dataset $\mathcal{D}$ – split into training and validation sets $\mathcal{D}_{train}$ and $\mathcal{D}_{valid}$, respectively – and a search space of Graph Neural Architectures $\mathcal{A}$, capable of generating a GNN with an architecture $a \in \mathcal{A}$ with its own set of hyperparameters Λ, the goal is to find the model with the highest expected accuracy $\mathcal{E}$ on $\mathcal{D}_{valid}$, when its parameters w^* are set on $\mathcal{D}_{train}$, setting the following bi-level optimization problem:

$$\operatorname*{argmax}_{a_\lambda \in \mathcal{A}, \lambda \in \Lambda, w^*} \mathcal{E}[(a_\lambda(w^*, \mathcal{D}_{valid}))]$$

$$\text{s.t. } w^* = \operatorname*{argmin}_w \mathcal{L}(a_\lambda(w, \mathcal{D}_{train})),$$

This section details the search spaces $\mathcal{A}$ previously defined for GraphNAS [9] and describes the evolutionary algorithm and the RL methods we evaluated in the context of GNN architecture search.

Table 2. Micro search space action and hyperparameters.

CNV	$GAT_{1,...,8}$, GCN, Cheb, SAGE, ARMA, SG, Linear, Zero
CMB	Add, Product, Concat
ACT	Sigmoid, tanh, elu, relu, linear
LR	$\{1 \times 10^{-2}, 1 \times 10^{-3}, 1 \times 10^{-4}\}$
DO	$\{0.0, 0.1, ..., 0.9\}$
WD	$\{0, 1 \times 10^{-3}, 1 \times 10^{-4}, 5 \times 10^{-4}, 1 \times 10^{-5}, 5 \times 10^{-5}\}$
HU	$2^i, i \in \{3, ..., 9\}$

4.1   Search Spaces

The two search spaces evaluated in this work, named by the authors in [9] as "Macro" and "Micro", are composed by different GNN layers, as detailed next.

Macro Search Space. The name "Macro" comes from the fact that architectures generated from this space always follow the same structure: each layer is composed by a multi-head attention mechanism ATT and the number of heads K, a choice of aggregator AGG, the output dimension DIM and an activation function ACT, in this order. The neighborhood sampling method is fixed as a first-order sampler, i.e. only direct neighbors of each node are sampled at each step.

Considering the definitions in Sect. 2, we have a new component here, which is the attention mechanism. As described by the authors in [20], an attention mechanism – implemented by the coefficients e_{ij}, is designed to attribute different importance value to the features of each of a node's neighbors. Such coefficients are calculated only for $j \in \mathcal{N}(i)$ for performance reasons (in order to avoid an $N \times N$ matrix), and in practice define the importance of node j's features over node i. They are implemented as a single-layer feed-forward neural network, and a range of options to this mechanism is available (see first column of Table 1). Multi-head attention is a way of having independent attention mechanisms over the node's features. It has been proven that concatenating the results of these independent mechanisms yields better results than using a single attention head [20].

Figure 2 presents the disposition of the actions. The number of multi-heads K can be merged with the attention mechanism ATT as they alter the same behavior. The output dimension DIM can also be merged with the activation function ACT.

Table 1 presents the options for each action on the layers. Considering the number of options for each action on the layers, the search space presents ($7 \times 6 \times 4 \times 7 \times 8$) = 9408 possibilities for each layer. According to the authors in [13], GNNs achieve the best overall results using architectures with 2 or 3 layers. Therefore in this paper the architectures have 2 layers, in a total of $9408^2 = 88,510,464$ architecture possibilities.

One important characteristic of this search space is that the hyperparameters of the GNNs, such as learning rate, dropout, weight decay are kept fixed. The learning rate is set to 0.005, the dropout to 0.6 and the weight decay to 5×10^{-4}.

Micro Search Space. The name "Micro" comes from the fact that architectures generated from this search space are composed by combining different convolution schemes, and do not follow a single fixed structure. The choice of actions in this space are: a convolutional layer CNV, a combination scheme CMB and an activation function ACT. The hyperparameters which can be tuned are: the learning rate LR, the dropout rate DO, the weight decay rate WD and the number of hidden units HU. In the options for CNV, the option $GAT_{1,...,8}$ means that there are 8 possible GAT convolutions, using 1 to 8 multi-heads attention.

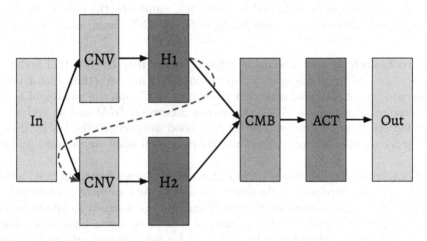

Fig. 3. Micro search space GNN architectures example

Figure 3 illustrates the types of architectures that can be generated from this space. The straight arrows represent one type of connectivity, where the input is fed to two separate convolutional layers and their outputs are fed to the combination layer. The dashed line represents the second type, when two convolutional layers are stacked before feeding the output to the combination layer. The full list of actions and hyperparameters for this space is presented in Table 2. Regarding the number of possibilities for each action and hyperparameter listed, there are $(15 \times 15 \times 3 \times 5 \times 3 \times 10 \times 5 \times 7) = 3,543,750$ architecture possibilities in this space.

Note that the architectures in the micro-space take advantage of convolutions. Graph convolution methods are classified mainly into two streams, both covered by the micro-search space: spectral-based and spatial-based methods [21]. Spectral methods [3,13] rely on spectral properties of the graph, by finding eigenvectors of the normalized graph Laplacian. This approach is limited because eigendecomposition is an expensive operation, eigenbasis are sensible to minimal graph perturbations and the learned filters do not generalize well to graphs of different structure (therefore they do not work well on inductive learning scenarios). Spatial-based methods [11,20] follow the message passing idea of traditional GNNs (also known as Recursive GNNs), in which a node's hidden representation is an input to its neighbors computation. These methods are scalable to large graphs and are more generalizable to various types of graphs (heterogeneous, directed, graphs which contain edge labels, etc.).

4.2   Search Methods

This section describes the two methods we apply to search the macro and micro search spaces described in the previous section: the evolutionary method and the reinforcement learning. We also describe the random search method that will be used as a baseline for the results.

Evolutionary Algorithm - Evolutionary methods are inspired by Darwin's theory of evolution, and evolve a set of individuals – which represent solutions to the problem at hand – for a number of iterations (also known as generations) [6]. From one iteration to the next, individuals are evaluated according to a fitness function, which assesses their ability to solve the problem. The value of fitness is used to probabilistic select the individuals that will undergo crossover and mutation operators, which are applied according to user-defined probabilities. We explore an evolutionary method inspired on the Aging Evolution method, described by Real et al. [16]. In this method, a population of individuals –i.e., a set of GNNs – is generated randomly by sampling options for each action in a layer, considering the number of layers specified. These GNNs are then trained in a training set and have their accuracy measured on a validation set. This value of accuracy is used to select an individual via tournament selection to generate a new offspring. The child individual is generated via mutation, which is uniform over the actions and replaces the selected action by a random option. The child individual is always added to the population and the oldest individual in the population (i.e., the individual that has been in the population for the highest number of iterations) is always removed (hence the name "Aging Evolution").

Reinforcement Learning - GraphNAS uses a LSTM (Long-Short Term Memory) network as a controller to generate fixed-length architectures, which act as GNN architecture descriptors and can be viewed as a list of actions. The accuracy achieved by the GNN in the validation dataset at convergence is used as the reward signal to the training process of the reinforcement learning controller. As the reward signal $\mathcal{R}$ is non-differentiable, a policy gradient method is used to iteratively update θ with a moving average baseline for reward to reduce variance.

Random Search - An initial random GNN is generated by sampling options from each action in a layer, for the specified number of layers. The GNN is trained and the accuracy on the validation set measured. This process is repeated for the specified number of iterations, storing the GNN with the highest accuracy.

5   Experimental Analysis

We assess the performance of the evolutionary algorithm (EA)[2], the reinforcement learning (RL) method and the random search (RS) on the transductive learning scenario, in a node classification task, over a set of 7 datasets in terms of accuracy and runtime, as detailed next. It is important to note that this work does not compare the architectures obtained by the optimization methods to hand-crafted ones, as that was already done in GraphNAS' paper [9].

5.1   Datasets

Table 3 presents the details of the datasets, as previously used in [19] and provided by `Pytorch Geometric`[3]. For all cases, we are dealing with a node classification task, where we use information from the nodes with known-labels to assign a class to nodes with unknown label (test set).

Table 3. Dataset characteristics.

Dataset (Abbrv.)	# Classes	# Features	# Nodes	# Edges
CORA (COR)	7	1433	2708	10556
Citeseer (CIT)	6	3703	3327	9104
Pubmed (MED)	3	500	19717	88648
Coauthor CS (CS)	15	6805	18333	163788
Coauthor physics (PHY)	5	8415	34493	495924
Amazon computers (CMP)	10	767	13752	491722
Amazon photo (PHO)	8	745	7650	238162

The first three datasets (**COR**, **CIT**, **MED**) are paper co-authorships networks, used previously in [13]. Nodes represent documents, and an edge between two documents means that one paper cited the other. Class labels represent subareas of machine learning [18]. Node features are sparse bag-of-words vectors.

CS and **PHY** are also co-authorship networks, based on the Microsoft Academic Graph from KDD Cup 2016. However, in these datasets nodes represent authors instead of papers, connected by an edge if they have co-authored a paper. Node features represent paper keywords for each author's papers. Class labels indicate the most active field of study for each author in the network.

CMP and **PHO** are segments of the Amazon co-purchase graph, where nodes represent products and edges are added between items frequently bought together. The nodes features are a bag-of-words representation of product reviews, and class labels represent the product category.

---

[2] Code available at: https://github.com/mhnnunes/nas_gnn.
[3] https://github.com/rusty1s/pytorch_geometric.

Table 4. Accuracies and execution times (in $\times 10^4$ seconds) of search methods.

		Macro		Micro	
		Accuracy	Time	Accuracy	Time
COR	**EA**	0.83 ± 0.007	0.75 ± 0.16	0.82 ± 0.005	1.73 ± 0.53
	RL	0.83 ± 0.003	1.45 ± 0.38	0.81 ± 0.001	2.42 ± 0.62
	RS	0.82 ± 0.003	0.96 ± 0.02	0.80 ± 0.009	1.20 ± 0.21
CIT	**EA**	0.75 ± 0.002	1.18 ± 0.10	0.71 ± 0.007	2.80 ± 0.72
	RL	0.73 ± 0.004	1.52 ± 0.42	0.68 ± 0.006	2.24 ± 0.08
	RS	0.73 ± 0.005	1.05 ± 0.03	0.69 ± 0.006	1.29 ± 0.04
MED	**EA**	0.82 ± 0.003	1.40 ± 0.37	0.82 ± 0.009	1.40 ± 0.09
	RL	0.80 ± 0.003	2.10 ± 0.14	0.76 ± 0.017	2.58 ± 0.28
	RS	0.85 ± 0.045	1.31 ± 0.02	0.80 ± 0.009	1.10 ± 0.18
CS	**EA**	0.98 ± 0.001	3.35 ± 0.78	0.99 ± 0.002	2.65 ± 0.48
	RL	0.95 ± 0.001	3.13 ± 0.11	0.97 ± 0.002	2.90 ± 0.34
	RS	0.97 ± 0.001	1.50 ± 0.03	0.99 ± 0.001	1.58 ± 0.05
PHY	**EA**	0.99 ± 0.002	4.21 ± 0.85	0.99 ± 0.000	1.53 ± 0.15
	RL	0.98 ± 0.001	3.34 ± 0.27	0.98 ± 0.001	2.01 ± 0.19
	RS	0.98 ± 0.001	2.08 ± 0.07	0.99 ± 0.001	1.11 ± 0.05
CMP	**EA**	0.91 ± 0.005	3.09 ± 0.49	0.93 ± 0.004	4.02 ± 1.94
	RL	0.90 ± 0.010	3.43 ± 0.21	0.92 ± 0.008	3.68 ± 0.27
	RS	0.89 ± 0.004	1.69 ± 0.07	0.92 ± 0.002	2.05 ± 0.07
PHO	**EA**	0.97 ± 0.002	2.48 ± 0.22	0.98 ± 0.004	1.66 ± 0.41
	RL	0.96 ± 0.005	3.65 ± 0.19	0.97 ± 0.002	1.88 ± 0.23
	RS	0.96 ± 0.002	1.82 ± 0.04	0.97 ± 0.002	1.08 ± 0.04

5.2 Experimental Setup

All search methods were executed for 1000 iterations in order to enable a fair comparison. In each iteration, a single GNN architecture is generated, trained on $\mathcal{D}_{train}$ and evaluated (in terms of accuracy) on $\mathcal{D}_{valid}$. The architecture with the highest validation accuracy is saved across iterations, and returned as the result of the optimization process. The generated architectures are trained using the following fixed hyperparameters for all search spaces and methods: minimizing cross-entropy loss using ADAM optimizer, initial learning rate of 0.005 and an early stopping strategy with a patience of 100 epochs.

Random search has only one parameter: the number of iterations. The reinforcement learning controller is trained using the same hyperparameters as described on GraphNAS' paper [9]: a one-layer LSTM with 100 hidden units, ADAM optimizer, learning rate at 3.5×10^{-4} and random initialization of weights. Aging Evolution has three main parameters: the population size, the tournament size k and the number of iterations n. The first parameter is related to the number of solutions evaluated during the search process, while the tournament size controls the convergence speed. The higher the value of k, the faster the algorithm converges. From all tested values ($\{100, 25\}, \{25, 2\}, \{100, 3\}$), the best results were achieved using the population size set to 100 and k set to 3.

The   dataset split between training, validation and testing sets was done in the same way as in the GraphNAS public code[4]: the last 1000 nodes are separated for validation and testing, split evenly between the two.

All experiments were repeated 5 times as the methods are non-deterministic. The experiments were run on a machine with a 16-core Intel(R) Xeon(R) Silver 4108 CPU @ 1.80GHz, 16GB DIMM DDR4 @ 2666 MHz RAM, and a NVIDIA GV100 [TITAN V] graphics card, with 12GB dedicated RAM.

5.3   Results

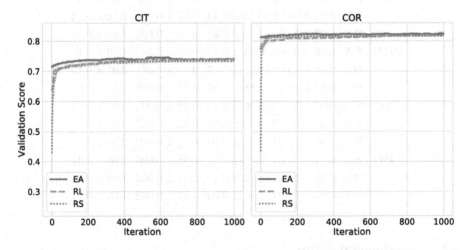

Fig. 4. Highest validation accuracy by iteration, for **CIT** and **COR** datasets, on the **Macro** search space.

Table 4 shows the results of accuracy and execution time for the   Macro and Micro search spaces,   at the end of the optimization process (after 1000 iterations). In terms of accuracy, the results obtained by the EA and RL methods are very similar to the ones obtained by the random search.   In terms of execution time, RS wins in most cases. The execution time for the search varies between 2 and 12 GPU hours.

Figure 4 presents the evolution of the highest validation accuracy value achieved by an GNN architecture across the iterations, by search method[5]. Each

---

[4] https://github.com/GraphNAS/GraphNAS.

[5] We present only the results for the **Macro** search space because the results for **Micro** are very similar.

line represents the mean validation score across all seeds, and the shaded area around it represents the standard deviation of this value. It is very clear that **all methods converge** (find a good performing architecture and plateaus) **within only a few iterations**. The fact that the EA already starts at a high value may be attributed to the population initialization process, depicted in Fig. 6.

It may seem counter-intuitive that we are using sophisticated methods to obtain results that can be also be achieved by a random search method, but as the authors in [2] have previously discussed, in large search spaces where many of the dimensions are irrelevant to the task at hand the random search can be as effective as more sophisticated methods. This problem is aggravated by the neutrality of the space, i.e., architectures in neighbour regions of the search space may differ in a few components but do not lead to a value of accuracy different from their neighbors [15]. Another stronger indicator of a neutral search space is the fact that many high quality individuals are generated in the initialization step, and evolution takes a minor part in improving them, as shown in Fig. 4.

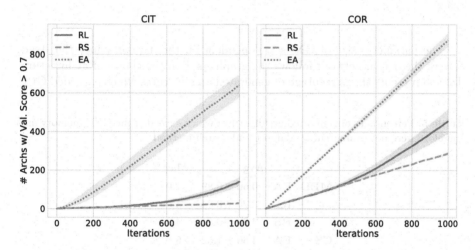

Fig. 5. Cumulative number of architectures with validation accuracy higher than threshold, for **CIT** and **COR** datasets, on the **Macro** search space.

Figure 5 presents the number of evaluated architectures with validation accuracy over 0.7, for **CIT** and **COR**, in the Macro search space. The 0.7 threshold was set because this value represents approximately the best accuracy value for **CIT** on the **Macro** search space. The pattern shown in the figure is consistent for **all datasets** in **both search spaces**. It shows that the EA tends to converge to a better region of the search space faster than the other two methods, thus evaluating more high quality architectures. Such tendency could be explained by the EA's selective pressure (driven by the tournament selection process), which makes the algorithm prioritize good individuals for mutation and evaluation.

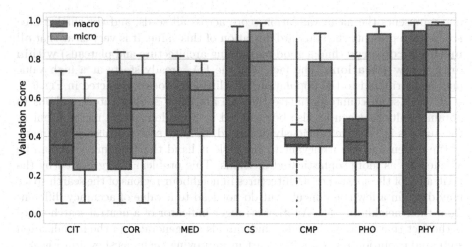

Fig. 6. Distribution of EA's initial population validation accuracies on both search spaces.

The parameter size of GNNs is dependent on the dataset (since the structure of the neural network follows the graph) and on the choice of architecture. Table 5 presents the percentage of generated architectures which exceeded GPU

Table 5. Percentages of generated architectures which exceeded the GPU memory and therefore were not evaluated, by dataset and search method

		Avg. %	Max %
MED	**EA**	0.60 ± 0.89	2.0
	RL	3.20 ± 0.84	4.0
	RS	2.80 ± 0.84	4.0
CS	**EA**	4.60 ± 1.52	6.0
	RL	10.20 ± 2.59	14.0
	RS	9.60 ± 1.52	11.0
PHY	**EA**	13.60 ± 1.82	**16.0**
	RL	41.80 ± 9.44	56.0
	RS	47.80 ± 0.45	48.0
CMP	**EA**	11.60 ± 2.61	14.0
	RL	47.00 ± 20.94	**81.0**
	RS	38.40 ± 1.67	41.0
PHO	**EA**	4.60 ± 2.70	9.0
	RL	20.80 ± 3.42	24.0
	RS	11.80 ± 1.48	14.0

memory, by each dataset and search method.[6] EA is consistently the search method for which the **smallest percentage** of generated architectures are too big for the GPU memory, with the highest value as 16%, while RL reaches 81% of all architectures being too large. This corroborates the findings of Real et al. [16] which state that Evolutionary Algorithms are able to find less complex but equally well performing architectures than RL.

6   Conclusions and Future Work

GNNs are able to achieve state-of-the-art performances in prediction tasks over networks. However, their design and optimization is currently hand-made and error prone. This paper compared the results of two NAS search methods – a reinforcement learning technique and an evolutionary algorithm – to a random search in the task of searching for architectures and hyperparameters for GNNs.

The three methods produced GNN architectures which achieved similar results in terms of accuracy when considering a set of 7 datasets and two architecture layer search spaces, with the random search being the fastest method followed by the evolutionary algorithm and reinforcement learning. Architectures generated by EA tend to fit in GPU memory, while the other methods generate oversized architectures in up to **80%** of cases. This shows that EA generates less complex structures while achieving a similar accuracy value to the other methods, corroborating the findings of Real et al. [16] for images.

In general, the results indicate that there are irrelevant dimensions to this task in the defined search spaces, which will require a more in-depth study of each of these spaces. Further, the neutrality of this space, i.e., the fact that neighbor solutions present different architectures but very similar results of accuracy make search even harder. As future work, we intend to perform a more in-depth investigation of the dimensions of the search space in order to identify those that may be irrelevant to search, as well as propose new search methods that may include mechanisms to try to avoid these neutral regions.

Acknowledgements. The authors would like to thank FAPEMIG (grant no. CEX-PPM-00098-17), MPMG (project Analytical Capabilities), CNPq (grant no. 310833/2019-1), CAPES, MCTIC/RNP (grant no. 51119) and H2020 (grant no. 777154) for the financial support.

References

1. Badrinarayanan, V., Kendall, A., Cipolla, R.: SegNet: a deep convolutional encoder-decoder architecture for image segmentation. TPAMI **39**(12), 2481–2495 (2017)
2. Bergstra, J., Bengio, Y.: Random search for hyper-parameter optimization. JMLR **13**, 281–305 (2012)

---

[6] The smallest datasets (**CIT** and **COR**) are not present in the table because none of the generated architectures for these datasets exceeded GPU memory.

3. Bruna, J., Zaremba, W., Szlam, A., LeCun, Y.: Spectral networks and locally connected networks on graphs. In: Bengio, Y., LeCun, Y. (eds.) ICLR 2014 (2014)
4. Cai, H., Chen, T., Zhang, W., Yu, Y., Wang, J.: Efficient architecture search by network transformation. In: AAAI 2018 (2018)
5. Defferrard, M., Bresson, X., Vandergheynst, P.: Convolutional neural networks on graphs with fast localized spectral filtering. In: Lee, D.D., Sugiyama, M., von Luxburg, U., Guyon, I., Garnett, R. (eds.) NeurIPS 2016, pp. 3837–3845 (2016)
6. Eiben, A., Smith, J.: Introduction to Evolutionary Computing. Springer, Heidelberg (2015). https://doi.org/10.1007/978-3-662-44874-8
7. Elshawi, R., Maher, M., Sakr, S.: Automated machine learning: state-of-the-art and open challenges. arXiv preprint arXiv:1906.02287 (2019)
8. Elsken, T., Metzen, J.H., Hutter, F.: Neural architecture search: a survey. JMLR **20**, 55:1–55:21 (2019)
9. Gao, Y., Yang, H., Zhang, P., Zhou, C., Hu, Y.: Graph neural architecture search. In: IJCAI 2020, pp. 1403–1409 (2020)
10. Gori, M., Monfardini, G., Scarselli, F.: A new model for learning in graph domains. In: Proceedings, 2005 IEEE International Joint Conference on Neural Networks, vol. 2, pp. 729–734. IEEE (2005)
11. Hamilton, W., Ying, Z., Leskovec, J.: Inductive representation learning on large graphs. In: NIPS 2017 (2017)
12. He, K., Zhang, X., Ren, S., Sun, J.: Deep residual learning for image recognition. In: CVPR 2016, pp. 770–778 (2016)
13. Kipf, T.N., Welling, M.: Semi-supervised classification with graph convolutional networks. In: ICLR 2017 (2017)
14. Liu, W., et al.: SSD: single shot multibox detector. In: Leibe, B., Matas, J., Sebe, N., Welling, M. (eds.) ECCV 2016. LNCS, vol. 9905, pp. 21–37. Springer, Cham (2016). https://doi.org/10.1007/978-3-319-46448-0_2
15. Pimenta, C.G., de Sá, A.G.C., Ochoa, G., Pappa, G.L.: Fitness landscape analysis of automated machine learning search spaces. In: Paquete, L., Zarges, C. (eds.) EvoCOP 2020. LNCS, vol. 12102, pp. 114–130. Springer, Cham (2020). https://doi.org/10.1007/978-3-030-43680-3_8
16. Real, E., Aggarwal, A., Huang, Y., Le, Q.V.: Aging evolution for image classifier architecture search. In: AAAI 2019 (2019)
17. Scarselli, F., Gori, M., Tsoi, A.C., Hagenbuchner, M., Monfardini, G.: The graph neural network model. In: IEEE TNN 2009 (2009)
18. Sen, P., Namata, G., Bilgic, M., Getoor, L., Galligher, B., Eliassi-Rad, T.: Collective classification in network data. AI Mag. **29**(3), 93 (2008)
19. Shchur, O., Mumme, M., Bojchevski, A., Günnemann, S.: Pitfalls of graph neural network evaluation. arXiv preprint arXiv:1811.05868 (2018)
20. Velickovic, P., Cucurull, G., Casanova, A., Romero, A., Liò, P., Bengio, Y.: Graph attention networks. In: ICLR 2018 (2018)
21. Wu, Z., Pan, S., Chen, F., Long, G., Zhang, C., Yu, P.S.: A comprehensive survey on graph neural networks. CoRR (2019)
22. Xu, K., Hu, W., Leskovec, J., Jegelka, S.: How powerful are graph neural networks? In: ICLR 2019 (2019). OpenReview.net
23. Zhang, M., Cui, Z., Neumann, M., Chen, Y.: An end-to-end deep learning architecture for graph classification. In: AAAI 2018 (2018)

24. Zhang, Z., Cui, P., Zhu, W.: Deep learning on graphs: a survey. In: TKDE 2020, p. 1 (2020)
25. Zhou, K., Song, Q., Huang, X., Hu, X.: Auto-GNN: neural architecture search of graph neural networks. arXiv preprint arXiv:1909.03184 (2019)
26. Zoph, B., Vasudevan, V., Shlens, J., Le, Q.V.: Learning transferable architectures for scalable image recognition. In: CVPR 2018 (2018)

People Identification Based on Soft Biometrics Features Obtained from 2D Poses

Henrique Leal Tavares[1]([✉]), João Baptista Cardia Neto[2], João Paulo Papa[1],
Danilo Colombo[3], and Aparecido Nilceu Marana[1]

[1] UNESP - São Paulo State University, Bauru, SP 17033-360, Brazil
{h.tavares,joao.papa,nilceu.marana}@unesp.br
[2] FATEC - São Paulo State Technological College, Catanduva, SP 15800-020, Brazil
joao.cardia@fatec.sp.gov.br
[3] Research and Development Center Leopoldo Américo Miguez de Mello
(CENPES/PETROBRÁS), Rio de Janeiro, RJ 21941-915, Brazil
colombo.danilo@petrobras.com.br

Abstract. An important challenge in the research field of Biometrics
is real-time identification, at a distance, in uncontrolled environments,
using low-resolution cameras. In such circumstances, soft biometrics can
be the only option. In this work, we propose two novel descriptor methods for biometric identification based on ensemble of anthropometric
measurements and joints heat-map of the person skeleton, captured
from video frames through state-of-the-art 2D poses estimation methods.
The proposed methods were assessed on a popular benchmark dataset,
CASIA Gait Dataset B, and obtained good results (85% and 89% of
rank-1 identification rates, respectively) with PifPaf 2D pose estimation
method.

Keywords: Biometrics · 2D pose estimation · Anthropometric
measurements · Joints heat-maps · People identification

1 Introduction

Biometrics is the science of establishing the identity of a person based on physical or behavioral attributes [9]. As biometric-based methods are more reliable
and difficult to fraud than the traditional identification methods based on possession (cards, documents, etc.) and knowledge (passwords, codes, etc.), they
have become increasingly used for human identification in different applications.

An important challenge in the area of Biometrics is the automatic identification carried out in real-time, in uncontrolled environments, using low-resolution
cameras, like CCTV cameras, installed at a distance in positions that are not
always favorable. In these types of scenarios, the use of traditional biometric features, such as iris, fingerprint, or even face, may be very difficult or unfeasible.
In such cases, the utilization of gait or soft biometrics can be the only option.

© Springer Nature Switzerland AG 2020
R. Cerri and R. C. Prati (Eds.): BRACIS 2020, LNAI 12319, pp. 318–332, 2020.
https://doi.org/10.1007/978-3-030-61377-8_22

Gait has gained interest especially because it requires low user interaction and, normally, the images needed for such type of subject identification can be low-resolution [4,5,7,15,17–19]. However, gait identification depends on motion, which is an important limitation. Besides, since it is a behavioral characteristic, it can be easily imitated by individuals that want to attack the biometric system. Soft biometrics are ancillary information (e.g. height, gender, skin color, hair color) easily distinguished at a distance but not fully distinctive when used individually in recognition tasks [16]. However, soft biometrics features can be very effective when used together with gait features. Gait and some soft biometric information can be extracted from videos through 2D poses. Fig. 1 shows the 2D outline of a human pose represented by the main parts of the human skeleton.

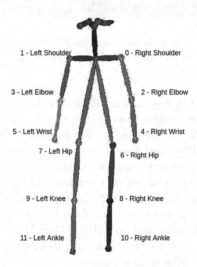

Fig. 1. Example of a 2D human pose represented by the main parts of the human skeleton. The labels refer to the 12 main joints of the human skeleton.

Recently, two methods for 2D human pose estimation from videos were proposed: OpenPose [6] and PifPaf [11]. OpenPose [6] utilizes Part Affinity Fields (PAFs) to learn how to associate parts of the body with individuals that are detected within an image. PifPaf [11] utilizes Part Intensity Field (PIF) to detect the body part of an individual and a Part Association Field (PAF) to associate parts of the body and build a whole human body. Figure 2 shows an example of a 2D human pose estimated by PifPaf. One can observe that the method was able to detect all 12 joint points of the human skeleton, and, consequently, their corresponding parts, like arms, legs and torso. Both methods, OpenPose and PifPaf, are able to estimate 2D poses from video in real time and, therefore, can be used for gait or soft biometric features extraction in biometric systems.

The goal of our work is to investigate if a soft biometric feature ensemble, composed by the lengths of the skeleton parts of the human body, related to

Fig. 2. Example of 2D human pose generated by PifPaf [11]. One can observe that the method was very precise in detecting the skeleton joints and their associated parts.

the twelve joints labeled in Fig. 1, has sufficiently discriminating information to identify a person. Another goal is to assess the use of joints heat-maps, outputted by OpenPose and PifPaf, as descriptors for gait recognition. As we intend to use methods like OpenPose and PifPaf to detect the parts of the human body skeleton, a secondary goal of our work, but equally important, is to analyse which method is more effective for our person identification application. Regarding the OpenPose method, we analysed two different implementation versions, one using Caffe Deep Learning Framework [10] and the other using Tensorflow framework [1]. Experimental results obtained on CASIA Gait Dataset B [13] showed that PifPaf leaded to better results and that both descriptor methods proposed in our work obtained good identification rates.

The rest of this paper is organized as follows: Sect. 2 gives a brief introduction to soft biometrics. Section 3 presents some related works. Section 4 presents details of the 2D pose estimation methods. Section 5 describes the proposed approach. Section 6 shows the experimental results, and Sect. 7 draws some conclusions of our work.

2 Soft Biometrics

Soft biometrics feature is defined as a characteristic that provides some type of discriminating information but not enough for assuring a subject identity [2]. There are two types of soft biometrics characteristics, discrete or continuous. A discrete characteristic is an intrinsic and more permanent trait from the subject's body, like skin tone and iris color, while a continuous characteristic is a trait that can change progressively over time, like height and weigh [2].

In general, soft biometrics are utilized together with hard features in order to increase the robustness of a biometric identification system [2]. For instance, a face-based biometric system will have its performance reduced in an environment where the presence of identical twins is frequent but, if soft biometrics features are used together, the performance of the biometric system can be maintained.

Contrary to the soft biometrics definition, some emerging works related to this subject have shown that, in many cases, soft biometrics information are discriminating enough. Analysing the works presented in Sect. 3 it is possible to see some pieces of evidence that point towards the possibility of using this type of feature on its own. Advances in Machine Learning, more specifically in deep learning, have had a positive impact in several areas. Surely, these advances will also help the biometric systems to learn the soft biometrics features that are most discriminatory for people identification.

3 Related Work

In this Section, some works related to ours are briefly presented. They were divided into two groups: gait-based and soft biometrics-based.

3.1 Gait-Based Methods

In Chao *et al.* [7] the proposed method is based on a network called GaitSet. This network utilizes sets of independent frames of gait to learn identity information. On the paper, it is noted that the method is robust to the permutation of frames and can integrate frames from different videos recorded in different scenarios. It achieved rank-1 accuracy of 95% on the CASIA Gait Dataset A on its best case and 62.5% on the worst scenario.

Ben *et al.* [5] had built a framework that utilizes tensor representation applied to cross-view gait recognition. There were three criteria utilized: Coupled Multi-Linear Locality-Preserved (CMLP) with the responsibility to preserve the tensorial manifold structure, Coupled Multi-Linear Marginal Fisher (CMMF) with the responsibility to encode intra-class compactness and inter-class separability, and Coupled Multi-Linear Discriminant Analysis (CMDA) with the responsibility to minimize the intra-class scatter and maximizes inter-class scatter. This work is also validated with the CASIA Gait Dataset A and has a rank-1 accuracy of 99% on the best case and 62% on the worst.

The work presented in [4] utilizes a couple patch alignment (CPA) to deal with changes in view, this is made for matching with different pairs of gaits. Each patch is made with a sample from the gait, its closest intra-class and inter-class neighbor. This is following by an objective function that balances the cross-view, intra-class, and inter-class variations. The results on the CASIA Gait Dataset A ranges from 48% to 100% of rank-1 recognition rate, depending on the utilized protocol.

3.2 Soft Biometrics-Based Methods

In [8] the authors propose a new fusion technique, a joint density distribution-based rank-score fusion, in order to combine rank and score information. Another interesting aspect of this work is that it evaluates the influence of distance depending on the soft biometric characteristic. Face, body, and clothing traits are utilized to build a representation for different subjects, several soft features are extracted from those different parts. More specifically, gender, age, height, weight, shoulder shape, hair color, hair length, neck length, humpback, and arm length are extracted from the body. The authors propose and utilize a new soft biometric database, in which the accuracy varies from 98.5% in close scenarios and 82.6% from a far scenario.

Ran, Rosenbush, and Zengh [12] utilize gender, body size, height, cadence, and stride as a set of characteristics to do the identification. They utilize two datasets for evaluating their method, the USF outdoor dataset, and the SET HD indoor dataset. The method achieved 80% of Genuine Accept Rate (GAR) at a False Accept Rate (FAR) of 0.05%.

In [3], the authors utilize a Microsoft Kinect sensor to extract the skeleton from the subjects and calculate several anthropometric measurements. The utilized features are left and right arms, left and right forearms, left and right legs, left and right thighs, thoracic spine, cervical spine, and height. For classification, the authors created a dataset with 8 subjects and trained several models to do classification: a multi-layer perceptron, a decision tree, a random forest, and a K-NN with $K = 1$. The work reported accuracy of, on average, 99%.

4   Pose Estimation

Human pose estimation has gained importance in recent years due to its great potential for use in many computer vision applications (e.g. human action recognition, human-computer interface). Its main objective is to localize joints in the body (e.g. elbow, knee) or parts (e.g. arm, legs) [14]. In this work, our interest is mainly focused on finding the main joints and parts of the human body and utilize joints heat-maps and parts lengths as biometric features for people identification. For this task, two state-of-the-art methods were assessed in our work, OpenPose [6] and PifPaf [11].

4.1   OpenPose

OpenPose was proposed in [6]. It mainly relies on a pipeline with multiple stages. It starts with a set of 2D confidence maps of the body part locations, then, in the next step, it generates a set of 2D vector fields of part affinities, which describes the affinities between two body parts. Figure 3 shows a block diagram of the OpenPose method. As one can see, the predictions and feature images in each stage are concatenated for the next stage. The four initial stages generates a set of part affinity fields and the last two stages are utilized to predict the confidence maps.

Finding body parts in an image utilizes the idea that, knowing that the image has a body, a part of it can be found in any pixel. The confidence maps are 2D representations that capture this heuristic. In a scenario where there is only one

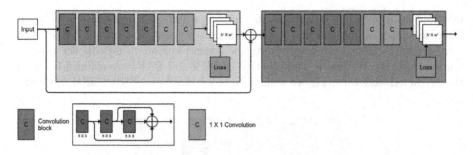

Fig. 3. A block diagram that illustrates the network architecture of the OpenPose multiple-stage CNN (block diagram inspired in [6]).

person in an image it is expected a single peak in the map, in other scenarios, there should be a peak per person. To associate different detected parts, a 2D vector with the encoded direction of one limb to another is utilized, the integral of a line that joins the current vector and other parts vector is calculated and utilized for joining parts.

4.2 TF-Pose Estimation

The TF-Pose Estimation[1] is an implementation of the OpenPose method using the Tensorflow framework [1]. It has some variation with changes in the network structure, which allows utilizing it in real-time on CPU or embedded devices.

We also utilized this implementation to verify if there is a difference in the observable behavior and if this impacts in any way the obtained result. In the Sect. 6 the results with both implementations of OpenPose method, original Caffe framework and TensorFlow framework, are listed.

4.3 PifPaf

In [11], Kreiss, Bertoni, and Alahi propose a method called PifPaf that focuses on estimating human poses in crowded images. PifPaf utilizes a shared ResNet with two head networks, one of the networks precisely estimates the location and size of a joint in the body, this is named Part Intensity Field (PIF). In order to estimate the association between the parts, other head networks are utilized, this is called Part Association Field (PAF).

The PIF can be defined as a structure that holds a scalar for the confidence in which that is a body part, a vector that points towards the closest body part, and one more scalar that has the estimated size of the joint. Since the confidence map of a PIF is coarse, it is necessary to join its vectorial part with a high-resolution confidence map.

For finishing the constructed predicted skeleton, it is necessary to build the PAFs. A PAF consists of two vectors specifying the parts that are associated with

---

[1] https://github.com/ildoonet/tf-pose-estimation.

two widths. In order to determine the vectors, a two-step algorithm is utilized. First, for determining the first vector, the closest joint of a part is found. The second step is needed for determining the second vector, to do this the ground truth is utilized. With this, the junction of PIFs and PAFs are utilized to build the predicted skeletons. Figure 4 shows a block diagram of the PifPaf method.

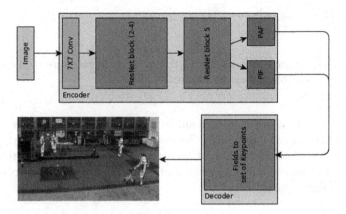

Fig. 4. A block diagram that illustrates the PifPaf network architecture. The encoder is responsible for extracting the fields from the original image and the decoder is responsible for transforming those fields into keypoints.

5   Proposed Methods

This work proposes a novel ensemble of soft-biometric features coupled with a pre-processing stage. Figure 5 presents a block diagram of our proposed method for people identification based on soft biometrics features obtained from 2D poses. As one can observe, our method has five main stages: background subtraction, pre-processing, pose estimation, feature extraction and, finally, person identification.

In the first stage, the input image is subtracted from the background image, thus removing the background and maintaining the foreground. In the second stage, we perform a morphological filtering on the foreground image resulting from the subtraction stage. More precisely, we apply an opening morphological operator in order to eliminate high frequency noise in the foreground areas of the image. In the third stage, we apply on the filtered images a 2D pose estimation method, such as OpenPose [6] or PifPaf [11], in order to obtain the skeleton of the person in the scene. In the fourth stage, the ensemble of soft biometrics features is extracted and a feature vector is generated. Our feature vector has dimension 8. The first feature is the length of right Humerus of the individual being identified, which is calculated by measuring the distance between joints 0 (right shoulder) and 2 (right elbow). The second feature is the length of left

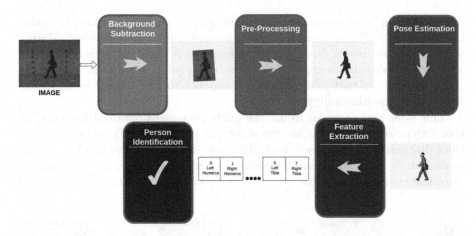

Fig. 5. Block diagram of the proposed method for people identification based on an ensemble of soft-biometric features obtained from 2D poses.

Humerus, which is calculated by measuring the distance between joints 1 (left shoulder) and 3 (left elbow). The third feature is the length of right Ulna, which is calculated by measuring the distance between joints 2 (right elbow) and 4 (right wrist), The fourth feature is the length of left Ulna, which is calculated by measuring the distance between joints 3 (left elbow) and 5 (left wrist), and so on. Finally, in the fifth stage, the feature vector obtained from the 2D skeleton is used to fed a classifier in order to identify the subject. In our work, we used a 1-NN classifier, and assessed two distance functions: Euclidean and City-Block.

In the target application for our method, the goal is to identify a person walking in a area under surveillance. So, the input information to our biometric system will be a sequence of video-frames captured by a surveillance camera (CCTV camera). So, we hypothesized that the biometric system will have in its database (gallery) at least one video per person which will be compared with the query video (probe). Therefore, in order to authenticate a subject's identity, the distance of the probe video to the gallery video associated to the claimed identity must be lower than a given threshold (in the authentication mode we have 1:1 comparison). For finding the identity of an unknown person, the probe video will have to be compared to all gallery videos (in the identification mode we have 1:n comparisons). In this case, the identity associated to the gallery video with the lowest distance to the probe video will be taken as the identity of the subject.

In order to compute the distance D between two videos, probe and gallery, that have m and n frames, respectively, the m feature vectors of the probe video are compared with the n feature vectors of the gallery video. Then, the distance D is taken as the mean of the k lowest frame distances (in our experiments the best results was obtained with $k = 11$).

Another important aspect of our method regards the occlusions of parts of the body that can occurs when capturing the video images. When occlusions occur, the methods of pose estimation will not be able to localize all joints. Actually, OpenPose and PifPaf returns a quality measure of each joint detected. Within our method, we used a function that determines a threshold of acceptance of a particular extracted feature, that is, we use a joint information only if the quality measure of that joint detection is superior to 70% (this threshold value was defined experimentally).

This work also proposes to use the joints heat-maps obtained from the 2D pose estimation methods as descriptor for gait recognition. Figure 6 presents a block diagram of the proposed method for gait recognition based on joints heat-maps features obtained from 2D poses. As one can observe, this method has three main stages: pose estimation, feature extraction and person identification.

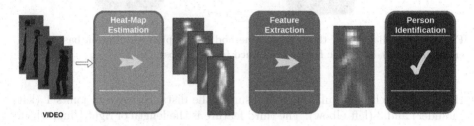

Fig. 6. Block diagram of the proposed method for gait recognition based on joint heat-map features obtained from 2D poses.

In the first stage, we apply on the input video a 2D pose estimation method, such as OpenPose [6] or PifPaf [11], in order to obtain the joints heat-maps from each frame from the video. In the second stage, a mean joints heat-map is calculated. Finally, in the third stage, the feature vector (mean joints heat-map) is used to fed a classifier in order to identify the subject. In this case, a ANN (Artificial Neural Network) classifier is used.

6   Experimental Results

In this Section, we present the experimental results obtained with the novel methods proposed for people identification based on the ensemble of anthropometric measurements and joints heat-maps obtained from 2D poses.

In our experiments we assessed the OpenPose (with its two implementations) and the PifPaf methods. These methods are described in Sect. 4. Besides, we assessed two distance functions (Euclidean and City-Block) when using the 1-NN classifier. For the ANN, we used a feedforward ANN (Artificial Neural Network) with an input layer, three hidden layers and an output layer, created by using Tensorflow [1] and Keras.

The proposed methods were assessed using the subset of videos captured at 90 degrees, available on CASIA Gait Dataset-B, which is described in Sect. 6.1.

6.1 Dataset

The dataset used in our experiments is a subset of CASIA Gait Dataset-B [13]. This dataset has several frames extracted from videos of 124 subjects from various angles and perspectives, backgrounds and previously classified silhouettes. Figure 7 shows two samples of frames of a video from this dataset.

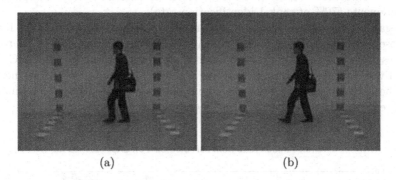

(a)   (b)

Fig. 7. Examples of images of a subject from two distinct videos from CASIA Gait Dataset-B, captured by two distinct cameras.

CASIA Gait Dataset-B is a very complete dataset. For a greater challenge in extracting characteristics, it has intra-class variations, that is, the same person was recorded with and without carrying a bag and with or without wearing a coat.

In our experiments we used 620 videos of the lateral view (view angle equal to 90 degrees) from the CASIA Gait Dataset-B, with the following information:

- The videos have 66 frames on average;
- From each frame it was detected 14 points of interest (joints of the estimated skeletons);
- For each person we used the videos with the three variations (person walking with no objects and no jacket, person carrying a bag and person wearing a jacket);
- Each variation has at least 2 shots;
- Right legs and arms were only possible to be captured in 36% of the frames due to the camera's angle.

6.2 Results

Figure 8 presents the Cumulative Match Curve (CMC) obtained with our proposed methods using the three pose estimation algorithms with the ANN and the 1-NN classifiers (with the Euclidean and City-Block distances). CMC is a measure commonly used to assess the performance of biometric systems that operate in identification mode (1:n comparisons).

Regarding the rank-1 identification rate, as one can see in Fig. 8, the best result was obtained with the PifPaf 2D pose estimation method with the ANN classifier and the joints heat-map features (89% of rank-1 identification rate). The second best result was also obtained by PifPaf, but with the 1-NN classifier, with Euclidean distance, and the anthropometric measurements (85% of rank-1 identification rate). The worst result was obtained with the TF-Pose-Estimation method, with 1-NN classifier, with City Block distance, and the anthropometric measurements (70% of rank-1 identification rate).

Regarding the overall performance, the best result was obtained by the PifPaf 2D pose estimation method with the ANN classifier and the joints heat-map features since this combination presented the highest area under the CMC. The overall worst result was obtained by the OpenPose 2D pose estimation method with the ANN classifier and the joints heat-map features, since this combination presented the lowest area under the CMC.

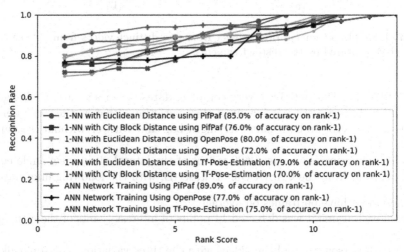

Fig. 8. Cumulative match curves obtained using three pose estimation methods (Pif-Paf, OpenPose and TF-Pose-Estimation), two biometric features (soft and gait), and three classifiers (ANN, 1-NN with Euclidean distance and 1-NN with City-Block distance).

In order to assess the performance of the proposed methods in a biometric systems that would operate in authentication mode, we also calculated the F1-Score. Table 1 shows the results obtained by using PifPaf as the 2D pose estimation method. The F1-Score was 89%, which can be considered a good result when taking into account that we are using gait biometric features in this experiment. OpenPose and Tf-Pose-Estimation methods obtained 12% and 14%, respectively, in loss of accuracy in relation to the PifPaf method.

Table 1. Precision, Recall and F1-Score values obtained by using PifPaf and ANN on the complete dataset.

	Precison	Recall	F1-Score
Accuracy			0.89
Macro avg	0.89	0.89	0.89
Weighted avg	0.89	0.89	0.89

Our last experiment was designed to simulate a scenario where it is necessary to identify a small group of known people (that is, a close set identification problem), but at a distance and in unconstrained conditions. This scenario could regard, for instance, a small set of employees being monitored while working on the factory floor of an industry. So, for that, we randomly chose ten individuals from the CASIA Gait Dataset B. For this experiment we used only the ANN classifier. We used 75% of the data available for the class for training the network and, consequently, 25% of the rest of the data for the evaluation of results. Figure 9 shows the CMC obtained in this experiment. As one can see, the best result jumped from 89% to 97% of rank-1 identification rate.

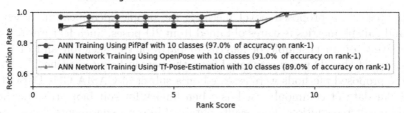

Fig. 9. Cumulative match curve obtained using the three pose estimation algorithms (PifPaf, OpenPose and TF-Pose-Estimation), with the ANN classifier, for 10 individuals.

The precision, recall and F1-Score measures for each individual obtained in this experiment by using the PifPaf 2D pose estimation and the ANN classifier are shown in Table 2. As one can see, the accuracy measure jumped from 89% to 96%. OpenPose and Tf-Pose-Estimation methods obtained 6% and 8%, respectively, in loss of accuracy in relation to the PifPaf method.

Table 2. Precision, Recall and F1-Score values obtained by using PifPaf and ANN on a subset of 10 individuals.

Individual	Precision	Recall	F1-Score
0	0.97	0.98	0.98
1	0.98	0.98	0.98
2	0.97	0.96	0.96
3	0.95	0.95	0.95
4	0.94	0.97	0.95
5	0.96	0.93	0.95
6	0.97	0.97	0.97
7	0.99	0.92	0.95
8	0.95	0.95	0.95
9	0.90	0.96	0.93
Accuracy			0.96
Macro avg	0.96	0.96	0.96
Weighted avg	0.96	0.96	0.96

7 Conclusions

In this paper we proposed novel methods for people identification based on soft and gait biometrics features obtained from 2D poses. Such features consist, respectively, of an ensemble of lengths of parts of the skeleton and of joints heat-maps of the 2D poses estimated from the person being identified.

The proposed methods were assessed on a subset of CASIA Gait Dataset B, a popular dataset commonly used as a benchmark for soft biometrics methods based on gait recognition. The results obtained in our experiments showed that the proposed methods were successful, mainly if is taken into account that we are using soft and gait biometrics in scenarios in which most traditional and hard biometric features will fail, or will not be possible to use. Our best results, 89% and 85% of rank-1 identification rates for gait and soft features, respectively, were reached by using the PifPaf method for 2D pose estimation. In a smaller subset randomly selected from the CASIA Gait Dataset B, the gait result jumped to 97%. In all experiments, PifPaf overcame both implementations of OpenPose method, when utilizing the same features, and the same classifier.

We did not compare our results with other works since we could not find other works that utilize the same dataset, the same experimental protocol, and just the same, or similar, anthropometric measurements (length of body's parts).

As our method is intended for scenarios where it is possible to record videos at a distance of people walking, a natural extension of our work is to fuse both anthopometric and gait features. This is one of our future work. Another extension of our work is to use a tracking method in order to be able to address the problem of identifying multiple persons present in the same scene.

Acknowledgments. This paper is one of the results of the ongoing Master's research being conducted by the first author on soft biometrics. This work has financial support of PETROBRÁS and is being developed at Recogna Laboratory - UNESP, campus of Bauru.

References

1. Abadi, M., et al.: TensorFlow: large-scale machine learning on heterogeneous systems, software available from tensorflow.org (2015). http://tensorflow.org/
2. Jain, A.K., Dass, S.C., Nandakumar, K.: Can soft biometric traits assist user recognition? (2004). https://doi.org/10.1117/12.542890
3. Araujo, R.M., Graña, G., Andersson, V.: Towards skeleton biometric identification using the microsoft kinect sensor. In: Proceedings of the 28th Annual ACM Symposium on Applied Computing, pp. 21–26 (2013)
4. Ben, X., Gong, C., Zhang, P., Jia, X., Wu, Q., Meng, W.: Coupled patch alignment for matching cross-view gaits. IEEE Trans. Image Process. **28**(6), 3142–3157 (2019). https://doi.org/10.1109/TIP.2019.2894362
5. Ben, X., Zhang, P., Lai, Z., Yan, R., Zhai, X., Meng, W.: A general tensor representation framework for cross-view gait recognition. Pattern Recogn. **90**, 87–98 (2019). https://doi.org/10.1016/j.patcog.2019.01.017. http://www.sciencedirect.com/science/article/pii/S0031320319300251
6. Cao, Z., Hidalgo, G., Simon, T., Wei, S., Sheikh, Y.: OpenPose: realtime multiperson 2D pose estimation using part affinity fields. CoRR abs/1812.08008 (2018). http://arxiv.org/abs/1812.08008
7. Chao, H., He, Y., Zhang, J., Feng, J.: GaitSet: regarding gait as a set for cross-view gait recognition. CoRR abs/1811.06186 (2018). http://arxiv.org/abs/1811.06186
8. Guo, B.H., Nixon, M.S., Carter, J.N.: Soft biometric fusion for subject recognition at a distance. IEEE Trans. Biometrics Behav. Identity Sci. **1**(4), 292–301 (2019)
9. Jain, A.K., Ross, A.A., Nandakumar, K.: Introduction to Biometrics. Springer, New Delhi (2011). https://doi.org/10.1007/978-0-387-77326-1
10. Jia, Y., et al.: Caffe: convolutional architecture for fast feature embedding. In: Proceedings of the 22nd ACM International Conference on Multimedia, pp. 675–678 (2014)
11. Kreiss, S., Bertoni, L., Alahi, A.: PifPaf: composite fields for human pose estimation. CoRR abs/1903.06593 (2019). http://arxiv.org/abs/1903.06593
12. Ran, Y., Rosenbush, G., Zheng, Q.: Computational approaches for real-time extraction of soft biometrics. In: 2008 19th International Conference on Pattern Recognition, pp. 1–4. IEEE (2008)
13. Yu, S., Tan, D., Tan, T.: A framework for evaluating the effect of view angle, clothing and carrying condition on gait recognition. In: 18th International Conference on Pattern Recognition (ICPR 2006), vol. 4, pp. 441–444 (2006). https://doi.org/10.1109/ICPR.2006.67
14. Sun, K., Xiao, B., Liu, D., Wang, J.: Deep high-resolution representation learning for human pose estimation. In: The IEEE Conference on Computer Vision and Pattern Recognition (CVPR), June 2019
15. Tian, Y., Wei, L., Lu, S., Huang, T.: Free-view gait recognition. PLOS ONE **14**(4), 1–24 (2019). https://doi.org/10.1371/journal.pone.0214389
16. Tome, P., Fierrez, J., Vera-Rodriguez, R., Nixon, M.S.: Soft biometrics and their application in person recognition at a distance. IEEE Trans. Inf. Forensics Secur. **9**(3), 464–475 (2014)

17. Bing Tong, S., zhuo Fu, Y., Fei Ling, H.: Cross-view gait recognition based on a restrictive triplet network. Pattern Recogn. Lett. **125**, 212–219 (2019). https://doi.org/10.1016/j.patrec.2019.04.010. http://www.sciencedirect.com/science/article/pii/S0167865518307475

18. Wang, Y., et al.: EV-GAIT: event-based robust gait recognition using dynamic vision sensors. In: The IEEE Conference on Computer Vision and Pattern Recognition (CVPR), June 2019

19. Zulcaffle, T.M.A., Kurugollu, F., Crookes, D., Bouridane, A., Farid, M.: Frontal view gait recognition with fusion of depth features from a time of flight camera. IEEE Trans. Inf. Forensics Secur. **14**(4), 1067–1082 (2019). https://doi.org/10.1109/TIFS.2018.2870594

Texture Analysis Based on Structural Co-occurrence Matrix Improves the Colorectal Tissue Characterization

Elias P. Medeiros[1]([⊠]), Daniel S. Ferreira[1,2]([⊠]), and Geraldo L. B. Ramalho[1]([⊠])

[1] Instituto Federal de Educação, Ciência e Tecnologia do Ceará, Fortaleza, CE, Brazil
{elias.paulino,daniels,gramalho}@ifce.edu.br
[2] Departamento de Engenharia de Teleinformática, Universidade Federal do Ceará, Fortaleza, CE, Brazil

Abstract. Colorectal cancer causes the deaths of thousands of people worldwide according to the World Health Organization. Automatic tissue recognition of histopathological images is essential for early disease diagnosis. Most research consists of employing texture descriptors to capture features that identify tumor samples. However, accurate multi-class classification is a challenge due to the complexity of colorectal tissue images. Recently, researchers have shown that the analysis of texture structural patterns degraded by image filtering provides valuable features for pre-diagnosis in several medical applications. Here we propose an approach to automatically classify eight types of colorectal tissues using Structural Co-occurrence Matrix. We carried on experiments on 5000 tissue patches from a public dataset to evaluate our algorithm, considering two scenarios: structural differences as a single descriptor, and combined with other characteristics. We found that our strategy improves the state-of-the-art, achieving, accuracy: 91.30%, precision: 91.41%, sensitivity: 91.31%, specificity: 98.76% e F1-score: 91.31%.

Keywords: Colorectal cancer · Structural co-occurrence matrix · Image classification

1 Introduction

Colorectal cancer is the term attributed to a pathology when a tumor is found in the large intestine or rectum. It can also be called colon or rectal cancer, depending on where the symptoms started [26]. Early detection of the disease is a vital strategy to find the tumor in the initial phase, thus increasing the chances of successful treatment. The presence of abnormalities caused by this cancer can be detected mainly by sigmoidoscopy and colonoscopy exams. Upon finding abnormalities in the examination, confirmation occurs through tissue biopsy of the suspected region, using the Hematoxylin-Eosin (H&E) staining technique [25]. Pathologists are responsible for diagnosing and studying tissue properties, such as structure, quantity, and shape of cells [9]. Figure 1 shows different examples

© Springer Nature Switzerland AG 2020
R. Cerri and R. C. Prati (Eds.): BRACIS 2020, LNAI 12319, pp. 333–347, 2020.
https://doi.org/10.1007/978-3-030-61377-8_23

of tissues that make up the colorectal structure. They have a wide range of lighting, stain intensity, and fabric textures. The diversity of characteristics and classes, makes the analysis process present some problems: (i) time spent; and (ii) biased nature. The pathologist's pre-existing knowledge guides manual analysis [14]. Therefore, to reduce the time of analysis and subjective interpretations, the use of CAD (Computer-Aided Diagnosis) systems can improve pathologists' productivity and increase confidence results.

Most CAD systems have the architecture divided into three parts: preprocessing, feature extraction, and classification [15,16,23]. In the context of feature extraction, several works aim to improve the classification performance by analyzing methods that represent, in the best way, important structures of the images of the problem addressed. Several feature extractors are explored in the literature regarding the construction of CADs for diagnostics in medical images. Narvàez F. et al. [15] used the curvelet multiresolution technique to extract texture resources, and Zerkine moments to extract the shape characteristics in breast images. PS e Dharun [19] explored Gray Level Co-occurrence Matrix (GLCM) texture information and shape (area, perimeter, and circularity) in the brain abnormality classification process by analyzing Magnetic Resonance Images (MRI) images. Filho et al. [22] compared the classification performance and extraction time of the methods GLCM, Hu moments, Statistical moments, Zernike's moments, Elliptic Fourier features, Tamura's features, Structural Cooccurrence Matrix (SCM), and Analysis of Human Tissue Densities (AHTD). They observed promising results of classification and extraction time of AHTD and SCM in analysis of lung and brain images.

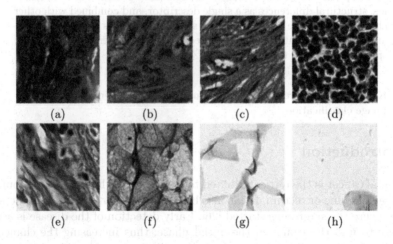

Fig. 1. H&E patches of colorectal tissue. (a) tumour epithelium, (b) simple stroma, (c) complex stroma, (d) immune cell conglomerates, (e) debris and mucus, (f) mucosal glands, (g) adipose tissue, and (h) background.

Most approaches for analyzing H&E histological images have two limitations: (i) they consider only two categories of tissue (tumour and stroma), which makes these approaches inappropriate for more heterogeneous parts of the tumor; or (ii) studies use private datasets, which impacts their reproducibility [12]. For example, Altunbay *et al.* [2] evaluated a private dataset during the application of graphs to mathematically represent normal, low-grade, and high-grade cancerous tissue structures.

Bianconi, Álvarez-Larrán e Fernández [3] used image characteristics based on visual perception (coarseness, contrast, directionality, line probability, and roughness) for discrimination of epithelial tissue and stroma on 1376 images. However, their dataset consists of samples with different resolution ranges (in pixels) per each class, which can lead to a bias in the classifier. On the other hand, Kalkan *et al.* [11] used information from GLCM and Gabor to separate the same tissue classes by analyzing 2000 images.

Kather *et al.* [12] introduced a new dataset with 5000 H&E images of colorectal cancer, including eight types of tissues. The authors used these images to compare the performance of a varied set of texture extractors (Gabor; GLCM; visual perception characteristics; Local binary patterns - LBP; and high and low order histogram statistics), in addition to a set of classifiers (k-Nearest Neighbor - kNN, Decision Tree, and SVM). The results were evaluated for characterization in two and eight types of tissue. Kather *et al.* [12] managed to improve state-of-the-art in the classification in two classes (epithelial-stroma), reaching 98.6% accuracy, associating characteristics of six extractors. As for eight types of tissues, the accuracy obtained was 87.4% with the union of information from five extractors. However, the approaches that achieved the best accuracy were those that demanded the longest extraction times. Before, no work has addressed the problem of separating H&E tissues from colorectal cancer into eight classes using only texture information. Some challenges remained open, such as: (i) improve accuracy result; (ii) decrease the feature extraction runtime of the approach and; (iii) identify which classes cause the most confusion when classifying the problem.

Wang *et al.* [28] proposed an algorithm called bilinear CNN (BCNN) for classification of histopathology images. The algorithm first decomposes the input into the H (Hematoxylin) and E (Eosin) components of the H&E images. BCNN is composed of two Convolutional Neural Networks (CNN) that are responsible for extracting characteristics from the decompositions. The outputs are combined by a bilinear pooling and classified with the Support Vector Machine (SVM). The results showed high efficiency of the method, but the authors selected a set of 1000 images from a universe of 5000 for evaluation. Rachapudi e Devi [20] also proposed a new CNN architecture. The complete model has sixteen convolutional layers, five layers of max-pooling, and a fully-connected layer. The network output is classified into eight classes.

The SCM is a generic structural analysis method proposed by Ramalho *et al.* [21] in a study of disease detection on medical images using rotation-invariant feature extractors. SCM was shown to be faster than most classic rotation-invariant feature extractors. Having the ability to capture structural degradation in images, SCM is a condensed global descriptor method useful to describe textures or region-guided structures that characterize marks of diseases in medical images. Previous works [17, 22, 23] have used a vector descriptor containing up to eight originally defined attributes computed from SCM giving it likeness power within classes.

Different from previous approaches, which explore classifying tissues using single descriptors, we propose to analyze the combination of them. Particularly, we are interested in investigating whether the structural analysis of the texture patterns provides valuable features to categorize colorectal tissue images, according to two scenarios: 1) SCM as a single descriptor, and 2) SCM combined with other state-of-the-art descriptors. Thus, we introduce SCM as a feature extractor on H&E images, addressing this method to classification into eight classes. This paper describes three contributions:

- **C#1**: Proposal of a SCM setup for extracting H&E image characteristics from colorectal tissue.
- **C#2**: Quantitative evaluation of SCM in comparison to state-of-the-art descriptors on H&E images.
- **C#3**: Ensemble of features that enhance the discrimination of colorectal tissues into different classes.

The remainder of this paper is organized as follows. Section 2 presents the SCM algorithm and discusses its application as a feature extractor for digital images. Section 3 introduces our methodology for colorectal tissue characterization, addressing the problem in two scenarios: single and combined descriptors. Section 4 presents the datasets and evaluation metrics. The experiment results are discussed in Sect. 5, and Sect. 6 draws our conclusions and future directions.

2   Structural Co-occurrence Matrix - SCM

SCM is a rotation-invariant features extraction technique for signals proposed by Ramalho *et al.* [21]. This method consists of a descriptor that enables the analysis of the relationship between low-level signal structures in 1D or 2D space. The information resulting is compacted in a 2D matrix. Figure 2 illustrates the computation of SCM.

Consider two bi-dimensional images f and g on the same domain and same spatial dimension. The SCM values are the frequency of the pair values satisfying the Eq. (1). For feature extraction applications, we make $g = f$. The k function provides an image transformation, usually either a high-pass or low-pass filter for the input signal. The goal of k is to evidence the structural difference between f and $k(g)$. The pixel intensity at position p is represented by f_p, and $k(g)_{p+d}$ given a d offset. For most feature extraction applications d is set to zero. Q is a quantization function, meaning it maps the pixel values of the image to the discrete domain where $Q(f)_p \in \{0, ..., N-1\}$ values. The 2D matrix of structural

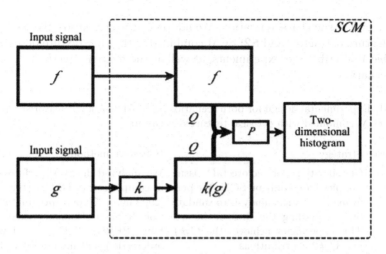

Fig. 2. Generic SCM model for two input signals. From Ramalho *et al.* [21]

co-occurrence of the signs $Q(f)$ and $Q(k(g))$, represented by $M = m_{ij}$ is obtained using the following expression:

$$m_{ij} = \#\{(i, j)|P(i, j), i = Q(f)_p, j = Q(k(g))_{p+d}\}, \tag{1}$$

where $\#\{\cdot\}$ represents the cardinality of the subset of pairs (i, j) that satisfies the property P. This property depicts the relationship between the discrete values $Q(f)_p$ and $Q(k(g))_{p+d}$ at the positions p e $p + d$, respectively. A set of attributes, derived from SCM, represent the structural information stored in M. These attributes are divided into three groups: statistical group, information group, and divergent group. In the statistical group are included the correlation (COR), inverse difference moment (IDM) and entropy (ENT). The divergent group presents the divergence of Kullback & Leibler (DKL) and complementary absolute difference (CAD). The information group presents Chi-square distance (CSD), Chi-square ratio (CSR) and mean absolute difference ratio (MDR) [21]. The attributes are often used as a feature vector to describe the input image f in most applications. SCM provides two marginal (marginal Y and marginal X) distributions given by the sum of rows and the sum of columns of M. The marginal histograms are equivalent to the histograms of the pixel values of the images f and g, respectively.

3   Methodology

We conducted experiments based on the approach described in Fig. 3. First, we converted the input RGB images to grayscale using the equation $GS = 0.298 * R + 0.5870 * G + 0.1140 * B$ [18], where GS is the grayscale image and R (Red), G (Green), B (Blue) are components of input image. Later, we extracted texture attributes using different techniques for investigating the impact of using SCM on colorectal tissue characterization. We addressed two scenarios: **S1:** SCM as a single feature extractor, and **S2:** SCM combined with other texture descriptors.

Table 1 describes our experiments, as well as the reference methods, to cover both scenarios.

Table 1. Experimental design for both scenarios: 1) SCM as a single feature extractor, and 2) SCM combined with classical texture descriptors.

Scenario	Strategy	Reference methods
1	Classification of colorectal tissue samples based only on SCM. We performed statistical analysis to validate SCM regarding the state-of-the-art. This experiment achieves the **C#1** and **C#2** contributions	We analyzed the SCM performance in comparison to GLCM [10]; LBP [29]; Gabor [7]; perceptual information from the human visual system (PERC) [3,27]; and Lower-order (HL) and higher-order (HH) histogram features [13]
2	Classification of colorectal tissue samples based on the combination of SCM and classic descriptors. We performed statistical tests to identify the potential descriptors for an ensemble solution based on SCM. This experiment leads to the **C#3** contribution	We referred the combination of five techniques called *best5* (PERC + HH + HL + LBP + GLCM). The *best5* was proposed by Kather *et al.* [12] for multi-class classification in colorectal tissue and corresponds to the state-of-the-art

We applied state-of-the-art machine learning algorithms to classify the colorectal image samples in our pipeline. We adopted the following algorithms: Decision Tree [24], KNN [8], and SVM with Linear kernel, Polynomial kernel (Poly) and Radial Basis Function kernel (RBF) [4,6]. For classification, the feature sets were divided into training (75%) and testing (25%). The process was repeated 1000x, always with random sampling for the training and test sets. Therefore, our findings for each descriptor are reasoned on the performance observation in 1000 iterations classification process.

4 Performance Assessment

4.1 Dataset

We carried out tests using a database containing 5000 image patches from different parts of colorectal cancer H&E exams. This dataset was introduced by Kather *et al.* [12] from the samples of the pathological archive of the University Medical Center Mannheim. The patches are labeled in eight types of tissue: epithelial tumour; simple stroma; complex stroma; immune cells; debris (including necrosis, bleeding, and mucus); normal mucous glands; adipose tissue and background. For each class, 625 non-overlapping patches were obtained, with dimensions 150px × 150px (74μm × 74μm), totaling 8 × 625 = 5000 images. Figure 1 shows some examples of the adopted database.

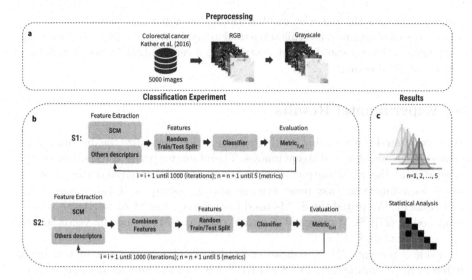

Fig. 3. A proposed approach to the colorectal cancer classification process. n stands for the different evaluation metrics employed. S1 and S2 correspond to our studied scenarios. (Color figure online)

4.2 Evaluation Metrics

We considered metrics based on True Positive (TP), True Negative (TN), False Positive (FP) and False Negative (FN) values to quantitatively assess our results. Specifically, we computed the accuracy (ACC), precision (PREC), sensitivity (SENS), specificity (SPEC), and F1-score as follows:

$$ACC = \frac{TP + TN}{TP + TN + FP + FN}, \tag{2}$$

$$PREC = \frac{TP}{TP + FP}, \tag{3}$$

$$SENS = \frac{TP}{TP + FN}, \tag{4}$$

$$SPEC = \frac{TN}{TN + FP}. \tag{5}$$

$F1\_score$ combines precision and recall (REC) to bring a unique number that indicates the general quality of the model. Recall has the same meaning as sensitivity (SENS).

$$F1\_score = 2 * \frac{PREC * REC}{PREC + REC}. \tag{6}$$

According to the proposed methodology (Fig. 3), we achieved a population of size N for each metric. Based on these values, we employed the one-way ANOVA statistical test [5] with the post-hoc Turkey test ($\alpha = 0.05$) [1] for each metric, reporting the pairwise combinations that differ significantly from each other. Our motivation to use this statistical analysis was to produce a systematic comparison between the different descriptors, revealing those whose results are potentially interesting.

5   Experimental Results

We investigated the following SCM parameters that maximize the discrimination performance on colorectal tissue images. Therefore, we perform tests varying the k filter; and the SCM output as a feature vector. We analyzed the following filters as k function: {low-pass: average and gaussian}, and {high-pass: sobel and laplace}. We tested the SCM-based feature vector as SCM attributes (ATT) proposed in [21]; SCM marginal X and Y (MX and MY); and the vectorization of the SCM matrix (LIST).

Table 2 presents the classification results of colorectal cancer images using different SCM parameters. The SVM RBF achieved the best results. Our experiments revealed that the low-pass filters in conjunction with the LIST lead to better characterization. Particularly, we show that the degradation of high frequencies of colorectal tissue images provides a valuable descriptor for pre-diagnosis in eight different classes. Therefore, we reported the average filter and LIST as our best SCM setup. For these parameters, we found ACC: 90.84%, PREC: 91.10%, SENS: 90.70%, SPEC: 98.69%, and F1-score: 90.79%. We adopted our best SCM parameters to analyze the scenarios S1 and S2.

Table 2. Classification results of H&E images of colorectal into eight tissue classes using different SCM parameters and SVM RBF classifier.

Filter	Output	ACC (%)	PREC (%)	SENS (%)	SPEC(%)	F1-score (%)
Average	LIST	**90.84 ± 1.47**	**91.10 ± 1.36**	**90.70 ± 1.44**	**98.69 ± 0.21**	**90.79 ± 1.39**
	MX	82.38 ± 1.49	82.48 ± 1.54	82.41 ± 1.48	97.48 ± 0.21	82.35 ± 1.51
	MY	80.18 ± 2.03	80.21 ± 2.16	80.14 ± 2.11	97.17 ± 0.29	80.09 ± 2.12
	ATT	77.92 ± 0.77	78.10 ± 1.06	77.92 ± 0.86	96.85 ± 0.11	77.87 ± 0.98
Gaussian	LIST	89.98 ± 0.69	90.08 ± 0.69	90.00 ± 0.70	98.57 ± 0.10	90.00 ± 0.69
	MX	82.84 ± 1.90	82.95 ± 2.07	82.80 ± 1.97	97.55 ± 0.27	82.82 ± 2.02
	MY	81.84 ± 1.43	81.91 ± 1.23	81.94 ± 1.24	97.41 ± 0.21	81.82 ± 1.30
	ATT	73.92 ± 2.67	73.99 ± 2.64	73.94 ± 2.62	96.28 ± 0.39	73.70 ± 2.64
Laplace	LIST	88.54 ± 1.45	88.75 ± 1.57	88.60 ± 1.49	98.36 ± 0.21	88.55 ± 1.46
	MX	80.96 ± 3.12	81.18 ± 2.92	80.98 ± 3.13	97.28 ± 0.44	80.89 ± 3.29
	MY	59.60 ± 1.90	61.80 ± 1.29	59.60 ± 1.39	94.23 ± 0.28	58.89 ± 1.56
	ATT	81.62 ± 1.64	81.98 ± 1.60	81.60 ± 1.64	97.37 ± 0.24	81.66 ± 1.62
Sobel	LIST	88.66 ± 1.08	88.79 ± 1.25	88.57 ± 1.11	98.38 ± 0.15	88.62 ± 1.17
	MX	82.04 ± 1.92	82.10 ± 1.93	82.04 ± 1.79	97.44 ± 0.28	81.99 ± 1.87
	MY	57.40 ± 3.24	59.39 ± 2.55	57.49 ± 3.10	93.92 ± 0.47	57.23 ± 3.23
	ATT	77.78 ± 2.28	77.73 ± 2.25	77.89 ± 2.13	96.83 ± 0.33	77.63 ± 2.22

5.1 S1 Analysis

S1 analyzes the performance of the SCM as a single descriptor. We found that
the SVM RBF classifier performs better with the surveyed descriptors, regardless of the evaluation metric. Figure 4 presents our quantitative evaluation. We
confirmed that the performance of *best5* is superior to that presented by each of
its single components, including Gabor. Our experiments revealed that the single
SCM performs better than *best5*, improving the state-of-the-art. For all extractors, SPEC was above 94%, showing the ability of the methods to identify true
negative cases. SCM contributed further, improving the performance on other
metrics (ACC, PREC, SENS, and F1-score). Thus, we argue that the modeling
of structural differences caused by a low-pass filter is a valuable descriptor for
the characterization of colorectal tissue images.

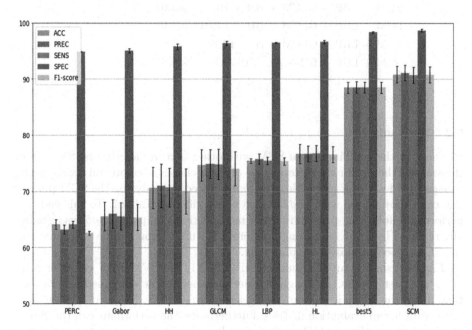

Fig. 4. Quantitative evaluation for the surveyed single descriptors. All values shown
are in (%).

Table 3. Average execution time for each feature extraction method.

Method	SCM	HL	HH	GLCM	PERC	LBP	Gabor
Time(s)	0.01	0.03	0.05	0.10	0.12	0.72	0.86

Table 3 describes the average runtime for all surveyed algorithms. We found
that SCM had the best performance, with a gain of 66.66% and 98.83% in feature

extraction runtime when compared to HL (second fastest algorithm) and Gabor (last fastest algorithm) methods, respectively. It is noteworthy that the SCM computational time is related to N value in the function Q of SCM (Eq. 1). In our experiments, we adopted N = 8. If N increases, consequently, the number of operations to compute the co-occurrence matrix and its attributes tends to increase considerably. We recommend to Ramalho et al. [21] for more details on the computational complexity of SCM.

Table 4. Combination of descriptors based on SCM.

Descriptors	Combined descriptor
SCM + GLCM + HH + HL + PERC	set_01
SCM + LBP + GLCM + HH + HL	set_02
SCM + LBP + GLCM + HH + PERC	set_03
SCM + LBP + GLCM + HL+ PERC	set_04
SCM + LBP + HH + HL + PERC	set_05

5.2 S2 Analysis

S2 analyzes the combination of SCM with other texture descriptors. Since that the state-of-the-art for characterization of eight classes of colorectal tissue corresponds to a descriptor ensemble, we are interested in analyzing the SCM performance in conjunction with *best5*, which is described in Table 1. To this end, we perform the following procedure: each method that composes *best5* was gradually replaced by SCM, proposing 5 different feature vectors. Table 4 summarizes the SCM-based proposed combinations.

Figure 5 shows our S2 results. Similar to S1, the SVM RBF classifier provides the best results. Although the performance of the combined descriptors that using the SCM is similar, we observed that all SCM-based sets outperform *best5* for all evaluation metrics. Furthermore, the performance of the single SCM and the combined SCM is close, signaling that the characteristics extracted by SCM are relevant in the combined feature vector. An interpretation here is that structural differences, encoded by SCM, are uncorrelated to the features extracted by other descriptors. Future works might address this finding towards the SCM interpretability.

Table 5 describes our quantitative results in terms of the metric averages. In this perspective, set_01 was our best combined SCM. However, the ANOVA and post-hoc Turkey statistical tests ($\alpha = 0.05$) did not identify statistical differences between set_01 and set_03 and set_04; set_03; set_03 and set_04; set_03 and set_04; and set_03 for the ACC, PREC, SPEC, SENS and F1-score metrics, respectively (Fig. 6). The proposed set_01 also had the advantage of being the fastest surveyed SCM-based combination. Since the SCM runtime significantly shorter than other surveyed techniques, we argue that SCM improved

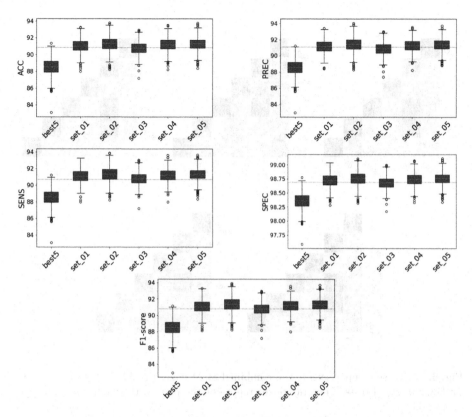

Fig. 5. Quantitative evaluation for SCM combined with classical texture descriptors. The horizontal line stands for the best single SCM result.

the performance of the ensemble of descriptors without relevantly impacting its computational time.

Figure 7 presents the confusion matrices for a colorectal tissue classification into eight classes. Our analysis was based on *set_*01, as it is our best performing algorithm. We observed that the discrimination between simple stroma and complex stroma is hard, due to the tissue's similarity. When comparing Fig. 7 (a) and Fig. 7 (b), *set_*01 contributed to decrease the error in all classes, mainly in classes simple stroma and complex stroma.

Finally, we analyzed the *set_*01 classification performance into eight classes of colorectal tissue, considering the state-of-the-art methods on our adopted dataset. Table 6 shows the result of the evaluation metrics for each method. We achieved results with ACC: 91.30%, PREC: 91.41%, SENS: 91.31%, SPEC: 98.76% and F1-score: 91.31% using 5000 images in the methodology. Our results are better when compared to the ACC of the Kather (2016) *et al.* [12] and Rachapudi e Devi (2020) [20] approaches. The ACC of the Wang *et al.* (2017) [28] approach was 1.4% higher; however, they used only 1000 images.

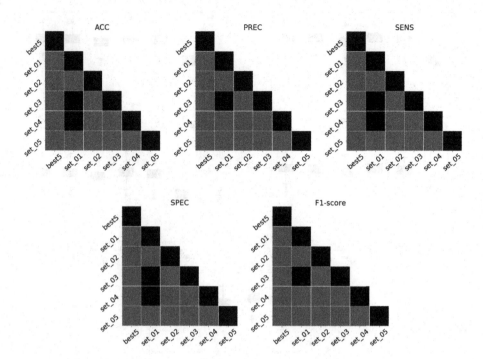

Fig. 6. Pairwise comparisons for all combined SCM using the ANOVA with post-hoc Turkey statistical tests. The gray boxes represent the pairs with significant difference at $\alpha = 0.05$.

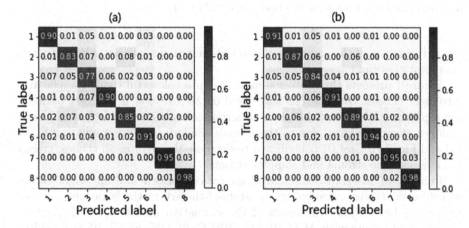

Fig. 7. Confusion matrices for colorectal tissue images classification. a) Results obtained from *best5* e b) *set_01*. The true labels and predicted labels are described as 1 - tumour epithelium, 2 - simple stroma, 3 - complex stroma, 4 - immune cell conglomerates, 5 - debris and mucus, 6 - mucosal glands, 7 - adipose tissue e 8 - background.

Table 5. Classification results of the H&E images of colorectal cancer into 8 classes using the SCM combined features.

Features	Classifier	ACC	PREC	SENS	SPEC	F1-score	Extraction time(s)
best5	KNN	82.2±0.84	82.57±1.08	82.29±1.06	97.45±0.11	82.13±1.12	1.04
	SVM Linear	87.46±1.04	87.49±1.01	87.47±1.05	98.20±0.14	87.44±1.06	
	SVM Poly	88.50±1.49	88.52±1.47	88.51±1.58	98.35±0.21	88.46±1.56	
	SVM RBF	**88.51±0.99**	**88.86±0.97**	**88.53±0.96**	**98.36±0.14**	**88.50±0.97**	
	Decison Tree	78.02±1.14	78.04±1.38	77.97±1.49	96.85±0.15	77.94±1.43	
set_01	KNN	87.78±1.20	87.92±1.28	87.82±1.28	98.25±0.16	87.78±1.28	0.33
	SVM Linear	89.72±0.81	89.84±0.70	89.77±0.79	98.53±0.11	89.75±0.77	
	SVM Poly	86.00±6.51	86.68±5.79	86.04±6.49	98.00±0.92	86.13±6.39	
	SVM RBF	**91.30±0.86**	**91.41±0.85**	**91.31±0.86**	**98.76±0.12**	**91.31±0.86**	
	Decison Tree	81.36±1.10	81.38±1.12	81.47±1.06	97.33±0.15	81.33±1.08	
set_02	KNN	87.20±0.85	87.65±0.98	87.26±0.91	98.17±0.11	87.24±1.09	0.93
	SVM Linear	89.52±0.58	89.72±0.82	89.54±0.77	98.50±0.07	89.58±0.81	
	SVM Poly	88.85±1.06	89.30±1.02	88.85±1.28	98.40±0.14	88.93±1.24	
	SVM RBF	**91.06±0.83**	**91.16±0.80**	**91.07±0.81**	**98.72±0.12**	**91.07±0.81**	
	Decison Tree	82.20±0.83	82.14±1.24	82.19±1.23	97.45±0.11	82.07±1.22	
set_03	KNN	87.58±1.47	87.98±1.45	87.61±1.53	98.22±0.21	87.60±1.50	1.02
	SVM Linear	89.32±0.66	89.44±0.67	89.32±0.69	98.47±0.09	89.31±0.67	
	SVM Poly	89.62±2.06	89.89±1.79	89.61±2.10	98.51±0.29	89.64±2.00	
	SVM RBF	**91.29±0.78**	**91.37±0.76**	**91.3±0.77**	**98.76±0.11**	**91.30±0.77**	
	Decison Tree	81.52±1.50	81.50±1.49	81.66±1.48	97.36±0.21	81.46±1.46	
set_04	KNN	87.56±1.07	87.97±1.23	87.56±1.22	98.22±0.14	87.58±1.20	1.00
	SVM Linear	89.74±0.79	89.92±0.88	89.73±0.81	98.53±0.11	89.75±0.84	
	SVM Poly	88.66±2.69	89.02±2.33	88.58±2.84	98.38±0.38	88.66±2.69	
	SVM RBF	**91.20±0.77**	**91.28±0.76**	**91.21±0.76**	**98.74±0.11**	**91.21±0.76**	
	Decison Tree	81.82±0.83	81.92±0.88	81.83±0.65	97.40±0.12	81.80±0.76	
set_05	KNN	86.96±0.30	87.37±0.49	86.97±0.34	98.13±0.04	87.02±0.36	0.95
	SVM Linear	88.96±0.93	89.15±0.93	88.99±0.86	98.42±0.13	89.01±0.88	
	SVM Poly	82.90±15.21	86.00±8.96	83.04±14.99	97.55±2.17	82.62±16.00	
	SVM RBF	**90.75±0.77**	**90.86±0.76**	**90.77±0.76**	**98.68±0.11**	**90.77±0.76**	
	Decison Tree	82.16±1.58	82.21±1.67	82.15±1.50	97.45±0.22	82.14±1.56	

Table 6. Comparison of the proposed approach with the state of the art for classification of the same image dataset. Only the standard deviations provided by the authors were presented.

Approach	Number of images	ACC (%)	PREC (%)	SENS (%)	SPEC (%)	F1-score (%)
This paper	5000	91.30 ± 0.86	91.41 ± 0.85	91.31 ± 0.86	98.76 ± 0.12	91.31 ± 0.86
Kather et al. (2016)	5000	87.40	–	–	–	–
Wang et al. (2017)	1000	92.60 ± 1.20	–	–	–	–
Rachapudi e Devi (2020)	5000	77.30	–	–	–	–

6   Conclusion and Future Directions

Here we studied state-of-the-art texture descriptors to categorize colorectal tissue samples into eight different classes, including normal and abnormal. We explored seven descriptors (SCM, HH, HL, PERC, Gabor, GLCM, LBP) in two scenarios:

isolated and in groups. Our experiments show that degrading high frequencies in the colorectal tissue images and encoding the resulting structural differences by SCM is valuable for pre-diagnosis into 8 classes. According to state-of-the-art, our approach enables gain of up to 3.93% in terms of accuracy when SCM is used as a single extractor, and 4.46%, when combined with others descriptors. Since the performance of single and combined SCM are close, we conclude that the texture information captured by SCM is relevant to achieving an accurate pre-diagnosis on colorectal tissues. Furthermore, our best SCM-based ensemble outperforms the state-of-the-art with up to 68.26% of the computational time.

Since we show the potential of SCM in applications with colorectal tissue images, future work should investigate the factors that make this structural co-occurrence matrix appropriate to this problem. Also, researches aimed at the development of CAD systems can benefit from the proposal of a descriptor that reaches a sensitivity of 91.31% and specificity of 98.76%, with satisfactory runtime to the clinical routine.

References

1. Abdi, H., Williams, L.J.: Newman-Keuls test and Tukey test. In: Encyclopedia of Research Design, pp. 1–11. Sage, Thousand Oaks (2010)
2. Altunbay, D., Cigir, C., Sokmensuer, C., Gunduz-Demir, C.: Color graphs for automated cancer diagnosis and grading. IEEE Trans. Biomed. Eng. **57**(3), 665–674 (2009)
3. Bianconi, F., Álvarez-Larrán, A., Fernández, A.: Discrimination between tumour epithelium and stroma via perception-based features. Neurocomputing **154**, 119–126 (2015)
4. Burges, C.J.: A tutorial on support vector machines for pattern recognition. Data Min. Knowl. Disc. **2**(2), 121–167 (1998)
5. Cardinal, R.N., Aitken, M.R.: ANOVA for the Behavioral Sciences Researcher. Psychology Press (2013)
6. Cortes, C., Vapnik, V.: Support-vector networks. Mach. Learn. **20**(3), 273–297 (1995)
7. Dunn, D., Higgins, W.E.: Optimal Gabor filters for texture segmentation. IEEE Trans. Image Process. **4**(7), 947–964 (1995)
8. Fisher, R.A.: On the mathematical foundations of theoretical statistics. Philos. Trans. Roy. Soc. London Ser. A Containing Papers Math. Phys. Char. **222**(594–604), 309–368 (1922)
9. Gurcan, M.N., Boucheron, L.E., Can, A., Madabhushi, A., Rajpoot, N.M., Yener, B.: Histopathological image analysis: a review. IEEE Rev. Biomed. Eng. **2**, 147–171 (2009)
10. Haralick, R.M., Shanmugam, K., Dinstein, I.H.: Textural features for image classification. IEEE Trans. Syst. Man Cybern. **6**, 610–621 (1973)
11. Kalkan, H., Nap, M., Duin, R.P., Loog, M.: Automated classification of local patches in colon histopathology. In: Proceedings of the 21st International Conference on Pattern Recognition (ICPR2012), pp. 61–64. IEEE (2012)
12. Kather, J.N., et al.: Multi-class texture analysis in colorectal cancer histology. Sci. Rep. **6**, 27988 (2016)

13. Malik, F., Baharudin, B.: The statistical quantized histogram texture features analysis for image retrieval based on median and Laplacian filters in the DCT domain. Int. Arab J. Inf. Technol. **10**(6), 1–9 (2013)
14. Mittal, H., Saraswat, M.: An automatic nuclei segmentation method using intelligent gravitational search algorithm based superpixel clustering. Swarm Evol. Comput. **45**, 15–32 (2019)
15. Narváez, F., Díaz, G., Poveda, C., Romero, E.: An automatic BI-RADS description of mammographic masses by fusing multiresolution features. Expert Syst. Appl. **74**, 82–95 (2017)
16. Panda, R.N., Baig, M.A., Panigrahi, B.K., Patro, M.R.: Efficient cad system based on GLCM & derived feature for diagnosing breast cancer. Int. J. Comput. Sci. Inf. Technol. **6**, 3323–3327 (2015)
17. Peixoto, S.A., Rebouças Filho, P.P.: Neurologist-level classification of stroke using a structural co-occurrence matrix based on the frequency domain. Comput. Electr. Eng. **71**, 398–407 (2018)
18. Pratt, W.K.: Digital Image Processing. A Wiley-Interscience Publication (1978)
19. PS, S.K., Dharun, V.: Extraction of texture features using GLCM and shape features using connected regions (2016)
20. Rachapudi, V., Devi, G.L.: Improved convolutional neural network based histopathological image classification. Evol. Intell., 1–7 (2020)
21. Ramalho, G.L.B., Ferreira, D.S., Rebouças Filho, P.P., de Medeiros, F.N.S.: Rotation-invariant feature extraction using a structural co-occurrence matrix. Measurement **94**, 406–415 (2016)
22. Reboucas Filho, P.P., Reboucas, E.D.S., Marinho, L.B., Sarmento, R.M., Tavares, J.M.R., de Albuquerque, V.H.C.: Analysis of human tissue densities: a new approach to extract features from medical images. Pattern Recogn. Lett. **94**, 211–218 (2017)
23. Rebouças Filho, P.P., et al.: Automatic histologically-closer classification of skin lesions. Comput. Med. Imaging Graph. **68**, 40–54 (2018)
24. Rokach, L., Maimon, O.: Top-down induction of decision trees classifiers-a survey. IEEE Trans. Syst. Man Cybern. Part C (Appl. Rev.) **35**(4), 476–487 (2005)
25. dos Santos, L.F.S., Neves, L.A., Rozendo, G.B., Ribeiro, M.G., do Nascimento, M.Z., Tosta, T.A.A.: Multidimensional and fuzzy sample entropy (SampEn Mf) for quantifying h&e histological images of colorectal cancer. Comput. Biol. Med. **103**, 148–160 (2018)
26. Society, A.C.: Colorectal Cancer Facts & Figures 2017–2019. American Cancer Society, Atlanta (2017)
27. Tamura, H., Mori, S., Yamawaki, T.: Textural features corresponding to visual perception. IEEE Trans. Syst. Man Cybern. **8**(6), 460–473 (1978)
28. Wang, C., Shi, J., Zhang, Q., Ying, S.: Histopathological image classification with bilinear convolutional neural networks. In: 2017 39th Annual International Conference of the IEEE Engineering in Medicine and Biology Society (EMBC), pp. 4050–4053. IEEE (2017)
29. Wang, L., He, D.C.: Texture classification using texture spectrum. Pattern Recogn. **23**(8), 905–910 (1990)

Unsupervised Learning Method for Encoder-Decoder-Based Image Restoration

Claudio D. Mello Jr$^{(\boxtimes)}$ (iD), Lucas R. V. Messias (iD);
Paulo Lilles Jorge Drews-Jr (iD), and Silvia S. C. Botelho (iD)

NAUTEC - Intelligent Robotics and Automation Group,
Center for Computational Science - C3, Federal University of Rio Grande - FURG,
Rio Grande, RS 96.203-000, Brazil
{claudio.mello,l.r.v.m,paulodrews,silviacb}@furg.br

Abstract. The restoration of a corrupted image is a challenge to computer vision and image processing. In hazy, underwater and medical images, the lack of paired images lead the state of the art to synthesize datasets. The Generative Adversarial Networks (GANs) are widely used in these cases. However, computational cost and training instability are current concerns. We present an unsupervised learning algorithm that does not requires paired dataset to train encoder-decoder-like neural network for image restoration. An encoder-decoder learn to represent its input data in a latent representation and reconstruct then in the output. During the training stage, our algorithm applies the encoder-decoder output image to a degradation block that reinforces its degradation. The degraded and input images are matched using a loss function. After the training process, we obtain a restored image from the decoder. We used ill-exposed images to evaluate and validate our algorithm.

Keywords: Unsupervised learning · Image restoration · Neural network

1 Introduction

Image restoration and image enhancement in the image processing has been received much attention from the researchers. Ill-exposition effects such as noise and over/under exposition are typical problems tackled to obtain image quality in many applications [1]. The over/under exposition occurs during image acquisition, when the signal is acquired by the sensor that presents limited acquisition range of light intensity. This situation produces the clipping phenomenon and interferes in bright regions of the scene, resulting in overexposure and in dark regions result in underexposure. In general, overexposure and underexposure are caused by poorly adjusted camera aperture, exposure time or gain [23].

Image restoration algorithms aim to reduce the artifacts in degraded images and to recover details, contrast and color. This is an important step in computing vision and applications like robotics, inspection, surveillance, recognition,

© Springer Nature Switzerland AG 2020
R. Cerri and R. C. Prati (Eds.): BRACIS 2020, LNAI 12319, pp. 348–360, 2020.
https://doi.org/10.1007/978-3-030-61377-8_24

etc. [19]. These algorithms must to tackle both spatially variant and invariant degradation [1]. However, recovering the image quality is an ill-posed problem given the image degradation is an irreversible process [15].

Convolutional Neural Networks (CNN)-based methodologies are widely used in the image restoration and enhancement-related works. Despite of the results, these works present complex network structures often using Generative Adversarial Networks (GANs) with moderate to high computational cost and hard training [12,14]. Although the GAN-based approach is unsupervised learning, the most of those works use supervised training strategy and require paired datasets containing images with and without specific features. Frequently, the datasets are synthesized to generate paired images [9,24,30].

Recently, unsupervised learning works using non-paired datasets have obtained great progress in image restoration [8], medical imaging problems [13,17,26] and audio denoising [16]. In these works are presented probabilistic approaches and encoder/decoder-type networks are used.

The Autoencoder (AE) is an encoder-decoder-like architecture and the most common in unsupervised deep learning. In general terms, AEs are neural networks which produce codifications from input data and they are trained so that their decodifications resemble of the inputs as closely as possible [4]. Based on AE's characteristic, we aim to restore an image that can be coded/decoded by AE or similar neural network. Our assumption establishes that the output (decodification) in the training stage can be properly degraded and applied to the loss function. As result we expect the neural network learn to restore the input degraded information. Thus, leading to correct the degradation imposed on training.

In this work, we present a new unsupervised training methodology to image restoration based on assumption described above. Our algorithm does not require paired dataset just a degradation function. The presented results are promising and are obtained for under and overexposed images generated via gamma variation. We restore ill-exposed RGB images and we compares the results obtained with supervised training using paired dataset.

2   Related Works

The application of the neural networks in image restoration increased in the recent years [7,11,28]. However, there are few proposals on CNN-based overexposed image restoration using unsupervised learning. In related works, the main issue tackled is the lack of paired datasets relating images with and without degradation. The main strategy adopt is synthesizing data and performing supervised training of the networks.

Although of recent advances, images containing high contrast scenes are difficult to be acquired. Images captured using conventional cameras operating in the visible light spectrum are commonly affected by artifacts and distortions due to excess or lack of light [22]. An image overexposure occurs when the sensor receives too much light, making it unable to differentiate the lighter parts of the image. The underexposure of an image may occur due to several factors, including insufficient lighting, short exposure time, or small aperture of the lens iris [21]. A method using a GAN to restore images ill-exposed is presented in [23]. Although the good looking results, the images show vibrant colors and unnatural-looking. High computational cost and instability of the GAN training convergence were observed.

A deep learning-based Single Image Contrast Enhancement (SICE) model is proposed in [3]. Three specialized networks are presented to correct the general image lightness, to restoring edges and high-frequency details and to combine the respective outputs.

Recently, similar approaches on dehazing and underwater image restoration were presented [7,9,24,30]. CNN-based structures using GANs and synthesizing datasets are adopted. An end-to-end image dehazing algorithm is presented in [7]. The model called FD-GAN uses a densely connected encoder-decoder as the generator. The main contribution is the novel Fusion-discriminator which integrates the high and low frequency information as additional priors and constraints into the learning procedure. The networks are trained using a synthetic dataset generated by the authors from the public datasets.

The approach described in [24] proposed the Underwater GAN (UWGAN) generate synthetic underwater style images. In the UWGAN training step, the generator network uses in-air image and depth map pairs based on improved underwater imaging model to output an underwater-style image. At the end of training, the generator is able to simulate an underwater-style image and a dataset is produced and used to training an Unet [18] architecture network.

Similarly, in [30] two GAN networks are employed to underwater image restoration. The UGAN proposed by [9] is trained to generate turbid images from clear ones. A synthetic dataset is produced and used to train another GAN. During inference, the generator network predicts clear images from blurry images as input.

The above mentioned works are supervised learning initiatives and they synthesize paired dataset to train the neural networks. On the other hand, unsupervised learning approaches no require paired images. This approach is adopted in several areas such as seismic analysis [5], medical imaging [13,26], image restoration [8] and audio applications [16]. In these works, probabilistic approaches are described with strategies applied to remove gaussian noise. Autoencoders or varational autoencoders are the chosen architectures to denoising task.

In this work, we present an unsupervised learning algorithm that requires only a dataset with non-paired images and a degradation function to train the neural network to perform image restoration.

3   Methodology

Synthesizing paired data is widely used to train neural networks to image restoration of ill-exposed images. These synthetic data are necessary to train the neural networks via supervised learning. However, we present an unsupervised learning algorithm and we do not use paired datasets. In this section, we describe our algorithm in details.

3.1 Proposed Algorithm

The most common architecture in unsupervised deep learning is the encoder-decoder. Many proposals need to compute costly optimization algorithms to encode a latent representation or sampling methods to reach a reconstruction (decode). Autoencoders capture both in their structure. With the advantage of training them become easier and faster [4].

In this work we present an unsupervised learning methodology for neural networks aiming to restore corrupted information. For this purpose, we applied to restoration of ill-exposed images. However, the method can be extended and evaluated to different contexts like audio denoising, sensor signal denoising, dehazing, underwater image restoration, etc. Fig. 1 describes our conceptual proposal. Essentially, an encoder-decoder network must reproduce its input information as precisely as according to its training. Therefore, a degraded input information, for example an image I, is coded by the encoder G_E generating a latent representation Y. The decoder G_D produces decoded and similar version $\hat{I}$ of the input image. In other words,

$$Y = G_E(I), \tag{1}$$

$$\hat{I} = G_D(Y). \tag{2}$$

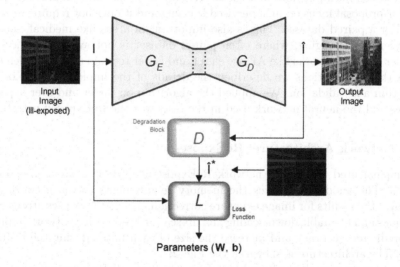

Fig. 1. The proposed image restoration architecture.

The core of our proposal is to insert a degradation step (a function or pipeline) D over training stage that simulate the effects to be removed on the encoder-decoder output image. Both input and degraded images are matched by loss function L. Our hypothesis is that, during the training stage, a proper degradation and the loss function minimization leads the encoder-decoder to perform an useful restoration of the input image. This restored image is presented at output $\hat{I}$.

In the training process, the *overdegradated* image $\hat{I}^*$ is the output image and it is matched to the input image using a loss function L. We assume $\theta = (\mathbf{W}, \mathbf{b})$, with the weight matrix $(\mathbf{W})$ and bias vector $(\mathbf{b})$, the parameters of the G_E and $\hat{\theta} = (\hat{\mathbf{W}}, \hat{\mathbf{b}})$, the parameters of the G_D. The hypothesis establishes that,

$$\hat{I}^* = D(\hat{I}), \tag{3}$$

and

$$\theta, \hat{\theta} = argmin[L(I, \hat{I}^*)]. \tag{4}$$

Thereby, minimizing the difference between I and $\hat{I}^*$ allow the image at decoder output be a restored version of input image (I'), thus

$$I' = D^{-1}(I). \tag{5}$$

A critical point is the degradation function or pipeline D (named degradation block) since the adequate specification allows and lead the network to learn what effects (artifacts) in image space that must be reduced. The correct specification can be complex and a challenging task. This conception can be applied to data from another nature. Similarly, the choose of the loss function is important and can be based on input data and characteristics of the restoration aimed.

Our proposal focus on unsupervised learning and it does not require an additional or a paired dataset. This is also important in areas like medical, seismic or underwater imaging, where often paired dataset is not available. Encoder-decoder neural networks like AE are easier and faster to train and to implement.

In the next sections we described the details of the implementation of the algorithm and validation. We applied the algorithm on under and over exposed images and the neural network used in the tests was the ReExposed-net [21].

3.2   Network Architecture: ReExpose-net

We implemented the neural network ReExpose-net [21] to restore ill-exposed images. This model minimizes the memory requirements of the network and improves the results for image exposure correction. The network uses atrous convolutions and trainable down-scaling and up-scaling layers. This network improve the prediction accuracy and increase the size of each batch during the training stage. The architecture is shown in the Fig. 2.

The convolutional block includes four 3×3 parallel atrous convolutional layers, with dilatation rates ranging from 2^0 up to 2^3. Thus, each convolutional block (Conv. Block) is able to cover 19 features in the input space using only nine trainable weights for each filter. Atrous convolutions provide context aggregation for each pixel by allowing the model to access a large region in the neighborhood, allowing us to reduce the amount of scaling layers in the network [21].

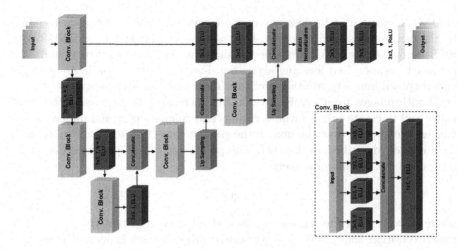

Fig. 2. ReExposed-net Architecture [21].

The learned down-scaling through strided convolutions perform better than pooling layers, especially in large saturated regions. For spatial up-scaling is used nearest neighbor interpolation followed by convolution. The Exponential Linear Unit (ELU) [6] activation function is used in the hidden layers. It speeds up the learning stage and leads to improve the image quality.

The downscale performed by the convolutional layers (stride = 2) and the upscale performed by the up sampling layers in the neural network identify the encoder-decoder aspect to the architecture.

3.3 Loss Function

The adopted loss function combines Structural Dissimilarity (DSSIM) [21] and Pixel-wise Euclidean Distance (L_2). DSSIM is based on SSIM [25], a similarity index calculated on various 3×3 window of an image. DDSIM provides great insight on image quality, but it does not able to account the pixel values in the exact position. The pixel values in ill-exposed regions in the image are closer the sensor limits. Then the loss function has the L_2 term and a element-wise multiplication reinforce the pixel values that are more likely to be affected by ill-exposure conditions. Assuming images I and $\hat{I}^*$ are in the representation interval $[0; 1]$, with an empirical constant $\lambda = 0.2$, loss function is shown in Eq. (6).

$$L(I, \hat{I}^*) = \lambda |0.5 - \hat{I}^*| \circ L_2(I, \hat{I}^*) + (1 - \lambda)DDSIM(I, \hat{I}^*), \qquad (6)$$

where the symbol $\circ$ represents element-wise multiplication.

3.4 Dataset and Training Parameters

We use the MIT-Adobe FiveK Dataset [2]. It features 5,000 images taken with SLR cameras by a set of different photographers. The images are made available in DNG file format, a lossless raw image format. All the information recorded by the camera sensor is preserved. For training, we converted the DNG file to RGB representation. We do not need paired images. We aimed to validate the proposed unsupervised learning algorithm. For this task, we evaluate over and underexposed image restoration problem. Thus, the degraded image dataset was build with images from FiveK dataset, but increasing or decreasing the luminance and to create bright or dark regions and simulate over and underexposed images. We applied over the images the gamma correction to produce bright or dark images, described by Eq. (7). The parameter $\gamma > 1$ produces darkening scenes and $\gamma < 1$ whitening scenes.

$$I = A.I_s{}^{\gamma}. \tag{7}$$

where $A = 1$, I_s is the dataset image and for overexposition the $\gamma = \frac{1}{3}$. For underexposition the $\gamma = 3$. The corrupted image dataset generated is used to train the neural network.

All trainable weights are initialized using the Glorot uniform initializer [10]. The model is trained using the Adam optimization algorithm with its default parameters. Training is stopped after 5000 batches are processed without yielding an improvement larger than 10^{-5}.

3.5 Degradation Block

Our methodology needs the degradation block D be able to reproduce the conditions of the input image degradation. Therefore, the degradation block is represented by the gamma factor, then:

$$D = A.I_s{}^{\gamma}, \tag{8}$$

$$\hat{I}^* = D(\hat{I}), \tag{9}$$

$$\hat{I}^* = A.\hat{I}^{\gamma}. \tag{10}$$

The training architecture diagram is indicated in the Fig. 3. The loss block represents the Eq. (6).

4   Experimental Results

In this section, we present the results obtained from image restoration algorithm based on unsupervised learning without paired dataset whose conceptual scheme is described in the methodology section. In addition, we shown a comparative between unsupervised and supervised learning via image quality metrics. The ReExpose-net is used to image processing in both cases.

The results are obtained for the ReExposed-net separately trained for under and overexposed images. We set $\gamma = 3.0$ to simulate underexposed images. The

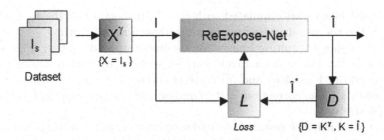

Fig. 3. The adopted architecture.

Fig. 4. Underexposed image restoration results. (a) Original Images, (b) Input images, (c) Output images (Restored).

overexposed images are simulated with $\gamma = \frac{1}{3}$. In both cases the training was performed in the unsupervised mode, i.e. without ground-truth.

Figure 4 shows resulting images for the underexposed context. Figure 4(a) corresponds the image from FiveK dataset without distortion shown to ease the comparison, Figs. 4(b) and (c) correspond the input image with distortion produced by gamma correction ($\gamma = 3$) and output image restored ($\hat{I}$ in the Fig. 3), respectively.

The neural network restored the underexposed images and the visual quality of the resulting images are is widely satisfactory. We can observe the improvement of regions with light colors in the scene. There is a slight change in the colors with enhancement of the red channel. These features are consequences of the gamma correction process, unsupervised learning, absence of ground-truth and of the ReExposed-net processing like highlighted in [21].

Figure 5 shows the resulting images for the overexposed images. Figure 5(a) correspond the images from FiveK dataset without distortion shown to ease the comparison, Figs. 5(b) and (c) correspond to the input image with distortion produced by gamma correction ($\gamma = \frac{1}{3}$) and output image, respectively. Our ReExposed-net restored the overexposed effect. The colors present changes with the red and the green channels are slightly attenuated. Dark regions in the no-distortion images are strongly darkened and present black areas in the resulting images. These areas in the input image are located near to the lower limit of the dynamic range. The overdegradation produces pixels values lower than zero and they are clipped. The content information is not recovered. These features are due to the gamma correction process, absence of ground-truth and the ReExposed-net processing. They can be perceived comparing Fig. 5(a) and Fig. 5(b).

In spite of color variations, the main result is the validation of the proposed unsupervised learning algorithm to restore degraded images. We use only one dataset and did not synthesize any ground-truth. The necessary information are the input information (with a specific noise or distortion), and a degradation pipeline or function. Also, it is important to highlight that the algorithm was implemented on a light neural network, whose performance was increased by the fact that the exta-degradation is an external function and it was not necessary to add on the network learning demands.

Tables 1 and 2 are show the quantitative results for under and overexposition cases, respectively. The values are obtained comparing both input and output images with the no-distortion image from FiveK dataset.

The indicated metrics are image quality metrics often utilized to evaluate image restoration methods. Specifically, Peak Signal-to-Noise Ratio (PSNR) [23], Mean Square Error (MSE) [23], Structural Similarity (SSIM) [25], Gradient Magnitude Similarity Deviation (GMSD) [27], CIEDE2000 [20], Feature Similarity and Feature Similarity with Chrominance (FSIM and FSIMc) [29].

We adopted a wide range of metrics to properly evaluate our method. We use pixel-wise metrics (PSNR, MSE), structure-related metrics (SSIM, GMSD), color metrics (CIEDE2000) and visual perception based metrics (FSIM, FSIMc).

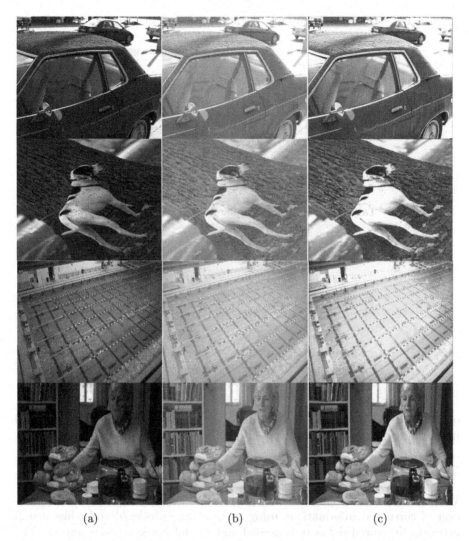

(a)   (b)   (c)

Fig. 5. Overexposed image restoration results. (a) Original Images, (b) Input images, (c) Output images (Restored).

From Table 1, all metrics point out that the neural network effectively performed the restoration of the underexposed images. There is a significant difference among metrics related pixel-wise, structure and color. However, the metrics related to visual perception produced close values.

Results for overexposed image restoration are shown in the Table 2. Although the most of metrics point out the best results for output image, the differences are smaller than underexposed case. In addition, the GMSD metric is smaller for the output image and values for visual perception metrics are close. These

Table 1. Quantitative results in underexposed images.

Metric	Input image	Output image
PSNR ↑	12.008	24.775
MSE ↓	4455.3	238.1
SSIM ↑	0.310	0.750
GMSD ↓	0.164	0.099
CIEDE2000 ↓	22.030	8.158
FSIM ↑	0.738	0.831
FSIMc ↑	0.717	0.813

Table 2. Quantitative results in overexposed images.

Metric	Input image	Output image
PSNR ↑	10.209	13.557
MSE ↓	6339.5	3008.2
SSIM ↑	0.658	0.781
GMSD ↓	0.075	0.104
CIEDE2000 ↓	26.582	23.589
FSIM ↑	0.926	0.921
FSIMc ↑	0.905	0.885

observations indicate that the performance of the neural network for overexposed image is lower than the underexposed one, assuming the mean of the dataset.

5   Conclusions and Future Works

We presented and validated a new unsupervised learning algorithm for restoration of corrupted informations using to train an encoder/decoder-like neural network. No paired datasets is needed and we did not synthesize anyone. The algorithm requires as input corrupted information and an adequate function or pipeline of additional degradation to be applied on output information. The input and output informations are matched in the loss function. Starting from no-corrupted images, we simulate under and overexposed images by gamma correction. The additional distortion required by algorithm also is performed by gamma correction. The results show the effectiveness of the method. However, a high degradation can limit the restoration. This was shown on underexposed image restoration results. The image restoration is related to effect produced by the degradation block and its incorrect definition or incorrect parameters can limit the output image quality. In further work, we will extend the analysis and the research to another image restoration context like underwater images. Furthermore, we intend to improve the presented degradation pipeline since the limitation in challenge situation.

References

1. Anwar, S., Barnes, N., Petersson, L.: Attention prior for real image restoration. arXiv:2004.13524v1 [cs.CV], April 2020
2. Bychkovsky, V., Paris, S., Chan, E., Durand, F.: Learning photographic global tonal adjustment with a database of input/output image pairs. In: IEEE CVPR, pp. 97–104, July 2011. https://doi.org/10.1109/CVPR.2011.5995413
3. Cai, J., Gu, S., Zhang, L.: Learning a deep single image contrast enhancer from multi-exposure images. IEEE Trans. Image Process. 1 (2018). https://doi.org/10.1109/TIP.2018.2794218
4. Charte, D., Charte, F., García, S., Del Jesus, M.J., Herrera, F.: A practical tutorial on autoencoders for nonlinear feature fusion: taxonomy, models, software and guidelines. Inf. Fusion 44, 78–96 (2018). https://doi.org/10.1016/j.inffus.2017.12.007
5. Chen, Y., Zhang, M., Bai, M., Chen, W.: Improving the signal-to-noise ratio of seismological datasets by unsupervised machine learning. Seismol. Res. Lett. (2019). https://doi.org/10.1785/0220190028
6. Clevert, D.A., Unterthiner, T., Hochreiter, S.: Fast and accurate deep network learning by exponential linear units (ELUs). arXiv preprint arXiv:1511.07289 (2015)
7. Dong, Y., Liu, Y., Zhang, H., Chen, S., Qiao, Y.: FD-GAN: generative adversarial networks with fusion-discriminator for single image dehazing. In: AAAI, pp. 10729–10736 (2020)
8. Du, W., Chen, H., Yang, H.: Learning invariant representation for unsupervised image restoration. arXiv:2003.12769v1, March 2020
9. Fabbri, C., Islam, M.J., Sattar, J.: Enhancing underwater imagery using generative adversarial networks. In: IEEE International Conference on Robotics and Automation (ICRA), pp. 7159–7165 (2018)
10. Glorot, X., Bengio, Y.: Understanding the difficulty of training deep feedforward neural networks. J. Mach. Learn. Res. Proc. Track 9, 249–256 (2010)
11. Gonçalves, L.T., Gaya, J.F.O., Drews-Jr, P.L.J., Botelho, S.S.C.: GuidedNet: single image dehazing using an end-to-end convolutional neural network. In: Conference on Graphics, Patterns and Images (SIBGRAPI), pp. 79–86 (2018)
12. Hashisho, Y., Albadawi, M., Krause, T., von Lukas, U.F.: Underwater color restoration using u-net denoising autoencoder. In: 2019 11th International Symposium on Image and Signal Processing and Analysis (ISPA), pp. 117–122 (2019)
13. Liu, J., Sun, Y., Eldeniz, C., Gan, W., An, H., Kamilov, U.S.: RARE: image reconstruction using deep priors learned without ground truth. IEEE J. Sel. Top. Sign. Proces. 14(6), 1–1 (2020)
14. Lucic, M., Kurach, K., Michalski, M., Gelly, S., Bousquet, O.: Are GANs created equal? a large-scale study. arXiv:1711.10337v4 (11 2018)
15. Mei, Y., et al.: Pyramid attention networks for image restoration. arXiv:2004.13824v1 [cs.CV] (04 2020)
16. Michelashvili, M., Wolf, L.: Audio denoising with deep network priors. arXiv:1904.07612v2, April 2019
17. Prakash, M., Lalit, M., Tomancak, P., Krull, A., Jug, F.: Fully unsupervised probabilistic noise2void. arXiv:1911.12291v2 [eess.IV], November 2019
18. Ronneberger, O., Fischer, P., Brox, T.: U-net: convolutional networks for biomedical image segmentation. ArXiv abs/1505.04597 (2015)

19. Ruiz-del-Solar, J., Loncomilla, P., Soto, N.: A survey on deep learning methods for robot vision. CoRR abs/1803.10862 (2018). http://arxiv.org/abs/1803.10862
20. Sharma, G., Wu, W., Dalal, E.: The ciede2000 color-difference formula: Implementation notes, supplementary test data, and mathematical observations. Color Res. Appl. **30**, 21–30 (2005). https://doi.org/10.1002/col.20070
21. Steffens, C.R., Drews-Jr, P.L.J., Botelho, S.S.C.: Deep learning based exposure correction for image exposure correction with application in computer vision for robotics. In: 2018 Latin American Robotic Symposium, 2018 Brazilian Symposium on Robotics (SBR) and 2018 Workshop on Robotics in Education (WRE), pp. 194–200 (2018)
22. Steffens, C.R., Messias, L.R.V., Drews-Jr, P.L.J., Botelho, S.S.C.: Can exposure, noise and compression affect image recognition? an assessment of the impacts on state-of-the-art convnets. In: 2019 Latin American Robotics Symposium (LARS), 2019 Brazilian Symposium on Robotics (SBR) and 2019 Workshop on Robotics in Education (WRE), pp. 61–66 (2019)
23. Steffens, C.R., Messias, L.R.V., Drews-Jr, P.L.J., Botelho, S.S.C.: CNN based image restoration: adjusting ill-exposed srgb images in post-processing. J. Intell. Roboti. Syst. **99**, 609–627 (2020)
24. Wang, N., Zhou, Y., Han, F., Zhu, H., Zheng, Y.: UWGAN: underwater GAN for real-world underwater color restoration and dehazing. arXiv preprint arXiv:1912.10269 (2019)
25. Wang, Z., Bovik, A.C., Sheikh, H.R., Simoncelli, E.P.: Image quality assessment: from error visibility to structural similarity. IEEE Trans. Image Process. **13**(4), 600–612 (2004)
26. Wu, D., Gong, K., Kim, K., Li, X., Li, Q.: Consensus neural network for medical imaging denoising with only noisy training samples. arXiv:1906:03639v1 (06 2019)
27. Xue, W., Zhang, L., Mou, X., Bovik, A.: Gradient magnitude similarity deviation: a highly efficient perceptual image quality index. IEEE Trans. Image Process. **23** (2013). https://doi.org/10.1109/TIP.2013.2293423
28. Zhang, H.-M., Dong, B.: A review on deep learning in medical image reconstruction. J. Oper. Res. Soc. Chin. **8**(2), 311–340 (2020). https://doi.org/10.1007/s40305-019-00287-4
29. Zhang, L., Zhang, L., Mou, X.: FSIM: a feature similarity index for image quality assessment. IEEE Trans. Image Process. **20**, 2378–2386 (2011). https://doi.org/10.1109/TIP.2011.2109730
30. Zhou, Y., Wang, J., Li, B., Meng, Q., Rocco, E., Saiani, A.: Underwater scene segmentation by deep neural network. In: UK-RAS19 Conference: Embedded Intelligence: Enabling & Supporting RAS Technologies, pp. 44–47, January 2019. https://doi.org/10.31256/UKRAS19.12

A Computational Tool for Automated Detection of Genetic Syndrome Using Facial Images

Eduardo Henrique Pais Pooch[(✉)], Thatiane Alves Pianoschi Alva, and Carla Diniz Lopes Becker

Universidade Federal de Ciências da Saúde de Porto Alegre (UFCSPA) - Rua Sarmento Leite, 245, Porto Alegre, Rio Grande do Sul, Brazil
{edupooch,thatiane,carladiniz}@ufcspa.edu.br

Abstract. Early diagnosis of genetic syndromes has a vital importance in the prevention of any potential related health problems. Down syndrome is the most common genetic syndrome. Patients with down syndrome have a high probability of developmental disorders, like Congenital Heart Disease, which is best treated when discovered in the early stages. These patients also have particular facial characteristics that are identified by geneticists in a physical exam. However, there is subjectivity in the professional analysis, which can lead to a late diagnosis, aggravating the patient's health condition. This paper proposes a software framework for the automatic detection of Down syndrome using facial features extracted from digital images, which could be used as a tool to help in the early detection of genetic syndromes. For training the machine learning model, we create a dataset gathering 170 pictures of children available on the internet. 50% of the pictures were of children with Down syndrome and the other 50% of healthy children. Then, we automatically identify faces and describe the images with facial landmarks. Next, we use two approaches for feature extraction. The first is a traditional computer vision approach using selected distances and angles and textures between the landmarks. The other, a deep learning approach using a Convolutional Neural Network to extract the features automatically. Then, the feature vector is fed to a Support Vector Machine with a linear kernel on both feature extraction approaches. We validate the results measuring the accuracy, sensitivity, and specificity of both feature extraction approaches using 10-fold cross-validation. The deep learning method resulted in an accuracy of 0.94, while the traditional approach achieved 0.84 of accuracy in our dataset. The results shows that the deep learning approach has a higher classification accuracy for this task, even with a small dataset.

Keywords: Genetic syndrome · Computer vision · Machine learning

© Springer Nature Switzerland AG 2020
R. Cerri and R. C. Prati (Eds.): BRACIS 2020, LNAI 12319, pp. 361–370, 2020.
https://doi.org/10.1007/978-3-030-61377-8_25

1   Introduction

Early detection of genetic syndromes in children has an important role in the treatment of patients and diseases. Precocious diagnosis helps to prevent critical health problems, such as congenital heart diseases and respiratory problems. Women in pregnancy routinely make ultrasound and blood tests. Other more invasive tests for prenatal diagnosis, such as amniocentesis and chorionic villus sampling, are required if there is a high risk of genetic syndromes After birth, the diagnosis is often based on physical variations or dysmorphology [1].

More than 50% of patients affected by genetic syndromes have particular facial characteristics [1], and there is a great subjectivity in the analysis and identification of dysmorphia as an identifier of a genetic syndrome, causing variation in the diagnostic decision of the physical examination among different professionals. After identifying the facial characteristics, the diagnosis can be confirmed by a cytogenetic examination of karyotype, fluorescence in situ hybridization (FISH), or molecular genetic tests, which perform DNA sequencing.

Down syndrome is the most well-known genetic syndrome with particular facial characteristics, and it is the main cause of congenital heart diseases [4]. This syndrome's diagnostic accuracy in newborns is around 64% [6,21], and can be lower in regions with precarious health conditions. Some other common syndromes associated with heart diseases are Turner, 22q11.2 deletion, Williams and Noonan [9]. After identifying congenital malformations, the newborn's cardiac condition needs to be evaluated and monitored; therefore, detecting these syndromes earlier may increase the chances of survival and the quality of life of these patients.

In the last few years, automatic methods to identify different genetic syndromes from digital images have been developed [1,23]. In countries where access to cytogenetic examinations is limited, the accuracy of an early diagnosis based in facial characteristics is critical. Some techniques of computer vision and machine learning have been explored for the task of a fast diagnosis in a significant number of genetic syndromes.

Zhao et al. [23] proposed a strategy based on machine learning to detect Down syndrome automatically. Geometric and texture features based on local binary patterns are extracted around the landmarks. The best performance achieved an accuracy of 0.95 using a Support Vector Machine (SVM) with a radial basis function kernel. In the study of Cerrolaza et al. [1], they presented a general framework for the detection of multiple genetic disorders, including Down syndrome, on a database of 145 facial pictures, combining geometrical and texture features. This study proposed the combination of morphological and local appearance features using local binary patterns, achieving a detection accuracy of 0.94. Dima et al. [3] uses facial recognition methods to extract the features and use SVMs and K-nearest neighbors classifiers in order to recognize the presence of Down syndrome. They use 50 down syndrome samples and achieve accuracy results ranging from 0.93 to 1.0 when using different datasets for the control group. Kumov et al. [11] explores 3D face reconstruction besides geometric and deep features to classify 8 genetic syndromes. They use a database of 1462 total samples and achieve an average accuracy of 0.92.

Deep convolutional neural networks (CNNs) are the state-of-the-art in image classification. CNNs use convolution for automated feature extraction and data classification, learning how to represent data with layers of abstraction [14]. The technique is being used in several medical image analysis applications with great results [15], making it a possible high accuracy diagnostic method for genetic syndromes.

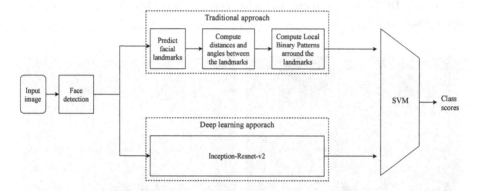

Fig. 1. The two proposed approaches in this paper.

In this paper, we propose a software implementation of two frameworks for automatic classification of Down syndrome, which are summarized in Fig. 1. The first is using geometric and texture features extracted from digital images to feed an SVM with a linear kernel based on previous works [1,23]. The second is a deep learning approach, using a pre-trained CNN model for feature extraction [22] and the same Linear SVM model for classification. We perform a comparison between the two approaches and demonstrate how this architecture can classify Down syndrome using facial images with high accuracy.

This study is divided into four sections. In Sect. 2, we present the suggested recognition methodologies. Section 3 presents the results of the data set simulations, comparison, and discussion of the efficiency of the methods. Finally, in Sect. 4, conclusions about the paper and indications of possible future work are presented.

2 Materials and Methods

2.1 Data Description

There is no publicly available dataset for this task. So, we created the dataset using 170 pictures of children available on the internet. Half of the images were of children with Down syndrome, and the other half was a control group. The samples of the Down syndrome group were collected from a gallery named Babies with Down Syndrome from a parenting website [18] and from the figures on the Kruszka et al. [10] paper, which presents Down syndrome in different ethnicities. The control group is a random sample manually selected from the children category on the ImageNet database [2]. The dataset used in this paper is composed of frontal images of children of both sexes, from 0 to 5 years of age, and from different ethnicities. The faces were automatically segmented from the images with a Python programming language script based on the Dlib-ml library [8] face detector. It is an object detector based on that uses a Histogram of Oriented Gradients (HOG) as a feature extractor and a linear classifier trained for face detection.

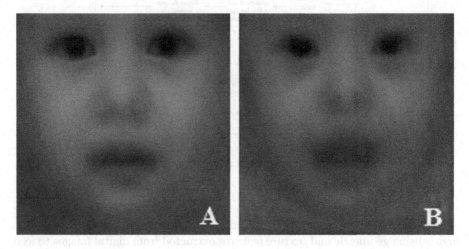

Fig. 2. Average image made of all the aligned samples from each group in the dataset. A: Control group; B: Down syndrome group.

2.2   Facial Landmarks

After segmenting the faces, the location of some facial landmarks were identified, and the segmented faces were aligned based on the eyes and mouth position to achieve a rotation normalization of the dataset and then all the images were resized to 150 × 150 pixels for scale normalization. We also used the facial landmarks to obtain a feature vector that can represent the particular facial characteristics of genetic syndromes, which was done by calculating some geometric relations between the landmarks and the texture around them. The facial landmarks were detected with Dlib's implementation of Kazemi et al. [7] pose estimator based on an ensemble of regression trees for face alignment and trained on the iBUG 300-W face landmark dataset [20] which can estimate the coordinates of 68 facial landmarks (see Fig. 2A).

2.3   Feature Extraction

Traditional Approach. The extracted geometric features were 9 Euclidean distances and 18 angles between the inner landmarks, which were based on the measures that geneticists use to assess particular facial characteristics and diagnose genetic syndromes (see Fig. 3). To improve the results, we used the local binary patterns (LBP) descriptor, a rotation, and grayscale invariant texture classification [17] based on pixel relations of the p surrounding points in a circularly symmetric neighborhood, in which the radius of the circle is equal to r. To compose the feature vector, 11 textures were extracted from the image in a 35 × 35 pixels region around some selected landmarks (see Fig. 3). For each region, the LBP representation with p = 12 and r = 2 was computed and used to build a normalized histogram counting the occurrence of each binary pattern, generating a p + 2 sized vector. Therefore, the feature vector is the concatenation of the distances, angles, and texture representations, forming a total of 181 values.

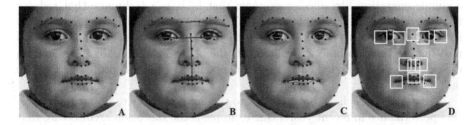

Fig. 3. A: Facial landmarks location; B: Computed distances between the points; C: Computed angles; D: Region of the extracted textures. (Image sample from the CAFE dataset [16]).

Deep Learning Approach. Deep-learning software attempts to simulate the brain process using a large array of neurons in an artificial neural network (ANN). The ANN learns, with high-level precision, to recognize patterns in digital representations of images and other data. Because of evolution and improvements in mathematical approach and powerful computers, computer scientists engineers can now model many more layers in ANN than ever before, creating the deep-learning networks (DLNs).

In the group of DLNs, we have CNNs, which are inspired in variants of multilayer perceptrons (MLPs). CNNs exploit spatially-local correlation by enforcing a local connectivity pattern between neurons of adjacent layers, proving to be very effective in areas such as image recognition and classification. The convolution layer's parameters consist of a set of learnable filters. Every filter is small spatially (along width and height). During the forward path, it is convolved each filter across the width and height of the input image and compute dot products between the entries of the filter and the input at any position. As we slide the filter over the width and height of the input image the activation map that gives the responses of that filter at every spatial position. So, the CNN will learn filters that activate when they found some type of features, such as an edge of some orientation on deep layers of the network, and each of them will produce a separate 2-dimensional activation map. These activation maps are produced along the depth dimension producing the output.

One of the first appearances of successful CNN applications in the literature was LeNet-5 [12] created for digit recognition in documents using the MNIST [13] dataset, which features 70,000 labeled handwritten digit images. The LeNet-5 architecture consists of two sets of convolution, activation and pooling, a fully-connected layer, an activation layer, another fully-connected layer containing a number of neurons equal to the number of output classes, and a softmax function for the classification. The convolution layers receive inputs, perform the convolution operation with a given kernel, and result in a feature map representing the characteristics extracted through the convolution [5]. The activation function is responsible for inserting non-linearity in the model. Pooling or subsampling reduces the dimensionality of the feature map by grouping the information using a simple operation like max or average; that is, it is a technique that reduces its size while maintaining information. The fully-connected layer combines the features extracted by the other layers to create the model. The last layer has a softmax activation function that shows the probability that the input is from each of the classes, and we choose the class with the highest probability as the output.

Very deep CNNs have been central to the largest advances in image recognition performance in recent years. One example is the Inception architecture that has been shown to achieve very good performance at relatively low computational cost [22]. In this work we studied and implemented the Inception-ResNet-v2 [22] to extract the image features automatically. Figure 4 shows the architecture of Inception-ResNet-v2. We use a model pre-trained on the ImageNet dataset and remove the softmax classifier in the last layer, keeping the 2048 features that the network extracted as the face's descriptor. We used a mini-batch size of 64 and trained for 35 epochs.

2.4   Classifiers

For the classification problem, a supervised learning method using linear classification was implemented using the Python programming language and the tools available on the Scikit-learn library [19]. Linear classification is a method of statistical classification, which will use the dataset to create a model that is able to separate the classes based on the samples feature values and deduce the class of a new unknown subject, in this case, syndromic or non-syndromic. To train the model, the presence or absence of Down syndrome is added to the feature vector. The trained and tested model is an Support Vector Machine with a linear kernel and a $C = 0.001$. The model optimizes a set of hyperplanes to separate the target classes based on the extracted features by maximizing the distance of the hyperplanes to the features of each class.

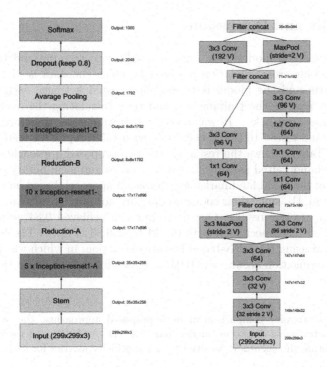

Fig. 4. Inception-ResNet-v2 architecture. Source: [22]

2.5 Evaluation Metrics

To evaluate performance, we split our dataset into two sets (train and test) and test the model on samples it has not seen during training. If the model infers that a sample is positive for down syndrome, it is a true positive (TP) outcome. If it was a wrong prediction and it was actually a negative sample, this prediction counts as a false positive (FP). If the model outputs negative for a sample that is really negative for down syndrome, it is a true negative (TN), and if it was actually a down syndrome (positive) case, it counts as a false negative (FN). The metrics used to evaluate performance are accuracy, sensibility, and specificity. Accuracy (Eq. 1) is the number of true outcomes over the total of samples. Sensitivity (Eq. 2) is the number of true positives over the total of positive samples, and specificity (Eq. 3) is the number of true negatives over the total of negative samples.

$$accuracy = \frac{TP + TN}{TP + TN + FP + FN} \tag{1}$$

$$sensitivity = \frac{TP}{TP + FN} \tag{2}$$

$$specificity = \frac{TN}{TN + FP} \tag{3}$$

3   Results and Discussion

To test the trained models, the data was split into training and testing sets using the k-folds cross-validation method, in which the data is divided into k different parts and the model is trained and tested k times, each time with one different part as the training set, and then calculating the average of the classification scores. A k = 10 was used for this validation, therefore the model was trained and tested 10 times, each time with 153 samples for training and 17 for testing. Table 1 shows the results of the two proposed approaches. The first one is using handcrafted features, distances, angles and local binary patterns on the regions of the facial landmarks, which was an approach based on the work of Cerrolaza et al. [1], since their code and data are not publicly available, we can't directly compare our results. Our first approach achieved 0.84 accuracy, 0.81 sensitivity and 0.86 specificity, with an AUC of 0.83 (Fig. 5). The best results were obtained using the CNN-based feature extraction, in which was possible to achieve a classification accuracy of 0.94, a sensitivity of 0.95, and a specificity of 0.92.

Table 1. Performance comparison of both proposed approaches, the one with the feature extractor using distances, angles and LBP (D+A+LBP) and the one using a CNN. Presenting the accuracy, sensitivity, and specificity metrics for each one.

	D+A+LBP	CNN
Accuracy	0.84	**0.94**
Sensitivity	0.81	**0.95**
Specificity	0.86	**0.92**

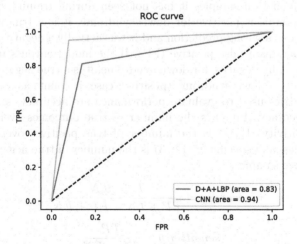

Fig. 5. Receiver Operating Characteristic (ROC) curve and area under curve of both proposed approaches traditional (D+A+LBP) and using convolutional neural networks (CNN). Vertical axis shows the models' true positive rate (TPR) and the horizontal axis the false positive rate (FPR).

4   Conclusion

Two fully automated framework for detecting Down syndrome were proposed, one using classic computer vision techniques based on previous works and another using a deep learning approach which achieved better results. In this work, it was possible to develop a model using a Linear SVM with a CNN feature extractor that is capable of identifying Down syndrome with 94% of accuracy using facial images, while the approach similar to the one presented by Zhao et al. [23], with some differences on the extracted features, such as a different selection of landmarks and Euclidean distances, instead of horizontal and vertical distances, showed an accuracy of 84% in our dataset. The results show that it is possible to build a software to assist the neonatal staff in identifying children that might have Down syndrome using state-of-the-art techniques in image classification, enabling an early detection that could improve the treatment of the Down syndrome conditions. A limitation of this study was the absence of a general database for this problem. We highlight the need for a public and well-controlled dataset for this task, in order to better develop and compare approaches in a closer to real-world scenario as well as evaluate possible bias. A mobile application powered by our model can improve the decision accuracy by having a protocol for image acquisition to standardize the submitted images, and present to the user a confidence metric for each classification. The code we created for this work can be used to create an application programming interface for a cloud-based mobile application. To evaluate the potentiality of this study, as future work, it is intended to test the methodology proposed in classifying other genetic syndromes, requiring an effort in gathering more samples of genetic syndrome patients.

References

1. Cerrolaza, J.J., Porras, A.R., Mansoor, A., Zhao, Q., Summar, M., Linguraru, M.G.: Identification of dysmorphic syndromes using landmark-specific local texture descriptors. In: 2016 IEEE 13th International Symposium on Biomedical Imaging (ISBI) (2016)
2. Deng, J., Dong, W., Socher, R., Li, L.J., Li, K., Li, F.-F.: ImageNet: a large-scale hierarchical image database. In: 2009 IEEE Conference on Computer Vision and Pattern Recognition, pp. 248–255. IEEE, June 2009
3. Dima, V., Ignat, A., Rusu, C.: Identifying down syndrome cases by combined use of face recognition methods. In: Balas, V.E., Jain, L.C., Balas, M.M. (eds.) SOFA 2016. AISC, vol. 634, pp. 472–482. Springer, Cham (2018). https://doi.org/10. 1007/978-3-319-62524-9_35
4. Ekure, E.N., Animashaun, A., Bastos, M., Ezeaka, V.C.: Congenital heart diseases associated with identified syndromes and other extra-cardiac congenital malformations in children in Lagos. West Afr. J. Med. **28**(1), 33–37 (2009)
5. Goodfellow, I., Bengio, Y., Courville, A.: Deep Learning. MIT Press, New York (2016). http://www.deeplearningbook.org
6. Hindley, D., Medakkar, S.: Diagnosis of down's syndrome in neonates. Arch. Dis. Child. Fetal Neonatal Ed. **87**(3), F220–1 (2002)

7. Kazemi, V., Sullivan, J.: One millisecond face alignment with an ensemble of regression trees. In: 2014 IEEE Conference on Computer Vision and Pattern Recognition (2014)
8. King, D.E.: Dlib-ml: a machine learning toolkit. J. Mach. Learn. Res. **10**(Jul), 1755–1758 (2009)
9. Ko, J.M.: Genetic syndromes associated with congenital heart disease. Korean Circ. J. **45**(5), 357–361 (2015)
10. Kruszka, P., et al.: Down syndrome in diverse populations. Am. J. Med. Genet. A **173**(1), 42–53 (2017)
11. Kumov, V., Samorodov, A.: Recognition of genetic diseases based on combined feature extraction from 2D face images. In: 2020 26th Conference of Open Innovations Association (FRUCT), pp. 1–7. IEEE (2020)
12. LeCun, Y., Bottou, L., Bengio, Y., Haffner, P.: Gradient-based learning applied to document recognition. Proc. IEEE **86**(11), 2278–2324 (1998)
13. LeCun, Y.: The MNIST database of handwritten digits (1998). http://yann.lecun. com/exdb/mnist/
14. LeCun, Y., Bengio, Y., Hinton, G.: Deep learning. Nature **521**(7553), 436–444 (2015)
15. Litjens, G., et al.: A survey on deep learning in medical image analysis. Med. Image Anal. **42**, 60–88 (2017)
16. LoBue, V., Thrasher, C.: The child affective facial expression (CAFE) set: validity and reliability from untrained adults. Front. Psychol. **5**, 1532 (2015)
17. Ojala, T., Pietikainen, M., Maenpaa, T.: Multiresolution gray-scale and rotation invariant texture classification with local binary patterns. IEEE Trans. Pattern Anal. Mach. Intell. **24**(7), 971–987 (2002)
18. Parenting: A special joy: babies with down syndrome galleries, December 2010. https://www.parenting.com/article/a-special-joy-babies-with-down-syndrome-galleries
19. Pedregosa, F., et al.: Scikit-learn: machine learning in Python. J. Mach. Learn. Res. **12**(Oct), 2825–2830 (2011)
20. Sagonas, C., Antonakos, E., Tzimiropoulos, G., Zafeiriou, S., Pantic, M.: 300 faces in-the-wild challenge: database and results. Image Vis. Comput. **47**, 3–18 (2016)
21. Sivakumar, S., Larkins, S.: Accuracy of clinical diagnosis in down's syndrome. Arch. Dis. Child. **89**(7), 691 (2004)
22. Szegedy, C., Ioffe, S., Vanhoucke, V.: Inception-v4, inception-resnet and the impact of residual connections on learning. CoRR abs/1602.07261 (2016). http://arxiv. org/abs/1602.07261
23. Zhao, Q., Rosenbaum, K., Okada, K., Zand, D.J., Sze, R., Summar, M., Linguraru, M.G.: Automated down syndrome detection using facial photographs. In: 2013 35th Annual International Conference of the IEEE Engineering in Medicine and Biology Society (EMBC), pp. 3670–3673. IEEE (2013)

Improving FIFA Player Agents Decision-Making Architectures Based on Convolutional Neural Networks Through Evolutionary Techniques

Matheus Prado Prandini Faria[1]([⊠]), Rita Maria Silva Julia[1], and Lídia Bononi Paiva Tomaz[2]

[1] Computer Science Department, Federal University of Uberlândia, Uberlândia, Brazil
matheusprandini.96@gmail.com, ritamariasilvajulia@gmail.com
[2] Computer Science Department, Federal Institute of Triângulo Mineiro, Uberaba, Brazil
ldbononi@gmail.com

Abstract. Convolutional Neural Network (CNN) is a fundamental tool in Deep Learning and Computer Vision due to its remarkable ability to extract relevant characteristics from raw data, which has been allowing significant advances in image classification tasks. One of the great challenges in using CNNs is to define an architecture that is suitable for the problem for which they are being designed. Thus, there is a big effort in many recent works to propose approaches to automatically define appropriate CNN architectures. Among them, the Convolutional Neural Network designed by Genetic Algorithm (CNN-GA) method stands out. As CNN-GA has only been validated in static scenarios involving image classification data sets, the main contributions of the present paper are the following: implementing an improved version of CNN-GA, named as Minimum CNN-GA (MCNN-GA), that automatically defines CNN architectures through a policy that minimizes the weigh vector dimensions and the classification error rate of the CNNs; implementing a set of imitation learning based agents that operate in a complex and dynamic scenario of FIFA game exploring distinct raw image representations for the environment at the input of CNNs designed according to the MCNN-GA approach. The performance of these agents were evaluated through their in-game score in tournaments against FIFA's engine. The results corroborate that the decision-making ability of such agents can be as good as human ability.

Keywords: Convolutional neural network · Genetic algorithm · Imitation learning

The authors thank CAPES for financial support.

R. Cerri and R. C. Prati (Eds.): BRACIS 2020, LNAI 12319, pp. 371–386, 2020.
https://doi.org/10.1007/978-3-030-61377-8_26

1   Introduction

Machine Learning (ML) has consolidated itself as one of the pillars for the advancement of Artificial Intelligence (AI) due to its success in investigating effective approaches to endow automatic agents with essential skills that make them able to solve modern life problems with as much autonomy as possible [20]. Among these problems, for example, autonomous systems, predictive systems, image recognition and digital games stand out [14]. In this way, ML allows for the conception of agents able to automatically learn to make decisions from experiences, instead of being previously programmed with built in knowledge to do this. In order to have such skill, the agents must count on powerful tools and adequate learning techniques that make them able to perceive the environment and to reprogram their own decision-making modules, so as to be capable of choosing actions that are compatible with such perception.

In this context, Deep Learning (DL) based agents highlight [12]. DL is an approach in which the agents seize the world (or environment) through a hierarchy of concepts [5]. Particularly, the DL based Convolutional Neural Networks (CNNs) - one of the key elements of Computer Vision (CV) - present excellent level of performance in tasks involving visual data, especially in image classification tasks [11,11]. That is why the CNNs have being very successfully used as tools that endow the agents with the capacity of perceiving the environment in which they operate and of acting according to this perception.

CV allows abstracting a high level of understanding from low-level information commonly known as raw images or raw pixels. Due to its simplicity, low-level information has been widely used to represent the environment in autonomous agents (for example, the current screenshot in player agents). Further, it is very suitable to be used as environment representation tool at the input of CNN-based agents.

In terms of supervised learning, the CNNs can be trained from a set of labeled instances so as to be apt to identify implicit behavior patterns in unknown instances. Due to its high generalization capacity, the appropriate choice of the CNN architecture depends on the characteristics of the task in which it will perform. This architecture is composed of elements called network structure hyperparameters, such as the number of layers and the number of neurons in each layer.

Most of the state-of-the-art works involving the construction of remarkable CNNs, such as AlexNet [11] and ResNet [6], relies on manually designed architectures carried out by specialists who have excellent domain knowledge from both this type of neural network and the problem investigated. In the context of digital games, CNNs are popularly used as a decision-making module in player agents both in the scenario of Imitation Learning (IL) [8] and Reinforcement Learning (RL) [9]. In [4], two autonomous agents are implemented with the purpose to replicate human behavior in the confrontation mode of the FIFA game. These agents perceive the environment through raw images and their decision-making modules are composed of CNNs trained by IL algorithms. It is important to note that the CNNs architecture was derived from the one used in [7], which was manually designed for navigation tasks. As these navigation tasks have simpler environment characteristics than the confrontation mode, the results presented in [4] can be improved through the use of a more suitable architecture for this problem. However, designing a new CNN manually from scratch can be a costly task even for an expert. Thus, it would be interesting to automate this process to build a more appropriate and efficient architecture.

Several proposals for automating the designing process of a CNN architecture have been made in the literature, divided into two main categories: based on evolutionary algorithms [13,18,23,24] and based on RL [1,25]. Among the former approach, Convolutional Neural Network designed by Genetic Algorithm (CNN-GA) highlights [23]. This method aims to evolve a population of CNN architectures to find the best one for a given image data set in terms of accuracy. It is noteworthy here that CNN-GA was validated only in a static scenario (evaluated on data sets), presenting a state-of-art performance in benchmark image classification tasks.

Motivated by such facts, in this paper the authors extend the approaches proposed in [23] and [4] by investigating an evolutionary approach based on CNN-GA to design optimized architectures used as decision-making modules for intelligent agents in the complex learning scenario of FIFA agents.

Thus, the main contributions here are: 1) Implementing a new version of CNN-GA, named Minimum CNN-GA (MCNN-GA), which improves the former with a new fitness function that combines the number of parameters (size of weights vector) and the classification error of a CNN architecture. In this sense, the term "Minimum" is due to the fact of minimizing the number of parameters of a CNN; 2) Proposing a set of IL-based FIFA agents whose decision-making architecture is automatically designed by MCNN-GA exploring two distinct raw images environment representations (with and without color information). These agents were evaluated in terms of in-game score playing against FIFA's engine on confrontation mode. The results showed the great potential that MCNN-GA has in building appropriate CNN architectures significantly lighter than those used in the agents developed in [4] (manually designed architectures).

The next sections are structured as following: Sect. 2 resumes the background; Sects. 3 and 4 describes, respectively, the MCNN-GA method and the player agents proposed in this paper; Sect. 5 presents the experiments and results; finally, Sect. 6 shows the conclusions and the future works.

2 Background

In neural networks projects (covering CNNs), one of the most important tasks is the definition of their architectures (for example, hyperparameters such as the number of layers and the number of neurons in each one). Thus, due to the huge number of possible combinations (large search space), the task of defining a good architecture manually is very costly. In this sense, recent works automate this process to build optimized architectures for a specific task, highlighting [23]. The mentioned work proposes the CNN-GA method that evolves a population of CNN architectures through genetic algorithm aiming to find the one with the best accuracy in an image classification task. It is noteworthy that CNN-GA obtained significant results in well-known benchmark image data sets, such as CIFAR-10 and CIFAR-100 [10]. As shown in Table 1, the present paper propose a new version named MCNN-GA that extends CNN-GA concerning the individual encoding and evaluation, limiting the size of an individual's architecture (in terms of blocks, which are composed of one or more layers) and developing a new fitness function that combines the classification error and the number of parameters (size of weights vector) of the CNN architecture. Besides, this new version uses the Stratified *K-fold cross-validation* [19] training strategy to better explore the distribution of the data increasing the probability of finding a higher quality decision-making CNN. Such a training strategy was adopted since it allows small data sets to be more appropriately exploited by the evolutionary method than Holdout [21] (originally used in CNN-GA). All the characteristics exposed in the referred table will be detailed in Sect. 3.

Table 1. Parallel between CNN-GA [23] and MCNN-GA

Approach	CNN-GA [23]	MCNN-GA
Individual encoding	Unlimited size	Limited size (maximum 10 blocks)
Blocks type	*Skip* and *Pooling*	*Skip* and *Pooling*
Fitness function	Maximize Accuracy	Minimize classification error and number of parameters
Training strategy	Holdout	Stratified *K-fold cross-validation*
Parent selection	Binary tournament	Binary tournament
Crossover	One-point	One-point
Mutation	Add, remove or modify a block	Add, remove or modify a block
Environmental selection	Binary tournament + Elitism	Binary tournament + Elitism
Experiments application	Image classification data sets	Dynamic Scenario of FIFA game

CNNs are commonly used in the context of player agents in both the IL and RL approaches, for example in [2,7,17]. In particular, [4] presents an investigation of two IL algorithms, named Direct Imitation (DI) and Deep Active Imitation (DAI), in confrontation mode of FIFA game. IL consists of a supervised DL training strategy where the agents learn by replicating human experts' behavior from demonstrations [8]. DI is one of the most trivial methods from IL approach aiming to produce a policy that maps observations to actions directly [15]. DAI extends DI through the use of active learning to improve the generalization of the policy obtained with the latter method [7]. So, [4] implemented two

IL-based agents (built according to DI and DAI) whose decision-making modules are composed of CNNs that process the current frame of the game. Both agents were evaluated according to the environment representation based on raw image without color information. As a result, the DAI-based agent obtained the best performance performing reasonably well compared to the human player used as a supervisor. It is important to note that the CNNs were retrieved from [7], which were manually designed for navigation tasks. In this way, the presented paper extends [4] improving the decision-making module with optimized CNNs architectures and investigating the behavior of IL-based agents with colored raw images environment representation, as shown in Table 2.

Table 2. Parallel between [4] and the present approach

Approach	[4]	Present paper
Learning strategy	IL based methods	IL based methods
Environment Representations	Grayscaled raw images	IL Grayscaled and colored raw images
CNNs design	Manually designed retrieved from [7]	Automatically designed through MCNN-GA
Experiments application	Confrontation mode	Confrontation mode

The FIFA game scenario named confrontation mode explores the situation involving two soccer players in yellow uniform controlled by an agent (human or autonomous) against two players in orange uniform manipulated by the game itself (FIFA's engine), as illustrated in Fig. 1.

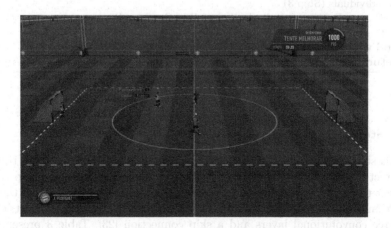

Fig. 1. Confrontation mode (Color figure online)

In this context, the agent's performance is measured by the in-game score. If the agent scores a goal, the in-game score increases by 1000 points. If FIFA's engine scores a goal, the in-game score decreases by 100 points. Thus, the objective is to achieve the highest possible in-game score within 45 s (duration of a game). It is necessary to emphasize that this scenario provides an excellent case study due to its properties (dynamic, multi-agent, unknown and stochastic [20]), which are major challenges in the learning process of a player agent.

3   MCNN-GA

This section presents the first contribution of the present paper: the implementation of MCNN-GA and its differences from GA-CNN. Algorithm 1 displays the structure of both methods. Basically, they try to find the best CNN architecture to classify an image classification data set through an evolutionary process consisting of the following steps: 1) Initialization of an arbitrary population; 2) Evaluation of individuals' fitness; 3) Generation of the offspring (new individuals); 4) Environmental selection.

Algorithm 1. Structure of CNN-GA and MCNN-GA

Input: size of the population N, number of generations T, the image data set for classification task.
   Output: best CNN architecture.

1: $P_0 \leftarrow$ Initialize a population with size N (Step 1)
2: $t \leftarrow 0$
3: **while** $t < T$ **do**
4:    Evaluate the fitness of each individual in P_t (Step 2)
5:    $Q_t \leftarrow$ Offspring generated by *Crossover* operation among the selected parent individuals (Step 3)
6:    $P_{t+1} \leftarrow$ Environmental selection of $P_t \bigcup Q_t$ (Step 4)
7:    $t \leftarrow t + 1$
8: **end while**
9: **return** Individual with the best fitness in P_t.

3.1   Step 1: Initializing Population

In this step, a population of predefined size individuals corresponding to CNN architectures is randomly initialized. Their architectures are made up of two distinct types of blocks: Skip and Pooling. The Skip block, inspired by the construction of the residual blocks (initially proposed in ResNet [6]), consists of two or three convolutional layers and a skip connection [23]. Table 3 presents the configuration used in the implementation of the Skip block. Note that the *Filter Size*, *Stride* and *Padding* hyperparameters are represented by constant values (retrieved from CNN-GA), while *Number of Filters* is the only one taken into account in the evolutionary process (F1 and F2). In the case of the present work, it can assume the following values: 8, 16, 32 or 64.

Table 3. Skip block hyperparameters configuration

	Number of filters	Filter size	*Stride*	*Padding*
conv1	F1	3×3	1×1	*same*
conv2	F2	3×3	1×1	*same*
conv3	F2	1×1	1×1	*same*

The *Pooling* block used in MCNN-GA has the *Filter Size* and *Stride* hyperparameters values equal to 2×2 (retrieved from CNN-GA). Thus, the only hyperparameter taken into account in the encoding process is the *pooling* operation type (P1). Such an operation can take one of the following values: *max pooling* or *average pooling*.

It is important to highlight that a great difference between MCNN-GA and CNN-GA is that, in the former, an individual is composed of a maximum of ten blocks, while there is no limit regarding the size of the architecture in the latter. The main reasons that justify this choice made by the present paper are: 1) the best architectures and, consequently, the best solutions found by CNN-GA in [23] had a maximum of nine blocks in their architectures; 2) ease of implementing a fitness function based on the number of parameters, since it is possible to compute the largest possible number of parameters a CNN architecture can assume based on the Skip and Pooling blocks configurations aforementioned; 3) execution time is a great weakness of evolutionary methods. Thus, limiting the maximum size of an individual is a way to speed up the execution of MCNN-GA.

3.2   Step 2: Evaluating Individuals

Algorithm 2 presents the process of evaluating individuals in a population. For each individual, two phases are performed: 1) CNN architecture decoding process (line 2); 2) Evaluation of its performance (lines 3 to 18).

In the first phase, the CNN architecture of the individual p is decoded (line 2). Throughout this process, the batch normalization operation followed by the *ReLu* activation function is added to the output of each convolutional layer in the *Skip* blocks, as done in CNN-GA. After this whole process, a fully connected layer - representing the output layer implemented with the *Softmax* activation function - is added to the end of such an architecture. The number of classes is determined by the problem addressed.

The second phase begins with the creation of the *partitions* used for training and testing the neural network using the stratified *k-fold cross-validation* method with K equal to five (line 3). In this way, it is guaranteed that the

---

Algorithm 2. Individuals evaluation

Input: Current Population P_t, image data set, MAX_PARAM.
Output: Population P_t evaluated.

1: **for** *each individual p in P_t* **do**
2:      $p.CNN \leftarrow$ CNN architecture decoding in p
3:      *partitions* $\leftarrow$ Stratified *k-fold cross-validation* $(K = 5)$
4:      *accuracy_sum* $\leftarrow 0$
5:      **for** *each iteration i in partitions* **do**
6:          Initializing $p.CNN$ with random parameters (weights vector)
7:          Training $p.CNN$ with training data
8:          $accuracy_i \leftarrow$ evaluating $p.CNN$ with test data
9:          *accuracy_sum* $\leftarrow$ *accuracy_sum* $+$ $accuracy_i$
10:     **end for**
11:     *dimension* $\leftarrow$ total number of parameters in $p.CNN$
12:     *dimension_fitness* $\leftarrow$ *dimension* / MAX_PARAM
13:     *accuracy_average* $\leftarrow$ *accuracy_sum* / K
14:     *error_fitness* $\leftarrow 1 -$ *accuracy_average*
15:     *p.fitness* $\leftarrow$ *dimension_fitness* $+$ *error_fitness*
16: **end for**
17: **return** P_t

---

architecture will be evaluated once with all the examples from the data set provided. Besides, the variable *accuracy_sum* (representing the sum of the accuracy obtained for each combination of *partitions*) is initialized equal to zero (line 4). For each iteration performed on *partitions*, the architecture is initialized with random parameters (line 6), trained with the training partitions for ten epochs and *Categorical Cross Entropy* loss function (line 7) and evaluated with the test partition (line 8). This evaluation generates a measure called $accuracy_i$ (accuracy referring to iteration i), which is added to *accuracy_sum* (line 11). Therefore, the individual's fitness is computed (lines 13 to 17).

The individual's fitness (line 18) corresponds to a linear combination between two terms defined in Eq. 1. TA_1 represents the fitness related to number of parameters of the architecture. TA_2 represents the performance of the architecture based on the classification error. Thus, F (fitness function) is a minimization problem in relation to TA_1 and TA_2, which are described below.

$$F = min(TA_1) + min(TA_2) = min(TA_1 + TA_2) \qquad (1)$$

– TA_1: the total number of parameters that make up the individual's architecture is set to *dimension* (line 13). Then, TA_1 (*dimension_fitness*) is defined based on the division of *dimension* by MAX_PARAM (line 14). MAX_PARAM is a constant value that represents the maximum number of parameters that a CNN architecture can have in the worst case (ten *Skip* Blocks with F_1 and F_2 equal to 64, corresponding to 3.474.500 parameters). Thus, TA_1 is better the smaller the *dimension* of an architecture.

– TA_2: the average accuracy obtained in *partitions* is calculated from *accuracy_sum* and K (number of *partitions*) and set to *accuracy_average* (line 15). Then, TA_2 (*fitness_error*) is defined based on the calculation of the average classification error (1 - *accuracy_average*) (line 16). Thus, TA_2 is better the smaller the error (a small error means the architecture has an excellent ability to deal with the data of the set provided).

3.3 Step 3: Generating the Offspring

The process of generating individual children consists of two phases: 1) parent individuals selection and execution of the *crossover* operation; 2) mutation operation executed on generated children individuals.

The first phase is performed using binary tournament [16] to select parent individuals and applying the one-point crossover operation [22] over each pair of selected parents to generate the offspring. The second phase can mutate the children through the following operations (as described in [23]): 1) Add a *Skip* block randomly; 2) Add a *Pooling* block randomly; 3) Remove a block randomly; 4) Modify a block configuration (hyperparameters) randomly. Finally, it is noteworthy here that the children are also limited to the number of ten blocks (excluding the output layer) in order to maintain the algorithm consistency.

3.4 Step 4: Environmental Selection

This process is carried out by combining the binary tournament and an elitist strategy. The best individual is automatically placed in the next population and the others are selected through binary tournament.

4 Implementation of the IL-Based Agents for FIFA

This section presents the second contribution of this paper: development of IL-based agents through distinct raw images environment representations and distinct ways of defining the CNN architecture. Such agents consist on two CNNs - named as *Movement CNN* and *Control CNN* - as defined in [4]. The former is responsible for the movement actions: *Go Left*, *Go Down*, *Go Right* and *Go Up* (named *movement actions*). and the latter controls the interactions involving the agent, the ball and the opponents: *Defending*, *Passing*, *Kicking* and *No Action* (named *control actions*). Table 4 summarizes the main characteristics of the agents proposed here.

Table 4. IL-based agents implemented for confrontation mode

Agent	Environment representation	CNN design
DI-Agent	Variable	Variable
DAI-Agent	Variable	Automatic

These agents present the following variations in relation to the environment representation and the CNNs designing process:

- *DI-Agent* with **Grayscaled Images** and **Manual**.
- *DI-Agent* with **Colored Images** and **Manual**.
- *DI-Agent* with **Grayscaled Images** and **Automatic**.
- *DI-Agent* with **Colored Images** and **Automatic**.
- *DAI-Agent* with **Grayscaled Images** and **Automatic**.
- *DAI-Agent* with **Colored Images** and **Automatic**.

The environment representation corresponds to the current frame of size 120 × 90. **Grayscaled Images** represents raw images without color information (that is, the current frame is a 120 × 90 × 1 matrix) and **Colored Images** represents raw images with color information (that is, the current frame is a 120 × 90 × 3 matrix). The CNN architecture can be manually designed (**Manual**), retrieved by the one used in [4] (composed of three Convolutional layer and Pooling layer pairs and two fully connected layers), or automatically designed through the M-CCN-GA method (**Automatic**).

It should be noted that *DI-Agent* takes into account demonstrations related to 100 games, producing a set of *movement actions* and *control actions* composed of 18000 and 3000 examples respectively. Both sets are balanced in the number of classes. The demonstrations were retrieved from real games in which the human agent to be mimicked faced the FIFA's engine and achieved an average in-game score of approximately 3000 (such supervisor is represented by one of the authors who has an advanced skill level of play in the confrontation mode). Besides, this agent varies in relation to environment representations and also in relation to the design of CNNs architectures. The *DAI-Agent* varies only with respect to state representations, being built from the *DI-Agent* versions that use MCNN-GA for automatic definition of the decision-making CNN architectures.

Considering the versions built from the CNNs architecture automatically designed, the MCNN-GA execution parameters are defined as following: population size equal to 10; number of generations equal to 10; *crossover* probability equal to 0.8 and mutation probability equal to 0.2.

Tables 5 and 6 present, respectively, the best architecture found by MCNN-GA for the *Movement CNN* and *Control CNN* decision-making modules with **Grayscaled Images**. The former consists of five blocks (11 layers), totaling 120.380 parameters. The latter consists of five blocks (9 layers), totaling only 20.348 parameters.

Similarly, Tables 7 and 8 present, respectively, the best architecture found by MCNN-GA for the *Movement CNN* and *Control CNN* decision-making modules with **Colored Images**. The former consists of ten blocks (22 layers), totaling 155.372 parameters. The latter consists of seven blocks (15 layers), totaling only 33.764 parameters.

It should be noted that, after finding the CNNs through the automatic method used here, all of them were trained with all the examples contained in their respective data sets. Finally, *DAI-Agent* variations were implemented using a 10% rate of active learning.

Table 5. Optimized *Movement CNN* for **Grayscaled Images**

Type	F1	F2	P1
Block 1 - *Skip*	64	8	–
Block 2 - *Pooling*	–	–	max pooling
Block 3 - *Skip*	16	8	–
Block 4 - *Pooling*	–	–	max pooling
Block 5 - *Skip*	4	32	–

Table 6. Optimized *Control CNN* for **Grayscaled Images**

Type	F1	F2	P1
Block 1 - *Skip*	16	8	–
Block 2 - *Pooling*	–	–	average pooling
Block 3 - *Pooling*	–	–	max pooling
Block 4 - *Skip*	32	16	–
Block 5 - *Pooling*	–	–	max pooling

Table 7. Optimized *Movement CNN* for **Colored Images**

Type	F1	F2	P1
Block 1 - *Skip*	8	64	–
Block 2 - *Pooling*	–	–	max pooling
Block 3 - *Skip*	32	16	–
Block 4 - *Skip*	32	16	–
Block 5 - *Skip*	64	16	–
Block 6 - *Skip*	64	32	–
Block 7 - *Pooling*	–	–	max pooling
Block 8 - *Pooling*	–	–	max pooling
Block 9 - *Skip*	32	16	–
Block 10 - *Skip*	8	64	–

Table 8. Optimized *Control CNN* for **Colored Images**

Type	F1	F2	P1
Block 1 - *Skip*	8	16	–
Block 2 - *Pooling*	–	–	average pooling
Block 3 - *Skip*	16	8	–
Block 4 - *Skip*	16	64	–
Block 5 - *Pooling*	–	–	max pooling
Block 6 - *Skip*	16	8	–
Block 7 - *Pooling*	–	–	max pooling

5 Experiments and Results

The experiments conducted here have as main objective validating MCNN-GA through *DI-Agent* and *DAI-Agent* in confrontation mode. Thus, two comparative analyses are performed: 1) evaluation of the impact that the architectures generated by MCNN-GA have on the performance of players' agents (Experiment 1); 2) evaluation of the environment representations described in the Sect. 4 through DAI (Experiment 2). These comparisons are essentially performed according to the in-game score. The experiments were executed in an architecture composed of a machine with a GPU Nvidia GeForce 940MX and 8 GB RAM.

5.1 Experiment 1: Evaluating MCNN-GA

The objective here is to evaluate the impact caused by MCNN-GA in the design of the decision-making modules for player agents. In this sense, a comparative analysis was made between *DI-Agent* variations that use the CNN architecture proposed in [7] (**Manual**) and those that use the architectures generated by

MCNN-GA (**Automatic**) based on the two representations studied here. The evaluation of the in-game score was performed through a tournament composed of 50 games between the variations and the FIFA's engine. It is emphasized that such an engine used to act against all variations had the same level of difficulty in the way as not to benefit any variation in particular.

Table 9 shows, respectively, the total mean (Mean) and the standard deviation (Std. Deviation) of the *DI-Agent* variations in the tournament made in terms of in-game score. The results indicate a very close performance between the manual and automatically designed variations. An *independent-samples t-test* [3] with a *significance level* ($\alpha = 0,05$) was conducted to validate the performance comparison using the software package SPSS.

The *t-test* is an inferential statistical test that determines whether there is a statistically significant difference between the means in two unrelated groups. In the case of this work, this method verifies if there is a superiority in terms of performance between the different variations compared. In this way, this method adopts a strategy of trying to prove the null hypothesis (here referred to as H_0). The null hypothesis is a statement of no effect or no difference and it is expected to be rejected by the experimenter. In this manner, the following H_0 was created: the **Manual** variations hold the same level of performance in terms of the in-game score as the **Automatic** variations.

Table 9. Results of *DI-Agent* variations in terms of the in-game score

Variation	Mean	Std. Deviation
Grayscaled Images and **Manual**	1100,00	1161,456
Grayscaled Images and **Automatic**	1298,00	1042,855
Colored Images and **Manual**	1752,00	1030,007
Colored Images and **Automatic**	1710,00	1009,395

As Table 10 states, the negative *t-value* (t(98) = -0,897) indicates that the in-game score mean for *DI-Agent* with **Grayscaled Images** and **Automatic** is slightly greater than the mean for *DI-Agent* with **Grayscaled Images** and **Manual** with *degrees of freedom* equals to 98. Regarding the **Colored Images** representation, the positive *t-value* (t(98) = 0,206) indicates that the in-game score mean for *DI-Agent* **Manual** is slightly greater than the mean for *DI-Agent* **Automatic** with *degrees of freedom* equals to 98. Since *p-value* is greater than the *significance level* ($\alpha = 0,05$) for both comparisons, the null hypothesis cannot be rejected for any of them. Therefore, these results suggest that the variations produced through MCNN-GA (**Automatic**) have the same performance level as those implemented according to [7] (**Manual**).

Table 10. *T-test* applied to the results of *DI-Agent* variations

Comparison	t-value	degrees of freedom	p-value
DI-Agent with **Grayscaled Images** and **Manual** × *DI-Agent* with **Grayscaled Images** and **Automatic**	-,897	98	,372
DI-Agent with **Colored Images** and **Manual** × *DI-Agent* with **Colored Images** and **Automatic**	,206	98	,837

The results obtained in this experiment validate the MCNN-GA method as an auxiliary tool in CNNs designing process in a complex game scenario (confrontation mode). The variations produced by the referred method obtained very competitive results compared to those that use manually designed architectures. The authors believe that agents had a performance far from the human level due to the limitation of DI learning strategy concerning generalization. Although the variations built from MCNN-GA have not significantly improved performance in confrontation mode, it is interesting to note that they are much lighter in terms of the number of parameters compared to the CNN architecture extracted from [7]. In fact, the latter architecture (defined manually) consists of 3.181.400 parameters, while the largest CNN built by MCNN-GA has only 155.372 parameters (approximately 20 times less). This has a direct impact on the memory (space complexity) needed to store the CNNs in a machine. It is expected that the smaller the network, the faster its inference, which can be crucial in choosing good decisions in real-time situations.

5.2 Experiment 2: Evaluating the Environment Representations

The objective here is to evaluate the decision-making quality of *DAI-Agent* variations - implemented from their respective *DI-Agent* versions that were produced by MCNN-GA (*DI-Agent* with **Grayscaled Images** and **Automatic** and *DI-Agent* with **Colored Images** and **Automatic**) - in terms of environment representation. In this sense, each *DAI-Agent* variation was submitted to a tournament consisting of 50 games against the FIFA's engine.

Table 11 shows, respectively, the total mean (Mean) and the standard deviation (Std. Deviation) of *DAI-Agent* variations in terms of in-game score. The results indicate a superior performance of the variation that considers the **Colored Images** representation over **Grayscaled Images**. An *independent-samples t-test* with a *significance level* ($\alpha = 0,01$) was conducted to validate the performance comparison. In this sense, the following null hypothesis H_0 was created: both variations have the same level of performance.

Table 11. Results of *DAI-Agent* variations in terms of the in-game score

Variation	Mean	Std. Deviation
Grayscaled Images and **Automatic**	1622,00	746,226
Colored Images and **Automatic**	2172,00	795,405

As Table 12 states, the negative *t-value* (t(98) = −3,566) indicates that the in-game score mean for the *DAI-Agent* with **Colored Images** is significant greater than the mean for the *DAI-Agent* with **Grayscaled Images** with *degrees of freedom* equals to 98. Since *p-value* = 0, 001 is lower than the *significance level* (α = 0, 01), the null hypothesis can be rejected. These results suggest that raw images with color information have a greater capacity to generate relevant features for player agents' decision-making compared to raw images without color information with a confidence interval of 99% in the confrontation mode.

Table 12. *T-test* applied to the results of *DAI-Agent*

Comparison	*t-value*	*degrees of freedom*	*p-value*
Grayscaled Images × **Colored Images**	−3,566	98	,001

The results obtained in this experiment allowed us to verify that the environment representation based on raw images with color information provides a better quality of perception of the current frame of the game and, consequently, provides better performance to a player agent in the confrontation mode than raw images without color information. It is necessary to emphasize that *DAI-Agent* with **Grayscaled Images** achieved similar performance to the agent developed in [4]. Thus, although a greater number of demonstrations and a higher active learning rate were used compared to the mentioned work, the results were not improved. Concerning *DAI-Agent* with **Colored Images**, its performance has improved considerably when using color information from images. In practical terms, it is the best agent generated in this paper and it demonstrates a performance very close to the human level in several executions, however, it does not have a consistent performance due to the stochastic character of the environment. These facts corroborate the limitations inherent to the representations used here (specially **Grayscaled Images**), which do not consider the sequence information of the environment in the decision-making process (only the current frame is processed by the agents' decision-making module).

6   Conclusion and Future Works

This paper investigated MCNN-GA as a tool to design optimized CNN architectures aiming to improve the decision-making of autonomous player agents.

The method showed great potential in generating light architectures with good quality in complex environments. However, the great weakness of MCNN-AG is related to the individuals evaluation based on the *k-fold cross validation* technique, which corresponds to a very expensive process in terms of computational time. The results also show the superiority in terms of performance of the agents implemented according to raw images with color information as environment perception. The authors believe that the results obtained here can be improved by the use of recurrent neural networks, which take into account in previously seen game frames in addition to the current one. The source code of this work is available at https://github.com/matheusprandini/MCNN-GA.

References

1. Baker, B., Gupta, O., Naik, N., Raskar, R.: Designing neural network architectures using reinforcement learning. In: 5th International Conference on Learning Representations (2017)
2. Buche, C., Even, C., Soler, J.: Autonomous virtual player in a video game imitating human players: the orion framework. In: 2018 International Conference on Cyberworlds, pp. 108–113 (2018)
3. Coleman, A.M.: A Dictionary of Psychology, 3rd edn. Oxford University Press, Oxford (2009)
4. Faria, M.P.P., Julia, R.M.S., Tomaz, L.B.P.: Evaluating the performance of the deep active imitation learning algorithm in the dynamic environment of FIFA player agents. In: 18th IEEE International Conference on Machine Learning and Applications (2019)
5. Goodfellow, I., Bengio, Y., Courville, A.: Deep Learning. MIT Press, New York (2016). http://www.deeplearningbook.org
6. He, K., Zhang, X., Ren, S., Sun, J.: Deep residual learning for image recognition. In: The IEEE Conference on Computer Vision and Pattern Recognition (2016)
7. Hussein, A., Elyan, E., Gaber, M.M., Jayne, C.: Deep imitation learning for 3D navigation tasks. Neural Comput. Appl. **29**, 389–404 (2018)
8. Hussein, A., Gaber, M.M., Elyan, E., Jayne, C.: Imitation learning: a survey of learning methods. ACM Comput. Surv. **50**, 21:1–21:35 (2017)
9. Justesen, N., Bontrager, P., Togelius, J., Risi, S.: Deep learning for video game playing. IEEE Trans. Games **12**, 1–20 (2020)
10. Krizhevsky, A., Nair, V., Hinton, G.: CIFAR-10 and CIFAR-100 (Canadian institute for advanced research). http://www.cs.toronto.edu/~kriz/cifar.html
11. Krizhevsky, A., Sutskever, I., Hinton, G.E.: ImageNet classification with deep convolutional neural networks. In: Pereira, F., Burges, C.J.C., Bottou, L., Weinberger, K.Q. (eds.) Advances in Neural Information Processing Systems 25, pp. 1097–1105. Curran Associates, Inc. (2012)
12. LeCun, Y., Bengio, Y., Hinton, G.E.: Deep learning. Nature **521**, 436–444 (2015)
13. Liu, H., Simonyan, K., Vinyals, O., Fernando, C., Kavukcuoglu, K.: Hierarchical representations for efficient architecture search (2018)
14. Liu, W., Wang, Z., Liu, X., Zeng, N., Liu, Y., Alsaadi, F.E.: A survey of deep neural network architectures and their applications. Neurocomputing **234**, 11–26 (2017)

15. Liu, Y., Gupta, A., Abbeel, P., Levine, S.: Imitation from observation: learning to imitate behaviors from raw video via context translation. In: 2018 IEEE International Conference on Robotics and Automation, pp. 1118–1125 (2018)
16. Miller, B.L., Miller, B.L., Goldberg, D.E., Goldberg, D.E.: Genetic algorithms, tournament selection, and the effects of noise. Complex Syst. **9**, 193–212 (1995)
17. Mnih, V., Kavukcuoglu, K., Silver, D., Rusu, A.A.: Human-level control through deep reinforcement learning. Nature **518**, 529–533 (2015)
18. Real, E., et al.: Large-scale evolution of image classifiers. In: Proceedings of the 34th International Conference on Machine Learning, Vol. 70, pp. 2902–2911 (2017)
19. Refaeilzadeh, P., Tang, L., Liu, H.: Cross-validation. In: Liu, L., Özsu, M.T. (eds.) Encyclopedia of Database Systems, pp. 532–538. Springer, Boston (2009). https://doi.org/10.1007/978-0-387-39940-9
20. Russell, S., Norvig, P.: Artificial Intelligence: A Modern Approach, 3rd edn. Prentice Hall Press, Upper Saddle River (2009)
21. Sammut, C., Webb, G.I.: Holdout evaluation. In: Encyclopedia of Machine Learning, pp. 506–507. Springer, Boston (2010). https://doi.org/10.1007/978-0-387-30164-8_369
22. Srinivas, M., Patnaik, L.M.: Genetic algorithms: a survey. Computer **27**, 17–26 (1994)
23. Sun, Y., Xue, B., Zhang, M., Yen, G.G., Lv, J.: Automatically designing CNN architectures using the genetic algorithm for image classification. IEEE Trans. Cybern. **50**(9), 1–15 (2020)
24. Xie, L., Yuille, A.: Genetic CNN. In: 2017 IEEE International Conference on Computer Vision, pp. 1388–1397 (2017)
25. Zoph, B., Le, Q.V.: Neural architecture search with reinforcement learning. In: International Conference on Learning Representations (2017)

Text Mining and Natural Language Processing

Authorship Attribution of Brazilian Literary Texts Through Machine Learning Techniques

Bianca da Rocha Bartolomei[(⊠)] and Isabela Neves Drummond

Institute of Mathematics and Computing, Universidade Federal de Itajubá (UNIFEI),
Itajubá, Brazil
{biancabartolomei,isadrummond}@unifei.edu.br

Abstract. Authorship attribution is the process of identifying the author of a particular document. This task has been performed by experts in the field. However, with the advancement of natural language processing tools and machine learning techniques, this activity has also been performed by computer systems. Authorship attribution has applicability from the detection of plagiarism and copyright to the resolution of forensic problems. There are several works on this subject in the English idiom, however those that consider texts in Portuguese are few. Therefore, this paper aims to study authorship attribution of texts of Brazilian literature. We carried out our experiments using Naïve Bayes and Random Forests methods, and for the feature extraction we considered Term Frequency - Inverse Document Frequency and Part of Speech techniques. The results showed that the Random Forests using as input the textual features extracted by Part of Speech presented the best cross-validation accuracy, although not the best runtime.

Keywords: Authorship attribution · Machine learning · Text mining

1 Introduction

The process of identifying who is the author of a particular document is called authorship attribution. The first studies involving this activity took place in the 19th century with the analysis of Shakespeare's plays [1]. The task was performed by a human specialist in the field. With the advancement of natural language processing tools and also many different machine learning approaches, this activity has been automated. Automatic models are able to extract features known as author's invariant, which are unconscious to those who write, but can define the authorship from a text.

The applications in this field are quite wide and affect several areas, starting with literary study, with the detection of plagiarism and copyright; criminal and civil laws; investigation of cyber crimes and computer forensics [2].

There are countless works about authorship attribution of texts, whether literary or not, however the amount of studies involving the Portuguese language

© Springer Nature Switzerland AG 2020
R. Cerri and R. C. Prati (Eds.): BRACIS 2020, LNAI 12319, pp. 389–402, 2020.
https://doi.org/10.1007/978-3-030-61377-8_27

is still quite small. In addition, when working with Portuguese idiom, some features should be considered like plural, gender, adverbs, diminutives and augments, nominal suffixes, verbal suffixes, accentuation, homographs and irregular verbs [3].

This paper describes the use of supervised machine learning techniques to assign authorship to literary texts. In this way, the authorship attribution is the process of identifying the most likely author of a given text, from a set of texts by several well-known authors. This is a classification problem, where each author represents a class. The attribution is made from the features belonging to the text to be classified [2]. Thus, the main point of the assignment are feature extraction and machine learning techniques.

This paper is organized as follows. In Sect. 2 we present some related works that were used as a basis for comparison. In Sect. 3 we show the steps performed in this work, which include the dataset, the summarizing method, the extracting features techniques, the features treatment, the models parametrization and the evaluation metrics used for the experiment. In Sect. 4 the results obtained are presented and compared. Finally, in Sect. 5, we present our conclusion and possible future works.

2   Related Works

There are several works on authorship attribution, which explore different aspects of the task, varying, for example, the way the features are extracted and the techniques used.

One of the main techniques is machine learning, which is divided into supervised and unsupervised models. The supervised models aim to forecast some value, which can be discrete or continuous. Conversely, unsupervised methods intend to group values in order to describe them [4]. Supervised approaches are the most widespread as can be seen in studies from Altheneyan and Menai [2], Mekala, Tippireddy and Bulusu [5] and Tamboli and Prasad [7]. However, there is still a small amount of studies on authorship attribution of texts in Portuguese. Therefore, it would be easier to compare results for the task in Brazilian texts with those present in the literature, if this work uses the most widespread methods, that are supervised machine learning techniques.

Considering classical machine learning techniques, we can find an amount of techniques employed for supervised tasks in the literature, including probabilistic models and decision trees.

The Naïve Bayes technique is a probabilistic method based on Bayesian Theorem. The Bayesian classifier considers a set of evidences, or features, X, and a set of hypotheses, or classes C, on which the probability of a class in the set C can be determined, given the set of X evidence [8]. The relation is represented mathematically by the Eq. 1.

$$P(C_i|X) = \frac{P(X|C_i) \times P(C_i)}{P(X)} \qquad (1)$$

This model was used by Altheneyan and Menai [2] for authorship attribution of Arabic texts. The authors compared the different implementations of the technique.

As for decision trees, one of the main techniques is Random Forests. This model can be defined as a set of trees, in which each one emerged from a random parameter. And each tree is worth a unit of vote in the most popular entry class x [9]. Random Forests were used for the authorship assignment task by Tamboli and Prasad [7], in which the model was compared with others, including Naïve Bayes. The same happens in the study by Mekala, Tippireddyand and Bulusu [5].

One of the most important steps in authorship attribution is the choice of techniques for extracting features, once there are several methods for this task. Altheneyan and Menai [2] considered number of words, frequency of elongations, among others. Mekala, Tippireddyand and Bulusu employed the Term Frequency - Inverse Document Frequency (TF-IDF) [5], which measures the importance of words in a given document considering the set of documents [6]. This technique consists of two parts: the first is TF, which is the number of times that a term appears in a document divided by the number of terms in the same document; and the second part is IDF, which is the logarithm of the division of the number of documents by the number of documents in which a given term appeared. Finally, these two parts are multiplied. When the final value of TF-IDF is high, it means that the evaluated word is important for the document and unusual in the set of documents [10]. In work by Tamboli and Prasad [7], it was used the Part of Speech (PoS) approach. What a PoS does is to label every word in a text as a morphological category, such as an adjective, pronoun or verb. This technique can be implemented in a supervised way or not. If supervised, then it is necessary to have a previously labeled *corpus*, on which training will be conducted to learn grammatical rules. One of the ways to implement a supervised PoS is stochastically, using the n-grams technique, which means, given any n, the label for a word is chosen from the probability calculation of the label to occur within the previous n labels [11].

Another aspect to be considered is the size of the documents analyzed, because it is not always necessary to use the entire document to identify the author, as shown by Dang [12]. In addition, the use of large amounts of text is limited by the experimental environment used. One of the solutions used is summarization. Summarization is responsible for reducing a text using some computational tool, in order to create a summary, in which the most important parts of the original document are concentrated [12]. One of the possible implementations is the TextRank algorithm, which is a ranking algorithm based on a graph that assigns importance to a vertex in a graph, using global information recursively computed from the entire graph [13].

Finally, it can be seen that in the studies [2,5,7], the main metric for evaluating the models is the model accuracy based on cross-validation.

3   Methodology

This section describes steps and artifacts for doing this work, represented by Fig. 1. The first activity was the selection of a Brazilian literary texts dataset. Subsequently, the process of summarizing the texts followed. With the dataset treated, the next step was the feature extraction, followed by the dataset balancing phase. Finally, the selected classification models were parameterized and the evaluation metrics were chosen, to proceed the experiments.

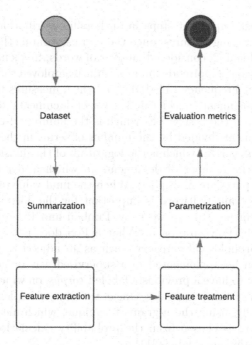

Fig. 1. Process flowchart

3.1   Dataset

The dataset used in the experiments proposed in this work is called Brazilian Portuguese Literature Corpus [14]. It was made available on the Kaggle platform, but all texts were taken from the Public Domain of the Brazilian Federal Government, and they are dated from 1840 to 1904. Altogether, there are 81 different texts by different authors, namely: Adolfo Caminha, Aluísio Azevedo, Bernardo Guimarães, Joaquim Manuel de Macedo, José de Alencar, Machado de Assis and Manuel Antônio de Almeida. The distribution of the number of texts by author can be seen in Fig. 2. The texts are in .*txt* extension.

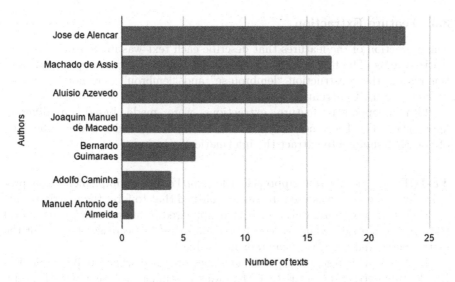

Fig. 2. Number of texts per author.

3.2 Summarization

One of the first steps to carry out this work was to summarize the literary texts, in order to minimize the computational cost of the feature extraction and machine learning models, thus reducing the runtime. Here we employed the TextRank graph-based algorithm [13]. Each vertex is a phrase of the text to be summarized. The graph is based on the idea of votes. So, when one vertex connects to another, a vote is computed for that vertex. In this way, the importance of a vertex is given by the number of vertex votes. The importance of a vertex also implies the weight of its vote. Thus, the main idea of this stage is to synthesize the most relevant sentences in each text [12]. For this, the procedure was divided into two parts, the pre-processing of the texts and the summarization itself.

When analyzing the texts individually, it was noticed the presence, in some, of headings with information about the sources that made them available. In addition, some of them contained page numbers and external links to the content of the text. Therefore, the first step was to remove these unwanted items using regular expressions. Since what matters for the authorship attribution is only the content of each text. With the texts ready to be summarized, the next step was to extract the sentences vectors for the sentences. The sentence vectors generated were placed in a similarity matrix, considering the cosine similarity, which is calculated from the normalized two-sentence scalar product [15]. Finally, the sentences were ranked according to their importance in the text and the ten most relevant sentences in each text were selected. The text Memórias Póstumas de Brás Cubas (Epitaph of a Small Winner) decreased in size from 352 KB to 4 KB, for example.

3.3 Feature Extraction

The extraction of the features that describe each text was made from the summarized texts. The first step was to separate verbs with enclitic pronouns. For this reason, the construction "lembrou-se" and "lembrou" were not considered different words. Punctuation was also removed.

Two independent feature extractions were made based on different approaches. The first one used TF-IDF, while the second relied on the Part of Speech technique to extract the information.

TF-IDF. In the TF-IDF approach, the *TfidfVectorizer* implementation provided by *scikit-learn* was used. It was established that to form the matrix resulting from this procedure, only words that appeared in at least 80% of the total texts in the dataset would be considered, in order to decrease the size of the matrix entry and avoid very small or null values.

Table 1 is a representation of some values obtained when applying the TF-IDF feature extraction to the text Memórias Póstumas de Brás Cubas(Epitaph of a Small Winner). The word "love" has a 0.00 relevance, on a scale ranging from 0 to 1, in the analyzed text. On the other hand, the word "house" has a representativeness of 0.19, being, therefore, more important for the text than the word "love". Also, the runtime of the algorithm was evaluated so that all texts in the dataset had their features extracted. The technique took 0.3755 s for the entire process to be completed.

Table 1. Memóórias Póstumas de Brás Cubas - TF-IDF feature extraction

	Amor (Love)	Casa (House)	Vida (Life)	⋯
Memórias Póstumas de Brás Cubas (Epitaph of a Small Winner)	0.0	0.19	0.02	⋯

PoS. For the PoS approach, the first step was to create a Bigram label with the *NLTK* library and train it with two datasets provided by the same library. The first dataset is Floresta, which consists of texts already labeled, and the second one is *MacMorpho*, composed by several journalistic texts extracted from *Folha de São Paulo* 1994.

The created tagger was used on the basis of Brazilian literary texts, evaluating each text and classifying each of its words morphologically, analyzing its syntactic structure. After the entire labeling process, the items of each morphological type were counted in each text. This is relevant, because one author may have as a characteristic of style a greater use of adjectives and adverbs, while another one uses a more direct style. Table 2 shows some of the morphological types and the number of times they appeared in the summarized text Memórias Póstumas de Brás Cubas (Epitaph of a Small Winner). It can be seen that for a considerable amount of nouns, there is a little use of adjectives modifying them, which can be an important feature to distinguish Machado de Assis from other authors. It should be noted that the runtime for the PoS algorithm was 85.0542 s.

Table 2. Memórias Póstumas de Brás Cubas - PoS feature extraction

	Adjective	Adverb	Noun	⋯
Memórias Póstumas de Brás Cubas (Epitaph of a Small Winner)	44	55	215	⋯

3.4 Feature Treatment

The machine learning model inputs are the two matrices generated by the feature extraction process. However, as shown in Fig. 2, the dataset is unbalanced, which means, the class sizes are different. The class that represents the author José de Alencar has more than 20 texts, while the class of the author Manuel Antônio de Almeida has only one text. This causes an overfitting problem in the model, which would tend to classify the texts as belonging to the classes with the greatest number of texts.

To solve the problem described, it is necessary to balance the classes. For this, there are two different approaches: subsampling and oversampling. In the subsampling, what is done is to reduce the amount of samples from the most populous classes [16]. Thus, there is a large reduction in the amount of data, which is not feasible for this work, since the number of texts is only 81. In oversampling, the goal is to increase the amount of sample in the classes with the lowest amounts [17], which is more suitable to the context.

The first measure taken was to generate six more summaries for the class of the author Manuel Antônio de Almeida, since there was only one text. Then, the next 10 most relevant sentences of his text were taken. This was done to avoid that all the data synthesized for this author were from the same summary. The same procedure was done for Adolfo Caminha, who won two more summaries, and Bernardo Guimarães, who increased the amount of texts in his class by one sample.

Then, the *imbalanced-learn* library was used to perform the oversampling. The technique used was *RandomOverSampler*, a simple strategy that generates new samples randomly, replacing the current available samples [20]. After the process, all classes presented 23 texts.

3.5 Parameterization

The two machine learning models chosen, Random Forest and Naïve Bayes, were implemented from *scikit-learn*. Before the experiments were performed, the models need to be parameterized.

It was found that for Naïve Bayes the parameter that most varied the results was *var_smoothing*. It is used to give more control over variances calculation [15]. Its possible values are in the range $[0, 1]$. For the model that uses the input extracted via TF-IDF, the *var_smoothing* of 0.02 was established. For the one who uses PoS, 0.001.

For Random Forest, the parameters were the number of trees and the maximum tree depth. For input with TF-IDF, the number of trees was 180 and the maximum depth was 35. For PoS, the values were 200 and 10, respectively.

3.6 Evaluation Metrics

The accuracy based on cross-validation is a well known metric for classification models evaluation. In cross-validation, any value k is chosen, so the dataset C is divided into k subsets of approximate sizes. k tests are performed, in which the classifier is trained with $k - 1$ subsets, excluding the C_k set, since it is used for the test [21]. In this work, k was established as 5 and it was used the *cross_val_score* implementation provided by *scikit-learn* [15].

Another way to evaluate the models was through confusion matrices, in which the results obtained from a prediction were placed. The lines were the expected classes and the columns were the prediction classes. The main diagonal corresponds to correct classification instances in a test dataset. A confusion matrix is interesting because it allows us to identify in which classes the errors occur [15].

In addition, the runtime is a factor of choice for the most suitable algorithms.

4 Experiments

This section gathers the results obtained from the experiments carried out with the *Brazilian Portuguese Literature Corpus* dataset. In all, four tests were performed. The first test was done with Naïve Bayes, using the features extracted by TF-IDF as input. The second one used Naïve Bayes with the features extracted by PoS. The third test was done with features extracted by TF-IDF, using the Random Forest model. Finally, the last test was the Random Forest model applied to input with features extracted by PoS.

It is important to note that all tests were performed in the same enviroment, which was Google Colab. This platform is a free Jupyter notebook environment where you can run Python code without the need for special configurations. Most importantly, the code is not executed on the user's machine, but in the cloud. It has a RAM with a capacity of around 12.72 GB and a disk of 48.97 GB [22].

4.1 Naïve Bayes

The confusion matrix of the classification process using the Naïve Bayes model with TF-IDF input is showed in Table 3. We can observe that the classes of the authors Aluísio Azevedo, Joaquim Manuel de Macedo and Adolfo Caminha were well classified. While most of the texts in class Manuel Antônio de Almeida were classified incorrectly, belonging to Bernardo Guimarães class.

Table 3. Naïve Bayes-TF-IDF confusion matrix

Real label \ Predicted label	Machado de Assis	José de Alencar	Aluísio de Azevedo	Joaquim M. de Macedo	Manuel A. de Almeida	Adolfo Caminha	Bernado Guimarães
Machado de Assis	3	2	0	1	0	0	0
José de Alencar	0	1	1	0	0	0	0
Aluísio de Azevedo	0	0	8	1	0	0	0
Joaquim Manuel de Macedo	0	1	0	3	0	0	0
Manuel Antônio de Almeida	0	0	0	0	1	0	2
Adolfo Caminha	0	0	1	0	0	6	0
Bernado Guimarães	0	1	0	0	0	0	1

For the test with Naïve Bayes using the input extracted via PoS, it is observed, through the confusion matrix in Table 4, that the classes of Manuel Antônio de Almeida, Adolfo Caminha and Bernado Guimarães were the ones that obtained the best results. On the other hand, most of the incorrect classifications occurred with texts by Machado de Assis, which were classified as José de Alencar.

4.2 Random Forests

The confusion matrix, represented in Table 5, shows the classification process using the Random Forest model with TF-IDF input. In general, the classification was correct, this can be seen by its main diagonal. Most of its errors are concentrated in the class of the author José de Alencar. His texts were mostly classified as belonging to Machado de Assis class.

The Table 6 refers to the confusion matrix obtained by the Random Forest classifier using the input extracted via PoS. It can be seen by its main diagonal that the model had a good classification. The only class in which classification errors appeared is the one that represents the author Machado de Assis.

Table 4. Naïve Bayes-PoS confusion matrix

Real label \ Predicted label	Machado de Assis	José de Alencar	Aluísio de Azevedo	Joaquim M. de Macedo	Manuel A. de Almeida	Adolfo Caminha	Bernado Guimarães
Machado de Assis	**2**	4	0	2	0	0	0
José de Alencar	0	**5**	3	0	0	0	0
Aluísio de Azevedo	1	0	**2**	1	0	0	1
Joaquim M. de Macedo	0	1	1	**2**	0	0	0
Manuel A. de Almeida	0	0	0	0	**4**	0	0
Adolfo Caminha	0	0	0	0	0	**3**	0
Bernado Guimarães	0	0	0	0	0	0	**2**

Table 5. Random Forest-TF-IDF confusion matrix

Real label \ Predicted label	Machado de Assis	José de Alencar	Aluísio de Azevedo	Joaquim M. de Macedo	Manuel A. de Almeida	Adolfo Caminha	Bernado Guimarães
Machado de Assis	**5**	0	0	0	0	0	0
José de Alencar	2	**1**	1	1	0	0	0
Aluísio de Azevedo	0	0	**4**	0	0	2	0
Joaquim M. de Macedo	0	0	0	**4**	0	0	0
Manuel A. de Almeida	0	0	0	0	**4**	0	0
Adolfo Caminha	0	0	0	0	0	**3**	0
Bernardo Guimarães	0	0	0	0	0	0	**6**

Table 6. Random Forest-PoS confusion matrix

Real label \ Predicted label	Machado de Assis	José de Alencar	Aluísio de Azevedo	Joaquim M. de Macedo	Manuel A. de Almeida	Adolfo Caminha	Bernado Guimarães
Machado de Assis	**5**	1	0	1	0	0	0
José de Alencar	0	**2**	0	0	0	0	0
Aluísio de Azevedo	0	0	**4**	0	0	0	0
Joaquim M. de Macedo	0	0	0	**4**	0	0	0
Manuel A. de Almeida	0	0	0	0	**8**	0	0
Adolfo Caminha	0	0	0	0	0	**3**	0
Bernado Guimarães	0	0	0	0	0	0	**5**

4.3   Discussion

In this work we compare the obtained results between them, since we do not have results in the literature using the same dataset. We also evaluate some results available in the literature for the same model, even using other textual datasets.

Table 7 aggregates the results obtained with the cross-validation technique for all the four tests alongside with its standard deviation and its runtime to complete the process.

Table 7. Naïve Bayes-PoS cross-validation

	Cross-validation	runtime
Naïve Bayes - TF-IDF	72% (+/− 13%)	0.0135 s
Naïve Bayes - PoS	60% (+/− 15%)	0.0204 s
Random Forest - TF-IDF	82% (+/− 14%)	1.0644 s
Random Forest - PoS	88% (+/− 17%)	1.6176 s

The results with Naïve Bayes using TF-IDF was better as we can see in Table 7. The cross-validation for Naïve Bayes with TF-IDF was 72% with a standard deviation of 13% and with PoS was 60% with a standard deviation of 15%. Analyzing the confusion matrices, we can observe that the most correct classes are those that have a smaller number of texts and, therefore, their synthesized data were quite concise in terms of stylometry. As for Naïve Bayes with TF-IDF the highest accuracy was for the author Aluísio Azevedo, which texts were correctly classified. In addition, Naïve Bayes' runtime with TF-IDF was faster than that of Naïve Bayes with PoS.

When comparing the results obtained by the tests with the Random Forests, we notice that the model using PoS presented a better result in the classification of the inputs, even though it has a higher runtime than the Random Forests with TF-IDF. As it can be observed in Table 7, the Random Forest test with TF-IDF had a cross-validation of 82% with standard deviation of 14%. The cross-validation for the test with PoS was 88% with 17% of standard deviation. Another important observation using Random Forests, both with PoS and TF-IDF, is that most of the wrong classifications happened with the classes that represent the authors José de Alencar and Machado de Assis.

In the work of Altheneyan and Menai [2], the authors achieved an accuracy of 82.30% for the attribution of authorship of Arabic texts with Naïve Bayes, as can be seen in Table 8. The feature extraction used by the authors is the aspect that can justify the difference between the results obtained in this work and those in the literature. Their work considered features such as total number of characters, total number of words, frequency of elongations, total number of scores, among others.

The results obtained with the Random Forest experiments were compared with the work *A Novel Document Representation Approach for Authorship Attribution* by Mekala, Tippireddy and Bulusu [5]. The authors used several approaches to extract features, including TF-IDF. Considering TF-IDF extration, the Random Forests model had the worst result of 69.45%.

Table 8. Comparison between results obtained and results in the literature

Models	TF-IDF	POS	Literature	
Naïve Bayes	72%	60%	82.30%	[2]
Random Forest	82%	88%	65.45% – 91.82%	[5]

5   Conclusion

This work presents a study of applying machine learning models for authorship attribution of Brazilian literary texts. The Naïve Bayes and Random Forests models were chosen based on their use by other similar works in the literature. In addition, each of them has been tested using two different feature extraction methods, TF-IDF and PoS. One of the dataset problems was the uneven distribution of texts by author. In order to obtain a balanced dataset, the oversampling technique was used, once the dataset is originally composed by 81 literary texts.

Considering the runtime, the TF-IDF extraction technique was faster than PoS, what is due to the simplicity of the TF-IDF approach, which is based on measuring the frequency of terms in *corpus* when compared to the PoS approach.

PoS approach needs a previous training, only then to be able to classify the terms of the texts morphologically, considering also previous terms.

Moreover, it was possible to verify that for Naïve Bayes, the feature extraction by TF-IDF provide a prediction model more accurated, with the advantage of a faster runtime.

For the Random Forests, the relationship between runtime and accuracy was more complex. Since the experiments carried out with the PoS feature extracted had a higher cross-validation score, 88%, but with a high runtime. In addition, it is important to consider the time of the PoS extraction itself. Which means that Random Forests model with PoS has higher computational cost.

Evaluating the four tests, it can be concluded that the Random Forests were the most suitable to the dataset, presenting a result close to that found in the literature for texts in other languages. Regarding the time, it must be considered that the application in question, authorship attribution, does not need the fastest possible time. Thus, opting for a higher accuracy over a faster runtime, justifies the choice of Random Forests with PoS as the model with the best performance.

Finally, we can mention as possible future works the comparison of other forms of features extraction. Also, it is important to evaluate other machine learning models applied to the authorship attribution of texts in Portuguese. Deep learning models could be evaluated in terms of accuracy and runtime, evaluating which cases are better options than the classic machine learning models studied in this work. Thus, for future works, other Portuguese textual datasets could also be used that exceed the literary scope and had a larger number of samples.

References

1. Elmanarelbouanani, S., Kassou, I.: Authorship analysis studies: a survey. Int. J. Comput. Appl. 86, December 2013. https://doi.org/10.5120/15038-3384
2. Altheneyan, A.S., Menai, M.E.B.: Naïve bayes classifiers for authorship attribution of Arabic texts. J. King Saud Univ. Comput. Inf. Sci. 26, 473–484 (2014). https://doi.org/10.1016/j.jksuci.2014.06.006
3. Orengo, V.M., Huyck, C.R.: A stemming algorithm for the Portuguese language, pp. 186–193 (2001). https://doi.org/10.1109/SPIRE.2001.989755
4. Faceli, K., Lorena, A.C., Gama, J., de Carvalho, A.C.P.L.F.: Inteligência Artificial: Uma Abordagem de Aprendizado de Máquina, Rio de Janeiro, LTC - Livros Técnicos e Cinetíficos Ltda. (2011)
5. Mekala, S., Tippireddy, R.R., Bulusu, V.V.: A novel document representation approach for authorship attribution. Int. J. Intell. Eng. Syst. 11, 261–270 (2018). https://doi.org/10.22266/ijies2018.0630.28
6. da Silva, C.L., Petry, L.M., Freitas, V., Dorneles, C.: Min. J. Ground Exploratory Anal. Newspaper Art. (2019). https://doi.org/10.1109/BRACIS.2019.00023
7. Tamboli, M.S., Prasad, R.: A robust authorship attribution on big period. Int. J. Electr. Comput. Eng. (IJECE) 9, 3167–3174 (2019). https://doi.org/10.11591/ijece.v9i4.pp3167-3174
8. Russell, S., Norvig, P.: Artificial Intelligence: A Modern Approach, 3rd edn. Elsevier Editora Ltd. (2009)

9. Breiman, L.: Random forests. Mach. Learn. **45**, 5–32 (2001). https://doi.org/10.1023/A:1010933404324

10. Lott, B.: Survey of keyword extraction techniques (2012)

11. Mannem, P., et al.: Parts of speech tagging for Indian languages: a literature survey. Int. J. Comput. Appl. 34 (2011). https://doi.org/10.5120/4119-5993

12. Dang, S.: A review of text mining techniques associated with various application areas. Int. J. Sci. Res. **4**, 2461–2466 (2015)

13. Mihalcea, R., Tarau, P.: TextRank: bringing order into texts. In: Proceedings of EMNLP-04 and the 2004 Conference on Empirical Methods in Natural Language Processing, July 2004

14. Tatman, R.: Brazilian Portuguese literature corpus, July 2017

15. Pedregosa, F., et al.: Scikit-learn: machine learning in Python. J. Mach. Learn. Res. **12**, 2825–2830 (2011)

16. Liu, A.Y.: The effect of oversampling and under sampling on classifying imbalanced text datasets (2004)

17. Corrêa, I., Drews-Jr, P., Souza, M., Garcia, V.: Supervised microalgae classification in imbalanced dataset (2016). https://doi.org/10.1109/BRACIS.2016.020

18. Cyran, K.A.: Machine learning approach to authorship attribution of literary texts (2007)

19. Aluísio, S., Pelizzoni, J., Marchi, A.R., de Oliveira, L., Manenti, R., Marquiafável, V.: An account of the challenge of tagging a reference corpus for Brazilian Portuguese. In: Mamede, N.J., Trancoso, I., Baptista, J., das Graças Volpe Nunes, M. (eds.) PROPOR 2003. LNCS (LNAI), vol. 2721, pp. 110–117. Springer, Heidelberg (2003). https://doi.org/10.1007/3-540-45011-4_17

20. Lemaître, G., Nogueira, F., Aridas, C.K.: Imbalanced-learn: a Python toolbox to tackle the curse of imbalanced datasets in machine learning. J. Mach. Learn. Res. **18**, 1–5 (2017)

21. Kohavi, R.: A study of cross-validation and bootstrap for accuracy estimation and model selection. In: Proceedings of the 14th International Joint Conference on Artificial Intelligence, vol. 2, pp. 1137–1143 (1995)

22. Pessoa, T., Medeiros, R., Nepomuceno, T., Bian, G., Albuquerque, V., Filho, P.: Performance analysis of Google colaboratory as a tool for accelerating deep learning applications, p. 1 (2018)

BERTimbau: Pretrained BERT Models for Brazilian Portuguese

Fábio Souza[1,3(✉)], Rodrigo Nogueira[1,2,3], and Roberto Lotufo[1,3] 🆔

[1] School of Electrical and Computer Engineering, University of Campinas, Campinas, Brazil
[2] University of Waterloo, Waterloo, Canada
[3] NeuralMind Inteligência Artificial, São Paulo, Brazil
{fabiosouza,rodrigo.nogueira,roberto}@neuralmind.ai
https://www.neuralmind.ai

Abstract. Recent advances in language representation using neural networks have made it viable to transfer the learned internal states of large pretrained language models (LMs) to downstream natural language processing (NLP) tasks. This transfer learning approach improves the overall performance on many tasks and is highly beneficial when labeled data is scarce, making pretrained LMs valuable resources specially for languages with few annotated training examples. In this work, we train BERT (Bidirectional Encoder Representations from Transformers) models for Brazilian Portuguese, which we nickname BERTimbau. We evaluate our models on three downstream NLP tasks: sentence textual similarity, recognizing textual entailment, and named entity recognition. Our models improve the state-of-the-art in all of these tasks, outperforming Multilingual BERT and confirming the effectiveness of large pretrained LMs for Portuguese. We release our models to the community hoping to provide strong baselines for future NLP research: https://github.com/neuralmind-ai/portuguese-bert.

1 Introduction

Transfer learning, where a model is first trained on a source task and then fine-tuned on tasks of interest, has changed the landscape of natural language processing (NLP) applications over the last years. The strategy of fine-tuning a large pretrained language model (LM) has been largely adopted and has achieved state-of-the-art performances on a variety of NLP tasks [7,27,29,48]. Aside from bringing performance improvements, transfer learning reduces the amount of labeled data needed for supervised learning on downstream tasks [10,24].

Pretraining these large language models, however, require huge amounts of unlabeled data and computational resources, with reports of models being trained using thousands of GPUs or TPUs and hundreds of GBs of raw textual data [18,29]. This resource barrier has limited the availability of these models, early on, to English, Chinese and multilingual models.

R. Cerri and R. C. Prati (Eds.): BRACIS 2020, LNAI 12319, pp. 403–417, 2020.
https://doi.org/10.1007/978-3-030-61377-8_28

BERT [7], which uses the Transformer architecture [44], among with its derived models, such as RoBERTa [18] and Albert [16], is one of the most adopted models. Despite having a multilingual BERT[1] model (mBERT) trained on 104 languages, much effort has been devoted on pretraining monolingual BERT and BERT-derived models on single languages, such as French [19], Dutch [6,45], Spanish [5], Italian [26], and others [2,13,21]. Even though it is unfeasable to train monolingual models for every language, these works are motivated by the superior performance and resource efficiency of monolingual models compared to mBERT.

Large pretrained LMs can be valuable assets especially for languages that have few annotated resources but abundant unlabeled data, such as Portuguese. With that in mind, we train BERT models for Brazilian Portuguese—which we nickname BERTimbau—using data from brWaC [46], a large and diverse corpus of web pages. We evaluate our models on three NLP tasks: sentence textual similarity, recognizing textual entailment, and named entity recognition. BERTimbau improves the state-of-the-art on these tasks over multilingual models and previous monolingual approaches, confirming the effectiveness of large pretrained LMs for Portuguese. We make BERTimbau models available to the community[2] on open-source libraries as to provide strong baselines for future research on NLP.

The paper is organized as follows: in Sect. 2, we present the related work. In Sect. 3, we briefly describe BERTimbau architecture and the pretraining procedure, such as the pretraining data, the vocabulary generation, and pretraining objectives. In Sect. 4, we describe the downstream tasks and datasets used to evaluate our models. Then, in Sect. 5, we describe our experiments and present and analyze our results. Lastly, we make our final remarks in Sect. 6.

2   Related Work

Our work encompasses three different topics which we discuss separately below.

2.1   Word Vector Representations

Word vector representations are a crucial component of many neural NLP models. Classic word embeddings [20,23] are static non-contextualized word-level representations that capture semantic and syntatic features using large corpora. More recently, contextual embeddings, such as ELMo [24] and Flair Embeddings [1], leverage the internal states of language models to extract richer word representations in context. These embeddings are used as features to task-specific models and consist of a shallow transfer learning approach, since downstream models have to be trained from scratch on each task of interest.

---

[1] https://github.com/google-research/bert/blob/master/multilingual.md.
[2] https://github.com/neuralmind-ai/portuguese-bert.

2.2   Deeper Transfer Learning

Deeper transfer learning techniques for NLP emerged by successfully fine-tuning large pretrained LMs with general purpose architectures, such as the Transformer [44], replacing task-specific models. Language modeling pretraining is shown to resemble a multitask objective that allows zero-shot learning on many tasks [28]. This pretraining stage benefits from diverse texts and can be further improved by additional pretraining with unlabeled data of downstream tasks' domains [9].

2.3   Word Representations for Portuguese

For the Portuguese language, recent works mainly explored and compared contextual embedding techniques. ELMo and Flair Embeddings trained on a large Portuguese corpora achieve good results on the named entity recognition task [3,36,37]. A comparison of ELMo and multilingual BERT using a contextual embeddings setup shows superior performance of Portuguese ELMo on semantic textual similarity task when no fine-tuning is used [32].

3   BERTimbau

Our approach closely replicates BERT's architecture and pretraining procedures with few changes. BERT consists of a Transformer encoder architecture [44] that is trained using a modified language modeling task called Masked Language Modeling (also known as Cloze task [41]), which we detail in Sect. 3.3. This design allows the pretraining of deep bidirectional word representations—in contrast to preceeding works that employ unidirectional LMs—and has many advantages over well established recurrent models, such as LSTM and GRU. By relying only on self-attention mechanisms instead of recurrence, Transformer models can see all input tokens at once and, hence, can model dependencies in long sequences in constant time while enabling much higher parallelization. For a detailed explanation of the Transformer architecture, we refer readers to Vaswani et al. [44].

We train BERTimbau models on two sizes: Base (12 layers, 768 hidden dimension, 12 attention heads, and 110M parameters) and Large (24 layers, 1024 hidden dimension, 16 attention heads and 330M parameters). The maximum sentence length is set to $S = 512$ tokens. We train cased models only since we focus on general purpose models and capitalization is relevant for tasks like named entity recognition [4,7].

3.1   Pretraining Data

For pretraining data, we use the brWaC [46] corpus, which contains 2.68 billion tokens from 3.53 million documents and is the largest open Portuguese corpus to date. On top of its size, brWaC is composed of whole documents and its methodology ensures high domain diversity and content quality, which are desirable features for BERT pretraining.

We use only the document body (ignoring the titles) and we apply a single post-processing step on the data to fix *mojibakes*[3] and remove remnant HTML tags using the `ftfy` library [40]. The final processed corpus has 17.5 GB of raw text.

3.2   Vocabulary Generation

We generate a cased Portuguese vocabulary of 30k subword units using the SentencePiece library [12] with the BPE algorithm [39] and 2,000,000 random sentences from Portuguese Wikipedia articles. The resulting vocabulary is then converted to WordPiece [38] format for compatibility with original BERT code.

3.3   Pretraining Objectives

Identical to BERT, we train our models on Masked Language Modeling (MLM) and Next Sentence Prediction (NSP) tasks. Each pretraining example is generated by concatenating two sequences[4] of tokens $\mathbf{x} = (x_1, \ldots, x_n)$ and $\mathbf{y} = (y_1, \ldots, y_m)$ separated by special [CLS] and [SEP] tokens as follows:

$$[\text{CLS}] \ x_1 \ \cdots \ x_n \ [\text{SEP}] \ y_1 \ \cdots \ y_m \ [\text{SEP}]$$

Given a sequence $\mathbf{x}$ of tokens from the corpus, in 50% of the time $\mathbf{y}$ is chosen as to form a contiguous piece of text, and 50% of the time $\mathbf{y}$ is a random sentence sampled from a distinct document of the corpus. Each example is then corrupted by randomly replacing 15% of the tokens of $\mathbf{x}$ and $\mathbf{y}$ by 1 of 3 options: a special [MASK] token with 80% probability, a random token from the vocabulary with 10% probability or, otherwise, keeping the original token. In this step we use whole word masking: if a word composed of multiple subword units is chosen to be corrupted, all of its subword units are also corrupted.

The corrupted sequences are used as inputs to BERT and the tokens' encoded representations are used as inputs for the pretraining tasks' classifiers. The NSP task takes the encoded [CLS] token and predicts if $\mathbf{y}$ is the original continuation of $\mathbf{x}$ or not. For the MLM task, each corrupted token has to be predicted back from the encoded masked positions.

Once pretrained, we fine-tune our models on each downstream task as described in Sect. 5.2. Next we will describe the evaluation tasks and datasets.

4   Evaluation Tasks and Datasets

We evaluate our models on 3 downstream NLP tasks: Sentence Textual Similarity (STS), Recognizing Textual Entailment (RTE), and Named Entity Recognition (NER). In the following sections we briefly define each task, the datasets, and the evaluation procedures.

---

[3] Mojibake is a kind of text corruption that occurs when strings are decoded using the incorrect character encoding. For example, the word "codificação" becomes "codificaÃ§Ã£o" when encoded in UTF-8 and decoded using ISO-8859-1.

[4] We use "sequence" and "sentence" interchangeably. A sentence is any contiguous span of text of arbitrary length.

4.1 Sentence Textual Similarity and Recognizing Textual Entailment

Sentence Textual Similarity (STS) is a regression task that measures the degree of semantic equivalence between two sentences in a numeric scale. Recognizing Textual Entailment (RTE), also known as Natural Language Inference (NLI), is a classification task of predicting if a given premise sentence entails a hypothesis sentence.

We use the ASSIN2 dataset [30], a shared task with 10,000 sentence pairs with STS and RTE annotations. The dataset is composed of 6500 train, 500 validation and 3000 test examples.

STS labels are continuous values in a scale of 1 to 5, where a pair of sentences with completely different meanings have label value of 1 and semantically equivalent sentences have value of 5. STS performance is evaluated using Pearson's Correlation as primary metric and Mean Squared Error (MSE) as secondary metric.

RTE labels are simply `entailment` and `non-entailment`. RTE performance is evaluated using macro F1-score as primary metric and accuracy as secondary metric.

4.2 Named Entity Recognition

Named Entity Recognition (NER) is the task of identifying text spans that mention named entities (NEs) and classifying them into predefined categories, such as person, organization, and location. We cast NER as a sequence labeling task that performs unified entity identification and classification. We use the IOB2 tagging scheme [42].

We use the Golden Collections of the First HAREM evaluation contests [35], First HAREM and MiniHAREM, as train and test sets, respectively, following previous works [4,34]. Both datasets contain multidomain documents annotated with 10 NE classes: Person, Organization, Location, Value, Date, Title, Thing, Event, Abstraction, and Other.

We employ the datasets on two distinct scenarios: a Total scenario that considers all 10 classes, and a Selective scenario that includes only 5 classes (Person, Organization, Location, Value, and Date). Table 1 presents some dataset statistics. We set aside 7% of First HAREM documents as a holdout validation set.

Table 1. Dataset statistics for the HAREM I corpora. The Tokens column refers to whitespace and punctuation tokenization.

Dataset	Documents	Tokens	Entities in scenario	
			Selective	Total
First HAREM	129	95585	4151	5017
MiniHAREM	128	64853	3018	3642

NER performance is evaluated using CoNLL 2003 [43] evaluation script,[5] that computes entity-level precision, recall, and micro F1-score on exact matches. In other words, precision is the percentage of named entities predicted by the model that are correct, recall is the percentage of corpus entities that were correctly predicted and F1-score is the harmonic mean of precision and recall.

Document Context and Max Context Evaluation for Token-Level Tasks. In token-level tasks such as NER, we use document context for input examples instead of sentence context to take advantage of longer contexts when encoding token representations from BERT. Following the approach of Devlin et al. [7] on the SQuAD dataset, examples longer than S tokens are broken into spans of length up to S using a stride of D tokens. Each span is used as a separate example during training. During evaluation, however, a single token T_i can be present in $N = \frac{S}{D}$ multiple spans s_j, and so may have up to N distinct predictions $y_{i,j}$. Each token's final prediction is taken from the span where the token is closer to the central position, that is, the span where it has the most contextual information. Figure 1 illustrates this procedure.

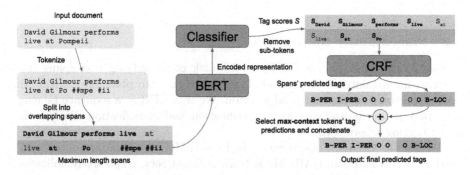

Fig. 1. Illustration of the proposed method for the NER task described in 4.2. Given an input document, the text is tokenized using WordPiece [47] and the tokenized document is split into overlapping spans of the maximum length using a fixed stride (=3, in the example). Maximum context tokens of each span are marked in bold. The spans are fed into BERT and then into the classification model, producing a sequence of tag scores for each span. The scores of subtoken entries (starting with ##) are removed from the spans and the remaining tags scores are passed to the CRF layer—if it is employed, otherwise the highest tag scores are used independently. The maximum context tokens are selected and concatenated to form the final predicted tags.

5   Experiments

In this section, we present the experimental setup and results for BERT pre-trainings and evaluation tasks. We conduct additional experiments to explore

---

[5] https://www.clips.uantwerpen.be/conll2002/ner/bin/conlleval.txt.

the usage of BERTimbau as a fixed extractor of contextual embeddings and also assess the impact of the long pretraining stage.

5.1 Pretrainings

Following Devlin et al. [7], models are pretrained for 1,000,000 steps. We use a learning rate of 1e−4, learning rate warmup over the first 10,000 steps followed by a linear decay of the learning rate over the remaining steps.

For BERTimbau Base models, the weights are initialized with the checkpoint of Multilingual BERT Base (discarding the word embeddings that are from a different vocabulary). We use a batch size of 128 and sequences of 512 tokens the entire training. This training takes 4 d on a TPU v3-8 instance and performs about 8 epochs over the pretraining data.

For BERTimbau Large, the weights are initialized with the checkpoint of English BERT Large (also discarding the word embeddings that are from a different vocabulary). Since it is a bigger model with longer training time, we follow the instructions of Devlin et al. [7] and use sequences of 128 tokens in batches of size 256 for the first 900,000 steps and then sequences of 512 tokens and batch size 128 for the last 100,000 steps. This training takes 7 d on a TPU v3-8 instance and performs about 6 epochs over the training data.

Note that in the calculation of the number of epochs, we are taking into consideration a duplication factor of 10 when generating the input examples. This means that under 10 epochs, the same sentence is seen with different masking and sentence pair in each epoch, which is effectively equal to dynamic example generation [18].

5.2 Fine-Tunings on Evaluation Tasks

To use BERTimbau on downstream tasks, we remove the MLM and NSP classification heads used during pretraining stage and attach relevant head(s) required for each task. Similar to pretraining, sentence-level tasks are performed on the encoded representation of the [CLS] special token, and token-level tasks use the encoded representation of each relevant token.

For all evaluation experiments, we use a learning rate schedule of warmup over the first 10% steps followed by linear decay of the learning rate over the remaining steps. Similar to pretraining, we use BERT's AdamW optimizer implementation with $\beta_1 = 0.9$, $\beta_2 = 0.999$ and L2 weight decay of 0.01. We perform early stopping and select the best model on the validation set of each task.

5.3 Sentence Textual Similarity and Recognizing Textual Entailment Results

For STS and RTE tasks, we attach two independent linear layers on top of BERTimbau and fine-tune the whole model in a multitask scheme using two losses: MSE loss for STS and cross-entropy loss for RTE. The final loss is the sum of both losses with equal weight.

We train BERTimbau Base with learning rate of 4e−5 and batch size 32 for 10 epochs, and BERTimbau Large with learning rate of 1e−5, batch size 8 for 5 epochs. We also train mBERT to compare it to BERTimbau models. mBERT is trained using learning rate of 1e−5 and batch size 8 for 10 epochs.

Table 2. Test scores for STS and RTE tasks on ASSIN2 dataset. We compare our models to the best published results. Best scores in bold. Reported values are the average of multiple runs with different random seeds. Star (⋆) denotes primary metrics. †: ensemble technique. ‡: extra training data.

Row	Model	STS		RTE	
		Pearson (⋆)	MSE	F1 (⋆)	Accuracy
1	mBERT + RoBERTa-Large-en (Averaging) [31] †	0.83	0.91	84	84.8
2	mBERT + RoBERTa-Large-en (Stacking) [31] †	0.785	0.59	88.3	88.3
3	mBERT (STS) and mBERT-PT (RTE) [33] ‡	0.826	0.52	87.6	87.6
4	USE+Features (STS) and mBERT+Features (RTE) [8]	0.800	**0.39**	86.6	86.6
5	mBERT+Features [8]	0.817	0.47	86.6	86.6
6	mBERT (ours)	0.809	0.58	86.8	86.8
7	BERTimbau Base	0.836	0.58	89.2	89.2
8	BERTtimbau Large	**0.852**	0.50	**90.0**	**90.0**

Results. Our results for both tasks are shown in Table 2. We compare our results to the best-performing submissions to official ASSIN2. All compared works employ mBERT or a Transformer-based architecture in their approaches. In the following paragraphs, we refer to each work using their corresponding row numbers in Table 2.

BERTimbau models achieve the best results on the primary metrics of both STS and RTE tasks, with the large model performing significantly better than the base variant. The previous highest scores (rows 1 and 2) for both STS Pearson's correlation and RTE F1 score are from ensemble techniques that combine mBERT fine-tuned on original ASSIN2 data and an English RoBERTa-Large fine-tuned on ASSIN2 data automatically translated to English. The averaging ensemble uses 2 models and the stacking ensemble uses 10 distinct fine-tuned models—5-fold stacking which results in 5 mBERT and 5 RoBERTa trained models. While this approach shows an interesting application of English models to Portuguese tasks, our BERTimbau models achieve higher performance using a single model and, hence, demand lower compute resources in both fine-tuning and inference stages.

Regarding our implementation using mBERT (row 6), it presents a lower performance compared to BERTimbau models, which highlights the benefits of Portuguese pretraining of BERTimbau. For STS task, we note that mBERT achieves the same MSE as BERTimbau Base, even though Pearson correlation is lower. Comparing it to other works' approaches, better performances are achieved using extra supervised training data and further pretraining of mBERT on Portuguese data (row 3), and also by combining it with hand-designed features (rows 4 and 5).

5.4   NER Task

To evaluate on NER, we attach a linear classifier layer on top of BERTimbau to predict the tag of each token independently. The model is trained using cross-entropy loss. We compute predictions and losses only for the first wordpiece of each token, ignoring word continuations, as depicted in Figure 1. Long examples are broken into spans using a stride of $D = 128$ as explained in Sect. 4.2.

It is common in NER for the vast majority of tokens not to belong to named entities (and have tag label "O"). To deal with this class imbalance, we initialize the classifier's bias term of the "O" tag with a value of 6 in order to promote a better stability in early training [17]. We also use a weight of 0.01 for "O" tag losses.

The model parameters are divided in two groups with different learning rates: 5e−5 for BERT model and 1e−3 for the classifier. We train the models for up to 50 epochs using a batch size of 16.

Since Linear-Chain Conditional Random Fields (CRF) [14] is widely adopted to enforce sequential classification in sequence labeling tasks [4,15,34], we also experiment with employing a CRF layer after the linear layer. We refer readers to Lample et al. [15] for a detailed explanation on CRF loss formulation and decoding procedure. Models with CRF are trained for up to 15 epochs.

In addition to BERTimbau Base and Large, we also train mBERT to compare monolingual versus multilingual model performances. mBERT is fine-tuned with the same hyperparameters.

When evaluating, we produce valid predictions by removing all invalid tag transitions for the IOB2 scheme, such as "I-" tags coming directly after "O" tags or after an "I-" tag of a different class. This post-processing step trades off recall for a possibly higher precision.

Results. The main results of our NER experiments are presented in Table 3. We compare the performances of our models on the two scenarios (total and selective) defined in Sect. 4.2.

Table 3. Results of NER task (Precision, Recall and micro F1-score) on the test set (MiniHAREM). Best results in bold. Reported values are the average of multiple runs with different random seeds. Star (⋆) denotes primary metrics.

Row	Architecture	Total scenario			Selective scenario		
		Prec.	Rec.	F1 (⋆)	Prec.	Rec.	F1 (⋆)
1	CharWNN [34]	67.2	63.7	65.4	74.0	68.7	71.2
2	LSTM-CRF [4]	72.8	68.0	70.3	78.3	74.4	76.3
3	BiLSTM-CRF + FlairBBP [36]	74.9	74.4	74.6	83.4	81.2	82.3
4	mBERT	71.6	72.7	72.2	77.0	78.8	77.9
5	mBERT + CRF	74.1	72.2	73.1	80.1	78.3	79.2
6	BERTimbau Base	76.8	77.1	77.2	81.9	**82.7**	82.2
7	BERTimbau Base + CRF	78.5	76.8	77.6	84.6	81.6	83.1
8	BERTimbau Large	77.9	**78.0**	77.9	81.3	82.2	81.7
9	BERTimbau Large + CRF	**79.6**	77.4	**78.5**	**84.9**	82.5	**83.7**

Table 4. NER performances (Precision, Recall and F1-score) on the test set (MiniHAREM) using BERTimbau as contextual embeddings in a feature-based approach. Star (⋆) denotes primary metrics.

Architecture	Total scenario			Selective scenario		
	Prec.	Rec.	F1 (⋆)	Prec.	Rec.	F1 (⋆)
mBERT + LSTM-CRF	74.7	69.7	72.1	80.6	75.0	77.7
BERTimbau Base + LSTM-CRF	78.3	73.2	75.6	84.5	78.7	81.6
BERTimbau Large + LSTM-CRF	77.4	72.4	74.8	83.0	77.8	80.3

BERTimbau outperforms the best published results, improving the F1-score by 3.9 points on the total scenario and by 1.4 point on the selective scenario. The BiLSTM-CRF+FlairBBP (row 3 of Table 3) model uses Portuguese Flair embeddings, which are contextual embeddings extracted from character-level language models. Interestingly, Flair embeddings outperform BERT models on English NER [1,7].

There is a large performance gap between BERTimbau and mBERT, which reinforces the advantages of monolingual models pretrained on multidomain data compared to mBERT, that is trained only on Wikipedia articles. This result is on par with other monolingual BERT works.

The CRF layer consistently brings performance improvements in F1 in all settings. However, F1 increases are pushed by a large boost in precision that is often associated with lower recall. It is worth noting that, without CRF, BERTimbau Large shows a close but inferior performance to the Base variant on the selective cenario. This result suggests that a more controlled fine-tuning scheme might be required in some cases, such as partial layer unfreezing or discriminative fine-tuning [25]—usage of lower learning rates for lower layers—, given that it is a higher capacity model trained on few data.

5.5 BERT as Contextual Embeddings

In this experiment, we evaluate BERTimbau as a fixed extractor of contextual embeddings that we use as features to train a downstream model on the NER task. In this feature-based approach, we train a BiLSTM-CRF model with 1 layer and 100 hidden units followed by a linear classifier layer for up to 50 epochs. Instead of using only the hidden representation of BERT's last layer, we sum the last 4 layers, following Devlin et al. [7]. The resulting architecture resembles the LSTM-CRF model [15] but using BERT embeddings instead of fixed word embeddings.

Results. We present the results on Table 4. Models of the feature-based approach perform significantly worse compared to the ones of the fine-tuning approach. The performance gap is found to be much higher than the reported values for NER on English language [7,25] and reaches up to 2 points on BERTimbau Base and 3.5 points on BERTimbau-Large. It is worth mentioning that BERTimbau models in this feature-based approach achieve better performances than a fine-tuned mBERT.

While BERTimbau Large is the highest performer when fine-tuned, we observe that it experiences performance degradation when used in this feature-based approach, performing worse than the smaller Base variant but still better than mBERT.

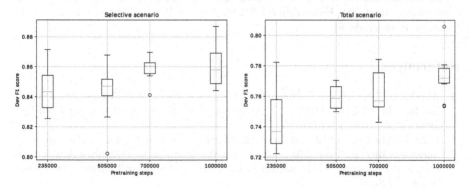

Fig. 2. Performance of BERTimbau Base on NER task using intermediate checkpoints of the pretraining stage. Scores reported on the validation set.

5.6 Impact of Long Pretraining

To assess the impact of long pretraining stage on the performance of downstream tasks, we repeat part of the NER fine-tuning experiment (Sect. 5.4) using intermediate checkpoints of BERTimbau Base pretraining procedure. We train BERT models (without CRF) using the checkpoints of steps 235k, 505k and 700k, which correspond to 23.5%, 50.5% and 70% of the complete pretraining of 1000k steps, respectively. All models are trained with the same hyperparameters and experimental setup described in Sect. 5.4.

The results are shown in Fig. 2. Performances on the downstream task increase non-linearly with pretraining steps, with diminishing returns as pretraining progresses. This is an expected result, as test performance of pretraining tasks are shown to follow a power law on the number of pretraining steps [11].

6 Conclusion

Our BERTimbau models are able to outperform previously published results and multilingual model equivalents on all evaluation tasks considered, on par with the results of similar works on other languages. We also demonstrate that they provide contextual embeddings that beat fine-tuned multilingual BERT on NER task, which is a lower compute alternative on limited resource scenarios. We hope that by releasing our models, others will be able to benchmark and improve the performance of many other NLP tasks in Portuguese. Experiments with more recent and efficient models, such as RoBERTa and T5, are left for future works.

Acknowledgements. R Lotufo acknowledges the support of the Brazilian government through the CNPq Fellowship ref. 310828/2018-0. We would like to thank Google Cloud for research credits.

References

1. Akbik, A., Blythe, D., Vollgraf, R.: Contextual string embeddings for sequence labeling. In: COLING 2018, 27th International Conference on Computational Linguistics, pp. 1638–1649 (2018)
2. Baly, F., Hajj, H., et al.: Arabert: transformer-based model for Arabic language understanding. In: Proceedings of the 4th Workshop on Open-Source Arabic Corpora and Processing Tools, with a Shared Task on Offensive Language Detection, pp. 9–15 (2020)
3. Castro, P., Felix, N., Soares, A.: Contextual representations and semi-supervised named entity recognition for Portuguese language, September 2019
4. Quinta de Castro, P.V., Félix Felipe da Silva, N., da Silva Soares, A.: Portuguese named entity recognition using LSTM-CRF. In: Villavicencio, A., et al. (eds.) PROPOR 2018. LNCS (LNAI), vol. 11122, pp. 83–92. Springer, Cham (2018). https://doi.org/10.1007/978-3-319-99722-3_9
5. Cañete, J., Chaperon, G., Fuentes, R., Pérez, J.: Spanish pre-trained BERT model and evaluation data. In: To Appear in PML4DC at ICLR 2020 (2020)

6. Delobelle, P., Winters, T., Berendt, B.: RobBERT: a Dutch RoBERTa-based language model. arXiv preprint arXiv:2001.06286 (2020)
7. Devlin, J., Chang, M.W., Lee, K., Toutanova, K.: BERT: pre-training of deep bidirectional transformers for language understanding. arXiv preprint arXiv:1810.04805 (2018)
8. Fonseca, E., Alvarenga, J.P.R.: Wide and deep transformers applied to semantic relatedness and textual entailment. In: Oliveira et al. [22], pp. 68–76. http://ceur-ws.org/Vol-2583/
9. Gururangan, S., Marasović, A., Swayamdipta, S., Lo, K., Beltagy, I., Downey, D., Smith, N.A.: Don't stop pretraining: adapt language models to domains and tasks. arXiv preprint arXiv:2004.10964 (2020)
10. Howard, J., Ruder, S.: Universal language model fine-tuning for text classification. In: Proceedings of the 56th Annual Meeting of the Association for Computational Linguistics (Volume 1: Long Papers), pp. 328–339 (2018)
11. Kaplan, J., et al.: Scaling laws for neural language models. arXiv preprint arXiv:2001.08361 (2020)
12. Kudo, T., Richardson, J.: SentencePiece: a simple and language independent subword tokenizer and detokenizer for neural text processing. arXiv preprint arXiv:1808.06226 (2018)
13. Kuratov, Y., Arkhipov, M.: Adaptation of deep bidirectional multilingual transformers for russian language. arXiv preprint arXiv:1905.07213 (2019)
14. Lafferty, J.D., McCallum, A., Pereira, F.C.N.: Conditional random fields: probabilistic models for segmenting and labeling sequence data. In: Proceedings of the Eighteenth International Conference on Machine Learning, ICML 2001, p. 282–289. Morgan Kaufmann Publishers Inc., San Francisco (2001)
15. Lample, G., Ballesteros, M., Subramanian, S., Kawakami, K., Dyer, C.: Neural architectures for named entity recognition. arXiv preprint arXiv:1603.01360 (2016). http://arxiv.org/abs/1603.01360, version 3
16. Lan, Z., Chen, M., Goodman, S., Gimpel, K., Sharma, P., Soricut, R.: Albert: a lite BERT for self-supervised learning of language representations. arXiv preprint arXiv:1909.11942 (2019)
17. Lin, T.Y., Goyal, P., Girshick, R., He, K., Dollár, P.: Focal loss for dense object detection. In: Proceedings of the IEEE International Conference on Computer Vision, pp. 2980–2988 (2017)
18. Liu, Y., et al.: Roberta: a robustly optimized BERT pretraining approach. arXiv preprint arXiv:1907.11692 (2019)
19. Martin, L., et al.: CamemBERT: a tasty French language model. arXiv preprint arXiv:1911.03894 (2019)
20. Mikolov, T., Chen, K., Corrado, G., Dean, J.: Efficient estimation of word representations in vector space. arXiv preprint arXiv:1301.3781 (2013)
21. Nguyen, D.Q., Nguyen, A.T.: PhoBERT: pre-trained language models for Vietnamese. arXiv preprint arXiv:2003.00744 (2020)
22. Oliveira, H.G., Real, L., Fonseca, E. (eds.): Proceedings of the ASSIN 2 Shared Task: Evaluating Semantic Textual Similarity and Textual Entailment in Portuguese, Extended Semantic Web Conference. No. 2583 in CEUR Workshop Proceedings (2020). http://ceur-ws.org/Vol-2583/
23. Pennington, J., Socher, R., Manning, C.: Glove: global vectors for word representation. In: Proceedings of the 2014 Conference on Empirical Methods in Natural Language Processing (EMNLP), pp. 1532–1543. Association for Computational Linguistics, Doha, Qatar, October 2014. https://doi.org/10.3115/v1/D14-1162. https://www.aclweb.org/anthology/D14-1162

24. Peters, M., et al.: Deep contextualized word representations. In: Proceedings of the 2018 Conference of the North American Chapter of the Association for Computational Linguistics: Human Language Technologies, Volume 1 (Long Papers), pp. 2227–2237 (2018)

25. Peters, M.E., Ruder, S., Smith, N.A.: To tune or not to tune? adapting pretrained representations to diverse tasks. In: Proceedings of the 4th Workshop on Representation Learning for NLP (RepL4NLP-2019), pp. 7–14 (2019)

26. Polignano, M., Basile, P., de Gemmis, M., Semeraro, G., Basile, V.: AlBERTo: Italian BERT language understanding model for NLP challenging tasks based on Tweets. In: Proceedings of the Sixth Italian Conference on Computational Linguistics (CLiC-it 2019), vol. 2481. CEUR (2019). https://www.scopus.com/inward/record.uri?eid=2-s2.0-85074851349&partnerID=40&md5=7abed946e06f76b3825ae5e294ffac14

27. Radford, A., Narasimhan, K., Salimans, T., Sutskever, I.: Improving language understanding with unsupervised learning. Tech. rep. OpenAI (2018)

28. Radford, A., Wu, J., Child, R., Luan, D., Amodei, D., Sutskever, I.: Language models are unsupervised multitask learners. OpenAI Blog 1(8), 9 (2019)

29. Raffel, C., et al.: Exploring the limits of transfer learning with a unified text-to-text transformer. arXiv preprint arXiv:1910.10683 (2019)

30. Real, L., Fonseca, E., Gonçalo Oliveira, H.: The ASSIN 2 shared task: a quick overview, pp. 406–412 (02 2020). https://doi.org/10.1007/978-3-030-41505-1_39

31. Rodrigues, R., da Silva, J., Castro, P., Felix, N., Soares, A.: Multilingual transformer ensembles for Portuguese natural language tasks. In: Oliveira et al. [22], pp. 27–38. http://ceur-ws.org/Vol-2583/

32. Rodrigues, R.C., Rodrigues, J., de Castro, P.V.Q., da Silva, N.F.F., Soares, A.: Portuguese language models and word embeddings: evaluating on semantic similarity tasks. In: Quaresma, P., Vieira, R., Aluísio, S., Moniz, H., Batista, F., Gonçalves, T. (eds.) PROPOR 2020. LNCS (LNAI), vol. 12037, pp. 239–248. Springer, Cham (2020). https://doi.org/10.1007/978-3-030-41505-1_23

33. Rodrigues, R., Couto, P., Rodrigues, I.: IPR: the semantic textual similarity and recognizing textual entailment systems. In: Oliveira et al. [22], pp. 39–47. http://ceur-ws.org/Vol-2583/

34. Santos, C.N.D., Guimaraes, V.: Boosting named entity recognition with neural character embeddings. arXiv preprint arXiv:1505.05008 (2015). https://arxiv.org/abs/1505.05008, version 2

35. Santos, D., Seco, N., Cardoso, N., Vilela, R.: HAREM: an advanced NER evaluation contest for Portuguese (2006)

36. Santos, J., Consoli, B., dos Santos, C., Terra, J., Collonini, S., Vieira, R.: Assessing the impact of contextual embeddings for Portuguese named entity recognition. In: 8th Brazilian Conference on Intelligent Systems, BRACIS, Bahia, Brazil, 15–18 October, pp. 437–442 (2019)

37. Santos, J., Terra, J., Consoli, B.S., Vieira, R.: Multidomain contextual embeddings for named entity recognition. In: IberLEF@SEPLN (2019)

38. Schuster, M., Nakajima, K.: Japanese and Korean voice search. In: 2012 IEEE International Conference on Acoustics, Speech and Signal Processing (ICASSP), pp. 5149–5152. IEEE (2012)

39. Sennrich, R., Haddow, B., Birch, A.: Neural machine translation of rare words with subword units. In: Proceedings of the 54th Annual Meeting of the Association for Computational Linguistics (Volume 1: Long Papers), Berlin, Germany, pp. 1715–1725. Association for Computational Linguistics, August 2016. https://doi.org/10.18653/v1/P16-1162. https://www.aclweb.org/anthology/P16-1162

40. Speer, R.: ftfy. Zenodo (2019). https://doi.org/10.5281/zenodo.2591652, version 5.5
41. Taylor, W.L.: "cloze procedure": a new tool for measuring readability. Journalism Q. **30**(4), 415–433 (1953)
42. Tjong, E.F., Sang, K., Veenstra, J.: Representing text chunks. In: Ninth Conference of the European Chapter of the Association for Computational Linguistics. Association for Computational Linguistics, Bergen, Norway, June 1999. https://www.aclweb.org/anthology/E99-1023
43. Tjong Kim Sang, E.F., De Meulder, F.: Introduction to the CoNLL-2003 shared task: Language-independent named entity recognition. In: Proceedings of the Seventh Conference on Natural Language Learning at HLT-NAACL 2003, pp. 142–147 (2003). https://www.aclweb.org/anthology/W03-0419
44. Vaswani, A., et al.: Attention is all you need. In: Advances in Neural Information Processing Systems, pp. 5998–6008 (2017)
45. Vries, W.D., Cranenburgh, A.V., Bisazza, A., Caselli, T., Noord, G.V., Nissim, M.: BERTje: a Dutch BERT model. arXiv preprint arXiv:1912.09582, December 2019
46. Wagner Filho, J., Wilkens, R., Idiart, M., Villavicencio, A.: The BRWAC corpus: a new open resource for Brazilian Portuguese, May 2018
47. Wu, Y., et al.: Google's neural machine translation system: bridging the gap between human and machine translation. arXiv preprint arXiv:1609.08144 (2016). http://arxiv.org/abs/1609.08144, version 2
48. Yang, Z., Dai, Z., Yang, Y., Carbonell, J., Salakhutdinov, R., Le, Q.V.: XLNet: generalized autoregressive pretraining for language understanding. arXiv preprint arXiv:1906.08237 (2019)

Deep Learning Models for Representing Out-of-Vocabulary Words

Johannes V. Lochter[1,2], Renato M. Silva[3(✉)], and Tiago A. Almeida[3]

[1] Department of Systems and Energy, University of Campinas (UNICAMP),
Campinas, São Paulo, Brazil
`johannes.lochter@facens.br`
[2] Smart Campus, Engineering College of Sorocaba (Facens),
Sorocaba, São Paulo, Brazil
[3] Department of Computer Science, Federal University of São Carlos (UFSCar),
Sorocaba, São Paulo, Brazil
`renatoms@dt.fee.unicamp.br, talmeida@ufscar.br`

Abstract. Communication has become increasingly dynamic with the popularization of social networks and applications that allow people to express themselves and communicate instantly. In this scenario, distributed representation models have their quality impacted by new words that appear frequently or that are derived from spelling errors. These words that are unknown by the models, known as out-of-vocabulary (OOV) words, need to be properly handled to not degrade the quality of the natural language processing (NLP) applications, which depend on the appropriate vector representation of the texts. To better understand this problem and finding the best techniques to handle OOV words, in this study, we present a comprehensive performance evaluation of deep learning models for representing OOV words. We performed an intrinsic evaluation using a benchmark dataset and an extrinsic evaluation using different NLP tasks: text categorization, named entity recognition, and part-of-speech tagging. Although the results indicated that the best technique for handling OOV words is different for each task, Comick, a deep learning method that infers the embedding based on the context and the morphological structure of the OOV word, obtained promising results.

Keywords: Natural language processing · Machine learning · Out-of-vocabulary words

1 Introduction

As the world became more digital, the amount of unstructured data available on the Internet has increased to the point where it has become unbearable to handle it using human resources. Natural language processing (NLP) tasks were

We gratefully acknowledge the support provided by the São Paulo Research Foundation (FAPESP; grants #2017/09387-6, #2018/02146-6), CAPES, and CNPq.

© Springer Nature Switzerland AG 2020
R. Cerri and R. C. Prati (Eds.): BRACIS 2020, LNAI 12319, pp. 418–434, 2020.
https://doi.org/10.1007/978-3-030-61377-8_29

developed to address it in an automatic way using computational resources. Some examples of these tasks are audio transcription, translation, assessment on text summaries, grading tests, and opinion mining [5].

For NLP tasks, a critical point is the computational text representation since there is no consensus on how to represent text proper ly using computational resources. The most classical text representation is bag-of-words. In this distributive representation, a vocabulary is collected in the training corpus and each sample[1] is represented by a vector where each element represents the occurrence or absence (1/0) of vocabulary terms in the document [16].

New text representation techniques have been studied due to known issues of bag-of-words representation: it loses word locality and fails to capture semantic and syntactic features of the words. To address these issues, other techniques were developed, such as the distributed text representation that learns fixed-length vector for each word, known as word embeddings. Using statistics of the context and the abundant occurrences of the words in the training corpus, learned word embeddings can capture the semantic similarity between words [19].

As each sample can have many words, a composition function is usually employed to encode all word embeddings into a single fixed-length representation per sample to satisfy the fixed-length input restriction on the most of the predictive models. Some representation techniques also encodes the position of a word in the sample, addressing the word locality issue [5].

The majority of predictive models for NLP tasks have their performance degraded when unknown words, which were not collected to build the vocabulary in the training phase or were discarded due to low frequency across the corpus, appear in the test. These words are called out-of-vocabulary (OOV) words and can degrade the performance of NLP applications due to the inefficiency of representation models to properly learn a representation for them.

In order to emphasize how an OOV word can hinder sentence comprehension, consider the following example originally written in "The Jabberwocky" by Lewis Caroll: "He went galumphing back". The nonce word "galumphing" was coined to mean "moving in a clumsy, ponderous, or noisy manner; inelegant". Since this word is an OOV, traditional models are not capable to handle it properly, ignoring it. The lack of representation for this word can restrict the predictive model capabilities to understand this sentence [1].

Handling OOV words in distributed representation models can be achieved with simple strategies. For instance, as OOV words have no word embedding representation learned, they can be ignored when the composition function is applied. This approach leads the predictive model to fit data without the knowledge of the absence of a word because it is unknown to the representation model. For such case, a random vector can be adopted for each OOV word or a single random vector can be adopted for all OOV words [31].

These simple strategies provide little or no information about unknown words to predictive models in downstream tasks. In order to enable a predictive model to use a vector representation for the unknown words, those words need to be

---

[1] In this study, we use the word sample to denote instance or text document.

replaced by a meaningful in-vocabulary word. For this specific task, most of the techniques available in literature fits in two groups: language models [27] and robust techniques capable of learning meaningful representation for OOV words using their structures or the context in which they appear [4,10].

In these two groups, there are several deep learning (DL) models. Some of them were developed to handle OOV, such as Comick [10] and HiCE [11], while evidence was found that pure neural architectures can also perform it, such as LSTM [18] and Transformer [29]. Some language models also had success in this task, such as RoBERTa [15], DistillBERT [24], and Electra [7].

Although several studies have shown that DL can be successfully applied in several NLP tasks, such as sentiment analysis [20], named entity recognition (NER), and part-of-speech (POS) tagging [11], there are few DL models for handling OOV words and no consensus on which one is the best. To fill that gap, we present a performance evaluation of state-of-the-art DL models considering different datasets and tasks that can be greatly affected by OOV words.

2   OOV Handling

For a text representation model unable to handle OOV words, when a word is unknown to this model, the word is ignored in sample context due to the incapacity of the model to generate a proper representation for it. This lack of information tends to degrade the performance of the predictive models as the number of OOV words per sample increases.

Simple replacement methods are straight forward solution to OOV handling, such as the replacement of every OOV word by the same random vector or a different random vector for each OOV word. There is also the zero-vector representation replacement which is suitable to inhibit activation through the neural network for that specific OOV word [31].

More elaborated methods for OOV handling also showed good results, although they were not able to learn new representations. For instance, Khodak et al. [12] proposed to average the embeddings of the closest words of a OOV word in the sample. Their proposed method failed to capture complex semantic relationship, but obtained better performance than ignoring the OOV words.

Some other methods are capable of handling OOV words by learning good enough representation to keep or enhance predictive model performance using OOV morphological structure and context information. In addition, some methods are able to predict the word of the vocabulary that is most similar to the OOV using a language model [11].

2.1   Approaches Based on the Word Context or Structure

According to Hu et al. [11], the representation methods able of learning new representations to handle OOV usually employ two different sources of information:

the morphological structure of the OOV words [4,10], or the context in which the OOV is inserted [11].

FastText is a popular distributed representation model that associates the morphological structure (subword) to the vector representation of the words. This feature enables FastText to handle OOV using its morphological learning capabilities. In this representation model, when an OOV word is found, a new representation is computed for it using its subwords [4].

Mimick [21], like FastText, also uses the word structure to represent an OOV. It is a deep learning model that uses every character of the OOV word to produce a vector representation. A neural network is fed with the characters of the word, and the target is a known representation for that word in a trained representation model. The objective function of the neural network minimizes the distance between the characters of the word and its known representation in the representation model. After Mimick is trained, a vector representation for an OOV word can be estimated using its characters [21].

The main drawback of methods that use only morphological information is their incapacity to handle OOV words that have different meanings in different contexts. When morphological information is the only source to generate a representation, an OOV word will always have the same representation regardless of the context in which it appears. To address it, a group of methods identified by Hu *et al.* [11] learns representation for an OOV word using context information, instead only morphological information.

Comick [10] was proposed using Mimick architecture as reference. This approach also feds the neural network with word embeddings from context. A composition function is applied to calculate a context vector using a window size of n words. The objective function averages context vector and characters vectors to minimize the distance between this composition and its known representation in the representation model. In this way, Comick differs from Mimick because it can generate different representations for an OOV when it has different meanings according to the context.

HiCE [11] is similar to Comick due to its capabilities to learn a representation for an OOV word using both morphological information and context. It is a deep architecture that finds a good enough representation for a word using the concept of few-shot learning. Using only few examples to feed a hierarchical attention network, HiCE found good enough representations for rare words, according to the high degree of confidence and the high Spearman correlation reported by Hu *et al.* [11]. The authors also proposed an adaptation model to extend the learned representations to different domains.

2.2 Deep Learning Based Language Models

Some language models are capable of predict the next word in a sentence while preserving the semantics of the original context due to their ability to learn semantic features from corpora [6]. While most traditional language models are statistical models, recent approaches uses deep learning (DL) architectures to perform neural network language model task with higher accuracy [27]. In the following, several DL language models are described.

LSTM. Recurrent neural networks are appropriate for sequence modeling due to their memory capability, which makes it possible to take into account all predecessor words, instead of a fixed context length as in feed-forward networks [18]. This architecture is hard to train because it suffers with the vanishing gradient problem while propagating gradient through the network. LSTM was proposed as a solution to the vanishing gradient problem using gates to define which features should be remembered across the network with a scaling factor fixed to one. The results obtained by Sundermeyer et al. [27] indicated that LSTM is better for sequence modeling in recognition systems than back-off models, which were considered state-of-the-art at the time.

Transformer. Introduced in Vaswani et al. [29] as an alternative to complex recurrent or convolutional neural networks in an encoder-decoder configuration. The transformer architecture is a simple network architecture based solely on attention mechanisms, dispensing with recurrence and convolutions entirely. As opposed to LSTM, which reads the text input sequentially (left-to-right or right-to-left), the transformer reads the entire sequence of words at once, which allows the model to learn the context of a word based on all of its surroundings. It is also efficient in terms of computational cost as it is more parallelizable, and it achieved state-of-the-art performance in many NLP tasks.

After the transformer architecture became available, many works were proposed using it as cornerstone (e.g., GPT-2 and BERT family). They use the principle of solely rely on attention mechanism for language modeling.

GPT-2. Defined by authors as "large transformer-based language model" [22], it was trained on a dataset of 8 million web pages to predict the next word, given all of the previous words within the text. The authors claim that GPT-2 is able to perform tasks as zero-shot, which means a task can be performed without any data of the specific domain.

Previously efforts on using transformer to language modeling relied on looking a text sequence from left-to-right or combined directions left-to-right and right-to-left in training phase due to the task of predicting next word. BERT (Bidirectional Encoder Representations from Transformers) [9] can have a deeper sense of language context and flow than single-direction language models because is bidirectionally trained. To make it possible, the authors of BERT redefined task on training phase as (1) masked model language (MLM), where instead to predict the next word, the model should predict any word masked (replaced by [MASK]) in the sample, and (2) predicting next sequence, where the model receives pairs of sentences as input and learns to predict if the second sentence is the subsequent one to the first sentence in the original document.

It is still an open challenge to train large models, such as BERT, and make them feasible for inference since their size is prohibitive in terms of computational cost. Considering that, many efforts have been reported in the literature to improve these models lowering the minimum requirements to run them, such as RoBERTa [15], DistillBERT [24], and Electra [7].

RoBERTa. Although BERT has shown state-of-the-art improvements in many NLP tasks in the last years, so far, it is expensive to process large corpus to train. RoBERTa [15] is a new approach to pretrain on BERT in a optimized way. Instead of a new architecture, RoBERTa is described as a new model which removes next sentence pretraining objective from original BERT and uses larger mini batches and learning rates to improve performance across many NLP tasks. The effectiveness of this new model is also related to a new dataset using public news article several times bigger than the one employed on BERT training.

DistillBERT. As described by Sanh *et al.* [24], DistillBERT is a smaller general-purpose language representation model which can be fine-tuned on specific domain tasks in NLP. On their work, the authors claim that knowledge distillation on pretraining can reduce BERT size by 40%, keeping 97% of language understanding capabilities, producing a model 60% faster.

Electra. While masked language models require a large amount of computational resources to train a model due to the expensive process of reconstruct masked original tokens, Electra [7] is an alternative more efficient that uses a replacement process with likely tokens instead the masking process from original BERT work to train the model. The results claimed by authors indicate lower costs on training with huge improvements to small models, leading to even better results than originally observed with BERT model. The authors also claim that for large models, such as RoBERTa, it uses less than 25% of computational resources, and can outperform them using the same amount of resources.

In this study, we present a performance evaluation of Comick, HiCE, LSTM, Transformer, GPT-2, RoBERTa, DistillBERT, and Electra in the task of handling OOV words. As these DL models have different strategies for handling OOVs, a comparative analysis may reveal the best strategy and present valuable insights for future research on the problem of handling OOV words. In the following, we present the experiments performed to evaluate these models. We performed an intrinsic evaluation using a benchmark dataset and an extrinsic evaluation using different NLP tasks.

3   Intrinsic Evaluation

To analyze the ability of DL models to find good representations for OOVs, we first performed an intrinsic evaluation using the benchmark Chimera dataset [14] that simulates OOV words in real-world applications. In this dataset, two, four, or six sentences are used to determine the meaning of a OOV, called chimera. This OOV is a nonword used to simulate a new word whose meaning is a combination of the meanings of two existing words. For each chimera, the dataset provides a set of six probing words and the human annotated similarities between the probing words and the chimeras [14]. Figure 1 presents an example of a chimera from the Chimera dataset.

To generate the vector representation of the sentences of the Chimera dataset, we used FastText word embeddings [4] trained on Twitter7 (T7) [30], a corpus of

Chimera: pirbin (a combination of **alligator** and **rattlesnake**)
Sentence 1: The outside section will also be used by PIRBINS, the rare L'hoest and Diana monkeys, cheetah and leopards. **Sentence 2:** But the kangaroo rat can hear the faint rustles of the PIRBIN's scales moving over the sand, and escape. **Sentence 3:** Blackmer and Culp are by now halfway across the swamp and have attracted the attention of several PIRBINS. **Sentence 4:** Faced with jewels, I sort of did a story and put a jungle PIRBINS into it.
Probe words: crocodile; iguana; gorilla; banner; buzzard; shovel
Human annotated similarities: crocodile (2.29); iguana (3.43); gorilla (2.14); banner (1.57); buzzard (2.71); shovel(1.43)

Fig. 1. Example of a Chimera defined by four sentences.

tweets posted from June 2009 to December 2009. We removed the retweets and empty messages of this corpus and selected only the English-language messages. At the end, we used the remaining 364,025,273 tweets to train the embeddings.

In this study, we infer an embedding for a given Chimera based on each sentence (two, four, or six sentences) that forms its meaning. If the model is unable to infer a vector for the chimera in a given sentence, the embedding of that chimera in that sentence is a vector of zeros. The final embedding for the chimera is the average of the vectors obtained for each sentence.

For each Chimera, we calculated the cosine similarity between the embeddings for the chimera and the probe words. Then, to evaluate the performance of a model for a given chimera, we calculated the Spearman correlation between the cosine similarities obtained by the model and the human annotations, as performed in other related studies (*e.g.,* Hu *et al.* [11], Khodak *et al.* [12]). In experiments where the model was unable to infer the embedding for a given chimera in any sentences, resulting in a zero vector for the embedding, we assigned a value of zero for the correlation (worst possible value). The overall performance of the model is the average Spearman correlation across all chimeras.

3.1   Baseline Methods

We compare the results of the DL models with the following baselines:

- **Oracle:** it is probably the best possible embedding for the Chimera, since it is the average vector of the embeddings of the two gold-standard words that compose the Chimera. Therefore, we considered the Spearman correlation between the results obtained by the Oracle and the human annotations as the upper bound performance.
- **Sum:** the vector of the Chimera in a sentence is the sum of the words embeddings of this sentence. Then, the final vector for the Chimera, is the average vector of the sentences that form the Chimera.
- **Average:** the vector of the Chimera in a sentence is the average of the embeddings of the words of this sentence. Then, the final vector for the Chimera, is the average vector of the sentences that form the Chimera.

3.2   Deep Learning Models

A chimera is a word that does not exist in the real-world. Therefore, in the intrinsic evaluation, it was not possible to perform experiments with techniques that analyze the morphological structure of the word, such as Comick. We present below the experimental settings for the evaluated DL models:

- **DistilBERT:** it was used the original model available in the `transformers`[2] library for Python 3, identified by the keyword '`distilbert-base-cased`'.
- **HiCE (context):** this model was trained using 10% of the T7 dataset using the default parameters of the official implementation[3]. However, the morphological information flag was set to `false`, since the OOVs in this experiment (chimeras) are not real-world words and, therefore, we can use only its context to infer its embedding.
- **GPT2:** it was used the original model available in the `transformers` library for Python 3, identified by the keyword '`gpt2-large`'.
- **Electra:** this language model was trained using 10% of the T7 dataset using the default parameters, except for the number of epochs, which was set to 10. The implementation for Electra model is available in the `transformers` library for Python 3.
- **LSTM:** this model was trained using 10% of the T7 dataset using default parameters, except for the number of epochs, which was set to 10. The implementation is available in the PyTorch[4] repository.
- **Transformer:** this model was trained using 10% of T7 dataset using default parameters, except for the number of epochs, which was set to 10. The implementation for Transformer architecture is available in the `transformers` library for Python 3.
- **RoBERTa:** it was used the original model available in the `transformers` library for Python 3, identified by the keyword '`roberta-base`'.

For all DL models, we used the default parameters because the computational cost to make an appropriate parameter selection is very high.

As the embeddings returned by the DL models are not in the same vector space of the word embeddings trained on T7, the models return a list of five candidate words to replace the OOV. The vector of the first candidate in the T7 vocabulary is used to represent the OOV.

3.3   Results

Table 1(a) shows the average Spearman correlation obtained on the Chimera dataset. The average ranking of each method is also presented. For each experiment, the method that obtained the best average Spearman correlation received

---

[2] Transformers. Available at https://huggingface.co/transformers, Accessed on October 10, 2020.

[3] HiCE. Available at https://github.com/acbull/HiCE, Accessed on October 10, 2020.

[4] PyTorch Github. Available at https://bit.ly/2B7LS3U, Accessed on October 10, 2020.

rank 1 and the worst one obtained rank 9 (nine techniques were evaluated). Therefore, the smaller the average ranking, the better the performance. The results are presented as a grayscale heat map, where the better the value, the darker the cell color. Bold values indicate the best result. Moreover, the methods are sorted by the average ranking.

To complement the analysis of the results, Table 1(b) presents the percentage of sentences where the evaluated DL model was able to find a vector for the chimera. The baseline methods (Average and Sum) always return a vector because they do not make any kind of prediction and, therefore, it was not necessary to include them in the table.

Table 1. Results obtained in the intrinsic evaluation.

(a) Average Spearman correlation.

| | | Avg. correlation | | | Avg. ranking |
		2 sent.	4 sent.	6 sent.	
Baselines	Oracle	0.40	0.40	0.43	- -
Baselines	Average	0.26	0.29	0.30	2.00
Baselines	Sum	0.26	0.28	0.30	2.33
Deep learning	DistilBERT	**0.27**	**0.31**	**0.37**	1.00
Deep learning	HiCE (context)	0.16	0.26	0.29	4.00
Deep learning	GPT2	0.15	0.22	0.20	5.00
Deep learning	Electra	0.06	0.17	0.16	6.00
Deep learning	LSTM	0.06	0.11	0.10	6.67
Deep learning	Transformer	0.05	0.04	0.10	7.67
Deep learning	RoBERTa	0.03	0.03	0.02	9.00

(b) Statistics of the OOVs.

		2 sent.	4 sent.	6 sent.
% of OOVs treated by the models	HiCE (context)	100.0	100.0	100.0
% of OOVs treated by the models	Electra	100.0	99.0	99.0
% of OOVs treated by the models	GPT2	98.0	97.0	97.0
% of OOVs treated by the models	LSTM	97.0	97.0	97.0
% of OOVs treated by the models	Transformer	97.0	97.0	97.0
% of OOVs treated by the models	DistilBERT	95.0	95.0	95.0
% of OOVs treated by the models	RoBERTa	9.0	10.0	10.0

In general, DistilBERT was the best technique for inferring embeddings for the chimeras. The average Spearman correlation of this tecnique was the closest to the upper bound performance (the one obtained by the Oracle) on the experiments with two, four, and six sentences.

The baseline approaches (Average and Sum) obtained, respectively, the second and third best performance. On the other hand, RoBERTa obtained the lowest Spearman correlation in all experiments, which also resulted in the worst average ranking. These results can be better understood when analyzed together with the statistics shown in Table 1(b). We can see that RoBERTa was able to infer embeddings for the Chimeras, in at most only 10% of the sentences.

HiCE was able to infer embeddings for the chimeras in 100% of the sentences, but obtained only the fourth best performance, in general. On average, its score was 26% lower than the one obtained by the best model (DistilBERT).

LSTM, Transformer, and GPT2 use only the context window to the left of the OOV to infer its embedding. Some chimeras are located in the first position of the sentence, leaving no context window to be used by these models. Therefore, in these sentences, they are not able to infer the vector for the chimera. Despite this, the percentage of chimeras handled by these models was still higher than DistilBERT and RoBERTa.

4   Extrinsic Evaluation

We conduct a comprehensive extrinsic evaluation of the DL models in three established tasks that can be greatly affected by OOV words:

- **text categorization:** we address the task of polarity sentiment classification in short and noise messages.
- **named entity recognition (NER):** this task seeks to locate and classify entities in a sentence.
- **part-of-speech (POS) tagging:** this task seeks to identify the grammatical group of a given word.

In all experiments, we used the same FastText word embeddings applied in the intrinsic evaluation to generate the vector representation of the documents.

Baseline Methods. We compare the results of the DL models with the following baselines:

- **Sum:** the OOV is represented by the sum of the embeddings of the words in the document.
- **Average:** the OOV is represented by the average of the embeddings of the words in the document.
- **Zero:** the OOV is represented by a vector of zeros.
- **Random:** all OOVs are represented by the same random vector generated at the beginning of the experiment.
- **FastText:** the OOV is represented by the vector obtained by the FastText [4].

Deep Learning Models. In the extrinsic evaluation, we performed experiments with the same DL methods used in the intrinsic evaluation using the same experimental settings described in Sect. 3.2. Additionally, we also evaluated models that use the morphological information of the OOV word to infer an embedding. We present below the experimental settings for these DL models:

- **HiCE:** this model was trained using 10% of the T7 dataset using the default parameters of the official implementation[5]. However, the morphological information flag was set to `true`, since, in this experiment, the word structure of the OOV words can be used to predict reliable embeddings.
- **Comick:** this model was trained using 10% of the T7 dataset using the default parameters on the private implementation obtained from Garneau *et al.* [10].

For all DL models, we used the default parameters because the computational cost to make an appropriate parameter selection is very high. Both HiCE and Comick infer the representation for the OOV in the same way as the other DL models we evaluated (see Sect. 3.2).

---

[5] HiCE. Available at https://github.com/acbull/HiCE, Accessed on October 10, 2020.

4.1   Text Categorization

In order to give credibility to the results and make the experiments reproducible, all tests were performed with the following real and public datasets: Archeage, Hobbit, and IPhone6 [17]; OMD (Obama-McCain *debate*) [25], HCR (*health care reform*) [26], Sanders [26], SS-Tweet (*sentiment strength* Twitter *dataset*) [28], STS-Test (Stanford Twitter *sentiment test set*) [2], and UMICH [17]. All sentences were converted to lowercase and they were processed using Ekphrasis [3], a tool to normalize text from social media.

Evaluation. The experiments were carried out with a Bidirectional LSTM. We built the LSTM using Keras[6] on top of TensorFlow[7]. All the documents were padded or truncated to 200 words. The OOVs that were not handled and paddings words were represented by a vector of zeros. We did not perform any parameter optimization for the neural network because the objective of this study is not to obtain the best possible result, but to analyze the ability of DL models to handle OOVs. The experiments were performed using stratified holdout validation with 70% of the documents in the training set and 30% in the test set. To compare the results, we employed the macro F-measure.

Results. Table 2(a) presents the macro F-measure. The average ranking of each method is also presented. For each dataset, the method that obtained the best macro F-measure received rank 1 and the worst one obtained rank 14 (fourteen methods were evaluated). The methods are sorted by the average ranking. Bold values indicate the best scores. Moreover, the scores are presented as a grayscale heat map, where the better the score, the darker the cell color.

To complement the analysis of the results, Table 2(b) presents the percentage of documents that have some OOV. Moreover, among the documents that have some OOV, Table 2(b) shows the percentage of them that had at least one OOV handled. The baseline methods (Zero, Random, Sum, and Average) always return a vector and, therefore, it was not necessary to include them in the table.

In general, none of the baseline techniques and DL models excelled in all datasets. Comick obtained the best score in three datasets (HCR, OMD, and UMICH) and the best average ranking, and Transformer obtained the best macro F-measure in four datasets (Archeage, Iphone, STS-Test, and UMICH). Two of these datasets, Iphone6 and Archeage, are the ones that have the highest percentage of documents with OOV. They are also the only datasets with more than 10% of documents with an OOV. Therefore, they are probably the datasets with the greatest degree of difficulty. Furthermore, RoBERTa was the technique that treated the lowest percentage of documents with OOVs. Despite this, RoBERTa obtained the fourth best average ranking among the DL methods.

---

[6] Keras. Available at https://keras.io/. Accessed on October 10, 2020.

[7] TensorFlow. Available at https://www.tensorflow.org/. Accessed on October 10, 2020.

Table 2. Experiments on text categorization.

(a) Macro F-measure obtained on text categorization.

		Macro F-measure									Avg.
		Archeage	HCR	Hobbit	IPhone6	OMD	SS-Tweet	STS-Test	Sanders	UMICH	ranking
Baselines	Zero	0.84	0.64	0.88	0.71	0.79	0.79	0.90	**0.86**	0.95	4.78
	Average	0.84	0.62	0.91	0.72	0.79	0.77	0.89	0.85	0.94	6.33
	Sum	0.84	0.65	0.87	0.67	**0.80**	0.64	0.89	**0.86**	**0.96**	6.44
	Random	**0.85**	0.67	0.83	0.72	0.77	0.76	0.89	0.82	0.95	8.00
	FastText	0.81	**0.68**	0.87	0.71	0.77	0.77	0.88	0.84	0.94	8.89
Deep learning	Comick	0.84	**0.68**	0.80	0.74	**0.80**	0.79	0.92	0.84	**0.96**	3.89
	Transformer	**0.85**	0.63	0.89	**0.76**	0.78	0.77	**0.93**	0.84	**0.96**	4.78
	GPT2	0.84	0.66	0.90	0.72	0.79	0.78	0.90	0.77	0.95	5.00
	RoBERTa	0.81	0.65	0.90	0.73	0.79	0.77	0.90	0.84	0.95	5.44
	Electra	0.83	0.64	0.84	0.71	0.79	0.77	0.90	**0.86**	0.95	6.11
	HiCE	0.82	0.61	0.90	0.73	0.78	0.78	**0.93**	**0.86**	0.94	6.22
	HiCE (context)	0.83	0.66	0.90	0.71	0.77	**0.80**	0.88	0.84	0.95	6.67
	LSTM	0.80	0.60	**0.92**	0.75	0.78	0.75	0.90	0.84	0.94	8.11
	DistilBERT	0.83	0.64	0.89	0.64	0.78	0.72	0.86	0.85	0.95	9.00

(b) Statistics of the OOVs.

		Archeage	HCR	Hobbit	IPhone6	OMD	SS-Tweet	STS-Test	Sanders	UMICH
	% of docs with OOV	51.0	4.0	6.0	61.0	3.0	5.0	3.0	9.0	2.0
% of docs treated by the models	HiCE	100.0	100.0	100.0	100.0	100.0	100.0	100.0	100.0	100.0
	HiCE (context)	100.0	100.0	100.0	100.0	100.0	100.0	100.0	100.0	100.0
	Comick	100.0	100.0	100.0	100.0	100.0	100.0	100.0	100.0	100.0
	Electra	99.0	100.0	100.0	100.0	100.0	95.0	90.0	100.0	95.0
	GPT2	91.0	97.0	93.0	95.0	88.0	93.0	90.0	93.0	79.0
	LSTM	91.0	97.0	93.0	94.0	88.0	92.0	90.0	93.0	79.0
	Transformer	91.0	97.0	93.0	94.0	88.0	92.0	90.0	93.0	79.0
	DistilBERT	75.0	45.0	76.0	89.0	75.0	65.0	80.0	70.0	79.0
	RoBERTa	27.0	43.0	34.0	17.0	39.0	33.0	20.0	39.0	47.0

DistilBERT, the best model in the intrinsic evaluation (Sect. 3.3), obtained the worst average ranking in the text categorization. One of the factors that may have contributed to the worsening of the result is the amount of OOVs that it was unable to handle. In the intrinsic evaluation, it was able to infer the vector for the chimera in 95% of the sentences, on average (Table 1(b)). However, on text categorization (Table 2(b)), on average, DistilBERT was able to infer vectors for OOVs in only 72% of the documents that had some OOV.

4.2 NER and POS Tagging

We performed experiments with three datasets commonly used in studies that address OOVs [11]: Rare-NER [8][8], Bio-NER [13][9], and Twitter-POS [23][10].

All sentences were converted to lowercase. All documents from all datasets are already provided separated by terms and each term has an associated label (entity or POS). Therefore, it was not necessary to perform tokenization.

---

[8] NER dataset with unusual and unseen entities in the context of emerging discussions.
[9] NER dataset with sentences that have technical terms in the biology domain.
[10] POS tagging dataset of Twitter messages.

Evaluation. The experiments with NER and POS tagging were carried out using a Bidirectional LSTM with a time distributed dense layer built using Keras on top of TensorFlow. All the documents were padded or truncated to 200 words. The OOVs that were not handled and paddings words were represented by a vector of zeros.

The experiments were performed using a holdout validation with 70% of the documents in the training set and 30% in the test set. To compare the results in the NER task, we employed the entity level F-measure. In the experiments with POS-tagging, we evaluated the prediction performance using the token accuracy.

Results. Table 3(a) presents the results obtained. For each dataset, the method that obtained the best F-measure (NER) or the best accuracy (POS-tagging), obtained rank 1, while the worst method obtained rank 14. The methods are sorted by the average ranking. In addition, Table 3(b) presents some statistics about the OOVs of the evaluated datasets.

Table 3. Experiments on NER and POS tagging.

(a) Performance on NER and POS tagging.

		NER (F-measure)		POS tagging (accuracy)	Avg. ranking
		Bio-NER	Rare-NER	Twitter-POS	
Baselines	FastText	**0.67**	**0.46**	**0.78**	1.00
	Average	0.63	**0.46**	0.76	1.67
	Random	0.62	0.45	0.74	5.00
	Zero	0.62	0.37	0.76	6.33
	Sum	0.60	0.44	0.74	7.33
Deep learning	LSTM	0.59	**0.46**	0.75	5.00
	Comick	0.62	0.44	0.73	6.67
	RoBERTa	0.62	0.42	0.73	7.33
	Transformer	0.56	0.45	0.75	7.67
	DistilBERT	0.58	0.39	0.76	8.00
	Electra	0.57	0.41	0.75	9.33
	GPT2	0.57	0.42	0.74	9.33
	HiCE (context)	0.61	0.39	0.73	10.00
	HiCE	0.58	0.42	0.71	10.67

(b) Statistics of the OOVs.

		Bio-NER	Rare-NER	Twitter-POS
	% of docs with OOV	85.0	69.0	85.0
% of docs treated by the models	Comick	100.0	100.0	100.0
	HiCE	100.0	100.0	100.0
	HiCE (context)	100.0	100.0	100.0
	GPT2	98.0	92.0	90.0
	LSTM	98.0	91.0	89.0
	Transformer	98.0	91.0	89.0
	DistilBERT	99.0	84.0	85.0
	Electra	93.0	82.0	85.0
	RoBERTa	23.0	67.0	74.0

FastText was the best technique in the experiments with NER and POS tagging. In the experiment with Bio-NER, the F-measure obtained by FastText was 5% higher than the second best score (Average) and 20% higher than the lowest score (Transformer). Furthermore, the average ranking of simple baselines techniques (Sum, Zero, Average, and Random) was higher than most DL methods. The reason may be that OOV words have little impact on the text of the evaluated datasets, for being rare words or have little semantic importance. We also believe that the noise from the documents in T7 (Twitter messages) may have affected the quality of the embeddings.

Among the DL models, LSTM obtained the best average ranking. Moreover, Comick repeated the good performance shown in the text categorization task, obtaining the second best average ranking in NER and POS tasks. This model, unlike most other DL models (except Hice) that we have evaluated, in addition to the context, analyzes the morphological structure of the OOVs. The results indicate that this characteristic may have benefited Comick in relation to the other DL models.

By the statistics shown in Table 3(b), we can note that the NER and POS tagging datasets have more OOVs than the text categorization datasets (Table 2(b)). In addition, NER and POS tagging can be more impacted by OOVs than the text categorization task, as almost all words are associated with a label (entity or POS). As in experiments with the text classification task, in the experiments with NER and POS tagging, RoBERTa was the DL model that treated the lowest percentage of documents with OOVS.

DistilBERT, the best method in the intrinsic evaluation (Sect. 3.3) and the second worst method in the text categorization (Sect. 4.1), obtained the second best performance in the POS tagging task, but was one of the worst methods in the experiments with NER.

5   Conclusions

The phenomenon of OOVs is a major problem in natural language processing tasks. Documents that have OOVs are usually incompletely represented by distributed text representation models. The lack of one or more words can significantly change the semantics of a sentence.

Distributed text representation models are not incremental and, therefore, the training process is performed only once due to the high computational cost demanded. As the model generated in the training is not updated over time, it is unable to deal with new words that were not seen during its training. Therefore, the more dynamic the communication becomes, the more OOVs appear, and the faster the model becomes obsolete.

In this paper, we presented a comprehensive performance evaluation of different DL models applied to handle OOVs. Among the evaluated models, DistilBERT, GPT2, Electra, LSTM, Transformer, and Roberta infer the embedding for a given OOV using approximation by the terms that appear next to the OOV in the sentence. Comick and HiCE, in addition to the context, use the morphological structure of the OOV for the inference.

To analyze these models, we performed an intrinsic evaluation using the benchmark Chimera dataset. We also performed an extrinsic evaluation with the text categorization task using nine public and well-known datasets for opinion polarity detection on Twitter messages, and with the tasks of NER and POS tagging, using three datasets frequently used in related studies.

There was no model that obtained the best performance in all evaluations. However, in general, Comick obtained a good performance in all extrinsic evaluation tasks, which resulted in higher average rankings than most other evaluated DL models. The ability of this model to analyze the morphological structure of OOVs and their context may have contributed to achieving superior performance, although Hice did not achieve the same success even having the same characteristics.

Considering each experiment more specifically, in the intrinsic evaluation, DistilBERT obtained the best performance, with a significant difference to other methods. In the text categorization task, in general, Comick was the best method to infer embeddings for OOVs. Finally, in NER and POS tagging, the best performance was obtained by one of the baseline techniques: FastText.

Based on the results, we can conclude that the task of inferring embeddings for OOVs generates different challenges for different evaluated scenarios. Therefore, we recommend that research on OOV handing techniques be addressed to specific tasks, increasing the chance of success.

We also noticed that for some scenarios with noisy texts (as in the datasets with Twitter messages) or sentences full of technical terms (as in the Bio-NER dataset), the context of OOVs and their morphological structure may not be enough to infer a good embedding. Therefore, we recommend using an architecture for OOV handling that combines the techniques analyzed in this study with simpler techniques based on spell checker and semantic dictionaries.

In future research, we also intend to analyze the techniques covered in this study in texts from other languages, such as Portuguese, and in other NLP tasks, such as knowledge base completion, stance detection, and question answering.

References

1. Adams, O., Makarucha, A., Neubig, G., Bird, S., Cohn, T.: Cross-lingual word embeddings for low-resource language modeling. In: Proceedings of the 15th Conference of the European Chapter of the Association for Computational Linguistics: Volume 1, Long Papers, pp. 937–947. Association for Computational Linguistics, Valencia, April 2017
2. Agarwal, A., Xie, B., Vovsha, I., Rambow, O., Passonneau, R.: Sentiment analysis of twitter data. In: Proceedings of the Workshop on Languages in Social Media (LSM 2011), pp. 30–38. Association for Computational Linguistics, Portland, June 2011
3. Baziotis, C., Pelekis, N., Doulkeridis, C.: DataStories at SemEval-2017 task 4: deep LSTM with attention for message-level and topic-based sentiment analysis. In: Proceedings of the 11th International Workshop on Semantic Evaluation (SemEval-2017), pp. 747–754. Association for Computational Linguistics, Vancouver, August 2017. https://doi.org/10.18653/v1/S17-2126
4. Bojanowski, P., Grave, E., Joulin, A., Mikolov, T.: Enriching word vectors with subword information. Trans. Assoc. Comput. Linguist. 5, 135–146 (2017)
5. Cho, K., et al.: Learning phrase representations using rnn encoder-decoder for statistical machine translation. In: Proceedings of the 2014 Conference on Empirical Methods in Natural Language Processing (EMNLP), pp. 1724–1734. Association for Computational Linguistics, Doha, October 2014
6. Chomsky, N.: Syntactic Structures. Mouton and Co., The Hague (1957)
7. Clark, K., Luong, M.T., Le, Q.V., Manning, C.D.: Electra: pre-training text encoders as discriminators rather than generators. In: International Conference on Learning Representations (2020)

8. Derczynski, L., Nichols, E., van Erp, M., Limsopatham, N.: Results of the WNUT2017 shared task on novel and emerging entity recognition. In: Proceedings of the 3rd Workshop on Noisy User-generated Text, pp. 140–147. Association for Computational Linguistics, Copenhagen, September 2017. https://doi.org/10.18653/v1/W17-4418

9. Devlin, J., Chang, M., Lee, K., Toutanova, K.: BERT: pre-training of deep bidirectional transformers for language understanding. CoRR abs/1810.04805 (2018). http://arxiv.org/abs/1810.04805

10. Garneau, N., Leboeuf, J.S., Lamontagne, L.: Predicting and interpreting embeddings for out of vocabulary words in downstream tasks. In: Proceedings of the 2018 EMNLP Workshop BlackboxNLP: Analyzing and Interpreting Neural Networks for NLP, pp. 331–333. Association for Computational Linguistics, Brussels, November 2018. https://doi.org/10.18653/v1/W18-5439

11. Hu, Z., Chen, T., Chang, K.W., Sun, Y.: Few-shot representation learning for out-of-vocabulary words. In: Proceedings of the 57th Annual Meeting of the Association for Computational Linguistics, pp. 4102–4112. Association for Computational Linguistics, Florence, July 2019. https://doi.org/10.18653/v1/P19-1402

12. Khodak, M., Saunshi, N., Liang, Y., Ma, T., Stewart, B., Arora, S.: A la carte embedding: cheap but effective induction of semantic feature vectors. In: Proceedings of the 56th Annual Meeting of the Association for Computational Linguistics (Volume 1: Long Papers), pp. 12–22. Association for Computational Linguistics, Melbourne, July 2018. https://doi.org/10.18653/v1/P18-1002

13. Kim, J.D., Ohta, T., Tsuruoka, Y., Tateisi, Y., Collier, N.: Introduction to the bio-entity recognition task at JNLPBA. In: Proceedings of the International Joint Workshop on Natural Language Processing in Biomedicine and Its Applications. JNLPBA 2004, pp. 70–75. Association for Computational Linguistics, Stroudsburg (2004)

14. Lazaridou, A., Marelli, M., Zamparelli, R., Baroni, M.: Compositional-ly derived representations of morphologically complex words in distributional semantics. In: Proceedings of the 51st Annual Meeting of the Association for Computational Linguistics (Volume 1: Long Papers), pp. 1517–1526. Association for Computational Linguistics, Sofia, August 2013

15. Liu, Y., et al.: Roberta: a robustly optimized BERT pretraining approach. CoRR abs/1907.11692 (2019)

16. Lochter, J., Pires, P., Bossolani, C., Yamakami, A., Almeida, T.: Evaluating the impact of corpora used to train distributed text representation models for noisy and short texts. In: 2018 International Joint Conference on Neural Networks (IJCNN), pp. 315–322, July 2018

17. Lochter, J., Zanetti, R., Reller, D., Almeida, T.: Short text opinion detection using ensemble of classifiers and semantic indexing. Expert Syst. Appl. **62**, 243–249 (2016)

18. Mikolov, T., Karafiát, M., Burget, L., Cernocký, J., Khudanpur, S.: Recurrent neural network based language model. In: INTERSPEECH, pp. 1045–1048 (2010)

19. Mikolov, T., Sutskever, I., Chen, K., Corrado, G., Dean, J.: Distributed representations of words and phrases and their compositionality. In: Proceedings of the 26th International Conference on Neural Information Processing Systems (NIPS 2013), pp. 3111–3119. Curran Associates Inc., Lake Tahoe (2013)

20. Ouyang, X., Zhou, P., Li, C.H., Liu, L.: Sentiment analysis using convolutional neural network. In: 2015 IEEE International Conference on Computer and Information Technology; Ubiquitous Computing and Communications; Dependable, Autonomic and Secure Computing; Pervasive Intelligence and Computing, pp. 2359–2364 (2015)

21. Pinter, Y., Guthrie, R., Eisenstein, J.: Mimicking word embeddings using subword RNNs. CoRR abs/1707.06961 (2017). http://arxiv.org/abs/1707.06961

22. Radford, A., Wu, J., Child, R., Luan, D., Amodei, D., Sutskever, I.: Language models are unsupervised multitask learners (2019)

23. Ritter, A., Clark, S., Mausam, Etzioni, O.: Named entity recognition in tweets: an experimental study. In: Proceedings of the 2011 Conference on Empirical Methods in Natural Language Processing, pp. 1524–1534. Association for Computational Linguistics, Edinburgh, July 2011

24. Sanh, V., Debut, L., Chaumond, J., Wolf, T.: DistilBERT, a distilled version of BERT: smaller, faster, cheaper and lighter (2019)

25. Shamma, D.A., Kennedy, L., Churchill, E.F.: Tweet the debates: understanding community annotation of uncollected sources. In: Proceedings of the First SIGMM Workshop on Social Media (WSM 2009), pp. 3–10. ACM, Beijing (2009). https://doi.org/10.1145/1631144.1631148

26. Speriosu, M., Sudan, N., Upadhyay, S., Baldridge, J.: Twitter polarity classification with label propagation over lexical links and the follower graph. In: Proceedings of the First Workshop on Unsupervised Learning in NLP (EMNLP 2011), pp. 53–63. Association for Computational Linguistics, Edinburgh (2011)

27. Sundermeyer, M., Schlüter, R., Ney, H.: LSTM neural networks for language modeling. In: INTERSPEECH (2012)

28. Thelwall, M., Buckley, K., Paltoglou, G.: Sentiment strength detection for the social web. J. Am. Soc. Inf. Sci. Technol. **63**(1), 163–173 (2012). https://doi.org/10.1002/asi.21662

29. Vaswani, A., et al.: Attention is all you need. In: Guyon, I., et al. (eds.) Advances in Neural Information Processing Systems 30, pp. 5998–6008. Curran Associates, Inc. (2017)

30. Yang, J., Leskovec, J.: Patterns of temporal variation in online media. In: Proceedings of the Fourth ACM International Conference on Web Search and Data Mining. WSDM 2011, pp. 177–186. ACM, New York (2011)

31. Yang, X., Macdonald, C., Ounis, I.: Using word embeddings in Twitter election classification. Inf. Retrieval J. **21**(2–3), 183–207 (2017). https://doi.org/10.1007/s10791-017-9319-5

DeepBT and NLP Data Augmentation Techniques: A New Proposal and a Comprehensive Study

Taynan Maier Ferreira[1,2]([✉]) [iD] and Anna Helena Reali Costa[1] [iD]

[1] Escola Politécnica, Universidade de São Paulo, São Paulo, Brazil
{taynan.ferreira,anna.reali}@usp.br
[2] Data Science Team, Itaú-Unibanco, São Paulo, Brazil

Abstract. Data Augmentation methods – a family of techniques designed for synthetic generation of training data – have shown remarkable results in various Deep Learning and Machine Learning tasks. Despite its widespread and successful adoption within the computer vision community, data augmentation techniques designed for natural language processing (NLP) tasks have exhibited much slower advances and limited success in achieving performance gains. As a consequence, with the exception of applications of back-translation to machine translation tasks, these techniques have not been as thoroughly explored by the wider NLP community. Recent research on the subject also still lacks a proper practical understanding of the relationship between data augmentation and several important aspects of model design, such as hyperparameters and regularization parameters. In this paper, we perform a comprehensive study of NLP data augmentation techniques, comparing their relative performance under different settings. We also propose Deep Back-Translation, a novel NLP data augmentation technique and apply it to benchmark datasets. We analyze the quality of the synthetic data generated, evaluate its performance gains and compare all of these aspects to previous existing data augmentation procedures.

Keywords: Data Augmentation · Natural Language Processing · Back-Translation · Machine learning

1 Introduction

Data Augmentation can be defined as any process of artificially creating new training data by applying class-preserving transformations to the original input data [5].

Partially supported by Itaú-Unibanco, CNPq (grants 25860/2016-7 and 530307027/2017-1), and CAPES (Finance Code 001). Any opinions, findings, and conclusions expressed in this manuscript are those of the authors and do not necessarily reflect the views, official policy or position of Itaú-Unibanco.

In harmony with Statistical Learning Theory, which states that discrepancy between training and generalization error shrinks as the number of training examples increases [9], Data Augmentation has successfully been used in the Machine Learning and Deep Learning communities to artificially inflate data for training and, as a consequence, obtain models with greater generalization power.

Its application by researchers range from Image Processing [18,25], Sound and Speech Recognition [1,21], and Time Series [27] to Natural Language Processing [15,24].

Specifically within the Video Processing community, Data Augmentation has been used successfully for several years now, being part of the training process of models related to some of the greatest achievements in Image Classification tasks, such as the AlexNet [17], All-CNN [23], and ResNet [11] models.

These remarkable accomplishments have led researchers to investigate the underlying theoretical principles governing Data Augmentation, trying to shed some light into its relationship to aspects such as model learning process, decision surface, etc. These researches show that Data Augmentation improve generalization by both increasing invariance and penalizing model complexity [5].

Data Augmentation also can be considered a form of implicit regularization, closely related to explicit regularization techniques such as Weight Decay and Dropout. In fact, the works of [30] and [16] indicate that, under certain circumstances, Data Augmentation and Dropout can be considered equivalent methods. Other studies, on the other hand, state that Data Augmentation exhibit superior performance in comparison to explicit regularization methods [12].

Despite unquestionable success in computer vision tasks, NLP research has not yet benefited as largely from data augmentation systems. General NLP tasks and challenges are often characterized by the low – or often unsuccessful – usage of data augmentation techniques. When analyzing the solutions proposed for some of the SemEval Tasks over the period of 2017–2019, e.g., we observe the following.

1. SemEval-2017 Task 5: there is no mention to the use of Data Augmentation methods by any of the participants [4];
2. SemEval-2018 Task 1: among 75 teams, only 2 teams acknowledge the use of some kind of Data Augmentation procedure [19];
3. SemEval-2019 Task 5: within 74 participants, only one of them indicates using some kind of Data Augmentation in her or his system [2].

We hypothesize that the low adoption of data augmentation techniques within the NLP community is a consequence of its primitive state, still in its infancy when compared to the advanced methods used in image processing tasks.

To address this research gap and provide practitioners and researchers general guidelines on its use, we conduct an in-depth investigation of NLP data augmentation techniques. We compare their output and relative performance under various settings and study their sensibility to different parameters. We also investigate the relationship between data augmentation, which is an implicit regularization technique, with an explicit regularization technique, namely the dropout procedure.

To further advance the state-of-the-art knowledge on the subject, we also present Deep Back-Translation, an unprecedented data augmentation technique for NLP tasks which stacks more intermediate layers of translation between the original and synthesized sentences. We apply Deep Back-Translation to benchmark datasets and compare its outcomes to results generated by previous existing methods. To the best of our knowledge, we are the first to propose and study such a technique for data augmentation.

The remainder of this paper is organized as follows. In Sect. 2 we describe the main and most recent proposals in the field of Data Augmentation for Natural Language Processing. Section 3 presents the new proposed method and our main objectives in this paper, followed by the Experimental Setup at Sect. 4. After presenting the main Results in Sect. 5, we summarize our main contributions and conclude the paper in Sect. 6.

2   Related Work

In this section we highlight several recent researches in the realm of NLP-specific Data Augmentation techniques that relate to the present work.

Easy Data Augmentation (EDA) is a technique first proposed by [26]. This method consists of applying a set of simple operations to the original text in order to generate new synthetic texts. The operations, all randomly applied according to the parameter α, which controls the percentage of words changed in any given sentence, are Synonym Replacement, Random Insertion, Random Swap and Random Deletion. Though maybe original in its proposal as a pure data augmentation technique, the use of this set of operations closely resembles the noise injection procedure proposed by [7] as an auxiliary task to improve neural machine translation models.

Considered crucial to neural machine translation tasks nowadays [10], Back-Translation (BT) is another method for generating auxiliary synthetic data. First introduced by [22], the term Back-Translation was initially conceived specifically within the context of machine translation, whereby monolingual data was leveraged by translating *Target Language* → *Source Language* (hence the term *Back*) in order to obtain additional training data for the *Source Language* → *Target Language* final translation task. The first implementation of Back-Translation as a data augmentation method for down-stream tasks seems to be the work of [28], which used Back-Translation to rephrase original sentences (i.e. generating paraphrases), producing extra data and obtaining state-of-the-art results on question answering tasks.

Supported by its remarkable success in neural machine translation tasks, there has been an emergence of numerous variations to the traditional Back-Translation method.

Iterative Back-Translation (IterativeBT), proposed in [13], is a process where models are successively trained using data Back-Translated by the previous model. This cyclical training yields generation of models which are able to improve at each iteration.

Noised Back-Translation (NoisedBT)[1], presented by [7] is a variation where noise is injected to the Back-Translated text. In the seminal paper, three types of noise are used: random deletion of words, random replacement of words and random swapping of words. Despite the fact that final noised sentences are not realistic, the authors argue that the superior performance obtained by noise injection could be attributed to the model becoming robust to reordering and substitutions occurring naturally on texts.

Tagged Back-Translation (TaggedBT) [3], heavily influenced by the works of [7] and [14], proposes another hypothesis for the superiority of noise injection in Back-Translation postulated in [7]: instead of increased text diversity, noise injection would instead benefit the final model by signaling which data is synthetic and which is original data.

Despite all of the proposed variations of Back-Translation, few researches have investigated how different choices and parameters of Back-Translation can affect its performance in down-stream tasks. Questions such as the impact of language used for translation (pivotal language), or even the effectiveness of EDA compared to BT, remain still open. The ablation studies in data augmentation performed by [28] seem to be the closest to partially address some of these open issues, though leaving the majority of the questions here proposed still unanswered.

Also, none of the Back-Translation variations take advantage of language translation at a greater level for general NLP Tasks. While IterativeBT is only applicable for Machine Translation tasks, TaggedBT and NoisedBT just add additional handcrafted information to the data. To the best of our knowledge, there has not been proposed any method that leveraged translations one step further when compared to traditional Back-Translation.

3   Deep Back-Translation

In this paper we contribute a new method for data augmentation, named Deep Back-Translation (DeepBT), which adds more layers of intermediate translations between the original text and the final paraphrase. Hence, using capital letters to represent original and final languages (which are always the same) and arrows to represent translations to languages L, while in the original Back-Translation we always have

$$A \rightarrow L \rightarrow A,$$

in DeepBT we could have n intermediate layers of translations:

$$A \rightarrow L_1 \rightarrow L_2 \rightarrow ... \rightarrow L_n \rightarrow A.$$

Figure 1 illustrates the difference with a 2-layer Deep Back-Translation.

For any given Back-Translation procedure (be it the Deep version or any other) we can define the concept of Multiplication Factor, which informs us by

---

[1] The term Noised Back-Translation was not used in [7], but coined by [3].

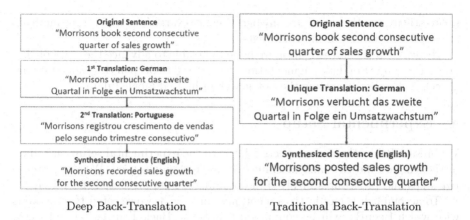

Fig. 1. Graphic representation of Deep- and traditional Back-Translation procedures. In DeepBT (a) we stack several intermediate layers (red) of translation. In traditional Back-Translation (b) there is always only one translation between original and synthesized text.

how much the available data has been multiplied by the data augmentation process.

Following the rationale of the Back-Translation method, which generates paraphrases with the same meaning and label as the original one, the Deep Back-Translation technique is created under the hypothesis that using several intermediate languages between the original and destination one could increase the difference between phrase and paraphrases, while still maintaining original meaning and label. With greater variability in training data we assume that we could reduce overfitting and achieve greater performance.

Therefore, we have three main objectives to be investigated in this paper, all addressing research gaps found in the NLP data augmentation literature.

First, we propose Deep Back-Translation, a new data augmentation technique. The development and evaluation of new NLP-specific data augmentation techniques is a relevant purpose since, as discussed before, it addresses the lack of new methods in this research area, in contrast to the continuous progress in data augmentation techniques in the computer vision community.

Second, and once again drawing inspiration from advances in the latest image processing research, we perform a systematic and comparative study of the main NLP data augmentation techniques. In particular, we aim at analyzing the relationship between data augmentation (an implicit regularization technique) to dropout (an explicit regularization procedure). This type of study has not yet been carried out within the NLP community, not even by the seminal papers that introduced the chosen methods.

Finally, we also want to address a research question related to the impact of linguistic styles in data augmentation techniques. Since NLP data augmentation methods rely, among others, on translation and the use of dictionaries, we want to investigate whether there are any performances variations when these methods are applied to formal or informal texts.

Hence, with this investigation we hope to be able to advance the state-of-the-art on the subject of data augmentation in NLP and to deepen the understanding of techniques that overcome the bottleneck of limited labeled data.

4   Experimental Setup

We present below the main aspects related to the experimental setup, introducing benchmark datasets and machine learning techniques used. We also summarize steps and procedures involved in obtaining results presented thereafter.

To accomplish the objectives mentioned in Sect. 3, we selected two methods with which DeepBT will be compared: traditional Back-Translation and EDA. These three data augmentation techniques were applied to the same benchmark datasets and using the same set of hyperparameters. Our baseline for comparing the outcomes will be the output of training without any data augmentation method and Mean Squared Error (MSE) was used as the performance metric in all experiments.

To address the first objective, we compare DeepBT to the other data augmentation procedures. To tackle the second objective we analyzed how each of the data augmentation techniques responded to varying dropout values. Finally, the third objective is reached by inspecting response of aforementioned data augmentation methods to each of the benchmark datasets.

Following [26] and others, data augmentation procedures were applied under varying dataset percentage usages (every decile from 10% to 100%), so as to assess the impact of data availability on results and conclusions.

4.1   Benchmark Datasets

We chose two datasets provided by the SemEval 2017 Task 5 challenge [4], in Tracks 1 and 2. Both of the datasets are used in NLP regression tasks within the domain of sentiment analysis.

The dataset of Track 1 (Microblog Messages) consists of messages collected in two different microblog platforms (StockTwits and Twitter) related to stock market events, discussion and assessments. The second dataset, associated to Track 2 (News Statements & Headlines), contains financial news headlines and texts crawled from sources on the Internet.

In both datasets, the label is a continuous sentiment score ranging from -1 (very negative) to $+1$ (very positive), with 0 being neutral. This sentiment score was labeled by domain experts to reflect the point of view of investors regarding negative, neutral or positive prospective trends for companies or stocks.

4.2   Data Augmentation and Preparation Procedure

Here we describe details about the investigated data augmentation methods and how the data was prepared.

Back-Translation and Deep Back-Translation. Since the focus of this paper is not related to the translation model itself and considering the high effort and computational resources required for training a translation language model, we chose to use a widely accepted translation API, namely the Google Cloud Translation API[2]. This allows for rapid and high quality translation for several different languages with minimal associated costs.

So as to use languages from different families, the languages chosen for both Deep- and traditional Back-Translation procedures were German, Russian and Portuguese. Thus, in choosing West Germanic (German), East Slavic (Russian) and Western Romance (Portuguese) languages, we hypothesize that this diversity could bring greater heterogeneity to the paraphrases generated by Back-Translation. For assessing the DeepBT method the experiments were carried out with 2-layer translation settings (e.g. *English → Russian → German → English*).

Easy Data Augmentation (EDA). The EDA method accepts two parameters to control its data augmentation process, namely α and n_{aug}. While the first controls the percentage of words in a sentence that are changed, the last is responsible for indicating the number of augmented sentences in the output. In the seminal paper [26], the authors propose general guidelines regarding optimal α and n_{aug} parameters to be used, depending on the size of the training dataset. For our experiments, to allow for best performance, we follow these guidelines, using $\alpha = 0.05$ and $n_{aug} = 8$ for both datasets.

The EDA procedure was applied using the original code made available by the authors [26].

Data Preparation. Well established NLP data preparation steps were equally employed on both datasets: stop words and punctuation removal, tokenization and lowercasing. In each sentence, the company or cashtag referred to were replaced with a generic token. All implemented models shared the same data preparation process.

The fact that we used original and synthetic data for training and validation in the cross-validation procedure required special attention as to how original and artificial data were distributed in the training phase. The presence, e.g., of original sentence in the training set and of the corresponding synthesized sentence in the validation set would entail the occurrence of data leakage. As a result, cross-validation was carefully designed so as to have original and synthesized sentences always together at the training or at the validations sets.

4.3   Machine Learning Model

We based our experiments in a Convolutional Neural Network (CNN). Choices regarding architecture, hyperparameters and feature representation were done

---

[2] https://cloud.google.com/translate/docs.

inspired by state-of-the-art models in sentiment analysis tasks [29] and the winning architectures of the SemEval 2017 Task 5 challenge [4,6].

The CNN architecture was composed of 2 convolutional layers followed by a single dense layer. Mean Squared Error (MSE) was picked as the loss function and *Adam* as the optimizer. The activation function for hidden and output layers where the *ReLU* activation function and the *tanh* function respectively. The "Wikipedia 2014 + Gigaword 5" pre-trained *GloVe* was used as our input word embedding [20]. To avoid drawing conclusions that are specific to some arbitrary chosen hyperparameter values [8] and seed, models were trained and results were averaged along the following values of hyperparameters: number of neurons in dense layer ({100, 150}), size of filters ({2, 3}) and dropout value ({0.0, 0.1, 0.2}).

5   Results

We begin showing the result of each method in the generation of artificial text. Next we present results related to each of our three main objectives. We end this section discussing the results and drawing conclusions from them.

5.1   Synthesized Data

Before diving into model results obtained by using each of the techniques, it is interesting to analyze synthesized data outputs of each of the methods so as to gain better insights into final model outcomes. Some examples are shown in Tables 1 and 2 for the Microblog Messages and for the News Statements & Headlines respectively.

Table 1. Synthesized Data generated by different data augmentation techniques in the Microblog Messages Dataset.

Original Sentence	EDA	Back-Translation	Deep Back-Translation
"watching for bounce tomorrow"	"watching for resile tomorrow"	"Watch out for jumping power tomorrow"	"I look forward to jumping tomorrow"
"Bad governance. not confident in core biz"	"in governance not confident bad core biz"	"Bad governance. not confident in core business"	"Bad governance. not confident in core business"
"#OwnItDon'tTradeIt"	"ownitdonttradeit"	"# OwnIt-Don'tTradeIt"	"# OwnItDon'tTradeIt"

We observe that, due to the random noise injected by EDA (random swap, random deletion, etc.) sentences generated by this method normally suffer from lack of correct grammatical or syntactical structure. Both Back-Translation methods, on the other hand, generally yield sentences with correct grammatical structure.

Table 2. Synthesized Data generated by different Data Augmentation techniques in the News Statements & Headlines respectively Dataset.

Original Sentence	EDA	Back-Translation	Deep Back-Translation
"Morrisons book second consecutive quarter of sales growth"	"Morrisons of second consecutive quarter book sales growth"	"Morrisons posted sales growth for the second consecutive quarter"	"Morrisons recorded sales growth for the second consecutive quarter"
"Britain's FTSE lifted by solid Kingfisher"	"britains ftse lifted united kingdom of great britain and northern ireland by solid kingfisher"	"Britain's FTSE lifted by solid kingfishers"	"British FTSE filmed by solid kingfishers"
"Brazil Vale says will appeal ruling to block assets for dam burst"	"says vale brazil will appeal ruling to block assets for dam burst"	"Brazil Vale will appeal to block assets for the dam breach"	"Brazil Vale Appeals Asset Lockout Dam Dam"

When comparing Back-Translation and DeepBT, we see that often the first is characterized by better capture of the true meaning of the sentence.

DeepBT may more easily loose the original ideas expressed by the original sentence, such as in *"I look forward to jumping tomorrow"* or in *"British FTSE filmed by solid kingfishers"*.

Interesting insights arise when analyzing the output of each method regarding formal (News & Headlines) and informal (Microblog Messages) texts. In the second example of the Microblog Messages Dataset we can see that, while EDA maintained the original term *biz*, both of the Back-Translation methods were able to identify this expression and output it as *business*. In contrast, the third example of this same dataset shows that none of the methods were able to capture the meaning of *#OwnItDon'tTradeIt* to generate paraphrases, probably due to the absence of spaces and the use of octothorpe sign (*hashtag*).

5.2   Performance

We now present the results for DeepBT for a variety of factors – such as percent of dataset used, multiplication factor and language of translation – and compare them to the results obtained by other data augmentation techniques.

We start comparing relative performance gain of DeepBT, BT and EDA against the baseline trained on 100% of the Microblog Messages dataset in Fig. 2a. Results show DeepBT achieving greater performance when compared to the traditional Back-Translation, though still inferior to EDA when smaller percentages of the dataset are made available. When the entire dataset is used in training, the three data augmentation techniques converge to similar results.

Figures 2b to 2d exhibit how each data augmentation technique's performance is affected by the Multiplication Factor parameter. We notice that BT and DeepBT yield superior results with greater Multiplication Factor, in contrast to what is observed in EDA. It is also interesting to notice how both translation-based methods start with much worse performance than the baseline with low percentages of dataset use, but rapidly respond to growing data availability.

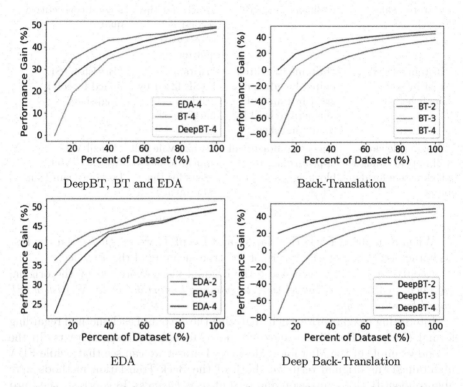

Fig. 2. Comparison of data augmentation techniques and response of each technique to varying Multiplication Factor in the Microblog Messages dataset. Performance gain is measured against the Baseline trained on 100% of the dataset. Labels are in the format "Method-Multiplication Factor".

Figure 3 presents models performance response for each of the chosen languages used for translation in DeepBT and traditional Back-Translation. We observe no performance distinction between any of the chosen languages used for translation purposes.

Table 3 presents how model performance is affected by the dropout hyper-parameter in both datasets. In contrast to the results obtained by [12] in image processing tasks, we obtained stable performance under varying dropout parameter.

An intriguing outcome of our experiments is that EDA had an average performance gain above 20% in all analyzed settings, far superior than the average

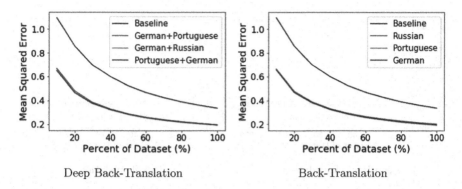

Deep Back-Translation   Back-Translation

Fig. 3. Comparison of models' performance for different languages choices for Back-Translation and Deep Back-Translation on the News Statements & Headlines dataset. Since we used a 2-layer DeepBT, in 3a each line represents a combination of two languages used in sequenced translation.

gains between 1% and 3% achieved by the original work proposing EDA [26]. We hypothesize that this could be attributable to difference in the characteristics of the dataset used in our experiments compared to the ones used in [26] (like linguistic style of the texts) and compare how each method responds to each dataset. The results are shown in Fig. 4.

While translation-based methods yielded similar results, regardless of the dataset, this was not the case for the EDA method. In the latter, we observed far better performance gains in the News Statements & Headlines dataset (formal language), 15%-20% higher than the outcomes obtained in the Microblog Messages dataset (informal language). Considering that the same methodology was applied to both datasets, which also have similar size, we hypothesize that this contrasting behavior can be attributable to greater stability of translation-based methods in comparison to EDA when facing different linguistic styles. Further experiments should be put forward to confirm this conjecture.

Table 3. Average MSE obtained by data augmentation techniques under different dropout values in the Benchmark Datasets.

Method	Microblog Messages			News Statements & Headlines		
	Dropout			Dropout		
	0.0	0.1	0.2	0.0	0.1	0.2
DeepBT	0.13 ± 0.02	0.12 ± 0.02	0.12 ± 0.02	0.21 ± 0.07	0.19 ± 0.06	0.19 ± 0.06
BT	0.15 ± 0.04	0.13 ± 0.03	0.13 ± 0.02	0.21 ± 0.07	0.18 ± 0.05	0.19 ± 0.06
EDA	0.12 ± 0.02	0.12 ± 0.02	0.12 ± 0.02	0.13 ± 0.02	0.13 ± 0.03	0.13 ± 0.02

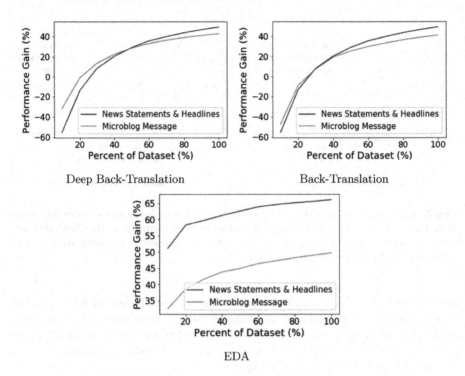

Deep Back-Translation   Back-Translation

EDA

Fig. 4. Comparison of performance gain brought by data augmentation techniques at each dataset. Translation-based methods showed grater homogeneity in outcomes when compared to EDA.

6   Conclusion

In this paper, we performed an in-depth study of data augmentation techniques and presented a new method, performing a systematic evaluation of it. We summarize below our main findings and contributions:

1. **Proposal of new Data Augmentation technique**: with Deep Back-Translation, we add to the state-of-the-art on NLP data augmentation techniques an unprecedented method. Despite its results being somewhat similar to traditional Back-Translation in most experimental setups, this new method yielded superior results in some scenarios, such as in low data availability settings. Furthermore, DeepBT showed greater stability among different datasets when compared to EDA.
2. **Assessing the impact of Dropout in various data augmentation techniques**: to the best of our knowledge, we are the first authors to perform an evaluation of the relationship between these explicit and implicit regularization techniques in a NLP task. This is extremely relevant since the combined use of implicit and explicit regularization techniques is commonplace among

practitioners, despite evidences from the computer vision community indicating that data augmentation could yield better results when used alone [12].

3. **Comparison of the effect of Data Augmentation techniques in Formal and Informal texts**: since NLP data augmentation techniques rely on auxiliary language processing tasks such as translation and synonym replacement, it is interesting to compare their relative behavior in texts with different levels of formality. The observed difference in performance underlines the importance of developing data augmentation techniques that are robust to linguistic preferences prevalent in informal texts, such as contractions, abbreviations, colloquialism, slang, among others.

With this paper, we hope to encourage further use of Back-Translation (and its variations) as auxiliary method for data augmentation in NLP tasks outside of the realm of neural machine translation, where it has already been heavily used.

As future work, we would like to assess Deep Back-Translation at a wider variety of settings, including classification datasets and tasks outside of the sentiment analysis domain. Also, we would like to broaden the scope of comparison of different NLP data augmentation techniques, including e.g. those involving the use of Generative Adversarial Networks. Finally, we would also like to explore data augmentation techniques that are not based on heuristic and handcrafted procedures, but rather learned in a machine learning framework so as to optimize the final model output.

References

1. Bao, F., Neumann, M., Vu, T.: CycleGAN-based emotion style transfer as data augmentation for speech emotion recognition. In: Proceedings of the Interspeech 2019, pp. 2828–2832, September 2019. https://doi.org/10.21437/Interspeech.2019-2293

2. Basile, V., et al.: SemEval-2019 task 5: multilingual detection of hate speech against immigrants and women in twitter. In: Proceedings of the 13th International Workshop on Semantic Evaluation, pp. 54–63. Association for Computational Linguistics, Minneapolis, June 2019. https://doi.org/10.18653/v1/S19-2007

3. Caswell, I., Chelba, C., Grangier, D.: Tagged back-translation. In: Proceedings of the Fourth Conference on Machine Translation (Volume 1: Research Papers), pp. 53–63. Association for Computational Linguistics, Florence, August 2019. https://doi.org/10.18653/v1/W19-5206

4. Cortis, K., et al.: SemEval-2017 task 5: fine-grained sentiment analysis on financial microblogs and news. In: Proceedings of the 11th International Workshop on Semantic Evaluation (SemEval-2017), pp. 519–535. Association for Computational Linguistics, Stroudsburg (2017). https://doi.org/10.18653/v1/S17-2089

5. Dao, T., Gu, A., Ratner, A., Smith, V., De Sa, C., Re, C.: A kernel theory of modern data augmentation. In: Chaudhuri, K., Salakhutdinov, R. (eds.) Proceedings of the 36th International Conference on Machine Learning. Proceedings of Machine Learning Research, vol. 97, pp. 1528–1537. PMLR, Long Beach, 09–15 June 2019. http://proceedings.mlr.press/v97/dao19b.html

6. Davis, B., Cortis, K., Vasiliu, L., Koumpis, A., Mcdermott, R., Handschuh, S.: Social sentiment indices powered by X-scores. In: ALLDATA 2016, The Second International Conference on Big Data, Small Data, Linked Data and Open Data, Lisbon, Portugal (2016)
7. Edunov, S., Ott, M., Auli, M., Grangier, D.: Understanding back-translation at scale. In: Proceedings of the 2018 Conference on Empirical Methods in Natural Language Processing, pp. 489–500. Association for Computational Linguistics, Brussels, October–November 2018. https://doi.org/10.18653/v1/D18-1045
8. Ferreira, T., Paiva, F., Silva, R., Paula, A., Costa, A., Cugnasca, C.: Assessing regression-based sentiment analysis techniques in financial texts. In: Anais do XVI Encontro Nacional de Inteligência Artificial e Computacional, pp. 729–740. SBC, Porto Alegre (2019). https://doi.org/10.5753/eniac.2019.9329, https://sol.sbc.org. br/index.php/eniac/article/view/9329
9. Goodfellow, I., Bengio, Y., Courville, A.: Deep Learning. The MIT Press, Cambridge (2016)
10. Graça, M., Kim, Y., Schamper, J., Khadivi, S., Ney, H.: Generalizing backtranslation in neural machine translation. In: Proceedings of the Fourth Conference on Machine Translation (Volume 1: Research Papers), pp. 45–52. Association for Computational Linguistics, Florence, August 2019. https://doi.org/10.18653/v1/ W19-5205
11. He, K., Zhang, X., Ren, S., Sun, J.: Deep residual learning for image recognition. In: 2016 IEEE Conference on Computer Vision and Pattern Recognition (CVPR), pp. 770–778 (2016). https://doi.org/10.1109/CVPR.2016.90
12. Hernández-García, A., König, P.: Further advantages of data augmentation on convolutional neural networks. In: Kůrková, V., Manolopoulos, Y., Hammer, B., Iliadis, L., Maglogiannis, I. (eds.) ICANN 2018. LNCS, vol. 11139, pp. 95–103. Springer, Cham (2018). https://doi.org/10.1007/978-3-030-01418-6_10
13. Hoang, V.C.D., Koehn, P., Haffari, G., Cohn, T.: Iterative back-translation for neural machine translation. In: Proceedings of the 2nd Workshop on Neural Machine Translation and Generation, pp. 18–24. Association for Computational Linguistics, Melbourne, July 2018. https://doi.org/10.18653/v1/W18-2703
14. Imamura, K., Fujita, A., Sumita, E.: Enhancement of encoder and attention using target monolingual corpora in neural machine translation. In: Proceedings of the 2nd Workshop on Neural Machine Translation and Generation, pp. 55–63. Association for Computational Linguistics, Melbourne, July 2018. https://doi.org/10. 18653/v1/W18-2707
15. Kobayashi, S.: Contextual augmentation: data augmentation by words with paradigmatic relations. In: Proceedings of the 2018 Conference of the North American Chapter of the Association for Computational Linguistics: Human Language Technologies, Volume 2 (Short Papers), pp. 452–457. Association for Computational Linguistics, New Orleans, June 2018. https://doi.org/10.18653/v1/N18-2072
16. Konda, K.R., Bouthillier, X., Memisevic, R., Vincent, P.: Dropout as data augmentation. arXiv abs/1506.08700 (2015)
17. Krizhevsky, A., Sutskever, I., Hinton, G.E.: ImageNet classification with deep convolutional neural networks. In: Proceedings of the 25th International Conference on Neural Information Processing Systems. NIPS 2012, vol. 1, pp. 1097–1105. Curran Associates Inc., Red Hook (2012)
18. Mikołajczyk, A., Grochowski, M.: Data augmentation for improving deep learning in image classification problem. In: 2018 International Interdisciplinary PhD Workshop (IIPhDW), pp. 117–122 (2018)

19. Mohammad, S., Bravo-Marquez, F., Salameh, M., Kiritchenko, S.: SemEval-2018 task 1: affect in tweets. In: Proceedings of The 12th International Workshop on Semantic Evaluation, pp. 1–17. Association for Computational Linguistics, New Orleans, June 2018. https://doi.org/10.18653/v1/S18-1001

20. Pennington, J., Socher, R., Manning, C.D.: GloVe : global vectors for word representation. In: Proceedings of the 2014 Conference on Empirical Methods in Natural Language Processing (EMNLP), pp. 1532–1543 (2014)

21. Salamon, J., Bello, J.P.: Deep convolutional neural networks and data augmentation for environmental sound classification. IEEE Signal Process. Lett. **24**(3), 279–283 (2017)

22. Sennrich, R., Haddow, B., Birch, A.: Improving neural machine translation models with monolingual data. In: Proceedings of the 54th Annual Meeting of the Association for Computational Linguistics (Volume 1: Long Papers), pp. 86–96. Association for Computational Linguistics, Berlin, August 2016. https://doi.org/10.18653/v1/P16-1009

23. Springenberg, J., Dosovitskiy, A., Brox, T., Riedmiller, M.: Striving for simplicity: The all convolutional net. In: ICLR (workshop track) (2015). http://lmb.informatik.uni-freiburg.de/Publications/2015/DB15a

24. Sugiyama, A., Yoshinaga, N.: Data augmentation using back-translation for context-aware neural machine translation. In: Proceedings of the Fourth Workshop on Discourse in Machine Translation (DiscoMT 2019), pp. 35–44. Association for Computational Linguistics, Hong Kong, November 2019. https://doi.org/10.18653/v1/D19-6504

25. Taylor, L., Nitschke, G.: Improving deep learning with generic data augmentation. In: 2018 IEEE Symposium Series on Computational Intelligence (SSCI), pp. 1542–1547 (2018)

26. Wei, J., Zou, K.: EDA: easy data augmentation techniques for boosting performance on text classification tasks. In: Proceedings of the 2019 Conference on Empirical Methods in Natural Language Processing and the 9th International Joint Conference on Natural Language Processing (EMNLP-IJCNLP), pp. 6382–6388. Association for Computational Linguistics, Hong Kong, November 2019. https://doi.org/10.18653/v1/D19-1670

27. Wen, Q., Sun, L., Song, X., Gao, J., Wang, X., Xu, H.: Time series data augmentation for deep learning: a survey (2020)

28. Yu, A.W., et al.: QANET: combining local convolution with global self-attention for reading comprehension. CoRR abs/1804.09541 (2018). https://arxiv.org/pdf/1804.09541

29. Zhang, Y., Wallace, B.C.: A sensitivity analysis of (and practitioners' guide to) convolutional neural networks for sentence classification. In: Proceedings of the 8th International Joint Conference on Natural Language Processing, pp. 253–263 (2017)

30. Zhao, D., Yu, G., Xu, P., Luo, M.: Equivalence between dropout and data augmentation: a mathematical check. Neural Netw. Off. J. Int. Neural Netw. Soc. **115**, 82–89 (2019)

Dense Captioning Using Abstract Meaning Representation

Antonio M. S. Almeida Neto[1]([✉]) [iD], Helena M. Caseli[1] [iD],
and Tiago A. Almeida[2]

[1] Department of Computing (DC), Federal University of São Carlos (UFSCar),
Rod. Washington Luis, km 235, São Carlos, SP, Brazil
`antoniomsaneto@gmail.com`, {`helenacaseli,talmeida`}`@ufscar.br`
[2] Department of Computer Science (DComp),
Federal University of São Carlos (UFSCar),
Rod. João Leme dos Santos, km 110, Sorocaba, SP, Brazil

Abstract. The world around us is composed of images that often need
to be translated into words. This translation can take place in parts,
converting regions of the image into textual descriptions what is also
known as dense captioning. By doing so, the information present in this
region is converted into words expressing the way objects relate to each
other. Computational models have been proposed to perform this task in
a similar way to human beings, mainly using deep neural networks. As the
same region of the image can be described in several different forms, this
study proposes to use the Abstract Meaning Representation (AMR) in
the generation of descriptions for a given region. We hypothesize that by
using AMR it would be possible to extract the meaning of the text and,
as a consequence, improve the quality of the sentences produced by the
models. AMR was investigated as a semantic representation formalism
evolving the so far proposed models that are based only on purely natural
language. The results show that the models trained with sentences in the
form of AMR led to better descriptions and the performance achieved
was superior in almost all evaluations.

Keywords: Dense captioning · Semantic representation · Abstract
Meaning Representation · Multimodal learning · Natural language
processing

1 Introduction

Natural language processing is the field of artificial intelligence that seeks to
automate and reproduce human behavior regarding specific linguistic knowledge
about a given language. The natural language generation aims to produce some
final, or partial, representation of a given language [16].

Dense captioning is a relatively new application among many that demands
natural language generation. In dense captioning, the regions of an image are

© Springer Nature Switzerland AG 2020
R. Cerri and R. C. Prati (Eds.): BRACIS 2020, LNAI 12319, pp. 450–465, 2020.
https://doi.org/10.1007/978-3-030-61377-8_31

converted into textual descriptions. Each description corresponds to the transformation of visual information into equivalent textual information in some natural language. This description can be defined as the production of a sentence that expresses the way objects relate to each other and their attributes in a given region of an image. An example of dense captioning is illustrated in Fig. 1[1].

Fig. 1. An example of dense captioning.   **Fig. 2.** Example of the problem of using natural language.

Currently, dense captioning is performed directly in the form of natural language, that is, the models are trained with sentences in pure natural language and their outputs are also sentences in natural language. This is a problem, as the number of sentences used for describing a scene in a data set grows each time a new example is added.

For example, the region marked in Fig. 2 (see footnote 1) has three different sentences describing it in the dataset: "a boy playing baseball"; "the boy who plays baseball"; "the boy baseball player". For a model based purely on natural language, these three sentences are considered three different examples, increasing the size necessary for training and requiring greater generalization.

To solve this problem, we propose to use a well-known semantic intermediate representation – Abstract Meaning Representation [2] –, so that sentences with the same meaning, but expressed using different words, could be represented in a single way. In this paper, we investigated different alternatives of using the AMR representations to generate image descriptions aiming at better preserving the relationship between the concepts represented in a region of an image.

The main contributions of this work are:

- The application of the main current semantic representation formalism, AMR, for the task of dense captioning. To the best of our knowledge it is the first time that AMR is used in a multimodal (visual) application;
- The construction of the first AMR dataset for dense captioning and possibly also the first AMR-image dataset;
- The corroboration of the hypothesis that the use of AMR sentences can benefit dense captioning compared to the use of pure natural language sentences;

---

[1] Image from Visual Genome Dataset [10].

– The verification of the limitation of the current evaluation measures (adapted from other tasks) applied in dense captioning, since they do not report the quality of the sentences regarding the regions of the image that they describe.

2   Related Work

The relevant related works in the literature combine visual and textual information for many applications such as information retrieval, image description and image captioning.

The work of Karpathy et al. [8] is located in the area of information retrieval and aims to retrieve the relevant images from a query in natural language (the opposite being also valid) through syntactic dependence information. Inspired by previous works, the authors observed that these dependency relationships provide information that is richer (especially the relationship between entities) and more efficient (from a computational point of view) than the words or bigrams themselves. The use of dependency relations, in contrast to the pure natural language, has provided better results.

The same authors proposed in [7] an alignment model between an image and its captions in order to learn how to generate descriptions for specific regions of an image, using a Bidirectional Recurrent Neural Network and Multimodal Recurrent Neural Network. The work aimed to produce descriptions for regions of the image. The proposed model is composed by RCNN [5], LSTM and an optimization method to reduce the number of selected regions. They also seek to combine and reduce the number of regions proposed by the RCNN. In their experiments, the authors reported 0.351 in BLEU-1.

In Johnson et al. [6], a model for generating dense captioning is developed using a Fully Convolutional Localization Network for the task of generating and understanding dense captioning, made up of CNN and RNN networks. However, a layer of dense localization has been inserted in the CNN to predict a set of regions of interest. This approach is similar to Faster-RCNN [17], except that the region of interest is replaced by a bilinear interpolation, the gradient back-propagation and also the regions of interest detected. In their experiments, the authors reported 5.39 in mAP (mean Average Precision).

More recently, Yang et al. [21] proposed more than one LSTM to generate the dense captioning: one for the context model (which is based on the whole image) and the other for modeling the sentence. This approach proved to be useful to distinguish similar elements based on the context. In their experiments, the authors reported 9.31 in mAP.

Currently, the method proposed by Yin et al. [22] is the state-of-the-art in dense captioning using sentences in natural language. This method generates the descriptions using information from the target region, neighboring and global regions (the whole image) in their respective LSTM, to be later combined. The authors reported a value of 10.51 in mAP.

The related works proposed so far generate dense captions based purely on natural language. However, as will be described in the next section, sentences that have the same meaning can be unified using an intermediate semantic representation. Therefore, the hypothesis assumed in this work is that using AMR can lead to better dense captions.

3   AMR

The main motivation for using a semantic representation in this work comes from Karpathy and Fei-Fei [7], in which the effectiveness of using syntactic relationships in multimodal information retrieval was verified, in contrast to the traditional approach of using the sentence in pure natural language. In such work, it was possible to observe that the use of the syntax helped in the recovery of the image from the text and vice-versa.

Based on the assumption that syntactic structures benefit the task of information retrieval, this work raised the hypothesis that semantics can similarly benefit the task of generating descriptions for regions of the image. To test this hypothesis we chose the Abstract Meaning Representation (AMR) [2] formalism.

AMR is a semantic representation formalism that has attracted great attention in recent years. It is a symbolic semantic representation for sentences, which aims to unify semantic tasks (e.g. named entity recognition, co-reference resolution and annotation of semantic relations), as happened with syntax tasks with the introduction of syntax banks [2][2]. The goal of AMR is to unify all these tasks in a simple, human-readable and easy to generate way. AMR supports predicate relations (arguments), including semantic roles (adapted from PropBank), enabling a large number of predicates, such as verbs, co-references, named entities, etc. [1].

To illustrate how AMR works, Fig. 3 shows an AMR representation for the three sentences used to describe the region in Fig. 2:

```
(p1 / play-01
  :ARG0 (b0 / boy))
  :ARG1 (b2 / baseball)
```

Fig. 3. AMR representation for the region in Fig. 2.

The transformation of a sentence in natural language to AMR is performed by Text-to-AMR parsers like [14]. The opposite process, which turns AMR into natural language is performed by AMR-to-Text parsers like the one proposed in [18]. The effectiveness of AMR in some NLP applications has already been proved for machine translation [19] and automatic summarization [11,13].

---

[2] The success of syntactic banks is due to the fact that unifying the various tasks in a single process allowed the use of a single tool. An example of a classic syntactic bank is the Penn Treebank.

4   Dense Captioning Using AMR

To verify the effectiveness of AMR in dense captioning, this work investigated several forms of using AMR representations as input for our baseline model proposed by Jhonson et al. [6], which is illustrated in Figure 4.

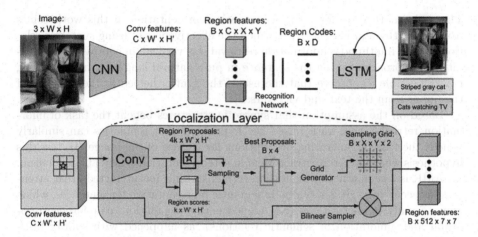

Fig. 4. Training process proposed by [6].

The model, called Fully Convolutional Localization Network, receives the image processed by CNN as input, which consists of 13 3×3 convolutional layers, interspersed with 5 parts of 2×2 max pooling. The last layer is removed to add a predator location layer like Regions of Interest (RoI), as in [17].

The RoI layer is a function of two inputs: the convolutional characteristics and the coordinates of the proposed regions. Gradients can be propagated for the characteristics, but not for the proposed regions. To get around this problem, they used bilinear interpolation. Bilinear interpolation is the extent of linear interpolation, in a regular grid of two variables. In general, polynomials are used to discover discrete points.

To generate the sentence in natural language, an LSTM was used. From the sequence of tokens $s_1, ...s_t$, the LSTM entry is $T + 2$ for word vectors $x_{-1}, x_0 x_1, ...x_T$, where $x_{-1} = CNN(I)$ is the encoded region and x_0 is the special start token (START).

As the model of [6] uses sentences in natural language for its training, those sentences with the same meaning but different syntactic structures are learned by the model independently. To solve this problem, instead of training our model to learn the syntax structures of the sentences, we trained it to learn the meanings of the sentences by converting the sentences to AMR representations and using them as input.

4.1   Proposed AMR Representations

Figure 5 illustrates the 4 forms of representations investigated in this work:

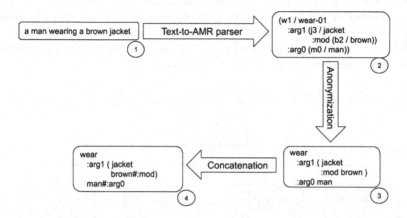

Fig. 5. The four sentence representations investigated in this work: (1) natural language (NL), (2) AMR, (3) anonymized AMR (AMR-Anon) and (4) concatenated anonymized AMR (AMR-Anon-Conc).

1. **Natural Language** (NL) – The first representation (identified by the number 1 in the figure) is the original natural language sentence available in the data set of Visual Genome (detailed in Sect. 5.1) . This is the representation adopted by Jhonson et al. [6] and, therefore, used as the baseline model in this work.
2. **Linearized AMR** (AMR) – The second form of representation is obtained by transforming the NL to AMR through the Text-to-AMR [14] parser and using the linearized form of the generated graph.
3. **Anonymized AMR** (AMR-Anon) – AMR can be simplified by means of an anonymization process, as specified in [9]. Anonymization aims to substantially reduce the complexity of linearized AMR graphs and provide better learning for models, especially those based on neural networks. To this end, it seeks to reduce the sparsity of words in the AMR, as frequently happen with named entities and quantifiers. Thus, the third way of representing sentences was obtained through the anonymization process of the sentences in linearized AMR.
4. **Concatenated Anonymized AMR** (AMR-Anon-Conc) – The fourth form of representation was obtained by concatenating concepts and arguments in the anonymized AMR in an attempt to maintain a stronger relationship between the argument of the verb (or attribute of a concept) and its immediately subsequent word. The concatenation was performed if the tag that represents an argument, starting with a colon (":"), is followed by a token that is not an open parenthesis ("("). This concatenation was delimited by the hashtag ("#").

4.2   Prediction of the Models

Figure 6 illustrates the process of predicting the descriptions of the regions of the images present in the test set. So, for example, from the image shown in Fig. 7, the sentences that were generated by each of the models are presented in Table 1, in which the column "predicted sentence" is the sentence as predicted by the model, and the column "converted to NL" represents the sentence predicted by the models after being converted to natural language.

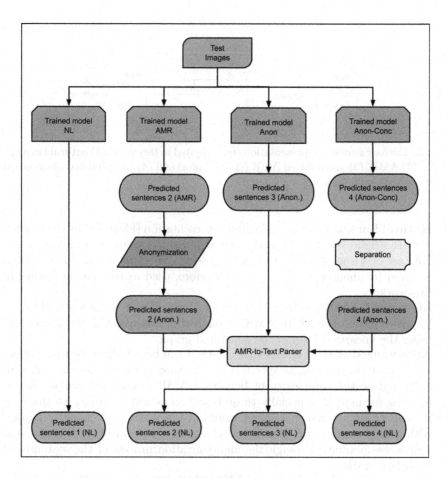

Fig. 6. Prediction process.

As can be seen in Table 1, the baseline sentence in natural language refers to one car being next to the other. The sentence predicted in AMR, in turn, is about the wheel on a bus that does not appear in the region of the image, that is, a wrong description given that context. The sentences generated by the models that use anonymized AMR, both in their linearized form (AMR-Anon) and in

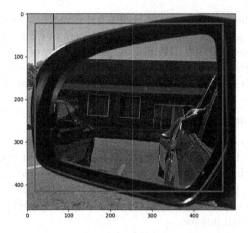

Fig. 7. Example from VG dataset [10].

the concatenated form (AMR-Anon-Conc), produced possible descriptions for the region of the image: one describing the car's windows and the other the parked black car.

Table 1. Sentences predicted for the region delimited in Fig. 7.

Model	Predicted sentence	Converted to NL
NL	a car on the side of a car	–
AMR	(w0/wheel: location (b1/bus))	the wheel on the bus
AMR-Anon	window: part-of car	windows of cars
AMR-Anon-Conc	park: arg1 (car black#:arg1-of)	the black car is parked

It is worth mention that in the sentences predicted by the models that used AMR, the relations between concepts/arguments are preserved in the predicted sentence, what allow the generation of richer descriptions in natural language.

5   Experiments and Results

To test our hypothesis and evaluate the effectiveness of AMR applied in dense captioning, we performed experiments as described in this section.

5.1   Dataset

The datased used in our experiments was the same as Johnson et al. [6]: the version 1.0 of Visual Genome (VG) [10]. This dataset was created from the intersection of other two sets: Yahoo Flickr Creative Commons 100 Million (YFCC100M) [20] and Microsoft Common Objects in Context (MSCOCO) [12].

The complete set has 108,077 images. The regions of an image can overlap, as long as their correspondent sentences are different. Each image has an average of 50 regions, and each description contains a maximum of 16 words.

5.2   Preprocessing

The same pre-processing steps and parameters from Johnson et al. [6] were used in this work. First, from the original set of 108,077 images, those containing less than 20 or more than 50 regions were removed to reduce the variation in the number of regions per image. This processing resulted in a total of 4,191,170 regions (one sentence for each region), reducing the set effectively used in the experiments to 91,610 images.

Following the parameters established by Johnson et al. [6], regions whose description sentence contained more than 10 tokens were discarded and tokens that occurred less than 15 times in the total set were replaced by UNK. Images for which all regions have been discarded have also been excluded. This process was carried out for all forms of representation of the sentence.

To allow reproducibility, we report in Table 2 all the preprocessing parameters applied in each form of representation of the sentences, as well as the size of the resulting set in terms of the number of images, sentences and vocabulary size[3]. It is worth mention that the resulting datasets differ according to each form of representation, since the transformation process (Text-to-AMR, anonymization, concatenation) can result in a greater or lesser number of tokens for the same sentence, what consequently impacts the frequency of occurrence.

Table 2. Preprocessing parameters

Representation	Parameters		Resulting set		
	Min. Freq.	Max. Length	Images	Sentences	Types/# Ocorr.
NL	15	10	91,610	4,168,246	12,514/20,915,384
AMR	15	10	90,193	989,079	4,782/6,752,935
AMR-Anon	15	10	91,607	3,135,352	6,038/18,275,781
AMR-Anon-Conc	15	10	91,609	3,548,212	18,601/17,173,659
AMR-Anon-Conc-2	15	20	91,609	4,165,814	21,928/25,106,949

Table 2 also presents another concatenated anonymized AMR, named concatenated anonymized AMR 2 (AMR-Anon-Conc-2). This representation follows the same idea as the concatenated anonymized AMR, but with the maximum sentence size limit increased to 20 in order to consider, in the dataset, a number of sentences closer to the baseline in natural language.

---

[3] The value before the slash indicates the total of unique tokens (types) and the value after the slash, the total number of occurrences of those tokens.

5.3   Results

The evaluation of the predicted sentences was carried out automatically in the entire test set (about 5,000 examples) using the measures mAP and SMATCH. It is important to say that, as stated by [23], BLEU [15] and METEOR [3] are not adequate to evaluate the language generation in multimodal domain. We also report a manual analysis of around 350 examples.

Mean Average Precision (mAP). The mAP is the main measure used to evaluate models for dense captioning in related works [6,21,22]. It is measured exactly on the sentences predicted by the models in relation to their respective references. Thus, the natural language model (baseline) was the only one in which the assessment took place in natural language. For the other models, evaluation occurred on their predicted sentences. So, the reference sentences (gold standard) used in the calculation of the mAP also went through the same processing of the sentences being evaluated, that is, they were converted to AMR, anonymized and concatenated, as needed.

The mAP results for the models are shown in Table 3. As can be noticed, the values obtained by the proposed models were higher than those obtained by the baseline. In [6], the authors report a mAP of 5.39 for the model trained in NL. In our experiment, carried out with the same parameter values of [6], we obtained a mAP of 4.824.

Table 3. mAP values for the sentences predicted by the models.

Model	mAP
NL	4.824
AMR	4.833
AMR-Anon	7.016
AMR-Anon-Conc	8.617
AMR-Anon-Conc-2	**8.978**

Among the models proposed in this work, the AMR model was the one with the lowest mAP value (4.83), while the concatenated anonymized AMR and concatenated anonymized AMR 2 models obtained the best results, 8.617 and 8.978, respectively, representing a gain of approximately 80% in relation to the baseline model. Thus, we can conclude that AMR in its concatenated anonymized form is effective in terms of mAP.

SMATCH. The AMR models proposed in this work were also evaluated using an appropriate measure for semantic representation: the SMATCH [4]. The SMATCH aims to verify the number of triples produced by converting the AMR into a format of propositional logic.

Figure 8 illustrates the process used to evaluate AMR sentences. In this figure, it is important to explain what are: reference, deanonymized and predicted sentences. The reference sentences are the test sentences (gold standard), in the same representation form used for training. Deanonymized sentences correspond to reference sentences transformed into natural language (for AMR models). For the baseline model, it represents the reference sentences, which were transformed into AMR and returned to natural language. The predicted sentences correspond to the predicted sentences by the models.

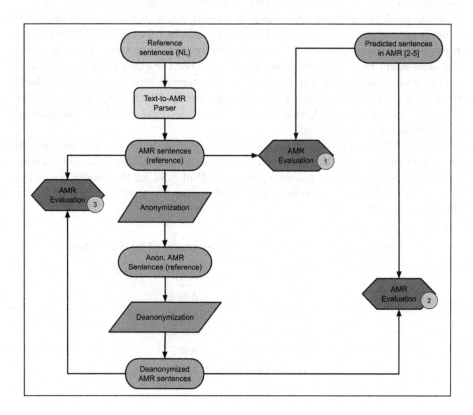

Fig. 8. AMR evaluation process.

Table 4 presents the results obtained by each of the models in their respective evaluation pairs, in which evaluations 1, 2, and 3 in Fig. 8 correspond to the evaluations "reference and predicted", "deanonymized and predicted", and "reference and deanonymized", respectively. The values of the column "reference and deanonymized" differ a little due to the training process in which the detection of region and the generation of the sentences can result in a different number of regions for the same image, in each model. However, it is noteworthy that the same test examples were used in all models.

Table 4. Results obtained using SMATCH

Model	Ref. & Pred.	Deanon. & Pred.	Ref. & Deanon.
NL	–	–	–
AMR	0.2021	0.1857	**0.8144**
AMR-Anon	**0.2058**	0.2432	0.7759
AMR-Anon-Conc	0.2039	**0.2472**	0.7748
AMR-Anon-Conc-2	0.1919	0.2337	0.7647

From the results in Table 4 it is possible to notice that the conversion from NL to AMR causes a significant error of about 20% (reference and deanonized evaluation), since comparing the reference sentence "with itself" after the deanonimization process, should result in a SMATCH of 1.0 and not just 0.76-0.81. The evaluation between the deanonymized and predicted sentences was slightly better than the references and predicted. This better result is justified by the fact that in the first evaluation pair, both sentences passed through the parsers and anonymizers, and consequently, accumulate noises from such tools.

Regarding the low values obtained for the SMATCH measure we can draw some conclusions. Although it takes into account the structure of the AMR graph and, thus, brings with it greater flexibility than the measures that compare natural language (such as METEOR), it is not the most suitable for dense captioning evaluation. This is because it is possible to generate a correct description for a given region that is completely different from the reference description in terms of the concepts used and the relationships between them. Thus, SMATCH was presented in this work only to complement the specific assessment of AMR, but not as an indication of the effectiveness of the model for dense captioning.

Human Evaluation. We also manually analyzed some natural language descriptions generated by the trained models. To do so, the outputs of the models that predict AMR have been transformed into natural language using the AMR-to-Text [18] parser. We chose to carry out the analysis of the sentences manually because automatic measures such as BLEU [15] and METEOR [3] are not adequate in this multimodal domain, given that they are sensitive to chosen words and word order.

For the evaluation, 400 examples were randomly selected. Of the total, 52 were ignored because the reference sentences (available in the VG dataset) do not adequately describe the specific region of the image (do not correctly describe the region or describe elements that are not in the region) or because they have a big intersection with another region of the same image also selected for evaluation. These problems come from the VG dataset. We evaluated the baseline model (NL) and the 3 best models according to mAP values: AMR-Anon, AMR-Anon-Conc and AMR-Anon-Conc-2.

For each test example, the predicted sentences were evaluated taking into account if the information they express was correct, partially correct or incorrect regarding the respective region. The model with the best sentence was also pointed out. Finally, the sentences predicted by the models trained with AMR were also classified as better, equal or worse than the baseline sentence.

Table 5. Human analysis of the sentences predicted by the models.

Model	Correct	Partially correct	Incorrect
NL	**145 (41.67%)**	58 (16.66%)	**145 (41.67%)**
AMR-Anon	104 (29.89%)	**78 (22.41%)**	166 (47.70%)
AMR-Anon-Conc	122 (35.06%)	61 (17.53%)	165 (47.41%)
AMR-Anon-Conc-2	109 (31.32%)	57 (16.38%)	182 (52.30%)

Table 5 presents the values of the manual analysis. The results demonstrate that the total number of correct and partially correct examples of the baseline (203 or 58.33%) were higher than the models AMR-Anon (182 or 52.30%), AMR-Anon-Conc (183 or 52.59%), and AMR-Anon-Conc-2 (166 or 47.70%). However, as already shown in the SMATCH evaluation, this can be due to the noise from the AMR-to-Text conversion.

Table 6. Comparison of predicted sentences in AMR in relation to the baseline.

Model	Better	Equal	Worse
AMR-Anon	**83 (23.85%)**	148 (42.53%)	117 (33.62%)
AMR-Anon-Conc	66 (18.96%)	**183 (52.59%)**	**99 (28.45%)**
AMR-Anon-Conc-2	70 (20.12%)	166 (47.70%)	112 (32.18%)

The results of the comparison with the baseline model are shown in Table 6. In this evaluation, for the AMR-Anon-Conc model, 71.55% (249 examples) of its sentences were better or equal to the baseline, against 67.82% (236 examples) of the AMR-Anon-Conc-2 model, and 66.38% (231 examples) for the AMR-Anon model.

An example that illustrates the need for human evaluation is presented in Fig. 7, for which the sentences predicted by each of the models were described in Table 1. Although all predicted sentences refer to different elements of the region, most of them are correct, since they correctly describe elements indeed present in the region.

From a more qualitative analysis of the data sample manually analyzed, it is possible to observe that the models trained in AMR suffer from the low frequency of some nouns (fence, balloons, tower, brisket and sign) which were wrongly described. On the other hand, the elements with greater frequency (buildings, curtains, snow, windows, shirts and ties) were better described. This problem was less harmful for the baseline model and we believe that this was due to the concatenation of the verb/concept AMR with the next word, which guarantees a stronger relationship, but also considerably decreases the frequency of the concatenated word.

6   Conclusions and Future Work

Dense captioning is a task traditionally performed in natural language. The models are trained with sentences in pure natural language and their outputs are also sentences in natural language. By doing so, sentences with the same meaning but different surface forms are considered different by these models, increasing the size necessary for training and requiring greater generalization.

To solve this problem, we proposed to use the AMR as a semantic intermediate representation, so that sentences with the same meaning, but expressed using different words, could be represented in a single way. In this paper, we investigated different alternatives of using the AMR representations to generate image descriptions aiming at better preserving the relationship between the concepts represented in a region of an image.

In our experiments, we observed that based on mAP values, AMR has benefits in relation to natural language in dense captioning, mainly the anonymized concatenated version.

In the evaluation with SMATCH, it is worth highlighting the impact of noise that the Text-to-AMR and AMR-to-Text conversion tools bring to the results (on average, 21%). This noise affected the performance also according to the human evaluation. We believe that this happened due to the noise brought by the AMR parsers since they were not trained on the same data set used in our experiments. The lots of noise produced by these tools impacted negatively the converted sentences.

However, despite the noise of the AMR tools, we can confirm our hypothesis that AMR indeed can have a good impact on dense captioning. It is mainly because the models can learn the meanings of the text instead of its syntax structure.

As a future work, we intend to cope with the noise from Text-to-AMR and AMR-to-Text parsers by training the [14] and [18] parsers in our dataset. We also aim at investigating the use of AMR in other dense captioning models, such as the state-of-the-art model from Yin et al. [22].

All our source code an datasets are available at: https://github.com/LALIC-UFSCar/densecap-amr.

Acknowledgements. This study was funded by the Coordenação de Aperfeiçoamento de Pessoal de Nível Superior - Brasil (CAPES) - Funding Code 001. The first author was funded by the grant #2018/1771510, CAPES. This research is also part of the MMeaning project, supported by São Paulo Research Foundation (FAPESP), grant #2016/13002-0.

References

1. Abend, O., Rappoport, A.: The state of the art in semantic representation. In: Proceedings of the 55th Annual Meeting of the Association for Computational Linguistics (Volume 1: Long Papers), vol. 1, pp. 77–89 (2017)
2. Banarescu, L., et al.: Abstract meaning representation for sembanking. In: Proceedings of the 7th Linguistic Annotation Workshop and Interoperability with Discourse, pp. 178–186 (2013)
3. Banerjee, S., Lavie, A.: METEOR: an automatic metric for MT evaluation with improved correlation with human judgments. In: Proceedings of the ACL Workshop on Intrinsic and Extrinsic Evaluation Measures for Machine Translation and/or Summarization, pp. 65–72. Association for Computational Linguistics, Ann Arbor, June 2005. https://www.aclweb.org/anthology/W05-0909
4. Cai, S., Knight, K.: Smatch: an evaluation metric for semantic feature structures. In: Proceedings of the 51st Annual Meeting of the Association for Computational Linguistics (Volume 2: Short Papers), pp. 748–752. Association for Computational Linguistics, Sofia, August 2013. https://www.aclweb.org/anthology/P13-2131
5. Girshick, R., Donahue, J., Darrell, T., Malik, J.: Rich feature hierarchies for accurate object detection and semantic segmentation. In: Proceedings of the 2014 IEEE Conference on Computer Vision and Pattern Recognition. CVPR 2014, pp. 580–587. IEEE Computer Society, Washington, DC (2014). https://doi.org/10.1109/CVPR.2014.81
6. Johnson, J., Karpathy, A., Fei-Fei, L.: DenseCap: fully convolutional localization networks for dense captioning. In: Proceedings of the IEEE Conference on Computer Vision and Pattern Recognition (2016)
7. Karpathy, A., Fei-Fei, L.: Deep visual-semantic alignments for generating image descriptions. In: The IEEE Conference on Computer Vision and Pattern Recognition (CVPR), June 2015
8. Karpathy, A., Joulin, A., Fei-Fei, L.: Deep fragment embeddings for bidirectional image sentence mapping. In: Proceedings of the 27th International Conference on Neural Information Processing Systems - Volume 2. NIPS 2014, pp. 1889–1897. MIT Press, Cambridge (2014). http://dl.acm.org/citation.cfm?id=2969033.2969038
9. Konstas, I., Iyer, S., Yatskar, M., Choi, Y., Zettlemoyer, L.: Neural AMR: sequence-to-sequence models for parsing and generation. In: Proceedings of the 55th Annual Meeting of the Association for Computational Linguistics (Volume 1: Long Papers), pp. 146–157. Association for Computational Linguistics, Vancouver, July 2017. https://doi.org/10.18653/v1/P17-1014, https://www.aclweb.org/anthology/P17-1014
10. Krishna, R., et al.: Visual genome: connecting language and vision using crowd-sourced dense image annotations. Int. J. Comput. Vision **123**(1), 32–73 (2017). https://doi.org/10.1007/s11263-016-0981-7
11. Liao, K., Lebanoff, L., Liu, F.: Abstract Meaning Representation for multi-document summarization. In: Proceedings of the 27th International Conference on Computational Linguistics, pp. 1178–1190. Association for Computational Linguistics, Santa Fe, August 2018. https://www.aclweb.org/anthology/C18-1101
12. Lin, T.-Y., et al.: Microsoft COCO: common objects in context. In: Fleet, D., Pajdla, T., Schiele, B., Tuytelaars, T. (eds.) ECCV 2014. LNCS, vol. 8693, pp. 740–755. Springer, Cham (2014). https://doi.org/10.1007/978-3-319-10602-1_48

13. Liu, F., Flanigan, J., Thomson, S., Sadeh, N., Smith, N.A.: Toward abstractive summarization using semantic representations. In: Proceedings of the 2015 Conference of the North American Chapter of the Association for Computational Linguistics: Human Language Technologies, pp. 1077–1086. Association for Computational Linguistics, Denver, May–Jun 2015. https://doi.org/10.3115/v1/N15-1114, https://www.aclweb.org/anthology/N15-1114

14. Lyu, C., Titov, I.: AMR parsing as graph prediction with latent alignment. In: Proceedings of the Annual Meeting of the Association for Computational Linguistics (2018)

15. Papineni, K., Roukos, S., Ward, T., Zhu, W.J.: BLEU: a method for automatic evaluation of machine translation. In: Proceedings of the 40th Annual Meeting on Association for Computational Linguistics. ACL 2002, pp. 311–318. Association for Computational Linguistics, Stroudsburg (2002). https://doi.org/10.3115/1073083.1073135

16. Reiter, E., Dale, R.: Building applied natural language generation systems. Nat. Lang. Eng. 3(1), 57–87 (1997)

17. Ren, S., He, K., Girshick, R., Sun, J.: Faster R-CNN: towards real-time object detection with region proposal networks. IEEE Trans. Pattern Anal. Mach. Intell. 39(6), 1137–1149 (2017). https://doi.org/10.1109/TPAMI.2016.2577031

18. Song, L., Zhang, Y., Wang, Z., Gildea, D.: A graph-to-sequence model for AMR-to-text generation. In: Proceedings of the 56th Annual Meeting of the Association for Computational Linguistics (ACL-18), Melbourne, Australia (2018)

19. Tamchyna, A., Quirk, C., Galley, M.: A discriminative model for semantics-to-string translation. In: Proceedings of the 1st Workshop on Semantics-Driven Statistical Machine Translation (S2MT 2015), pp. 30–36. Association for Computational Linguistics, Beijing, July 2015. https://doi.org/10.18653/v1/W15-3504, https://www.aclweb.org/anthology/W15-3504

20. Thomee, B., et al.: YFCC100M: the new data in multimedia research. Commun. ACM 59(2), 64–73 (2016). https://doi.org/10.1145/2812802

21. Yang, L., Tang, K., Yang, J., Li, L.J.: Dense captioning with joint inference and visual context. In: IEEE Conference on Computer Vision and Pattern Recognition (CVPR), July 2017

22. Yin, G., Sheng, L., Liu, B., Yu, N., Wang, X., Shao, J.: Context and attribute grounded dense captioning. In: The IEEE Conference on Computer Vision and Pattern Recognition (CVPR), June 2019

23. Yu, L., Poirson, P., Yang, S., Berg, A.C., Berg, T.L.: Modeling context in referring expressions. In: Leibe, B., Matas, J., Sebe, N., Welling, M. (eds.) ECCV 2016. LNCS, vol. 9906, pp. 69–85. Springer, Cham (2016). https://doi.org/10.1007/978-3-319-46475-6_5

Can Twitter Data Estimate Reality Show Outcomes?

Kenzo Sakiyama, Lucas de Souza Rodrigues,
and Edson Takashi Matsubara(✉)

Universidade Federal de Mato Grosso do Sul - FACOM, Campo Grande, Brazil
kenzosakiyama@gmail.com, lucas.rodrigues@ifms.edu.br,
edsontm@facom.ufms.br
http://www.facom.ufms.br/

Abstract. People's opinions can impact the real world in many different ways. The election of politics, the sales of products, stock market prices, and consumer habits are just a few examples. However, exploring this relationship between people's opinions and real-world events requires data from both sides, which is usually expensive and hard to obtain. In this study, on one side, we address this problem by extracting data from Twitter, and on the other side, the real-world outcomes of a reality show. We carefully select a reality show that uses the audience's opinion to define the elimination of participants. This relationship brings an interesting case of a causal relationship between audience opinion and real-world events. From Twitter, we obtained simple features, such as the counts of likes, retweets, followers, specific hashtags along with sentiment analysis counts obtained from a fine-tuned BERT. From the TV show, we obtained the eliminated candidate and the percentage of audience rejection of the eliminated candidate. To answer the question posed in the title, we empirically evaluate eleven standard machine learning algorithms using the collected features. The models were able to achieve 88.23% of accuracy to predict the eliminated candidate in the reality show.

Keywords: Natural Language Processing · Sentiment analysis · Tweets

1 Introduction

Currently, getting opinions from people on issues of public utility, products, or services is a difficult task, as it involves financial costs, software tools, and field staff for data collection. In recent years, researchers have faced this challenge with AI techniques capable of extracting information from the Internet, optimizing complex tasks at a low operating cost. With the advancement of Natural Language Processing (NLP) and machine learning techniques, the use of sentiment analysis allows us to extract information from social networks, microblogs, and websites, which allows us to assess the public's sense of something. Recent

R. Cerri and R. C. Prati (Eds.): BRACIS 2020, LNAI 12319, pp. 466–482, 2020.
https://doi.org/10.1007/978-3-030-61377-8_32

work in machine learning uses sentiment analysis to predict future scenarios in various areas such as economics, politics and health.

The use of social networks to estimate the outcomes of the real-world event has become an active research area in recent years [2]. The relationship of Twitter data and real-world effects can be applied in many different ways. The assumption generally used in many research papers assumes that Twitter data can reflect real-world effects. Stock market moods [5], political striking news detector [21], sentiment analysis of movie reviews [1], detect people who are at a suicidal risk [7], enterprise outcomes [14] are just a few examples.

This work relates messages on social networks to produce information about the assessment of public sentiment about television programs. In a similar way to the previously cited works, our goal is to use Twitter data to study a real world problem. Our study brings a simple approach with the use of machine learning to predict participants potentially eliminated in each round on a reality show. The results reflect the opinion of the real world when comparing the official results with those obtained in our work.

The strength of our purpose is the simplicity of the features and algorithms. We have used from simple features, such as counts of hashtags, to more sophisticated features like sentiment analysis. We systematically evaluate the models and the features. Using empirical results, we make the following contributions:

1. Bidirectional Encoder Representations (BERT [11]) based sentiment analysis: we fine-tuned a BERT based sentiment analysis model to identify positive, neutral, and negative tweets of the reality show participants. The number of positive, neutral, and negative created a set of sentiment analysis features. We used these features to improve the predictions of reality show outcomes.
2. Predicting the participant rejection: in this experiment, our goal was to evaluate the prediction of the participant rejection employing mean squared error, mean absolute error, correlation, and the plots of prediction and real values. For numerical stability, we performed ten times 17-cross validation for the algorithms tested.
3. Predicting the participant elimination: during the reality show, the participant with a higher percentage of rejection was eliminated. Due to the restricted number of training examples, we performed 17-cross validation to evaluate whether the rejection as a ranking problem could correctly predict the candidate elimination.
4. Datasets: this study produced two datasets, one with the tabulated results of the candidate elimination during the reality show named SENTCOUNTBBB, and another dataset pointing to the tweets used for sentiment analysis named TWEETSENTBBB. Both datasets can be downloaded at https://github.com/liafacom/bbb2020.

This work is organized as follows. In Sect. 2, we briefly describe the related works. Sect. 3 describes the methodology, Sect. 4 shows our experimental results. Section 5 presents our final remarks.

2   Related Work

There is a series of works related to use social networks and pre-training general language representations, and we briefly review the most widely-used approaches in this section.

2.1   Sentiment Analysis

Sentiment Analysis is a process of determining whether a text is positive, negative, or neutral. According to Mantyla et al. [15], sentiment analysis is one of the fastest-growing research areas in computer science. Mantyla et al. [15] investigated the evolution of the analysis of feelings in several sources: newspapers, tweets, photos, and chats. In his experiments, several methods of data analysis were found and divided into three groups: machine learning, natural language processing, and sentiment analysis specific methods. The authors state that areas of application of sentiment analysis are on the rise, and new sentiment analysis techniques will become a standardized part of many services and products.

Other works highlight the use of sentiment analysis to study the polarity of feelings [25], a kind of diffusion of feeling related to textual information extracted from the social network. They are mechanisms to assist the sentiment analysis process and describe an increase of between 5% to 8% in the tasks of feeling classification related, for example, to tweets.

Competitions of machine learning problems are import references to the state of the art algorithms. The International Workshop on Semantic Evaluation (SemEval) [20] has a continuous series of Contests of semantic computational systems. In the 2017 track of sentiment analysis, shallow convolutional neural network and Bi-LSTM were amongst the top methods. In 2018, the Bidirectional Encoder Representations from Transformers (BERT) from Google AI Language draw attention from the Natural Language Processing community when it obtained eleven state-of-the-art results of General Language Understanding Evaluation (GLUE) benchmark [24]. After that, many BERT variations were proposed and still are among the state-of-the-art in 2020.

2.2   Machine Learning and Social Network Analysis

Recently in Brazil, there were elections for President with data analysis on social networks to assess each candidate and the feelings of voters. The work [21] proposes a detector of breaking news events based on the time series of the number of positive, negative, and neutral tweets obtained from a sentiment analysis classifier about presidential candidates. This study describes the use of domain adaptation and convolutional neural network (CNN) for sentiment analysis [13]. The results show that the sentiment analysis classifier reaches 74% accuracy for the three classes (positive, neutral, negative).

Another technique applied with machine learning widely used in social networks are traditional methods or pre-processing of text data using a distributed representation of words and phrases to classify tweets [19]. The work makes use of Long Term Memory Networks (LSTM) and Convolutional Neural Networks (CNNs) with promising results for accuracy up to 81%.

In more sensitive areas such as economics, a sentiment analysis tool is essential for the financial market to give consumers feedback. Consumer sentiment directly impacts stock prices and global market expectations. The technique applied at work [17] uses WORD2VEC and N-gram to analyze the public's feelings in tweets through supervised learning. Analysis and correlation is made between the movements of a company's stock market and the feelings in tweets.

2.3 Monitoring Social Media Research

Monitoring on social networks has become essential for government entities, large corporations, and global companies. Recent work evaluates the public's reaction to the measures adopted in a country, companies, or TV programs.

The work of [16] proposes the use of tweets to obtain information on the assessment of public sentiment towards television programs in Indonesia. The work uses the Naive Bayes classification algorithm to analyze tweets.

Some studies use data from Twitter messages related to elections [6]. Its proposal relates to the use of programs developed by companies for sentiment analysis, the study predicts the chances of a party winning, using the public's opinion.

Another poll in the political field took place in the last elections in the USA [8], Donald Trump and Hillary Clinton used social networks to promote their campaigns and polarize their ideas to followers impacting the election results.

3   Methodology

In this section, we describe the steps that we adopted to analyze the tweets related to the 2020 edition of a reality show in Brazil and to predict the most likely participant to be eliminated during the program. Later, compare the results obtained with the real outcomes of the reality show. We believe that the viewers' vote can be estimated looking at their tweets using sentiment analysis. The Fig. 1 shows the steps of our methodology to estimate participants to be eliminated from the reality show.

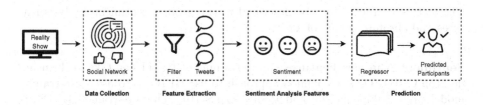

Fig. 1. Sequence of steps for predicting participants

Below we present Sects. 3.1, 3.2, and 3.3, where we describe step by step the pipeline showed in the Fig. 1: data collection, feature extraction, and the prediction of an eliminated candidate.

3.1 Data Collection

Since the Twitter text data are unstructured, we pre-process the tweets in a tabular format. We collected text from Twitter using the API TWINT [26]. TWINT is an advanced Twitter scraping tool written in Python that allows for scraping tweets from Twitter profiles without using Twitter's official API.

The literature shows different ways to perform tokenization and text processing of tweets. In this work, we used the text processor and tokenizer implemented by the winner of SemEval-2017 Task 4 'Sentiment Analysis in Twitter' [3]. The proposed pipeline by the winners include word segmentation, word normalization, and sentiment-aware tokenization. In this work, we used only the word processing pipeline, to process words with emphasis, dates, times, currencies, emoticons emojis, acronyms and censored words. We preserve usernames and remove information such as date, URLs, numbers, percentages. The tool is also able to unzip hashtags: (#ForaBabu to Fora Babu).

We collected a tweet corpus using queries with pre-determined hashtags [12] and usernames of each participant in the reality show TV. The use of usernames in the queries was possible due the fact that in the 2020 edition of the reality show every participant has its own Twitter account, maintained by the TV producers. The github repository of the paper source code lists the hashtags and mentions used to build our Corpus.

After a few tasks and internal voting during the week, the reality show defines a set of participants to compose a list of possible candidates to eliminate (usually three participants). Then, the audience of the reality show vote in the candidate to be eliminated. The candidate with a higher rejection is eliminated from the program. As soon as this list of candidates was defined, we started the data collection procedure with the hashtags of these candidates until the day of the eliminated participant's disclosure. Altogether 17 collections were carried out, matching all the 17 eliminations of the reality show summing up to 3.485.394 tweets. We named this dataset as TWEETSENTBBB.

3.2 Feature Extraction

The feature extraction can be described in two groups: sentiment analysis features and absolute counts of Twitter.

Sentiment Analysis Features: a full training of BERT can take weeks using desktop GPUs. Thus, a common approach to use BERT is to use a pre-trained model from Wikipedia [23] and BooksCorpus [10]. This pre-trained BERT is generic to solve many different NLP tasks by adding a classification head (fully connected layer) on top of the model. In this fashion, we trained a classification head on top of BERT using the TweetSentBR [4] tweet corpus. This Corpus consists of tweets related to popular Brazilian TV shows and generated during the exhibition of the shows. Table 1 shows the class distribution of the dataset.

Table 1. Statistics of TweetSentBR

Dataset	Total	Positive	Neutral	Negative
TweetSentBR	12.312	5.542(45%)	3.136(%25)	3.634(%29)

From Table 1, we observe a class imbalance in the dataset. Since the minority class has an expressive amount of examples (25%), we decided to not modify the dataset further.

The tweets collected by Twint contain information such as: date of creation, quantity of likes and retweets(shares), username, hashtags, mentions besides the text present in the tweet. We classified the tweets in positive, neutral and negative using BERT-NEURALMIND.

Absolute Counts of Twitter: for each candidate in elimination, we counted the total positive, neutral and negative tweets and the total number of likes and retweets related to the candidate. From the quantities, we extracted statistics such as percentage of positive, neutral and negative tweets related to each candidate and the percentages to these categories related to the total number of collected tweets. We also counted the number of tweets published in each day before the elimination results, the amount of likes and retweets each tweet has and used these quantities as features.

As mentioned before, each candidate has its own Twitter account. Since the number of followers in the social media is related to the participant popularity, we decided to keep track of the number of followers of each one of the participants. We updated the number of followers every 15 days.

Finally, analysing the tweets we noticed that the hashtags "*#Fica{Candidate}*" and "*#Fora{Candidate}*" are the most present hashtags in the tweets. They are equivalent to "*#Stay{Candidate}*" and "*#Out{Candidate}*" respectively, and play an important role in the dispute in Twitter. Hence, we decide to count the total amount of each of these hashtags related to each candidate and use the quantities as features.

Final Feature Set: the final feature set with description of each feature is shown in the Table 2. At the end of the feature extraction, we created a small data set with 51 examples from the 17 eliminations where, in average, 3 participants were nominated. This dataset is named SENTCOUNTBBB.

Figure 2 shows the Pearson correlations of the features, where we can highlight four regions: (1) top left corner where `positive`, `negative` and `neutral` shows correlation above 0.87, (2) the correlation of individual percentage, (3) the

Table 2. SENTCOUNTBBB features and their descriptions

Feature	Description
positive	Quantity of positive tweets related to the candidate
neutral	Quantity of neutral tweets related to the candidate
negative	Quantity of negative tweets related to the candidate
positive_individual_pct	Percentage of tweets classified as positive in the tweets related to the candidate
neutral_individual_pct	Percentage of tweets classified as neutral in the tweets related to the candidate
negative_individual_pct	Percentage of tweets classified as negative in the tweets related to the candidate
positive_global_pct	Percentage of tweets classified as positive and related to the candidate, in all tweets collected in the elimination
neutral_global_pct	Percentage of tweets classified as neutral and related to the candidate, in all tweets collected in the elimination
negative_global_pct	Percentage of tweets classified as negative and related to the candidate, in all tweets collected in the elimination
day1	Quantity of tweets related to the cadidate published in the first day of elimination
day2	Quantity of tweets related to the cadidate published in the second day of elimination
day3	Quantity of tweets related to the cadidate published in the third day of elimination
likes	Total quantity of likes in all tweets related to the candidate
retweets	Total quantity of retweets (shares) in all tweets related to the candidate
followers	Quantity of followers of the candidate in a near period
fica	Quantity of hashtags "#$Fica\{CandidateName\}$"
fora	Quantity of hashtags "#$Fora\{CandidateName\}$"

correlation of `positive`, `negative`, `neutral`, `day1`, `day2` and `day3`, and (4) the correlation of `likes` and `retweets`. We need to remind that each example of the dataset represents a candidate nominated for elimination. The first region indicates that the number of `positive`, `negative` and `neutral` are proportionally similar. Looking at the dataset, we can see that some candidates with a lower number of followers receive a lower number of tweets. The second region indicates that the number of positive and negative tweets per candidate are inversely proportional, which indicates that when people write positive tweets, they avoid writing negative tweets and vice versa. The neutral tweets are more correlated to positive than negative tweets. The third region indicates that people tend to tweet more messages on `day2` and `day3` than `day1`. The forth region shows that likes and retweets are highly correlated, indicating that when a person like a tweet, it is very likely that this person will also retweet it.

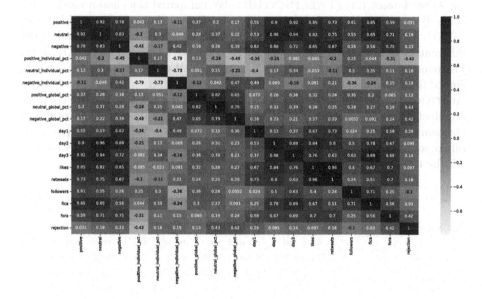

Fig. 2. Pearson correlation of the extracted features.

3.3   Prediction

The reality show of analysis defines the elimination according to the number of votes of the audience. The candidate with the highest percentage of votes is eliminated. This percentage represents our target value. Therefore, we modeled the problem as a regression problem. We evaluated different regression models using the extracted features described in the previously section.

4 Experimental Evaluation

We conducted three sets of experiments with the following purposes: finding a good sentiment analysis model, predicting the participant rejection as a regression problem, and predicting the eliminated candidate.

4.1 Finding a Good Sentiment Analysis Model

In sentiment analysis, our goal is to use a state-of-the-art classifier to determine the sentiment of a given text. We decided to use BERT as our text classifier. In specific, we wanted to use a BERT model that was fine-tuned in the Brazilian Portuguese language to perform the task. Hence, we chose the BERT-NEURALMIND [22] as the model of the sentiment analysis of tweets. We compared the BERT-NEURALMIND model with several results obtained from a work on the same dataset [21] (TWEETSENTBR). We compared the chosen model with Naive Bayes (Gausian and Multinomial Naive Bayes), Support Vector Machines (SVM), and K-Nearest Neighbour (KNN) from sklearn [18]. These models use different text representations such as BOW (Bag of Words), TF-IDF(Term Frequency - Inverse Document Frequency), and WORD2VEC and FASTTEXT embeddings. We also compared the BERT-NEURALMIND with the BERT-BASE-MULTILIGUAL-CASED from Google [11]. From a simple KNN to a more complex architecture using transformers(BERT), we aimed to evaluate models of different complexities. Table 3 shows the results in terms of the chosen metrics, the difference in the metrics is statically significant according to analysis of variance. All the metrics were obtained using 10-fold-cross validation.

Table 3. Cross-validation results on TWEETSENTBR

Algorithm represen.	Accuracy mean ± std	precision mean ± std	recall mean ± std	F1-macro mean ± std
knn bow	0.574 ± 0.012	0.550 ± 0.016	0.528 ± 0.013	0.529 ± 0.014
mnb bow	0.638 ± 0.013	0.610 ± 0.016	0.601 ± 0.014	0.597 ± 0.014
svm bow	0.647 ± 0.013	0.621 ± 0.015	0.614 ± 0.015	0.614 ± 0.015
knn tfidf	0.611 ± 0.009	0.592 ± 0.012	0.562 ± 0.011	0.565 ± 0.013
mnb tfidf	0.634 ± 0.006	0.638 ± 0.012	0.571 ± 0.007	0.563 ± 0.009
svm tfidf	0.647 ± 0.013	0.620 ± 0.015	0.611 ± 0.014	0.612 ± 0.140
knn avg-fasttext	0.590 ± 0.010	0.571 ± 0.015	0.548 ± 0.012	0.531 ± 0.013
gnb avg-fasttext	0.552 ± 0.021	0.548 ± 0.021	0.548 ± 0.021	0.538 ± 0.021
svm avg-fasttext	0.659 ± 0.006	0.635 ± 0.007	0.628 ± 0.006	0.628 ± 0.006
TextCNN word2vec	0.660 ± 0.009	0.633 ± 0.012	0.629 ± 0.011	0.630 ± 0.012
TextCNN fasttext	0.684 ± 0.009	0.659 ± 0.012	0.656 ± 0.009	0.656 ± 0.010
bert-multilingual-cased	0.659 ± 0.011	0.635 ± 0.013	0.632 ± 0.014	0.632 ± 0.014
bert-neuralmind	**0.720 ± 0.017**	**0.698 ± 0.017**	**0.696 ± 0.018**	**0.697 ± 0.017**

Recent studies [9] have analyzed the attention heads of the BERT model, showing that they are capable of capturing syntactic and semantic information, beyond the context, to build word representations. Thus, BERT uses a more representative model than the other text representations evaluated. According to our empirical results, BERT-NEURALMIND model outperformed the other well-known representation methods, BOW and TF-IDF, and even the word embeddings. BERT works with larger sequence lengths than the used TEXTCNN model, and outperformed TEXTCNN accuracy in 3.6%. Running a unpaired t-test between TextCNN fasttext and bert-neuralmind, the obtained p-value is less than 0.0001. Therefore we can reject the null-hypothesis with 95% confidence.

It is also worth mentioning that the BERT-BASE-MULTILIGUAL-CASED model had worse performance than the fine-tunned version of the model and the TEXTCNN with FASTTEXT embeddings generated with tweets. This fact shows the positive impact of fine-tunning the original model in a specific language corpus.

Analyzing the obtained results, BERT-NEURALMIND absolute values in accuracy, precision, recall and f1-macro outperformed all the other classifiers evaluated. Hence, we chose the BERT-NEURALMIND as the model in charge of generating sentiment analysis related features (shown in Table 2) to our work.

4.2 Predicting the Participant Rejection as a Regression Problem

A well-designed choice usually avoids convoluted solutions and first evaluate simple methods. Therefore, our first attempt to deal with the problem proposed in this paper considered standard machine learning algorithms provided by sklearn [18]. We tested the following regression models: Linear, Ridge, Lasso, Elastic Net, SVR (Support Vector Regressor), KNN (K-Nearest Neighbours), SGD (Stochastic Gradient Descent) and Random Forest. We also experimented with three ensembles based in the previously listed models. The first ensemble is based on the Ada Boost method and uses Decision Trees as an estimator. The second one uses an ensemble based on bagging and uses SVR as its estimator. Finaly, the third one is based on averaging the predictions of individual models to generate the final prediction (voting ensemble).

The models hyper-parameters were defined using grid-search. We performed 10 repetitions of 17-cross validation and averaged the metrics results for numeric stability. We used 17 folds to match the 17 eliminations of the reality show. Table 4 shows the results using mean squared error (MSE) and mean absolute error (MAE). The models were sorted in descending order of MSE. Analysing the results from Table 4, we noted that the difference in the metrics is not statistically significant according to analysis of variance. The lowest MAE and MSE were obtained using random_forest.

Figure 3 shows the random_forest predictions versus the true rejection rate. The Pearson correlation between these two variables is 0.767. Before 0.7 on the x-axis, the predictions are in general close to the true values. After 0.7, on the right side of side of the figure, the model the prediction can miss the correct value by large margin.

Table 4. Average of mean squared error (MSE) and mean absolute error (MAE) of ten repetitions of 17-cross validation. Lower values mean a better result

Model	MSE (mean, stddev)	MAE (mean, stddev)
linear_regression	(0.150, 0.247)	(0.260, 0.164)
lasso	(0.090, 0.061)	(0.257, 0.097)
elastic_net	(0.086, 0.051)	(0.252, 0.086)
knn	(0.083, 0.060)	(0.233, 0.100)
sgd	(0.079, 0.053)	(0.236, 0.084)
ridge	(0.072, 0.050)	(0.227, 0.086)
ensamble2 (bagging, svr)	(0.057, 0.047)	(0.194, 0.079)
ensamble3 (voting of svr, knn and rigde)	(0.056, 0.040)	(0.193, 0.076)
svr	(0.047, 0.035)	(0.176, 0.070)
ensamble1 (adaboost, dt)	(0.043, 0.051)	(0.143, 0.090)
random_forest	**(0.032, 0.034)**	**(0.134, 0.072)**

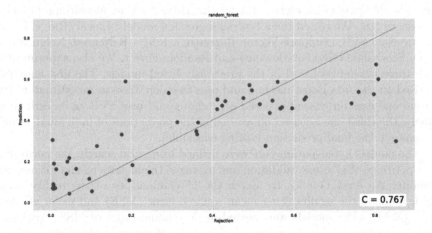

Fig. 3. `random_forest` prediction (y-axis) versus true rejection rate (x-axis)

4.3 Predicting the Eliminated Candidate

We performed another experiment using a 17-cross validation procedure to use a data sampling strategy closer to the real world problem. For this procedure, we enumerated each elimination from 1 to 17. Each elimination has, on average, 3 participants nominated for elimination and we used these examples

Table 5. Grid-Search parameter range and chosen parameters for all the models. The models used in the `ensamble3` were chosen based in the models with lowest MSE in the Table 4

Model	Grid-Search Parameters	Choosen Parameter
random_forest	estimators: from 100 to 10000 criterion: MSE and MAE	estimators: 200 criterion: MSE
ensamble1	estimators: from 3 to 10000 learning rate: from 0.05 to 1 loss: linear, square and exponential	estimators: 5 learning rate: 0.2 loss: linear
ensamble2	estimators: from 2 to 19	estimators: 15
svr	kernel: linear, poly, RBF and sigmoid degree (for the poly kernel only): from 2 to 7 C: from 0.3 to 0.95 episilon: from 0.05 to 0.8 gamma: from 0.1 to 4	kernel: RBF C: 4 episilon: 0.1 gamma: 2.1
knn	neighbors: from 1 to 11 metric: Minkowski and Chebyshev p (for the Minkowski metric only): from 1 to 7	neighbors: 3 metric: Minkowski p: 2
lasso	alpha: from 1×10^{-5} to 7	alpha: 0.01
elastic_net	alpha: from 1×10^{-5} to 5 l1-ratio: from 0 to 1	alpha: 0.1 l1-ratio: 0
ridge	alpha: from 0.01 to 17	alpha: 0.5
sgd	loss: Squared, Huber, Epsilon Insensitive and Squared Epsilon Insensitive penalty: l1, l2, elastic net alpha: from 1×10^{-6} to 1 l1-ratio: from 0 to 1 epsilon: from 0.05 to 0.4 lr schedule: constant, optimal, invscaling and adaptive	loss: Episilon Insensitive penalty: l1 alpha: 0.0001 l1-ratio: 0.7 epsilon: 0.15 lr schedule: constant

(participants) as test sets. For an elimination N, we trained our regression model in the other N - 1 eliminations and used the N-elimination as a test set. Note that this is not leave-out-out, the dataset has 51 examples, and on a leave-one-out procedure uses 50 examples to train and 1 to test, this is not what we did. In our experimental setup, we split the folds into 17 folds, one fold for each elimination (Table 5).

The candidate with the highest prediction value output is predicted to be eliminated. We compare each prediction with the real eliminated candidate to determine the amount of hits of each model. We performed 10 repetitions of the described procedure and averaged the results.

We counted the number of hits in each elimination. The distribution of hits by elimination is shown in the Fig. 4. We can visualize that the eliminations with the worst performance of the models were the 5th, 14th and the 16th ones. Note that none of the models was able to predict the correct candidate in the 5th elimination. Table 6 shows the participants of these eliminations, focusing on features not related to sentiment analysis.

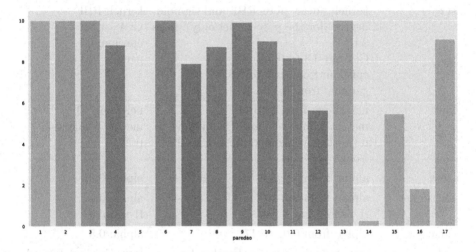

Fig. 4. This figure shows how many times each model predicted the correct N-elimination in the 10 runs, on average. One model can predict the correct elimination of the N-elimination a maximum of 10 times.

Table 6. Eliminations with worst performance. Candidates in bold were eliminated.

Elimination	Candidate	Likes	Retweets	Followers	Fica	Fora	Rejection
5	Flayslene	244.036	42.579	104.138	283	2.872	17.64%
5	Felipe Prior	703.621	137.701	204.031	798	7.069	29.27%
5	**Bianca Andrade**	136.880	18.673	411.880	608	1918	53.09%
14	Mari	857.323	89.995	269.733	1.998	2.469	3.88%
14	Babu	3.730.971	617.153	638.797	4.318	20.326	41.33%
14	**Gizelly**	984.954	140.133	313.866	2.279	17.902	54.78%
16	Babu	756.662	51.204	735.315	3.229	3.266	4.58%
16	Manu Gavassi	2.777.914	287.943	4.916.179	15.957	27.271	41.26%
16	**Mari**	1.492.459	174.831	344.116	6.314	25.433	54.16%

In the Table 6, the candidates with more engagement of the public (`likes` and `retweets`) are also the candidates with more hashtags `fora`. From Fig. 2 we can observe that these features are highly correlated with each other (Pearson correlation of 0.7). As mentioned before, the hashtags "$\#Fora\{Candidate\}$" are a huge indicator of the opinion of a Twitter user.

To further investigate the eliminations 5th, 14th and 16th, Table 7 shows the predicted eliminated candidates in the first repetition of 17-fold cross validation procedure for the five methods with most hits: (`random_forest`, `svr`, `ensamble1`, `ensamble1`, `ensamble2` and `ensamble3`) from the previously evaluation.

Analysing the Table 7, we can conclude that the candidates with the most engagement and hashtags "Fora" were also the most likely to be predicted as eliminated by the studied models. Therefore, we can conclude that these features show some of the bias present in the Twitter data, hindering in same cases the capability of generalization of the used models.

Finally, by counting the number of hits of each model in each repetition, we observed that the `ensamble2` model was able to predict the correct elimination of 15 eliminations out of 17 (on average). The wrong predictions observed in the ten repetitions of the model were the 5th and the 14th elimination, already discussed. Therefore, the model achieved 88.23% accuracy in the task of predicting the eliminated candidates.

Table 7. Table showing the predictions in the 5th, 14th and 16th eliminations for the models `random_forest`, `svr`, `ensamble1`, `ensamble1`, `ensamble2` and `ensamble3`.

Elimination	Model	Prediction	Eliminated
5	random_forest	Felipe Prior	Bianca Andrade
5	svr	Felipe Prior	Bianca Andrade
5	ensamble1	Felipe Prior	Bianca Andrade
5	ensamble3	Felipe Prior	Bianca Andrade
5	ensamble2	Felipe Prior	Bianca Andrade
14	random_forest	Babu	Gizelly
14	svr	Babu	Gizelly
14	ensamble1	Babu	Gizelly
14	ensamble3	Babu	Gizelly
14	ensamble2	Babu	Gizelly
16	random_forest	Manu Gavassi	Mari
16	svr	Mari	Mari
16	ensamble1	Manu Gavassi	Mari
16	ensamble2	Mari	Mari
16	ensamble3	Manu Gavassi	Mari

5   Conclusion

This study aimed to evaluate empirically the use of machine learning algorithms to predict the outcome of reality shows using features extracted from Twitter. First, we evaluate the algorithms as a regression problem that predicts the rejection of the candidates. However, in the reality-show, the candidate with a higher percentage of rejection was eliminated from the program. The use of MAE and MSE, where not enough to answer the question posed in this paper. Therefore, we continue to evaluate the problem using a 17-cross validation strategy and ranking the candidates according to their rejection. Using this approach, we were able to evaluate the hit rate of the elimination during the 17 rounds of candidates' elimination.

To the best of our knowledge, this is the first study to approach the causal effect of sentiment analysis obtained from BERT and counting features from Twitter in reality shows. A Pearson correlation of 0.767 for `random_forest` and information of tweets collected before the effect in the reality show, this study shows an experimental evaluation of a cause-effect relation using social network mapped to real-world events. Bear in mind that the Twitter public represents only a fraction of the total reality show public. Hence, the set of analysis may not be representative enough for the analysis.

The results indicate that `ensamble2` can correctly predict the eliminated candidate with an accuracy of 88.23% using the proposed feature set of nine sentiment analysis features and eight features based on counts of tweets and hashtags. Based on the conducted experimental evaluation, it seems feasible to estimate the outcomes of reality shows using Twitter data.

The caveat of this study is the coarse granularity of the dataset, which generated a small dataset of only 51 examples. Therefore, further studies will be necessary to evaluate models based on deep learning to determine whether it is possible to achieve similar or better results with a more end-to-end learning strategy without all pre-processing steps used in this study. Another caveat of the proposal is that in the first round of elimination, we do not have access to the other sixteen eliminations, to train a model. This study would be useful only in the last elimination of the program. However, it is possible to use the dataset produced in this work to train eliminations on further versions of the reality show, when the reality show setup is similar to the 2020 version.

References

1. Amolik, A., Jivane, N., Bhandari, M., Venkatesan, M.: Twitter sentiment analysis of movie reviews using machine learning techniques. Int. J. Eng. Technol. **7**(6), 1–7 (2016)
2. Anber, H., Salah, A., Abd El-Aziz, A.: A literature review on Twitter data analysis. Int. J. Comput. Electr. Eng. **8**(3), 241 (2016)

3. Baziotis, C., Pelekis, N., Doulkeridis, C.: DataStories at SemEval-2017 task 4: deep LSTM with attention for message-level and topic-based sentiment analysis. In: Proceedings of the 11th International Workshop on Semantic Evaluation (SemEval-2017), pp. 747–754. Association for Computational Linguistics, Vancouver, August 2017

4. Bertini Brum, H., das Graças Volpe Nunes, M.: Building a sentiment corpus of tweets in Brazilian Portuguese. arXiv preprint arXiv:1712.08917 (2017)

5. Bollen, J., Mao, H., Zeng, X.: Twitter mood predicts the stock market. J. Comput. Sci. **2**(1), 1–8 (2011)

6. Bose, R., Dey, R.K., Roy, S., Sarddar, D.: Analyzing political sentiment using Twitter data. In: Satapathy, S.C., Joshi, A. (eds.) Information and Communication Technology for Intelligent Systems. SIST, vol. 107, pp. 427–436. Springer, Singapore (2019). https://doi.org/10.1007/978-981-13-1747-7_41

7. Braithwaite, S.R., Giraud-Carrier, C., West, J., Barnes, M.D., Hanson, C.L.: Validating machine learning algorithms for Twitter data against established measures of suicidality. JMIR Ment. Health **3**(2), e21 (2016)

8. Buccoliero, L., Bellio, E., Crestini, G., Arkoudas, A.: Twitter and politics: evidence from the US presidential elections 2016. J. Mark. Commun. **26**, 114–88 (2020)

9. Clark, K., Khandelwal, U., Levy, O., Manning, C.D.: What does BERT look at? An analysis of BERT's attention. arXiv preprint arXiv:1906.04341 (2019)

10. Davies, M.: Google books corpora, February 2020. https://www.english-corpora. org/googlebooks/#. Accessed 27 Apr 2020

11. Devlin, J., Chang, M.W., Lee, K., Toutanova, K.: BERT: pre-training of deep bidirectional transformers for language understanding. In: Proceedings of the 2019 Conference of the North American Chapter of the Association for Computational Linguistics: Human Language Technologies, Volume 1 (Long and Short Papers), pp. 4171–4186 (2019)

12. Giannoulakis, S., Tsapatsoulis, N.: Evaluating the descriptive power of Instagram hashtags. J. Innov. Digit. Ecosyst. **3**(2), 114–129 (2016)

13. Hu, B., Lu, Z., Li, H., Chen, Q.: Convolutional neural network architectures for matching natural language sentences. In: Advances in Neural Information Processing Systems, pp. 2042–2050 (2014)

14. Lim, S., Tucker, C.S.: Mining Twitter data for causal links between tweets and real-world outcomes. Expert Syst. Appl. X **3**, 100007 (2019)

15. Mäntylä, M.V., Graziotin, D., Kuutila, M.: The evolution of sentiment analysis–a review of research topics, venues, and top cited papers. Comput. Sci. Rev. **27**, 16–32 (2018)

16. Mulyani, E.D.S., Rohpandi, D., Rahman, F.A.: Analysis of Twitter sentiment using the classification of Naive Bayes method about television in Indonesia. In: 2019 1st International Conference on Cybernetics and Intelligent System (ICORIS), vol. 1, pp. 89–93. IEEE (2019)

17. Pagolu, V.S., Reddy, K.N., Panda, G., Majhi, B.: Sentiment analysis of Twitter data for predicting stock market movements. In: 2016 International Conference on Signal Processing, Communication, Power and Embedded System (SCOPES), pp. 1345–1350. IEEE (2016)

18. Pedregosa, F., et al.: Scikit-learn: machine learning in Python. J. Mach. Learn. Res. **12**, 2825–2830 (2011)

19. Reddy, D.M., Reddy, N.V.S.: Twitter sentiment analysis using distributed word and sentence representation. arXiv abs/1904.12580 (2019)

20. Rosenthal, S., Farra, N., Nakov, P.: SemEval-2017 task 4: sentiment analysis in twitter. arXiv preprint arXiv:1912.00741 (2019)

21. Sakiyama, K.M., Silva, A.Q.B., Matsubara, E.T.: Twitter breaking news detector in the 2018 Brazilian presidential election using word embeddings and convolutional neural networks. In: 2019 International Joint Conference on Neural Networks (IJCNN), pp. 1–8. IEEE (2019)

22. Souza, F., Nogueira, R., Lotufo, R.: Portuguese named entity recognition using BERT-CRF. arXiv preprint arXiv:1909.10649 (2019)

23. Wales, J.D.: Wikipedia, February 2020. https://www.wikipedia.org/. Accessed 27 Apr 2020

24. Wang, A., Singh, A., Michael, J., Hill, F., Levy, O., Bowman, S.R.: Glue: a multitask benchmark and analysis platform for natural language understanding. arXiv preprint arXiv:1804.07461 (2018)

25. Wang, L., Niu, J., Yu, S.M.: SentiDiff: combining textual information and sentiment diffusion patterns for Twitter sentiment analysis. IEEE Trans. Knowl. Data Eng. **32**, 2026–2039 (2019)

26. Zacharias, C., Poldi, F.: GitHub - twintproject/twint: an advanced Twitter scraping & OSINT tool written in Python that doesn't use Twitter's API, allowing you to scrape a user's followers, following, tweets and more while evading most API limitations, February 2020. https://github.com/twintproject/twint. Accessed 27 Apr 2020

Domain Adaptation of Transformers for English Word Segmentation

Ruan Chaves Rodrigues$^{(\boxtimes)}$, Acquila Santos Rocha,
Marcelo Akira Inuzuka, and Hugo Alexandre Dantas do Nascimento

Institute of Informatics, Federal University of Goiás, Goiânia, Brazil
ruanchaves93@gmail.com, acquila@discente.ufg.br
{marceloakira,hadn}@inf.ufg.br

Abstract. Word segmentation can contribute to improve the results of natural language processing tasks on several problem domains, including social media sentiment analysis, source code summarization and neural machine translation. Taking the English language as a case study, we fine-tune a Transformer architecture which has been trained through the Pre-trained Distillation (PD) algorithm, while comparing it to previous experiments with recurrent neural networks. We organize datasets and resources from multiple application domains under a unified format, and demonstrate that our proposed architecture has competitive performance and superior cross-domain generalization in comparison with previous approaches for word segmentation in Western languages.

Keywords: Word segmentation · Transformer architecture · Knowledge distillation

1 Introduction

Word segmentation is a low-level task in natural language processing. It can be defined as the activity of introducing proper word boundaries between the characters in a sentence. Depending on how the task is defined, these boundaries may introduce space between words, if they are absent (e.g. *computerscience* becomes *computer science*), or between morphemes that belong to the same word (e.g. *sunglasses* becomes *sun glasses*).

Word segmentation is distinct from procedures such as stemming or abbreviation expansion, in which non-delimiter characters can be added, eliminated or changed. It is also not the same as the detection of segmentation candidates, that is, to determine whether a sequence of characters should be segmented or not.

The word segmentation task is fundamental in Chinese natural language processing. There are no spaces between words in day-to-day Chinese usage, and understanding where word boundaries are located is a process left to the reader. As a consequence, there is considerable disagreement between Chinese speakers about where such boundaries should eventually be inserted. In fact,

© Springer Nature Switzerland AG 2020
R. Cerri and R. C. Prati (Eds.): BRACIS 2020, LNAI 12319, pp. 483–496, 2020.
https://doi.org/10.1007/978-3-030-61377-8_33

Sproat et al. [20] indicate that there is only a 75% agreement between segmentation judgements, and that incompatible segmentation criteria are adopted even among datasets compiled by professional linguists [4, 7].

Such a complex linguistic situation turns word segmentation into a prerequisite for training word-based models for the Chinese language. On the other hand, due to the natural presence of word spacing, word segmentation is not necessary for most natural language processing tasks in Western languages. As a result, word segmentation for those languages is less intensely researched than it is for Chinese, and there are significantly less resources available for it. Nevertheless, even for Western languages, there are still several use cases and linguistic domains in which word spacing is not present, and word segmentation becomes a fundamental requirement before downstream natural language tasks can be performed.

The purpose of this work is to advance the state-of-the-art research in word segmentation for Western languages, while focusing on the English language as a case study. Our main contribution is to evaluate the performance of a compact Transformer architecture, more specifically the Transformer$_{MINI}$ model trained by Turc et al. from a BERT$_{BASE}$ model [21] through a procedure known as Pre-trained Distillation (PD). We provide state-of-the-art results for word segmentation on the WMT17 English news dataset, and we also found our approach to be competitive with previous models for identifier splitting under certain experimental settings, as demonstrated on Sect. 5.

Another contribution of our research is to propose an unified format for word segmentation datasets, and to organize resources for multiple application domains, such as Wikipedia and newspaper word segmentation, hashtag splitting and identifier splitting.

The next sections are organized in the following way: in Sect. 2, we provide a historical background for the word segmentation problem in Western languages. We also present its current application domains, and we review previous works on word segmentation for both Chinese and Western languages. Next, in Sect. 4, we detail the experimental setup, including our datasets and our chosen models. Then, in Sect. 5, we examine the results obtained by Transformer$_{MINI}$ in domain-specific and cross-domain settings. Finally, we conclude in Sect. 6 with a discussion about the main aspects of our work and suggestions for future studies.

Our code is made available at https://github.com/ruanchaves/BERT-WS.

2   Contextualization

In this section, we provide the motivation behind the study of word segmentation. Next, we provide the historical background and the application domains which are relevant for Western languages; then we look into model architectures used for word segmentation, and reflect about the distinct aspects and trade-offs that must be considered in our architectural choices.

2.1   Motivation

Before going through the literature for the word segmentation task, a reasonable question to be posed is whether we actually need to perform it in the first place. Xiaoya Li et al. [11] demonstrate that recent character-based deep learning architectures pretrained for the Chinese language, such as Chinese BERT, are capable of achieving benchmarks either superior or equivalent to word-based models on Chinese natural language processing tasks.

However, Wang et al. [22] prove that character-based models, such as Chinese BERT, are significantly vulnerable to character-level adversarial attacks, indicating that word segmentation may still be necessary for applications where robustness is a fundamental requirement. Therefore, the application domains and model architectures for word segmentation still remain relevant and worthy of our attention.

2.2   Historical Background

As a historical note, it should be said that word spacing was not a widespread practice in the Western world before the 12th century, as most manuscripts adopted the *scriptura continua* writing style of classical Latin, which did not have any form of separation between words. Saenger [17] describes the emergence of word spacing as being closely tied to deep cultural changes that took place in the Western world during the Late Middle Ages. Therefore, word segmentation techniques may be useful while dealing with sources from centuries when it was not an usual practice to insert spaces between words, as often is the case with historical Latin manuscripts [16].

Automatic word segmentation became increasingly in demand after the Internet and personal computers became widespread in the Western world. Social media hashtags, source code identifiers, email and URL addresses are common examples of contemporary text formats which do not include spaces between words.

2.3   Application Domains

In the field of software engineering, identifier splitting (IS) is a common task that can be useful for code clone detection or code summarization [10]. Another emerging context is *hashtag splitting* (HS), which can improve the performance of sentiment analysis [13] and event detection [14] pipelines. Other applications include scenarios where word spacing was lost or corrupted for some reason, as it can often happen while converting files between text and PDF formats. It is also common for optical character recognition systems to fail to recognize spaces between words in handwritten or low resolution scans [8].

Although word spacing is widespread in modern Western languages, there are still some Western languages where it is common practice to assemble long compound words. An example would be the Dutch legal term *arbeidsongeschik-theidsverzekering*, which means "Total permanent disability insurance" and is formed by the concatenation of *arbeid* (labour), *ongeschiktheid* (inaptitude) and *verzekering* (insurance). This feature can become a problem in Neural Machine Translation (NMT), specially on its word alignment (WA) subtask, in which matches between words of a certain language pair are identified. In this context, word segmentation becomes necessary, and it is often referred to by the term *compound splitting* [8, 15].

2.4   Related Work

Recurrent neural networks and Transformer architectures are the prevalent approaches for word segmentation. One of the main practical distinctions between Transformer architectures and standard recurrent networks such as LSTM is that, for the latter, computation of states cannot occur in parallel, significantly limiting its decoding speed. Such scalability concerns are particularly important for word segmentation, which often must be performed over massive amounts of raw data, before other preprocessing steps can take place [7].

Previous research by Huang et al. [7] and Yang [23] has demonstrated that the majority of the information that is relevant to word segmentation is concentrated on the inferior layers of BERT [7]. This insight often leads to approaches in which most superior layers of a large BERT model are pruned with only a small impact on model accuracy for word segmentation tasks [7].

An alternative to pruning a large model is to train a lightweight model from scratch. Duan et al. have performed experiments with Gaussian-masked Directional (GD) Transformer architectures, which benefit from the insight that most relevant information for word segmentation is highly localized, rather than spread over the context of an entire sentence [7]. In addition, Ma et al. [12] demonstrate that a simple Bi-LSTM model trained according to the best practices can outperform Transformer architectures which were not pretrained specifically for word segmentation.

Another important discussion is whether word segmentation should be treated as a language modeling task, where word segmentation candidates are found through a heuristic search algorithm and ranked according to their estimated likelihood [6], or as a character-based sequence labeling task, where individual characters are given labels according to their position in a word. [10]

Labeling schemes formulate word segmentation as a sequence to sequence translation task, in which characters are "translated" into their positional labels [18]. A common labeling scheme is known as *BMES*. In this scheme, the first character of a word is labeled as B (begin), the intermediate characters are labeled as M (middle), and the final character is labeled as E (end). Words with a single character can be labeled as S (single).

Dealing with word segmentation as a sequence labeling task is particularly interesting within the context of the Chinese language, as individual characters are morphemes in and of themselves. For Western languages, characters have no individual meaning, and alternative approaches to word segmentation may be considered in place of or in combination with sequence labeling [25].

Table 1 outlines the main experiments considered during our research. Jiechu Li et al. [10] and Doval et al. [6] investigated distinct word segmentation approaches for Western languages, and their datasets are considered in our experiments. Furthermore, the architecture proposed by Jiangping Lei [9] is adapted to our Transformer$_{MINI}$ model, as described in Sect. 4.

Table 1. Previous word segmentation approaches considered in this work. Here we list the models and the experimental settings found on their original papers: their approach to word segmentation, and the languages and domains of their datasets.

Models	Approach	Languages	Domains
Lei [9]	Character-based sequence labeling task with BERT$_{BASE}$ for Chinese [5]	Chinese	News from The People's Daily (人民日报) newspaper
Li et al. [10]	Character-based sequence labeling task with CNN-BiLSTM-CRF	English	Source code identifiers
Doval et al. [6]	Language modeling task with LSTM and Beamsearch	English, Spanish, German, Turkish, and Finnish	2016 news datasets from WMT17 [2] and the Sentiment140 tweets dataset [1]

Jiangping Lei [9] and Jiechu Li et al. [10] follow the same task paradigm. Word segmentation is targeted as a character-based sequence labeling task, similar to what has been the usual approach deployed by state-of-the-art models for Chinese Word Segmentation. On the other hand, the architecture proposed by Doval et al. [6] deals with word segmentation as a language modeling task.

3   Evaluated Models

During our experiments, we evaluated three models for word segmentation: Transformer$_{MINI}$ [21], CNN-BiLSTM-CRF [10] and a LSTM architecture designed by Doval et al. [6].

The Transformer$_{MINI}$ model (also known as BERT-Mini) utilized throughout our experiments was pretrained by Turc et al. [21] and released on the official BERT repository[1]. The pretraining and distillation steps of Transformer$_{MINI}$, illustrated on Fig. 1, were carefully described by Turc et al. [21]. A Transformer model, after being exposed to a preliminary pretraining step with unlabeled language model data, is submitted to knowledge distillation with unlabeled transfer data produced by a teacher model, BERT$_{BASE}$.

The resulting Transformer$_{MINI}$ student model is then fine-tuned by ourselves in distinct ways for each one of our specific tasks, as described on Sect. 5. We

---

[1] https://github.com/google-research/bert.

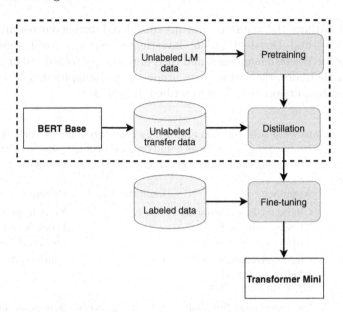

Fig. 1. An illustration of the training procedure of the Transformer$_{MINI}$ model utilized in our experiments. The pretraining and distillation steps were performed by Turc et al. [21], and then, during our experiments, the resulting model was individually fine-tuned for each one of our word segmentation tasks, as described on Sect. 5.

fine-tuned Transformer$_{MINI}$ to the character-based sequence labeling task with labeled data.

We evaluate Transformer$_{MINI}$ as a component to the word segmentation architecture proposed by Jiangping Lei[2] [9]. In this architecture, all characters were labeled according to the *BMES* scheme described under Subsect. 2.4.

In the case of CNN-BiLSTM-CRF, we simply report the benchmarks published by Jiechu Li et al. [10], and compare them to what can be achieved by Transformer$_{MINI}$.

For the LSTM architecture, we chose to utilize a neural network with three layers and 1,500 neurons per layer[3] which has already been trained by Doval et al. [6]. Although the model was tested on new datasets, no changes were made to the model and to the parameters recommended on its original paper.

4   Experimental Setup

Our main goal during our experiments was to measure how Transformer$_{MINI}$ compared against previous state-of-the-art approaches for word segmentation in Western languages. For this purpose, we collected multiple datasets across

---

[2] https://github.com/jiangpinglei/BERT_ChineseWordSegment.
[3] https://www.grupocole.org/software/VCS/segmnt/models/.

distinct linguistic domains. After putting these datasets under a unified format, we evaluated Transformer$_{\text{MINI}}$ and two distinct state-of-the-art models on their word segmentation tasks.

4.1 Collected Datasets

The datasets were collected from several sources, considering a monolingual app-roach centered on the English language. As a result, we built a multi-domain corpora divided into four groups, one for each application domain: Wikipedia word segmentation, news word segmentation, identifier splitting and hashtag splitting. For both of these sources we splitted all the data into sentences, from which we removed all special characters such as blog mentions or SGML tags[4].

We evaluate Transformer$_{\text{MINI}}$ regarding its domain-specific and cross-domain performance, comparing it to the current state-of-the-art models for each one of these aspects.

The domain-specific performance was measured through the datasets that closely match the domain in which Transformer$_{\text{MINI}}$ has been pre-trained: the News (WMT17) and Wikipedia (2019/11/30) datasets. On the other hand, cross-domain performance experiments were performed with datasets meant for iden-tifier splitting, which are expected to be farther away from the original linguistic domain of our pretrained Transformer$_{\text{MINI}}$ model.

Wikipedia (2019/11/30). We fetched the English Wikipedia dump on 2019/11/30, and then used Wikiextractor[5] to collect all articles. Next we took a random set of this articles and applied the pre-processing protocol, described earlier in this section, resulting in 892, 521 sentences.

News (WMT17). We used the same source employed by Doval et al. [6], the monolingual English news dataset from the 2016 WMT17 shared task[6], from which we got a set of 10 million sentences.

Hashtags (Shaikh_2019). For the hashtag splitting domain, we used the dataset made available online by Mohammed Shaikh[7] with approximately 700, 000 hashtags. It should be noticed that, for monolingual evaluation, there is a certain amount of noises in this dataset, since a small percentage of the hashtags are not written in English.

---

[4] Standard Generalized Markup Language is a metalanguage through which it is pos-sible to define markup languages for documents.

[5] https://github.com/attardi/wikiextractor.

[6] http://www.statmt.org/wmt17/.

[7] https://github.com/afmdnf/hashtag-lstm.

Identifier Splitting. Jiechu Li et al. [10] use the nomenclature 'oracles' to refer to identifier splitting datasets. Their work considered four frequently used oracles, namely *Binkley*, containing 2, 664 samples, *BT11*, with 21, 122 samples, *Jhotdraw*, with 974 samples and *Lynx*, with 3, 085 samples. To conduct the experiments for the identifier splitting task, we used all aforementioned oracles except Lynx, since it demands an abbreviation expansion step prior to the word segmentation task.

4.2 Format Unification

A common approach to build a word segmentation dataset is to join entire sentences into a single string, while disregarding delimiter characters, as it was done by Doval et al. [6]. In order to avoid undesirable biases and to maintain real-world applicability, we chose a different approach: instead of join the entire sentence, we join only a sub-sequence of words from sentence. All datasets were structured as a spreadsheet of tab-separated values (TSV). Although the construction process varies according to each domain, we formatted the datasets, so that all spreadsheets contain the same columns. The name and meaning of each column are described in Table 2.

In order to split the texts into sentences, we use the NLTK PunktSentenceTokenizer routine[8]. In this paper, we consider a token to be the word unit separated by a delimiting character. Only spaces were considered as delimiting characters, and all other punctuation which was not detected to be a sentence boundary was treated as a single token.

In order to fit the Wikipedia (2019/11/30) and News (WMT17) datasets in our standard format, starting from a random position of each sentence, we chose a random sequence of tokens, from which all the space character were removed. For example, consider the sentence 'Mass strandings are infrequent' and suppose we have randomly chosen a sequence of 3 tokens (*token_size*), starting from position 5 of the sentence (*joined_position*). According to this values the resulting *joined_tokens* for this sentence is 'strandingsareinfrequent'.

For each sample on the hashtags (Shaikh_2019) dataset, we placed the original hashtag on the *joined_tokens* field and its splitted version on *original_token*. We also converted all identifier splitting oracles to our standard format, following a similar procedure. For both Hashtags (Shaikh_2019) and identifier splitting datasets, we left the *sentence* field blank, because the joined tokens aren't associated with any surrounding sentence context.

5 Results

In our experiments for domain-specific performance, we fine-tuned Transformer$_{\text{MINI}}$ on the News dataset, and then we tested it on the test subsets for the News and Wikipedia datasets. Furthermore, we also performed experiments to measure cross-domain performance on identifier splitting datasets.

---

[8] https://www.nltk.org/_modules/nltk/tokenize/punkt.html.

5.1 Domain-Specific Performance

We decided to compare Transformer$_{MINI}$ to the neural network trained by Doval et al. [6] specifically for the News dataset. As it can be seen on Table 3, the results achieved by Transformer$_{MINI}$ were significantly superior to the neural approach of Doval et al., and the performance gap between the models further increased after they were tested on Wikipedia.

This shows that, compared to a previous state-of-the-art model, Transformer$_{MINI}$ not only achieved higher benchmarks, but was capable of maintaining them across closely related datasets.

5.2 Cross-Domain Performance

In this section we evaluate the cross-domain performance of Transformer$_{MINI}$. We describe results for a preliminary experiment on hashtag splitting, and then we proceed to an in-depth study of how Transformer$_{MINI}$ compares to other solutions for identifier splitting.

Hashtag Splitting. Although we could not find a systematic benchmark on the hashtag splitting dataset, its original source presented a BiLSTM baseline which was trained on the available samples, reaching a F1-score of 0.754. After fine-tuning Transformer$_{MINI}$ on the same samples, it reached a F1-score of 0.823.

Identifier Splitting. Here we compare a CNN-BiLSTM-CRF model which has been specifically trained for identifier splitting [10] with Transformer$_{MINI}$ models which were only fine-tuned over distinct identifier splitting datasets.

We start from the experiments proposed by Jiechu Li et al. [10]. His results are reported on Table 4. In this experiment, the model is trained on each one of the datasets, and then tested on all the others. For instance, it can be seen on Table 4 that, after training on the augmented Binkley dataset (Binkley$_{AUG}$), the CNN-BiLSTM-CRF model was able to achieve an accuracy of 0.760 on the augmented BT11 dataset (BT11$_{AUG}$).

An important aspect of their experiments is that, when used as training sets, all identifier splitting datasets (Binkley, BT11 and Jhotdraw) were augmented with 5000 identifiers that were manually created by the authors. In their work, they are called *fake identifiers*. We demonstrate on Table 5 that 65% of the augmented Binkley dataset and 83% of the augmented Jhotdraw dataset is made of fake identifiers.

As these fake identifiers were not made publicly available by the authors, we have no means to assert their quality or their impact on the final results achieved by CNN-BiLSTM-CRF. Therefore, we decided to perform the same experiments with Transformer$_{MINI}$ while replacing the augmented datasets by the original datasets, without considering any sort of data augmentation technique.

The Jhotdraw dataset is remarkably distinct from the Binkley and BT11 datasets, as it has many identifiers that bear no resemblance to patterns usually

Table 2. A description of the unified format adopted for all of our datasets. When there is no original context for the joined tokens, the *sentence* and *sentence_len* fields will be left blank.

Field	Description	Example
joined_tokens	Sequence of joined tokens	'strandingsareinfrequent'
original_token	Correct splitting of *joined_tokens*	'strandings are infrequent'
sentence	Original sentence with all the delimiting characters	'Mass strandings are infrequent'
joined_position	Position where *joined_tokens* begins at sentence	5
token_size	Number of joined tokens in *joined_tokens*	3
sentence_len	Number of tokens from the original sentence	4

Table 3. Domain-specific performance of Transformer$_{\text{MINI}}$

Model	Dataset					
	News			Wikipedia		
	Recall	Accuracy	F1	Recall	Accuracy	F1
Transformer$_{\text{MINI}}$ [21]	0.96	0.96	0.96	0.95	0.94	0.94
Doval et al. [6]	0.84	0.82	0.83	0.82	0.69	0.74

found on general-domain texts. Single character domain-specific terms are relatively more common, as in the case of the identifier "frelativey", that should be correctly segmented as "f relative y".

We believe these features have contributed to the low accuracy of the results obtained by Transformer$_{\text{MINI}}$ on the Jhotdraw datasets on Table 4. A low accuracy is obtained either by fine-tuning the model on Jhotdraw and testing on the other datasets (i.e., the results on the *Jhotdraw* row), or by testing on Jhotdraw after fine-tuning either on Binkley or BT11 (i.e., the results on the *Jhotdraw* column).

Compared to the Transformer$_{\text{MINI}}$ model, CNN-BiLSTM-CRF was able to surpass its results on Jhotdraw through the addition of fake identifiers. However, in one case, its results were inferior to Transformer$_{\text{MINI}}$: CNN-BiLSTM-CRF achieved an accuracy of 0.760 on the BT11 dataset after being trained on the augmented Binkley dataset, and Transformer$_{\text{MINI}}$ achieved an accuracy of 0.856 after being trained on the original Binkley dataset.

As fake identifiers constitute the large majority of the BT11$_{\text{AUG}}$ and Jhotdraw$_{\text{AUG}}$ datasets, we believe the results reported on Table 4 have more to say about the fake identifiers than about the evaluated datasets themselves.

Table 4. Cross-domain performance (i.e., accuracy of splitting) of CNN-BiLSTM-CRF [10] on the augmented datasets.

Model	CNN-BiLSTM-CRF		
Training	Evaluation		
	Binkley	BT11	Jhotdraw
Binkley$_{AUG}$	–	0.760	0.655
BT11$_{AUG}$	0.833	–	0.796
Jhotdraw$_{AUG}$	0.896	0.936	–

Table 5. Composition of the augmented datasets utilized by Li et al. [10] during his cross-domain performance experiments.

Dataset	Total	Original	Fake	Percentage of fake identifiers
Binkley$_{AUG}$	7663	2663	5000	65%
BT11$_{AUG}$	26122	21122	5000	19%
Jhotdraw$_{AUG}$	5974	974	5000	83%

It is also a fact that these fake identifiers were not able to consistently improve the accuracy of the CNN-BiLSTM-CRF results. Therefore, future work on the CNN-BiLSTM-CRF architecture proposed by Jiechu Li et al. [10] should perform ablation studies to carefully evaluate the impact of data augmentation techniques on model performance (Table 6).

Table 6. Cross-domain performance (i.e., accuracy of splitting) of Transformer$_{MINI}$[21] on the original datasets.

Model	Transformer$_{MINI}$		
Training	Evaluation		
	Binkley	BT11	Jhotdraw
Binkley	–	0.856	0.189
BT11	0.785	–	0.358
Jhotdraw	0.356	0.250	–

6 Conclusion

Pretrained distillation [21] of general-purpose Transformer models has proven itself to be a strategy that is both competitive and scalable in comparison with previous state-of-the-art results for English word segmentation. The quality of our results was found to be directly correlated with how close the evaluated datasets were to the linguistic domain in which the Transformer model has been pretrained.

Our end goal is to eventually develop a framework for word segmentation that can achieve consistent results across multiple languages and application domains. Therefore, we plan to expand and diversify our evaluated tasks, as well as to consider recent techniques for domain adaptation. As we are seeking to develop a lightweight architecture, inexpensive domain adaptation techniques are particularly desirable, such as the approach suggested by Ye et al. [24].

All of our collected datasets were automatically constructed from randomly selected sentences, either by ourselves or their original authors. Unfortunately, as we are dealing with a relatively unexplored field, we could not find any datasets that had a careful linguistic justification for their design choices, and most of them did not even have a standard train, dev and test split.

Therefore, in order to avoid undesirable biases, we believe that dataset construction should be considered as a relevant research direction for future work on word segmentation for Western languages. A notable example is that hashtags datasets which are built from random sentences are usually highly imbalanced in favor of noun-based hashtags [3], with verb-based hashtags being significantly less represented. Such linguistic concerns should be carefully considered while building datasets to evaluate the performance and the reliability of word segmentation systems.

Other directions include looking into the feasibility of deploying Conditional Random Fields in combination with the Transformer architecture, as has already been done for named entity recognition tasks [19] and Chinese word segmentation [23]. Furthermore, as we are primarily focused on Western languages, it is also particularly relevant to compare the quality of the results that can be obtained by tackling word segmentation as a language modeling task or as a character-based sequence labeling task.

References

1. Sentiment140 - a Twitter sentiment analysis tool (2009). http://help.sentiment140.com/for-students
2. EMNLP 2017 - second conference on machine translation (WMT17) (2017). http://www.statmt.org/wmt17/. Accessed 17 May 2020
3. Celebi, A., Ozgur, A.: Segmenting hashtags and analyzing their grammatical structure. J. Assoc. Inf. Sci. Technol. **69** (2017). https://doi.org/10.1002/asi.23989
4. Chen, X., Shi, Z., Qiu, X., Huang, X.: Adversarial multi-criteria learning for Chinese word segmentation (2017)
5. Devlin, J., Chang, M.W., Lee, K., Toutanova, K.: BERT: pre-training of deep bidirectional transformers for language understanding. arXiv preprint arXiv:1810.04805 (2018)
6. Doval, Y., Gómez-Rodríguez, C.: Comparing neural- and N-gram-based language models for word segmentation. J. Assoc. Inf. Sci. Technol. **70**(2), 187–197 (2018). 10/gfs6rd, https://doi.org/10.1002/asi.24082, 105

7. Huang, W., Cheng, X., Chen, K., Wang, T., Chu, W.: Toward fast and accurate neural Chinese word segmentation with multi-criteria learning. arXiv:1903.04190 [cs], March 2019. http://arxiv.org/abs/1903.04190, 97.9

8. Inuzuka, M.A., Rocha, A.S., Nascimento, H.A.D.: Segmentation of words written in the Latin alphabet: a systematic review. In: Quaresma, P., Vieira, R., Aluísio, S., Moniz, H., Batista, F., Gonçalves, T. (eds.) PROPOR 2020. LNCS (LNAI), vol. 12037, pp. 291–302. Springer, Cham (2020). https://doi.org/10.1007/978-3-030-41505-1_28

9. Lei, J.: Bert 系列（五）——中文分词实践 f1 97.8%(附代码). https://jiangpinglei.git hub.io/BERT系列（五）——中文分词实践-F1-97-8-附代码 (2019). Accessed 17 May 2020

10. Li, J., Du, Q., Shi, K., He, Y., Wang, X., Xu, J.: Helpful or not? An investigation on the feasibility of identifier splitting via CNN-BiLSTM-CRF (2018)

11. Li, X., Meng, Y., Sun, X., Han, Q., Yuan, A., Li, J.: Is word segmentation necessary for deep learning of Chinese representations? arXiv preprint arXiv:1905.05526 (2019)

12. Ma, J., Ganchev, K., Weiss, D.: State-of-the-art Chinese word segmentation with Bi-LSTMs. In: Proceedings of the 2018 Conference on Empirical Methods in Natural Language Processing, pp. 4902–4908. Association for Computational Linguistics, Brussels, October–November 2018. https://doi.org/10.18653/v1/D18-1529, https://www.aclweb.org/anthology/D18-1529

13. Maddela, M., Xu, W., Preoţiuc-Pietro, D.: Multi-task pairwise neural ranking for hashtag segmentation (2019)

14. Morabia, K., Bhanu Murthy, N.L., Malapati, A., Samant, S.: SEDTWik: segmentation-based event detection from tweets using Wikipedia. In: Proceedings of the 2019 Conference of the North American Chapter of the Association for Computational Linguistics: Student Research Workshop, pp. 77–85. Association for Computational Linguistics, Minneapolis, June 2019. https://doi.org/10.18653/v1/N19-3011, https://www.aclweb.org/anthology/N19-3011

15. Pan, Y., Li, X., Yang, Y., Dong, R.: Morphological word segmentation on agglutinative languages for neural machine translation (2020)

16. Rhodes, D.: Conditional random field Latin word segmenter (2013)

17. Saenger, P.: Space Between Words: The Origins of Silent Reading. Stanford University Press, Palo Alto (1997)

18. Shi, X., Huang, H., Jian, P., Guo, Y., Wei, X., Tang, Y.-K.: Neural Chinese word segmentation as sequence to sequence translation. In: Cheng, X., Ma, W., Liu, H., Shen, H., Feng, S., Xie, X. (eds.) SMP 2017. CCIS, vol. 774, pp. 91–103. Springer, Singapore (2017). https://doi.org/10.1007/978-981-10-6805-8_8

19. Souza, F., Nogueira, R., Lotufo, R.: Portuguese named entity recognition using BERT-CRF. arXiv:1909.10649 [cs], February 2020. zSCC: NoCitationData[s0]

20. Sproat, R.W., Shih, C., Gale, W., Chang, N.: A stochastic finite-state word-segmentation algorithm for Chinese. Comput. Linguist. **22**(3), 377–404 (1996). https://www.aclweb.org/anthology/J96-3004

21. Turc, I., Chang, M.W., Lee, K., Toutanova, K.: Well-read students learn better: on the importance of pre-training compact models (2019)

22. Wang, B., Pan, B., Li, X., Li, B.: Towards evaluating the robustness of Chinese BERT classifiers (2020)

23. Yang, H.: BERT meets Chinese word segmentation (2019)

24. Ye, Y., Zhang, Y., Li, W., Qiu, L., Sun, J.: Improving cross-domain Chinese word segmentation with word embeddings. In: Proceedings of the 2019 Conference of the North, pp. 2726–2735 (2019). https://doi.org/10.18653/v1/N19-1279, http://arxiv.org/abs/1903.01698, arXiv: 1903.01698
25. Zhang, Y., Clark, S.: Syntactic processing using the generalized perceptron and beam search. Comput. Linguist. **37**(1), 105–151 (2011). https://doi.org/10.1162/coli_a_00037, https://www.aclweb.org/anthology/J11-1005

Entropy-Based Filter Selection in CNNs Applied to Text Classification

Rafael Bezerra de Menezes Rodrigues$^{(\boxtimes)}$ ⓘ, Wilson Estécio Marcílio Júnior ⓘ, and Danilo Medeiros Eler ⓘ

São Paulo State University-UNESP, Presidente Prudente, SP 19060-900, Brazil
{rafael.rodrigues,wilson.marcilio,danilo.eler}@unesp.br

Abstract. Filter selection in convolutional neural networks aims at finding the most important filters in a convolutional layer, with the goal of reducing computational costs and needed storage, as well as understanding the networks' inner workings. In this paper we propose an entropy-based filter selection method that ranks filters based on the mutual information between their activations and the output classes using validation data. Our proposed method outperforms using filters' absolute weights sum by a large margin, allowing to regain better performance with fewer filters.

Keywords: Filter pruning · Convolutional neural networks · Mutual information

1 Introduction

Deep neural networks commonly have millions of parameters, demanding high computational cost and storage when deploying a model. This makes it difficult to take advantage of their state-of-the-art performance in low-resource settings, such as embedded systems. In order to tackle this issue, network pruning methods have been developed with the aim of reducing the number of parameters in neural network's layers, whether for fully connected layers [2] or, more recently, convolutional layers [3]. The network pruning pipeline consists of three steps [5]: train an overly parameterized model; prune it according to a defined process; and fine-tune the model. Besides reducing the size of the network and amount of computation, elimating irrelevant and redundant features leads to better understanding of data and might improve classification performance [1].

Feature selection methods can be broadly divided into three groups: Filter, Wrapper and Embedded methods [1]. Filter methods rank the features according to some criteria, such as Pearson correlation between each feature and the output classes to discover relevant features, or between pairs of features to find redundant variables. They are computationally light and help to avoid overfitting, but might end up with a suboptimal set of features due to redundant variables. Wrapper methods use a different strategy, in which the quality of a subset of features is measured according to the performance of a classifier [1]. They search the

© Springer Nature Switzerland AG 2020
R. Cerri and R. C. Prati (Eds.): BRACIS 2020, LNAI 12319, pp. 497–510, 2020.
https://doi.org/10.1007/978-3-030-61377-8_34

feature space, sequentially or heuristically, while trying to maximize the objective function of a classifier, thus obtaining better sets of features. An instance of heuristic Wrapper methods are Genetic Algorithms, where chromosome bits can represent if a feature is included or not in the subset and individuals' aptitudes are measured according to the classifier's performance. For large classifiers, such as deep neural networks, training a new classifier for each candidate solution is impracticable. Embedded methods try to compensate the drawbacks of both previous methods by incorporating feature selection in the training process.

In this work we propose an entropy-based filter selection method for convolutional layers, which ranks filters in a layer according to the mutual information between their activations and the output labels using validation data. Mutual information is used to capture how discriminative of the output classes each filter is, thus allowing the elimination of irrelevant filters. In our experiments the proposed method outperforms ranking filters by their absolute weights sum [3], allowing to regain better performance with fewer filters. Filter pruning was applied to the last convolutional layer only, however we show that it can drastically reduce its output size, which also reduces the connections to its next layer.

The experiments were conducted using the character-level Convolutional Neural Network (CNN) presented by Zhang et al. in [12] applied on two text classification datasets. First, in order to evaluate the proposed method versus using filters' absolute weights sum, a full CNN is trained and filters from the last convolutional layer are ranked according to different metrics, then subsets with the k-most important filters (pruning) are used to train new classifiers (fine-tuning). We also compare pruned and fine-tuned models performance against training the smaller architectures from scratch, where pruning and fine-tuning yielded better results on the smallest dataset, however for the biggest dataset pruning and fine-tuning served most as an architecture search paradigm, corroborating the findings in [5].

This paper is structured as follows. Section 2 introduces the CNN used in our experiments, the applied mutual information theory, and the absolute weights sum pruning technique. Section 3 presents our method and how we summarize a filter's activations as a single value. Our experiment is outlined in Sect. 4 along with the details of both used datasets. Section 5 presents the results and discussion, followed by Sect. 6 with our conclusions and future work.

2   Background

2.1   Character-Level CNNs for Text Classification

Language models tend to have large vocabularies, including up to millions of unique tokens, which can lead to very sparse representations or large weights matrices for storage when using dense representations, such as word embeddings. Working at the character-level requires building alphabets in the tens or hundreds of characters, consequently reducing the amount of space needed to deploy a model.

Each character can be represented as an one-hot vector, an array with the length of the model's alphabet filled with zeros and with value equal to one at the index of the character. Thus, a text sequence becomes a matrix, from which convolutional neural networks can learn to identify patterns from raw signals [12], i.e., sequences of one-hot encoded characters. The input of each convolutional layer is sweeped in a single dimension, left to right, instead of two dimensions as usual when the input is an image.

The CNN used in our experiments is the same as presented by [12], whose architecture is shown in Table 1, with the only exception being the input size, limited to 312 characters. Our feature selection approach aims at reducing the number of filters in the last convolutional layer, which reduces the size of the flatten layer and consequently the connections between it and the first dense layer, the largest weights matrix in this network.

Table 1. Character-Level CNN presented by [12], with 312 characters at the input. The size of the alphabet is denoted by m, and n is the number of classes. Dropout layers were ommited.

Layer	Trainable parameters	Output size
Input	0	$312 \times m$
Convolution, 256 filters, kernel $7 \times m$	$256 * 7 * m + 256$	306×256
Max-pooling, 3×256	0	102×256
Convolution, 256 filters, kernel 7×256	459,008	96×256
Max-pooling, 3×256	0	32×256
Convolution, 256 filters, kernel 3×256	196,864	30×256
Convolution, 256 filters, kernel 3×256	196,864	28×256
Convolution, 256 filters, kernel 3×256	196,864	26×256
Convolution, 256 filters, kernel 3×256	196,864	24×256
Max-pooling, 3×256	0	8×256
Flatten	0	2048
Dense, 1024 neurons	2,098,176	1024
Dense, 1024 neurons	1,049,600	1024
Dense, n neurons	$n * 1025$	n

The output from the last max-pooling layer represents the embedding of an input text, in which each filter from the last convolutional layer contributes with eight values. Each value corresponds to how much a given filter was activated by some portion of the text.

2.2 Mutual Information

Mutual information is one of many feature selection methods, for which the common goal is to remove irrelevant and redundant features from an input following some methodology [1]. It can be used to analyze each input feature individually in regard to the output classes, producing a metric that determines their ability to decrease the uncertainty of a given class, i.e., if the feature is good to discriminate between the classes. In our experiments we used Sci-Kit Learn's [7] implementation of mutual information, which follows the process described in [8], with number of neighbors equals to three.

To determine the mutual information of a given variable, first the algorithm finds the distance between each datapoint and the third closest from the same class. In a following step it counts how many datapoints, considering now all classes, are within the same distance radius. The more datapoints from other classes are within the same radius, the less discriminative is the feature, i.e., less relevant for classification. Figure 1 shows the radius around a chosen blue datapoint. On the left a good variable is spotted since there are no green points inside the radius, however on the right several green points are within the neighborhood of the chosen blue datapoint, meaning this variable does not discriminate classes as well as the variable on the left.

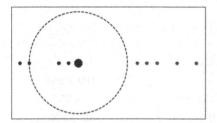 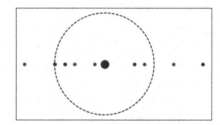

Fig. 1. Mutual information example: a variable that is a great discriminator on the left, and a not so good one on the right

Other common feature selection techniques belonging to the Filter methods are the Pearson Correlation and the Analysis of Variance (ANOVA). Pearson correlation is able to capture only linear depencies between variables [1], thus it was not considered to incorporate our method. And ANOVA needs its inputs to have normal distributions [9], which is not the case in our experiments given the Shapiro-Wilk test obtained. ANOVA also needs equal standard deviation between variables' groups, which is also not true in our case.

Besides finding relevant filters according to the output classes, mutual information can also be applied between each pair of filters, where high dependence means redundant features. However, as the number of filters grow, the problem becomes intractable. Therefore other methods are needed to compare filters pairwise, which will be discussed in Sect. 6 along with our conclusions.

2.3   Absolute Weights Sum

Li et al. present in [3] a data-free filter selection method that uses absolute weights values to determine filter importance. For a given convolutional layer, its filters are sorted according to their absolute weights sum. The bigger the sum, the more important the filter is. The same authors report experiments with robust CNNs, such as VGG-16 and ResNet-110, where they were able to reduce computation operations by a third while maintaining accuracy close to the original models.

Computing filters' absolute weights sum takes less than a second in our setting, while to determine all filters' mutual information according to the output classes takes around 40 s. Our entropy-based method is slower to run, however, as shown in our results, classwise information given by mutual information can help to select better filters and regain accuracy more rapidly.

3   Entropy-Based Filter Selection Method

Figure 2 shows the application of our entropy-based filter selection method over the last convolutional layer of a CNN. First a large CNN is trained and used as a feature extractor over the training and validation data, and the output of the last max-pooling layer is obtained for each input example. We name these outputs as texts' embeddings. Then a filter ranking method is applied using validation embeddings only, and the k-best filters are selected to downsize the embeddings. Finally, a smaller classifier, including only the fully connected layers, is trained from scratch.

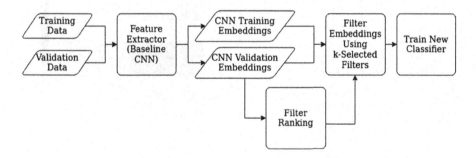

Fig. 2. Data-driven pipeline to rank filters and train a new classifier

Instead of pruning the final convolutional layer and freeze convolutionals layers' weights, we chose to extract embeddings for all texts and select activations from the k-best filters to speedup retraining the new classifiers. This way we were able to skip all convolutional operations required to train the fully connected layers while selecting several different amounts of filters. The scheme depicted in Fig. 2 can be easily adapted to obtain activations for any layer of interest, use the activations to sort filters by their importance, prune the layer with the best filters, and retrain the new model as desired.

To evaluate the importance of each filter using its mutual information relative to the output classes, it is necessary to summarize all its generated values into a single one. As detailed in Sect. 2.1, each filter in the final convolutional layer contributes with eight features from the input text, therefore a method to unify these values must be defined.

Empirically, we found that the leftmost feature from each filter can be used to summarize its importance. For that, we trained a character-level CNN as presented by [12], with 1024 neurons in the dense layers, using the AG News dataset[1], a common benchmark for text classification which is presented in more details in Sect. 4.1. Then applied mutual information for each individual feature, i.e., each one of the 2048 features was treated as a different variable.

Left side of Fig. 3 shows a histogram with the importance of each feature, where it can be seen that most features are not important to discrimate between the classes while few features are highly important. The right side of Fig. 3 shows a heatmap where the rows are eight selected filters, the columns are the eight features generated by each filter, and the values are the importance of each feature treated individually. Seven filters were randomly selected and filter 13 was included because it has the most important feature. As one can see, the importance of a feature extracted by a given filter diminishes along the text in most cases. We argue that this happens because text is read from left to right and some texts are right-padded with empty characters to match the input size, therefore filters are not activated with the same frequency in all positions over the text.

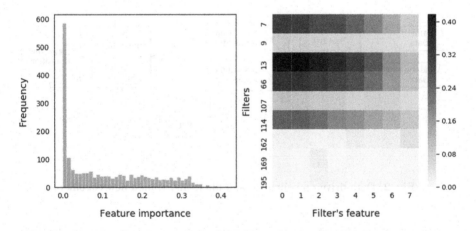

Fig. 3. Individual feature importance. On the left a histogram with all filter importances, on the right a heatmap with features from eight selected filters

Although in some cases, such as filters 13 and 162, the first feature is not the most important, the difference between the most important feature and the first feature is very small for all filters. Another compelling reason to use only the first feature, instead of some aggregated value as their mean, is the fact that it

---

[1] http://groups.di.unipi.it/~gulli/AG_corpus_of_news_articles.html.

could put in disadvantage classes with smaller texts, since the rightmost features from each filter would be less activated for them.

Besides using the leftmost activation to summarize a filters' activations, we also tried to use the mean of the values as well as the maximum. Using the mean resulted in slightly worse performance when compared to the leftmost activation, while using the maximum value yielded similar performance. The results are ommited for brevity. Throughout the paper the experiments use the leftmost activation only.

In this work our method is demonstrated using character-level CNNs, which apply convolutions in a single dimension. However, it can also be applied over 2D or 3D convolutional layers, as long as each filter's activations can be summarized in the form of a single value.

4   Experiments

This section presents the text classification datasets used in our experiments, then describes the process followed to validate the proposed filter selection method.

4.1   Datasets

AG News. The AG News is a well known benchmark dataset for text classification including news articles labeled with their respective category. We used the same subset of samples as [12] containing the four most frequent classes, as shown in Table 2. Original data is divided in training and test sets. We randomly selected 10% of the training samples as validation samples while preserving class balance. As valid characters we use the same 70 characters reported by [12].

Table 2. AG news selected categories and dataset size

Category	Training samples	Validation samples	Test samples
Business	27000	3000	1900
Sci/Tech	27000	3000	1900
Sports	27000	3000	1900
World	27000	3000	1900
Total	108000	12000	7600

DSL Shared Task. The other text classification goal in our experiments is discriminating between similar languages, where the model needs to diferentiate between language variaties such as brazilian and european portuguese or canadian and european french. The datasets used come from the DSL Shared Tasks

[10], a competition that occurred from 2014 to 2017[2] aiming at the development of language detectors.

As validation and test sets we used only the dataset from 2017, with 2000 and 1000 samples for each class, respectively, restricting our model to the 14 languages in that year's competition. As training data we used training samples from all previous competitions, while removing duplicates. Table 3 shows the languages and variaties with their corresponding number of samples in the training set. No method to compensate data imbalance was applied.

To build the alphabet, which consists of the networks known characters, we chose the 70 most common characters in each language variety, as long as they appear at least 10 times in the training set, considering data from 2017 only. Non-printable characters and the blank space were also removed, as done by [12]. The final alphabet for this dataset has 172 characters.

Table 3. DSL shared task's languages and experiment dataset size

Language group	Language variety	Training samples
Serbo-Croatian	Bosnian	69416
	Croatian	71040
	Serbian	68616
Austronesian	Indonesian	71186
	Malay	67887
Portuguese	Brazilian	69209
	European	71525
Spanish	Argentine	38663
	Castilian	69437
	Peruvian	17972
French	Canadian	17959
	Hexagonal	36000
Persian	Iranian	17958
	Dari (Afghanistan)	17964
Total		704832

4.2 Experiment Outline

Our experiment is composed by two parts: retraining new classifiers using features from the k-best filters; and comparing pruned and fine-tuned models against retraining the pruned architecture from scratch. Two different baseline CNNs, with 256 and 1024 neurons in the dense layers, were trained on both datasets, resulting in four different baselines.

---

[2] All four competition datasets are available at: http://ttg.uni-saarland.de/resources/ DSLCC/.

In the first part, for each baseline, we use the trained model to extract embeddings from training and validation text inputs. Then, for each validation embedding, we select the leftmost activation from each filter, as detailed in Sect. 3, and compute the mutual information between each variable – filter – and the output classes. To train new classifiers, we identify the k-most important features and select all the features from those filters to reduce our training and validation embeddings. Note that new classifiers are composed by dense layers only.

To compare our approach versus using absolute weights sum to select the filters, we sort filters according to each metric and train new classifiers using the same value of k. For the DSL Shared Task dataset k = 8, 16, 32, 64, 128, 256, and for the AG News we also used k = 4, since this dataset has fewer target classes.

As pointed out by [6], discriminating between languages is hardest between similar languages, such as brazilian and european portuguese, in comparison to classifying between distinct languages like portuguese and spanish. Following this idea, for the DSL Shared Task dataset, we also sort filters by their importance intragroups, i.e., apply the mutual information process for each one of the six language groups.

In order to apply the intragroup approach and compare it to the other methods, we need to define a process to pick filters in groups of 8 elements, so to match the value of k selected filters in each round. Table 4 shows the 5 most important filters, identified by their index in the layer, for each language group as well as for all languages. At each addition of filters we choose one filter from each group and two filters that are good discriminators in general, without repetition. The filters used when k = 8 are highlighted in blue, while filters in red are also included when k = 16. Note that the filter 152 was not added in the second round for the persian language since it had been added first for serbo-croatian languages. When not considering the intragroup information, the process is to simply pick the k-best filters sorted according to all languages–last row in Table 4.

Table 4. Intragroup filter selection example. Each group's list contains all 256 filters.

Linguistic group	Filters sorted by importance				
Serbo-Croatian	152	6	105	201	27
Austronesian	45	36	73	213	21
Portuguese	166	77	28	20	235
Spanish	170	98	22	228	30
French	25	48	224	96	62
Persian	18	152	59	118	32
All	240	155	122	207	132

5 Results and Discussion

In this section, baseline CNNs with 1024 neurons in the fully connected layers will be called Baseline 1024. The same goes to baseline CNNs with 256 neurons.

5.1 Comparing Filter Selection Methods

The goal in this part of the experiment is to determine which filter selection method is best at regaining the baseline's accuracy using fewer filters. Figure 4 shows the results with the AG News dataset for Baseline 256 on the left and Baseline 1024 on the right. Using mutual information allowed to regain better performance using fewer filters in both cases. In the case of Baseline 1024, sorting filters by mutual information allowed the retrained classifier to obtain 98.43% of the baseline f1-score using only four filters, while reducing the network size by 43.98%. On the other hand, sorting by weights absolute mean, the retrained classifier achieved only 83.34% of the original score under the same condition.

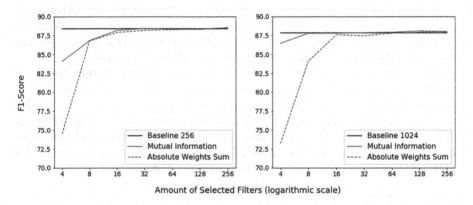

Fig. 4. F1-scores for retrained models on the AG news dataset. On the left results for Baseline 256, on the right Baseline 1024

Table 5 presents the f1-scores for retrained classifiers starting from Baseline 256 in more details, with network's total parameters and network reduction after downsizing the number of features using k-selected filters. Non-trainable parameters, which corresponds to characters one-hot embeddings, were ommited. One can see that classifier's original accuracy was fully regained using 32 filters, which results in 23.36% less parameters, and did not get significant improvement by adding more filters than that.

To get a second glance at the proposed filter selection method, the same process was applied over the DSL Shared Task dataset. Figure 5 shows the results, which also include intragroup mutual information as detailed in Sect. 4.2. Both mutual information methods outperformed weights absolute mean, with some leverage to the intragroup approach when using fewer filters. For Baseline 1024,

Table 5. Retrained classifiers sorting filters by mutual information for the AG news dataset

Model	F1-Score	Total parameters (millions)	Dense parameters (millions)	Total parameters reduction (%)
Baseline 256	88.37	1.964	0.591	–
4 Filters	84.08	1.447	0.075	26.28
8 Filters	86.81	1.456	0.083	25.87
16 Filters	88.16	1.472	0.100	25.05
32 Filters	88.41	1.505	0.132	23.36
64 Filters	88.31	1.570	0.198	20.03
128 Filters	88.31	1.701	0.329	13.35
256 Filters	88.51	1.964	0.591	0

mutual information allowed to regain performance close to original using 32 filters, while the same result was obtained by absolute weights sum when considering 128 filters, that results in 16.67% less parameters when using our method.

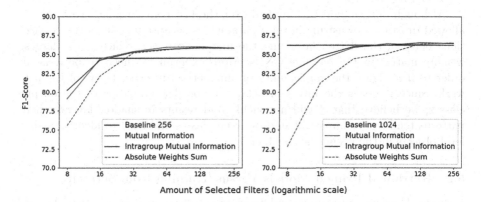

Fig. 5. F1-scores for retrained models on the DSL shared task dataset. On the left results for Baseline 256, on the right Baseline 1024

Another interesting fact is that retraining Baseline 256 with 64 filters or more allowed the fine-tuned model to achieve greater performance than its original model, reaching the same performance as Baseline 1024. This means that the features were there for Baseline 256's dense layers, but somewhere along the training process they could not handle the features well. Training new fully connected layers from scratch while using ready embeddings since the first epoch allowed the new classifiers to consistently achive better results.

In Table 6 the results for retraining from Baseline 1024 are reported in more details. One can see that using a model with only two thirds of the original parameters (64 filters) the network regained original performance. The reason why total parameter reduction is considerably greater for Baseline 1024 is that it has four times more neurons than Baseline 256 in the dense layers, escalating the number of connections.

Table 6. Retrained classifiers sorting filters by intragroup mutual information for the DSL shared task dataset

Model	F1-score	Total parameters (millions)	Dense parameters (millions)	Total parameters reduction (%)
Baseline 1024	86.16	4.717	3.162	–
8 Filters	82.42	2.685	1.131	43.07
16 Filters	84.77	2.751	1.196	41.68
32 Filters	85.98	2.882	1.327	38.90
64 Filters	86.17	3.144	1.589	33.34
128 Filters	86.49	3.668	2.114	22.23
256 Filters	86.41	4.717	3.162	0

Although our method was applied over the last convolutional layer only, it allowed drastical downsizing in total parameters because it reduces the connections between the flatten layer and the first dense layer, which is the largest weights matrix in the baseline CNNs. More experiments need to be done in order to investigate the applicability of our method in other layers of the network. Since it needs the filters in a layer to be discriminative of the output classes, we believe that it will not yield good results in shallow layers, where patterns tend to be simpler, shared between classes and used to identify higher level objects in the deeper layers [4,11].

5.2 Retrained Pruned Models Versus Training from Scratch

Liu et al. [5] argue that pruning and fine-tuning large models is valid to find a better architecture, but these retrained models cannot get better performance than retraining the pruned architecture from scratch. In this section we retrain from scratch some selected architectures and compare them with the best retrained models.

For the AG News dataset we chose 16 as the best number of filters. We trained two models following the pruned architecture with 16 filters in the last convolutional layer and compare them against the baselines and the retrained models with same number of filters. Table 7 presents the results, where the best smaller model is the pruned and fine-tuned one using 256 neurons at the dense layers, beating the same architecture trained from scratch by 1.03. Although this seems to go against the findings by [5], the same authors also point out that it can happen for small datasets. The difference between fine-tuning and training from scratch is even more evident when using 1024 neurons.

Table 7. Comparing baseline models, fine-tuned models, and pruned architectures trained from scratch for the AG news dataset

Model	Number of filters	Epochs until convergence	F1-score	Total parameters (millions)
Baseline 1024	256	11	87.84	4.717
Pruned and fine-tuned 1024	16	6	87.78	2.751
Pruned 1024 trained from scratch	16	30	85.53	2.751
Baseline 256	256	13	88.37	1.964
Pruned and fine-tuned 256	16	8	88.16	1.472
Pruned 256 trained from scratch	16	14	87.13	1.472

The same process was applied to the DSL Shared Task dataset, however considering the best number of filters as 64, with results shown in Table 8. In this case retraining the pruned architecture from scratch achieved the best results, with f1-score equals 86.23. Although the fine-tuned model from baseline 1024 got similar results, it has nearly twice as much parameters. This result over a larger dataset corroborates the findings by [5], that training the pruned architecture from scratch yields better results.

Table 8. Comparing baseline models, fine-tuned models, and pruned architectures trained from scratch for the DSL shared task dataset

Model	Number of filters	Epochs until convergence	F1-score	Total parameters (millions)
Baseline 1024	256	9	86.18	4.717
Pruned and fine-tuned 1024	64	4	86.17	3.144
Pruned 1024 trained from scratch	64	9	85.71	3.144
Baseline 256	256	7	84.46	1.964
Pruned and fine-tuned 256	64	7	85.64	1.570
Pruned 256 trained from scratch	64	10	86.23	1.570

Tables 7 and 8 also present the epochs that each model took to converge, where convergence means to achieve the lowest error over the validation set. Fine-tuning takes fewer epochs than the baseline to train, or the same amount for the case of Baseline 256 on the DSL Shared Task dataset, while training the pruned atchitecture from scratch usually takes more epochs or the same amount.

6   Conclusion

In this work we proposed an entropy-based filter selection method to reduce convolutional layers that outperforms using absolute weights sum by a large margin, with the drawback that it might not be applied to shalow convolutional layers.

Although we demonstrated it using a character-level CNN, which performs convolutions on a single dimension, it can also be used in CNNs with 2D or 3D convolutions as long as one filter's activations can be summarized as a single value.

According to [1], features that individually are not good discriminators between classes can be of use when combined with some other feature, therefore as future work the importance of subsets of filters can also be investigated. Another improvement in the proposed method would be to prioritize filters that are not correlated to any of the already selected filters, thus not adding redundant filters until all types of filters are selected. In order to do this, the filter clustering method proposed by [4] can be used to identify groups of filters, then select the most important filter from each group.

It also becomes interesting to use our technique to identify relevant filters and focus on them to understand network's predicitions, since our method finds good discriminator between classes.

Acknowledgements. This work was supported by São Paulo Research Foundation–FAPESP (Grant Numbers #2018/17881-3 and #2018/25755-8).

References

1. Chandrashekar, G., Sahin, F.: A survey on feature selection methods. Comput. Electr. Eng. **40**(1), 16–28 (2014)
2. LeCun, Y., Denker, J.S., Solla, S.A.: Optimal brain damage. In: Advances in Neural Information Processing Systems, pp. 598–605 (1990)
3. Li, H., Kadav, A., Durdanovic, I., Samet, H., Graf, H.P.: Pruning filters for efficient convNets. arXiv preprint arXiv:1608.08710 (2016)
4. Liu, M., Shi, J., Li, Z., Li, C., Zhu, J., Liu, S.: Towards better analysis of deep convolutional neural networks. IEEE Trans. Visual Comput. Graph. **23**(1), 91–100 (2016)
5. Liu, Z., Sun, M., Zhou, T., Huang, G., Darrell, T.: Rethinking the value of network pruning. arXiv preprint arXiv:1810.05270 (2018)
6. Malmasi, S., Dras, M.: Language identification using classifier ensembles. In: Proceedings of the Joint Workshop on Language Technology for Closely Related Languages, Varieties and Dialects, pp. 35–43 (2015)
7. Pedregosa, F., et al.: Scikit-learn: machine learning in Python. J. Mach. Learn. Res. **12**, 2825–2830 (2011)
8. Ross, B.: Mutual information between discrete and continuous data sets. PLOS ONE **9**(2), 1–5 (2014)
9. St, L., Wold, S., et al.: Analysis of variance (ANOVA). Chemometr. Intell. Lab. Syst. **6**(4), 259–272 (1989)
10. Zampieri, M., et al.: Findings of the VarDial evaluation campaign (2017)
11. Zeiler, M.D., Fergus, R.: Visualizing and understanding convolutional networks. In: European Conference on Computer Vision, pp. 818–833. Springer (2014). https://doi.org/10.1007/978-3-319-10590-1_53
12. Zhang, X., Zhao, J., LeCun, Y.: Character-level convolutional networks for text classification. In: Advances in Neural Information Processing Systems, pp. 649–657 (2015)

Identifying Fine-Grained Opinion and Classifying Polarity on Coronavirus Pandemic

Francielle Alves Vargas[✉][iD], Rodolfo Sanches Saraiva Dos Santos[✉],
and Pedro Regattieri Rocha[✉]

Institute of Mathematical and Computer Sciences,
University of Sao Paulo, São Carlos, Brazil
{francielleavargas,rodolfosanches,pedro.regattieri.rocha}@usp.br

Abstract. In this paper, we explore the fine-grained opinion identification and polarity classification tasks using twitter data on the COVID-19 pandemic in Brazilian Portuguese. We trained machine learning-based classifiers using a few different methods and tested how well they performed different tasks. For polarity classification, we tested a cross-domain strategy in order to measure the performance of the classifiers among different domains. For fine-grained opinion identification, we provide a taxonomy of opinion aspects and employed them in conjunction with machine learning methods. Based on the obtained results, we found that the cross-domain data improved the results of the polarity classification. For fine-grained opinion identification, the use of a domain taxonomy presented competitive results for the Portuguese language.

Keywords: Fine-grained opinion mining · Natural language processing · Computational social science.

1 Introduction

The Novel Coronavirus (2019-nCoV) pandemic is definitely a new experience for the world and its human inhabitants, as people all collectively grapple with what this global pandemic may mean to them. Around the world, the Covid-19 pandemic drastically altered people's way of life, and unlike during pandemics in the past, people now use social networks to express their opinions on different aspects of the crisis. The extraction, classification and summarization of user opinions on the web during a pandemic of this magnitude provides relevant data for both public and private agencies so they may take deliberate action while confronting the social crisis. Additionally, opinion mining systems with a focus in the pandemic may be employed to combat misinformation. For example, understanding the public's opinion about a given issue may provide reliable information about the specific topics which are being discussed.

The area of extraction, classification and summarization of subjective text on the web is called opinion mining, also known as sentiment analysis. According

R. Cerri and R. C. Prati (Eds.): BRACIS 2020, LNAI 12319, pp. 511–520, 2020.
https://doi.org/10.1007/978-3-030-61377-8_35

to [20], opinion mining or sentiment analysis aims to automatically extract and process relevant opinions, providing useful information to the interested reader. There are usually three granularity levels for opinion mining [9]: the (i) document, (ii) sentence and (iii) aspect levels. At the document level, all of the opinions expressed in the document are weighted. At this level, a full document is rated as either positive, negative, or neutral. At the sentence level, the opinion expressed in each sentence of the document is determined. At the most refined level, in order to figure out exactly the object properties that the user liked or disliked, an aspect-based analysis is carried out. For example, in the sentence "It is cheap but the screen resolution is not good, the "price" aspect is well evaluated, while the "screen resolution" is not. According to [9], aspects represent properties or parts of entities that are evaluated by users, in opinionated texts, as in comments on websites and blogs on the web.

As claimed by [18], aspect-based opinion mining (also know as fine-grained opinion mining) consists of 4 main tasks: (i) explicit and implicit aspect extraction; (ii) aspect clustering; (iii) polarity extraction; and (iv) opinion summarization. The aspects of the evaluated object may be explicitly or implicitly expressed. Explicit aspects are those that are cited in the text, e.g., "screen resolution" in the previous example. Implicit aspects, on the opposite, must be inferred and are usually signaled by clue terms. It is the case for the "price" aspect in the example, that was signaled by the word "cheap". In the aspect clustering task, groups of similar aspects must be found, as the user may employ different words to refer to the same aspect, such as "value", "cost" and "investiment" to refer to the "price" aspect. For polarity extraction, the expressed sentiments in relation to the cited aspects must be identified and extracted. In our example, one must identify that the "price" shows a positive polarity, while the "screen resolution" shows a negative one. Lastly, in the opinion summarization step, a general overview of the opinions of an object of interest must be produced.

In this paper, we explore the fine-grained opinion identification and polarity classification tasks of twitter data related to Covid-19 pandemic in Portuguese language. In order to achieve this, we trained machine learning-based classifiers using different strategies for each task. For polarity classification, we used different user review data sets on different domains to evaluate the performance of classifiers among the domains. As for fine-grained opinion identification, we built a taxonomy of aspect in order to improve our machine learning approach. Based on the obtained results, we found that the cross-domain strategy improves the polarity classification, and the utilization of a domain taxonomy presents competitive results for opinion mining aspect extraction task for the Portuguese language.

The remainder of this paper is organized as follows. In Sect. 2, we present some related work. In the Sect. 3, we show an overview of data. In Sect. 4, we detail the experiments that were performed. In Sect. 5, we report the results of out evaluation. In Sect. 6, we present our final remarks.

2 Related Work

Aspect extraction is an aspect-based sentiment analysis task, which consists of the automatic information extraction in subjective texts. The first one study related to aspect extraction task was proposed in [8] and [9], who also introduced the distinction between explicit and implicit aspects.

For aspect extraction task, we observed on the literature two main approaches: statistic and knowledge-based approaches. On knowledge approach (also called linguistic rules approach), in [9], proposed a word-based translation model (WTM) in order to find the association between opinion aspects and polarity. In this proposal, we considered noun/noun phrases as potential aspects and adjectives as their polarity. In [12], developed a parser responsible to extract explicit and implicit aspect in product reviews. The authors used a set of syntactic rules with a focus on subject noun and sentences, which do not have subject noun. Additionally, they employed rules related to position among modifier of noun. In [19], a new tree nucleus function to model the phrase dependency trees was proposed. First one, constructed phrase dependency tree from the results of chunking and dependency parsing. The second step, extracted candidate product features and candidate opinion expressions. In the last step, performed the extracting of relations among product features and opinion expressions. For Portuguese, which is the language of interest in this paper, in [17], [16], [2] and [7], proposed methods for clustering and extraction of aspects in product reviews using machine learning and lexicon-based approaches.

For polarity classification, a wide range of work's literature present different methods, and the most of them employ machine learning and lexicon-based methods. In general, the proposals of area make use mainly of nouns, verbs, adjectives and adverbs [14], [11] to identify the polarities positive, negative or neutral. For the Portuguese language, in [1] and [7], proposed lexicon-based methods for polarity classification in product reviews. In [4], the authors proposed a new corpus of twitter data on TV show domain, and they used machine learning methods.

To the best of our knowledge, no previous works exists on fine-grained opinion mining on the COVID-19 for the Portuguese language.

3 Data Overview

We select two different corpus, ReLi [6] and TweetSentBr [4] dataset of the literature and create a new one, which we referred by OPCovid-BR. We describe in brief overview on each dataset below.

3.1 ReLi [6]

The ReLi corpus consists of 1.600 book reviews, which are in Portuguese language. As an example, one may find the following review: *Ótimo livro, bem diferente do que eu imaginava. Apesar de antigão, éuma leitura gostosa, com*

a linguagem bem moderna. Um livro adolescentes, de aqueles momento foda-se.
"Amazing book, totally different which I imagined. Although old, it is a pleasant reading, with very modern language. A teenagers book, from those fuck time.". According to [15], it is possible to notice the challenges in dealing with such texts. They are usually marked by orality and informality, orthographic, grammar errors, as well bad language.

3.2 TweetSentBr [4]

TweetSentBr is a sentiment corpus for Brazilian Portuguese manually annotated with 15.000 sentences (17.166 tokens) on TV show domain. An example is describe to follow: *A pior vilã dos desenhos animados está hoje no programa.* "The worst cartoon villain is on the show today". The sentences were labeled in three classes (positive, neutral and negative) by seven annotators, following literature guidelines for ensuring reliability on the annotation.

3.3 OPCovid-BR[1]

We also create a new dataset, which consists of 600 twitters on the Covid-19 pandemic domain. We developed a twitter API and extracted twitters using the follow key term search: "coronavirus". The OPCovid-Br were annotated by three annotators, with concordance among annotators equal to 82,77 %. It is annotated with the binary document polarity (positive and negative) and fine-grained opinion (explicit aspects) for each twitter. An overview on the OPCovid-BR dataset is displayed in the Table 1 and 2.

Table 1. OPCovid-BR overview: polarity labels.

Polarity	Total of Twitters
Negative	300
Positive	300
Total	600

OPCovid-BR consists of a corpus of twitter data on coronavirus pandemic. There are 300 tweets with positive labels and 300 tweets with negative labels, as well as labels for fine-grained opinion identification.

For polarity classification, each twitter was labeled in positive or negative. An example is shown in Fig. 1.

For fine-grained opinion task, each tweet was annotated with at most three and at least one aspects. According to [10], in the fine-grained opinion mining (also called aspect-based opinion mining), it possible identify preciously what is being said about the opinion target. For example, when any consumer user

---

[1] https://github.com/francielleavargas/OPCovid-BR.

Table 2. OPCovid-BR overview: fine-grained opinion labels.

Categories	Total of aspects
Political & Social	132
Health	109
Localization	49
Economy	32
Education & Science	28
Total	350

Twitter 1	Twitter 2
não eh possível que o mundo esteja errado sobre o coronavírus e sobre suas precauções e que o mundo todo queira derrubar um zé ninguém de um país emergente. (negative)	o brasil tem 14.026 pessoas que já se curaram do coronavírus. uma boa notícia em meio a tantas negativas. graças a deus. (positive)

Fig. 1. Polarity annotation of twitter data on the COVID-19.

evaluate a smartphone, in fact, this user is evaluating proprieties of the product as for example, battery, price, touchscreen, size, design, etc. This proprieties are denominated "aspects", which according [9], represent properties or parts of entities that are evaluated by users in subjective texts, as in comments on websites and blogs on the web. An example of is shown in Fig. 2.

Twitter 1	Twitter 2
wilson witzel com coronavírus - recorde de mortes no rj e br - 90% das utis ocupadas em hospital referência municipal - 70% das uti ocupadas na rede estadual - hospitais de campanha não estão prontosc- estudo prevê isolamento intermitente até 2022. vamos levar a sério?	não há vacina. o distanciamento social é a única medida para evitar a contaminação [por coronavírus] no planeta! nós multiplicamos por dez os casos de covid-19 em 15 dias na ilha de são luís, diz @.

Fig. 2. Fine-grained opinion identification of twitter data on the COVID-19.

Note that, for Twitter 1 the aspects identified were: "wilson witzel (wilson witzel)", "mortes" (deaths), "UTI ocupadas" (occupied ICU), "hospitais de campanha" (field hospital), "isolamento intermitente" (intermittent insulation). In the Twitter 2, the aspects identified were "vacina" (vaccine), "distanciamento social" (social distance), "contaminação" (contamination), "casos de covid-19" (covid-19 cases), "são luis" (são luis).

We also provide a new taxonomy composed by opinion aspects, which we categorized in five main groups: economic, education-science, localization, political-social and health. The partial overview of the taxonomy is shown in Fig. 3.

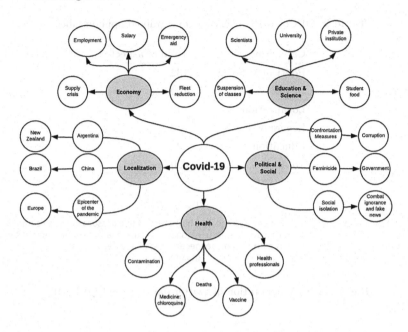

Fig. 3. Partial overview of the taxonomy for fine-grained opinion mining.

It is noticeable that for each identified fine-grained opinion, we categorize them along with your respective group. As for example: "vaccine" and "deaths" integrated the health category; "government", "social isolation" and "feminicide" integrated the political category.

4   Experiments

We carry out a wide variety of experiments for polarity and fine-grained opinion classification. We used Python 3.6 and tweepy, rdflib, beautifulSoup, spacy and skit-learn libraries for implementation.

For features extraction, we employed different algorithms for polarity classification and fine-grained opinion identification tasks. We opted the features extraction algorithm for polarity classification task proposed in [4]. Nonetheless, for fine-grained opinion classification, we proposed a new algorithm shown in Algorithm 1, which to extract features using our taxonomy of aspects.

For the construction of machine learning models, we used the ReLi [6], Tweet-SentBR [4] and OPCovid-19 datasets. We choose Naive Bayes [3], Decision Tree [13] and SVM [5] classifiers. For SVM, we performed three different training using the following metrics: (i) SVM 1 train, the kernel is equal to poly with and degree equal to 2; (ii) SVM 2, we used the linearSVC;, and (iii) SVM 3, the kernel is equal to sigmoid.

For our experiments, we split our data to train (70%), validation (15%) and test (15%). For validation, we has been applied the 10-fold cross-validation.

Algorithm 1. Features extraction for fine-grained opinion classification

---

Ensure: Lemmatization (twitters data and taxonomy of aspects)
1: F = frequency of aspects
2: **function** CLASSIFY(twitter)
3:     $F \leftarrow [0,0,0,0,0]$ ▷ In order: economy(0), education-science(1), location(2), political-social(3), health(4)
4:     **for** aspect in annotated aspects **do**
5:         **if** aspect in twitter **then**
6:             I ← index of the aspect class in the frequency variable
7:             $F[I] \leftarrow F[I]$ + aspect class weight
8:         **end if**
9:     **end for**
10:     **return** class with highest value in frequency variable
11: **end function**

Table 3. Precision

N.	Tasks	Classification	Decision Tree	Naive Bayes	SVM1	SVM2	SVM3
1		ReLi	0.53	0.52	0.60	0.54	0.50
2		TweetSentBr	0.58	0.53	0.55	0.53	0.55
3	Task 1	OPCovid-BR	0.51	0.56	0.50	0.59	0.48
4		ReLi + TweetSentBr	0.55	0.51	0.55	0.52	0.54
5		ReLi + TwitterSent + OPCovid-BR	0.54	0.52	0.59	0.57	**0.60**
5		Political& Social	0.70	0.46	0.74	0.68	0.68
6		Economy	1.00	0.50	0.00	0.00	0.00
7	Task 2	Education&Science	0.50	0.08	0.00	0.00	0.00
8		Localization	0.54	0.58	0.44	0.54	0.50
9		Health	0.85	0.93	0.73	0.75	0.72
10		**weighted avg**	**0.78**	0.76	0.65	0.66	0.64

5 Evaluation

Wee have computed the traditional machine learning evaluation measures of Accuracy, Precision, Recall, F1-Score. The results are showed in Table 3, 4 and 5.

Surprisingly, cross-domain data improved the results for polarity classification task (See task 1 in Tables 3, 4 and 5), which they show better performance using the SVM classifiers. For fine-grained opinion classification task (See task 2 in Tables 5, 6 and 7), the results using a domain taxonomy show to be competitive with Naive Bayes classifier in terms of accuracy, with Decision Tree classifier in terms of recall and with SVM classifiers in terms of F1-score.

Table 4. Recall

N.	Tasks	Classification	Decision Tree	Naive Bayes	SVM1	SVM2	SVM3
1		ReLi	0.49	0.48	0.60	0.46	0.49
2		TweetSentBr	0.58	0.52	0.56	0.53	0.58
3	Task 1	OPCovid-BR	0.51	0.53	**0.63**	0.60	0.49
4		ReLi + TweetSentBr	0.56	0.49	0.56	0.52	0.57
5		ReLi + TwitterSent + OPCovid-BR	0.54	0.53	0.54	0.56	0.62
5		Political& Social	0.68	0.68	0.45	0.61	0.61
6		Economy	0.17	0.33	0.00	0.00	0.00
7	Task 2	Education&Science	0.67	1.00	0.00	0.00	0.00
8		Localization	0.74	0.58	0.42	0.37	0.32
9		Health	0.84	0.46	0.91	0.90	0.87
10		**weighted avg**	**0.76**	0.53	0.70	0.72	0.69

Table 5. F1-score

N.	Tasks	Classification	Decision Tree	Naive Bayes	SVM1	SVM2	SVM3
1		ReLi	0.49	0.49	0.49	0.42	0.50
2		TweetSentBr	0.58	0.52	0.55	0.53	0.54
3	Task 1	OPCovid-BR	0.51	0.54	0.50	0.59	0.49
3		ReLi + TweetSentBr	0.55	0.50	0.55	0.52	0.54
4		ReLi + TwitterSent + OPCovid-BR	0.54	0.53	0.55	0.56	**0.60**
5		Political& Social	0.69	0.55	0.56	0.64	0.64
6		Economy	0.29	0.40	0.00	0.00	0.00
7	Task 2	Education&Science	0.57	0.15	0.00	0.00	0.00
8		Localization	0.62	0.58	0.43	0.44	0.39
9		Health	0.84	0.62	0.81	0.82	0.79
10		**weighted avg**	**0.76**	0.58	0.66	0.69	0.66

6   Final Remarks

Opinion mining or sentiment analysis is an important step to understanding what people on the internet think, especially when people around the world are thrown into disarray, such as during the 2019-nCoV pandemic. Extracting and classifying fine-grained opinion and polarity may provide insight into the public that both governmental and private agencies can use to make informed decisions while combating the virus. In this paper we have explored the fine-grained opinion identification and polarity classification of twitter data on the COVID-19 pandemic in Brazilian Portuguese. We trained machine learning models in order to classify the opinion and polarity classes. We also used two other data sets that had different domains (i.e. not the Coronavirus Pandemic) to measure the performance of cross domain polarity classifiers. Based on the obtained results, we found that the cross-domain strategy improved the results of polarity classification and we also found that we generally had the best performance using the SVM [5] algorithm. As for the fine-grained opinion classification, the usage of a domain taxonomy presented favorable results when applied to fine-grained opinion classification in Portuguese language.

Acknowledgments. The authors are grateful to CAPES and CNPq for supporting this work.

References

1. Avanço, L., Nunes, G.M.V.: Lexicon-based sentiment analysis for reviews of products in Brazilian Portuguese. In: Proceedings of the Brazilian Conference on Intelligent Systems, pp. 277–281. São Carlos, Brazil (2014)
2. Balage Filho, P.P., Pardo, T.A.S.: Aspect extraction using semantic labels. In: Proceedings of the 8th International Workshop on Semantic Evaluation, Dublin, Ireland, pp. 433–436 (2014)
3. Bayes, T.: An essay towards solving a problem in the doctrine of chances. Phil. Trans. Royal Soc. London **53**, 370–418 (1763)
4. Brum, H., Volpe Nunes, M.d.G.: Building a sentiment corpus of tweets in Brazilian Portuguese. In: Proceedings of the 11th International Conference on Language Resources and Evaluation). European Language Resources Association (ELRA), Miyazaki, Japan, May 2018
5. Cristianini, N., Ricci, E.: Support Vector Machines, pp. 928–932. Springer, Boston (2008)
6. Freitas, C., Motta, E., Milidiú, R., Cesar, J.: Vampiro que brilha... rÁ ! desafios na anotação de opinião em um córpus de resenhas de livros. In: Anais do XI Encontro de Linguística de Corpus, pp. 1–13. São Carlos, Brazil (2012)
7. Freitas, L.A., Vieira, R.: Ontology based feature level opinion mining for Portuguese reviews. In: Proceedings of the 22nd International Conference on World Wide Web, pp. 367–370. Association for Computing Machinery (2013)
8. Hu, M., Liu, B.: Mining opinion features in customer reviews. In: Proceedings of the 19th National Conference on Artificial Intelligence (2004)
9. Liu, B.: Sentiment Analysis and Opinion Mining. Morgan & Claypool Publishers, 1st edn. (2012)
10. Liu, B., Hu, M., Cheng, J.: Opinion observer: analyzing and comparing opinions on the web. In: Proceedings of the 14th International Conference on World Wide Web, pp. 342–351. Association for Computing Machinery, New York (2005)
11. Pang, B., Lee, L., Vaithyanathan, S.: Thumbs up? sentiment classification using machine learning techniques. In: Proceedings of the Conference on Empirical Methods in Natural Language Processing, Stroudsburg, USA, pp. 79–86 (2002)
12. Poria, S., Cambria, E., Gui, C., Gelbukh, A.: A rule-based approach to aspect extraction from product reviews (2014)
13. Rokach, L., Maimon, O.: Decision Trees, vol. 6, pp. 165–192, January 2005
14. Turney, P.D.: Thumbs up or thumbs down?: Semantic orientation applied to unsupervised classification of reviews. In: Proceedings of the 40th Annual Meeting on Association for Computational Linguistics, pp. 417–424. Association for Computational Linguistics, Stroudsburg (2002)

15. Vargas, F.A., Pardo, T.A.S.: Clustering and hierarchical organization of opinion aspects: a corpus study. In: Proceedings of the 14th Meeting of Linguistics of Corpus and 9th Brazilian School of Computational Linguistics pp. 342–351. Rio Grande do Sul, Brazil (2017)

16. Vargas, F.A., Pardo, T.A.S.: Aspect clustering methods for sentiment analysis. In: 13th International Conference on the Computational Processing of Portuguese, pp. 365–374. Canela, RS, Brazil (2018)

17. Vargas, F.A., Pardo, T.A.S.: Linguistic rules for fine-grained opinion extraction. In: Workshop Proceedings of the 14th International AAAI Conference on Web and Social Media, pp. 01–06. Association for the Advancement of Artificial Intelligence (2020)

18. Wu, C.W., Liu, C.L.: Ontology-based text summarization for business news articles. In: Computers and Their Applications, pp. 389–392. ISCA (2003)

19. Wu, Y., Zhang, Q., Huang, X., Wu, L.: Phrase dependency parsing for opinion mining. In: Proceedings of the 2009 Conference on Empirical Methods in Natural Language Processing, pp. 1533–1541. Association for Computational Linguistics, Singapore (2009)

20. Zhao, L., Li, C.: Ontology based opinion mining for movie reviews. In: Karagiannis, D., Jin, Z. (eds.) KSEM 2009. LNCS (LNAI), vol. 5914, pp. 204–214. Springer, Heidelberg (2009). https://doi.org/10.1007/978-3-642-10488-6_22

Impact of Text Specificity and Size on Word Embeddings Performance: An Empirical Evaluation in Brazilian Legal Domain

Thiago Raulino Dal Pont[1]([✉]), Isabela Cristina Sabo[2], Jomi Fred Hübner[1], and Aires José Rover[2]

[1] Department of Automation and Systems, Federal University of Santa Catarina, Florianópolis, SC 88040-900, Brazil
thiagordalpont@gmail.com, jomi@das.ufsc.br
[2] Department of Law, Federal University of Santa Catarina, Florianópolis, SC 88040-900, Brazil
isabelasabo@gmail.com, aires.rover@ufsc.br

Abstract. Word embeddings is a text representation technique capable of capturing syntactic and semantic linguistic patterns and of representing each word as an n-dimensional dense vector. In the domain of legal texts, there are trained word embeddings in languages like English, Polish, and Chinese. However, to the best of our knowledge, there are no embeddings based on Portuguese (Brazilian and European) legal texts. Given that, our research question is: does the specificity and size of the text corpus used for a word embedding training contribute to a more successful classification? To answer the question, we train word embeddings models in the legal domain with different levels of specificity and size. Then we evaluate their impact on text classification. To deal with the different levels of specificity, we collect text documents from different courts of the Brazilian Judiciary, in hierarchical order. We used these text corpora to train a word embeddings model (GloVe) and then had then evaluated while classifying processes with a deep learning model (CNN). In a context perspective, the results show that in word embeddings trained on smaller corpora sizes, text specificity has a higher impact than for large sizes. Also, in a corpus size perspective, the results demonstrate that the greater the corpus size in embeddings training, the better are the results. However, this impact decreases as the corpus size increases until a point where more words in the corpus have little impact on the results.

Keywords: Word embeddings · Legal corpora · GloVe · Text classification · Convolutional Neural Network

T. R. Dal Pont and I. C. Sabo—This research was supported by grants from CNPq (National Council for Scientific and Technological Development) and CAPES (Coordination for the Improvement of Higher Education Personne).

R. Cerri and R. C. Prati (Eds.): BRACIS 2020, LNAI 12319, pp. 521–535, 2020.
https://doi.org/10.1007/978-3-030-61377-8_36

1 Introduction

Text classification is an important part of Text Mining (TM) and Natural Language Processing (NLP) and it has been applied in many contexts [9,19,27]. Since texts are unstructured data, we can to transform them into a structured format so that it is possible to perform supervised learning using an available classifier, such as Support Vector Machines (SVM) or Convolutional Neural Networks (CNN) [18].

One of the many methods to represent text in a structured format is the Vector Space Model (VSM), where each document is represented through a numerical vector. This representation can be created using the Bag of Words (BOW) model where the vector values may be frequencies of each word in the text, generating sparse and high-dimensional representations [4]. New models of VSM representations have been proposed recently, such as Word2Vec [22], GloVe [24], and FastText [7]. They use a technique known as word embeddings. It represents each word as an n-dimensional dense vector, capable of capturing syntactic and semantic linguistic patterns [36]. To learn word embeddings, an unsupervised technique, such as GloVe, is applied to a large text corpus. Its size and the covered subjects have an impact on the quality of the embeddings and its performance in text classification. Commonly, word embeddings models use texts from several contexts. However, embeddings trained using only texts related to the classification task may get better results [20].

In the domain of legal texts, there are trained word embeddings in languages like English [10] and Polish [29]. However, to the best of our knowledge, there are no embeddings based on Portuguese legal texts. We have searched the main knowledge bases (Scopus, IEEE Xplore, ACM DL and Web of Science) for papers published in the last ten years. Nevertheless, Portuguese embeddings in multi-genre texts were already investigated [15,25].

Therefore, our research question is: does the specificity and size of the text corpus used for a word embedding training contribute to a more successful classification? To answer the question, we train word embedding models in the legal domain with different levels of specificity and size. Then, we evaluate their impact on classification. To deal with the different levels of specificity, we collect texts from different courts of the Brazilian Judiciary, in hierarchical order. These text corpora are used to train a word embeddings model (GloVe) and then are evaluated while classifying processes with a deep learning model (CNN).

The motivation for our research is the lack of academic work about word embeddings applied to legal texts in Portuguese (Brazilian and European). Moreover, few papers evaluate the influence of the level of specificity and size of the data set on classification [20]. Our contribution is thus a better understanding of the impact of specificity and the size of the text corpus in word embeddings. Although the results presented here are focused on the legal and Portuguese context, the results and methods can be reused in other applications.

This paper is organized as follows: In Sect. 2, we present some concepts on Brazilian Judiciary organization, word embeddings and text classification. In Sect. 3, we expose some works about word embeddings applied in different

Portuguese domain, and in other languages legal domain. In Sect. 4, we describe the methodology and strategies used in the experiments. In Sect. 5, we show and discuss the results. Finally, there are concluding remarks and new perspectives of work in Sect. 6.

2   Background

To contextualize our dataset and the models used in this work, in the following sections we present relevant concepts on the hierarchy of the Brazilian Judiciary, as well as the tasks of text representation and text classification.

2.1   Courts of Brazilian Judiciary

In order to may correct errors of the judges and also guarantee the non-conformity of the losing party about unfavourable judgments, modern legal systems enshrine the principle of double a degree of jurisdiction. That means that the losing party has the possibility of obtaining a new judgment. For this, all Brazilian Judiciary have higher and lower courts. Above all of them are the Federal Supreme Court (STF), the highest level of the Brazilian Judiciary [13].

According to Brazilian Federal Constitution [1], the Judiciary is composed of: a) Federal Supreme Court (STF); b) Superior Court of Justice (STJ); c) Federal Regional Courts (TRFs) and Federal judges; d) Superior Labor Court (TST), Labor Regional Courts (TRTs) and Labor judges; e) Superior Electoral Court (TSE), Electoral Regional Courts (TREs) and Electoral judges; f) Superior Military Court (STM) and Military judges; g) State Courts (TJs) and State judges. Also, Federal and State Courts can, within their jurisdiction, create the Special Courts (JECs and JEFs), which are responsible for judging local less complex cases. We illustrate this organization in Fig. 1.

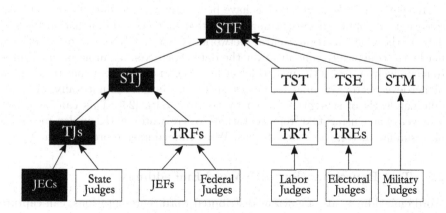

Fig. 1. Brazilian Judiciary organization chart

In our evaluation step, we classify JECs judgments about air transport failures, which belongs to Consumer Law. This legal subject is under the jurisdiction of the State Courts. Therefore, as highlighted in Fig. 1, JECs are submitted to TJs, which in turn are submitted to STJ and STF. So, we used judgments from TJs, STJ and STF to training the word embeddings model.

2.2   Text Classification with CNN

In the text classification task, we use data to construct a model that learns to relate its features to one of the class labels preset. For a given test instance for which the class is unknown, the training model is used to predict a class label for this instance [4].

To evaluate if the prediction is correct (true positives and true negatives) or wrong (false positives and false negatives), we use some metrics, such as accuracy, precision, recall, and F1 score [3]:

- **Accuracy** is the fraction of test instances in which the predicted value matches the ground-truth value.
- **Precision** is the percentage of instances predicted to belong to the positive class that was correct.
- **Recall** is the percentage of ground-truth positives that have been recommended as positives.
- **F1** score is the harmonic mean between the precision and the recall.

When dealing with multiclass datasets we can measure the F1 score using the macro F1 score. It first calculates the metric for each class and then takes the average of them. In this way, the metric considers the performance equally in each class, surpassing any class imbalances. In contrast, there are other average methods such as micro and weighted F1 score however, these variations do not take class imbalance into account [35].

Recently, deep learning models have been proven to be effective in text classification. A popular deep learning model that attracts more attention for text data is the Convolutional Neural Networks (CNNs). CNNs use convolutional masks to sequentially convolve over the data. For texts, a simple mechanism is to recursively convolve the nearby lower-level vectors in the sequence to compose higher-level vectors. Similar to images, such convolution can naturally represent different levels of semantics shown by the text data [23]. This can be better achieved when using text representations where words with similar meanings have similar vectors, as it occurs with Word Embeddings representation [6].

2.3   Text Representation with Word Embeddings

Word embeddings, also known as distributed word representations, can capture both the semantic and syntactic information of words while representing them as n-dimensional dense vectors [20].

These representations are generated from large unlabeled corpora through a training process that varies among existing algorithms. In Word2Vec Skipgram for instance, the representations are generated and modified as it tries to predict surrounding words in a phrase, based on the current word [22]. On the other hand, GloVe creates a co-occurrence matrix containing the frequencies of words in different contexts. Then it applies a dimensionality reduction technique to produce final representations [24].

Word embeddings have been used in many NLP tasks beyond text classification. These tasks include clustering [3], text summarization [5], and many others.

3   Related Work

After a systematic review, we select six works related to ours: a) two about word embeddings applied in multi themes in Portuguese (Brazilian and European), b) two about word embeddings applied in legal theme in several languages, and c) two about text classification in STF.

Hartmann et al. [15] evaluated different word embedding models trained on a sizeable Portuguese corpus (1,395,926,282 tokens in total). They trained 31 word embedding models using FastText, GloVe, Wang2Vec and Word2Vec, and evaluated them intrinsically on syntactic and semantic analogies and extrinsically on POS tagging and sentence semantic similarity tasks. The results obtained from intrinsic and extrinsic evaluations were not aligned with each other, contrary to what they expected. GloVe produced the best results for syntactic and semantic analogies, and the worst, together with FastText, for both POS tagging and sentence similarity.

Rodrigues et al. [25] evaluated different word representation models on semantic similarity tasks, trained on a Portuguese dataset provided by a workshop (10,000 sentences). They used word embeddings (Word2Vec and FastText) and deep neural language models (ELMo and BERT). The results indicated that ELMo language model was able to achieve better accuracy than any other pretrained model which has been made publicly available for the Portuguese language. They also demonstrate that FastText skip-gram embeddings can have a significantly better performance on semantic similarity tasks.

Chalkidis and Kampas [10] trained a word embeddings model on a large legal corpus from various public legal sources in English (UK legislation, European legislation, Canadian legislation, Australian legislation, English-translated legislation from EU countries, English-translated legislation from Japanese, US Supreme Court decisions and US Code), which sums up to a total of approximately 492,000,000 tokens. They trained based on the Word2Vec and Skip-gram model, instead of the most recent FastText. They justified that Word2Vec is reported to provide better semantic representation than FastText, which tends to be highly biased towards syntactic information.

Smywiński-Pohl et al. [29] trained word embeddings models (Word2Vec and GloVe) to find out which of them is best suited for establishing the correspondence between Polish legal and extra-legal terminology. The corpora are composed of text data collected from two databases: a) National Corpus of Polish, which includes texts of different genres, such as novels, transcripts of parliamentary speeches and newspaper articles, which sums up to a total of 2,591,817,208 tokens; b) judgments from Polish Supreme Court, Polish Constitutional Tribunal, Polish common courts, Polish National Chamber of Appeal and Polish administrative courts, which sums up to a total of 4,076,628,858 tokens. The results showed the superiority of Word2Vec CBOW negative sampling variant in their problem.

Finally, Correia da Silva et al. [28] and Braz et al. [8], representing a national project developed at the STF, classified different types of judgments using deep learning models (Bi-LSTM and CNN) and obtained satisfactory results. However, the published papers suggest that they used a model of word embeddings already trained for the task of text representation.

In our survey, we did not find publications concerned with the training of word embeddings in Portuguese legal texts. Therefore, with this paper, we plan to contribute in this direction.

4   Experiments

To answer our research question, we build different embeddings (varying the specificity and size of the text corpus used to train them) and evaluate their performance in a classification problem. The classification problem concerns specific texts in the area of air transport services. We expect that more specific embeddings require smaller corpus size than general embeddings.

In this section, we explain the pipeline of this work, starting from corpora construction for word embeddings training and also the dataset used for text classification. Then, we describe the embedding training steps and the classification model used to evaluate our embeddings.

4.1   Corpus Construction

The following sections describe the steps[1] we followed for the construction of the text corpora used for (i) training the different word embeddings we want to evaluate as well as (ii) the corpus considered for the text classification.

Concerning embeddings training, the first step is to obtain the collection of legal documents from the court web portals, followed by raw text extraction from these documents. To enable us to evaluate the specificity influence of these legal corpora, we divided it into two contexts: related to general legal texts and related to air transport services text. We also collected texts from other general topics (not related to legal domains) that are already compiled and freely available. Having the corpora for legal and miscellaneous contexts, we applied

---

[1] Code and Word Embeddings available at https://github.com/thiagordp/embeddings_in_law_paper.

some processing steps to remove noise from texts. To evaluate the influence of corpus size in embeddings training, we divided these three corpora into smaller pieces based on word count.

Concerning the classification task, the construction of the corpora is based on JEC processes related to air transport services. We are thus interested in the quality of the classification in this specific domain and, of course, the impact of the specificity of the embeddings in this specific distribution problem.

Legal Context Corpus for Embeddings Training. To train the embeddings it is required large text corpora to be able to get good embeddings. However, in the Brazilian Portuguese language, we could not find any dataset available on the Internet containing enough legal text corpora for our purposes. Thus, we had to build our legal corpora.

Our main sources of legal text are Brazilian courts platforms. We collected judgments from the webpages of STF [30], STJ [31], and TJ-SC (State Court of Santa Catarina) [34]. We also collected judgments from the JusBrasil portal containing processes related only to failures on air transport service from all TJs (State Courts) of Brazil [16]. Table 1 shows the number of processes acquired and word count for each Tribunal:

Table 1. Acquired process from courts for embeddings training

Source	Collegial judgments	Individual judgments	Subtotal	Word count
STF	64,779	118,910	183,689	294,937,185
STJ	101,141	0	101,141	312,687,450
TJ-SC	989,964	662,535	1,652,499	3,060,212,814
TJs (JusBrasil)	34,239	0	34.239	78,138,337
		Total	1,971,568	3,745,975,786

After downloading all processes, most of them in PDF and Rich Text Format (RTF) formats, we extracted raw texts from these files. We did not apply Optical Character Recognition (OCR) in scanned PDF documents, due to time limits to finish the experiments, so only digital PDFs were accounted with RTF files in Table 1.

With the extracted texts, we applied some pre-processing steps, as discussed further in this section. Then we built the legal text corpora containing all the processes related to all law subjects, which we call *general* legal text corpora in this work. Using this base, we created another text corpora whose processes are related only to air transport and consumer law, and we call it *air transport* legal text corpora.

Global Context Corpus Acquisition. To be able to compare how good embeddings trained with legal texts perform against those created with all kinds of texts, we also created other corpora from a variety of sources. Thus, we searched for free available textual datasets. In this work, we call these texts as *global* context texts. Table 2 shows all the global text datasets used. Then we apply some preprocessing steps, as will be described further in this section.

Table 2. Global context corpora

Dataset	Documents	Word count	Source
Wikipedia Dump in Portuguese	1,014,713	303,622,360	[2]
Brazilian Literature Books	169	37,848,783	[33]
Old Newspapers	617,627	26,441,581	[32]
Folha de São Paulo News	165,641	74,594,367	[21]
HC News Corpus	494,128	27,170,063	[12]
Blogspot Posts	2,181,073	696,657,915	[26]
Wikihow Instructions	786,283	22,471,312	[11]
Total	5,259,634	1,188,806,381	

Legal Context Corpus for Text Classification. To evaluate each of the trained embeddings, we used a set of judgments from the JEC located at the Federal University of Santa Catarina (JEC/UFSC), which is related only to failures on air transport services (Consumer Law). In these processes, the consumer claims for compensation for material or moral damages against an airline company due to failures in its services. We extracted nearly one thousand judgments, divided into four class labels, corresponding to 26%, 10%, 62%, and 2% of this dataset, respectively:

1. Well-founded: The consumer wins the lawsuit.
2. Not founded: The consumer loses the lawsuit.
3. Partly founded: The consumer wins part of the lawsuit (for example, when he/she plead a greater compensation than the assigned value by the judge).
4. Dismissed without prejudice: The consumer makes a procedural error (for example, when he/she indicate as a defendant the wrong airline company). So the consumer can file a new lawsuit.

Before the text classification task, we applied some preprocessing steps as discussed further in this section and then created three subsets of processes, the training, validation and test sets, corresponding to 70%, 15%, and 15% of the dataset. In these sets, all the classes are distributed proportionally. We used the training and validation sets during the training of the classification model. Then, we evaluated this model with the test set.

Corpus Processing. After text extraction from the documents, we applied some pre-processing steps, which are required before training the embeddings or text classification. The first of them was the conversion to lower case. Then punctuation marks were removed, as well as special characters and some symbol characters. We removed stopwords neither apply stemmization or lemmatization, following the literature [22, 24].

In relation to our three corpora used in embeddings training, which comprising 3.7 billion, 100 million and 1.19 bilion words for *general, air transport* and *global* corpora, respectively, we created others based on them according to the following smaller corpora sizes, considering the word count: 1,000; 10,000; 50,000; 100,000; 200,000; 500,000; 1,000,000; 5,000,000; 10,000,000; 25,000,000; 100,000,000; 500,000,000; 750,000,000 and 1,000,000,000.

We choose these corpora sizes to be able to compare the variation on evaluation metrics while increasing corpora size. For the air transport context, we could not embrace all these sizes due to limited corpora available. The largest sub-base had 100 million words for this context.

Finally, each of these smaller corpora was used to train one different word embeddings representation.

4.2 Embeddings Training

In this work, we chose GloVe representation due to its good results in many NLP tasks, including text classification, and also for its training time which is significantly less than other techniques like Word2Vec and FastText [24]. In terms of GloVe parameters, we kept most of the default values, except for windows size, training iterations, and vector size, which were set to 5, 100, and 100, respectively. With these values, we achieved better results in text classification.

Considering the corpus sizes described in Sect. 4.1 and the parameters above described, we trained 15 representations for *general* and *global* contexts bases. For *air transport* context base, we trained 11 embeddings.

4.3 Embeddings Evaluation in Legal Text Classification

To evaluate the GloVe embeddings representations, we applied each of them to the task of text classification on judgments from JEC/UFSC. Also, we used Convolutional Neural Networks as a classification model based on the literature [17]. Figure 2 illustrates this model.

This CNN takes into account the order of the words by stacking the corresponding embeddings for each word as they occur in the text. Them it applies multiple convolutional masks with different dimensions that correspond to the red and yellow contours in Fig. 2. Mask widths are equal to word embedding size while the heights can vary. In this context, mask height can be related to the idea of N-Grams, since they embrace multiple embeddings at the same time. In the original model, these heights were set to 3, 4, and 5. We added one more mask of height 2, which increased classification metrics. Also, we set to 10 the number

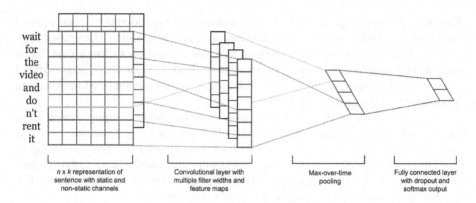

wait
for
the
video
and
do
n't
rent
it

n x k representation of   Convolutional layer with   Max-over-time   Fully connected layer
sentence with static and   multiple filter widths and   pooling   with dropout and
non-static channels   feature maps   softmax output

Fig. 2. CNN model for text classification [17]

of masks for each of these sizes, without affecting our results, but decreasing the required training time.

In this work, we applied each of the embeddings trained in conjunction with the CNN described in the classification of JEC/UFSC judgments, where Out of Vocabulary (OOV) words are replaced by an vector of random values. Thus, we trained and tested 41 models. Furthermore, due to the stochastic nature of neural networks training methods [14], each of these models was trained and tested 200 times and the resulting evaluation metrics were averaged.

Finally, we compare the performance in classification using Accuracy and Macro F1-Score.

5   Results Analysis and Discussion

In the following sections, we present and discuss our results for text classification using trained embeddings for *global*, *general*, and *air transport* contexts with multiple corpus training sizes.

5.1   Experimental Results

Following the steps presented in Sect. 4, we trained all 41 word embeddings representations for GloVe.

To illustrate how these embeddings behave, in Fig. 3, we used Principal Component Analysis (PCA) to create a projection in two dimensions of a set of words from *general* context embedding trained with 1 billion words.

Using each embedding, we trained and tested CNNs for text classification in JEC/UFSC judgments. These two steps were repeated 200 times, and the evaluation metrics were averaged for each group of repetitions.

In Fig. 4 and 5, we present the results, for accuracy and F1-Score, respectively, from test data applied to each CNN model. These results are related to embeddings trained with *general*, *air transport*, *global* texts. The x-axis denotes the

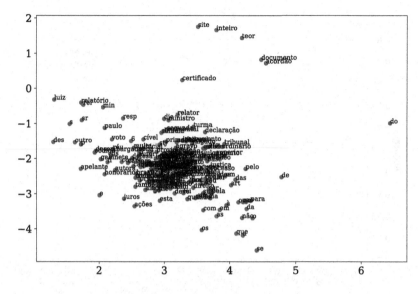

Fig. 3. Word embeddings projection

corpus sizes used to train the embeddings, while the y-axis represents accuracy or F1-Score. Each data point represents the average of the evaluation metric, after 200 train and test repetitions using each specific embedding.

5.2 Discussion from Context Perspective

In this section, we will consider the first part of our research question: Does the specificity of the corpora in embeddings training contribute to the quality of the classification?

In terms of accuracy, when we compare *global* against others (Fig. 4), we have that higher text specificity leads to better results, for most of the corpus sizes used for embeddings training. Furthermore, when comparing *general* and *air transport* curves, there is a significant difference in accuracy only for the lowest and highest x-values. However, in terms of F1-Score, as shown in Fig. 5, our observations change, once *general* and *air transport* curves have a similar shape. Also, for the highest corpus sizes, *general* and *global* curves converge to similar values of F1-Score. We believe that these differences in accuracy and F1-Score emerge from the fact that our dataset to text classification is imbalanced, once the former does not take this fact into account, while the latter does. However, this result still requires further investigation.

In general, we can note that for smaller corpora size for embeddings training, text specificity has a more impact than for large sizes.

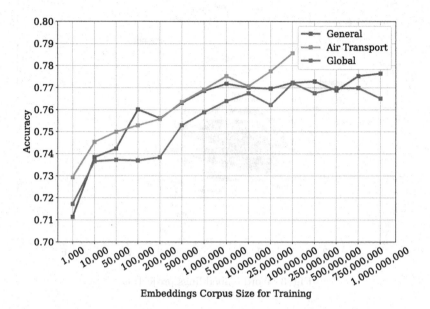

Fig. 4. Accuracy for test set from CNN model

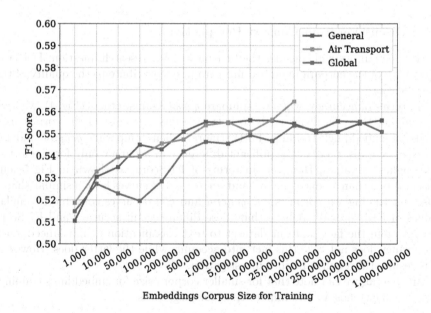

Fig. 5. Macro F1-Score for test set from CNN model

5.3 Discussion from Corpus Size Perspective

In this section, we will consider the second part of our research question: How does the corpus size contribute to the embedding quality?

When we observe both accuracy and F1-Score measures from Fig. 4 and 5, it is clear the tendency for improvement while increasing corpus size. However, the metrics converge with the largest corpus sizes. There are two exceptions. The first one occurs with smaller values of corpus sizes for *global* curve, as it decreases in F1-Score measures. The second corresponds to the last data point in *air transport* curves. The former can happen we the classifier performs poorly for some classes while gets better in others. The latter may indicate that those curves could improve if we had more significant corpus sizes related to that context.

In general, we can note that the greater the corpus size in embeddings training, the better are the results. However, this impact decreases as the corpus size increases until a point where more words in the corpus have little impact on the results.

6 Conclusion and Future Work

The research allowed us to learn more about the behaviour of word embedding models in different variations of the text in the legal domain, such as the specificity (context) and the corpus size.

In the context of legal documents in Portuguese, we concluded that there is more assertiveness when the trained text resembles the text that we have to classify. This behaviour does not occur with the size of the training corpus, because when you reach a certain amount of words, the results suggest stability.

Despite the moderate gain in accuracy with the specific texts as test set (air transport curve), we consider this result relevant because it shows that the use of billions of tokens, as in previous works, does not bring great contributions. Therefore, the specificity of the text set impacts more positively on the classification task than the size of the text set.

Finally, the results presented in this work cover only CNNs. Thus, we intend to check how our conclusions fit when using other classification and representation techniques, although we believe the results would not change significantly. In our future work, we intend to use as classification models SVM, LSTM, Attention Mechanisms etc. Also, we plan to create and experiment new legal datasets to the classification task. We also aim to use other word embeddings models, such as Word2Vec and FastText, as well as other text representation approaches, like BERT, ELMo etc. Finally, we would experiment word embeddings in other tasks, such word analogies and word similarity.

References

1. Brazilian Federal Constitution (1988). http://www.planalto.gov.br/ccivil_03/constituicao/constituicao.htm

2. Ptwiki dump progress on 20191120 (2019). http://wikipedia.c3sl.ufpr.br/ptwiki/20191120/

3. Aggarwal, C.C.: Machine Learning for Text, 1st edn. Springer, Cham (2018). https://doi.org/10.1007/978-3-319-73531-3

4. Aggarwal, C.C., Zhai, C. (eds.): Mining Text Data, 27th edn. Springer, Boston (2012). https://doi.org/10.1007/978-1-4614-3223-4

5. Alami, N., Meknassi, M., En-nahnahi, N.: Enhancing unsupervised neural networks based text summarization with word embedding and ensemble learning. Expert Syst. Appl. **123**, 195–211 (2019)

6. Aubaid, A.M., Mishra, A.: Text classification using word embedding in rule-based methodologies: a systematic mapping. TEM J. **7**(4), 902–914 (2018)

7. Bojanowski, P., Grave, E., Joulin, A., Mikolov, T.: Enriching word vectors with subword information. Trans. Assoc. Comput. Linguist. **5**, 135–146 (2016)

8. Braz, F.A., et al.: Document classification using a Bi-LSTM to unclog Brazil's supreme court. In: NeurIPS Workshop on Machine Learning for the Developing World (ML4D), 8 December 2018

9. Cardoso, E.F., Silva, R.M., Almeida, T.A.: Towards automatic filtering of fake reviews. Neurocomputing **309**, 106–116 (2018)

10. Chalkidis, I., Kampas, D.: Deep learning in law: early adaptation and legal word embeddings trained on large corpora. Artif. Intell. Law **27**(2), 171–198 (2019). https://doi.org/10.1007/s10506-018-9238-9

11. Chocron, P., Pareti, P.: Vocabulary alignment for collaborative agents: a study with real-world multilingual how-to instructions. In: Proceedings of the Twenty-Sixth International Joint Conference on Artificial Intelligence, IJCAI-18, International Joint Conferences on Artificial Intelligence Organization, pp. 159–165, July 2018

12. Christensen, H.: HC Corpora (2016). https://web.archive.org/web/20161021044006/http://corpora.heliohost.org/

13. Cintra, A.C.d.A., Grinover, A.P., Dinamarco, C.R.: Teoria geral do processo. Malheiros (2011)

14. Cohen, P.R.: Empirical Methods for Artificial Intelligence. MIT Press, Cambridge (1995)

15. Hartmann, N., Fonseca, E., Shulby, C., Treviso, M., Rodrigues, J., Aluisio, S.: Portuguese word embeddings: evaluating on word analogies and natural language tasks (Section 3), August 2017

16. JusBrasil: JusBrasil. Conectando pessoas à justiça (2020). https://www.jusbrasil.com.br/home

17. Kim, Y.: Convolutional neural networks for sentence classification. In: Proceedings of the 2014 Conference on Empirical Methods in Natural Language Processing (EMNLP), vol. 2017-January, pp. 1746–1751. Association for Computational Linguistics, Stroudsburg, September 2014

18. Kowsari, K., Meimandi, J., Heidarysafa, M., Mendu, S., Barnes, L., Brown, D.: Text classification algorithms: a survey. Information **10**(4), 150 (2019)

19. Kumar, G.R., Mangathayaru, N., Narasimha, G.: Intrusion detection using text processing techniques. In: Proceedings of the The International Conference on Engineering & MIS 2015 - ICEMIS 2015. ACM Press (2015)

20. Lai, S., Liu, K., He, S., Zhao, J.: How to generate a good word embedding. IEEE Intell. Syst. **31**(6), 5–14 (2016)

21. Marlessonn: News of the Brazilian newspaper (2019). https://www.kaggle.com/marlesson/news-of-the-site-folhauol

22. Mikolov, T., Chen, K., Corrado, G., Dean, J.: Efficient estimation of word representations in vector space. In: 1st International Conference on Learning Representations, ICLR 2013 - Workshop Track Proceedings, pp. 1–12, January 2013
23. Peng, H., et al.: Large-scale hierarchical text classification with recursively regularized deep graph-CNN. In: Proceedings of the 2018 World Wide Web Conference, WWW 2018, pp. 1063–1072. International World Wide Web Conferences Steering Committee, Republic and Canton of Geneva (2018)
24. Pennington, J., Socher, R., Manning, C.: Glove: global vectors for word representation. In: Proceedings of the 2014 Conference on Empirical Methods in Natural Language Processing (EMNLP), vol. 19, pp. 1532–1543. Association for Computational Linguistics, Stroudsburg (2014)
25. Rodrigues, R.C., Rodrigues, J., de Castro, P.V.Q., da Silva, N.F.F., Soares, A.: Portuguese language models and word embeddings: evaluating on semantic similarity tasks. In: Quaresma, P., Vieira, R., Aluísio, S., Moniz, H., Batista, F., Gonçalves, T. (eds.) PROPOR 2020. LNCS (LNAI), vol. 12037, pp. 239–248. Springer, Cham (2020). https://doi.org/10.1007/978-3-030-41505-1_23
26. Santos, H., Woloszyn, V., Vieira, R.: BlogSet-BR: a Brazilian Portuguese blog corpus. In: Proceedings of the Eleventh International Conference on Language Resources and Evaluation (LREC 2018). European Language Resources Association (ELRA), Miyazaki (2018)
27. Sheikhalishahi, S., Miotto, R., Dudley, J.T., Lavelli, A., Rinaldi, F., Osmani, V.: Natural language processing of clinical notes on chronic diseases: systematic review. JMIR Med. Inform. **7**(2), e12239 (2019)
28. da Silva, N.C., et al.: Document type classification for Brazil's supreme court using a convolutional neural network. In: 10th International Conference on Forensic Computer Science and Cyber Law (ICoFCS), Sao Paulo, Brazil, October 2018
29. Smywiński-Pohl, A., Lasocki, K., Wróbel, K., Strzałta, M.: Automatic construction of a polish legal dictionary with mappings to extra-legal terms established via word embeddings. In: Proceedings of the Seventeenth International Conference on Artificial Intelligence and Law - ICAIL 2019. ACM Press (2019)
30. STF: Supremo Tribunal Federal (2020). http://portal.stf.jus.br/
31. STJ: STJ - Jurisprudência do STJ (2020). https://scon.stj.jus.br/SCON/
32. Tan, L.: Old newspapers (2020). https://www.kaggle.com/alvations/old-newspapers
33. Tatman, R.: Brazilian literature books (2017). https://www.kaggle.com/rtatman/brazilian-portuguese-literature-corpus
34. TJSC: Jurisprudência Catarinense - TJSC (2020). http://busca.tjsc.jus.br/jurisprudencia/
35. Uysal, A.K.: An improved global feature selection scheme for text classification. Expert Syst. Appl. **43**, 82–92 (2016)
36. Wang, S., Zhou, W., Jiang, C.: A survey of word embeddings based on deep learning. Computing **102**(3), 717–740 (2019)

Machine Learning for Suicidal Ideation Identification on Twitter for the Portuguese Language

Vinícios Faustino de Carvalho, Bianca Giacon, Carlos Nascimento, and Bruno Magalhães Nogueira(✉)

Federal University of Mato Grosso do Sul, Campo Grande MS, Brazil
bruno@facom.ufms.br

Abstract. Suicidal ideation is one of the main predictors of the risk of suicide attempt and can be described as thoughts, ideas, planning, and desire to commit suicide. Fast detection of such ideation in early stages is essential for effective treatment. Many expressions of suicidal ideation can be found in publications in social networks, especially by young people. Previous works explore the automatic detection of suicidal ideation in social networks for the English language using machine learning algorithms. In this work, we present the first exploration of machine learning algorithms for suicidal ideation detection for the Portuguese language. We compared three classifiers in Twitter data: SVM, LSTM, and BERT (multilingual and Portuguese). Results suggest that BERT is effective for suicidal ideation identification in Portuguese data, achieving 79% of F1 score and less than 9% false negative score.

Keywords: Machine learning · Suicidal ideation · Natural language processing · Social networks

1 Introduction

Suicide is a serious public health problem and a significant cause of death around the world. According to data from the World Health Organization (WHO) [29], about 800,000 people commit suicide each year. In Brazil, 10,000 people die as a result of suicide per year, resulting in a rate of 5.5 suicides per 100,000 people in 2015 [29].

According to [42], suicidal ideation can be understood as a set of thoughts, ideas, planning, and desires about killing themselves. Also, suicidal ideation is one of the main predictors to suicide attempts.

Suicide prevention is heavily dependent on surveillance and monitoring of suicide attempts or self-harm trends and also recurring subjects of these individuals [36]. As stated by [20], early detection and treatment are the most effective ways to prevent suicide attempts.

Younger people tend to use the Internet as a place to debate, to seek help, and to discuss information related to depression and suicide. This interaction

R. Cerri and R. C. Prati (Eds.): BRACIS 2020, LNAI 12319, pp. 536–550, 2020.
https://doi.org/10.1007/978-3-030-61377-8_37

occurs in social network platforms, such as Twitter, Facebook, and blogs. On the other hand, suicidal ideation frequently appears in publications made by users in social networks [32].

Automated solutions are essential for fast-tracking suicidal-related content in the massive amount of publications produced continuously in social networks. These solutions typically use machine learning [25] and natural language processing [22]. The identification of suicidal ideation is treated as a text classification task where the class value indicates whether a publication is suicidal-related.

Most of the existing work on automated suicidal ideation detection focuses on applying of classical machine learning algorithms on Twitter data, like Support Vector Machines (SVM), Decision Trees, Naïve Bayes, Neural Networks, and Logistic Regression [9,27]. Some recent works on suicidal ideation detection also use deep learning approaches, which achieved state-of-the-art performance in many text classification tasks [16]. For example, Long Short-Term Memory networks (LSTM) were used to suicide ideation detection in the work of Birjali et al. [5] and Sawhney et al. [34]. Deep learning algorithms-Convolutional Neural Networks (CNN) and LSTM-were also used to detect depression and Post-traumatic stress disorder (PTSD) on Twitter data [28], as depression is one of the main risk factors of suicide [42].

All the actual work listed before focus on texts written in English. There is no work that explores the automatic detection of suicidal ideation in Portuguese as far as we are concerned. Moreover, recent advances in machine learning brought models that stood out as the state of the art in various tasks. In particular, a more recent algorithm that achieved high performance in natural language processing tasks is BERT [13], a language model which stands for Bidirectional Encoder Representations from Transformers. One of BERT's essential features is that the model can be fine-tuned with just one additional output layer, which allows us to get excellent performance in various tasks, including text classification. No work in the literature fine-tunes BERT for suicidal ideation detection.

Thus, our objective is to identify suicidal ideation on Twitter publications written in Brazilian Portuguese in the present work. We hypothesize that machine learning algorithms, especially deep learning-based algorithms, are useful in this task. We monitored Twitter using Portuguese keywords generated by the translation of English keywords extracted from other work [14,27]. Domain experts labeled these data. The resulting dataset was presented to three different classifiers: BERT, LSTM, and SVM. We tested two different versions of BERT: one pretrained with a multilingual corpus and one fine-tuned with a Brazilian Portuguese corpus. Results indicate that the version of BERT fine-tuned with Brazilian Portuguese corpus outperformed the compared algorithms.

In summary, the main contributions of this paper are:

- A first exploration on the automatic machine learning-based detection of suicidal ideation on the Portuguese language;
- The usage of recent deep learning-based approaches for suicidal ideation detection. As far as we are concerned, our work is the first that uses BERT, a

state-of-the-art classifier in many text classification tasks, for suicidal ideation detection;
- A vocabulary in Portuguese for suicidal ideation monitoring in social networks, validated by domain specialists, as well as automated tools for data collection and analysis.

The remainder of this paper is organized as follows: In Sect. 2, we discuss a background around the main themes of this work; comments about previous related work is presented in Sect. 3; In Sect. 4, we describe the experimental methodology of this work; The results and discussion are presented in Sect. 5; and, in Sect. 6, our conclusions and future works are discussed.

2   Background

In this section, we present the main concepts discussed in this work to understand the problem better. Also, we present and discuss the techniques employed in the study.

2.1   Suicidal Ideation

Suicide is considered a worldwide public health problem and the second leading cause of death in the world among young people aged 15 to 29 years. In Brazil, only in the period between 2011 and 2015, the youth suicide rate increased by 12%, being the fourth-largest cause in this group [8]. Furthermore, it is estimated that for each suicide that occurs, there are at least 20 attempts [40].

There is not only one definition acceptable of suicide, but it implies a conscious desire to die and a clear notion of the consequences of act [3]. Suicide behavior is frequently divided into three categories: suicide attempt, consummated suicide, and suicidal ideation [4]. Suicidal ideation is one of the main predictors of the risk of suicide attempt and can be described as thoughts, ideas, planning, and desire to commit suicide.

According to [30], younger people are more vulnerable to pro-suicide online communication due to easy access to digital media and teenagers' inherent characteristics. Internet, along with social media, can make access to pro-suicide content easier. Also, younger people are exposed to cyberbullying, which requires extra attention to this matter.

Online text monitoring can anticipate or even detect suicide risk conversations, which can directly help people in need of help, saving lives [33]. Thus, fast detection and detection in early stages are essential for an effective treatment.

Automated solutions meet this need since they can rapidly and effectively analyze large amounts of data in a reasonable time. In Brazil, however, suicidal ideation is studied mainly using questionnaires applied to a target group, without automated tools, like reported in [7,38] and [3]. Thus, there is a latent need for computational solutions to help professionals in this task. Our work explores text classification for this end, which is explained in the next section.

2.2 Text Classification

Text classification is one of the tasks in natural language processing. The objective of this task is to classify documents into a fixed number of predefined categories [21]. The main application of text classification is the automatic document categorization [35]. Text classification is also often used in sentiment analysis, exploring many levels of granularity and different ranges of sentiment. The most common scenario is to classify texts as positive, negative, or neutral, according to the sentiment expressed in the document [1].

To perform text classification, the first step is to convert the text into a representation that can be used to machine classification. An approach widely used is the bag of words (BoW). Bag of words is an approach for feature extraction in texts, where the words in the document are used to form a histogram, and each word counts as a feature. Despite its simplicity and good results in a large variety of applications, BoW is not adequate for small text data [15].

Besides this BoW approach, recent advances in NLP brought the popularization of word embedding approaches [39]. Word embeddings are real vectors distributed in a multidimensional space induced through unsupervised learning. Each vector dimension of a specific word represents a characteristic of it. This is done to capture semantic, syntactic, or morphological properties of the word in question [11].

One of the most popular word embedding model is Word2Vec [24]. This model improves the quality of learned words and phrases representations. Another advantage of methods like this is that the vectors can be somewhat meaningfully combined using vector addition. With this approach, it is possible to pre-train word embeddings with unlabeled data. Then we can use this representation in a classifier without the overhead of training the representation every time [31].

After obtaining a proper word representation, classification algorithms can be used to learn and predict class labels. Classical machine learning algorithms like Support Vector Machines and Multilayer Perceptron (MLP) are commonly used and achieve good results in various applications [35]. Deep learning algorithms, however, brought a new perspective in the state-of-the-art results in text classification. In this paper, we use LSTM, a Recurrent Neural Network, and BERT, a Transformer network. These architectures are briefly explained in the next sections.

2.3 LSTM

Recurrent neural networks (RNN) are neural networks designed for sequential data, exploring the sequential dependency among attributes. Attributes dependency is especially important in text processing since semantic properties can be discovered by considering the context in which a word occurs [2]. The architecture of an RNN is substantially different from feed-forward architectures. Instead of operating only with the input, RNNs also use an intern state that traces what has already been processed [6]. This characteristic leads RNN's to perform better than feed-forward networks in temporal sequences and data that the context information is important in the result.

One type of RNN is the Long Short-Term Memory (LSTM) [17]. LSTMs are designed to mitigate one of the most common problems in training RNNs: the vanishing and exploding gradients. LSTMs have fine-grained control over the data written to its long-term memory [2]. This control enables the neural network to use more context information to compute the output [17]. In an LSTM unit, there are gates used to control the update of long and short term information, making the model learn which information is more useful for that unit [19].

2.4   BERT

Bidirectional Encoder Representations from Transformers (BERT) is a language model algorithm using bidirectional transformers [13]. To better understand how BERT works, it is necessary to understand the encoders and transformers.

An encoder is a recurrent neural network with one recurrent unit (LSTM or Gated Recurrent Unit (GRU)) unit for each input element. It allows us to encode a variable input size into a vector with a fixed dimension size [10]. Analog to encoders, decoders convert a vector with a fixed dimension into a variable number of outputs. For each output, there is an RNN unit, like the encoders [10].

The transformer is another crucial concept to understand BERT. A transformer neural network architecture is composed of a set of encoders and a set of decoders. This is known as a sequence-to-sequence architecture, in which an input sequence is mapped into an output sequence [2]. Both encoders and decoders use attention mechanisms. Transformer architectures achieved state of the art in many translation tasks, such as English-to-German and English-to-French translation [41].

BERT is a transformer-based algorithm designed to pre-train deep bidirectional representations from the unlabeled text. The BERT model can be fine-tuned with just one additional output layer. This model achieved the state-of-the-art performance in a wide range of tasks in natural language processing by the time of its publication [13].

BERT's model architecture is a multi-layer bidirectional Transformer encoder, with the implementation based on [41]. The framework has two steps: (i) pre-training; and (ii) fine-tuning. The pre-training step uses unlabeled text and masks it to make the process bidirectional. The input embeddings of BERT are the sum of the token embeddings, segmentation embeddings, and position embeddings. Fine-tuning BERT to a task consists of adding the specific inputs and outputs of the task and fine-tune all the parameters end-to-end [13].

3   Related Work

Some related works use machine learning algorithms to detect suicidal ideation in social networks, like [5,9,27,34], and [20].

In [27], different classification algorithms were compared, such as SVMs and Logistic Regression (LGR), to classify Twitter texts. Their approach was to collect data on Twitter based on a vocabulary of suicide ideation. The data was labeled in three different levels of concern. Then the data was split into two sets with different distributions of the classes. The SVM model performed better over LGR and achieved 76% of precision with both datasets combined. A limitation discussed by the authors is that the reliability of the classifier due to the search terms may not be the ideal and that the data used have no other context information, like age, gender, and location.

Using a more significant number of classes to annotate the data, [9] uses seven classes, listing different contexts that suicide-related texts can appear. They used three different sets, each one with its own feature sets. Their best results were achieved with random forest, obtaining a precision value in all classes of 73.2%.

In [5], authors also employ machine classification for suicide prediction in social networks. The procedures follow the idea of defining a vocabulary of terms to search on social networks, and then to annotate the data to proceed with the classification. They collected tweets and annotated them into two categories with and without risk with a distribution of 67% and 33%, respectively. Among the tested algorithms, Sequential Minimal Optimization (SMO) performed better to identify tweets with the risk of suicide with an F-score of 89.3%.

Deep learning models were used in [34] in order to classify tweets. They used, among classic machine learning methods, an LSTM for classification. Nevertheless, the algorithm that performed better was Random Forest with better precision (84.2%), accuracy (85.8%), and F1 score (84.4%). Different feature sets were used, but the better performance was obtained using all features. Also, this work did not use any distributed representation for the data, like word2vec.

Ji et al. [20] also compared several machine learning models, including LSTM with word2vec word embedding. However, the best results came from XGBoost for accuracy (95.71%), recall (96.68%), and F1-score (95.83%). The deep learning model only got better precision (97.86%). They used the Reddit Dataset, where the posts are usually bigger than on Twitter since tweets have a little character limitation. As in [34], the best results in [20] were also achieved when employing data derived features, such as topic probabilities and linguistic features, which implies in an extra data preprocessing effort.

Other works do not directly focus on detecting suicidal ideation, but identifying depression in messages shared on Twitter. In Nadeem's work [26], classical machine learning algorithms were also explored in this task, such as decision trees and Naive-Bayes classifiers, in addition to the SVM and LGR as mentioned earlier. The data used was not collected by the authors; they used the CLPsych dataset that contains tweets about Major Depression Decease. In this work, the better algorithm tested was Logistic Regression, with an F1-score of 84%.

Orabi et al. [28] used different architectures of convolutional neural networks and recurrent neural networks to detect depression in Twitter messages. They also used the CLPsych dataset to classify depression and PTSD. In the classifiers, a word embedding approach was used to enhance the representation of the texts.

Among the different architectures of deep learning used, CNN with a global max-pooling performed better, with an F1-Score of 86.967%.

To identify the connectivity and communication of suicidal users on Twitter, [12] provides graph representations of the users' behavior on the platform. This graph suggests that the spread of suicidal ideation content is more significant than reported in other works and indicates a high level of information spread in these users' communications.

The present work has a similar approach to [27] in the matter of data collection and annotation process. However, our objective is to identify suicide ideation in tweets written in Brazilian Portuguese. As far as we are concerned, this is the first work that performs automatic detection of suicidal ideation in Portuguese. We compare state-of-the-art classifiers (LSTM and BERT), which were not previously used in this task, with the classic machine learning algorithm (SVM) as a baseline. Our experimental methodology is presented in the next section.

4   Experimental Methodology

The procedure in this work can be divided into three main tasks: data collection, data annotation, and data preprocessing and automated classification with machine learning algorithms, as shown in Fig. 1. Each of these tasks is discussed in the next sections.

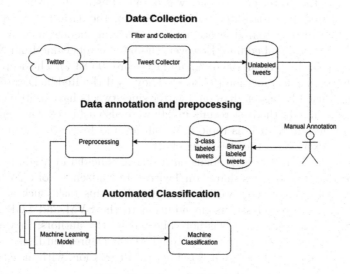

Fig. 1. System architecture.

4.1 Data Collection

Twitter implements a public API[1] that provides access to get tweets and use the platform with programming languages. Automatizing data collection is desirable because it allows the collection of tweets at any time for an undefined amount of time. To collect tweets from the stream, one must provide a list of terms the will be tracked.

In order to collect data from Twitter API, one must define a set of keywords to search for in messages published in their platform. An adequate vocabulary with domain-related words and sentences must be chosen to assure quality and sufficient data. In this work, the vocabulary to monitor in tweets was defined by merging different vocabularies from two studies on suicidal ideation detection [14, 27]. We translated these words and sentences from English to Portuguese as we were monitoring messages written in Brazilian Portuguese. Finally, we validated our vocabulary with domain specialists to filter the most useful expressions of suicidal thought detection.

The final vocabulary contains phrases like *"I don't wanna live anymore"*, *"tired of living"*, *"I wish I was dead"*, *"I want to kill myself"*, as well as other similar expressions. The full vocabulary is available on the project repository[2]. A Python program, named Tweet-Collector[3], was developed to wrap the tracking of tweets with Tweepy and provide a reliable continuous collection of tweets. From 18th February 2020 to 11th March 2020, more than 15,000 tweets were collected. It is strongly relevant to highlight that no information about the user is stored, assuring data anonymization in our analysis.

4.2 Data Annotation

The machine learning algorithms used in this work are supervised learning methods, which means that the data must be labeled in order to the algorithms work. Out labeling process relies on the intervention of a domain specialist. To avoid overload in the specialist annotation work, from the collected tweets, we randomly selected a set of 2446 tweets to be labeled.

Each tweet was initially labeled according to three classes, as suggested in other previous work on suicidal ideation detection: "Safe to ignore"; "Possibly concerning" and "Strongly concerning". The more detailed definition of each level of concerning is displayed as follows:

- *Strongly concerning*: The text shows the author's real intention to commit suicide. For example, "I want to kill myself" not conditioned to other factors that can exempt the seriousness of the statement, such as "I will kill myself if I forget my bring my lunch";
- *Possibly concerning*: If the text doesn't fit in the other two levels. It indicates that there is no certainty that it is either safe or concerning, so it is classified as a middle level.

---

[1] https://developer.twitter.com/.

[2] https://github.com/viniciosfaustino/suicide-detection.

[3] https://github.com/viniciosfaustino/tweet-collector.

- *Safe to ignore*: The tweet contains no evidence that the person is having suicidal thoughts, besides containing one of the monitored expressions present in our vocabulary.

A second approach was also made, dividing the problem only by two classes: "Safe to ignore" and "Possibly concerning". Here, "Possibly concerning" includes every tweet that may contain suicidal ideation, independent if it is strong or not. Our domain specialist suggested this division. According to our specialist, the first step in helping someone is to identify those who are in need. Thus, in practical terms, it is more important to identify whether there is a concern about suicide in the given text than graduate it in levels.

The distribution of classes in both datasets is shown in Table 1. It is possible to notice that when considering two classes, the resulting dataset is almost balanced. When dividing the dataset into three classes, there is an unbalanced distribution of the elements.

Table 1. Classes distribution in our assembled datasets.

# of classes	# of Examples	Safe to ignore	Possibly concerning	Strongly concerning
3	2446	1265	480	701
2	2446	1265	1181	–

4.3   Data Preprocessing and Classification

In our experiments, different machine learning models were used for this task, and each one has different preprocessing procedures. The chosen models are BERT, LSTM, and SVM. Also, we used only textual features, like a bag of words and word embeddings, with no extra information. The parameters were empirically selected during experimentation.

Only punctuation was removed on the BERT application, but meaningful characters as hashtags and emojis were kept. For LSTM, besides removing punctuation, it was also converted to lower case. For SVM, punctuation was removed, converted to lower case, and converted to a word2vec representation using the pre-trained word embeddings dataset from [18] using a skip-gram model with 50 dimensions.

To guarantee the generalizability of the classifiers, we split the dataset into ten folds using Stratified K-Fold implemented on Scikit-Learn. Resulting folds were saved separately, so all the models could perform on the same data, to ensure a fair comparison between the models.

Twelve Transformer blocks compose the architecture of the BERT model used. Its hidden size is 768 and 12 self-attention heads, in a total of 110M parameters. We used two different pre-trained $BERT_{BASE}$, the BERT-base-multilingual-cased, and one that was fine-tuned in Brazilian Portuguese, BERT-base-Portuguese-cased [37]. Since we are using pre-trained models, the task left is to fine-tune, which is made by adding a fully-connected layer output in the model, unfreeze all layers and train the model in the task. We used five cycles to fine-tune both models, the multilingual, and the Brazilian Portuguese one. The optimizer used was AdamW.

The LSTM approach is based on [23], and it uses an LSTM as a language model attached to an LSTM classifier. The first part learns a representation for the text input, as an encoder, and the last one is responsible for the classification itself. The language model is an LSTM with three RNN with dropout layers, one Linear, and one RNN with Dropout, with 853,328 parameters. The language model is the input for the classifier itself, that is another LSTM with three RNN with dropout layers, one BatchNorm, one Dropout, one Linear, one ReLU, one BatchNorm, one Dropout, and one Linear, with 62,652 parameters. The language model is fed with the input data, and its output goes to the classifier. The optimizer for both is ADAM, and five cycles were used to train the model.

The SVM model used is the built-in implementation of Scikit-Learn, and the parameters were chosen by random search. We set the search with the kernels *linear*, *poly*, and *rbf*. To the regularization parameter, C, we set a uniform distribution between 2 and 10, and the kernel coefficient, *gamma*, was another uniform distribution between 0.1 and 2.

Since we have two datasets with distinct class sizes, the random search was used for each dataset, using a ten-fold split. The best parameters for the binary class dataset were kernel: *rbf*, C: *9.0127*, gamma: *0.7293*, and for the 3-class dataset, kernel: *rbf*, C: *9.0332*, gamma: *1.0312*.

All models were evaluated both in 2-classes and 3-classes datasets. Results and their discussion are shown in next section.

5   Results and Discussion

The results of the compared algorithms in both datasets can be observed in Table 2. It is possible to observe that BERT fine-tuned using a Brazilian Portuguese dataset achieved the best performance in all scenarios, followed by multilingual BERT. This is expected behavior since BERT outperformed other state-of-the-art classifiers in many tasks. Also, the significant difference between Portuguese and multilingual approaches shows the importance of fine-tuning BERT, considering the target language. Since our task relies on tweets, there are many language-specific expressions to be considered that are not properly represented in the multilingual approach.

The difference in performance achieved by the classifiers when facing two and three classes also stands out. The binary approach consists of a more straightforward problem than the multiclass approach, which should justify some differences in classifiers' performance. However, the scale of the difference showed how hard it is to differentiate a "possibly concerning" message from the other scales of concern. To illustrate this, in Table 3, we present the confusion matrix achieved by the best classifier in the multiclass scenario-BERT-Portuguese. For the "possibly concerning" class, BERT-Portuguese present 65% of accuracy, which is

Table 2. Results for accuracy, precision, recall and F1.

Model	Classes	Accuracy	Precision	Recall	F1
BERT-multil.	2	0.7382 ± 0.0472	0.7417 ± 0.0429	0.7390 ± 0.0456	0.7370 ± 0.0487
BERT-port.	2	$\mathbf{0.7886 \pm 0.0279}$	$\mathbf{0.7909 \pm 0.0254}$	$\mathbf{0.7887 \pm 0.0276}$	$\mathbf{0.7879 \pm 0.0283}$
LSTM	2	0.6461 ± 0.0294	0.6517 ± 0.0279	0.6468 ± 0.0303	0.6425 ± 0.0332
SVM	2	0.7049 ± 0.0289	0.7053 ± 0.0292	0.7033 ± 0.0291	0.7032 ± 0.0292
BERT-multil.	3	0.6377 ± 0.0208	0.5511 ± 0.0421	0.5287 ± 0.0264	0.5153 ± 0.0364
BERT-port.	3	$\mathbf{0.7041 \pm 0.0225}$	$\mathbf{0.6458 \pm 0.0280}$	$\mathbf{0.6470 \pm 0.0293}$	$\mathbf{0.6446 \pm 0.0283}$
LSTM	3	0.5838 ± 0.0212	0.5080 ± 0.1303	0.4592 ± 0.0274	0.4254 ± 0.0308
SVM	3	0.6304 ± 0.0320	0.5611 ± 0.0621	0.5217 ± 0.0340	0.5119 ± 0.0373

much worse than the accuracy presented in other classes. Since "possibly concerning" can be considered as an intermediate category between strong in need for attention and non-suicidal patients, considering this class may lead classifiers to misclassification. In practice, considering this intermediate class may be harmful for adequate detection of patients for specialist intervention.

Table 3. Confusion matrix for multiclass classification with BERT-Portuguese.

	Safe to ignore	Possibly concerning	Strongly concerning
Safe to ignore	**1044**	155	66
Possibly concerning	157	**198**	125
Strongly concerning	54	135	**512**

Another aspect to be observed in the multiclass approach is the high rate of false negatives (i.e., "possibly concerning" and "strongly concerning" messages considered as "safe to ignore" and/or "possibly concerning", respectively). In a scenario where specialists need to detect patients in need of help as soon as possible, false-negative results must be avoided the most. In the multiclass approach, false-negative rate is about 14%, while in the binary approach, this rate drops to 9%, as can be observed in Table 4. This meets the recommendation of our domain specialist to consider just two classes. Since a suicidal patient needs intervention despite his/her level of suicidal tendencies, it is more important to detect those in need of intervention accurately.

Table 4. Confusion matrix for binary classification with BERT-Portuguese.

	Safe to ignore	Possibly concerning
Safe to ignore	**1016**	250
Possibly concerning	240	**941**

Comparing our results with other state-of-the-art results in suicidal detection in Twitter messages written in English [20,34], it is possible to observe that F1 score achieved in our experiments is, in average, 5% lower. First, it is important to analyze that such work employed much larger datasets. Also, the best results in these work employ other data derived features, such as topic probabilities and linguistic features. The usage of such features implies extra data preprocessing and data annotation effort, which can be time-consuming. Finally, the difference in complexity and subtlety of English and Portuguese languages also contributes to this difference.

6   Conclusion and Future Work

Fast-tracking of suicidal ideation is essential for the effective treatment of those in need. With the popularization of social networks, various suicidal ideation manifestation may be observed in publications, especially on Twitter. Automated solutions may help in the fast identification of these publications.

In this work, we present the first machine learning-based approach for suicidal ideation in Brazilian Portuguese. We explored Twitter publications that were captured using a set of keywords validated by domain specialists. We explored three different classification algorithms: SVM, BERT, and LSTM. As suggested by specialists, we compared the performance of these algorithms in two scenarios, considering two and three levels of suicidal risk, respectively. Results suggest that BERT is effective for suicidal ideation identification, presenting high accuracy and low false-negative rates in both scenarios. The binary scenario, however, was especially important for practical solutions. Since our solution is based on a pre-trained model, fine-tuning for new data is particularly fast. This makes our model adequate for the rapid incorporation of new data and model update, which is vital for a fast suicidal ideation detection.

In future work, we intend to explore data derived features, such as topic probabilities and linguistic features. Despite being time-consuming, these features improved classification performance in work on English-written Twitter publications. Such features would be used along with other classifiers, such as Random Forest and XGBoost. Also, we intend to expand our labeled dataset by incorporating other data annotated by a domain specialist. We hypothesize that, by incorporating more training data, BERT would achieve classification performance comparable with results achieved by state-of-the-art algorithms in this task in English data. Finally, the next step of our work is to incorporate our model as a bot to monitor and detect suicidal ideation in comments made on the official channels of our university in social networks.

References

1. Agarwal, A., Xie, B., Vovsha, I., Rambow, O., Passonneau, R.J.: Sentiment analysis of twitter data. In: Proceedings of the Workshop on Language in Social Media (LSM 2011), pp. 30–38 (2011)

2. Aggarwal, C.C.: Neural Networks and Deep Learning. Springer, Cham (2018). https://doi.org/10.1007/978-3-319-94463-0
3. Araújo, L.D.C., Vieira, K.F.L., Coutinho, M.D.P.D.L.: Ideação suicida na adolescência: um enfoque psicossociológico no contexto do ensino médio. Psico-USF 15(1), 47–57 (2010)
4. Berman, A.L., Silverman, M.M., Bongar, B.M.: Comprehensive Textbook of Suicidology. Guilford Press, New York (2000)
5. Birjali, M., Beni-Hssane, A., Erritali, M.: Machine learning and semantic sentiment analysis based algorithms for suicide sentiment prediction in social networks. Procedia Comput. Sci. 113, 65–72 (2017)
6. Boden, M.: A guide to recurrent neural networks and back propagation. The Dallas project (2002)
7. Borges, V.R., Werlang, B.S.G.: Estudo de ideação suicida em adolescentes de 15 a 19 anos. Estudos de Psicologia (Natal) 11(3), 345–351 (2006)
8. Brasil, Ministério da Saúde, S.D.V.E.S.: Perfil epidemiológico das tentativas e óbitos por suicídio no brasil ea rede de atenção à saúde. Bol Epidemiol 48 (2017)
9. Burnap, P., Colombo, W., Scourfield, J.: Machine classification and analysis of suicide-related communication on Twitter. In: Proceedings of the 26th ACM Conference on Hypertext and Social Media, pp. 75–84 (2015)
10. Cho, K., Van Merriënboer, B., Bahdanau, D., Bengio, Y.: On the properties of neural machine translation: encoder-decoder approaches. arXiv preprint arXiv:1409.1259 (2014)
11. Collobert, R., Weston, J., Bottou, L., Karlen, M., Kavukcuoglu, K., Kuksa, P.: Natural language processing (almost) from scratch. J. Mach. Learn. Res. 12, 2493–2537 (2011)
12. Colombo, G.B., Burnap, P., Hodorog, A., Scourfield, J.: Analysing the connectivity and communication of suicidal users on Twitter. Comput. Commun. 73, 291–300 (2016)
13. Devlin, J., Chang, M.W., Lee, K., Toutanova, K.: BERT: pre-training of deep bidirectional transformers for language understanding. In: Proceedings of the 2019 Conference of the North American Chapter of the Association for Computational Linguistics: Human Language Technologies, pp. 4171–4186 (2019)
14. Du, J.: Extracting psychiatric stressors for suicide from social media using deep learning. BMC Med. Inform. Decis. Mak. 18(2), 43 (2018)
15. Goldberg, Y.: Neural Network Methods in Natural Language Processing. Morgan and Claypool Publishers, San Rafael (2017)
16. Goodfellow, I., Bengio, Y., Courville, A.: Deep Learning. The MIT Press, Cambridge (2016)
17. Graves, A.: Long short-term memory. In: Supervised Sequence Labelling with Recurrent Neural Networks, pp. 37–45. Springer (2012). https://doi.org/10.1007/978-3-642-24797-2_4
18. Hartmann, N., Fonseca, E., Shulby, C., Treviso, M., Rodrigues, J., Aluisio, S.: Portuguese word embeddings: evaluating on word analogies and natural language tasks. arXiv preprint arXiv:1708.06025 (2017)
19. Hochreiter, S., Bengio, Y., Frasconi, P., Schmidhuber, J., et al.: Gradient flow in recurrent nets: the difficulty of learning long-term dependencies (2001)
20. Ji, S., Yu, C.P., Fung, S.F., Pan, S., Long, G.: Supervised learning for suicidal ideation detection in online user content. Complexity 2018 (2018)

21. Joachims, T.: Text categorization with support vector machines: learning with many relevant features. In: Nédellec, C., Rouveirol, C. (eds.) ECML 1998. LNCS, vol. 1398, pp. 137–142. Springer, Heidelberg (1998). https://doi.org/10.1007/BFb0026683

22. Manning, C.D., Schütze, H.: Foundations of Statistical Natural Language Processing. MIT Press, Cambridge (1999)

23. Merity, S., Keskar, N.S., Socher, R.: Regularizing and optimizing LSTM language models. arXiv preprint arXiv:1708.02182 (2017)

24. Mikolov, T., Sutskever, I., Chen, K., Corrado, G.S., Dean, J.: Distributed representations of words and phrases and their compositionality. In: Burges, C.J.C., Bottou, L., Welling, M., Ghahramani, Z., Weinberger, K.Q. (eds.) Advances in Neural Information Processing Systems 26, pp. 3111–3119. Curran Associates, Inc. (2013)

25. Mitchell, T.M.: Machine Learning, 1st edn. McGraw-Hill Inc, New York (1997)

26. Nadeem, M.: Identifying depression on twitter. arXiv preprint arXiv:1607.07384 (2016)

27. O'dea, B., Wan, S., Batterham, P.J., Calear, A.L., Paris, C., Christensen, H.: Detecting suicidality on Twitter. Internet Interv. 2(2), 183–188 (2015)

28. Orabi, A.H., Buddhitha, P., Orabi, M.H., Inkpen, D.: Deep learning for depression detection of Twitter users. In: Proceedings of the Fifth Workshop on Computational Linguistics and Clinical Psychology: From Keyboard to Clinic, pp. 88–97 (2018)

29. Organization, W.H., et al.: Preventing suicide: a global imperative. World Health Organization (2014)

30. Pereira, C.C.M., Botti, N.C.L.: O suicídio na comunicação das redes sociais virtuais: revisão integrativa da literatura. Revista Portuguesa de Enfermagem de Saúde Mental 17, 17–24 (2017)

31. Qi, Y., Sachan, D.S., Felix, M., Padmanabhan, S.J., Neubig, G.: When and why are pre-trained word embeddings useful for neural machine translation? arXiv preprint arXiv:1804.06323 (2018)

32. Recupero, P.R., Harms, S.E., Noble, J.M.: Googling suicide: surfing for suicide information on the internet. J. Clin. Psychiatry (2008)

33. Robertson, L., Skegg, K., Poore, M., Williams, S., Taylor, B.: An adolescent suicide cluster and the possible role of electronic communication technology. Crisis (2012)

34. Sawhney, R., Manchanda, P., Singh, R., Aggarwal, S.: A computational approach to feature extraction for identification of suicidal ideation in tweets. In: Proceedings of ACL 2018, Student Research Workshop, pp. 91–98 (2018)

35. Sebastiani, F.: Machine learning in automated text categorization. ACM Comput. Surv. 34(1), 1–47 (2002)

36. Sher, L.: Preventing suicide. QJM 97(10), 677–680 (2004)

37. Souza, F., Nogueira, R., Lotufo, R.: Portuguese named entity recognition using BERT-CRF. arXiv preprint arXiv:1909.10649 (2019)

38. Souza, L.D.D.M., et al.: Ideação suicida na adolescência: prevalência e fatores associados. Jornal Brasileiro de Psiquiatria 59(4), 286–292 (2010)

39. Turian, J., Ratinov, L., Bengio, Y.: Word representations: a simple and general method for semi-supervised learning. In: Proceedings of the 48th Annual Meeting of the Association for Computational Linguistics, pp. 384–394. Association for Computational Linguistics (2010)

40. Vasconcelos-Raposo, J., Soares, A.R., Silva, F., Fernandes, M.G., Teixeira, C.M.: Níveis de ideação suicida em jovens adultos. Estudos de psicologia 33(2), 345–354 (2016)

41. Vaswani, A., et al.: Attention is all you need. arXiv preprint arXiv:1706.03762 10 (2017)
42. Werlang, B.S.G., Borges, V.R., Fensterseifer, L.: Fatores de risco ou proteção para a presençade ideação suicida na adolescência. Interam. J. Psychol. **39**(2), 259–266 (2005)

Pre-trained Data Augmentation for Text Classification

Hugo Queiroz Abonizio$^{(\boxtimes)}$ ⓘ and Sylvio Barbon Junior ⓘ

State University of Londrina (UEL), Londrina, Brazil
{hugo.abonizio,barbon}@uel.br

Abstract. Data augmentation is a widely adopted method for improving model performance in image classification tasks. Although it still not as ubiquitous in Natural Language Processing (NLP) community, some methods have already been proposed to increase the amount of training data using simple text transformations or text generation through language models. However, recent text classification tasks need to deal with domains characterized by a small amount of text and informal writing, e.g., Online Social Networks content, reducing the capabilities of current methods. Facing these challenges by taking advantage of the pre-trained language models, low computational resource consumption, and model compression, we proposed the *PRE-trained Data AugmenTOR* (PREDATOR) method. Our data augmentation method is composed of two modules: the Generator, which synthesizes new samples grounded on a lightweight model, and the Filter, that selects only the high-quality ones. The experiments comparing Bidirectional Encoder Representations from Transformer (BERT), Convolutional Neural Networks (CNN), Long Short-Term Memory (LSTM) and Multinomial Naive Bayes (NB) in three datasets exposed the effective improvement of accuracy. It was obtained 28.5% of accuracy improvement with LSTM on the best scenario and an average improvement of 8% across all scenarios. PREDATOR was able to augment real-world social media datasets and other domains, overcoming the recent text augmentation techniques.

Keywords: Data augmentation · Text classification · Online social networks

1 Introduction

Data augmentation techniques have been successfully applied in machine learning models to improve their generalization capacity. It is a common strategy to avoid overfitting the training data, mainly on scenarios of data scarcity and situations where labeled examples are expensive. Since the performance of machine

The authors would like to thank the financial support of the National Council for Scientific and Technological Development (CNPq) of Brazil - Grant of Project 420562/2018-4 - and Fundação Araucária.

R. Cerri and R. C. Prati (Eds.): BRACIS 2020, LNAI 12319, pp. 551–565, 2020.
https://doi.org/10.1007/978-3-030-61377-8_38

learning models is highly correlated with the amount and the quality of the data used during its training, low-data scenarios become a challenge for practitioners [13].

Several techniques have been proposed and evaluated for image data [30], but the field of textual data augmentation is still incipient. Simple transformations, such as flipping, cropping, and other image manipulations, are often label-preserving on image classification tasks [3,18], but this assumption does not hold for text data. Changing words order or removing some parts of a sentence might change its whole semantic, resulting in low-quality samples and negatively impacting the performance.

In recent years, different text transformation strategies have been proposed. Varying from synonyms replacements [17,33], paraphrasing through translation models [23] and text generation using language models [16], however no gold standard technique has been developed yet. A recent method, entitled Easy Data Augmentation (EDA) [28], has been proposed combining synonym replacement with other simple methods such as random deletion and random swap of words. Those methods were found to increase the accuracy of classification on small datasets.

An often employed technique is the Back-Translation (BT), which works by making a round-trip translation using a secondary language. Given a sentence written in a language L_a and a translation system between L_a and a different language L_b, the BT approach firstly translates the sentence from L_a to L_b and then translates it back to the original L_a language, generating a slightly different sentence. Previous work demonstrated that this approach leads to better results on Neural Machine Translation [23] and reading comprehension [32]. BT was also applied to low-resource text classification [24], yielding an improvement in classification accuracy using different secondary languages.

Most recent work has proposed the use of pre-trained language models [1,19], leveraging transfer learning for improving text generation capabilities when synthesizing new samples. However, those pre-trained models increase the computational requirements of classification pipelines, in contrast to simple sample transformations initially proposed. Beyond its resources requirements, those approaches can be prohibitively expensive and have raised a concern regarding the energy efficiency of those models [22,26].

In contrast with dictionary-based approaches, such as EDA, those pre-trained language models can deal better with noisy text coming from Online Social Networks (OSN). OSN texts are characterized by an informal writing style and the presence of Internet slangs [14], which leads to frequent out-of-vocabulary words in this scenario. Language model-based approaches, on the other hand, can learn to reproduce the dataset writing style and pre-trained models leverage a priori knowledge to extend its generation capabilities [20].

Tackling the challenges of text augmentation in different and recent domains, we present the *Pre-trained Data Augmentor* (PREDATOR), a novel method for textual data augmentation that combines the high performance achieved by transfer learning of pre-trained models approaches with lower computational resource consumption. The simplicity of our methods is grounded on simple pre-trained models obtained by model compression [9]. PREDATOR works by synthesizing new high-quality samples to improve classification performance, particularly on small datasets, proving its effectiveness even on noisy social media datasets. We evaluated our method in three different datasets (*SST-2*, *AG-NEWS*, and *CyberTrolls*) from different media sources, using four different classifiers (Bidirectional Encoder Representations from Transformer, Convolutional Neural Networks, Long Short-Term Memory, and Multinomial Naive Bayes) and comparing the results with two other techniques present in the literature (Easy Data Augmentation and Back-Translation). The results demonstrated that PREDATOR increased the accuracy of all classifiers, achieving an average of 8% improvement in accuracy and a maximum of 28.5% on the best scenario, and statistical analysis demonstrated that its performance is similar to the real data to increase the dataset. The code is publicly available at https://github.com/hugoabonizio/predator.

The remainder of this paper of structured as follows: Sect. 2 describes the proposed method, detailing the pipeline and required parameters. Section 3 presents the evaluated datasets, the classification algorithms, and the methods compared with ours. In Sect. 4 we discuss the results regarding the amount of augmentation with all combinations of classifiers and datasets, and compare our method with two other widely adopted techniques on the same augmentation amount setups. Finally, on Sect. 5 we conclude and discuss future work.

2   Proposed Method

The main idea of our proposal regards the joint contribution of a text generator module and a filter module leading a two-step sample synthesizing approach. Thus, boosted by semi-supervised classification model, our method delivers a robust text augmentation method.

The PREDATOR architecture is composed of two modules: the Generator and the Filter, as shown in Fig. 1. These modules are responsible for synthesizing new samples and filtering high-quality ones, respectively. The first one, the Generator, is based on a language model [2], i.e., a model trained to predict the probability distribution for next tokens for a given context until it reaches a stop condition. This module is responsible for learning to generate new data corresponding to original classes while increasing its variability. The Filter module uses a text classifier trained on the original dataset towards selecting the high-quality new samples, i.e., accept as augmented samples only the synthetic samples in which the classifier has a high confidence of belonging to one of the given classes.

Figure 1 illustrates the main steps of our method pipeline. The first step is to initialize the Filter module by training its classifier to predict new sample labels based on the original dataset. Then, the Generator is trained to learn to synthesize new samples based on the original sentences by fine-tuning its language model. With both modules initialized, for each iteration of the method,

PRE-trained **D**ata Augmen**TOR**

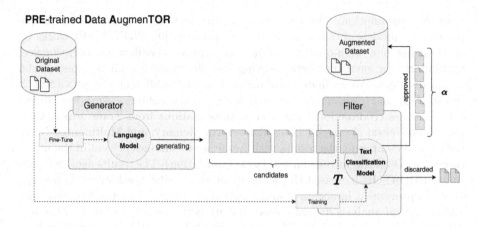

Fig. 1. Overview of proposed approach: PRE-trained Data AugmentOR

the generated samples are filtered, discarding low-quality sentences. The selected synthetic samples are accumulated until it reaches a previously defined number of augmented samples.

Among the several recently developed language models, we propose the usage of DistilGPT2 [21] on the Generator module. DistilGPT2 is a compressed version of GPT-2 obtained through knowledge distillation [9], becoming two times faster and having 33% fewer parameters than the smallest version of original GPT-2 with a minimal reduction in performance. This reduction makes the process of fine-tuning and posterior text generation much faster and reproducible with lower resources when compared to prior works.

The fine-tuning step may vary depending on the dataset, especially when its content is very different from the original corpus DistilGPT2 was trained. However, since DistilGPT2 was trained using OpenWebTextCorpus [8], a very diverse corpus, experimental results indicate that fine-tuning for only one epoch was enough to generate high-quality texts for augmenting the target dataset, even with a noisy dataset collected from social media interactions.

Given the language model fine-tuned in the given dataset domain, different methods can be employed to generate new texts. The previous work attached the class labels to condition the generation of text. Our proposed approach differs from previous [1] by simply concatenating three random samples from the target class using the language model input, i.e., given random samples s from from a target class, and a separator token already included on model vocabulary SEP, the input is given by s_1SEPs_2SEPs_3SEP. Thus, the following generation maintains the characteristics of the target class. The generation is done by sampling the probability of the language model, which, in contrast with beam-search and greedy decoding, generates higher-quality and more diverse texts [11]. The decoding strategy used in PREDATOR is top-k sampling [7], with $k = 40$.

After Generator, the next step is the selection of those new synthesized samples and the imputation of its class by a classifier trained on the original dataset. This process is performed by the Filter module for avoiding low-quality samples in the final augmented dataset, essential to leverage high accurate outcomes. Current state-of-the-art classification models are often large Transformer-based classifiers [29], which makes them too resource-hungry. Therefore they are not well suited to be directly applied to a pipeline of data augmentation. Thus, Sanh et al. [21] developed DistilBERT, a compressed model obtained through knowledge distillation which results in a 40% smaller model than original BERT, while being 60% faster and achieving 97% of its original performance. Therefore, DistilBERT is used as the classifier for the Filter module, maintaining a competitive performance, and meeting the requirements of computational resources.

PREDATOR requires two main hyperparameters: α as the augmentation rate, determining how many new samples need to be synthesized by the Generator module, and T as the threshold confidence for the Filter module. The α parameter controls the increase of the augmented dataset regarding the original sample size, i.e., given a dataset with n samples per class and an α of 0.25, the resulting augmented dataset will have $1.25n$ samples for each class. In this work, we conduct a more in-depth experiment with different values for α, showing its behavior on different datasets.

The T parameter controls the flexibility of the Filter, determining whether it is stricter or more flexible on the sample quality selection. Quality refers to the predictive power regarding classification task considering the given samples. With a more strict T value, the module only selects the samples that its classifier predicts the class with the most confidence. In contrast, a lower value of T might approve samples that the classifier has more uncertainty. This value represents a trade-off since a higher value makes it difficult for the Generator to synthesize enough samples, making it necessary to do more iterations to satisfy α requirement. On the other hand, lower values can lead to low-quality augmented dataset due to noisy samples.

Another aspect that needs to be assessed is the variability sought by data augmentation methods. With a high T, only samples with high certainty will be selected, i.e., only samples very similar to the original dataset will be included in the augmented dataset, which is a suboptimal result. Since the objective of data augmentation is to enrich the dataset with different samples without losing its class nature, a certain amount of uncertainty is required. Preliminary experiments showed that a value of 0.7 (default value) for T is a reasonable default since it is a balance on prediction confidence. However, different values of T for different scenarios might be explored in future works, especially on datasets with a higher class carnality.

2.1 PREDATOR Augmentation Example

Table 1 shows examples of original samples and synthesized to illustrate the learning of writing style from the original dataset and the class preserving of the generated text. Besides writing style, the text length also tends to be similar, even though it was not directly enforced.

Table 1. Examples of original and synthesized samples for each dataset showing samples from the same class.

Dataset	Real samples	Augmented samples
AG-NEWS	*Apple cuts prices and improves products. Apple introduced a range of new machines on Tuesday, as it gears up for the annual christmas shopping season. As part of the launch, it cut the price of its entry-level iBook G4 notebook computer and boosted chip speed across the line*	*Apple Shares. A European market leader says heś ready to share it. Apple stock was high Tuesday on the eve of its annual earnings*
CyberTrolls	*i hate that! That sucks!!!*	*oh man that sucks!! LOL*
SST-2	*The jokes are flat, and the action looks fake*	*... even some jokes are off the charts and probably just too out of place*

3 Materials and Methods

3.1 Datasets

We evaluated the method on three different text classification datasets. *SST-2* (Stanford Sentiment Treebank)[1] [25], a classic dataset for sentiment classification on movie reviews widely applied as benchmark [16,19,28], with two classes: positive and negative. *AG-NEWS*[2] [33], another common benchmark dataset, but for topic classification task [12,31] composed of news belonging to four classes: world, sports, business and science/technology. *CyberTrolls*[3] is a more recent dataset for the task of aggressiveness detection in social networks, with examples of two classes: cyber-aggressive and non cyber-aggressive. *CyberTrolls* presents a challenge for text mining given the noisy characteristics of this media. With those three datasets we test our method with representations of different written styles to compare its behavior on more formal and more informal data sources.

The experiments were conducted to reproduce a small data scenario, where the classifier is trained on a restricted number of samples, and data augmentation performance is compared with the subsampled set. Previous work has simulated low-data regime setting by subsampling original datasets [13,19,28],

---

[1] https://nlp.stanford.edu/sentiment/.
[2] http://groups.di.unipi.it/~gulli/AG_corpus_of_news_articles.html.
[3] https://www.kaggle.com/dataturks/dataset-for-detection-of-cybertrolls.

becoming a common practice when evaluating augmentation techniques. Thus, for each dataset, 100 samples per class were subsampled and, since it is a non-deterministic process, this procedure was repeated ten times to average the final results. Those subsample of the dataset are assigned as the original performance on experiments, and all augmentation was made based on them. It is important to emphasize that only train sets were subsampled, validation and test sets were kept the same as originals.

3.2 Text Classification Algorithms

We conducted the experiments using four different classification models: Bidirectional Encoder Representations from Transformers (BERT) [5], Convolutional Neural Networks (CNN) [15], Long Short-Term Memory (LSTM) [10], and Multinomial Naive Bayes (NB) [27]. Those classifiers were selected to represent different categories, having NB as a classic text classifier often compared as a baseline method [27], LSTM, and CNN as common deep learning classifiers [33], and BERT representing the most recent progress achieving state-of-the-art on numerous NLP tasks including text classification.

3.3 Augmentation Methods

Since the text data augmentation is still an emergent topic, several methods have been proposed, but there is still no de facto standard technique. The two most applied techniques found in the literature are synonyms replacement and BT. EDA is an extension of synonyms replacing, introducing simple text transformations that were successfully applied to other works. Therefore, we compared PREDATOR with those two widely applied augmentation techniques (EDA and BT), observing the boosting on performance on different domains.

The first compared technique was BT, where each sentence of the original dataset is translated into a different language and then translated back to the source. This method requires two models: a model to translate from source language to a target one, and the inverse model. Among the alternatives of models, we conducted the experiments using the models proposed by Edunov et al. [6], a Transformer model made publicly available[4].

The second compared technique was EDA, which the code is publicly available[5]. The hyperparameters used to generate new samples were the recommended default, generating 9 new samples for each sample in the original datasets.

4 Results and Discussion

In this section, we expose two perspectives to evaluate the proposed method. First, it is discussed the augmentation ratio and classification performance from

---

[4] https://pytorch.org/hub/pytorch_fairseq_translation/.

[5] https://github.com/jasonwei20/eda_nlp.

the original size to nine augmented outcomes using three different datasets (*AG-NEWS*, *CyberTrolls*, and *SST-2*) and four classification algorithms (BERT, CNN, LSTM, and NB). The second perspective supports the comparison between PREDATOR and current text augmentation methods (EDA and BT) using the original textual resource (Truth) with the same amount of samples for each method.

4.1   Augmentation Capabilities

The main goal when using augmentation methods, independent of problem, is to improve the predictive performance. We performed 9 augmentation rates (0.1, 0.25, 0.50, 1.0, 1.5, 2.0, 3.0, 6.0 and 9.0) on *AG-NEWS*, *CyberTrolls* and *SST-2* datasets. Figure 2 shows the accuracy of different augmentation rates across all four different classification algorithms (BERT, CNN, LSTM and NB).

A prominent boosting in performance was obtained by LSTM on *AG-NEWS* (first column and third row in Fig. 2) since the original accuracy of 71% reach 84% when augmented. In the same combination of algorithm and dataset, we can observe an improvement of the model quality grounded in the reduction on the accuracy standard deviation, highlighted by the performance shadowed mark.

The overall accuracy improvement between the original size and the maximum augmentation (9x) across all scenarios is exposed in Table 2. The results were grouped by dataset with the biggest improvement highlighted in bold and the smallest underlined. As the previous case presented, LSTM obtained the biggest improvement in all scenarios. BERT provided small improvement on *AG-NEWS* and *CyberTrolls* and CNN on *SST-2*.

Table 2. Augmentation results grouped by datasets for each classifier, highlighting the biggest improvements in bold and the smallest improvements underlined.

Dataset	Algorithm	Avg. Improvement	Original Acc
AG-NEWS	BERT	+0.7%	87.2%
AG-NEWS	CNN	+2.2%	83.1%
AG-NEWS	LSTM	**+28.5%**	65.4%
AG-NEWS	NB	+2.8%	80.0%
CyberTrolls	BERT	+6.8%	63.5%
CyberTrolls	CNN	+0.7%	65.2%
CyberTrolls	LSTM	**+10.0%**	58.7%
CyberTrolls	NB	+3.5%	62.8%
SST-2	BERT	+2.0%	81.5%
SST-2	CNN	+12.9%	63.8%
SST-2	LSTM	**+17.9%**	59.8%
SST-2	NB	+8.6%	63.4%

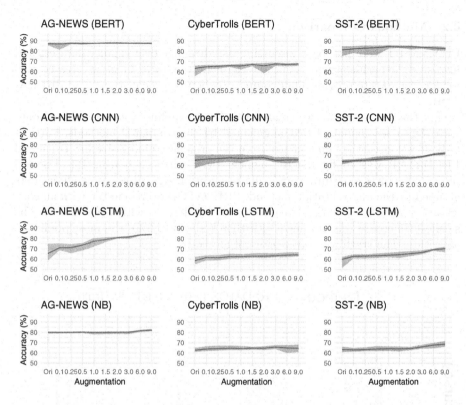

Fig. 2. Performance obtained from different algorithms and datasets using the original size and different augmented sources with PREDATOR.

It is important to note that improvement depends on the original performance, since LSTM was the algorithm that took the most advantage of PREDATOR, but obtained the lower average performance in comparison to the other algorithms. The high improvement on LSTM results indicates it tends to overfit the training data when using a small number of samples, thus the variances introduced by augmentation act as a regularizer, highly improving its performance. Conversely, BERT presented the smallest difference when using our augmentation, but high classification performance.

The higher average performance increases were achieved in *SST-2* and *AG-NEWS* datasets, traditional benchmarks with cleaned texts. *SST-2* is composed of movie reviews and *AG-NEWS* is composed of news articles, which tend to have a more formal writing style. On the other hand, the lowest average increase happened in *CyberTrolls* dataset, which is composed of highly noisy texts, containing emojis and specific social media expressions. Even though this writing style might be more difficult for the language model to reproduce, the fine-tuning step and the proposed input seed strategy was proved effective in writing style conditioning and robust on noisy datasets.

4.2   Methods Comparison

EDA and BT were proposed discussing its advantage over smaller datasets and intensively experiment with a specific augmentation rate. Thus, we created particular scenarios to provide a fair comparison between the methods and PREDATOR.

For the first scenario, we created the Truth baseline from the original data with the same amount of text augmented by the compared methods. Particularly, it was used the triple of size from the original training size for Truth and PREDATOR was configured to augment twice when comparing BT. Figure 3 presents boxplots of accuracy with all classification algorithms using all datasets grouped by the augmentation method. PREDATOR overcome BT with all classification algorithms, with a greater accuracy difference with CNN (2.3%). The smallest difference was obtained with BERT about (0.2%), the most predictive algorithm. The truth was superior to all methods.

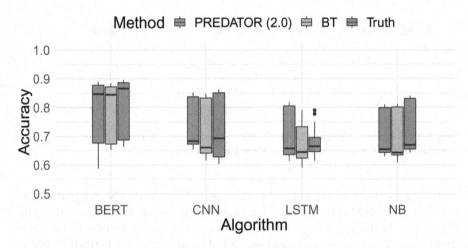

Fig. 3. Accuracy comparison among augmentation methods (PREDATOR and BT) and truth dataset. The boxplots were computed with the three datasets (*AG-NEWS*, *CyberTrolls* and *SST-2*) grouped by classification algorithms.

In order to observe a performance superiority of a method, we evaluated the statistical significance using the Friedman test, with a significance level of $\alpha = 0.05$. The null hypothesis is that the augmentation methods are similar. Anytime the null hypothesis is rejected, the Nemenyi post hoc test is applied, stating that the performance of a pair of methods are significantly different if their corresponding average ranks differ by at least a critical distance value. When multiple methods are compared in this way, a graphic representation can be used to represent the results with the Critical Difference Diagram, as proposed by Demšar [4]. This analysis is shown in Fig. 4, where it is possible to conclude that Truth and PREDATOR are similar, PREDATOR and BT are similar, and Truth and BT are statistically different. Thus, we can claim that our proposal is statistically similar to the usage of the original data.

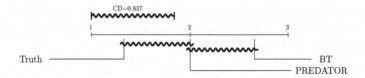

Fig. 4. Comparison of the accuracy values obtained by augmentation methods (PREDATOR and BT) and truth with the nemenyi test. Groups that are not significantly different ($\alpha = 0.05$ and $CD = 0.83$)

In the second scenario, PREDATOR was compared to EDA, one of the most recent proposals. A situation to the first scenario was found. Truth, the real data, obtained the best performance followed by PREDATOR and EDA. In this scenario, PREDATOR augmented the original dataset by nine to match the same amount of Truth and EDA. The more significant difference between the proposed method and EDA was using NB, about 6.3% accuracy, as Fig. 5 shows. Again, BERT achieved the smallest accuracy difference, an average of 0.1%.

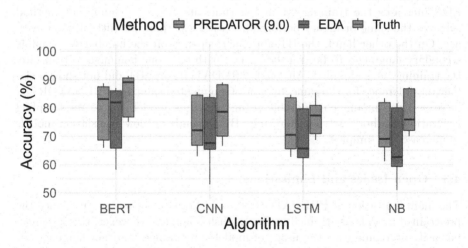

Fig. 5. Accuracy comparison among augmentation methods (PREDATOR and EDA) and truth dataset. The boxplots were computed with the three datasets (*AG-NEWS*, *CyberTrolls* and *SST-2*) grouped by classification algorithms

Using the same statistical assumptions of the first scenario, we employed Friedman and Nemenyi test to compare PREDATOR, EDA, and Truth. As Fig. 6

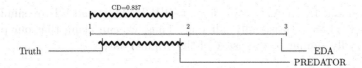

Fig. 6. Comparison of the accuracy values obtained by augmentation methods (PREDATOR and EDA) and truth with the nemenyi test. Groups that are not significantly different ($\alpha = 0.05$ and $CD = 0.83$)

exposes, it is possible to conclude that Truth and PREDATOR are similar and statistically different from EDA. Thus, we can affirm that our proposal produces synthetic data able to support the training of a text classification model capable of obtaining results statistically similar to the usage of real data. Therefore, PREDATOR generates samples that lead to superior performance than EDA.

The results reveal that the PREDATOR approach is an effective method for improving the performance of the classifiers, resulting in similar or greater performance than its alternatives. It also proves to be robust to noise on the text and informal writing, improving the accuracy of the model on a real-world social media dataset. Other methods, such as EDA, depend on a fixed dictionary to work, causing out-of-vocabulary issues on rare words, neologisms, and internet slang. This explains the poor performance of EDA on *CyberTrolls* dataset and shows better handling of these issues by our method.

Another aspect is the amount of added samples to the training set. While EDA increases the training set in ten times its size by default, our method achieves the same or higher performance with less than one-fifth of it on average. On the other hand, the BT approach depends on the number of available secondary languages to be translated, i.e., with only one language to translate, the training set is doubled. Although PREDATOR and BT did not show a significant difference for the same amount of augmented samples, PREDATOR can increase this amount considerably using the same model due to its sampling generation method. At the same time, BT depends on new translation models to increase the samples.

4.3   Open Issues and Limitations

The main limitation of PREDATOR is its Anglophone-centric nature since the pre-trained models are trained in the English language. However, this issue can be easily overcome with the usage of models pre-trained on other languages or even multilingual models on its modules. The PREDATOR architecture enables the change of its kernel models, making it possible to be applied to different languages or taking advantage of newer classifiers and language models in the future.

The minimum amount of samples required to apply the method successfully is not yet clear. Too few amounts of samples can not be enough to train an effective Filter module since it depends on its classifier quality. More experiments need to be taken in order to define the optimal hyperparameters for each different scenario.

5   Conclusion

We presented PREDATOR, a novel data augmentation technique for text classification leveraged by transfer learning using a language model to synthesize new high-quality samples. Our proposed method was experimentally compared with two widely adopted methods across three datasets using four text classifiers. The results show that PREDATOR is effective with either cleaner benchmark datasets and noisy real-world datasets. Besides achieving a better average performance than the compared methods, it is statistically similar to using the original dataset with the same amount of data.

A natural progression of this work is to analyze the behavior of different kernels, using different language models and classifiers to improve its applicability to lower-resourced languages. Another possible area of future research would be to expand the experiments with newer language model-based approaches, assessing its resources consumption and complexity.

References

1. Anaby-Tavor, A., et al.: Not enough data deep learning to the rescue. arXiv preprint arXiv:1911.03118 (2019)
2. Bengio, Y., Ducharme, R., Vincent, P., Jauvin, C.: A neural probabilistic language model. J. Mach. Learn. Res. **3**(2), 1137–1155 (2003)
3. Cubuk, E.D., Zoph, B., Mane, D., Vasudevan, V., Le, Q.V.: Autoaugment: learning augmentation policies from data (2019).https://arxiv.org/pdf/1805.09501.pdf
4. Demšar, J.: Statistical comparisons of classifiers over multiple data sets. J. Mach. Learn. Res. **7**, 1–30 (2006)
5. Devlin, J., Chang, M.W., Lee, K., Toutanova, K.: BERT: pre-training of deep bidirectional transformers for language understanding. In: Proceedings of the 2019 Conference of the North American Chapter of the Association for Computational Linguistics: Human Language Technologies, Vol. 1 (Long and Short Papers), pp. 4171–4186. Association for Computational Linguistics, Minneapolis, Minnesota, June 2019. https://doi.org/10.18653/v1/N19-1423, https://www.aclweb.org/anthology/N19-1423
6. Edunov, S., Ott, M., Auli, M., Grangier, D.: Understanding back-translation at scale. In: Proceedings of the 2018 Conference on Empirical Methods in Natural Language Processing, pp. 489–500. Association for Computational Linguistics, Brussels, Belgium, Oct-Nov 2018. https://doi.org/10.18653/v1/D18-1045, https://www.aclweb.org/anthology/D18-1045
7. Fan, A., Lewis, M., Dauphin, Y.: Hierarchical neural story generation. In: Proceedings of the 56th Annual Meeting of the Association for Computational Linguistics (Volume 1: Long Papers), pp. 889–898. Association for Computational Linguistics, Melbourne, Australia, July 2018. https://doi.org/10.18653/v1/P18-1082, https://www.aclweb.org/anthology/P18-1082

8. Gokaslan, A., Cohen, V.: Openwebtext corpus (2019). http://Skylion007.github.io/OpenWebTextCorpus

9. Hinton, G., Vinyals, O., Dean, J.: Distilling the knowledge in a neural network. In: NIPS Deep Learning and Representation Learning Workshop (2015). http://arxiv.org/abs/1503.02531

10. Hochreiter, S., Schmidhuber, J.: Long short-term memory. Neural Comput. **9**(8), 1735–1780 (1997). https://doi.org/10.1162/neco.1997.9.8.1735

11. Holtzman, A., Buys, J., Du, L., Forbes, M., Choi, Y.: The curious case of neural text degeneration. arXiv preprint arXiv:1904.09751 (2019)

12. Howard, J., Ruder, S.: Universal language model fine-tuning for text classification. In: Proceedings of the 56th Annual Meeting of the Association for Computational Linguistics (Volume 1: Long Papers), pp. 328–339. Association for Computational Linguistics, Melbourne, Australia, July 2018. https://doi.org/10.18653/v1/P18-1031, https://www.aclweb.org/anthology/P18-1031

13. Hu, Z., Tan, B., Salakhutdinov, R.R., Mitchell, T.M., Xing, E.P.: Learning data manipulation for augmentation and weighting. In: Wallach, H., Larochelle, H., Beygelzimer, A., dÁlché-Buc, F., Fox, E., Garnett, R. (eds.) Advances in Neural Information Processing Systems, vol. 32, pp. 15764–15775. Curran Associates, Inc. (2019). http://papers.nips.cc/paper/9706-learning-data-manipulation-for-augmentation-and-weighting.pdf

14. Igawa, R.A., et al.: Account classification in online social networks with lbca and wavelets. Inform. Sci. **332**, 72–83 (2016)

15. Kim, Y.: Convolutional neural networks for sentence classification. In: Proceedings of the 2014 Conference on Empirical Methods in Natural Language Processing (EMNLP), pp. 1746–1751. Association for Computational Linguistics, Doha, Qatar, October 2014. https://doi.org/10.3115/v1/D14-1181, https://www.aclweb.org/anthology/D14-1181

16. Kobayashi, S.: Contextual augmentation: data augmentation by words with paradigmatic relations. In: Proceedings of the 2018 Conference of the North American Chapter of the Association for Computational Linguistics: Human Language Technologies, Vol. 2 (Short Papers), pp. 452–457. Association for Computational Linguistics, New Orleans, Louisiana, June 2018. https://doi.org/10.18653/v1/N18-2072, https://www.aclweb.org/anthology/N18-2072

17. Kolomiyets, O., Bethard, S., Moens, M.F.: Model-portability experiments for textual temporal analysis. In: Proceedings of the 49th Annual Meeting of the Association for Computational Linguistics: Human Language Technologies: Short Papers, Vol. 2, pp. 271–276. HLT '11, Association for Computational Linguistics, USA (2011)

18. Krizhevsky, A., Sutskever, I., Hinton, G.E.: Imagenet classification with deep convolutional neural networks. In: Advances in Neural Information Processing Systems, pp. 1097–1105 (2012)

19. Kumar, V., Choudhary, A., Cho, E.: Data augmentation using pre-trained transformer models. arXiv preprint arXiv:2003.02245 (2020)

20. Petroni, F., et al.: Language models as knowledge bases. In: Proceedings of the 2019 Conference on Empirical Methods in Natural Language Processing and the 9th International Joint Conference on Natural Language Processing (EMNLP-IJCNLP), pp. 2463–2473. Association for Computational Linguistics, Hong Kong, China, November 2019. https://doi.org/10.18653/v1/D19-1250, https://www.aclweb.org/anthology/D19-1250

21. Sanh, V., Debut, L., Chaumond, J., Wolf, T.: Distilbert, a distilled version of bert: smaller, faster, cheaper and lighter. arXiv preprint arXiv:1910.01108 (2019)

22. Schwartz, R., Dodge, J., Smith, N.A., Etzioni, O.: Green ai. ArXiv abs/1907.10597 (2019)
23. Sennrich, R., Haddow, B., Birch, A.: Improving neural machine translation models with monolingual data. In: Proceedings of the 54th Annual Meeting of the Association for Computational Linguistics (Volume 1: Long Papers), pp. 86–96. Association for Computational Linguistics, Berlin, Germany, Auguest 2016. https://doi.org/10.18653/v1/P16-1009, https://www.aclweb.org/anthology/P16-1009
24. Shleifer, S.: Low resource text classification with ulmfit and backtranslation. arXiv preprint arXiv:1903.09244 (2019)
25. Socher, R., et al.: Recursive deep models for semantic compositionality over a sentiment treebank. In: Proceedings of the 2013 Conference on Empirical Methods in Natural Language Processing, pp. 1631–1642 (2013)
26. Strubell, E., Ganesh, A., McCallum, A.: Energy and policy considerations for deep learning in NLP. In: Proceedings of the 57th Annual Meeting of the Association for Computational Linguistics, pp. 3645–3650. Association for Computational Linguistics, Florence, Italy, July 2019. https://doi.org/10.18653/v1/P19-1355, https://www.aclweb.org/anthology/P19-1355
27. Wang, S., Manning, C.D.: Baselines and bigrams: simple, good sentiment and topic classification. In: Proceedings of the 50th Annual Meeting of the Association for Computational Linguistics: Short Papers, vol. 2, pp. 90–94. Association for Computational Linguistics (2012)
28. Wei, J., Zou, K.: EDA: easy data augmentation techniques for boosting performance on text classification tasks. In: Proceedings of the 2019 Conference on Empirical Methods in Natural Language Processing (EMNLP-IJCNLP), pp. 6382–6388. Association for Computational Linguistics, Hong Kong, China, November 2019. https://doi.org/10.18653/v1/D19-1670, https://www.aclweb.org/anthology/D19-1670
29. Wolf, T., et al.: Huggingface's transformers: state-of-the-art natural language processing. ArXiv abs/1910.03771 (2019)
30. Wong, S.C., Gatt, A., Stamatescu, V., McDonnell, M.D.: Understanding data augmentation for classification: when to warp. In: 2016 International Conference on Digital Image Computing: Techniques and Applications (DICTA), pp. 1–6. IEEE (2016)
31. Yogatama, D., Dyer, C., Ling, W., Blunsom, P.: Generative and discriminative text classification with recurrent neural networks. arXiv preprint arXiv:1703.01898 (2017)
32. Yu, A.W., Dohan, D., Luong, T., Zhao, R., Chen, K., Le, Q.: Qanet: combining local convolution with global self-attention for reading comprehension (2018). https://openreview.net/pdf?id=B14TlG-RW
33. Zhang, X., Zhao, J., LeCun, Y.: Character-level convolutional networks for text classification. In: Proceedings of the 28th International Conference on Neural Information Processing Systems, Vol. 1, pp. 649–657. NIPS'15, MIT Press, Cambridge, MA, USA (2015)

Predicting Multiple ICD-10 Codes from Brazilian-Portuguese Clinical Notes

Arthur D. Reys[1,2]([✉]), Danilo Silva[1], Daniel Severo[2], Saulo Pedro[2],
Marcia M. de Sousa e Sá[3], and Guilherme A. C. Salgado[2]

[1] Federal University of Santa Catarina, Florianópolis, Brazil
danilo.silva@ufsc.br
[2] 3778 Healthcare, Belo Horizonte, Brazil
{arthur.reys,severo,saulo.pedro,guilherme}@3778.care
[3] Syrian-Lebanese Hospital, São Paulo, Brazil
marcia.sa@hsl.org.br
https://3778.care/

Abstract. ICD coding from electronic clinical records is a manual, time-consuming and expensive process. Code assignment is, however, an important task for billing purposes and database organization. While many works have studied the problem of automated ICD coding from free text using machine learning techniques, most use records in the English language, especially from the MIMIC-III public dataset. This work presents results for a dataset with Brazilian Portuguese clinical notes. We develop and optimize a Logistic Regression model, a Convolutional Neural Network (CNN), a Gated Recurrent Unit Neural Network and a CNN with Attention (CNN-Att) for prediction of diagnosis ICD codes. We also report our results for the MIMIC-III dataset, which outperform previous work among models of the same families, as well as the state of the art. Compared to MIMIC-III, the Brazilian Portuguese dataset contains far fewer words per document, when only discharge summaries are used. We experiment concatenating additional documents available in this dataset, achieving a great boost in performance. The CNN-Att model achieves the best results on both datasets, with micro-averaged F1 score of 0.537 on MIMIC-III and 0.485 on our dataset with additional documents.

Keywords: ICD coding · Clinical notes · Natural language processing · Multi-label classification · Neural networks

1 Introduction

Throughout the stay of a patient in a hospital, a series of documents are written about their situation, including symptoms, clinical evolution, diagnoses and medical history. After the release of a patient, medical coders analyze their documentation and assign to that stay a list of codes based on the International Classification of Diseases (ICD), a standard system maintained by the World

R. Cerri and R. C. Prati (Eds.): BRACIS 2020, LNAI 12319, pp. 566–580, 2020.
https://doi.org/10.1007/978-3-030-61377-8_39

Health Organization [24,25]. Those codes identify a variety of clinical information, which is useful for billing purposes, health plan communication and organizing databases for research and statistical analysis [12].

Currently the ICD coding process is manually performed by specifically trained coders. The granularity of the coding system makes differences between similar codes very subtle. Moreover, much of the information in clinical records comes in unstructured free text and the language used is specific to the medical field, containing abbreviations, ambiguous terms and typos. Together, those factors make manual coding an expensive, time consuming and error-prone task.

The development of machine learning models over free text from Electronic Health Records (EHR) for automated ICD coding has been discussed for over two decades [15]. Recently, models based on natural language processing techniques using advanced neural networks have shown relevant performance improvements [18,22]. However, most of these works involve English-language data. To the best of our knowledge, only [6,8,9,23,32] have considered a Portuguese-language dataset. Except for [8], which focuses on a different task of coding the causes of death from death certificates, and [23], which aims at predicting groups of oncology ICD codes from pathology reports, all others use small datasets to predict a limited set of ICD codes. Also, none provide comparisons with accessible datasets.

In this work, we consider the problem of automatically assigning multiple diagnostic ICD codes to a patient stay based on Brazilian Portuguese free-text clinical notes, considering all available codes. Specifically, we develop and compare Logistic Regression (LR), Convolutional Neural Network (CNN), Recurrent Neural Network (RNN) and CNN-based attention models with optimized hyperparameters. We present a case study based on data from Syrian-Lebanese Hospital, a Brazilian hospital in São Paulo, where we intend to deploy our best performing model in order to support the ICD tagging process. Additionally, we provide results for the publicly available English-language dataset MIMIC-III (Medical Information Mart for Intensive Care) [13,14], where we outperform previous work among models of the same families and the current state of the art.[1]

2 Background

2.1 Previous Work

In the ICD coding task, researchers often have to decide which codes will be the target of the study. While some works consider all types of ICD codes [36], others use a limited amount of ICD codes [32] or limit the scope to Diagnoses ICD codes [19]. This is done mainly because of differences in datasets and the large class imbalance observed in the majority of them. As free text inputs for this specific task, most works use discharge summaries, as they condense

---

[1] Code for MIMIC-III is available at https://github.com/3778/icd-prediction-mimic.

information about a patient stay in a single document [22]. However, [8] and [36] have experimented using additional documents.

The structure of the ICD system is used to develop a hierarchical approach to assist predictions in [2] and [28]. A method based on ICD co-occurrence is proposed in [33]. In [5], overlaps between ICD descriptions and words in documents compose a rule-based method. More prominently, works use machine learning models such as SVM (Support Vector Machine) [28], Naive Bayes [20,26] and kNN (k-Nearest Neighbors) [30].

Convolutional Neural Networks (CNN) have been widely used in the literature, achieving good results in the ICD coding task [17,19,22]. The advantage of this architecture over more traditional machine learning models (such as LR and SVM) is its capability of capturing local contextual features [19]. Recurrent Neural Networks have also been used due to their ability to associate information in longer contexts than CNNs [1,2,10]. In particular, LSTM (Long Shortterm Memory) and GRU (Gated Recurrent Unit) recurrent networks capture information within a large contextual window. These approaches have achieved improvements over older machine learning models, as free text usually have high complexity and their comprehension rely on local and global semantic relations between terms and sequences.

In addition to neural networks, innovative models include ensemble of different architectures [35,36] and *per-label* attention mechanisms [18,22]. Per-label attention consists of weighing a base representation of documents differently for each ICD code. In the specific task of this work, including only Diagnoses ICD codes, [22] appears to hold the current state of the art.

Due to the limited availability of public EHRs, most works focus on MIMIC [14], a freely accessible dataset in English language. Works aimed at EHRs in Portuguese are rare and use different private data sources [6,8,9,23,32]. Among these, [6] shows how an hierarchical approach can improve results in the ICD coding task. An approach based solely on structured data is presented in [9]. In [32], a CNN with self-taught GloVe embeddings is presented to predict a small set of possible codes from free text, while a cost-sensitive learning approach is implemented to overcome class imbalance. These works use relatively small datasets and focus on few codes. In turn, [23] and [8] use large collections of data. In [23], SVM is used to predict groups of topographical and morphological oncology ICD codes from pathology reports, in a one-*versus*-all approach. Finally, [8] uses a recurrent neural network with attention to predict ICD codes corresponding to death causes from death certificates and related documents. However, oncology ICD groups and death causes ICD codes still comprise smaller sets than diagnostic codes, while pathology reports and death certificates have significant structural, semantic and lexical differences from clinical notes such as discharge summaries.

2.2 Feature Extraction

Training a computational model over free text requires some kind of feature extraction method. Among different methods, some encode whole documents into vectors, without regard to the order of the words. This is called a Bag-of-Words (BoW) representation, with the most popular approach being TF-IDF (Term Frequency – Inverse Document Frequency) [31]. Others generate latent vector representations of words, as Word2Vec [21], GloVE [27] and Fast-Text [3], allowing documents to be represented as a sequence of word vectors. More enhanced methods at word level include ELMo [29] and BERT [7]. In these methods, the same word can be mapped to different vectors, depending on their surrounding context. Other methods include character-level and paragraph-level representations [16, 37].

In this work we use TF-IDF features for Logistic Regression and Word2Vec for the neural networks. Hence, a detailed description of these methods is given.

Term Frequency – Inverse Document Frequency. TF-IDF [31] aims to reflect the importance of a word in a document, given a *corpus*. Based on a BoW model, a document is converted into a multi-hot encoding of words contained in it, based on vocabulary constructed from the *corpus*. Then, words are given weights based on their importance for each document. The importance of a word increases proportionally to the number of times it appears in that document (term frequency) and inversely proportional to the total of documents that contain it (inverse document frequency).

Word2Vec. Word2Vec [21] is a representation model that takes into account order and context of words in documents. The inputs are tokenized texts and the model builds a vocabulary associating words to correspondent fixed-dimension vectors. Tokens in the *corpus* are projected into a multi-dimensional space, allowing identification of interdependent relations between different terms, through cosine similarity.

Word2Vec is composed of a single hidden layer neural network. Two methods can be applied in the embedding training: CBoW (Continuous Bag-of-Words) and Skip-gram. In CBoW, a word is predicted from a limited amount of words that precede and succeed it. The context words are converted into BoW features, losing local ordering information between them. In Skip-gram, the task is to predict, from a given word, a limited amount of words around it. In this case, the order of the context words influences the network projection, as nearby words receive higher weights. The latent word representations resulting from Word2Vec training can be loaded as an embedding layer in neural network based models. An embedding layer is a mapping of discrete input variables (e.g. tokens representing words) to corresponding vector representations.

3 Datasets and Preprocessing

3.1 MIMIC-III Dataset

The MIMIC-III dataset—the third revision of MIMIC, v1.4—is a publicly accessible English-language dataset that includes numerous tables relative to patients in Beth Israel Deaconess Medical Center, in the United States [14]. Each admission of a patient to the hospital is associated to several documents, as well as to an ordered list of ICD codes, using the Diagnoses ICD-9-CM (where CM stands for Clinical Modification) coding system, at the most specific level (i.e. subcategories).

Table 1. Statistics of document types in MIMIC-III and HSL datasets.

Dataset	Unique patients	Admissions	Total documents	Avg. words per sample[a]
MIMIC-III[b]	41127	52722	52722	1327.5
HSL-S	51298	77005	77005	94.6
HSL-E	50899	76159	919713	1483.0
HSL-A	42153	59249	63423	155.4
HSL-SEA	51298	77005	1060141	1730.4

[a]Concatenation of all documents corresponding to the same admission.
[b]Only discharge summaries.

As the majority of related works, only free text discharge summaries were selected, totaling 52722 hospital admissions from 41127 unique patients. We found a total of 6918 unique ICD codes associated with these documents.

We perform light preprocessing on the input texts, removing date/hour patterns, special characters and applying lowercase. The same data split as in [22] was used, consisting of 47719 samples in the training set, 1631 in the validation set, and 3372 in the test set. In this split, no patient is listed in more than one subset.

3.2   HSL Dataset

The HSL dataset contains de-identified documents linked to patients from the Syrian-Lebanese Hospital (HSL). Collected between 2016 and 2018, texts are written in Brazilian Portuguese. The dataset includes different types of documents in free text. Each document has a hospital admission ID from which different documents can be linked. We removed all admissions that did not have a linked discharge summary, totaling 77005 admissions from 51298 unique patients. Each admission has a list of ICD codes tagged by professional medical coders using Diagnoses ICD-10 codes at the most specific level (i.e. subcategories). We found 5360 unique codes in the dataset.

Initially, we selected only discharge summaries (S), with each admission containing a single document. This set is referenced as HSL-S. However, after further analysis, we decided to include additional free text documents which were numerously available, in particular: clinical developments (E) and anamnesis/physical exams (A). Unlike discharge summaries, a wide range of types E and A documents are attached to each admission, from none to several texts. Table 1 shows the total of documents per type and the unique admissions and patients linked to

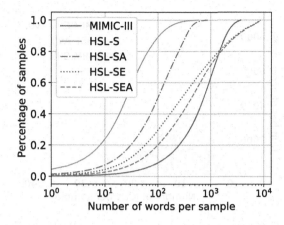

Fig. 1. Word count per sample cumulative distribution on all datasets.

these documents. These additional documents were concatenated to type S documents with the same admission ID, to form the input text for each sample. This better reproduces the human coding process that takes place at Syrian-Lebanese Hospital, where coders observe all documents of an admission to determine the correspondent ICD codes. We refer to the dataset with concatenated types S, E and A documents as HSL-SEA.

Text preprocessing is done in the same way as in MIMIC-III. We split data ensuring no patient was present in more than one subset, totaling 69309 samples in the training set; 2313 in the validation set; and 5383 in the test set.

3.3   Comparison Between Datasets

Besides language, the presented datasets have some relevant differences. Figure 1 shows the cumulative distribution of word count per sample in all datasets, after text preprocessing. Table 1 presents differences in the number of documents selected for each dataset. As also shown in Table 1, MIMIC-III discharge summaries have a much larger average of words per sample than HSL-S. The concatenation of S, E and A documents to HSL-SEA result in an average closer to MIMIC-III. From these statistics, we can assert that MIMIC-III discharge summaries contain, objectively, far more data than HSL-S, while having a closer average and distribution to HSL-SEA. Also, by looking at random samples, we noticed more detailed and well written texts in MIMIC-III.

The ICD coding systems adopted by the datasets are also different. While MIMIC-III uses a Clinical Modification of ICD-9, HSL uses the newer ICD-10.

Figure 2 shows histograms of the number of ICD codes per sample for both datasets. While maximum, minimum and standard deviation are similar, the average number of ICD codes per sample is lower in HSL. Note that the classes are extremely imbalanced in both datasets, as some examples show in Table 2.

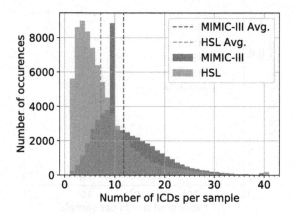

Fig. 2. ICD count per sample histograms on MIMIC-III and HSL.

We also note that 4.47% and 4.48% of the ICD codes contained in the test sets are not present in the training sets, respectively, in MIMIC-III and HSL.

Table 2. Percentage of samples tagged with the 1st, 10th, 100th and 1000th most frequent ICD codes on MIMIC-III and HSL.

Dataset	MIMIC-III	HSL
1st	38.02%	34.37%
10th	11.67%	10.71%
100th	2.23%	1.26%
1000th	0.15%	0.06%

4   Methods

In this section we present the evaluation metrics and models used in this work.

4.1   Evaluation Metrics

We used popular metrics for multi-label tasks, namely, F1, precision and recall, all micro-averaged over different classes [19]. Micro-averaging presents a more representative result considering the large and imbalanced sets of classes, and is indeed used in most works that do not limit the number of ICD codes.

Micro-averaged precision and recall are defined, respectively, as

$$P_{\text{micro}} = \frac{\sum_{c=1}^{C} \sum_{n=1}^{N} y_n^c \hat{y}_n^c}{\sum_{c=1}^{C} \sum_{n=1}^{N} \hat{y}_n^c}, \qquad R_{\text{micro}} = \frac{\sum_{c=1}^{C} \sum_{n=1}^{N} y_n^c \hat{y}_n^c}{\sum_{c=1}^{C} \sum_{n=1}^{N} y_n^c}, \qquad (1)$$

where C is the number of classes, N is the number of samples and y_n and $\hat{y}_n$ are, respectively, true and predicted vectors with C binary entries, each indicative of a class c in a sample n. The F1 score (higher is better) is defined as the harmonic mean between precision and recall.

Each model (with the exception of the Constant model) outputs, per sample, a vector with C real-valued entries between 0 and 1 corresponding to the confidence of prediction for each class. In particular, if the model was trained on only $C' < C$ classes, we assign $\hat{y}_n^c = 0$, $\forall n$, for the remaining classes c not seen by the model. In order to compute the above metrics, we analyze a range of thresholds to binarize outputs, selecting the best one for each model based on F1 in the validation set.

4.2 Models

The models developed in this work are described below. We used the Keras[2] framework with Tensorflow[3] backend in all implementations. Models were trained to a maximum of 10 epochs. Instead of applying Early Stopping, after each epoch we computed F1 in the validation set. When training was over, we restored weights corresponding to the epoch with the best result. For our study we used an AWS EC2 virtual machine with 8 vCPUs and a NVIDIA T4 GPU.

Constant (Top-k). The objective of this baseline model is to determine whether the performance of real models is greater than that of an implementation which does not use the input texts.

The Constant model predicts a constant list of k ICD codes for all samples. The ICD codes selected are the k most occurring in the training set. The parameter k was optimized to obtain the best F1 in the validation sets, resulting in $k = 15$ for MIMIC-III and $k = 8$ for HSL.

Logistic Regression. In the LR model, we convert the multi-label problem into a set of binary classification problems, one for each class. The inputs of the LR model are TF-IDF features computed over each dataset.

TF-IDF was implemented using Scikit-learn[4]. Stopwords were removed from the preprocessed texts, using default Portuguese and English stopwords from Natural Language Toolkit[5]. Maximum vocabulary size was fixed to the 20000 most frequent words.

The hyperparameters of the LR were optimized via Grid Search considering different optimizers, learning rates from 0.0001 to 0.1 in multiples of 10 and L2 regularizer parameters from 0 (no regularization) to 10, also in multiples of 10. The final model uses Adam optimizer with learning rate 0.1 and all other

---

[2] http://keras.io/.

[3] https://tensorflow.org/.

[4] https://scikit-learn.org/.

[5] https://www.nltk.org/.

optimizer parameters set to default values. No regularization is performed. Each training epoch took 50 s for MIMIC-III and 60 s for HSL-S and HSL-SEA.

Convolutional Neural Network. The CNN implemented in this work consists of an embedding layer loaded with Word2Vec word vectors, followed by a single one-dimensional convolutional layer and Batch Normalization [11]. On the output, a Global Average Pooling operation precedes a fully connected layer with as many units as the number of classes for each dataset. It is based on the implementation of [22], but with some modifications: our tests showed that removing Dropout, adding Batch Normalization and increasing kernel size from 4 to 10 improved results, as well as performing Global Average Pooling instead of Global Max Pooling. The layers and respective parameters are shown in Table 3. We used Adam optimizer with learning rate 0.001 for MIMIC-III and 0.003 for HSL-SEA.

Table 3. Architectures and parameters for the neural network models.

CNN	GRU	CNN-Att
Input	Input	Input
Embedding (size 300)	Embedding (size 300)	Embedding (size 300)
Conv1D (500 filters, kernel 10, tanh)	GRU (500 units, tanh)	Conv1D (500 filters, kernel 10, tanh)
Batch Normalization	Batch Normalization	Batch Normalization
GlobalAveragePooling1D	GlobalAveragePooling1D	Attention
Output (sigmoid)	Output (sigmoid)	Output (sigmoid)

Given that the CNN model involves Batch Normalization, it is mandatory for the inputs to have fixed sizes [11]. However, samples have a large variation in number of words, as shown in Fig. 1. To ensure a fixed-length input, texts with fewer words than needed were padded with padding tokens by the end, while texts with more than the maximum of words had their end truncated. The padding token points to a null vector in the embedding layer. Observing the distribution of text sizes among the datasets, the fixed-length of the inputs was set to: 2000, for the MIMIC-III dataset; 300, for HSL-S; and 4000, for HSL-SEA.

Word2Vec vectors were trained using Gensim[6]. The embeddings were self-trained due to the specificity of the Brazilian Portuguese clinical language, containing medical terms, abbreviations and acronyms [32]. Words appearing in less than 10 samples were not considered. We experimented vector lengths between 100 and 600, CBoW and Skip-gram implementations, and whether stopwords should be removed. These parameters were optimized for the HSL-S dataset,

---

[6] https://radimrehurek.com/gensim/.

resulting in vectors with length 300, Skip-gram training algorithm and stop-words not being removed.

Each epoch took 310 s when training for MIMIC-III and 820 s for HSL-SEA.

Recurrent Neural Network. The RNN model consists of an embedding layer loaded with Word2Vec word vectors, followed by a GRU layer. Then, Batch Normalization and Global Average Pooling are performed. In the output we define a fully connected layer with as many units as the number of classes for each dataset. As in the CNN model, the samples were processed to fit in a fixed-length input, in this case to allow faster training on the GPU. In this work, we used GRU layers for their better results over traditional RNNs, while keeping a simpler architecture (and being more quickly trainable) than LSTMs [4].

Three base architectures were tested: the first one is such as shown in Table 3; the second has an extra GRU layer; the last has a bidirectional GRU instead of a common GRU. The first architecture yielded the best results, so each parameter was then individually optimized from this base architecture.

Among the optimized parameters are: optimizer; learning rates from 4e−4 to 1e−2 in steps of 1e−4; masking of padding tokens, to avoid their influence on model predictions; sample weighting inversely proportional to the number of true ICD codes; fine-tuning of the embedding layer; and Pooling methods. Adam optimizer with 8e−4 learning rate resulted in improvements in F1, so as fine-tuning the embedding layer. Average Pooling proved to be greatly superior than Max Pooling. The final architecture is shown in Table 3. This model is referred simply as GRU in the next sections.

The GRU model uses the same Word2Vec vectors trained for the CNNs. Training times per epoch were 268 s for MIMIC-III and 785 s for HSL-SEA.

Convolutional Neural Network with Attention. The CNN model with Attention (CNN-Att) is based on the current state of the art CAML (Convolutional Attention for Multi-Label Classification) [22], with some modifications. The model is also similar to our conventional CNN model, with the only difference that the Global Pooling is replaced by a *per-label* attention mechanism (which computes a separate context vector for each label as a weighted average of the input sequence) and each fully-connected sigmoid output unit takes as input only its corresponding context vector. The attention operation is a scaled dot-product [34] and uses a separate trainable target vector for each label (see [22] for details).

Compared to the original CAML model, we removed Dropout from the embedding layer, which in our initial experiments did not seem to improve the performance, and added Batch Normalization after the convolutional layer, since it typically allows for a faster convergence of training. We increased the number of filters in the convolutional layer from 50 to 500. These modifications improved metrics in our tests. Also, to allow faster convergence, we scheduled the learning rate to start at 0.001 in the first two epochs, and only then decrease to 0.0001. Table 3 presents the architecture and parameters used in the CNN-Att.

Following our other neural network models, we used Word2Vec word embeddings, and the samples were processed to fit a fixed-length input (see the CNN model subsection). Training the CNN-Att took 1600 s per epoch for MIMIC-III and 3700 s per epoch for HSL-SEA.

5   Results and Discussion

We trained our models using MIMIC-III and HSL datasets. This section shows achieved results and comparisons, as well as experiments regarding additional documents in HSL.

Table 4. Performance of different models on MIMIC-III dataset. Entries with no citation brackets correspond to our models.

Model	Threshold	F1	Precision	Recall
Constant	–	0.192	0.188	0.196
LR [22]	–	0.242	–	–
flat-SVM [19]	–	0.253	0.635	0.158
LR	0.19	0.406	0.425	0.388
CNN [22]	–	0.402	–	–
CNN [19]	–	0.399	0.440	0.366
CNN	0.30	0.423	0.467	0.387
Bi-GRU [22]	–	0.393	–	–
GRU	0.32	0.468	0.543	0.412
CAML [22]	–	0.524	–	–
CNN-Att	0.28	**0.537**	0.590	0.492

5.1   MIMIC-III Results

Table 4 shows the results obtained for all models on the MIMIC-III test set. As baselines for comparison, we also present results from other works in the literature.

The Constant model achieves very poor results, as expected. Our LR with optimized hyperparameters greatly outperforms similar LR [22] and SVM [19] linear models, presented as baselines in these works. This suggests that these models were underfitting due to lack of hyperparameter optimization; indeed, we noticed that the LR from [22] used a default L2 regularization parameter of 1, while we adopted no L2 regularization. The F1 achieved by our LR is comparable to CNN implementations with Word2Vec features found in [19] and [22], while our CNN shows an improvement over these models. The GRU returns significant improvements over all previous models, as well as over a similar model presented in [22]. Finally, the CNN-Att outperforms all other models, including the original CAML [22].

5.2 HSL Results

For the HSL dataset, we first selected only discharge summaries (HSL-S), to allow a more direct comparison with MIMIC-III, which uses only this type of document. As HSL-S and MIMIC-III are very different datasets, we did not expect identical results. Even so, when training the LR model, the results we obtained were much lower than expected, namely, an F1 of 0.316, which is 20% below that of MIMIC-III.

These results, as well as the fact that HSL-S has a considerably lower average of words per sample than MIMIC-III, lead our study to experiment with other documents available in HSL. We trained the LR model on different combinations of concatenated documents: types S and A; types S and E; and types S, A and E (refer to Sect. 3.2 for an explanation of each document type). Table 5 presents metrics computed over the validation set. Clearly, adding documents to discharge summaries—thus increasing average words per sample—shows improvements in metrics, with a large increase in F1 when using HSL-SEA.

Table 5. Validation metrics of LR model trained over HSL considering different concatenated document types.

Documents	Threshold	F1	Precision	Recall
S	0.26	0.316	0.320	0.312
S and A	0.25	0.347	0.359	0.336
S and E	0.27	0.357	0.382	0.336
S, E and A	0.25	**0.367**	0.390	0.346

Considering the outcomes of these experiments, we then trained all models on HSL-SEA. As CNN and RNN are sensitive to the order of concatenation of documents, we experimented orders S-A-E and S-E-A. We adopted the latter one, as it achieved slightly better results. Compared to HSL-S, we achieved consistently better results when using HSL-SEA, for all models. Metrics on the HSL-SEA test set are shown in Table 6. Once more, the CNN is slightly superior than the LR, while the GRU model shows improvements over both of those models. The CNN-Att model presents again the best results, significantly ahead of all other models.

Note that each model on HSL-SEA achieves a performance comparable to (up to about 10% below) that same model on MIMIC-III. This is evidence that HSL-SEA has comparable quality to MIMIC-III discharge summaries for ICD code prediction.

Table 6. Performance of different models for HSL-SEA dataset.

Model	Threshold	F1	Precision	Recall
Constant	–	0.203	0.183	0.228
LR	0.25	0.368	0.400	0.340
CNN	0.26	0.374	0.386	0.363
GRU	0.29	0.441	0.508	0.390
CNN-Att	0.29	**0.485**	0.543	0.438

6   Conclusion

This work presented a study on automated ICD coding from free text, using four learning models trained on two datasets. For MIMIC-III, we reproduced and improved results of similar models in the literature, outperforming the state of the art on the prediction of diagnosis codes from discharge summaries. Results show that using a CNN with per-label attention outperforms conventional CNN, GRU and LR models, attaining a Micro-F1 of 0.537.

For the HSL dataset, we observed that using only discharge summaries was insufficient to achieve results similar to MIMIC-III. Besides the different coding system, word count statistics and detail levels in documents may explain the loss in performance. After concatenating additional documents found in HSL, we observed a significant improvement. Again, the best performance was achieved by our optimized CNN-Att model, with a Micro-F1 of 0.485.

We believe our best model trained on HSL could be suited to assist medical coders using clinical records in Brazilian Portuguese, allowing for gains in efficiency and a decrease in errors in the manual ICD tagging process. We are working towards the deployment of a pilot trial to test the usefulness of the model and better understand its limitations in a practical setting.

Acknowledgments. The authors would like to thank Ricardo Giglio, Dr. Mauro Cardoso, Dr. Flávio Amaro, and Marcio Gregory for helpful discussions, as well as 3778 Healthcare and Syrian-Lebanese Hospital for their support of this research.

References

1. Ayyar, S.: Bear don't walk IV, O.: Tagging patient notes with ICD-9 Codes. In: Proceedings of the 29th NIPS (2016)
2. Baumel, T., et al.: Multi-label classification of patient notes a case study on ICD code assignment. In: AAAI Workshops (2017)
3. Bojanowski, P., et al.: Enriching word vectors with subword information. TACS **5**, 135–146 (2016)
4. Chung, J., et al.: Empirical evaluation of gated recurrent neural networks on sequence modeling. In: Proceedings of the NIPS 2014 Workshop on Deep Learning (2014)

5. Crammer, K., et al.: Automatic code assignment to medical text. In: Proceedings of the Workshop on BioNLP 2007, p. 129 (2007). https://doi.org/10.3115/1572392. 1572416

6. de Lima, L.R.S., Laender, A.H.F., Ribeiro-Neto, B.A.: A hierarchical approach to the automatic categorization of medical documents. In: Proceedings of the 7th CIKM, pp. 132–139 (1998). https://doi.org/10.1145/288627.288649

7. Devlin, J., et al.: BERT: pre-training of deep bidirectional transformers for language understanding. In: Proceedings of the NAACL-HLT 2019 (2019)

8. Duarte, F., et al.: Deep neural models for ICD-10 coding of death certificates and autopsy reports in free-text. J. Biomed. Inform. 80, 64–77 (2018). https://doi.org/10.1016/j.jbi.2018.02.011

9. Ferrão, J., et al.: Using structured EHR data and SVM to support ICD-9-CM coding. In: Proceedings of the 2013 IEEE ICHI, pp. 511–516 (2013). https://doi.org/10.1109/ICHI.2013.79

10. Huang, J., Osorio, C., Sy, L.W.: An empirical evaluation of deep learning for ICD-9 code assignment using MIMIC-III clinical notes. Comput. Methods Prog. Biomed. 177, 141–153 (2019). https://doi.org/10.1016/j.cmpb.2019.05.024

11. Ioffe, S., Szegedy, C.: Batch normalization: accelerating deep network training by reducing internal covariate shift. In: Proceedings of the 32nd ICML, vol. 37 (2015)

12. Jensen, P.B., Jensen, L.J., Brunak, S.R.: Mining electronic health records: towards better research applications and clinical care. Nat. Rev. Genet. 13, 395–405 (2012). https://doi.org/10.1038/nrg3208

13. Johnson, A., Pollard, T., Mark, R.: The MIMIC III clinical database (2016). https://doi.org/10.13026/C2XW26

14. Johnson, A.E.W., et al.: MIMIC-III, a freely accessible critical care database. Sci. Data 3, 160035 (2016). https://doi.org/10.1038/sdata.2016.35

15. Larkey, L.S., Croft, W.B.: Automatic assignment of ICD9 codes to discharge summaries. Tech. rep. University of Massachusetts, Amherst, MA (1995)

16. Le, Q.V., Mikolov, T.: Distributed representations of sentences and documents. In: Proceedings of the 31st ICML (2014)

17. Li, C., et al.: Convolutional neural networks for medical diagnosis from admission notes. arXiv:1712.02768 [cs] (2017)

18. Li, F., Yu, H.: ICD coding from clinical text using multi-filter residual convolutional neural network. In: Proceedings of he 34th AAAI Conference on Artificial Intelligence (2020)

19. Li, M., et al.: Automated ICD-9 coding via a deep learning approach. IEEE/ACM Trans. Comput. Biol. Bioinform. 16, 1193–1202 (2019). https://doi.org/10.1109/TCBB.2018.2817488

20. Medori, J., Fairon, C.: Machine learning and features selection for semi-automatic ICD-9-CM encoding. In: Proceedings of the NAACL HLT 2010 Second Louhi Workshop on Text and Data Mining of Health Documents, Los Angeles, California, USA, pp. 84–89. Association for Computational Linguistics, June 2010

21. Mikolov, T., et al.: Efficient estimation of word representations in vector space. In: Proceedings of the ICLR Workshop (2013)

22. Mullenbach, J., et al.: Explainable prediction of medical codes from clinical text. In: Proceedings of the 2018 NAACL-HLT, vol. 1, pp. 1101–1111 (2018). https://doi.org/10.18653/v1/N18-1100

23. Oleynik, M., Patrão, D.F.C., Finger, M.: Automated classification of semi-structured pathology reports into ICD-O using SVM in Portuguese. Stud. Health Technol. Inform. 235, 256–260 (2017)

24. WHO Organization: International Classification of Diseases: [9th] Ninth Revision, Basic Tabulation List with Alphabetic Index. World Health Organization (1978)
25. WHO Organization: ICD-10: international statistical classification of diseases and related health problems: tenth revision. World Health Organization (2004)
26. Pakhomov, S.V.S., Buntrock, J.D., Chute, C.G.: Automating the assignment of diagnosis codes to patient encounters using example-based and machine learning techniques. JAMIA **13**, 516–525 (2006). https://doi.org/10.1197/jamia.M2077
27. Pennington, J., Socher, R., Manning, C.: Glove: global vectors for word representation. In: Proceedings of the 2014 EMNLP, pp. 1532–1543. Association for Computational Linguistics (2014). https://doi.org/10.3115/v1/D14-1162
28. Perotte, A., et al.: Diagnosis code assignment: models and evaluation metrics. JAMIA **21**, 231–237 (2014). https://doi.org/10.1136/amiajnl-2013-002159
29. Peters, M.E., et al.: Deep Contextualized Word Representations. In: Proceedings of the 2018 NAACL-HLT. vol. 1. Association for Computational Linguistics (2018). https://doi.org/10.18653/v1/N18-1202
30. Ruch, P., et al.: From episodes of care to diagnosis codes: automatic text categorization for medico-economic encoding. In: Proceedings of the AMIA Annual Symposium, pp. 636–640 (2008)
31. Salton, G., Buckley, C.: Term-weighting approaches in automatic text retrieval. Inf. Process. Manage. **24**, 513–523 (1988). https://doi.org/10.1016/0306-4573(88)90021-0
32. dos Santos, A.B.V., Gumiel, Y.B., Carvalho, D.R.: Using deep convolutional neural networks with self-taught word embeddings to perform clinical coding. Iberoamerican J. Appl. Comput. **8**, 10–27 (2018)
33. Subotin, M., Davis, A.R.: A method for modeling co-occurrence propensity of clinical codes with application to ICD-10-PCS auto-coding. JAMIA **23**, 866–871 (2016). https://doi.org/10.1093/jamia/ocv201
34. Vaswani, A., et al.: Attention is all you need. In: Proceedings of the 31st NIPS, Long Beach, California, USA, pp. 6000–6010. Curran Associates Inc. (2017)
35. Xie, P., Xing, E.: A neural architecture for automated ICD coding. In: Proceedings of the 56th ACL, vol. 1, pp. 1066–1076. Association for Computational Linguistics (2018). https://doi.org/10.18653/v1/P18-1098
36. Xu, K., et al.: Multimodal machine learning for automated ICD coding. In: Proceedings of the 4th Machine Learning for Healthcare Conference (2019)
37. Zhang, X., Zhao, J., LeCun, Y.: Character-level convolutional networks for text classification. In: Proceedings of the 28th NIPS, vol. 1, pp. 649–657 (2015)

Robust Ranking of Brazilian Supreme Court Decisions

Jackson José de Souza$^{(\boxtimes)}$ and Marcelo Finger

Department of Computer Science, Institute of Mathematics
and Statistics, part of University of Sao Paulo, Sao Paulo, SP, Brazil
{jackson,mfinger}@ime.usp.br

Abstract. This work studies quantitative measures for ranking judicial decisions by the Brazilian Supreme Court (STF). The measures are based on a network built over decisions whose cases were finalized in the Brazilian Supreme Court between 01/2001 and 12/2019, obtained by crawling publicly available STF records. Three ranking measures are proposed; two are adaptations of the PageRank algorithm, and one adapts Kleinberg's Algorithm. All are compared with respect to agreement on top 100 rankings; we also analyze each measure robustness based on self-agreement under perturbation.

We conclude that all algorithms show the network of citations is highly robust under perturbation. Both versions of PageRank, even if producing different rankings, achieved robustness results which are indistinguishable via statistical tests; Kleinberg's algorithm achieves more promising results to rank leading cases at the top, but it does need more research to achieve this goal.

Keywords: Legal scores · Importance scores · Authority scores · PageRank algorithm · STF · Brazilian Supreme Court · Robustness · Complex networks · Information extraction

1 Introduction

This work investigates the development of NLP-based, network analysis tools aiming at understanding the impacts of Brazilian Supreme Court decisions. At the moment there is no widely accepted and used tool, but recent developments in computer science and artificial intelligence allow us to put forward a few proposals. In this spirit, this work is a step in the discussion of the desired properties of quantitative feasible measures as the robustness of Brazilian Supreme Court decisions network and the authority of decisions in it as a measure of relevance to improve the knowledge of Brazilian Supreme Court functioning and the impact of its decisions in the judicial system.

The Brazilian Supreme Court (*Supremo Tribunal Federal*, STF) is of utmost importance to Brazilian society. It decides about matters that maintain the peaceful coexistence among different parts of society such as right to same-sex

© Springer Nature Switzerland AG 2020
R. Cerri and R. C. Prati (Eds.): BRACIS 2020, LNAI 12319, pp. 581–594, 2020.
https://doi.org/10.1007/978-3-030-61377-8_40

marriage and unconstitutionality of Brazilian media law. In times of political turmoil STF plays a major role imposing limits on questionable discretionary acts of legislative and executive powers. Legal scholars are interested to analyze the influence these and other decisions judged by STF imposes on future cases that deal with similar matters. Most of these decisions usually are leading cases, i.e., "a judicial decision that first definitively settled an important legal rule or principle. Such important case is used as guidance by lawyers and judges who face similar issues later" [6].

STF has several attributions that opens it to receive an overwhelming number of cases every year, which makes it impossible to do such analysis only by human effort. STF is the constitutional court; it is the original jurisdiction of cases such as those involving congressmen [19]; and it is also the court of last resort in Brazilian law. This last attribution has a negative impact on the performance of Brazilian Supreme Court [8], generating a high demand for lawsuits and appeals.

In the last years STF has judged more than 80,000 cases annually. By contrast, such a demand provided an impulse for the public availability of electronic information on the court decisions, which is the starting point for this work.

Recently, STF has made its judicial acts freely available on the web, including the court's decisions. We crawled that material to extract and process decision data in text, written in Portuguese, to build a network of citations only with decisions judged by STF. Based upon such complex network and the concept of authority scores, we propose several *decision ranking measurements* built on PageRank [15] and Kleinberg's [11] ranking algorithms. Our goal is to investigate the level of *agreement between measurements*, their *robustness* in respect to the network, and whether most of best ranked decisions in the measurements are *leading cases*. The latter goal is a desirable ranking for legal scholars. For robustness, it is expected that the rankings be preserved under addition/removal of some small random number of nodes, otherwise the ranking is too unstable to be useful.

The results of experiments ran in this work are supposed to reflect STF everyday activities; the proposed measures analyze all decision data judged between 01/2001 and 12/2019 and decisions cited by them. Thus, as a first result of this work the best ranked decisions are those which deal with matters that reach STF most often and that come from the biggest litigants as government ones. Leading cases and those which attract most attention of legal scholars and the press, such as controversial matters or fundamental constitutional matters because they usually are not much cited by cases that reach STF most often.

The proposed measurements are an exploratory attempt to provide quantitative support for claims about Brazilian Supreme Court decisions, and should be considered among the first steps toward more complex analyses of decision structure. The results achieved are promising, as they put in evidence a few important characteristics of STF, having a potential to produce social impact. The present work provides a platform on which more complex natural language processing may be performed, such as legal argument extraction and evaluation of the impact of specific laws in the society.

This paper is structured as follows. We start by analyzing related work on metrics and algorithms for quantitative and qualitative ranking measurements of legal decisions in Sect. 2. Then, we describe how STF decision data was extracted and preprocessed in Sect. 3. We proceed by describing the modeling of a network of STF decisions in Sect. 4, based on which we adapt node ranking algorithms to create measurements of decision ranking. The agreement of those rankings is statistically analyzed, and then we describe a set of robustness tests on those rankings. Results obtained are discussed in Sect. 5 and we conclude on the compliance of the proposed measurements to the desired properties of agreement and robustness.

2 Related Work

Algorithms for discovering authoritative nodes in complex networks, which receive a large number of references, and hub nodes, which refer to several nodes, were proposed by Kleinberg [11]. A few networks built on the U.S. and Europe Supreme Court decisions have been studied in the last decades [1,9,14,20]. Both works [1,9] have found that Kleinberg's algorithm leads to scores that usually meet the evaluation of legal experts on relevant decisions, in which relevant *hub decisions* are those which cite many relevant authoritative decisions; similarly, relevant *authoritative decisions* are those cited by many relevant hubs. The analysis on the *in via incidentale* rulings of the Italian Constitutional Court (ICC) [1] identified decisions that lost relevance due to the definitive resolution of a matter settled in the ruling; its decision network topology was scale-free according to a power law.

Another approach was proposed by [14] with good results using closeness metrics, which calculates the distance between decisions, such as *proximity prestige* and *generalized core*, but in their judgement the best results were achieved with "Marc in-degree", a metric that only takes into account incoming citations. On the same line, sink distance metrics combined with the single-linkage hierarchical clustering algorithm produced more accurate and more interpretable clusterings in the work done by [3]. In contrast, [14] and [20] did not achieve good or meaningful importance scores with PageRank algorithm [15], i.e. the relative importance of the referring case does not seem to predict the relevance of the cases it refers to. The results obtained by [14] using in-degree HITS showed low correlation between their results and the number of publications in specialized magazines and the number of citations in literature, too.

There are some studies about the Brazilian Supreme Court, but none of them address the topic of authority scores nor the Supreme Court decisions citations' network. The FGV Rio Law School[1] has been publishing quantitative analysis in the form of articles and reports since 2010 about the Brazilian Supreme Court. Usually, the reports address questions aiming at fostering debate about the court's activity through queries that can be answered with basic statistics such as "what are the heaviest users of the court?", "which justices take more

---

[1] https://portal.fgv.br/en.

time to judge preliminary injunctions[2]?", "which types of cases take more time to have a final decision and which ones are judged fastest?" [7]. The articles published by FGV Rio Law School about the Brazilian Supreme Court address more specific issues like "Is it short the time taken to request to view a case in comparison with U.S. Supreme Court?" [10] and "What affect more courts cohesion, the court's workload or differences between the justices' personalities?" [12].

3   Data Extraction and Preparation

The Brazilian Supreme Court (STF) is composed by eleven magistrates, one which is the President of the court elected for a two-year term. There are two collegiate decision-making bodies: the Plenary, composed by all eleven magistrates; and the two Panels of the court, composed by five magistrates each except the President. The type of decisions pronounced by these decision-making bodies are called "acórdão". Also, decisions may be pronounced by only one magistrate of the court and such type of decision is called "monocratic" (single magistrate). In this work we will analyze data of the former decision type.

The data were extracted from cases' entries found in STF jurisprudence search engine [5]. An entry contains summarized information about one decision organized in sections. The entry header contains the decision code, petition type, name of magistrate-rapporteur[3] and decision date. Other two entry sections of interest to this work are "Note" (*Observação*), and "Decisions in the same direction" (*Acórdãos no mesmo sentido*).

The decision-making process adopted by collegiate decision-making bodies in STF is the seriatim. In this model each judge writes his/her own opinion citing jurisprudence to support it. The decision is the result of opinions followed by the majority of judges in the case and it is written by a judge that followed the majority. An entry's "Note" section contains all jurisprudent decisions cited, even ones that do not support it. This section also contains other data meant to provide more information for legal researchers about the case; decisions are also mentioned in this section for other reasons. We contacted the court's jurisprudence sector staff responsible for case entries registration to understand the criteria adopted for insertions in this section, looking for information to correctly extract the decisions cited by judges in a case; the rules created for extracting decisions cited required several consulting rounds, as the jurisprudence sector staff uses multiple patterns to list them.

A high number of decisions are decided by STF and, naturally, many of them share same matter and content. In this context, matter is the subject of a dispute in a case as same-sex marriage. Content of a decision is the expression of

---

[2] In Brazil preliminary injunctions are petitions which have temporary and urgent features. It aims to avoid likely violations of constitutionally guaranteed rights or irreparable harm in the absence of preliminary relief caused by delay in the legal process.

[3] A magistrate-rapporteur is an STF judge assigned at the start of a case the task of marshalling the arguments of parties [16].

a judgement that settles finally and authoritatively matters in dispute in a case. The content states which order was given, how it should be served and it cites the decisions foundations as the prevailing thesis and the laws that supports it. Therefore, decisions that share same matter and content are those whose subject in dispute and order given are fundamentally the same.

To increase efficiency, the court groups cases in bulks by similarity of matter and decides them in the same session, issuing the same decision to all cases in a batch. Not all such decisions are registered in jurisprudence search engine, so as to improve its performance and to improve legal researchers experience when filtering through similar cases. If there already exists an entry whose decision shares the same matter and content of that batch, all decision in the batch are registered in the section "Decisions in the same direction" (*Acórdãos no mesmo sentido*). Otherwise, a decision in the batch is chosen for a new entry, and the remaining ones are registered in the "Decisions in the same direction" section. The decisions inserted in this section will be called *similar decisions*.

This study is based on 103,168 STF decisions case entries judged between 01/2001 and 12/2019 using STF jurisprudence search engine [5]; data extracted and parsed was stored in a MongoDB database as semi-structured data.

4   Decision Network Modeling

A decision is the result of a judgement whose goal is to resolve a specific matter and many decisions finalized in STF may have a binding effect for future cases that deal with same matter. So, it is important to study how decisions are made and how precedents propagate influence. Our main goal is to propose a measure that enables us to rank STF decisions by their relevance and to find out if the network of citations is robust. For that, we have to build the network of citations using STF decisions and execute ranking algorithms over the network. The ranking orders the decisions by their relevance in the network and, therefore, their degree of influence over other decisions. Therefore, the decision network is meant to be a network of decision influence.

In the process of building the network of citations, consider nodes as decisions and edges as citations to decisions cited in magistrates opinions during judgement. Let N be the number of decisions, each decision represented by a node A_i, $i = 1, ..., N$. If the entry for process A_i mentions n_i decisions $A_{i_1}, ... A_{i_{n_i}}$, we create n_i edges connecting A_i to each cited decision. The cited decisions are meant to be precedents for decision that cite it, but as mentioned in Sect. 3 that may not always be the case. However, as we will consider all cited decisions as precedents, as we cannot distinguish which citations are part of opinions that followed the majority of magistrates in each case. Another issue concerning equating citation with precedent is that even decisions cited in opinions followed by the majority of magistrates may cite overruled decision due to a change of court's position concerning a particular matter. Even under those circumstances, the cited decisions are relevant because they contribute to foster the decision's prevailing arguments, and can be considered as influences.

For each similar decision S_k^i of A_i, let we also create edges to link S_k^i to $A_{i_1}, \ldots A_{i_{n_i}}$, given that each similar decision shares same matter and content of decision A_i. Although entries were not created for similar decisions, they were also judged like the one for which the entry was created. We assume similar decisions cite the exact same decisions as entry ones, even if this has not been confirmed by a search in the *complete case files*. But including them to network of citations contributes to consolidate court's jurisprudence about the matter at issue in these decisions.

4.1   Network Node Ranking Algorithms

PageRank. PageRank is an algorithm created for ranking website pages in a network using a relevance metric for such pages. It outputs a probability distribution, i.e. a probability of reaching each page (node) in the network, that represents the likelihood of a person randomly clicking on links to get to a specific page. PageRank works on the idea that a page pointed by many other pages may be more relevant than those pointed by only a few pages; furthermore, the relevance of a page increases if the pages pointing to it are also relevant. We present the original PageRank model and algorithm, and then an adapted version of it.

We start with a simplified example. Consider a directed network in which a node a is pointed by nodes b and c. The value $PR(a)$, where $PR(a)$ is the PageRank value of node a, is the sum of $PR(b)/N_b$ and $PR(c)/N_c$, where N_b and N_c are the number of nodes pointed by nodes b and c. The original PageRank model in (1), PR_1, calculates the value of $PR_1(a)$ as the summation of PR_1 of each node $b \in L_a$, where L_a are a's incoming links, i.e. the set of nodes that point to a, divided by the number of nodes pointed by b, N_b. Thus, the relevance of b is split between all N_b pages pointed by b and page a gets a $1/N_b$ fraction of it. To guarantee that no node has $PR_1(a) = 0$, for any a in the network, we multiply the sum by weight p, which is a damping factor, and add the factor $\frac{1-p}{N}$ to the model. The value of p used here is 0.85 as in [4]. The idea behind the damping factor is that there is a chance of getting stuck in a page that has no link to other pages. So, the damping factor considers the possibility of jumping from one page without links to another one. In the modeling adopted here this probability is $1 - p$. Thus, we have the following model.

$$PR_1(a) = p \sum_{b \in L_a} \frac{PR_1(b)}{N_b} + \frac{1-p}{N} \tag{1}$$

The computation of the PageRank of all nodes is iterative; the algorithm initially assigns the same value to each node in the decision network and iteratively updates the value of each node a calculating the value $PR_1(a)$ in each iteration until they converge under certain criteria. The convergence criteria used in this case is the Euclidean distance, given by (2), between one iteration, PR_1^m, and the previous one, PR_1^{m-1}, where L is the set of nodes in the decision network. When this difference is less than a precision ϵ, which for this algorithm in this work is 10^{-8}, the algorithm stops.

$$\sqrt{\sum_{a \in L}(PR_1(a)^m - PR_1(a)^{m-1})^2} < \epsilon \qquad (2)$$

The adapted version of PageRank model, PR_2, modifies the original model by replacing the rank computation given in (1) by that given by Eq. (3). The motivation for this adaptation lies in a point raised by STF experts pointing that the "relevance" of a decision should not be reduced if it has a great number of citations of precedence, thus avoiding the division of $PR_2(b)$ by N_b. The resulting ranking model is computed by:

$$PR_2(a) = p \sum_{b \in L_a} PR_2(b) + \frac{1-p}{N} \qquad (3)$$

Kleinberg's HITS. Kleinberg's algorithm, also called Hyperlink-Induced Topic Search (HITS), was designed to find most relevant pages as an answer for broad search topics in the context of the web. Many relevant pages do not contain the query string used in a search, so they are not retrieved in a search; to find these pages the links present in the retrieved pages are explored. The idea is that there is a relationship between two types of pages in this context, the *hub* pages and the *authoritative* pages. Authoritative pages are those which are most relevant to the initial query, usually have a large number of incoming links and there is a considerable overlap in the set of pages retrieved in the search that point to them; an authoritative page needs not be among the retrieved pages. Hub pages are those, among retrieved pages in the initial query, that point to, i.e. have links to, authoritative pages and also there is an overlap of retrieved pages that are pointed by them. So, a good hub is a page that points to many good authorities; a good authority is a page that is pointed to by many good hubs. In this work, the pages are the decisions and the retrieved pages are the decisions present in the network of citations.

To obtain the authoritative and hub decisions we have to calculate decisions' authority and hub scores. Consider a set of decisions large enough to contain most of authoritative decisions and hub decisions and that each decision has an non-negative *authority weight* $x^{\langle a \rangle}$, where a is a decision, and an non-negative *hub weight* $y^{\langle a \rangle}$.

$$x^{\langle a \rangle} = \sum_{b \in L_a} y^{\langle b \rangle} \qquad (4)$$

$$y^{\langle a \rangle} = \sum_{a \in L_b} x^{\langle b \rangle} \qquad (5)$$

HITS initially assigns the same value to each decision and iteratively updates the authority weight and the hub weight of decision a in each iteration as follows, until they converge under certain criteria. The convergence criteria used in this case is the sum of absolute difference of hub weights between the iteration m

and the previous one, $m - 1$, given by (6), where L is the set of decisions in the network of citations.

$$\sum_{a \in L} |x_m^{\langle a \rangle} - x_{m-1}^{\langle a \rangle}| < \epsilon \tag{6}$$

When this difference is less than a precision ϵ, which for this algororithm in this work is 10^{-8}, the algorithm stops. The authority and hub scores are the authority and weights obtained after the algorithm converged.

4.2 Computing Metrics of Decisions Relevance and Their Robustness

Based on the output of PageRank models PR_1 and PR_2 and Kleinberg's algorithm, we build for each algorithm output a list of the top 100 most relevant decisions which we call *Top100Decisions*. It was designed to evaluate the decision network robustness obtained by all 3 algorithms with the understanding that a high degree of agreement among those measurements indicates that they are capturing in the higher levels of relevance, a similar notion of decision authority.

This notion of relevance-by-agreement gives us a "static" view of robustness. To obtain a "dynamic" view of robustness, we have to apply some form of perturbation to the network. The "dynamic" view of robustness is designed to find out if the list of most relevant decisions, which in this case is the *Top100Decisions*, changes significantly among multiple runs when the decision network is altered by removing various sets of decisions from it. If the list of *Top100Decisions* among multiple runs does not change dramatically the decision network is found to be robust.

To evaluate the robustness an algorithm is run multiple times, a percentage of nodes is randomly removed from the network at the beginning of each experiment run and the *Top100Decisions* are re-evaluated after that perturbation[4]. The nodes removed are inserted back in the network in the end of each run, and the random perturbation is repeated in the following experiment runs. We call this test *Top100Decisions perturbation*. To ensure that the *Top100Decisions perturbation* of all algorithms and perturbation levels are comparable we remove the exact same decisions from the network for all algorithms, i.e. the same perturbation is applied to all algorithms in a certain run; we perform 10 runs for all algorithms outputs. The test is run for 3 levels of perturbation by removing 10%, 20% and 30% of nodes before building the *Top100Decisions* list[5].

The way we measure robustness is as follows: we compute *Top100Decisions* 10 times on all the 280,144 decisions in the network, the first one without perturbations and the other nine with the same random perturbation level, e.g. 10% perturbation. We then consider a decision as "agreed upon" if it reaches

---

[4] The removal of decisions is done before building the network; if a decision entry is removed, all similar decisions related to this decision are also removed.

[5] The source code for the scraper, the experiments run and the results analysis are available at https://github.com/jacksonjos/analise-juridica.

80%-threshold, that is, the decisions is present in at least 8 of 10 running trials. The standard computation, which runs the PageRank algorithm on all decisions in the network, we call the *Top100Decisions perturbation all*. To evaluate if PR_1 and PR_2 *Top100Decisions perturbation all* are really different algorithms we verify the intersection of decisions of *Top100Decisions perturbation all* metric and we employ the Chi-squared hypothesis test over the resulting table of all *Top100Decisions perturbation all* executions made for PR_1 and PR_2 models.

5 Results

There exists a few metrics to specific network topologies to evaluate robustness like those in [2,17], but in this work we propose another one to evaluate if a network is robust or not. Considering the centrality of *Top100Decisions perturbation* metric and that it can take values between 0 and 100, we fix the threshold of 50 as a criteria to determine if the network is robust or not. So, when an *Top100Decisions perturbation* metric is above 50 the network is robust; it is weak, otherwise. The results show the decision network is robust in respect to the *Top100Decisions perturbation all* metric for the perturbation levels 10% and 20% for all algorithms and robust for perturbation level 30% for PageRank models PR_1 and PR_2 as can be seen at Table 1 because the *Top100Decisions perturbation* is closer to 100 than to zero.

Table 1. *Top100Decisions perturbation all* results for each perturbation level for PR_1 and PR_2 PageRank models and Kleinberg algorithm.

Top100Decisions perturbation all metric for each algorithm			
Perturbation level	PR_1	PR_2	Kleinberg
10%	93	73	90
20%	78	54	56
30%	62	59	16

Those results indicate that the network of citations is highly robust, which can be explained by that fact the network of citations has a scale-free [13] topology with a power-law degree distribution with $\gamma = 2.00$ as we can see at Fig. 1 and Fig. 2. The network is scale-free when the probability of node degrees in a network, i.e. number of edges per node, decays as the degree k increases and that is of the form $P_{deg}(k) = \alpha k^{-\gamma}$ This means there are few nodes with a high number of edges, which can be verified in Fig. 2.

To compare the dynamic robustness revealed by the increasing levels of perturbation between the measurements of PR_1 and PR_2 models we use a hypothesis test whose main purpose is to compare if two distributions with some properties are statistically different. The hypothesis test expects that two experiments following the same probability distribution will be similar unless there is some

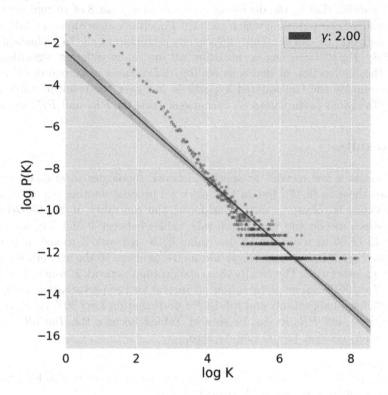

Fig. 1. Probability of node degree vs normalized frequency of node degree (log scale, both).

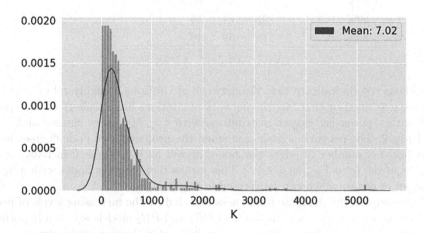

Fig. 2. Distribution of node degrees in the decision network.

Table 2. Intersection of decisions obtained by *Top100Decisions perturbation all* intersection for all combinations of algorithms and perturbation levels.

Top100Decisions perturbation all metric intersection between algorithms results			
Perturbation level	PR_1 vs PR_2	PR_1 vs Kleinberg	PR_2 vs Kleinberg
10%	43	5	11
20%	38	3	9
30%	33	2	2

reason, e.g. the data, that proves otherwise. We call this premise the null hypothesis, H_0, which is rejected in favor of the alternative hypothesis, H_1, if the data indicates that both experiments are statistically different. It is important to note that the impossibility to prove the alternative hypothesis does not prove that the null hypothesis is true. This just means that was not possible to disprove it.

In this work we use the Chi-squared hypothesis test, χ^2-test, to compare the dynamic robustness between PR_1 and PR_2 measurements indicated by *Top100Decisions perturbation* metrics results at Table 1 which follow the χ^2-distribution. We calculate the p-value, the significance level, that indicates the probability that the null hypothesis, i.e. the similarity of distributions, is true. If the value is smaller, for example, than 0.05 it suggests that the null hypothesis must be rejected in favour of the alternative hypothesis. In this work the p-value adopted to reject the null hypothesis is 0.05 for both *Top100Decisions perturbation* metrics. The hypothesis test for the *Top100Decisions perturbation all* metric obtained a p-value of 0.45 which implies that we can not reject the hypothesis that both PageRank model versions, PR_1 and PR_2, retrieve the same decisions ranking.

Other indicator that may contribute to the analysis is the intersection of decisions in the *Top100Decisions perturbation all* metric between both PR_1 and PR_2 models. We can see the intersection of decisions between PageRank models PR_1 and PR_2 is lower than 50, but they don't change much between the different levels of perturbation suggesting as can be seen at Table 2 they may not be different models. Checking the scores of *Top100Decisions* in an execution of PageRank algorithm without removing decisions from the network for both models we notice that while PR_1 model PageRank scores decrease smoothly for the 100 first positions. When looking at PR_2 model PageRank scores they decrease abruptly even in the 100 first positions and also decrease in steps, i.e., some decisions share the same score in each score value decrease creating a chart that looks like stairs. Nevertheless, we do not have strong evidence PageRank models PR_1 and PR_2 are different and produce distinct rankings. With respect to similarity of rankings produced by Kleinberg and PageRank algorithms the intersection of decisions between Kleinberg algorithm and PageRank models PR_1 and PR_2 for the *Top100Decisions perturbation all* metric is low enough to assume Kleinberg decisions ranking is different of rankings obtained by PageRank models PR_1 and PR_2.

The last goal of this work is to evaluate if the algorithms studied in this work successfully ranked at the top the leading cases, i.e. if the algorithms have meaningful authority scores. We asked legal scholars to take a look at *Top100Decisions perturbation all* 10% perturbation level output for all three algorithms and they informed that PageRank algorithms PR_1 and PR_2 ranked best many decisions that deal with procedural issues, i.e. decisions the judges do not even get to decide about the merits of the case. The rankings may have this result because theses decisions were picked by some magistrate in STF to be cited in other decisions to support the application of some legal principles and they happen to be cited a lot because the legal principles it contain are commonly cited. Kleinberg's algorithm turns out to have a more interesting result. Some few leading cases were found in the *Top100Decisions perturbation all* 10% perturbation level output and some decisions are related to social security matters that reach STF very often, many of these ones share the same content which is not desirable for legal scholars. However, it would be desirable to find more leading cases about different areas of law and without recurrence what is a direction for improvement. In both cases, similar decisions may have an amplifier factor because they multiply the number of citations pushing some decisions to the top of the rank, even if they do not settle important legal principles.

6   Contributions and Future Work

In this work we found the modification in the original PageRank model for calculating the probability of reaching each decision in the decision network, as suggested by STF experts, did produce a significantly different ranking for the *Top100Decisions Perturbation all* test. Locally, each measure was computed differently, but due to network structure, globally yield statistically indistinguishable robustness results.

Considering the ranking produced by the PageRank algorithm, it does not produce an expected relevance ranking, according to STF experts. Kleinberg's algorithm suggests with some improvements may retrieve leading cases, but it requires more research to do so. We conclude that the algorithms employed achieved a good quantitative measure, but not a qualitative one because they not measure the importance or impact of the decisions on other decisions but how much decisions support most common matters that reach the court. This particular finding confirms some claims of [18] that relevant decisions responsible to improve and develop the law and of most interest for STF experts, like those concerning constitutional matters, are not in dense regions of the decision network because they are not used as binding precedents and, therefore, are usually cited by just a few other decisions.

A important evolution of this work is to develop tools that analyze the full content of the decision, identifying arguments in it for and against the decision. Also work alongisde with STF experts to create separate network of citations per area of law and obtain authority scores for each network to retrieve leading cases.

Acknowledgments. We thank Mayara C. Melo and Alessandro Calò for developing previous work that made possible to push this research forward; Dr. Luís Matricardi for helping a lot on understanding concepts of law and properties of STF decisions and suggesting readings regarding these subjects; Felipe Farias and the STF jurisprudence sector staff for helping with questions about the data in cases entries. Also, we thank Dr. Juliano Souza de Albuquerque Maranhão (Faculty of Law of the University of São Paulo), president of Lawgorithm Association for Research in Artificial Intelligence and Law, and Dr. Jorge Alberto Araújo de Araújo for all help analyzing the lists of top decisions studied in this work to find out if they contain leading cases and for ideas of analyses and future works.

This study was financed in part by the Coordenação de Aperfeiçoamento de Pessoal de Nível Superior - Brasil (CAPES) - Finance Code 001. M. Finger was partly supported by Fapesp, processes 20/06443-5 (SPIRA), 19/07665-4 (C4AI) and 14/12236-1 (Animals).

References

1. Agnoloni, T., Pagallo, U.: The case law of the Italian constitutional court, its power laws, and the web of scholarly opinions. In: Proceedings of the 15th International Conference on Artificial Intelligence and Law, pp. 151–155. ACM (2015). http://doi.acm.org/10.1145/2746090.2746108

2. Albert, R., Jeong, H., Barabási, A.L.: Error and attack tolerance of complex networks. Nature **406**(6794), 378 (2000). https://doi.org/10.1038/35019019

3. Bommarito II, M.J., Katz, D.M., Zelner, J.L., Fowler, J.H.: Distance measures for dynamic citation networks. Phys. A **389**(19), 4201–4208 (2010). https://doi.org/10.1016/j.physa.2010.06.003

4. Brin, S., Page, L.: The anatomy of a large-scale hypertextual web search engine. Comput. Netw. ISDN Syst. **30**(1–7), 107–117 (1998). https://doi.org/10.1016/S0169-7552(98)00110-X

5. Pesquisa de jurisprudência : STF - Supremo Tribunal Federal. http://stf.jus.br/portal/jurisprudencia/pesquisarJurisprudencia.asp

6. Definitions, U.L.: Leading case law and legal definition (2020). https://definitions.uslegal.com/l/leading-case/

7. Falcão, J., Cerdeira, P., Arguelhes, D.: I relatório do supremo em números-o múltiplo supremo. Revista de Direito Administrativo **262**, 399–452 (2013)

8. Falcão, J., Hartmann, I.A., Chaves, V.P.: III Relatório Supremo em números: o Supremo e o tempo (2014)

9. Fowler, J.H., Jeon, S.: The authority of supreme court precedent. Soc. Netw. **30**(1), 16–30 (2008). https://doi.org/10.1016/j.socnet.2007.05.001

10. Hartmann, I.A., dos Santos Junior, F.A.C., Silva, F.A., Appel, O., et al.: Pedidos de vista no tribunal superior eleitoral. REI Revista Estudos Institucionais 3(2), 1074–1111 (2017)

11. Kleinberg, J.M.: Authoritative sources in a hyperlinked environment. J. ACM **46**(5), 604–632 (1999). https://doi.org/10.1145/324133.324140

12. Nunes, J.L., de Oliveira Chaves Filho, L., et al.: Explicando o dissenso: uma análise empírica do comportamento judicial do supremo tribunal federal e da suprema corte dos estados unidos. REI Revista Estudos Institucionais 2(2), 899–931 (2016)

13. Onnela, J.P., et al.: Structure and tie strengths in mobile communication networks. Proc. Nat. Acad. Sci. **104**(18), 7332–7336 (2007). https://doi.org/10.1073/pnas.0610245104. https://www.pnas.org/content/104/18/7332

14. van Opijnen, M.: Citation analysis and beyond: in search of indicators measuring case law importance. In: JURIX , vol. 250, pp. 95–104 (2012). https://doi.org/10. 3233/978-1-61499-167-0-95
15. Page, L., Brin, S., Motwani, R., Winograd, T.: The PageRank citation ranking: bringing order to the web. Technical report, Stanford InfoLab (1999)
16. Stewart, W.J.: Collins Dictionary of Law. HarperCollins Publishing, London (2006)
17. Schneider, C.M., Moreira, A.A., Andrade, J.S., Havlin, S., Herrmann, H.J.: Mitigation of malicious attacks on networks. Proc. Natl. Acad. Sci. **108**(10), 3838–3841 (2011). https://doi.org/10.1073/pnas.1009440108
18. Vojvodic, A.d.M.: Precedentes e argumentação no Supremo Tribunal Federal: entre a vinculação ao passado e a sinalização para o futuro. Ph.D. thesis, Universidade de São Paulo (2012)
19. Wikipedia: Supreme court. http://en.wikipedia.org/w/index.php?title=Supreme %20court&oldid=779636937. Accessed 11 May 2017
20. Winkels, R., de Ruyter, J., Kroese, H., et al.: Determining authority of Dutch case law. Front. Artif. Intell. Appl. **235** (2011). https://doi.org/10.3233/978-1-60750-981-3-103

Semi-Supervised Sentiment Analysis of Portuguese Tweets with Random Walk in Feature Sample Networks

Pedro Gengo[1(✉)] and Filipe A. N. Verri[2]

[1] Data Science Team, Itaú Unibanco, São Paulo, SP, Brazil
pedro.gengo.lourenco@gmail.com
[2] Computer Science Division, Aeronautics Institute of Technology (ITA),
São José dos Campos, SP, Brazil
verri@ita.br

Abstract. Nowadays, a huge amount of data is generated daily around the world and many machine learning tasks require labeled data, which sometimes is not available. Manual labeling such amount of data may consume a lot of time and resources. One way to overcome this limitation is to learn from both labeled and unlabeled data, which is known as semi-supervised learning. In this paper, we use a positive-unlabeled (PU) learning technique called Random Walk in Feature-Sample Networks (RWFSN) to perform semi-supervised sentiment analysis, which is an important machine learning that can be achieved by classifying the polarity of texts, in Brazilian Portuguese tweets. Although RWFSN reaches excellent performance in many PU learning problems, it has two major limitations when applied in our problem: it assumes that samples are long texts (many features) and that the class prior probabilities are known. We leverage the technique by augmenting the data representation in the feature space and by adding a validation set to better estimate the class priors. As a result, we identified unlabeled samples of the positive class with precision around at 70% in higher labeled ratio, but with high standard deviation, showing the impact of data variance in results. Moreover, given the properties of the RWFSN method, we provide interpretability of the results by pointing out the most relevant features of the task.

Keywords: Sentiment analysis · Semi-supervised classification · Positive-unlabeled learning · Random walk

1 Introduction

Traditionally, machine learning tasks are divided in two categories: supervised learning, tasks whose input data are labeled, or unsupervised learning, when data are unlabeled. However, a third paradigm, called semi-supervised learning,

---

The original version of this chapter was revised: the title has been corrected. The correction to this chapter is available at https://doi.org/10.1007/978-3-030-61377-8_47

© Springer Nature Switzerland AG 2020, corrected publication 2020
R. Cerri and R. C. Prati (Eds.): BRACIS 2020, LNAI 12319, pp. 595–605, 2020.
https://doi.org/10.1007/978-3-030-61377-8_42

combines labeled and unlabeled data and take advantage of this combination [13]. The study of techniques in this paradigm is extremely important because data has been generated at an increasing rate and, in many applications, manual labelling is expensive and time-consuming.

One particular task is sentiment analysis, which is also called opinion mining. The goal of sentiment analysis is to extract sentiments and opinions from natural language text using computational methods [7]. A more specific task is polarity classification, which consists of classifying texts in the following classes: positive, negative and neutral. Methods used in this task range from machine learning algorithms to lexical or distant approach [3–5,10,11].

We model the task of polarity classification as a problem where we have few labeled examples of the positive class and all others are unlabeled. This problem is called positive-unlabeled (PU) learning, and is an inner class problem of semi-supervised learning. The goal of PU learning is to label all unlabeled input samples at once (transductive learning) or to construct a function that discriminates positive and negative samples (inductive learning) [9].

One technique of PU learning is Random Walk in Feature-Sample Networks (RWFSN) [12], which is a graph-based technique with steps: *a)* Convert the dataset into a sparse binary representation. *b)* Create a bipartite graph where samples and features are the vertices, and edges are the connection between a sample and a feature. *c)* Perform a random walk process over the graph, applying a scaling factor(constant) at labeled samples. *d)* Use the limiting distribution of the Markov chain to calculate the positive-class confidence of unlabeled sample. *e)* Order the unlabeled samples by their positive-class confidence and, with knowledge of the positive-class prior probability, classify the unlabeled samples.

A limitation with this technique, showed at [12], is that the classification step depends on the assumption of knowing the positive-class prior probability and this information may not be known in real-world problems.

In this paper, we adapted a dataset of Brazilian Portuguese tweets to be able to use it in the PU learning technique described to classify unlabeled samples. We also proposed a modification of the technique, using a validation set to choose the threshold of positive-class confidence without the need of knowing the prior probabilities.

The rest of this paper is organized as follows. In Sect. 2, we present the proposed modification of the technique to deal with the problem of unknown prior probability. Section 3 describes the preprocessing steps used to treat tweets of the dataset. Finally, Sects. 4 and 5 show our results and conclude this paper.

2   Model Description

The general idea of the model proposed in [12] is that it receives a dataset where each sample is either a positive or negative sample and only a few positive samples are labeled. Using the prior probability P^+ and the positive-class confidence, it classifies the unlabeled data.

In the following subsections, we explain the steps of the learning algorithm.

2.1   Construction of the Feature-Sample Network

The *Feature-Sample Network*, $\mathcal{G}$, is a bipartite complex network whose edges associates samples and features of the dataset $\mathcal{D}$. So, the vertices are samples and features, and an edge exists only where we have the presence of the feature in a sample.

We can construct this complex network defining the vertex set $\mathcal{V}$ like $\{v_1, ..., v_N, v_{N+1}, ..., v_{N+M}\}$, where N represents the number of samples and M the number of features in the dataset. Besides that, an edge exists only between samples and features. So, we can define the adjacency matrix of this graph as:

$$A = \begin{bmatrix} 0 & X \\ X^T & 0 \end{bmatrix} \text{ ,where X is the dataset.}$$

An import condition is that $\mathcal{G}$ has a single connected component. If it is false, only the largest connected component will be considered.

2.2   Modeling of the Random Walk Process

We want to reach the stationary distribution π for an irreducible Markov Chain. So, we model the transition matrix P to guarantee the existence and uniqueness of this limiting distribution.

The limiting distribution of a random walk is reached independently of the initial conditions if the Markov chain is ergodic. To satisfy this requirement, we model the transition matrix P as

$$p_{ij} = \frac{w_{ij} \nu_j}{\lambda \nu_i},$$

where the matrix $W = (w_{ij})$ has elements

$$w_{ij} = \begin{cases} 1, & \text{if i} = \text{j} \\ \beta a_{ij}, & \text{if } x_i \text{ is a positive sample} \\ a_{ij}, & \text{otherwise,} \end{cases}$$

ν is the eigenvector associated with the leading eigenvalue λ of the matrix W and β is a hyperparameter called scaling factor, and a_{ij} is the element of the adjacency matrix of the bipartite graph.

2.3   Estimation of the Positive-Class Confidence

The transition matrix P describes a system with a limiting distribution π reached independently of the initial setting. We expected that the limiting probabilities related to positive samples are greater than the ones associated with negative samples.

So, we estimate the positive-class confidence $f(x_i) = \pi_i$ for all unlabeled samples. The stochastic vector $\pi = (\pi_i)$ is a eigenvector associated with the leading eigenvalue of matrix P^T.

Algorithm 1 illustrates the steps of RWFSN method[1].

---

[1] Available on https://github.com/pedrogengo/RWFSN.

Algorithm 1: RWFSN

Data: Dataset with samples as binary feature vectors, β
Result: Positive-class confidence of unlabeled data ($f(\boldsymbol{x}_i)$)
Initialization;
Create the adjacency matrix A;
Remove all disconnected elements;
Create the matrix $W = (w_{ij})$;
Do the spectral decomposition of the matrix W;
$\lambda \leftarrow$ leading eigenvalue of matrix W;
$\nu \leftarrow$ eigenvector associated with λ;
Calculate the transition matrix P;
Do the spectral decomposition of the matrix P^T;
$\pi \leftarrow$ eigenvector associated with the leading eigenvalue of P^T;
$f(\boldsymbol{x}_i) \leftarrow \pi$;
end

2.4   Classification of the Unlabeled Samples

Differently from the original model, where the classification of unlabeled samples is made by using the prior probability of positive class to determine the threshold to classify as either positive or negative, we propose a new way to perform this classification.

Based on the idea of validation set, widely used in supervised learning, we propose to split a fraction of the labeled data, once it was shown that the model works well with few labeled samples, and use this split as an unlabeled data. So, after run the Algorithm 1, we could choose the threshold ($f_{validation}^n$) based on this samples that we know and were not scaled by β remaining on the same scale as unlabeled data.

The predicted class $c(\boldsymbol{x}_i)$ of an unlabeled sample $\boldsymbol{x}_i$ regarding the validation data is

$$c(\boldsymbol{x}_i) = \begin{cases} +1, & \text{if } f(\boldsymbol{x}_i) > f_{\text{validation}}^n \\ -1, & \text{otherwise.} \end{cases}$$

In Fig. 1, we observe the proposed method to select the threshold to classify the unlabeled data.

3   Dataset and Preprocessing

To perform sentiment analysis of Portuguese tweets, the TweetSentBR dataset [2] was used, which contains 15047 tweets ID's and their classifications. The tweets were related to Brazilian program shows. Of the total, 44% are positive, 26% are neutral and 29% negatives. However, during the scrapping using Twitter's API and the tweet's ID as keys, some of tweets were not found. Therefore, the collected dataset is formed by 11610 samples, where 45% are positive, 25% are neutral and 30% negative.

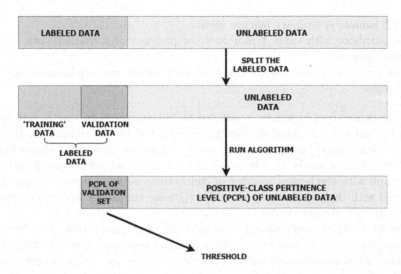

Fig. 1. Proposed change to use a validation set to choose better the threshold of positive-class confidence. We split the labeled data in two parts: training set and validation set. The validation set is treated as unlabeled data. Thus, we can choose the threshold based on positive-class confidence of validation set, which we known that are positive samples.

When we analyzed some of positive, neutral and negative tweets, we observed that, in some cases, the neutral tweets had a subtle difference from the other classes, as we can see in this neutral example, which has an positive emoji (heart):

Example. Consegui achar os ep de hoje do Master chef ♡ foi o ep 11 né gente? #MasterChefBR

Thus, we decided to use only the positive and negative classes in our experiments once their difference is stronger. In Table 1, we can see the class distribution over the dataset used in the experiments.

Table 1. Class distributions of data used

Class	Frequency	Total
Positive	0.606	5232
Negative	0.394	3400
Total		**8632**

RWFSN technique requires that the input data satisfy 3 requirements, which are:

- Each sample is a binary feature vector;
- An attribute with value 1 indicates the presence of a characteristic of that data instance;
- The similarity of two samples depends only on the number of shared characteristics.

Tweets are short messages, restricted to 280 characters in length, but by the time the dataset was labeled, they were restricted to 140 characters. Because of this short length, people use acronyms, emoticons and other characters to substitute formal words [1]. The problem with that is we need to create a bipartite graph with the largest number of connections between positive class examples and with the use of acronyms, abbreviations, emoticons and others, similar tweets did not present shared characteristics.

So, to satisfy the requirements and solve the mentioned problem it was necessary to preprocess the dataset to achieve a binary representation with the largest numbers of connections (less sparse), relevant features and viable dimensionality.

The steps performed to tokenize and to preprocess the data were:

1. A binary feature which indicates the presence of an exclamation mark in the tweet (1) or not (0) was created. (flag_exclamation)
2. A binary feature which indicates the presence of a question mark in the tweet (1) or not (0) was created. (flag_question)
3. All characters were converted to lowercase.
4. HTML entities were replaced by their character and ASCII escape sequences were replaced by a blank space.
   Example: "\n" → " "
          "<" → "<"
5. Accents were removed.
6. Usernames (an @ symbol followed by up to 15 characters) were replaced by the tag __user__.
7. Hashtags (a # symbol followed by any sequence of characters) were replaced by the tag __hashtag__.
8. A binary feature which indicates the presence of words with sequence of repeated characters was created. (flag_n_letter)
9. A sequence of repeated characters was replaced by only one character. In this context of modeling the problem as a bipartite graph we wanted to have the largest number of connections. So, we would like to reduce this words with repeated characters to an unique form, in order to have fewer variations of the same word, resulting in more connections.
   Example: "ammooooooooo" → "amo"
10. Punctuation marks were removed.
11. A binary feature which indicates the presence of positive emojis in unicode format was created. To do so, we used a dictionary of emoji's sentiment [6]. So, we iterated the tweet and if an emoji was found we check its polarity magnitude. If this polarity was positive (greater than 0.1), this feature was 1, else 0 (flag_pos_emoji).

12. As in the previous topic, a binary feature which indicates the presence of negative emojis was created (flag_neg_emoji).
13. We performed a data augmentation by inserting root words or infinitive verb tense based on *thesaurus* Delaf Unitex [8] as the following procedures: creating a key-value structure to store the root words and their similar words or infinitive verb tense (we treated the words with repeated characters in this dictionary to follow the preprocessing done with tweets); inserting to this dictionary some acronyms and abbreviations frequently used, which were detected by tweet's analysis; ultimately, we iterated all tweets and inserted the root word or the infinitive verb tense of each word right after it.

The following tweet is used to demonstrate how was the output after preprocessing.

Original tweet:
Vou ficar tonta @pefabiodemelo #conversacombial #MasterChefBR
Preprocessed tweet:
vou ir ficar ficar tonta tonto _ _user_ _ _ _hashtag_ _ _ _hashtag_ _

After preprocessing the data, we tokenized the tweets and used bag of words (BoW) approach to represent the data as a binary vector. When we tokenized the data, we obtained 10986 different tokens. We selected only the 199 most frequent features because we achieved better results and reduced the sparsity of the matrix.

4   Results and Discussion

This section is divided in three subsections. In the first one, we compare the polarity classification of Portuguese tweets using RWFSN and a baseline model based on the neighborhood graph. In the second one we present the results using the proposed method (validation set) to choose the threshold. And in the last one we present a discussion about interpretability of the model and relevant features.

4.1   Performance Comparison of Polarity Classification

We compare the RWFSN model with a baseline semi-supervised method. Such baseline method is based on the neighborhood graph and we choose this method because it uses the same classification mechanism providing a fair comparison. We use the prior probability to perform the classification step in order to prove that the method can be extended to the task of polarity classification of tweets.

The baseline method is based on the construction of a k-NN graph of the dataset, where distance between two sample was done by using the Jaccard index. We calculated the positive-class confidence as

$$f^{k-NN}(\boldsymbol{x_i}) = (\min_j l_{ij})^{-1},$$

where l_{ij} is the length of the shortest path between vertices v_i and v_j. The positive class pertinence of a sample is inversely proportional to the shortest distance from the associated vertex to any labeled vertex.

To compare the performance between models, we observed the accuracy and the F-score, in order to compile precision and recall. We tested each ratio with 3 different samples and reported the average of metrics. The results are shown at the Table 2 with the respective best parameters.

Table 2. Comparison between the methods. the best parameter is shown.

Labeled ratio	RWFSN		k-NN graph	
	Accuracy (β)	F-score	Accuracy(k)	F-score
1%	0.5203 (17)	0.6035	**0.5792 (30)**	**0.6376**
5%	**0.5819 (14)**	**0.65175**	0.5779 (30)	0.6373
10%	**0.6154 (9)**	**0.6759**	0.6045 (30)	0.6668

As we can see in Table 2, as the labeled sample ratio increases the RWFSN performance surpasses the baseline model. Although the difference was not large, this behavior suggests that RWFSN can exploit the label information more efficiently than the KNN graph.

4.2   Choice of Threshold Using Validation Set

We also perform tests using a validation set, which corresponds to 20% of the size of labeled set, to choose the threshold instead of using the prior probability, which is unknown in most of real-world problems. With the validation set we needed to choose a specific positive-class confidence of the samples in this set. So, in order to be conservative, we choose the higher positive-class confidence, because there is an idea that unlabeled samples that have confidence greater than the largest confidence of the validation set are probably also positive.

To evaluate the performance of this choice we used as metric the precision, which tell us how many examples were really positive of what we predict positive, and F-Score. We also performed a Receiver Operating Characteristic (ROC) Curve analysis using all the positive-class confidences of the validation set, in order to evaluate the performance with different choices of positive-class confidence in validation set.

We used different labeled ratio with the respective optimal β, which has been found empirically and we tested in 10 differents samples for each ratio. The average of precision and F-Score and its standard deviation achieved are presented at Table 3 and the ROC Curves are presented at Fig. 2.

As we can see in Table 3, we verified that the precision increases with more labeled data and the F-Score decreases. Another important result is about the impact of variance due to the sample used, which can be verified by the order

Table 3. Results using the higher positive-class confidence of the validation set for different labeled ratio

Labeled ratio	Optimal β	Precision	F-Score
1%	17	0.63 ± 0.02	0.12 ± 0.10
5%	16	0.65 ± 0.09	0.04 ± 0.03
10%	9	0.70 ± 0.04	0.01 ± 0.01

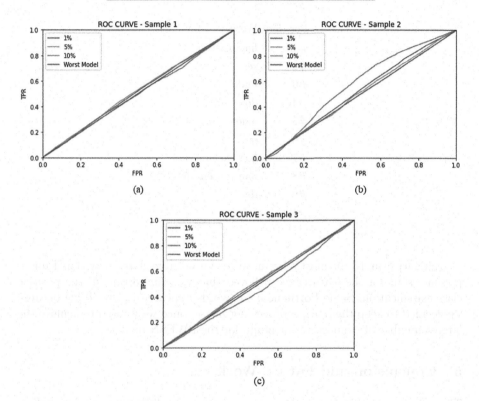

Fig. 2. ROC curves using different labeled ratio and different samples. Each curve color represents one different labeled ratio. As we can see in (a), (b) and (c) we obtained different results for each of the samples, showing the impact of the variance on the results.

of the standard deviation in precision. In the Fig. 2 we can corroborate what is verified at Table 3, as we show different curves that depend on the sample used.

4.3   Relevant Features

One important aspect of the RWFSN is that the states associated with each relevant features of the positive class has high stationary probabilities. So, if we look inside the positive-class confidence of features set, we can see which features are relevant to the positive class. With that, we can extend this technique to classify the polarity of words in a specific context or make a feature selection.

At Table 4, we can see the 10 most relevant features associated with the positive class using 5% of labeled samples and β equals to 16.

Table 4. Top 10 most relevant features using 5% of labeled samples and β equals to 16.

Position	Word
#1	flag_n_letter
#2	ser
#3	lindo
#4	do
#5	amor
#6	que
#7	flag_exclamation
#8	flah_pos_emoji
#9	de
#10	como

Not surprisingly, the most relevant words for positive class, showed in Table 4, presents some of the features we believed that were important to the positive class classification. Some Portuguese stopwords appeared in this Table because we decided to keep them at preprocessing phase, once we wanted to achieve the largest number of connections at graph for the RWFSN model.

5   Conclusion and Future Work

The polarity classification task is very important in different scenarios such as product reviews and analysis of social media content. However, many machine learning methods suffers from the need of a large amount of labeled data during their training phases. In this paper, we proposed a preprocessing approach of the Portuguese tweets dataset to use this dataset in RWFSN, as well as a change in the classification mechanism of the model.

The results shown that model and classification mechanism proposed can be used to perform this task, but the issue of matrix sparsity should be further explored and it is important to take care of the sample used and about the distribution of your dataset, due to the impact of the data variance in the results. Moreover, this model has higher interpretability, which can be explored in other tasks such as word polarity detection. However, the limitation of the parameter β can impair the use of the model in this task.

As future works, we intend to explore ways to reduce the matrix sparsity by using sentiment lexicon for Portuguese and compare the presented model with traditional machine learning techniques. We also intend to explore a way to find the optimal β and use the model presented for active learning techniques.

Acknowledgment. This work was supported by Itaú-Unibanco.

Any opinions, findings, and conclusions expressed in this manuscript are those of the authors and do not necessarily reflect the views, official policy or position of Itaú-Unibanco.

References

1. Agarwal, A., Xie, B., Vovsha, I., Rambow, O., Passonneau, R.: Sentiment analysis of twitter data. In: Proceedings of the Workshop on Languages in Social Media, LSM 2011, pp. 30–38. Association for Computational Linguistics, Stroudsburg (2011). http://dl.acm.org/citation.cfm?id=2021109.2021114
2. Brum, H.B., das Graças Volpe Nunes, M.: Building a sentiment corpus of tweets in brazilian portuguese (2017). CoRR abs/1712.08917 http://arxiv.org/abs/1712.08917
3. Corrêa Jr, E.A., Marinho, V.Q., Santos, L.B.D., Bertaglia, T.F.C., Treviso, M.V., Brum, H.B.: Pelesent: Cross-domain polarity classification using distant supervision (2017)
4. Dos Santos, C., Gatti, M.: Deep convolutional neural networks for sentiment analysis of short texts. In: Proceedings of COLING 2014, the 25th International Conference on Computational Linguistics: Technical Papers, pp. 69–78 (2014)
5. Go, A., Bhayani, R., Huang, L.: Twitter sentiment classification using distant supervision. CS224N Proj. Rep. Stanford **1**(12), 2009 (2009)
6. Kralj Novak, P., Smailović, J., Sluban, B., Mozetič, I.: Sentiment of emojis. PLoS ONE **10**(12), e0144296 (2015). https://doi.org/10.1371/journal.pone.0144296
7. Liu, B.: Sentiment Analysis and Opinion Mining, pp. 1–135. Cambridge University Press, New York (2015)
8. Muniz, M.C.M.: A construção de recursos lingüístico-computacionais para o português do brasil: o projeto de unitex-pb. São Carlos (2004)
9. Muñoz-Marí, J., Bovolo, F., Gómez-Chova, L., Bruzzone, L., Camp-Valls, G.: Semisupervised one-class support vector machines for classification of remote sensing data. IEEE Trans. Geosci. Remote Sens. **48**(8), 3188–3197 (2010)
10. Pak, A., Paroubek, P.: Twitter as a corpus for sentiment analysis and opinion mining. LREC **10**, 1320–1326 (2010)
11. Taboada, M., Brooke, J., Tofiloski, M., Voll, K., Stede, M.: Lexicon-based methods for sentiment analysis. Comput. Linguist. **37**(2), 267–307 (2011)
12. Verri, F.A.N., Zhao, L.: Random walk in feature - sample networks for semi-supervised classification. In: 5th Brazilian Conference on Intelligent Systems Random, pp. 235–240 (2016). https://doi.org/10.1109/BRACIS.2016.41
13. Zhu, X., Goldberg, A.B.: Introduction to semi-supervised learning. Synth. Lect. Artif. Intell. Mach. Learn. **3**(1), 1–130 (2009). https://doi.org/10.2200/S00196ED1V01Y200906AIM006

The Use of Machine Learning
in the Classification of Electronic Lawsuits:
An Application in the Court of Justice
of Minas Gerais

Adriano Capanema Silva[(⊠)] [iD] and Luiz Cláudio Gomes Maia[iD]

Faculty of Business Sciences – FACE, FUMEC University, Belo Horizonte, Brazil
acapanema@fumec.edu.br, luiz.maia@fumec.br

Abstract. With the abundance of electronic lawsuits already implemented throughout Brazil, courts have a valuable source of information in text format that constitute attractive bases for the application of Artificial Intelligence (AI) and machine learning (ML). In this research, supervised learning approaches were explored for the automatic classification of types of documents in electronic court proceedings of the Court of Justice of Minas Gerais (TJMG). The methodology is composed of cross-validation within the specific corpus of the legal domain, comparing traditional classifiers and more recent methods based on neural networks and deep learning models, using Glove word vectors generated for the Portuguese Language and Convolutional Neural Network (CNN). This work achieved high precision in the results and if implemented in the courts it can provide significant savings in financial and human resources, allowing lawsuits classification activities, currently done manually by employees, to be performed in seconds by the machine. The result of this experiment shows that the hit rates for the CNN and SVM classifiers exceed 93% and is considered a high result. Based on the assumption that Glove brings extra semantic resources that can help in classifying texts from court proceedings, this work demonstrates Glove's effectiveness by showing that a CNN with Glove surpasses SVM.

Keywords: Glove · Artificial Intelligence · Machine learning · Text classification · Electronic lawsuits

1 Introduction

Artificial Intelligence (AI) can be applied to the legal area, through solutions that help meet the exponential growth of society's demands for justice and providing people to be released to perform intellectual work, since the adoption of automated instruments that performing repetitive tasks can contribute to a more efficient and faster judiciary. In this scenario, the information contained in the Electronic Judicial Process (PJE) database, a system that allows the practice and monitoring of the procedural act, is perceived as a relevant data source for the application of AI techniques. With the implementation of the PJE in several Brazilian courts, the virtualization of Justice maintains a historical

© Springer Nature Switzerland AG 2020
R. Cerri and R. C. Prati (Eds.): BRACIS 2020, LNAI 12319, pp. 606–620, 2020.
https://doi.org/10.1007/978-3-030-61377-8_43

trend of growth, and the number of electronic documents available in the courts has been growing exponentially. This creates an opportunity - until then non-existent - for exploring a huge amount of unstructured data with AI.

This work presents a proof of concept using machine learning techniques for automatic classification of electronic lawsuits. It was defined the model that presents the best result in the light of the configurations adequate to the corpus of electronic lawsuits of the Court of Justice of Minas Gerais (TJMG).

The TJMG has a project to digitize millions of civil lawsuits that reach the Court on appeal, from the district that already have the PJE system in place. These lawsuits will be judged electronically and returned to the counties in the same way, that is, in a virtual way. For the project in question, a team of employees has allocated to carry out the classification of the types of procedural documents (Initial Petition, Petition, Order, Sentence, Decision, etc.). Each lawsuit - often with more than 200 pages – analyzed is to define which type of document each procedural part refers to, this being a time consuming, expensive job and with an average time spent of 50 min to classify the parts of a single lawsuit. In the TJMG collection in February/2020 there were 2,050,597 civil physical lawsuits with the potential to be scanned and classified automatically. Figure 1 shows the time it would take to classify manually all of these lawsuits.

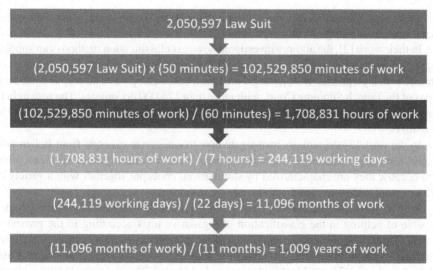

Fig. 1. Estimated time for manual classification of the TJMG physical lawsuit collection in February/2020.

This work presents not only a proof of concept, but also investigated which machine learning algorithm achieved the best performance in the analysis of the content of electronic lawsuits. Machine learning models have been developed using the database of the PJE system that can be reused by judicial systems. With the use of these automated models, the possibility of saving up to 1,708,831 h of work by the TJMG team designated to manually classify the procedural documents is envisaged, as shown in Fig. 1.

Different learning algorithms were explored to train classification models and used to classify electronic lawsuits in defined classes of text, comparing learning algorithms in order to define the one that presents the best classification result for electronic lawsuit corpus of TJMG. Support Vector Machine (SVM), Naïve Bayes e AdaBoost classifiers were analyzed and tested in addition to the application of more recent methods based on neural networks and deep learning models using the incorporation of Glove words generated for the Portuguese Language and Convolutional Neural Network (CNN). The tasks performed, seen as supervised learning, made the automated identification of the label "type of procedural part" of lawsuits based on model training and suggested probability. Comparisons were made using metrics - accuracy, recall, F1 score and training times and classification - disseminated in the academic context and relevant to the topic [1].

The remainder of this paper is organized as follows. Section 2 discusses the work related to the classification of legal documents. Section 3 describes the proposed methodology. Section 4 presents the experiments and results obtained. Finally, Sect. 5 concludes with final observations and guidelines for future work.

2   Related Work

Seven articles were analyzed and chosen, adequate to the proposed work and which make sense to the scope of this experiment.

In their work [2], the authors investigate how text classification methods can support legal professionals, presenting experiments that apply machine learning techniques to predict with high precision which legal area each case belongs to. The corpus used decisions of the French Supreme Court, with more than 126,000 documents. The researchers developed a medium probability method that combines the output of several SVM classifiers.

In the work [3] the authors investigate how to categorize excerpts from Italian normative texts. They categorize fragments of legal texts, considering them a difficult problem because they are characterized by summarized concepts, together with a variety of phrases used to denote complicated sentence structures.

In their work [4], the authors address the exploration of automatic methodologies capable of helping in the classification of legislative texts according to the provision model. The article presents a module capable of classifying fragments of legislative texts into types of provision. Two machine learning methodologies (Naive Bayes and Multiclass Support Vector Machine) were used and tested. The experiences give evidence of satisfactory results.

In their work [5] the authors denote that the traditional classification of documents uses information retrieval techniques, such as continuous bag-of-words or tf-idf, widely used in natural language processing. With the introduction of word2vec by Google, a new approach to document representation emerges. The work is carried out under the assumption that word2vec brings semantic resources that help in the classification of text.

In the work [6], the authors show studies related to the use of deep learning for binary classification of legal documents. Specifically, they carried out experiments to compare results of deep learning with the results obtained using SVM algorithm on legal subject data sets. The results showed that CNN performed better with a higher volume of training data and is an adequate method for classifying text in the legal domain.

In their work [7], the authors propose a system for predicting areas of law ("Legal areas"). The work uses a data set of sentences from the Supreme Court of Singapore (comprising 31 labels from relevant legal areas) to comparatively study the performance of text classification approaches for classification of legal areas. Newer models are compared to traditional statistical models. The experiment compares legal text classification techniques and explores more deeply how document scarcity and length affect performance.

In the work [8], the authors perform an extreme text classification with several labels, which refers to the task of marking documents with relevant labels from an extremely large data set, usually containing thousands of labels (classes). Were explored various types of neural networks based on RNN and CNN classifiers.

Table 1. Summary of previous studies on the classification of judicial documents.

Algorithm	Accuracy	Size	Language	Class
SVM	96%	126,000	French	8
NCD	85%	70	Italian	24
BCN	80%	70	Italian	24
SVM	88%	582	Italian	11
NB	82%	582	Italian	11
Word2vec	74%	18,000	English	10
CNN	82%	25,000	English	2
SVM	78%	25,000	English	2
BERT	60.70%	6,227	English	31

Table 1 summarizes the performance of these previous studies on the automatic classification of court documents.

3   Methodology

In this section, we present the methodology for classifying electronic lawsuits. Section 3.1 describes the proposed model. Section 3.2 presents the data and, finally, Sect. 3.3 presents the performance metrics used to evaluate the quality of the proposal of this work.

3.1   Stages

In order to evaluate the performance of the different algorithms used in the classification of electronic lawsuits (classification with traditional Machine Learning, classification with Neural Networks and classification with Glove word vectors), were applied the following steps:

1. *Data collection*: Stage that involves the extraction of the content of electronic legal proceedings and the label "type of documents".
2. *Pre-processing*: Performing the tasks of the pre-processing phase in a correct order, as the sequence in which these changes are made can affect the result.
3. *Selection of characteristics*: Step that analyzes the electronic lawsuits of the data set in its textual form and extracts its characteristics generating a numerical representation of each text.
4. *Class balancing*: The data sets already reduced are balanced in order to balance the contribution of each class in the training of the model.
5. *Battery of tests*: The data set is shuffled and divided and the parts are destined to the training and classification stages.
6. *Training subset*: It consists of the first part of the reduced and scrambled data set that are used to train the machine learning algorithms.
7. *Model*: It involves the construction of an artifact resulting from the training of the learning model. This training step involves dividing the data set into k-parts.
8. *Parameter optimization*: It is an iterative step performed in the training in order to select the hyper-parameters that will generate the best results. The Randomsearch technique was used to optimize the parameters of the models [9].
9. *Subset of tests*: It consists of the second part of the reduced and scrambled data set that are used to test the machine learning algorithms.
10. *Evaluation*: The model is evaluated by classifying the subset of tests and the result of each metric is calculated.
11. *Extraction of results*: It is performed after testing the models. The main results obtained are F1 score, precision and recall.
12. *Comparison of results*: Comparison of results to obtain the best model.

3.2 Dataset

The evaluative studies of the classifiers used the database of the PJE system, composed of legal proceedings from the 1st instance[1] of the TJMG.

Documents attached by users (lawyers) in image format were disregarded and those typed directly into the system or attached in PDF format were considered.

The initial corpus consists of 3,000,853 electronic lawsuits, which were distributed in court between 2012 and January/2020. A process consists of procedural pieces[2], which are documents that have an average of 5,400 tokens (words) and are significantly larger than classic texts, such as news articles, commonly found in data sets used to compare machine learning models in classifying text.

This work aims to create a model that allows the automatic classification of electronic lawsuits and to make the experiments feasible, it was considered only part of the initial

---

[1] The first instance is the first hierarchical jurisdiction, i.e., the first body of Justice to which the citizen must address a dispute resolution request.

[2] Legal term referring to the manifestations in processes such as: Initial Petition, Contestation (defense), Embargos, Sentence.

corpus. To this end, the judicial processes of a specific District[3] were selected. Thus, it is possible to achieve the amount of data necessary to carry out the training of the machine learning models and, at the same time, to make possible an adequate infrastructure to the volume of data handled.

The initial corpus showed inconsistency in the quality of data by that of users (lawyers), often filling in information about the process in an erroneous way, among which we can highlight the filling in the field "type of legal document" (type of document). In order to minimize this inconsistency in data quality and, thus, avoid training the algorithms with incorrect data, some rules were applied to the document extraction flow:

- Selection only of documents that have content - To avoid the use of processes registered in the PJE system and that do not contain procedural documents.
- Selection of documents with a size greater than 3,000 bytes only - To avoid the use of documents in which the user does not fill in the content of the procedural document via the PJE text editor, and only informs a small text, such as: "Follow document attached"

After the application of these extraction rules and the selection of data for a specific region, a final data set was created, with 111,343 documents to be used in the experiments of this work.

Treatment of Unbalanced Data. It can be seen that, based on electronic legal proceedings, the number of types of documents is unbalanced. "Initial Petitions" are much more numerous than the documents of "Sentence", "Order" and "Decision". In this situation, the classification model generated could be skewed, that is, it would tend to classify the new data with the class that has more examples, in the case "Initial Petition".

To deal with the problem of unbalanced data, the undersampling method was used, which consists of randomly reducing the examples of the majority class. Thus, the 111,343 documents were evenly distributed among the classes Initial Petition, Sentence, Order and Decision.

3.3 Performance Metrics

Works that involve comparisons of any nature require the definition of which is the subject of evaluation, as well as from which point of view it will be evaluated. The metrics used in the evaluation of the models of this experiment were Precision, Revocation, Score F1, Time of training and Time of classification.

4 Experiments

4.1 Classification with Traditional Machine Learning Techniques

The Naïve Bayes, AdaBoost and SVM classifiers were used to classify electronic court cases. The labeled data set was divided into two parts: training and testing, divided as

---

[3] Judicial circumscription, under the jurisdiction of one or more judges of law.

follows: 70% for training (77,940 documents) and 30% for tests (33,403 documents). The test data set was also used to adjust the parameters of the models. The test data set was used to decide the best values for hyperparameters. The k-fold method (k = 10) was used to partition the data set and then perform the tests for each of the partitions. From the results, the average was calculated and the final result of each algorithm was obtained. In this way, a more realistic result was achieved.

Word Frequencies with TF-IDF. As it is not possible to work directly with text when using machine learning algorithms, the conversion of texts into numbers was performed, that is, the documents were converted into vectors of numbers. The data were tokenized and, to calculate the word frequencies, the TF-IDF method (Inverse frequency of documents) was used, which reduces the words that appear repeatedly in the documents. The TF-IDF depicts word frequency scores and highlights the most interesting, frequent in a document, but not between documents. Finally, a vocabulary was created with 317,633 terms from the database of lawsuits with their respective inverse document frequencies.

Pre-Processing. The NLTK[4] and Scikit-learn[5] libraries were used to create a stopwords list. 203 stopwords were used, among which 170 were detected in the set of lawsuits. The stop words are repeated 67,249,028 times in the data set. The removal of the stop words did not affect the F1 Score of the SVM classifiers and improved the performance of the other classifiers. It was found that stopwords are very frequent in the data, and therefore have been removed to emphasize words that make sense for the legal business, thus, the algorithms obtained better classification results. By removing unnecessary words, performance gains were obtained by processing less data.

Results. Table 2 presents the results of the traditional classification of the type of document of the electronic lawsuits. The numbers in bold represent the best performance.

Table 2. Ranking of traditional classifiers in the electronic lawsuit dataset.

Classifier	Precision	Recall	Score F1	Time (minutes)
SVM (One-vs-One)	**0.9340**	**0.9340**	**0.9340**	**14.75**
SVM (One-vs-Rest)	0.9338	0.9338	0.9338	20.95
Naive Bayes	0.8558	0.8558	0.8558	0.85
AdaBoost	0.8000	0.7974	0.7974	1,198.80

- SVM, in both strategies (One-Vs-One and Once-Vs-Rest), performed better than the other classifiers.

---

[4] The Natural Language Toolkit is a set of libraries and programs for symbolic and statistical processing of natural language written in the Python programming language.

[5] Open source machine learning library for the Python programming language.

- SVM (One-Vs-One) was considered the winning algorithm, among the traditional classifiers, as it proved to be slightly superior to SVM (One-Vs-Rest), reaching the value of 0.9340 for all performance measures, while the (One-Vs-Rest) reached the value of 0.9338.
- However, the SVM classifier with the One-Vs-Rest strategy consumed less training time, 14.75 min versus 20.95 min for the best strategy (One-Vs-One).
- Naive Bayes overall performance was weak compared to SVM. One reason for this poor performance may be the dependency feature on the data.
- Observing the results, the AdaBoost algorithm had the worst performance, demonstrating to be more vulnerable to randomness and making a lower success rate in relation to the other classifiers. One explanation would be that AdaBoost focuses on training examples classified incorrectly, making the Boosting technique more vulnerable to overfitting.
- The AdaBoost classifier had a classification time vastly superior to the other classifiers, spending 1,198.80 min to train, while the second classifier that used more time to train - SVM (One-vs-One) - consumed 20.95 min.

4.2 Classification with Deep Learning Techniques

In this section, it has described a set of experiments, using CNN, to classify automatically types of documents in electronic lawsuits.

CNN Architecture. CNN have been used to classify the types of documents in electronic lawsuits. A CNN has been implemented which has an incorporation, a convolution, a maxpooling and a fully connected layer similar to the one highlighted in [10]. This simple CNN-random architecture demonstrated a good performance in the classification of electronic lawsuits. In addition, the performance of random CNN have been compared with CNN architectures using Glove word vectors in Portuguese. Deeper models using Glove outperformed the simple CNN model. Figure 2 shows the architecture of the CNN model. In the Embedding layer, each batch of data has 150 records and the maximum size considered for documents in a lawsuit was the first 1,200 words. The data are been represented in a $150 \times 1,200$ matrix. Each of the words in the records is been replaced by a dense vector of size 100, resulting in a $150 \times 1,200 \times 100$ matrix

In the convolution layer, 512 filters of sizes one, two, three, four and five (2,560 in total) have been used. The elementary multiplication of each of the filters with the vector representation of a record creates a resource map. For example, given the document type "Sentence", the elementary multiplication of the filter with size 2 with each word in the document type leads to a resource map with 1,200 scalar values. As we have 512 filters, we have 512 resource maps with different sizes.

Then, the resulting 2,560 resource maps were fed into the pool layer, which extracts the maximum value from each of the resource maps. Then, the grouped resources are concatenated to create a single resource map. Some of the features are eliminated at random (set to zero) to avoid over-adjustment as suggested in [11]. Finally, the remaining resources were fed into the fully connected layer that has a sigmoid function, which assigns an independent probability to each class. This process is been repeated for each batch.

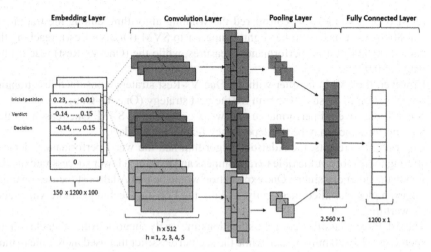

Fig. 2. CNN's architecture for classifying court cases.

A comparison was made between a CNN-Random that learns word vectors during the supervised classification process with CNN-Glove-Static-Portuguese, which uses pre-trained word vectors from a large corpus in an unsupervised way, as described below.

CNN-Random. The word incorporation is initialized with a dense vector of 100 dimensions that are updated during the training process. The values of the vector are initialized in the range [−0.25, 0.25] and are displayed randomly from a uniform distribution.

CNN-Glove-Portuguese. The words are initialized with pre-trained Glove vectors [12] from a large corpus of Brazilian Portuguese and European Portuguese, from different sources and genres. This data set has 1,395,926,282 words in Portuguese. Of the 337,086 words present in the documents of electronic legal proceedings, 92,398 (27%) were found in the Portuguese language data set. The words in the text were replaced by their incorporations through a lookup table. If the incorporation of a word were not available, the word vector would start randomly in the range [−0.25, 0.25]. We also compared the CNN-Glove-Static-Portuguese that does not update word vectors during training with a CNN-Glove-Dynamic-Portuguese that updates word vectors during training.

Hyper-Parameter Adjustment. CNN's performance is highly dependent on its hyperparameters. One of the crucial stages of deep learning and working with neural networks is the optimization of hyper-parameters. The models used in this work have a large number of parameters to adjust and choose. We chose to use the Random Search optimization technique. This method is widely used in the application of the deep learning model [13] and uses random combinations of parameters using specific values, which allows faster processing, as it does not exhaustively explore all possible values of hyper-parameters. The CNN parameters explored for this classification of lawsuits, used cross-validation (k = 10) and number of interactions = 5, totaling 50 model trainings in order to find the ideal values. Table 3 shows the standard hyperparameters used and Table 4 shows the hyperparameters processed through Random Search.

Table 3. Default values of hyperparameters used in CNN

Hyper-parameters	Value
Activation Function	ReLu
Mini lot size	150
Strid Size	1
Pooling	Max

Table 4. CNN hyper-parameter ranks.

Hyper-parameters	Value
Learning rate	[0,0001, 1]
Number of Filters	[32, 64, 128, 512]
Drop Out	[0,1, 0,5]
Kernel size	[3, 5, 7]

To process the optimization of hyperparameters using Random Search, we divide the data into a set of tests and a training set. 30% of court documents were randomly selected for the test suite (33,403 records). The training set included the remaining 77,940 records. Cross-validation was used 10 times in the training set to adjust the hyperparameters, which contributed to the deep learning models consuming a high training time.

Next, CNN was tested using the best hyper-parameters identified in the test suite. The values of the hyper-parameters that obtained the best F1 Score for CNN-Glove-Portuguese (static and dynamic) are show in the Table 5.

Table 5. Final hyper-parameters used in cnn-glove-portuguese.

Hyper-parameters	Valor
Activation function	ReLu
Mini lot size	150
Stride Size	1
Pooling	Max
Learning Rate	0,001
Number of filters	512
Drop Out	0,5
Kernel size	5

One problem with training neural networks is in choosing the number of training times to use. Too many times can lead to over-tuning of the training data set, while too few times can result in a misfit model. In this work, the "early stop" method was used, which allows specifying a number of training periods and interrupting it when the model's performance stops improving in the validation data set.

Results. The model that obtained the best performance, among the deep learning algorithms, was the CNN-Glove-Static-Portuguese. The evaluation of this model in the set of tests presented is through the calculations of the Accuracy metric (Fig. 3) and the Loss metric (Fig. 4).

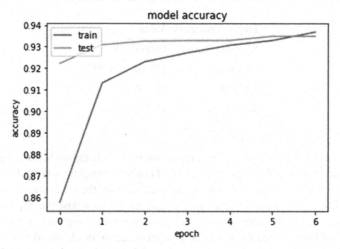

Fig. 3. Accuracy of the best performing deep learning model - CNN-Glove-Static-Portuguese.

Table 6 shows the performance of the CNN classifiers.

- The CNN-Glove-Static-Portuguese model achieved the best performance, however it obtained a slightly higher performance compared to the other models.
- The expectation was that pre-trained word vectors would improve CNN's performance. However, CNN-Glove-Dynamic-Portuguese with dynamic word vectors performed slightly less than CNN with static Glove word vectors. One reason is that only 27.1% of the words in the data are found in the Glove database and random word vectors replace the remaining 13%. Some of the words that were not found are abbreviations. Nevertheless, Glove's pre-trained word vectors outperformed the other CNN-Random and SVM classifiers.
- The word vectors trained with the Glove algorithm (CNN-Glove-Static-Portuguese and CNN-Glove-Dynamic-Portuguese) surpassed the SVM, which shows that ready-made vector resources (pre-trained) bring extra semantic resources that collaborate in the tasks of text classification and machine learning.
- CNN-Random did not outperform SVM, although it is a more complex classifier.

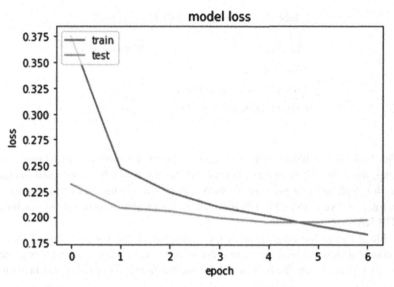

Fig. 4. Loss of the best performing deep learning model CNN-Glove-Static-Portuguese.

Table 6. Result of classification of lawsuits with convolutional neural networks.

Classifier	Accuracy	Precision	Recall	Score F1	Loss
CNN-Random	0.9337	0.9348	0.9323	0.9336	0.2026
CNN-Glove-Dynamic-Portuguese	0.9345	0.9366	0.9324	0.9345	0.2006
CNN-Glove-Static-Portuguese	**0.9349**	**0.9370**	**0.9329**	**0.9350**	**0.1970**

- Regardless of the classifier, the Score F1 exceeded 0.93, which is presumed to be a great result. Document data from electronic lawsuits is long, averaging 5,400 words per record. The classification of a long text can be considered more accessible than a short text. In addition, common resource selection and weighting methods work best on long texts. Another issue to take into account is the significant amount of data involved in this work. Previous studies on the classification of court documents (Table 1) that used machine learning algorithms obtained results of accuracy that varied from 49.2% to 96%. When comparing the results of this work with the results of the previous researched studies, it was found that the classification results of this work are similar to the performance of the work with better performance and superior to the results of all the others.

Execution Time. Table 7 shows the execution time of CNN models. Deep learning models generally employ longer runtime than classic machine learning algorithms.

Table 7. Execution time for CNN classifiers.

Classifier	Time (hours)
CNN-Random	2.60
CNN-Glove-Dynamic-Portuguese	2.83
CNN-Glove-Static-Portuguese	1.40

The Scikit-learn library was used to implement the models with classic machine learning, the TensorFlow library [14] for the execution of CNN codes and the Gensim library for application of pre-trained word vectors (Glove) for the Portuguese language, on an Intel (R) Xeon (R) CPU E5-4610 V2 @ 2.30 GHz (16 processors) machine with 164 GB RAM.

Table 8 shows the time spent in the execution of the optimization of the hyper-parameters, which is a lengthy process because it is necessary to adjust a large number of parameters to choose the best model for the classification of electronic lawsuits.

Table 8. Execution time for optimization of CNN parameters.

Optimization method	Time (hours)
Random Search	163.00

The source code used in the experiments of this work is available in a public repository[6].

5 Conclusion and Future Work

TJMG has millions of physical lawsuits that are digitized and classified manually. This is an expensive job and an automatic classification of electronic lawsuits would be of great importance for the institution. In this scenario, the present study uses machine learning techniques to propose and build a proof of concept and takes as a question to investigate which supervised machine learning algorithm has the best performance for the current problem of automatic classification of electronic lawsuits.

Traditional machine learning algorithms like SVM, Naïve Bayes, and AdaBoost have been compared with current algorithms that use deep learning and incorporation of Glove words generated for the Portuguese language, through CNN. The main results obtained were evaluated under the metrics F1 score, precision and recall. To train the classifiers, 111,343 electronic court documents were handled and distributed in the TJMG and duly labeled. As each document is associated with a certain label (Initial Petition, Sentence, Order, Decision), we worked with the multiclass classification.

---

[6] https://github.com/adrianocapanema/ClassificationOfElectronicLawsuitsAnApplicationInTh eTJMG.

The results showed accuracy in predicting the types of documents in court proceedings, with the Score F1 of all CNN and SVM classifiers exceeding 93%. In the implementation of traditional classifiers, it was found that SVM had a better performance than other traditional classifiers. The implementation of recent classifiers showed that the CNN-Glove-Static-Portuguese model achieved the best performance in comparison to the other CNN models.

The word vectors trained with the Glove algorithm (CNN-Glove-Static-Portuguese and CNN-Glove-Dynamic-Portuguese) outperformed the other classifiers, which demonstrates that ready-made vector resources (pre-trained) bring extra semantic resources that collaborate in the tasks classification of legal proceedings.

Considering that CNN-Glove-Static-Portuguese obtained the best result among the analyzed algorithms with 93.7% precision, it is concluded that such model, presents itself as the best option for the classification of texts in the legal domain.

It is believed that the automation provided by the techniques used in this research promotes both a scientific and practical contribution within the scope of the judiciary, and can collaborate to increase productivity, saving time and financial resources. We also understand that the AI models created in this study can be used not only by the TJMG, but also by courts throughout Brazil.

Future works could compare the performance of the CNN models trained in this research with other word vectors in the Portuguese language such as Bert, Word2Vec and FastText. Considering that a reduced number of words from the judicial data was found in the Glove database, it is proposed to adapt word vectors generated in the Portuguese language to the legal domain including the forensic corpus as one of the corpora used to train the word vectors, the in order to expand the semantic representation of the models.

References

1. Evaluation: from Precision, Recall and F-measure to ROC, Informedness, Markedness and Correlation (2011)
2. Sulea, O.-M., Zampieri, M., Malmasi, S., Vela, M., Dinu, L.P., van Genabith, J.: Exploring the Use of Text Classification in the Legal Domain (2017)
3. Mastropaolo, A., Pallante, F., Radicioni, D.P.: Legal documents categorization by compression. In: Proceedings of the Fourteenth International Conference on Artificial Intelligence and Law - ICAIL'13, p. 92. ACM Press, Rome (2013)
4. Francesconi, E., Passerini, A.: Automatic classification of provisions in legislative texts. Artif. Intell. Law 15, 1–17 (2007). https://doi.org/10.1007/s10506-007-9038-0
5. Lilleberg, J., Zhu, Y., Zhang, Y.: Support vector machines and Word2vec for text classification with semantic features. In: 2015 IEEE 14th International Conference on Cognitive Informatics & Cognitive Computing (ICCI*CC), pp. 136–140. IEEE, Beijing (2015)
6. Wei, F., Qin, H., Ye, S., Zhao, H.: Empirical study of deep learning for text classification in legal document review. In: 2018 IEEE International Conference on Big Data (Big Data), pp. 3317–3320 (2018). https://doi.org/10.1109/BigData.2018.8622157
7. Howe, J.S.T., Khang, L.H., Chai, I.E.: Legal Area Classification: A Comparative Study of Text Classifiers on Singapore Supreme Court Judgments (2019). arXiv:1904.06470 [cs]
8. Chalkidis, I., Fergadiotis, M., Malakasiotis, P., Aletras, N., Androutsopoulos, I.: Extreme Multi-Label Legal Text Classification: A case study in EU Legislation (2019). arXiv:1905.10892 [cs]

9. Pedregosa, F., et al.: Scikit-learn: machine learning in Python. J. Mach. Learn. Res. **12**, 2825–2830 (2011)
10. Kim, Y.: Convolutional neural networks for sentence classification (2014). arXiv preprint arXiv:1408.5882
11. Agarwal, A., Negahban, S., Wainwright, M.: A simple way to prevent neural networks from overfitting. Ann. Stat. **40**, 1171–1197 (2012)
12. Hartmann, N., Fonseca, E., Shulby, C., Treviso, M., Rodrigues, J., Aluisio, S.: Portuguese Word Embeddings: Evaluating on Word Analogies and Natural Language Tasks (2017). arXiv: 1708.06025 [cs]
13. Bergstra, J., Bengio, Y.: Random search for hyper-parameter optimization. J. Mach. Learn. Res. **13**, 281–305 (2012)
14. Abadi, M., et al.: Tensorflow: A system for large-scale machine learning. In: Symposium on Operating Systems Design and Implementation, pp. 265–283 (2016)

Towards a Free, Forced Phonetic Aligner for Brazilian Portuguese Using Kaldi Tools

Ana Larissa Dias(✉)(iD), Cassio Batista(iD), Daniel Santana(iD), and Nelson Neto(iD)

Institute of Exact and Natural Sciences, Federal University of Pará,
Augusto Corrêa 1, Belém 66075–110, Brazil
larissa.engcomp@gmail.com, {cassiotb,nelsonneto}@ufpa.com,
daniel.santana.1661@gmail.com

Abstract. Phonetic analysis of speech, in general, requires the alignment of audio samples to its phonetic transcription. This task could be performed manually for a couple of files, but as the corpus grows large it becomes unfeasibly time-consuming, which emphasizes the need for computational tools that perform such speech-phonemes forced alignment automatically. Therefore, due to the scarce availability of phonetic alignment tools for Brazilian Portuguese (BP), this work describes the evolution process towards creating a free phonetic alignment tool for BP using Kaldi, a toolkit that has been the state of the art for open-source speech recognition. Five acoustic models were trained with Kaldi and tested in phonetic alignment, where the evaluation took place in terms of the phone boundary metric. The results show that its performance is similar to some Kaldi-based aligners for other languages, and superior to an outdated phonetic aligner for BP based on HTK toolkit.

Keywords: Phonetic alignment · Acoustic modeling · Kaldi · Brazilian Portuguese

1 Introduction

In order to analyze the prosodic features of speech sounds, it is becoming increasingly expected for linguists to take into account a large amount of data, often including several hours of recorded speech. With the advancement of local and cloud storage infrastructures, the biggest problem facing today linguists is not keeping the data safe, but generating its annotation, which includes utterance, word, syllable, and phoneme segmentations. Moreover, the analysis of the prosodic structure of speech very often requires the alignment of the speech recording with a phonetic transcription of the speech. Research into natural prosody generation for speech synthesis is an example of an issue that needs phonetically-annotated data with a high level of precision [3].

Nevertheless, it is obvious that transcribing and aligning several hours of speech by hand is very time-consuming, even for experienced phoneticians. Thus,

R. Cerri and R. C. Prati (Eds.): BRACIS 2020, LNAI 12319, pp. 621–635, 2020.
https://doi.org/10.1007/978-3-030-61377-8_44

several approaches have been applied to automate this process, some of them brought from the automatic speech recognition (ASR) domain. Before the emergence of deep neural networks (DNNs) as the state of the art for most of the current machine-learning tasks, the most widely explored phonetic alignment technique was to use hidden Markov models (HMM) in a forced-alignment mode.

In this context, automatic alignment tools such as P2FA [40], Prosodylab-aligner [16], EasyAlign [14], SPPAS [5], and Train&Align [7] have been developed and released. In addition to a phonetic dictionary, all these tools rely on the acoustic modeling of the language via HMMs, supported by a widespread HMM-based speech recognition toolkit, called HTK [39]. Such aligners provide the user with pre-trained speaker-independent models of each language, or can instead train models of each phoneme (monophone models) or group of phonemes (triphone models) directly on the corpus to be aligned. Then, these models are used to align an audio file with its phonetic transcription.

The Montreal Forced Aligner (MFA) [26] is an example of a system for speech-text alignment that uses a standard architecture combining HMMs and Gaussian mixture models (HMM-GMM), adapted from existing Kaldi recipes [34], which offers advantages over the HTK toolkit underlying most existing aligners. MFA is a 29-language (Brazilian Portuguese included), multilingual update to the English-only Prosodylab-aligner [16], and maintains its key functionality of training on new data, as well as incorporating improved architecture (triphone GMMs and speaker adaptation), which also offers the possibility of using DNN-based acoustic models for forced alignment. MFA performs well relative to its predecessor, generally resulting in more accurate word and phone boundaries.

Another Kaldi-based forced aligner is Gentle [29], which is available either as a graphical user interface in a web browser, programmatic HTTP requests, or as a Python library. Gentle is built on top of Kaldi's time-delay neural networks (TDNN) chain model [32]. Currently, Gentle performs forced alignment only on English data and it does not appear to be an academic work since no publications have been found. Therefore, to the best of our knowledge, there is no work regarding Gentle's performance compared to the others currently available automatic alignment tools.

Regardless of the technique adopted, phonetic alignment resources for Brazilian Portuguese (BP) are still scarce. In order to mitigate this gap, our first effort contributed with an automatic phonetic alignment tool for BP [37], consisting of grapheme-to-phone (G2P) converter, syllabification system and HMM-based acoustic models trained over the HTK toolkit. As usual, tests comparing the automatic versus manual segmentations were performed. An extra comparison was made with EasyAlign [14], which was the only other aligner that supported BP at that moment. It was observed that the tools achieved equivalent behaviors, considering two metrics: boundary-based and overlap rate.

Therefore, this work proposes an update to [37] by providing a free phonetic aligner for BP using Kaldi tools. Assuming Kaldi is pre-installed as a dependency, the proposed aligner works fine under Linux environments via command line, but also provides a graphical interface as a plugin to Praat [6], a free software package for speech analysis in phonetics. Multi-tier TextGrid files are created on-the-fly using phonetic and syllabic annotations from dictionaries constructed from a list of 200,000 most-frequent words in BP [12], but G2P and syllabification systems are also available as a package to be called for missing words [19]. All tools provided in the context of this work are being released as open-source under the MIT license [17].

Some intra- and inter-evaluation procedures were performed, the former considering all acoustic models trained within the Kaldi's default GMM and DNN pipeline, while the latter applied the HTK-based aligner from [37] over the same dataset for the sake of a fair comparison. The similarity measure is given in terms of the absolute difference, within a tolerance threshold, between the forced alignments with respect to manual ones, which is called phonetic boundary [26].

The remainder of the paper is organized as follows. Section 2 presents details regarding the phonetic alignment process using Kaldi. Sections 3 and 4 shows the experiments conducted over a hand-aligned evaluation dataset and the results achieved with respect to the phone boundary metric, respectively. Finally, Sect. 5 shows the conclusion and plans for future work.

2   Methodology

This Section describes the process to perform forced alignment using Kaldi tools, as well as details concerning the training of acoustic models.

2.1   Training Dataset

To build an effective acoustic model (AM), a relatively large amount of labeled data is required. The Brazilian Portuguese corpora used to train AMs with Kaldi consists of seven data sets summarized in Table 1. The data sets contain audio files in an uncompressed, linear, signed PCM (namely, WAVE) format, and are sampled at 16 kHz with 16 bits per sample. It is worth noting that Constitution and Consumer Protection Code corpora share the same speaker. Also, due to

Table 1. Audio corpora used to train acoustic models.

Dataset	Ref.	Hours	Words	Speakers
LapsStory	[28]	5h:18m	8,257	5
LapsBenchmark	[28]	0h:54m	2,731	35
Constitution	[31]	8h:58m	5,330	1
Consumer Protection Code	[31]	1h:25m	2,003	1
Spoltech LDC	[24]	4h:19m	1,145	475
West Point LDC	[25]	5h:22m	484	70
CETUC	[35]	144h:39m	3,528	101
Total		170h:51m	14,518	687

the foreign words excess amidst the corpus, the actual number of speakers in West Point was reduced.

2.2   Acoustic Models

Our acoustic models were trained by adapting a recipe created in [4], originally based on Kaldi's Wall Street Journal (WSJ) recipe, to Kaldi's Resource Management (RM) recipe [11]. The deep-learning-based training approach in Kaldi actually uses the HMM-GMM training as a pre-processing stage. Figure 1 shows the pipeline to training a HMM-DNN acoustic model based on HMM-GMM triphones using Kaldi. In the front-end, the acoustic waveforms from the training corpus are windowed at every 25 ms with 10 ms of overlap, being encoded as a 39-dimension vector: 12 Mel frequency cepstral coefficients (MFCCs) [8] using C0 as the energy component, plus 13 delta (Δ, first derivative) and 13 acceleration ($\Delta\Delta$, second derivative) coefficients are extracted from each window.

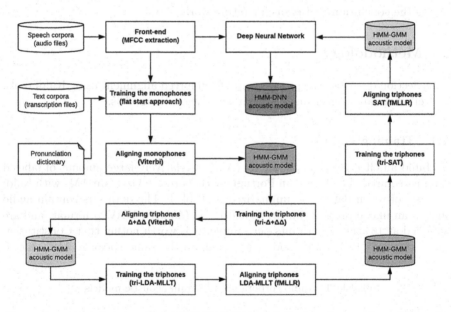

Fig. 1. Stages for training a hybrid HMM-DNN, triphone-based acoustic model on the top of HMM-GMM AMs. Scripts are based on Kaldi's RM recipe, and the DNN in particular was trained following the **nnet2** setup.

The flat-start approach models 39 phonemes (38 monophones plus one silence model) as context-independent HMMs, using the standard 3-state left-to-right HMM topology with self-loops. At the flat-start, a single Gaussian mixture models each individual HMM with the global mean and variance of the entire training data. Also, the transition matrices are initialized with equal probabilities.

The Viterbi training is the algorithm [38] used by Kaldi to re-estimate the models at each training step. Likewise, in order to allow training algorithms to improve the model parameters, Viterbi alignment is applied after each training step. Subsequently, the context-dependent HMMs are trained for each triphone, first with the delta and after with the acceleration coefficients. Each triphone is represented by a leaf on a decision tree. Eventually, leaves with similar phonetic characteristics are then tied/clustered together.

The next step is the linear discriminant analysis (LDA) [9] combined with the maximum likelihood linear transform (MLLT) [15]. The LDA technique takes the feature vectors and splices them across several frames, building HMM states with a reduced feature space. Then, a unique transformation for each speaker is obtained by a diagonalizing MLLT transform. On top of LDA+MLLT features, a speaker normalization that uses feature-space maximum likelihood linear regression (fMLLR) as alignment algorithm is applied [10].

The last step of the HMM-GMM training step is the speaker adaptive training (SAT) [1,2]. The SAT technique is applied on top of the LDA+MLLT-fMLLR features performing adaptation and projecting training data into a speaker normalized space. This way, by becoming independent of specific training speakers, the acoustic model generalizes better to unseen testing speakers [27].

The HMM-DNN model is obtained as a final-stage AM by using the neural network to model the state likelihood distributions as well as to input those likelihoods into the decision tree leaf nodes [20]. In short terms, the network input are groups of feature vectors and the output is given by the aligned state of the HMM-GMM system for the respective features of the input. The number of HMM states in the system also defines the DNN's output dimension [23].

For specifications and parameters regarding the DNN training, the reader is referred to [4]. The only major modification was to switch from a script called `train_pnorm_fast.sh` to `train_pnorm_simple2.sh`, because it is an updated version of the former script and uses the 'online' preconditioning, which is faster especially on GPUs. Therefore, a NVIDIA Titan Xp card was used to speed up the neural network training procedure.

2.3   Kaldi Forced Phonetic Alignment

The forced phonetic aligner proposed in this work uses Kaldi, a toolkit that is under active development and provides state-of-the-art algorithms for many assorted speech recognition tasks, including stable neural-network frameworks. Our aligner has also been developed as a plugin for Praat, a popular speech analysis software among linguists, which aims to ensure a user-friendly interface requiring only a few manual steps in the process.

The plugin's interface was developed in Praat's programming language— Praat Scripting. As input, it requires an audio file, the corresponding orthographic transcription and the path to Kaldi's root directory. Following a successful alignment, a multi-level annotation TextGrid file can be load into Praat. The scripts developed to handle the actual steps of the phonetic alignment process are written in Python. Besides, Kaldi provides standardized Bash scripts,

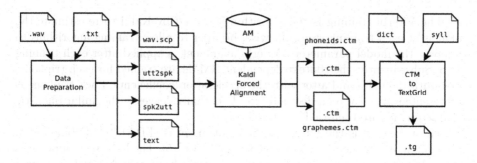

Fig. 2. The phonetic alignment process with its main inputs and outputs.

which wrap the Kaldi C++ executables tools that are employed in our proposed aligner. Figure 2 summarizes the whole phonetic alignment process using Kaldi.

For data preparation, the first step consists in checking whether there are any new words in the input data that were not part of the acoustic model training corpora. A Python script searches for the input data's words in the pronunciation dictionary (or lexicon), which contains each word of the training corpora with its phonemic pronunciation. Then, if any word is not found, the script uses the grapheme-phoneme conversion module from a natural language processing (NPL) tool [18] to extend the lexicon with each new word along with its respective phonemic pronunciation. Next, Kaldi requires the creation of some files that contain information regarding the specifics of the audio file and its transcription: `text`, `wav.scp`, `utt2spk`, and `spk2utt`. Our proposed aligner also rely on scripts that create these files automatically on the fly.

Kaldi forced alignment block itself performs several steps for obtaining the time-marked conversation (CTM) files, which contains a list of phonemes with both its start time and duration. First, Kaldi scripts are used to extract and normalize cepstral features (MFCCs) from time-domain audio data. Then the forced alignment step, that employs the aforementioned pre-trained acoustic models, is computed by Kaldi using Viterbi beam search algorithm [21].

In Kaldi, an alignment is a representation of the HMM states sequence taken by the Viterbi's alignment of an utterance [33]. In order to compute the forced alignment, Kaldi calls the `compile-train-graphs` tool, which creates a finite-state transducer (FST) binary graph from the lexicon's orthographic representation expanded into acoustic model features (phone-level HMM states). Next, the binary graph is read by either `gmm-align-compiled` or `nnet-align-compiled`, depending on the features of the acoustic model, HMM-GMM or HMM-DNN, to produce state-level alignments in a binary format compressed by `gunzip`. Lastly, `ali-to-phones` and `get-train-ctm` tools are employed to read the compressed file in order to obtain `phoneids.ctm` and `graphemes.ctm` output files, respectively, which contain a time-aligned, readable transcription of an utterance at the phoneme and grapheme level, individually.

The last block of the phonetic alignment process handles the conversion of both CTM files to a Praat's TextGrid (`.tg`), a text file containing the alignment information. For this task, a phonetic (`dict`) and a syllabic separation (`syll`) dictionaries were created based on a list of the most frequent words for Brazilian Portuguese provided by FrequencyWord project [12], which uses the OpenSubtitles.org [30] data. The GNU Aspell [13], a free spell checker, was used to discarding unsuited words from the FrequencyWord's list, resulting in a list of 200,000 most-frequent words that generated the phonetic and the syllabic dictionaries for BP, available at [19]. The phonetic alphabet used is the International Phonetic Alphabet (IPA), an alphabetic system of phonetic notation. Therefore, CTM files are read by a Python script that in the conversion process uses the `dict` and `syll` dictionaries generating the TextGrid file as output.

Finally, the aligner offers the option to promptly display the current resulting TextGrid in the Praat interface or to proceed to align a new audio file. Figure 3 shows the Praat's TextGrid editor displaying an audio file waveform followed by its spectrogram and its aligner's resulting multi-tier TextGrid containing five tiers: phonemes, syllables, words, phonetic transcription and orthographic transcription, respectively.

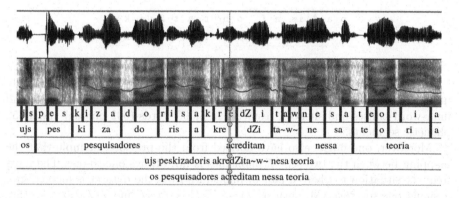

Fig. 3. A example of a resulting TextGrid, displayed by Praat's TextGrid editor, including five tiers: phonemes, syllables, words, phonetic transcription and orthographic transcription, respectively.

3 Experiments

This Section describes the experiments conducted in order to evaluate how good is the alignment estimates provided by Kaldi, as well as the characteristics of the dataset used as ground-truth during the measurements.

3.1 Evaluation Dataset

In these experiments, the automatic alignment was estimated on the basis of the manual segmentation. The original dataset used for assessing the accuracy of the phonetic aligner is composed of 200 utterances spoken by a male speaker, in a total of 7 min and 58 s of hand-aligned audio. Praat's TextGrid files, whose phonetic timestamps were manually adjusted by a phonetician, are available alongside audio and text transcriptions.

Nevertheless, apart from the fact that this corpus uses a different set of phonemes rather than the ones understood by acoustic models (AMs), it also has some problems of phonetic mismatches between phonemes and syllabic TextGrid tiers, phonemes among graphemes, and lots of cross-word phonemes between words, which makes the mapping between both phoneme sets challenging, given that the grapheme-to-phoneme (G2P) software used during AM training handles only internal-word conversion [36].

The example below shows the phonetic transcription for the phrase "*para a informática*" given by the original dataset (top) and the acoustic model (bottom), which then becomes "*parinformática*" in the former due to cross-word rules that suppress vowel sounds altogether. Phrases like this have been therefore discarded, since the boundary between words is lost.

p	a	4	i~	n	f		o h/	m	a	t S	i	k	a
p	a	r	a		a	i~	f	o	R	m	a	tS	i	k	a

In the end, 31 audios were removed from the dataset, which left 169 only utterances remaining (6 min and 42 s). The filtering also discarded intra- and inter-word pauses and silences, resulting in 1,099 words (602 unique) and 4,652 phonetic segments.

Moreover, one might also have noticed from the previous example that the mapping between the two sets of phonemes is not always one-to-one. The most frequent situation is where a pair of phonemes from the dataset is merged into a single one for the AM, such as /i~/ /n/ → /i~/ and /t/ /S/ → /tS/. However, a single phoneme can also be less frequently split to one more, such as /u/ /S/ → /u/ /j/ /s/.

To deal with this irregular mapping, we used the Many-to-Many alignment model (m2m-aligner) software [22] in the core of a pipeline that converts the original TextGrid from the evaluation dataset to a TextGrid that is compatible with the AM's phonetic dictionary (or lexicon), as shown in Fig. 4 for the phrase "*os jardins*".

The m2m-aligner works in an unsupervised fashion, using an edit-distance-based algorithm to align two different (unaligned) strings from a file in the news format, in order for them to share the same length. As this algorithm works based on frequency counts (e.g., how many times phonemes /d/ and /Z/ are merged to /dZ/), all 169 TextGrid files from our evaluation dataset, represented as short .tg, are used to compose a single news file.

Additionally, the news file created by our tg2news.py script is composed by the phonemes of a pair of words rather than isolated words, in order to avoid the effects of the cross-word boundaries. An example of the set of phonemes of

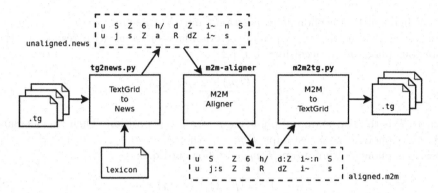

Fig. 4. TextGrid mapping pipeline.

two sentences within a single **news** file is shown in Table 2. In practice, the file is composed of the last two columns, separated by a tabular '\t' character, so every other phoneme token is separated by a single space. The string mapping is finished when the m2m-aligner provides a one-to-one mapping in a file that joins some phonemes together through a colon ':' character (see Fig. 4).

Table 2. Example of a single **news** file with phonemes from two out of 169 TextGrid files for sentences *"a questão foi retomada"* and *"os jardins são lindos"*.

Word pair	Dataset phonemes	AM phonemes
a questão	6 k e s t a~ w~	a k e s t a~ w~
questão foi	k e s t a~ w~ f o j	k e s t a~ w~ f o j
foi retomada	f o j h/ e t o~ m a d 6	f o j R e t o m a d a
os jardins	u S Z a h d Z i~ n S	u j s Z a R dZ i~ s
jardins são	Z a h/ d Z i~ n S s a~ w~	Z a R dZ i~ s s a~ w~
são lindos	s a~ w~ l i~ n d u S	s a~ w~ l i~ d u s

Finally, as the m2m-aligner provides the mapping for phonemes, our script m2m2tg.py provides the timestamps calculations prior to creating the converted TextGrid file. Table 3 shows how the phonetic timestamps, in milliseconds, are mapped accordingly.

Table 3. Timestamps' conversion for phrase *"os jardins"*.

000	114	175	268	375	445	496	571	714	752
u	S		Z	6	h\	d	Z	i~	n	S
u	j	s	Z	a	R	dZ		i~		s

| 000 | 057 | 114 | 175 | 268 | 375 | | 496 | | 714 | 752 |

When splitting a phone p at position t into N phonemes $p'_i, p'_{i+1}, \ldots, p'_{i+n}$, with $n = N - 1$, first we calculate a boundary offset b_o with respect to the previous phone p_{t-1}, according to Eq. 1:

$$b_o = \frac{T(p_t) - T(p_{t-1})}{N}, \tag{1}$$

where T is the timestamp of phoneme p at a given position. Then we calculate the new phonemes' timestamps $T(p'_{i+j})$, where $j = [0, 1, \ldots, n]$, based on the previous phoneme timestamp $T(p_{t-1})$, following Eq. 2:

$$T(p'_{i+j}) = T(p_{t-1}) + (i + j) \times b_o. \tag{2}$$

For merging N phonemes $p_t, p_{t+1}, \ldots, p_{t+n}$ into one p'_i, on the other hand, one simply needs to consider the timestamp of the last phoneme $T(p_{t+n})$.

3.2   Phone Boundary

Kaldi's alignment output is stored into a time-marked conversation (CTM) file, which contains a list of phonemes together with this start time with respect to the audio timing, and its duration, both in seconds. After CTM has been parsed to TextGrid, a script then reads two TextGrids (the one generated by Kaldi alignment pipeline, and the other one mapped from the original evaluation dataset to the AM phone set) in order to compute the phone boundary.

Phonetic boundary simply considers the absolute difference between the ending time of both phoneme occurrences [26]. The calculation is performed five times, one for each acoustic model, and takes place over all 169 utterances from the evaluation dataset.

Besides, for the sake of comparison, the phonetic boundary was also computed a sixth time over the alignments provided by an outdated HTK-based phonetic aligner that works for Brazilian Portuguese [37]. The phoneme set of the HTK-based aligner is the same used for our Kaldi AMs, so there has been no trouble using the same scripts to calculate the boundaries, given the phonemes are the same and are not position-shifted.

4   Results

Figure 5 shows histograms of the distribution of the difference between manual and forced alignments (phone boundary), on logarithmic scale, for HTK-based and all Kaldi-based acoustic models. The dashed, vertical line is at 5 ms just for localization purposes. All distributions are right-skewed, which is expected and also in conformity with the results for the English version of the MFA aligner [26]. All our Kaldi AMs apparently behave similarly, but one can see that HTK-based aligner is a little bit more to right-sided, meaning its timestamps differ more from the ground-truth evaluation dataset than all Kaldi-based forced alignments.

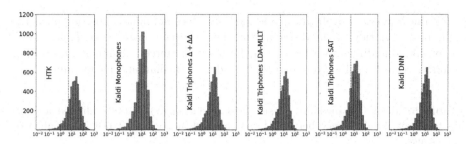

Fig. 5. Histogram of phone boundaries between ground-truth hand-aligned and forced alignments by both HTK and Kaldi.

Table 4. Cumulative percentage below a tolerance threshold, in milliseconds, of the differences between forced aligned audio and ground-truth (hand aligned) phonemes.

Toolkit/model	Tolerance (ms)			
	<10 (%)	<25 (%)	<50 (%)	<100 (%)
HTK-based [37]	33.95	65.73	86.40	96.54
Kaldi monophones	45.57	83.89	**96.71**	99.39
Kaldi triphones Δ+ΔΔ	**48.36**	**85.35**	**96.71**	**99.71**
Kaldi triphones LDA-MLLT	47.66	83.82	96.53	**99.71**
Kaldi triphones SAT	46.62	83.03	96.08	99.55
Kaldi DNN	46.49	82.65	96.15	99.66

The situation is verified in Table 4, which shows the cumulative distribution of phone boundaries with respect to some tolerance thresholds, in milliseconds. Roughly only 1% of phoneme tokens aligned by Kaldi acoustic models are off the 100 ms tolerance, against 4% of tokens aligned by HTK. In fact, approximately 96% of phonemes were under the 50 ms tolerance when aligned by Kaldi AMs, considering an average of all models.

Table 4 also highlights the best results in bold. Although the triphone-Δ+ΔΔ model holds the 48.36% and 85.35% most accurate alignments under 10 ms and 25 ms, respectively, this does not strictly mean it is the overall best model. According to the results, no Kaldi AM is the true "winner", since the first two tolerances of all triphone-based models contain a distribution close to 46% and 83% of tokens, respectively. The DNN model is also pretty close to those values, and the monophone model, although it is the worst among the Kaldi AMs, still performs relatively better than HTK.

It is important to mention that, despite the results from [37] had been calculated over the same evaluation dataset, the filtering process described in Sect. 3 resulted in a different number of files that undergo the alignment process: 169 vs. 181. Results, however, are compatible to the ones from [37], and differ at most by a 2.5% tolerance factor.

Table 5 shows both the mean and median values in milliseconds, as well as the standard deviation of the distributions. The comparison HTK vs. Kaldi shows a considerable difference of ~5–11 ms between the forced alignments from both toolkits regarding the ground-truth, considering median and mean values.

Table 5. Mean (μ), median and standard deviation (σ) of the evaluated aligners in terms of the difference between forced aligned and ground-truth (hand aligned) phonemes.

Toolkit/model	μ (ms)	median (ms)	σ
HTK-based [37]	26.043	15.961	32.378
Kaldi monophones	15.233	11.196	16.327
Kaldi triphones $\Delta+\Delta\Delta$	**14.438**	**10.357**	15.178
Kaldi triphones LDA-MLLT	14.726	10.577	**15.095**
Kaldi triphones SAT	15.359	10.834	16.314
Kaldi DNN	15.306	10.904	15.864

Among the Kaldi AMs, the highest mean and median values are found in triphone-SAT and monophone models, respectively. The most accurate alignments, on the other hand, are once more given by the triphone-$\Delta+\Delta\Delta$ model. However, once again there is no great difference among AMs, as both mean and median values are within a pretty close range (~14–15 ms and ~10–11 ms, respectively).

5   Conclusion

This paper presented what appears to be the first effort towards creating a free, forced phonetic aligner for Brazilian Portuguese (BP) using Kaldi tools. The proposed aligner works either via command line (Linux) or in a graphical interface as a plugin to Praat. Up-to-date phonetic and syllabic dictionaries create over a list of 200,000 most-frequent words for BP [19] are also provided, as well as standalone grapheme-to-phoneme and syllabification systems for handling missing words outside the dictionaries [18].

For evaluation, a comparison among the Kaldi-based acoustic models trained with an updated version of the scripts from [4] was performed, as well as a comparison to an outdated HTK-based aligner from [37].

Results regarding the absolute difference between forced and manual aligned utterances (phone boundary metric) showed that the HTK-based aligner performed worse when compared to any of the Kaldi-based models. A possible reason for such bad a result might be that the version of HTK shipped with the aligner from [37] uses Baum-Welch algorithm for training HMMs while Kaldi uses Viterbi training [4].

Among Kaldi models, triphone-Δ+$\Delta\Delta$ stands out as being virtually the best one. However, with just a $\sim$1–3% difference in tolerance, and $\sim$1 ms difference in both mean and median values, we cannot tell whether it is significant enough to classify one model into being better than the others, as they appear pretty close at glance. We would rather prefer to state that the linear sequence of model training does not result in lower errors regarding phonetic boundaries as it resulted in lower error rates for speech recognition [4].

As future work, we will perform a somehow "biased" experiment of using the same evaluation dataset for training the acoustic models with Kaldi, as also reported in [26]. MFA supports such feature, and their experiments show that retraining on the dataset to be aligned often improves (or at least it barely hurts) alignment accuracy relative to using acoustic models pre-trained on a larger dataset.

Gentle, on the other hand, already supports Kaldi's nnet3 setups, performing on the top of a pre-trained TDNN chain model, which may give us a reason to update again the recipe from [4] from nnet2 to nnet3 using either Mini-Librispeech or Aspire Kaldi default recipes. As for MFA, a DNN framework based on Kaldi's nnet2 is already operational, but still unstable and may not give a better result than the alignments produced by the standard HMM-GMM pipeline according to the MFA's documentation.

Furthermore, although our proposed aligner can be used as a plugin to Praat, we plan in the future to make it portable to MFA or Gentle under the same licensing, as to avoid open-source competition. Both software codebases are more well documented and well maintained, seem to be up-to-date, and may potentially cover a more broad community.

Acknowledgment. We gratefully acknowledge NVIDIA Corporation with the donation of the Titan Xp GPU used for this research. The authors also would like to thank CAPES and CNPq research funding agencies, and Federal University of Pará (UFPA) under Edital n° 06/2019 – PIBIC/PROPESP for the financial support.

References

1. Anastasakos, T., McDonough, J., Makhoul, J.: Speaker adaptive training: a maximum likelihood approach to speaker normalization. In: 1997 IEEE International Conference on Acoustics, Speech, and Signal Processing, vol. 2, pp. 1043–1046 (1997)
2. Anastasakos, T., Mcdonough, J., Schwartz, R., Makhoul, J.: A compact model for speaker-adaptive training. In: Proceedings of the ICSLP, pp. 1137–1140 (1996)
3. Batista, C., Cunha, R., Batista, P., Klautau, A., Neto, N.: Utterance copy in formant-based speech synthesizers using LSTM neural networks. In: 2019 8th Brazilian Conference on Intelligent Systems (BRACIS), pp. 90–95, October 2019. https://doi.org/10.1109/BRACIS.2019.00025
4. Batista, C., Dias, A.L., Sampaio Neto, N.: Baseline acoustic models for Brazilian Portuguese using Kaldi tools. In: Proceedings of IberSPEECH, pp. 77–81 (2018). https://doi.org/10.21437/IberSPEECH.2018-17

5. Bigi, B., Hirst, D.: Speech phonetization alignment and syllabification (SPPAS): a tool for the automatic analysis of speech prosody. In: Proceedings of Speech Prosody, pp. 1–4, May 2012. https://www.isca-speech.org/archive/sp2012/papers/sp12_019.pdf

6. Boersma, P., Weenink, D.: Praat: doing phonetics by computer (version 6.1.15) [computer program] (2020). https://www.fon.hum.uva.nl/praat/

7. Brognaux, S., Roekhaut, S., Drugman, T., Beaufort, R.: Train&align: a new online tool for automatic phonetic alignment. In: IEEE Workshop on Spoken Language Technology, pp. 416–421 (2012). https://doi.org/10.1109/SLT.2012.6424260

8. Davis, S., Mermelstein, P.: Comparison of parametric representations for mono-syllabic word recognition in continuously spoken sentences. IEEE Trans. Acoust. Speech Signal Process. **28**(4), 357–366 (1980). https://doi.org/10.1109/TASSP.1980.1163420

9. Duda, R.O., Hart, P.E., Stork, D.G.: Pattern Classification, 2nd edn. Wiley Interscience, Hoboken (2000)

10. Gales, M.J.F.: Maximum likelihood linear transformations for hmm-based speech recognition. Comput. Speech Lang. **12**(2), 75–98 (1998). https://doi.org/10.1006/csla.1998.0043

11. GitHub: Kaldi speech recognition toolkit (2018). https://github.com/kaldi-asr/kaldi

12. GitHub: Frequencywords (2020). https://github.com/hermitdave/FrequencyWords

13. GitHub: GNU Aspell (2020). https://github.com/GNUAspell/aspell

14. Goldman, J.P.: EasyAlign: an automatic phonetic alignment tool under Praat. In: Proceedings of Interspeech, pp. 3233–3236 (2011). https://archive-ouverte.unige.ch/unige:18188

15. Gopinath, R.A.: Maximum likelihood modeling with Gaussian distributions for classification. In: IEEE International Conference on Acoustics, Speech and Signal Processing, ICASSP, vol. 2, pp. 661–664, May 1998. https://doi.org/10.1109/ICASSP.1998.675351

16. Gorman, K., Howell, J., Wagner, M.: Prosodylab-aligner: a tool for forced alignment of laboratory speech. Can. Acoust. **39**(3), 192–193 (2011). https://jcaa.caa-aca.ca/index.php/jcaa/article/view/2476

17. Grupo FalaBrasil: Ferramentas para alinhamento fonético em português brasileiro (2020). https://gitlab.com/fb-align/

18. Grupo FalaBrasil: NLP: Gerador de ferramentas para processamento de linguagem natural (2020). https://gitlab.com/fb-nlp/nlp-generator

19. Grupo FalaBrasil: Recursos prontos para processamento de linguagem natural em português brasileiro (2020). https://gitlab.com/fb-nlp/nlp-resources

20. Guiroy, S., Cordoba, R., Villegas, A.: Application of the Kaldi toolkit for continuous speech recognition using hidden-Markov models and deep neural networks. In: Proceedings of IberSPEECH 2016, pp. 187–196 (2016). https://iberspeech2016.inesc-id.pt/wp-content/uploads/2017/01/OnlineProceedings_IberSPEECH2016.pdf

21. Huang, X., Acero, A., Hon, H.W.: Spoken Language Processing: A Guide to Theory, Algorithm, and System Development, 1st edn. Prentice Hall PTR, Upper Saddle River (2001)

22. Jiampojamarn, S., Kondrak, G., Sherif, T.: Applying many-to-many alignments and hidden Markov models to letter-to-phoneme conversion. In: Human Language Technologies 2007: The Conference of the North American Chapter of the Association for Computational Linguistics; Proceedings of the Main Conference, Rochester, New York, pp. 372–379. Association for Computational Linguistics, April 2007. http://www.aclweb.org/anthology/N/N07/N07-1047
23. Kipyatkova, I., Karpov, A.: DNN-based acoustic modeling for Russian speech recognition using Kaldi. In: Ronzhin, A., Potapova, R., Németh, G. (eds.) SPECOM 2016. LNCS (LNAI), vol. 9811, pp. 246–253. Springer, Cham (2016). https://doi.org/10.1007/978-3-319-43958-7_29
24. LDC: CSLU: Spoltech Brazilian Portuguese version 1.0 (2018). https://catalog.ldc.upenn.edu/LDC2006S16
25. LDC: West point Brazilian Portuguese speech (2018). https://catalog.ldc.upenn.edu/LDC2008S04
26. McAuliffe, M., Socolof, M., Mihuc, S., Wagner, M., Sonderegger, M.: Montreal forced aligner: trainable text-speech alignment using Kaldi. In: Proceedings of Interspeech, pp. 498–502, August 2017. https://doi.org/10.21437/Interspeech.2017-1386
27. Miao, Y., Zhang, H., Metze, F.: Speaker adaptive training of deep neural network acoustic models using I-vectors. IEEE/ACM Trans. Audio Speech Lang. Process. **23**(11), 1938–1949 (2015)
28. Neto, N., Patrick, C., Klautau, A., Trancoso, I.: Free tools and resources for Brazilian Portuguese speech recognition. J. Braz. Comput. Soc. **17**(1), 53–68 (2010). https://doi.org/10.1007/s13173-010-0023-1
29. Ochshorn, R.M., Hawkins, M.: Gentle forced aligner [computer program] (2020). https://github.com/lowerquality/gentle
30. opensubtitles.org: Opensubtitles (2020). https://www.opensubtitles.org/
31. PCD Legal: PCD legal: Acessível para todos (2018). http://www.pcdlegal.com.br/
32. Povey, D.: Chain models (2020). https://kaldi-asr.org/doc/chain.html
33. Povey, D.: Kaldi documentations (2020). https://kaldi-asr.org/doc/index.html
34. Povey, D., et al.: The Kaldi speech recognition toolkit. In: IEEE 2011 Workshop (2011)
35. PUC-Rio: Centro de estudos em telecomunicações (CETUC) (2018). http://www.cetuc.puc-rio.br/
36. Siravenha, A., Neto, N., Macedo, V., Klautau, A.: Uso de regras fonológicas com determinação de vogal tônica para conversão grafema-fone em Português Brasileiro. In: 7th International Information and Telecommunication Technologies Symposium (2008)
37. Souza, G., Neto, N.: An automatic phonetic aligner for Brazilian Portuguese with a Praat interface. In: Silva, J., Ribeiro, R., Quaresma, P., Adami, A., Branco, A. (eds.) PROPOR 2016. LNCS (LNAI), vol. 9727, pp. 374–384. Springer, Cham (2016). https://doi.org/10.1007/978-3-319-41552-9_38
38. Viterbi, A.: Error bounds for convolutional codes and an asymptotically optimum decoding algorithm. IEEE Trans. Inf. Theory **13**(2), 260–269 (1967). https://doi.org/10.1109/TIT.1967.1054010
39. Young, S., Ollason, D., Valtchev, V., Woodland, P.: The HTK Book. Cambridge University Engineering Department, Version 3.4 (2006)
40. Yuan, J., Liberman, M.: Speaker identification on the SCOTUS corpus. J. Acoust. Soc. Am. **123**(5), 3878–3881 (2008). https://doi.org/10.1121/1.2935783

Twitter Moral Stance Classification Using Long Short-Term Memory Networks

Matheus Camasmie Pavan[iD], Wesley Ramos dos Santos[iD],
and Ivandré Paraboni[✉][iD]

University of São Paulo (EACH-USP), Av Arlindo Bettio 1000, São Paulo, Brazil
{matheus.pavan,wesley.ramos.santos,ivandre}@usp.br

Abstract. In Natural Language Processing, stance detection is the computational task of deciding whether a piece of text expresses a favourable or unfavourable attitude (or stance) towards a given topic. Stance detection may be divided into two subtasks: deciding whether a piece of text conveys any stance towards the target topic and, once we have established that the text does convey a stance, determining its polarity (e.g., favourable or unfavourable) towards the target. Both tasks - hereby called stance recognition and (stance) polarity classification - are the focus of the present work. Taking as a basis a corpus of 13.7k tweets in the Brazilian Portuguese language, and which conveys stances towards five moral issues (abortion legislation, death penalty, drug legalisation, lowering of criminal age, and racial quotas at universities), we compare a number of long short-term memory (LSTM) and bidirectional LSTM (BiLSTM) models for stance recognition and polarity classification. In doing so, the two tasks are addressed both independently and as a joint model. Results suggest that the use of BiLSTM models with attention mechanism outperform the alternatives under consideration, and pave the way for more comprehensive studies in this domain.

Keywords: Natural Language Processing · Stance classification · Morality · Sentiment analysis

1 Introduction

Natural language text of the kind found in, e.g., social media, conveys a wide range of sentiment-related information that provide us with a glimpse on social trends, public opinion and others. Motivated by these opportunities, the task of stance detection has emerged as a popular research topic in Natural Language Processing (NLP) and related fields [12].

Stance detection is generally understood as the computational task of deciding whether a piece of text expresses a favourable or unfavourable attitude (or stance) towards a given target topic [13]. Stance detection is related to, but distinct from standard sentiment analysis in that positive and negative sentiments may reflect either favourable or unfavourable stances towards the target [12].

© Springer Nature Switzerland AG 2020
R. Cerri and R. C. Prati (Eds.): BRACIS 2020, LNAI 12319, pp. 636–647, 2020.
https://doi.org/10.1007/978-3-030-61377-8_45

For instance, given the target topic 'vaccination', a sentence as in 'it is painful, but it will protect you against this horrible disease' expresses a negative feeling (which would possibly be recognised as such by a sentiment analysis system), but it expresses a favourable stance towards this particular topic.

Stance detection is usually divided into two subtasks: first, there is the question of deciding whether a piece of text conveys any stance towards the target topic at all, or whether it is simply neutral in this respect. Second, once we have established that the text does convey a certain stance, there is the issue of determining its polarity (e.g., favourable or unfavourable) towards the target.

Both tasks - hereby called stance recognition and polarity classification - are illustrated below, using the target topic 'vaccination' as an example.

Stance recognition:

- *The doctor said that they ran out of vaccines*
  (no stance towards vaccination)
- *Measles is an entirely vaccine-preventable disease*
  (a stance towards vaccination)

Polarity classification:

- *Vaccination prevents you from catching measles*
  (a stance in favour of vaccination)
- *Yeah, but by catching the disease you become immune anyway*
  (a stance against vaccination)

Both stance recognition and polarity classification are well-known tasks in English natural language processing [1,5,6,9,12,13,18,19], and have been the focus of a recent shared task competition based on a Twitter corpus [12] for which a wide range of shallow and deep learning methods have been considered. However, there is still relatively few studies based on other domains, and even less so based on other languages.

Based on these observations, our own work addresses the issue of stance detection in the Brazilian Portuguese language using a Twitter corpus of stances towards five moral issues: abortion legislation, death penalty, drug legalisation, lowering of criminal age, and racial quotas at universities. More specifically, we shall focus on the use of a number of long short-term memory network (LSTM) [7] architectures, which have been shown to obtain positive results in a large number of tasks both in NLP and related fields, and which may be potentially useful to the tasks at hand.

The present work makes use of standard and bidirectional LSTM (BiLSTM) models with and without attention mechanism. These models are to be applied to both stance recognition and polarity classification, and the two tasks are to be addressed both independently and as a joint model. In doing so, our goal is to compare alternative LSTM strategies to more traditional (e.g., shallow) methods to determine which, if any, is more suitable to the present setting.

The rest of this paper is organised as follows. Section 2 reviews recent work in stance detection for the English language. Section 3 presents the bulk of the current work: the Twitter corpus taken as the basis of our experiments (Sect. 3.1), computational models (Sect. 3.2), evaluation (Sect. 3.3) and results (Sect. 4). Section 5 presents final remarks and points to future work.

2   Related Work

The work in [1] is among the first to approach the computational recognition of stance in text, which was implemented by analysing a corpus of 4873 posts in on-line discussion forums. The dataset is made up of 14 topics, ranging from entertaining to moral issues. The classification of favourable/unfavourable stances obtained an accuracy of up to 69%, outperforming a unigram baseline model that obtained up to 60% accuracy. Complementary results also suggest that taking into account the context of the dialogue might be a helpful strategy.

Stance recognition in on-line discussion forums is also the focus of the work in [6], which analyses the effects of size and quality of training data, model complexity, features expressiveness, and the use of extra-linguistic constraints on the task accuracy. The dataset under consideration was created from an on-line debate forum, and labelled as *for* or *against* stances by considering four popular domains: abortion, gay rights, Obama, and the legalisation of marijuana. The experiments provide a number of contributions on how to build models of this type, and on which kinds of knowledge to consider.

In the context of the SemEval-2016 competition [11], 19 participating systems joined in the task of supervised stance recognition in a dataset made up of tweets in the English language[1]. The training corpus, described in detail in [11], contains 2914 tweets about five topics (Hillary Clinton, climate change, atheism, feminism, and abortion.). The corpus comprises, on average, 583 tweets per topic, but classes are generally unbalanced. On average, there is a 25.8% of positive stances and a 47.9% of negative stances.

For the supervised stance recognition task at SemEval-2016 [12], the work in [19] accomplishes the best overall performance. The model employs a recurrent neural network (RNN) with features learned by distant supervision from large unlabelled corpora. Word- and phrase-based embedding representations are computed by using Word2vec skip-gram [10], and then taken as an input to learn sentence representations with the aid of a hashtag prediction model. In addition to that, sentence vectors are optimised by using labelled examples taken from the training data.

Also a participating system at SemEval-2016, the work in [18] presents a convolutional neural network (CNN) model that substitutes the prediction of maximal validation accuracy for a voting scheme with further improvements, including the training of an individual model for each of the target topics in the corpus. This strategy obtained the second best overall results of the shared task stance classification task.

---

[1] Additional teams also participated in the unsupervised track of the competition, which is presently not discussed.

Following the public release of the SemEval-2016 stance corpus, a number of additional studies have attempted to improve results even further. For instance, the work in [9] addresses the use of rules about political alliances and enmities as a means to improve the overall accuracy in the recognition of political stances. The proposed model consists of a stance classifier with added semantic features representing world knowledge rules, and it was found to be superior to the best-performing participant systems in the previous shared task when applied to the recognition of political stances.

The work in [13] carries out a post-hoc analysis of the SemEval-2016 stance classification task. Unlike the overall winner of the competition in [19], the work presents a much simpler and yet more accurate model. The proposal makes use of linear SVM classifiers and features computed from the training portion of the corpus, including word and character n-gram counts, and word embedding models computed from an external data set. In order to further improve the classifiers, the models make use of additional unlabelled data obtained from distant supervision.

The work in [2] introduced an architecture based on two bidirectional-encoded LSTM networks to address the stance classification task. In this approach, the first neural network receives as an input the sequence of words representing the topic. The output of this network is then taken as the initial state for the second model, whose inputs are tweet word sequences. The work also presents a weakly supervised learning method that is shown to improve results for the task.

Finally, the work in [5] investigates the reproducibility of existing approaches to stance recognition by using both neural and shallow classification models alike. To this end, six models were considered, including CNNs, RNNs, SVMs and BERT [4] classifiers. All models were validated by using both the SemEval-2016 stance corpus, and a set of health-related on-line news articles. The study presents a comparative analysis of different approaches and discusses their main shortcomings.

3   Current Work

Beyond the scope of the SemEval challenge and accompanying data, computational approaches to stance recognition are relatively scarce, and even more so in the case of the Brazilian Portuguese language. In what follows we will focus on the computational task of recognising stances towards moral issues, which are ubiquitous in Brazilian Twitter text.

The present study addresses two main tasks - moral stance recognition and moral stance polarity classification - based on tweets that make reference to certain topics of this kind. As discussed in Sect. 1, stance recognition is presently defined as the binary classification task of deciding whether a given piece of text conveys any attitude towards a given topic or not, and polarity classification is defined as the binary task of deciding whether a given stance shows a favourable or unfavourable attitude towards a given topic. Both issues will be dealt with separately and also as a joint task, and in doing so we shall focus on the use of long short-term memory (LSTM) and bidirectional LSTM (BiLSTM) models with and without attention mechanism.

3.1 Data

The present work makes use of a corpus of Twitter texts in the Brazilian Portuguese Twitter language described in [15]. The corpus conveys over 13.7k moral stances towards five topics: abortion legislation, death penalty, drug legalisation, lowering of criminal age, and racial quotas at universities. As discussed in [15], messages were selected by searching Twitter for certain key words (e.g., 'quotas' etc.). For each topic, an initial set of 7000k tweets was selected for manual revision and annotation. Annotation was performed by three judges and validated by a fourth in case of conflict.

Corpus labelling consisted of assigning a positive/negative label to all instances that unequivocally expressed a stance on the target topic, and by assigning the 'other' label to any instance that did not meet these criteria. Thus, instances labelled as 'other' comprise texts that do contain a key word of interest, but which do not convey any obvious stance about the target topic, and which are therefore regarded as noise[2].

As a means to keep a minimum balance between for/against stances towards each topic, annotation proceeded until at least 300 instances of each of the three classes were obtained, or until reaching the end of the data. The 'other' class, however, still remains several times larger than the for/against classes in all topics. The corpus class distribution is summarised in Table 1. For further details, we report to [15].

Table 1. Class distribution.

Topic	for	against	other
Abortion	300	444	2630
Criminal age	303	300	1493
Death penalty	861	304	1578
Drugs	395	241	1542
Racial quotas	300	424	2656
Overall	2159	1713	9899

---

[2] For instance, 'death penalty for people who litter the streets' does not convey a genuine stance towards the issue of death penalty.

3.2 Models

For both stance and polarity recognition tasks, we consider three strategies based on Long-Short Term Memory networks: standard Long-Short Term Memory (LSTM), standard Bidirectional LSTM (BiLSTM), and BiLSTM with an attention mechanism (BiLSTM+att). This section discusses the implementation of these models, and their variations under consideration.

In a series of pilot experiments, we performed grid search over development data, and considered a large number of parameters and network configurations. These alternatives included varying the size of the embedding layer (100, 128, 200, 256), the numbers of cells in the recurrent layer (ranging from 2 to 20), and the number of recurrent layers (up to 2). In what follows we shall focus on the best overall resulting models only.

The standard LSTM models take as an input sequences of up to 30 words, and use a 100-dimensional embedding layer for stance detection and a 128-dimensional embedding layer for polarity prediction followed by a single recurrent layer. This consists of four LSTM cells in the stance recognition task, and two LSTM cells in polarity classification. A dense layer connects the recurrent output to the final binary output.

Similarly to the LSTM architecture, our BiLSTM models take as an input sequences of up to 30 words, and use a 256-dimensional embedding layer for stance detection and a 128-dimensional for polarity prediction followed by a single bidirectional recurrent layer. This consists of four LSTM cells in each direction for both stance recognition and polarity classification. Finally, a dense layer connects the recurrent output to the final binary output.

In addition to the standard BiLSTM architecture, we also implemented BiLSTM+att by including a multi-head attentional mechanism [17]. This consists of a 128 or 256 embeddings layer, 16 or 128 LSTM units per recurrent layer, 8 or 32 to attention model depth, and 2 or 4 attention heads. Multi-head attention is illustrated in Fig. 1 and further discussed below.

Multi-head attention was implemented as follows. From an input consisting of two tensors in the form $[batch\_size, lstm\_layer\_size, seq\_length]$ representing the state and the output of every cell at every point in time, two parameters - model depth md and number of heads nh - are selected so that model density d is defined as $d = md/nh$. Next, the input to the attention mechanism is fed into three dense layers whose number of neurons is equal to the model density d. Each of these layers generates an item from the key set (k), query (q), and value (v). Once these values are generated, q and k are multiplied matrix-wise, and the result is scaled and normalised through a softmax layer, and then the output of the softmax layer is multiplied matrix-wise by v. The process is encapsulated into a head represented by the light blue area in Fig. 1. Finally, the results of each head are concatenated and a dense layer generates the output of the attention mechanism.

The two main architectures - with and without attention mechanism - are summarised in Fig. 2.

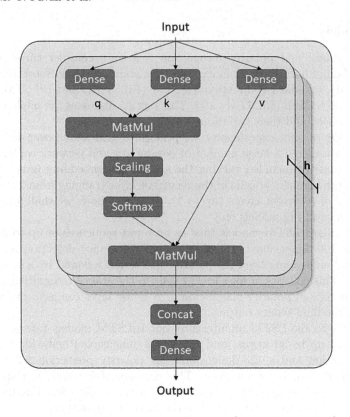

Fig. 1. Multi-head attention mechanism. (Color figure online)

Our three LSTM-based models - LSTM, BiLSTM and BiLSTM+att - are to be compared against two baseline systems that make use of L2-regularised logistic regression. The baseline models consist of TF-IDF word unigram and variable-length (2 to 8) character n-grams, in both cases using univariate feature selection with ANOVA f-value as the score function. A majority class baseline is also included for illustration purposes.

3.3 Experiments

In the case of the LSTM models, we considered a range of values for the following network hyper parameters: learning rate (from 0.0001 to 0.1), dropout probability (from 0.3 to 0.8), and maximum number of epochs for early stopping (from 0 up to 20). All LSTM and BiLSTM models were trained using 10-fold cross-validation data in batches of 100 observations. The learning rate is set at 0.01 with a 0.5 dropout probability and using early stopping with a maximum of 5 epochs. The training was limited to 20 epochs.

For the logistic regression baseline models, we considered C values in the 1e−3 to 1e−1 range, and tolerance values in the 1e−5 to 1e−3 range. For the

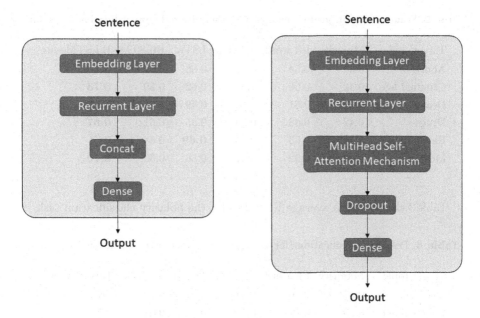

Fig. 2. Plain LSTM (left) and LSTM with attention mechanism (right.)

word-based models, the k parameter for univariate feature selection was searched in the 25000 to 3000 range at 300 intervals. For the character-based models we considered the 22000 to 13000 range at 1000 intervals.

The evaluation of the present LSTM classifiers and baseline systems was carried out by performing three kinds of experiment. In the first two, stance and polarity recognition tasks are assessed individually. In the third and more comprehensive experiment, by contrast, the two models are combined as a single task in which the first classifier attempts to recognise whether the text expresses a stance towards the target topic or not and, if so, the second classifier is invoked so as to determine its polarity. This is intended to provide insights not only on each individual classifier, but also to consider a more realistic scenario in which decisions made during stance recognition actually affect the outcome of the downstream polarity classifier.

4   Results

This section presents results of our three experiments, that is, for the individual stance recognition and polarity classification tasks, and for joint stance and polarity classification. In all cases, best results for each target topic are highlighted. Table 2 shows macro average F1 scores for the stance recognition task.

From these results we notice that, with the exception of standard LSTM, neural models generally outperform the simpler Logistic Regression alternatives, and that the use of the attention mechanism in BiLSTM+att is best of all in all tasks.

Table 2. Stance recognition (for+against vs. other) 10-fold cross validation F1 results.

Target topic	Majority	LR.word	LR.char	LSTM	BiLSTM	BiLSTM+att
Abortion	0.44	0.54	0.56	0.66	0.61	**0.70**
Criminal age	0.42	0.56	0.61	0.62	0.59	**0.73**
Death penalty	0.37	0.51	0.56	0.49	0.50	**0.72**
Drugs	0.42	0.48	0.53	0.55	0.61	**0.67**
Racial quotas	0.44	0.53	0.58	**0.69**	0.68	0.68
Mean	0.42	0.52	0.57	0.56	0.60	**0.70**

Table 3 shows macro average F1 scores for the polarity classification task.

Table 3. Polarity classification (for vs. against) 10-fold cross validation F1 results.

Target topic	Majority	LR.word	LR.char	LSTM	BiLSTM	BiLSTM+att
Abortion	0.37	0.70	**0.76**	0.71	0.69	0.75
Criminal age	0.33	0.72	**0.78**	0.66	0.69	0.76
Death penalty	0.43	0.66	0.70	0.70	0.70	**0.78**
Drugs	0.38	0.65	**0.71**	0.63	0.68	**0.71**
Racial quotas	0.37	0.70	**0.79**	0.66	0.68	0.71
Mean	0.38	0.69	**0.75**	0.67	0.69	0.74

In the case of the polarity classification task, we notice that mean results for LR.char and BiLSTM+att are similar, with a minimal advantage for the LR.char method.

Finally, Table 4 shows macro average F1 scores for joint stance and polarity classification.

Table 4. Joint stance and polarity classification 10-fold cross validation F1 results.

Target topic	Majority	LR.word	LR.char	LSTM	BiLSTM	BiLSTM+att
Abortion	0.29	0.33	0.28	0.45	0.46	**0.47**
Criminal age	0.28	0.34	0.35	0.47	0.48	**0.73**
Death penalty	0.24	0.39	0.41	0.45	0.50	**0.74**
Drugs	0.28	0.29	0.29	0.39	0.42	**0.47**
Racial quotas	0.29	0.33	0.34	0.41	0.41	**0.42**
Mean	0.28	0.34	0.33	0.43	0.45	**0.57**

Once again, the use of attention in BiLSTM+att outperforms all alternatives under consideration. We notice however that these results are on average lower than those previously obtained by the individual stance recognition and polarity classification models. The difference is mainly due to the observation that the joint task is considerably harder since as classification errors percolate from the first task to the second, that is, any error produced by either of the two classifiers counts as an overall error in the present analysis.

4.1 Discussion

Generally speaking, the use of bi-directional LSTM models with attention produces the best results in all three experiments reported in the previous section. However, compared with the best non-LSTM alternative, the much simpler LR.char model, we notice that the benefits of the BiLSTM+att approach vary considerably.

The BiLSTM+att approach keeps well ahead of LR.char in the stance recognition task (experiment 1), and this advantage is carried on to the results of the joint stance and polarity classification task (experiment 3). However, the polarity classification task (experiment 2) shows that results for both LR.char and BiLSTM+att are essentially similar, and in that case there is little motivation for using the more computationally expensive neural model.

There are at least two possible explanations for the differences across tasks. First, we notice that stance recognition in experiment 1 has been presently modelled as a ternary classification task, whereas polarity classification is a binary task. Second, recall that stance classes (for/against/other) are heavily imbalance (cf. Table 1), whereas the polarity class is not. The rate of samples that present some stance is on average 28.1%. Although, for polarity labels, the rate for stances against the topic is on average 44.2%. In other words, stance recognition is essentially more complex than polarity classification in both problem definition and class distribution.

The present experiments suggest that more complex text classification problems such as moral stance recognition may make explicit some of the limitations of models based on regularised logistic regression that rely on word or character features. Bi-directional LSTMs with attention (and to a lesser extent even some of our simpler LSTM-based alternatives), by contrast, seem to be generally more suitable for problems of this kind.

5 Final Remarks

This work has discussed the issue of stance detection in Brazilian Portuguese using a Twitter corpus of 13.7k stances towards topics of moral nature. We presented a number of LSTM and BiLSTM models addressing stance recognition and polarity classification individually, and also as a joint task. Our current results suggest a strong preference for using bi-directional LSTMs with attention mechanism over the alternatives under consideration, and pave the way for more comprehensive studies in this domain.

As future work, we intend to expand the current dataset by adding both more instances and more topics as a means to fully benefit from the current methods, and to provide more robust results in the joint task. These results, in our present experiments, may haven been particularly hindered by class imbalance and lack of data.

Possible extensions and applications of the current methods are numerous. In particular, we envisage the related issue of reputation classification [14] (e.g., deciding whether a piece of text information improves or damages the perceived reputation of an entity such as a company, public figure etc.), and using stance classification as a means to improve NLP tasks such as author profiling [8,16] and author identification [3], among others. These are also left as future work.

Acknowledgements. This work has been supported by the University of São Paulo PRP grant nr. 668/2018.

References

1. Anand, P., Walker, M., Abbott, R., Tree, J.E.F., Bowmani, R., Minor, M.: Cats rule and dogs drool!: classifying stance in online debate. In: Proceedings of the 2nd Workshop on Computational Approaches to Subjectivity and Sentiment Analysis (ACL-HLT 2011), pp. 1–9. Association for Computational Linguistics, Portland (2011)
2. Augenstein, I., Rocktaschel, T., Vlachos, A., Bontcheva, K.: Stance detection with bidirectional conditional encoding. In: Proceedings of the 2016 Conference on Empirical Methods in Natural Language Processing, pp. 876–885. Association for Computational Linguistics, Austin (2016)
3. Custódio, J.E., Paraboni, I.: EACH-USP ensemble cross-domain authorship attribution. In: Working Notes Papers of the Conference and Labs of the Evaluation Forum (CLEF-2018), Avignon, France, vol. 2125 (2018)
4. Devlin, J., Chang, M.W., Lee, K., Toutanova, K.: BERT: pre-training of deep bidirectional transformers for language understanding. In: 2019 Conference of the North American Chapter of the Association for Computational Linguistics: Human Language Technologies, pp. 4171–4186. Association for Computational Linguistics, Minneapolis (2019). https://doi.org/10.18653/v1/N19-1423
5. Ghosh, S., Singhania, P., Singh, S., Rudra, K., Ghosh, S.: Stance detection in web and social media: a comparative study. In: Crestani, F., et al. (eds.) CLEF 2019. LNCS, vol. 11696, pp. 75–87. Springer, Cham (2019). https://doi.org/10.1007/978-3-030-28577-7_4
6. Hasan, K.S., Ng, V.: Stance classification of ideological debates: data, models, features, and constraints. In: Proceedings of the International Joint Conference on Natural Language Processing, pp. 1348–1356. Association for Computational Linguistics, Nagoya (2013)
7. Hochreiter, S., Schmidhuber, J.: Long short-term memory. Neural Comput. 9(8), 1735–1780 (1997)
8. Hsieh, F.C., Dias, R.F.S., Paraboni, I.: Author profiling from facebook corpora. In: 11th International Conference on Language Resources and Evaluation (LREC-2018), ELRA, Miyazaki, Japan, pp. 2566–2570 (2018)

9. Lai, M., Hernández Farías, D.I., Patti, V., Rosso, P.: Friends and enemies of Clinton and Trump: using context for detecting stance in political tweets. In: Sidorov, G., Herrera-Alcántara, O. (eds.) MICAI 2016. LNCS (LNAI), vol. 10061, pp. 155–168. Springer, Cham (2017). https://doi.org/10.1007/978-3-319-62434-1_13

10. Mikolov, T., Sutskever, I., Chen, K., Corrado, G.S., Dean, J.: Distributed representations of words and phrases and their compositionality. In: Burges, C.J.C., Bottou, L., Welling, M., Ghahramani, Z., Weinberger, K.Q. (eds.) Advances in Neural Information Processing Systems, vol. 26, pp. 3111–3119. Curran Associates, Inc. (2013)

11. Mohammad, S.M., Kiritchenko, S., Sobhani, P., Zhu, X., Cherry, C.: A dataset for detecting stance in tweets. In: Proceedings of 10th Edition of the the Language Resources and Evaluation Conference (LREC-2016), Portoroz, Slovenia (2016)

12. Mohammad, S.M., Kiritchenko, S., Sobhani, P., Zhu, X., Cherry, C.: Semeval-2016 task 6: detecting stance in tweets. In: Proceedings of the International Workshop on Semantic Evaluation, San Diego, California, USA (2016)

13. Mohammad, S.M., Sobhani, P., Kiritchenko, S.: Stance and sentiment in tweets. ACM Trans. Internet Technol. Argum. Soc. Media 17(3), 23 (2017)

14. Rantanen, A., Salminen, J., Ginter, F., Jansen, J.: Classifying online corporate reputation with machine learning: a study in the banking domain. Internet Res., November 2019. https://doi.org/10.1108/INTR-07-2018-0318

15. dos Santos, W.R., Paraboni, I.: Moral stance recognition and polarity classification from twitter and elicited text. In: Recents Advances in Natural Language Processing (RANLP-2019), Varna, Bulgaria, pp. 1069–1075 (2019). https://doi.org/10.26615/978-954-452-056-4_123

16. Silva, B.B.C., Paraboni, I.: Learning personality traits from Facebook text. IEEE Lat. Am. Trans. 16(4), 1256–1262 (2018). https://doi.org/10.1109/TLA.2018.8362165

17. Vaswani, A., et al.: Attention is all you need. In: Guyon, I., et al. (eds.) Advances in Neural Information Processing Systems, vol. 30, pp. 5998–6008. Curran Associates, Inc. (2017)

18. Wei, W., Zhang, X., Liu, X., Chen, W., Wang, T.: pkudblab at SemEval-2016 task 6: a specific convolutional neural network system for effective stance detection. In: Proceedings of the International Workshop on Semantic Evaluation, San Diego, California, USA (2016)

19. Zarrella, G., Marsh, A.: MITRE at SemEval-2016 task 6: transfer learning for stance detection. In: Proceedings of the International Workshop on Semantic Evaluation, San Diego, California, USA (2016)

A Study on the Impact of Intradomain Finetuning of Deep Language Models for Legal Named Entity Recognition in Portuguese

Luiz Henrique Bonifacio[1(✉)], Paulo Arantes Vilela[1,2], Gustavo Rocha Lobato[2], and Eraldo Rezende Fernandes[1]

[1] Universidade Federal de Mato Grosso do Sul, Campo Grande, Brazil
`luiz.bonifacio@ufms.br`
[2] Ministério Público do Estado de Mato Grosso do Sul, Campo Grande, Brazil

Abstract. Deep language models, like ELMo, BERT and GPT, have achieved impressive results on several natural language tasks. These models are pretrained on large corpora of unlabeled *general domain* text and later supervisedly trained on downstream tasks. An optional step consists of finetuning the language model on a large *intradomain* corpus of unlabeled text, before training it on the final task. This aspect is not well explored in the current literature. In this work, we investigate the impact of this step on named entity recognition (NER) for Portuguese legal documents. We explore different scenarios considering two deep language architectures (ELMo and BERT), four unlabeled corpora and three legal NER tasks for the Portuguese language. Experimental findings show a significant improvement on performance due to language model finetuning on intradomain text. We also evaluate the finetuned models on two general-domain NER tasks, in order to understand whether the aforementioned improvements were really due to domain similarity or simply due to more training data. The achieved results also indicate that finetuning on a legal domain corpus hurts performance on the general-domain NER tasks. Additionally, our BERT model, finetuned on a legal corpus, significantly improves on the state-of-the-art performance on the LeNER-Br corpus, a Portuguese language NER corpus for the legal domain.

Keywords: Natural language processing · Named entity recognition · Deep learning

1 Introduction

Transfer learning has achieved impressive results in several areas. More specifically, in Natural Language Processing (NLP), deep language models have been trained on large amounts of raw text and transferred to improve performance on different downstream tasks. Models like ULMFiT [12], ELMo [16], BERT [7] and

R. Cerri and R. C. Prati (Eds.): BRACIS 2020, LNAI 12319, pp. 648–662, 2020.
https://doi.org/10.1007/978-3-030-61377-8_46

GPT [20] learn language features at different levels by deriving powerful textual representations, which are useful to solve several tasks. Usually, such representations are incorporated in models that are supervisedly trained to perform specific tasks. Those language models (LMs) are usually trained on general-domain corpora, i.e., corpora comprising texts from several sources involving different domains. Although general-domain LMs have achieved great success on considerable tasks, some domains present highly specific language styles that may limit the benefits of such LMs. In such cases, one optional step is to finetune the language model on this intradomain data, before supervisedly training on the final task. This idea was successfully explored in some previous work [12,13,19] but is generally neglected in current literature.

The legal domain is a particular case in which raw text is vast but labeled data is usually scarce; which is specially true for non-English languages like Portuguese. In this work, we explore the impact of finetuning deep LMs in legal documents before supervisedly training NER models on legal labeled corpora. We consider BERT- and ELMo-based LMs trained on three general-domain corpora: the 104 largest Wikipedias, the Portuguese Wikipedia, and the brWaC corpus (a 2.7-billion-token Portuguese corpus). We train legal-domain LMs by finetuning general-domain LMs on the Acórdãos-TCU corpus[1], which comprises judgment reports from TCU (the Federal Court of Accounts in Brazil).

We evaluate the proposed general- and legal-domain LMs on three NER tasks: *(a) HAREM*: documents from different sources annotated with general entities [10]; *(b) LeNER-Br*: judgment reports from different Brazilian Courts and some law texts manually labeled with general and legal entities [14]; and *(c) DrugSeizures-Br*: initial petitions regarding drug seizures from the Ministério Público de Mato Grosso do Sul[2]. The data used in the latter scenario was collected during the development of this work. Experimental results show that legal-domain LMs outperform general-domain LMs on scenario (b), while hurt performance on scenario (a). This clearly indicates that finetuning LMs on intradomain data is beneficial in this case.

We also consider an additional hypothesis represented by scenario (c). The DrugSeizures-Br is a corpus composed of legal documents as well, but its language style is different from the ones found in LeNER-Br and Acórdãos-TCU. When considering scenario (c), we could not find significant performance improvement from using legal- instead of general-domain LMs. In fact, in some cases, the legal-domain finetuning hurts performance. That is intradomain LM finetuning is beneficial when the used corpus is similar enough to the final task. In other words, some domains may present such strong language variations that intradomain finetuning is undermined.

Lastly, after finetuning the BERT-based LM on intradomain data, we achieve new state-of-the-art performance on the LeNER-Br corpus. Before describing our work in details, we discuss some related work in the next section.

---

[1] https://github.com/netoferraz/acordaos-tcu.
[2] The public agency for law enforcement and prosecution of crimes in the Brazilian state of Mato Grosso do Sul.

2 Related Work

NER in legal domain has been explored in different languages. Angelidis et al. [2] applied deep learning models to solve NER on Greek legislation. Badji [3] applied a rule-based system to identify legal entities in Spanish and English texts from Twitter and news articles. Similarly, Dozier et al. [9] introduced a NER system to identify judges, attorneys, jurisdictions and courts in legal documents. This system is based on context rules and statistic models.

Regarding the Portuguese language, most of the papers on general-domain NER evaluate its models on the HAREM corpus [8,17] which comprises documents from several sources involving different domains. Souza et al. [22] recently proposed a start-of-the-art model for HAREM, using a BERT-based model. The BERT-based LM was finetuned on brWaC corpus, a large corpus for Portuguese language. For the final task, a CRF layer was used for token classification. Besides the results, one key contribution of this work is the provision of pretrained BERT models. Castro et al. [6] explored transfer learning of ELMo-based models on three different Portuguese-language NER corpora. One important contribution of their work consists of two ELMo-based LMs for Brazilian Portuguese: one trained on Wikipedia and the other on brWaC corpus.

In our work, we employ two deep LMs: ELMo and BERT. ELMo is a contextualized word representation architecture based on character convolutions and a two-layer biLSTM. During the LM pretraining, it computes word representations as functions of the entire sentence, combining (although independently) both forward and backward LSTM hidden states. Parameters for token representation are shared while LSTM parameters are split for both directions. The resulting word embedding is a linear combination of the intermediate biLSTM layers. When supervisedly training on the final task, these layer weights can be adjusted.

BERT (Bidirectional Encoder Representations from Transformers) is a sentence encoder based on the Transformer architecture [23]. Unlike previous LMs, BERT is able of computing the encoding of a word considering both left and right contexts at the same time. This novel bidirectional conditioning is achieved by means of a novel masked language training procedure. It also considers a second pretraining task named Next Sentence Prediction (NSP), allowing the model to incorporate multi-sentence context. Nowadays there are pretrained BERT LMs available for different languages.

Although finetuning LMs may be beneficial when working with specific domains or languages, just a few papers deal with this method in the literature. Rother and Rettberg [21] trained an ULMFiT model for hate speech identification on German tweets. The language model was pretrained on German Wikipedia articles and finetuned on German tweets. The perplexity obtained for finetuned language model was three points lower than the original ULMFiT model. The classifier achieved F1-score of 80%.

More recently, LMs have made its way through restrict domain areas, as proposed by Hakala and Pyysalo [11] in which a BERT model is applied to biomedical NER. The same domain is tackled on Alsentzer et al. [1], where BERT models were trained on large publicly available datasets. Some recent works have focused on comparing pretrained multilingual models with monolingual ones. CamemBERT [15], for instance, is a BERT-based model that achieved state-of-the-art results on several French tasks. AlBERTo [18], also a BERT-based model, achieved state-of-the-art results for different Italian classification tasks.

3 Methodology

Our methodology can be split into three main components: (i) a deep language model pretrained on a general-domain corpus; (ii) finetuning of this LM on legal domain corpus; and (iii) supervised training on a NER task. Baselines were obtained by skipping component (ii), i.e., using the models from component (i) directly on component (iii). The assessment of component (ii) is the main object of this study. In the following, we describe each component variation explored in this work.

3.1 General-Domain Language Models

We consider general-domain language models based on two architectures: ELMo and BERT. Regarding the ELMo architecture, we used two general-domain LMs: Wikipedia and brWaC. The former was pretrained on the Portuguese Wikipedia, and the latter was pretrained on the brWaC corpus. Both models are available in the official ELMo website[3] and were provided by de Castro [4,5]. We also considered two general-domain BERT LMs. The first one is the official BERT Multilingual model, pretrained on the 100 largest Wikipedias, including Portuguese. When we started developing this work, there was no BERT LM pretrained exclusively on Portuguese texts. Moreover, pretraining a BERT-based LM from scratch requires great computational power, since the Masked Language Model training is a data intensive procedure [7]. For this reason, we decided to finetune the BERT Multilingual model on the Portuguese Wikipedia in order to derive a second general-domain BERT LM, but now finetuned on Portuguese text. We present basic statistics for the Portuguese Wikipedia and the brWaC corpus in Table 1. The Wikipedia corpus was built from a Portuguese Wikipedia dump. After collecting the data, all hyperlinks, reference citations and HTML

Table 1. Basic statistics for the unlabeled corpora.

Corpus	Docs	Sentences	Tokens
Wikipedia	934k	13,900k	1,400mi
brWaC	3,530k	145,000k	2,680mi
Acórdãos-TCU	298k	9,000k	912mi

---

[3] https://allennlp.org/elmo.

tags were removed. We also removed duplicated punctuation and empty articles. The spaCy[4] library was used to all preprocessing steps. The brWaC corpus comprises more than 60 million pages collected from Brazilian domains on the internet. More details about this corpus can be found in the original paper by Wagner Filho et al. [24].

The general-domain LMs described above are used to derive our four baselines. We use these LMs to train and evaluate NER models directly on the supervised tasks. Our aim is to uncover the circumstances in which the intradomain finetuning outperforms these baselines.

3.2   Legal-Domain Language Models

In order to derive legal-domain language models, we finetuned two LMs described above: BERT Mutilingual and ELMo brWaC. Both models were finetuned on the Acórdãos-TCU corpus[5] which comprises judgments from the Tribunal de Contas da União (the Brazilian Federal Court of Accounts) during the period from 1992 to 2019. Basic statistics for this corpus are presented in Table 1.

3.3   NER Tasks

Transferring a deep language model to perform a final task consists of plugging a task-specific layer (or sequence of layers) on top of the underlying language model. For NER tasks, the general approach consists of using the word-level representations provided by the underlying LMs as input to a token classification model. When supervisedly training on final task, the weights of the LM layers can be adjusted or not.

Regarding the ELMo-based LMs, we used the NER model provided in the AllenNLP library[6]. This model consists of a CRF layer on top of a biLSTM. The LM itself is not updated during training, however, ELMo includes a trainable weight vector to compute a linear combination of the underlying LM layers. For BERT-based models, all LM parameters are updated during supervised training. The BERT-based NER models consist of a simple token classification layer on top of the word-level representations provided by the LM. This is the standard approach regarding BERT for NER tasks since, in the original BERT paper, no significant improvement was observed when a CRF layer was plugged on top of the token classifier.

We make use of three NER corpora in this work in order to assess the considered models. In Table 2, we depict basic statistics of these corpora. HAREM is a general-domain corpus, while LeNER-Br and DrugSeizures-Br fits in the legal domain. In the following, we detail each of these corpora.

---

[4] https://spacy.io/.

[5] https://www.kaggle.com/ferraz/acordaos-tcu.

[6] https://allennlp.org/.

Table 2. Basic statistics for the NER corpora.

Corpus	Docs	Sentences	Tokens
HAREM	257	8k	156k
LeNER-Br	70	10k	318k
DrugSeizures-Br	6,218	118k	6,400k

HAREM is a Portuguese-language NER corpus, which was developed in two shared tasks. This corpus is manually labeled with ten types of named entities and is usually divided in two versions: total and selective scenario. Per-entity statistics of HAREM for the total scenario are presented in Table 3. The development set was built by randomly selecting 10% of the training set. In the selective scenario, the original entity types are grouped in five super-types: Person, Organization, Location, Value and Date. In Table 4, we present entity frequencies for this scenario.

Table 3. Named entity frequencies in the HAREM corpus for the total scenario.

Entity type	Train	Dev	Test
Person	1,010	20	832
Time	404	32	361
Location	1,129	108	877
Organization	885	40	625
Value	422	41	326
Title	189	7	190
Thing	128	7	170
Event	125	3	57
Abstraction	372	34	228
Other	37	3	28
Overall	4,701	295	3,694

LeNER-Br is a Brazilian Portuguese corpus composed of 70 legal documents, of which 66 documents came from different Brazilian Courts and the remaining four are legislation documents. Besides the four traditional entities, two additional classes related to the legal domain were annotated: Legislação (Laws) and Jurisprudência (previous decisions regarding legal cases). In Table 5, we present basic statistics regarding this corpus. The corpus is split into three subsets: 50 documents for training, 10 documents for validation and 10 documents for testing.

DrugSeizures-Br comprises 6,218 petitions filed from 2015 to 2019 by Ministério Público do Estado de Mato Grosso do Sul regarding seizures of illegal

Table 4. Named entity frequencies in the HAREM corpus for the selective scenario.

Entity type	Train	Dev	Test
Person	1,010	20	832
Time	404	32	361
Location	1,129	108	877
Organization	885	40	625
Value	422	41	326
Overall	3,850	241	3,021

Table 5. Named entity frequencies in the LeNER-Br corpus.

Entity type	Train	Dev	Test
Person	1,525	310	233
Time	1,334	234	192
Organization	2,400	561	501
Location	611	109	47
Law	1,920	397	378
Legal case	1,104	207	185
Overall	8,894	1,818	1,536

drugs. Entities of interest covering 25 types were extracted from petition metadata. This data is not aligned to the petition text like in a traditional NER corpus. We applied an approximate string matching algorithm to find entity occurrences within the petition texts, thus deriving a traditional labeled NER corpus. In Table 6, we present basic statistics regarding this corpus for the total scenario. Inspired by the HAREM corpus, we grouped the 25 entity types into 6 super-types (Location, Person, Time, Organization, Drug and Other), as emphasized in Table 6. We then generated a corpus representing a selective scenario by using five of these super-types (we dropped the Other super-type). Basic statistics of the selective scenario are presented in Table 7.

All these datasets are labeled using the IOB tagging scheme. This scheme is based on token classification and considers three labels for each entity type. These labels indicate that a specific token is the beginning of an entity (B), inside an entity (I) or outside any entity (O). Therefore, models have to assign to each token a label.

4   Experimental Evaluation

We report on several experiments to highlight the impact of intradomain LM finetuning for legal named entity recognition. The metrics were computed using

Table 6. Named entity frequencies in the DrugSeizures-Br corpus for the total scenario.

Entity *Super-Type*/Type	Train	Dev	Test
Location			
Street name	5,386	807	1,556
Street number	2,799	424	789
Neighborhood	4,762	769	1,426
City	18,412	2,737	5,477
Addr. complement	202	51	97
Origin city	3,927	529	1,130
Origin state	5,066	670	1,324
Target city	3,764	460	1,030
Target state	7,283	966	2,118
Person			
Indicted	26,304	3,351	7,671
Active witness	8,977	945	1,601
Witness	4,346	654	862
Passive witness	443	64	86
Investigated	1,242	201	362
Victim	1,180	173	433
Author	3,162	446	834
Defendant	1,186	122	218
Attorney	67	1	3
Time			
Time	746	95	227
Date	715	145	222
Organization			
Prosecutor	140	16	31
Authority	3,042	490	815
Drug			
Name	13,711	2,089	3,862
Other			
Penal norm	5,938	870	1,825
Drug quantity	3,128	403	848
Overall	125,928	17,478	34,847

Table 7. Named entity frequencies in the DrugSeizures-Br corpus for the selective scenario.

Entity type	Train	Dev	Test
Person	46,907	5,957	12,070
Time	1,461	240	449
Location	51,601	7,413	14,947
Organization	3,182	506	846
Drug	13,711	2,089	3,862
Overall	116,862	16,205	32,174

the seqeval Python library[7]. This is a well-tested implementation of the official CoNLL-2002 evaluation script[8]. This metric is an entity-level evaluation metric, i.e., a predicted entity is considered correct only if its exactly span and type is correct[9]. All performances reported in this section are F1-score averages over five runs along with the respective standard deviations. All metrics were computed on the corresponding test set for each considered NER task.

For BERT-based models, we used the WordPiece tokenizer and preserved letter case. We finetuned BERT-based LM for one epoch on Wikipedia and five epochs on Acórdãos-TCU. For both BERT finetuning procedures, Masked LM and NSP unsupervised tasks were used. The rate for masked tokens was set to 0.15. The batch size was 8 and the maximum sequence length 256. Those parameters were shared for the two models. Regarding the final task supervised training, the batch size was set to 16, due to memory and processing restrictions. The learning rate was set to $5 \cdot 10^{-5}$ and the dropout rate to 0.1. The supervised training phase performed 10 epochs over the training data. All hyper-parameters were tuned using the development sets. All experiments with BERT-based models were performed by means of the Transformers library[10].

ELMo-based LM was finetuned just for one epoch on Acórdãos-TCU. As the pretrained LM used was the brWaC, comprising the biggest corpus used in this work, we decided to finetune this LM just for one epoch. Considering the supervised training on NER task, we used AllenNLP library to train and evaluate these models. Some parameters were kept the same for the three evaluated models, namely batch size (32), dropout (0.5) and training epochs (75). Early stopping criteria was used in case the training did not improve development F1-score for 25 epochs. The learning rate was set to 10^{-3} for Wikipedia-pretrained model and to $5 \cdot 10^{-3}$ for the remaining ones.

---

[7] https://github.com/chakki-works/seqeval.

[8] https://www.clips.uantwerpen.be/conll2002/ner/bin/conlleval.txt.

[9] The results presented in the LeNER-Br paper are based on the token-level evaluation, which is not standard in the literature and provides much higher numbers.

[10] https://github.com/huggingface/transformers.

4.1 Results and Discussion

First we examine the impacts of BERT-based LM finetuning evaluating it on LeNER-Br corpus. The experimental results from pretrained and finetuned-based BERT models are provided in Table 8. It compares the results for the three BERT models: Multilingual (general), Wikipedia (general) and Acórdãos (specific). The model finetuned in the intradomain corpus outperforms our baselines. The Acórdãos-based model achieves the highest overall F1-score. If we consider the standard deviation of the Wikipedia model, there is a slight difference when compared to Multilingual. Regarding entity types, we can observe that the higher performance increase is related to Law entity, up to 2 points on the Acórdãos model. On the other hand, Time, Local and Legal Case entities results show a small drop on the same model evaluation.

We obtained interesting findings for the HAREM task, as show in Table 9. The Acórdãos finetuning considerably decreased the overall performance in more than 2 points. Unlike for the LeNER-Br task, to finetune on the legal domain has harmed model performance. This is kind of expected since we are finetuning

Table 8. BERT results on LeNER-Br.

Entity type	Multilingual	Wikipedia	Acórdãos
Person	91.38 ± 0.87	91.70 ± 1.25	93.66 ± 0.21
Time	94.43 ± 0.39	94.48 ± 1.32	92.39 ± 1.82
Location	68.67 ± 1.89	68.65 ± 3.36	65.78 ± 2.26
Organization	84.76 ± 0.57	85.25 ± 0.66	86.43 ± 0.49
Law	93.40 ± 1.23	93.44 ± 2.05	95.48 ± 0.47
Legal case	84.04 ± 0.63	84.17 ± 1.22	83.26 ± 0.73
Overall	88.81 ± 0.29	88.90 ± 1.72	89.39 ± 0.08

Table 9. BERT results on HAREM (total scenario).

Entity type	Multilingual	Wikipedia	Acórdãos
Person	78.44 ± 0.16	78.12 ± 0.19	75.49 ± 0.17
Time	90.26 ± 0.32	88.63 ± 0.47	88.67 ± 1.14
Location	81.22 ± 0.41	81.80 ± 0.36	80.51 ± 0.22
Organization	75.58 ± 0.13	76.03 ± 0.52	70.24 ± 0.98
Value	78.96 ± 1.23	79.01 ± 0.65	76.95 ± 0.32
Title	48.97 ± 0.76	49.24 ± 1.03	44.02 ± 3.18
Thing	38.90 ± 1.78	40.02 ± 1.96	40.00 ± 2.74
Event	42.46 ± 2.30	42.58 ± 2.57	34.67 ± 1.23
Abstraction	48.61 ± 3.13	49.65 ± 1.78	45.05 ± 2.65
Other	15.98 ± 5.45	15.63 ± 3.25	12.64 ± 4.23
Overall	73.48 ± 0.21	73.85 ± 0.67	71.03 ± 0.48

the model on a specific domain before applying it on a general domain. As this is a general-domain corpus, it did not take advantages from Acórdãos-TCU language model finetuning. Entities as Organization and Title were substantially affected on Acórdãos performance. On the other hand, Wikipedia model has outperformed the baseline model. Wikipedia is a general-domain corpus and the finetuning process might have benefited the LM to capture generic features from language. For HAREM selective scenario, the same behavior was observed on results. The best overall is from Wikipedia model. Once again, Multilingual and Wikipedia results were close, as show on Table 10.

Table 10. BERT results on HAREM (selective scenario).

Entity type	Multilingual	Wikipedia	Acórdãos
Person	79.20 ± 1.18	80.03 ± 2.02	75.91 ± 0.32
Time	89.88 ± 0.71	90.01 ± 0.89	88.22 ± 0.27
Location	79.41 ± 0.25	79.85 ± 0.74	80.13 ± 0.06
Organization	72.32 ± 0.18	72.29 ± 0.63	71.29 ± 0.15
Value	78.34 ± 0.58	78.66 ± 1.12	75.84 ± 0.54
Overall	79.39 ± 0.35	79.51 ± 0.54	77.72 ± 0.26

DrugSeizures-Br results for total scenario are reported in Table 11. In this scenario, we noticed that about four entity classes results are zero. Some aspects of DrugSeizures-Br corpus may explain this results: a minor number of annotations for the considered entities, poor quality of annotation or resulting noise from the string matching algorithm. The model generalization ability may be affected by this reason. Considering the selective scenario for DrugSeizures-Br results are show in Table 12. We can see that grouping the entities in super-types is effective when considering the NER task. In this case, the results from Acórdãos model are minimally inferior compared to Wikipedia model. We evaluated three different ELMo models: Wikipedia, brWaC and Acórdãos. Wikipedia model corresponds to the pretrained weights publicly available on Allennlp page (See footnote 3). brWaC model was provided by de Castro [4] and Acórdãos model was derived from finetuning the brWaC pretrained model on Acórdãos-TCU corpus. Table 13 shows that Acórdãos model improved F1 overall performance by 2.2 and 0.75 over Wikipedia and brWaC models.

Although the results are lower than those achieved by BERT-based models, it shows that the finetuning step is beneficial when considered intradomain data. We report on Table 14 the best results for LeNER-Br corpus. LeNER-Br original results were improved by ELMo and BERT after language model finetuning, as reported on Sect. 3. Our BERT-based LM finetuned on Acórdãos-TCU corpus has improved the state-of-the-art results by 3.97 points.

Table 11. BERT results on DrugSeizures-Br (total scenario).

Entity *Super-Type*/Type	Multilingual	Wikipedia	Acórdãos
Location			
Street name	61.81 ± 0.40	62.74 ± 0.98	58.80 ± 0.71
Street number	64.04 ± 0.16	63.25 ± 0.14	62.67 ± 0.19
Neighborhood	56.13 ± 0.21	52.25 ± 0.78	59.44 ± 0.32
City	86.74 ± 0.40	87.54 ± 0.74	86.39 ± 0.23
Addr. compl.	13.54 ± 11.8	10.36 ± 16.6	5.97 ± 5.32
Origin city	68.60 ± 0.26	67.95 ± 0.55	69.29 ± 0.34
Origin state	56.41 ± 0.44	56.58 ± 0.39	57.39 ± 0.39
Target city	51.27 ± 0.32	52.03 ± 0.69	52.72 ± 0.16
Target state	60.39 ± 0.38	60.89 ± 0.47	62.15 ± 0.14
Person			
Indicted	88.21 ± 0.21	88.05 ± 0.32	86.60 ± 0.14
Active witness	41.74 ± 0.23	42.22 ± 0.87	38.85 ± 0.42
Witness	19.38 ± 2.50	15.65 ± 7.58	22.03 ± 1.43
Passive witness	0.00 ± 0.00	0.00 ± 0.00	0.00 ± 0.00
Investigated	0.00 ± 0.00	0.00 ± 0.00	2.27 ± 0.22
Victim	30.62 ± 0.27	28.65 ± 2.26	23.29 ± 1.94
Author	67.77 ± 0.58	67.36 ± 1.04	66.20 ± 0.91
Defendant	0.00 ± 0.00	0.00 ± 0.00	0.00 ± 0.00
Attorney	0.00 ± 0.00	0.00 ± 0.00	0.00 ± 0.00
Time			
Time	93.42 ± 0.35	95.25 ± 0.25	92.78 ± 0.14
Date	85.80 ± 0.62	85.79 ± 0.73	84.18 ± 0.18
Organization			
Prosecutor	86.47 ± 1.25	87.22 ± 1.48	88.29 ± 0.20
Authority	92.70 ± 0.42	92.87 ± 0.32	91.50 ± 0.29
Drug			
Name	89.23 ± 0.79	89.55 ± 1.07	90.57 ± 0.93
Other			
Penal norm	94.68 ± 2.53	92.15 ± 1.96	94.07 ± 0.74
Drug quantity	75.98 ± 1.31	75.88 ± 1.96	74.61 ± 0.35
Overall	73.83 ± 1.55	72.98 ± 1.27	72.78 ± 0.89

Table 12. BERT results on DrugSeizures-Br (selective scenario).

Entity type	Multilingual	Wikipedia	Acórdãos
Person	81.70 ± 0.31	81.68 ± 0.69	81.94 ± 0.47
Time	87.99 ± 0.11	88.39 ± 0.18	82.13 ± 0.51
Location	76.80 ± 0.61	76.92 ± 2.36	77.94 ± 0.65
Organization	92.60 ± 0.58	93.62 ± 0.44	91.41 ± 0.64
Drug	91.63 ± 0.17	91.72 ± 0.21	89.71 ± 0.32
Overall	80.94 ± 0.07	81.12 ± 0.18	81.04 ± 0.35

Table 13. ELMo results on LeNER-Br.

Entity type	Wikipedia	brWaC	Acórdãos
Person	95.61 ± 0.87	95.85 ± 0.76	98.28 ± 0.41
Time	89.41 ± 1.78	91.17 ± 0.89	93.77 ± 1.05
Location	72.95 ± 2.01	75.04 ± 1.99	75.21 ± 1.13
Organization	85.15 ± 0.57	86.00 ± 0.82	86.54 ± 0.74
Law	87.64 ± 0.93	89.14 ± 0.82	88.08 ± 1.18
Legal case	75.74 ± 3.92	80.23 ± 3.88	80.81 ± 1.14
Overall	86.21 ± 0.68	87.66 ± 1.01	88.41 ± 0.41

Table 14. Comparison of our best models with the previous state-of-the-art results on LeNER-Br. The BERT and ELMo models were based on the LMs finetuned on the Acórdãos corpus. The LSTM+CRF model was proposed on LeNER-Br paper [14].

Entity type	LSTM+CRF	BERT	ELMo
Person	80.83 ± 1.83	93.66 ± 0.21	98.28 ± 0.41
Time	90.12 ± 2.28	92.39 ± 1.82	93.77 ± 1.05
Location	64.81 ± 3.05	65.78 ± 2.26	75.21 ± 1.13
Organization	84.37 ± 0.62	86.43 ± 0.49	86.54 ± 0.74
Law	93.01 ± 0.61	95.48 ± 0.47	88.08 ± 1.18
Legal case	81.22 ± 1.65	83.26 ± 0.73	80.81 ± 1.14
Overall	85.42 ± 0.61	89.39 ± 0.08	88.41 ± 0.41

5 Conclusion

In this paper, we present a study regarding the impact of intradomain finetuning of deep language models for named entity recognition on the legal domain. We have explored different Portuguese NER corpora. By means of the legal domain LeNER-Br corpus, we demonstrated how intradomain LM finetuning considerably improve performance. Additionally, a new state-of-the-art result was achieved for this corpus. On the other hand, the same finetuned model, when applied to the general domain HAREM corpus, presented inferior results compared to general-domain models. Moreover, for the DrugSeizures-Br corpus, which also comprises legal documents, general-domain models outperformed legal-domain models. This is probably due to the fact that the Acórdãos-TCU corpus, used for intradomain LM finetuning, presents a quite different language style from the DrugSeizures-Br corpus. We are now creating a larger corpus with the same style as DrugSeizures-Br. We plan to evaluate the LM finetuning procedure on this corpus, in order to validate our findings.

References

1. Alsentzer, E., et al.: Publicly available clinical BERT embeddings. CoRR, abs/1904.03323 (2019). http://arxiv.org/abs/1904.03323
2. Angelidis, I., Chalkidis, I., Koubarakis, M.: Named entity recognition, linking and generation for Greek legislation. In: Proceedings of JURIX 2018 (2018)
3. Badji, I.: Legal entity extraction with NER systems, June 2018. http://oa.upm.es/51740/
4. de Castro, P.V.Q.: Aprendizagem profunda para reconhecimento de entidades nomeadas em domínio jurídico. Master's thesis, Programa de Pós-graduação em Ciência da Computação (INF) (2019). http://repositorio.bc.ufg.br/tede/handle/tede/10276. Instituto de Informática - INF (RG)
5. Quinta de Castro, P.V., Félix Felipe da Silva, N., da Silva Soares, A.: Portuguese named entity recognition using LSTM-CRF. In: Villavicencio, A., et al. (eds.) PROPOR 2018. LNCS (LNAI), vol. 11122, pp. 83–92. Springer, Cham (2018). https://doi.org/10.1007/978-3-319-99722-3_9
6. de Castro, P.V.Q., da Silva, N.F.F., da Silva Soares, A.: Contextual representations and semi-supervised named entity recognition for Portuguese language. In: Proceedings of IberLEF@SEPLN 2019 (2019)
7. Devlin, J., Chang, M.-W., Lee, K., Toutanova, K.: BERT: pre-training of deep bidirectional transformers for language understanding (2018). http://arxiv.org/abs/1810.04805
8. do Amaral, D.O.F., Vieira, R.: NERP-CRF: uma ferramenta para o reconhecimento de entidades nomeadas por meio de conditional random fields. Linguamática **6**, 41–49 (2014)
9. Dozier, C., Kondadadi, R., Light, M., Vachher, A., Veeramachaneni, S., Wudali, R.: Named entity recognition and resolution in legal text. In: Francesconi, E., Montemagni, S., Peters, W., Tiscornia, D. (eds.) Semantic Processing of Legal Texts. LNCS (LNAI), vol. 6036, pp. 27–43. Springer, Heidelberg (2010). https://doi.org/10.1007/978-3-642-12837-0_2
10. Freitas, C., Mota, C., Santos, D., Oliveira, H.G., Carvalho, P.: Second HAREM: advancing the state of the art of named entity recognition in Portuguese. In: Proceedings of LREC 2010 (2010)
11. Hakala, K., Pyysalo, S.: Biomedical named entity recognition with multilingual BERT. In: Proceedings of the 5th Workshop on BioNLP Open Shared Tasks, Hong Kong, China, November 2019, pp. 56–61. Association for Computational Linguistics (2019). https://doi.org/10.18653/v1/D19-5709. https://www.aclweb.org/anthology/D19-5709
12. Howard, J., Ruder, S.: Fine-tuned language models for text classification. CoRR, abs/1801.06146 (2018). http://arxiv.org/abs/1801.06146

13. Lample, G., Conneau, A.: Cross-lingual language model pretraining. CoRR, abs/1901.07291 (2019). http://arxiv.org/abs/1901.07291
14. Luz de Araujo, P.H., de Campos, T.E., de Oliveira, R.R.R., Stauffer, M., Couto, S., Bermejo, P.: LeNER-Br: a dataset for named entity recognition in Brazilian legal text. In: Villavicencio, A., et al. (eds.) PROPOR 2018. LNCS (LNAI), vol. 11122, pp. 313–323. Springer, Cham (2018). https://doi.org/10.1007/978-3-319-99722-3_32
15. Martin, L., et al.: CamemBERT: a tasty French language model. In: Proceedings of the 58th Annual Meeting of the Association for Computational Linguistics. Association for Computational Linguistics (2020). https://doi.org/10.18653/v1/2020.acl-main.645
16. Peters, M.E., et al.: Deep contextualized word representations. CoRR, abs/1802.05365 (2018). http://arxiv.org/abs/1802.05365
17. Pirovani, J., Oliveira, E.: Portuguese named entity recognition using conditional random fields and local grammars. In: Proceedings of LREC 2018, May 2018
18. Polignano, M., Basile, P., de Gemmis, M., Semeraro, G., Basile, V.: AlBERTo - Italian BERT language understanding model for NLP challenging tasks based on tweets. In: CLiC-it (2019)
19. Radford, A., Narasimhan, K., Salimans, T., Sutskever, I.: Improving language understanding by generative pre-training (2018)
20. Radford, A., Wu, J., Child, R., Luan, D., Amodei, D., Sutskever, I.: Language models are unsupervised multitask learners (2019)
21. Rother, K., Rettberg, A: ULMFiT at GermEval-2018: a deep neural language model for the classification of hate speech in German tweets. In: Proceedings of the GermEval 2018 Workshop, September 2018
22. Souza, F., Nogueira, R., Lotufo, R.: Portuguese named entity recognition using BERT-CRF. arXiv:1909.10649 (2020)
23. Vaswani, A., et al.: Attention is all you need. CoRR, abs/1706.03762 (2017). http://arxiv.org/abs/1706.03762
24. Wagner Filho, J.A., Wilkens, R., Idiart, M., Villavicencio, A.: The brWaC corpus: a new open resource for Brazilian Portuguese. In: Proceedings of the Eleventh International Conference on Language Resources and Evaluation (LREC 2018), Miyazaki, Japan, May 2018. European Language Resources Association (ELRA) (2018). https://www.aclweb.org/anthology/L18-1686

Correction to: Semi-Supervised Sentiment Analysis of Portuguese Tweets with Random Walk in Feature Sample Networks

Pedro Gengo and Filipe A. N. Verri

Correction to:
Chapter "Semi-cupervised Sentiment Analysis of Portuguese Tweets with Random Walk in Feature Sample Networks"
in: R. Cerri and R. C. Prati (Eds.): *Intelligent Systems*,
LNAI 12319, https://doi.org/10.1007/978-3-030-61377-8_42

Inadvertently the authors of this chapter released it without correcting an error in the title. This has now been corrected and the corrected title reads: "Semi-Supervised Sentiment Analysis of Portuguese Tweets with Random Walk in Feature Sample Networks".

The updated version of this chapter can be found at
https://doi.org/10.1007/978-3-030-61377-8_42

Author Index

Printed in the United States
By Bookmasters